Graphing

Symmetry to y-axis: $f(-x) = f(x)$

Symmetry to origin: $f(-x) = -f(x)$

Graphs of $y = f(x)$ and $y = f(-x)$ are reflections in the y-axis.

Graphs of $y = f(x)$ and $y = -f(x)$ are reflections in the x-axis.

A graph of $y = f(x) + c$ is translated c units vertically with respect to a graph of $y = f(x)$.

A graph of $y = f(x - c)$ is translated c units horizontally with respect to a graph of $y = f(x)$.

Graphs of a function and its inverse function are reflections in the line $y = x$.

Zeros of Polynomial Function

$F(x) = a_nx^n + a_{n-1}x^{n-1} + \cdots + a_1x + a_0$

Rational zeros are of the form p/q where p is a factor of a_0 and q is a factor of a_n.

The number of zeros equals the degree of the polynomial.

Properties of Logarithms

$y = \log_b x$ if and only if $x = b^y$

$\log_b xy = \log_b x + \log_b y$

$\log_b \frac{x}{y} = \log_b x - \log_b y$

$\log_b x^p = p \log_b x$

$\log_b b = 1, \ \log_b 1 = 0$

$\log_b x = \frac{\log x}{\log b}$

$b^{\log_b x} = x$

Area and Volume Formulas

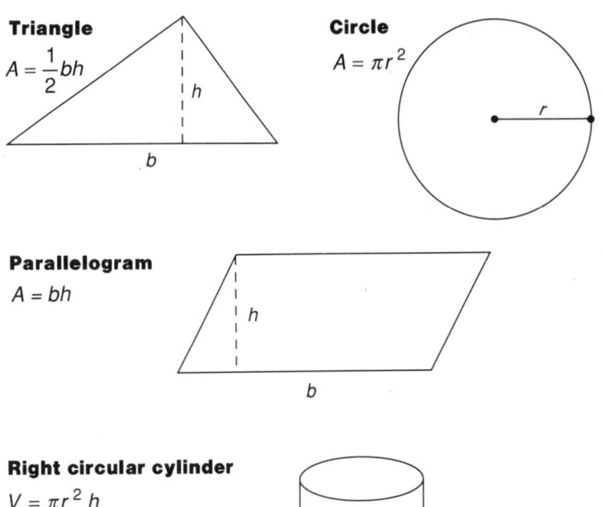

Triangle

$A = \frac{1}{2}bh$

Circle

$A = \pi r^2$

Parallelogram

$A = bh$

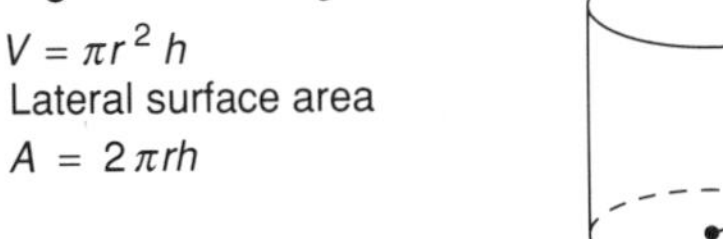

Right circular cylinder

$V = \pi r^2 h$

Lateral surface area

$A = 2\pi rh$

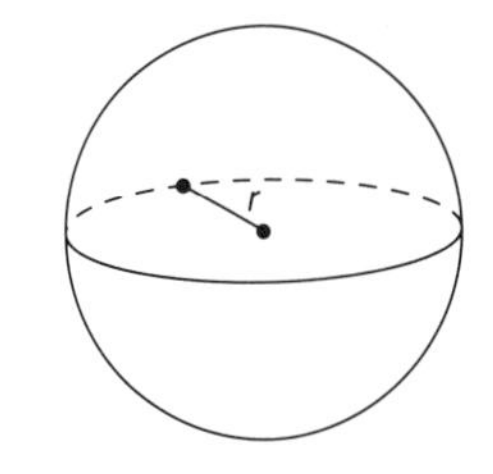

Sphere

$V = \frac{4}{3}\pi r^3$

$A = 4\pi r^2$

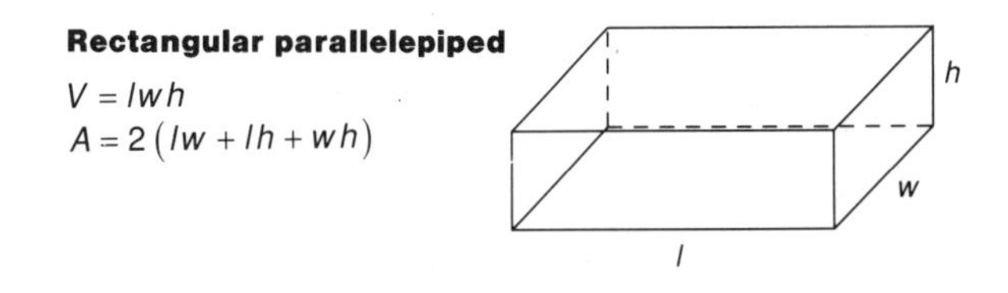

Rectangular parallelepiped

$V = lwh$

$A = 2(lw + lh + wh)$

Right circular cone

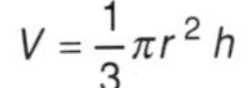

$V = \frac{1}{3}\pi r^2 h$

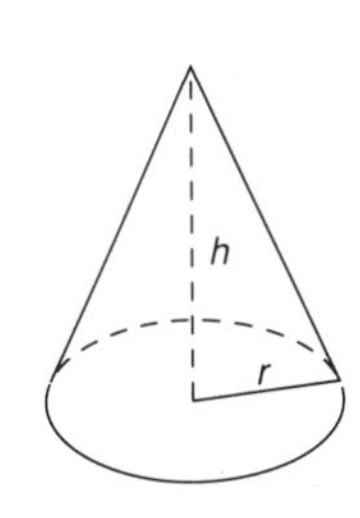

Precalculus

Murray Gechtman
Los Angeles Pierce College

Wm. C. Brown Publishers

Book Team

Editor *Earl McPeek*
Developmental Editor *Janette S. Scotchmer*
Production Editor *Jane E. Matthews*
Designer *K. Wayne Harms*
Art Editor *Mary E. Swift*

Wm. C. Brown Publishers

President *G. Franklin Lewis*
Vice President, Publisher *George Wm. Bergquist*
Vice President, Operations and Production *Beverly Kolz*
National Sales Manager *Virginia S. Moffat*
Group Sales Manager *Vincent R. Di Blasi*
Assistant Vice President, Editor in Chief *Edward G. Jaffe*
Executive Editor *Earl McPeek*
Marketing Manager *John W. Calhoun*
Advertising Manager *Amy Schmitz*
Managing Editor, Production *Colleen A. Yonda*
Manager of Visuals and Design *Faye M. Schilling*
Production Editorial Manager *Julie A. Kennedy*
Production Editorial Manager *Ann Fuerste*
Publishing Services Manager *Karen J. Slaght*

WCB Group

President and Chief Executive Officer *Mark C. Falb*
Chairman of the Board *Wm. C. Brown*

Cover photo by Mark Tomalty/Masterfile

Copy Editor *Barbara A. Bonnet*

Library of Congress Catalog Card Number: 90–80921

ISBN 0–697–11773–1

Printed in the United States of America by Wm. C. Brown Publishers, 2460 Kerper Boulevard, Dubuque, IA 52001

10 9 8 7 6 5 4 3 2 1

To my wife Lucille, my sons Joe and Dan, and my daughter Risa

Brief Contents

Chapter 1 Equations and Inequalities *1*
Chapter 2 Functions and Graphing *71*
Chapter 3 Systems of Linear Equations *131*
Chapter 4 Polynomial Functions *169*
Chapter 5 Rational Functions *205*
Chapter 6 Exponential and Logarithmic Functions *226*
Chapter 7 Trigonometric Functions *256*
Chapter 8 Trigonometric Identities and Equations *312*
Chapter 9 Applications of the Trigonometric Functions *344*
Chapter 10 Polar Coordinate Graphs and Parametric Equations *382*
Chapter 11 The Conic Sections *406*
Chapter 12 Vectors *470*
Chapter 13 Induction, Series, and the Binomial Theorem *504*
Appendix A Tables *537*
Appendix B Answers to Odd-Numbered Problems *555*

Contents

Preface *xvii*

Chapter 1 **Equations and Inequalities** 1

1.1 Linear Equations *1*
Sets *1*
Replacement and Solution Sets *2*
Equivalent Equations *3*
Linear Equations *5*
Mathematical Models *5*

1.2 Exponents *11*
Integral Exponents *12*
Exponents in the Form $1/n$ *14*
Exponents in the Form m/n *15*
Rational Number Exponents *16*
Radical Notation *17*

1.3 Factoring *22*
Common Monomial Factors *22*
Trinomials of the Form $ax^2 + bx + c$ *23*
Difference of Two Squares *25*
Sums and Differences of Cubes *25*
Factoring by Grouping *26*
Other Kinds of Factoring *26*

1.4 Complex Numbers *28*
Conjugates and Quotients in C *30*
Subsets of C *31*

1.5 Quadratic Equations *33*
Completing the Square *34*
The Quadratic Formula *36*
Discriminant and Roots *36*
Applications *38*

1.6 Inequalities *42*
Generating Equivalent Inequalities *43*
Intervals *46*
Sign Graphs *47*
1.7 Equations Solvable as Linear or Quadratic Equations *52*
Equations in Quadratic Form *54*
1.8 Absolute Value in Equations and Inequalities *57*
Absolute Value Properties *57*
Solving $|y| = k$, $|y| < k$, and $|y| > k$ *59*
1.9 Absolute Value and Error *63*
Chapter 1 Review 66
Chapter 1 Test 1 70
Chapter 1 Test 2 70

Chapter 2 **Functions and Graphing** 71

2.1 Functions *71*
Relations *72*
Functions *72*
Functions Represented by Equations *73*
Function Notation *76*
Graphs *76*
Increasing and Decreasing Functions *78*
Constructing Functions *79*
2.2 Linear Functions *83*
Slope-Intercept Form *85*
General Form *87*
Lines Parallel to the Coordinate Axes *88*
Parallel and Perpendicular Lines *89*
Distance and Midpoint Formulas *91*
Some Graphs *93*
2.3 Quadratic Functions *97*
Finding the Vertex of a Parabola *98*
Max-Min Problems *100*
The Difference Quotient *101*
2.4 Graphing Techniques *104*
Symmetry *105*
Intercepts *106*
Domain and Range *107*
New Graphs from Old: Reflections *109*
Translations *111*
Stretchings and Contractions *112*

2.5 Algebra of Functions *115*
Composition of Functions *116*
Inverse Functions *119*
Inverses and Composition *121*
Graphing Inverse Functions *122*
Chapter 2 Review 127
Chapter 2 Test 1 129
Chapter 2 Test 2 129

Chapter 3 **Systems of Linear Equations** 131

3.1 2×2 Linear Systems *131*
Test for Consistency *132*
The Point of Intersection of Nonparallel Lines *134*
Linear Combinations *135*
Applications *136*
3.2 3×3 Systems *141*
3×3 Linear Systems *141*
Linear Combinations *142*
Applications *145*
3.3 Elementary Transformations and Matrices *148*
Elementary Transformations *148*
Matrices *150*
Echelon Form *152*
Consistency *154*
3.4 $M \times N$ Linear Systems *156*
Applications *159*
Chapter 3 Review 165
Chapter 3 Test 1 168
Chapter 3 Test 2 168

Chapter 4 **Polynomial Functions** 169

4.1 The Arithmetic Operations on Polynomials *169*
The Arithmetic of Polynomials *170*
Synthetic Division *171*
4.2 Zeros of Polynomial Functions I *174*
Roots That Are Not Integers *177*
4.3 Zeros of Polynomial Functions II *179*
Finding Rational Zeros *181*
Shortening the Search for Rational Zeros *183*
4.4 Graphing Polynomials *187*

4.5 Approximating Irrational Zeros *197*
Iteration *197*
Evaluating Polynomials with a Calculator *199*
Chapter 4 Review 201
Chapter 4 Test 1 204
Chapter 4 Test 2 204

Chapter 5 **Rational Functions** 205

5.1 Graphing Rational Functions I *205*
Vertical Asymptotes *206*
Horizontal Asymptotes *208*
Graphs *211*
5.2 Graphing Rational Functions II *216*
5.3 Partial Fractions *219*
Chapter 5 Review 224
Chapter 5 Test 1 225
Chapter 5 Test 2 225

Chapter 6 **Exponential and Logarithmic Functions** 226

6.1 Exponential Functions *226*
Defining b^x When x Is Irrational *228*
Properties of Irrational Exponents *230*
6.2 The Natural Base e *234*
Population Growth *234*
Radioactive Decay *237*
Compound Interest *237*
6.3 Logarithmic Functions *241*
Evaluating $\log_b x$ *242*
Properties of Logarithmic Functions *244*
6.4 Logarithmic and Exponential Equations *247*
Some Applications *249*
Chapter 6 Review 253
Chapter 6 Test 1 255
Chapter 6 Test 2 255

Chapter 7 **Trigonometric Functions** 256

7.1 Periodic Functions *256*
Periodicity *260*
7.2 Radian Measure *264*
Circular Arcs and Radian Measure *264*
Arc Length *266*
Quadrant Position of Terminal Side *267*

7.3 The Trigonometric Functions 269
Evaluating the Trigonometric Functions for $\pi/2 < \theta < 2\pi$ 273
7.4 Some Properties of the Trigonometric Functions 278
More Identities 280
Evaluating the Trigonometric Functions over $(-\infty,\infty)$ 281
Cofunctions and Tables 284
A Summary of the Section's Identities 286
7.5 Graphs of the Trigonometric Functions 287
The Basic Sine Curve 287
The Basic Cosine Curve 288
The Basic Tangent Curve 289
Graphs of the Reciprocal Functions 290
Graphs of $y = A f[B(x + C)]$ 291
Sums of Trigonometric Functions 295
7.6 The Inverse Trigonometric Functions 298
Evaluating $\sin^{-1} x$, $\cos^{-1} x$, and $\tan^{-1} x$ 300
Some Relationships 301
$\text{Csc}^{-1} x$, $\text{Sec}^{-1} x$, $\text{Cot}^{-1} x$ 303
Chapter 7 Review 308
Chapter 7 Test 1 310
Chapter 7 Test 2 311

Chapter 8 **Trigonometric Identities and Equations** 312

8.1 Identities 312
Verifying That an Equation Is an Identity 313
Sum and Difference Formulas 315
8.2 More Identities 320
Double Angle Formulas 321
Half Angle Formulas 323
Evaluating Expressions That Contain Inverse Trigonometric Functions 325
8.3 Trigonometric Equations 328
Equations with Inverse Trigonometric Functions 332
An Application 333
8.4 Numerical Techniques 337
Chapter 8 Review 341
Chapter 8 Test 1 343
Chapter 8 Test 2 343

Chapter 9 **Applications of the Trigonometric Functions** 344

9.1 Angular and Linear Velocity 344
9.2 Right Triangle Trigonometry 349
9.3 The Sine and Cosine Laws 356
Law of Sines 356
Law of Cosines 359

9.4 Vectors *364*
Magnitude of Vectors *365*
Vector Addition *365*
Vector Applications *366*
9.5 Trigonometric Models of Periodic Phenomena *372*
Beats *373*
Modulation Theory *374*
Chapter 9 Review 377
Chapter 9 Test 1 380
Chapter 9 Test 2 381

Chapter 10 **Polar Coordinate Graphs and Parametric Equations** 382

10.1 Polar Coordinates *382*
Polar and Rectangular Equations *384*
10.2 Polar Coordinate Graphs *389*
Symmetry Tests *390*
10.3 Parametric Equations *395*
10.4 Complex Roots and deMiovre's Theorem *399*
Polar Form of a Complex Number *399*
Product of Complex Numbers *400*
Chapter 10 Review 404
Chapter 10 Test 1 405
Chapter 10 Test 2 405

Chapter 11 **The Conic Sections** 406

11.1 Circles *408*
Circles Determined by Three Conditions *410*
Parametric Equations *413*
11.2 Parabolas *414*
Translated Parabolas *416*
11.3 Ellipses *423*
Translated Ellipses *426*
Parametric Equations *427*
11.4 Hyperbolas *430*
Asymptotes *433*
Translated Hyperbolas *435*
Parametric Equations *438*
11.5 Rotation of Axes *440*
Identifying Conics *447*
11.6 Eccentricity and Polar Coordinate Forms *449*
Polar Coordinate Forms of the Conics *454*

11.7 Tangents and the Reflecting Property of the Conics *458*
The Angle between Two Lines *459*
Tangents *460*
Reflecting Property of the Parabola *463*
Reflecting Property of the Ellipse and the Hyperbola *464*
Chapter 11 Review 466
Chapter 11 Test 1 468
Chapter 11 Test 2 469

Chapter 12 **Vectors** 470

12.1 Vector Algebra *470*
Scalar Multiplication *471*
Properties of Vector Addition and Scalar Multiplication *473*
Basis Vectors *473*
12.2 The Dot Product *477*
The Angle between Two Vectors *478*
The Distance from a Point to a Line *480*
Properties of the Dot Product *482*
12.3 Parametric Equations of a Line *485*
Subdivision of a Line Segment *488*
12.4 Lines and Planes in Three Dimensions *492*
A Three-Dimensional Coordinate System *492*
Vectors in 3-Space *493*
Lines in 3-Space *494*
Planes in 3-Space *497*
Chapter 12 Review 500
Chapter 12 Test 1 502
Chapter 12 Test 2 503

Chapter 13 **Induction, Series, and the Binomial Theorem** 504

13.1 Mathematical Induction *504*
13.2 Sequences, Series, and Summation *510*
Series *511*
Sigma Notation *512*
Properties of Σ Notation *512*
13.3 Arithmetic and Geometric Series *515*
Arithmetic Series *516*
Geometric Sequences *517*
Geometric Series *518*
Infinite Geometric Series *520*

13.4 Counting and the Binomial Theorem *524*
Permutations of n-Member Sets *525*
r-Member Subsets of n-Member Sets *527*
Combinations *528*
The Binomial Theorem *530*
Chapter 13 Review 533
Chapter 13 Test 1 535
Chapter 13 Test 2 535

Appendix A Tables 537
Appendix B Answers to Odd-Numbered Problems 555
Index 675

Preface

Precalculus is written for students who have completed intermediate algebra and trigonometry and expect to take the standard three-semester lower-division calculus course. It is designed to help students bridge the gap in mathematical sophistication between high school intermediate algebra and college calculus.

APPROACH

The elementary functions, their graphs, and mathematical modeling are the core of the text. From the examples in the discourse to the exercise sets, the student is engaged more with analytical activities than mechanical manipulation and computation.

ABOUT THIS BOOK

This precalculus text does not simply consist of topics that have been thrown together from intermediate algebra, college algebra, and trigonometry; it demands more of the student than just rehashing material he/she has seen before. Familiar topics are set in new and different settings. The student is drawn deeper into the subject matter by the easily readable discourse, the detailed examples, and the exercise sets.

1. The basic intermediate algebra topics are reviewed in chapter 1. Reviewing inequalities and absolute value in sections 1.6 and 1.8 is recommended.
2. Chapter 2 contains a detailed discussion of functions and their graphs and a number of interesting applications. This is the core chapter in the book. The student is asked to apply the ideas in this chapter over and over again throughout the text.
3. The systems of linear equations are placed in chapter 3 so that the techniques for solving linear systems can be used throughout the text to generate polynomial approximations of some of the elementary functions. The techniques are also used in chapter 5 to construct partial fraction decompositions. There are some unusual applications of linear systems in section 3.4.
4. Chapter 4 contains a development of "Theory of Equations." Iteration is introduced in section 4.5 as a means of approximating irrational roots of polynomial equations. The technique is used in later chapters to approximate roots of logarithmic, exponential, and trigonometric equations. The calculator plays a vital role in these root approximations.
5. In chapter 5, the student is informally introduced to the notion of a limit in the discussions on the graphing of rational functions. The emphasis placed on the construction of vertical, horizontal, and oblique asymptotes forces the student to think about the process of "going to infinity." Some unusual applications of rational functions are also included in this chapter.

6. Exponential functions are introduced in the context of population growth. Base e is derived in section 6.2 through a pseudo-limit process. The log function is developed as the inverse of the exponential function. The exercise sets in this chapter are filled with applications.
7. Throughout the introduction, development, and application of the trig functions in chapters 7–9, it is emphasized that trig functions help us model periodic phenomena. Graphing and applications are emphasized throughout these chapters. The treatment of the inverse trig functions is more detailed and complete than the student will find in most precalculus texts.
8. A thorough development of polar coordinate graphing is given in chapter 10. The chapter also contains a novel but careful treatment of parametric equations.
9. Both trigonometry and algebra are employed at a fairly sophisticated level in the study of the conic sections. For this reason, a reasonably complete discussion of the conic sections is included in chapter 11. The material in this chapter is demanding because it is a compilation of many of the ideas in the earlier chapters. There are a large number of interesting applications in the exercise sets.
10. Vector methods in two and three dimensions are introduced in chapter 12. They are used for generating equations for lines in the plane and in 3-space and for generating equations of planes in 3-space.
11. In chapter 13 the student is shown how to construct mathematical induction proofs. Counting techniques are used to generate the binomial coefficients. Applications of geometric and arithmetic series in the exercise sets are also included in this chapter.

FEATURES

○ Graphing is emphasized as a way of "seeing" a function and its properties (chapter 2). With the introduction of each new function, the ideas introduced in chapter 2 are reinforced throughout the text. Graphs are used to help the student see and understand complicated relationships, such as those that exist between members of a family of lines (section 3.1), or the conic sections (chapter 11). The student learns that analysis, rather than mechanical point-plotting, produces representative and meaningful graphs (sections 4.4 and 5.1). Graphing is also used to find approximate solutions to problems that either cannot be solved algebraically or to find algebraic solutions that are very difficult to obtain (section 2.1, exercise 61; section 4.4, exercise 39).

- Mathematical modeling is integrated throughout the book. Starting with the simple design model in section 1.1, through the maximum-minimum models in section 2.3, the growth and decay models in chapter 6, the periodic models in chapters 7 and 9, the orbital models in chapter 11, and ending with the series models in chapter 13, the student is asked to construct and interpret mathematical models of natural phenomena.
- Methods for solving systems of linear equations are introduced early (chapter 3) so that those techniques can be used throughout the book. The equation of a curve through a given set of points is determined in section 4.4, exercise 36. A given function is approximated by a polynomial function in section 6.4, exercise 50. Partial fraction decompositions are constructed with the aid of systems of linear equations in section 5.3.
- Important ideas are introduced with concrete examples. In chapter 2, functions are introduced after the relationship between a student's test grade and the amount of time that the student studies for a test is discussed. Exponential functions and the natural base are introduced (chapter 6) in the context of population growth. The study of trig functions begins in chapter 7 with an attempt to write an expression that describes the periodic motion of the piston of a pump.
- New ideas grow out of previously discussed material and in turn reinforce ideas and methods that came earlier in the text. As an example, the development of exponential functions in sections 6.1 and 6.2 relies on the student learning the properties of rational exponents in section 1.2, the notion of increasing and decreasing functions in section 2.1, and the discussion of the difference quotient in section 2.3. This spiralling of an idea also shows up in the exercise sets. (See section 2.1, exercise 55; section 2.3, exercise 41; and section 8.3, exercise 47.)
- Calculator use is required throughout the text for computing (sections 7.6 and 8.1), for solving equations by iteration or interpolation, (section 4.5, exercises 47 and 48; and sections 6.4 and 8.4), and for establishing horizontal and vertical asymptotes (section 5.1).
- The notion of a limit is confronted often, but gently, throughout the text. The difference quotient, average, and marginal changes are discussed in chapter 2 in relation to the graphs of certain population growth and cost functions. The behavior of polynomial graphs as $|x|$ assumes larger and larger values is taken up in chapter 4. Asymptotic behavior of graphs of rational functions (chapter 5) occasions a number of discussions of the limit process both in the discourse and in the exercise sets. The notion of limits is important when writing an expression for population growth (section 6.2) and when showing how the natural base comes into being in population growth expressions.

- The examples and exercise sets move the student away from memorization and computation toward analysis and therefore, from high school mathematics to the threshhold of calculus (see section 2.1, exercises 55–61; or section 4.4, exercises 41–47).
- The last one or two sections in many chapters include material that is more demanding than the material in the rest of the chapter. These sections can be skipped without loss of continuity.
- $\varepsilon - \delta$ methods are masked in the discussion of absolute value and error in section 1.9.
- Systems of linear equations are solved by matrix methods in sections 3.3 and 3.4. Section 3.4 also contains a discussion of $m \times n$ linear systems and their application in circuit analysis and traffic flow analysis.
- Iteration is introduced in section 4.5.
- Partial fractions are discussed in section 5.3.
- Interpolation is used to approximate roots of trig equations in section 8.4.
- In section 9.5, certain periodic phenomena are modeled by trig functions.
- Some instructors may choose to leave out sections 10.3 on parametric equations and 10.4 on complex roots and deMoivre's Theorem.
- Unlike many precalculus texts, this book includes material on polar forms of the conic sections (section 11.5) and tangents to the conic sections (section 11.6).
- The text includes 3,620 exercises.

PEDAGOGY

Precalculus includes a consistent, logical pedagogical system designed to motivate concepts and reinforce learning.

- The chapter introduction provides an overview of the chapter and highlights the significance of topic treatments.
- The Example/Solution format relates theory with real applications. Solutions are provided in a step-by-step format.
- The boxed material highlights salient points for easier referencing by the student.
- Selected illustrations are provided on a grid background to help the student with his/her graphing.
- The exercise sets at the end of each section are divided into (**A**) and (**B**) subsets. The (**A**) exercises are designed to follow the text's discourse so that a student can practice using certain formulas or applying certain definitions. The (**B**) exercises, which expand on ideas in the text, generally require a reasonable amount of analysis. Many of the (**B**) exercises ask the student to construct an informal proof. The instructor may desire to assign the (**B**) exercises as take-home problems.

- End-of-chapter material includes: *Major Points for Review,* which outlines section features, major points, and terms for the student to review; *Review Exercises,* which cover the concepts presented in the chapter; and *Chapter Tests 1 and 2,* which are designed as fifty-minute to one hour practice tests that are meant to bracket between easy and difficult a typical one hour chapter exam.
- Answers to odd-numbered exercises are found at the end of the book in Appendix B so that the student can receive immediate feedback on his/her work.

COURSE OPTIONS

The following are suggested syllabi for precalculus courses using this text.

15 wks; 5 mtg/wk

Section	Hours
1.6,1.8,1.9	3
2.1–2.5	5
3.1–3.4	4
4.1–4.5	5
5.1–5.3	3
6.1–6.4	4
7.1–7.6	6
8.1–8.4	4
9.1–9.5	5
10.1–10.4	4
11.1–11.7	7
12.1–12.4	4
13.1–13.4	4
5 tests	5
Review	12
Total hours	75

15 wks; 3 mtg/wk

Section	Hours
1.6,1.8	2
2.1–2.5	5
4.1–4.4	4
5.1,5.2	2
6.1–6.4	4
7.1–7.6	6
8.1–8.3	3
9.1–9.3	3
11.1–11.4	4
13.1–13.4	4
3 tests	3
Review	5
Total hours	45

2–10 wk Quarters; 3 mtg/wk

1st Quarter		2nd Quarter	
Section	Hours	Section	Hours
1.6,1.8	2	8.1–8.4	4
2.1–2.5	5	9.1–9.4	4
3.1,3.2	2	10.1–10.3	3
4.1–4.4	4	11.1–11.6	6
5.1,5.2	2	13.1–13.4	4
6.1–6.4	4	3 tests	3
7.1–7.6	6	Review and/or optional material	6
3 tests	3	Total hours	30
Review	2		
Total hours	30		

SUPPLEMENTARY MATERIALS

Precalculus will be accompanied by an *Instructor's Manual,* which will have answers to all exercises and chapter tests, two master tests for each of the chapters, an answer key to the master tests, and a set of transparency masters. For the student, *Precalculus* has a *Student's Solutions Manual,* which provides solutions to selected problems. *TestPak 3.0*, a computerized testing service, is available to the adopting instructor. It consists of a bank of 1,500 questions organized by chapter, section, objective, and difficulty level. The instructor may generate his/her own tests from the diskettes.

Other supplementary material includes the *Test Item File*, a printout of the *TestPak* so that the instructor may select from the 1,500 test questions or use the Wm. C. Brown (800) service to generate tests. *QuizPak* is a quizzing program that allows students to take quizzes on randomly chosen questions from selected chapters. *GradePak* allows instructors to create a file of grades on as many as 250 students per file.

TestPak 3.0, QuizPak, and *GradePak* are available for IBM, Apple, and Macintosh personal computers. For a full description of the supplementary material to *Precalculus,* you may contact your Wm. C. Brown sales representative.

ACKNOWLEDGMENTS

I want to thank my longtime friend, Dr. Murray Geller of the Jet Propulsion Laboratory, for reading the manuscript and offering many helpful suggestions. He also provided me with a number of interesting exercises. Thanks also to longtime friend and Pierce College colleague, Irving Drooyan, for his support and encouragement.

The manuscript was reviewed by a number of colleagues around the country, including William Peirce, Cape Cod Community College; Leland J. Fry, Kirkwood Community College; Fredric W. Tufte, University of Wisconsin, Platteville; Herbert E. Kasube, Bradley University; Annamarie Langlie, North Hennepin Community College; Larry J. Bowen, University of Alabama; Terry Tiballi, North Harris County College; Chris Kolaczewski, University of Akron; John F. Keating, Massasoit Community College; John Hansen, Armstrong State College. Their comments and suggestions had a major impact on the final form of this book. I thank them for their help.

Finally, I want to thank J. Earl McPeek, executive editor at Wm. C. Brown Publishers, for keeping in touch; Jan Scotchmer, senior developmental editor, for hanging in there through the tough times; and Jane Matthews, production editor, for putting it all together.

Murray Gechtman

Chapter 1

Equations and Inequalities

The review material that accounts for most of this chapter has been included here because of its importance in the development of the mathematical ideas and techniques that are at the heart of a precalculus course. You must be able to factor quickly and correctly. You must recognize linear and quadratic equations and know how to solve them. You must be able to work easily with rational exponents.

The chapter includes topics that have traditionally been difficult for algebra students: word problems, linear and quadratic inequalities, and absolute value. Success in this precalculus course depends, very much, on how well you master (or have mastered) the material in this chapter.

1.1 Linear Equations

We shall need to review some set terminology and set notation before proceeding to solve equations.

Sets

Recall that a set is a well-defined collection of distinct objects, each of which is said to be a member of, or an element of, the set. Capital letters such as A, B, N, etc., are used to name sets. In fact, N shall name the set of natural numbers, 1,2,3, . . . *, throughout the text.

A set will often be indicated by listing its members between braces. Thus, the set that includes the first five even natural numbers is $\{2,4,6,8,10\}$. We shall also use set builder notation to specify a set. As an illustration, $\{n|n \in N\}$ is read, "The set of all elements *n,* such that *n* is a natural number." The symbol ε denotes membership in a set. The set of all odd natural numbers can be specified as $\{2n - 1|n \in N\}$. The expression in braces is read, "The set of all elements, $2n - 1$, such that n is a natural number."

*The ellipsis symbol, . . . , means "and so forth."

If every element of set A is also an element of set B, then A is a **subset** of B. If A is a subset of B, and B contains at least one element that is not in A, then A is a **proper subset** of B. The set of odd natural numbers is a proper subset of the set of natural numbers.

The set that contains no elements is called the **empty set,** or the **null set.** It is identified by the symbol $\emptyset$. Two sets are equal if they contain the same members.

Replacement and Solution Sets

The set of all possible number replacements for a variable in an equation is called the **replacement set** for the variable. Unless otherwise stated, the replacement set is assumed to be a subset of the set of real numbers. We shall use **R** to name the set of real numbers.

Example 1 Determine the replacement set for the variable in each equation.

a. $\dfrac{2y + 3}{y - 2} = 5$

b. $\dfrac{x}{x - 1} - \dfrac{3}{2x + 5} = 1$

Solution The replacement set for the variable in an equation includes all the real numbers for which each term in the equation denotes a real number.

a. A fraction does not define a real number if its denominator is 0. Hence, 2 cannot be in the replacement set for y. We can specify the replacement set of all real numbers except 2 as $y \neq 2$.

b. The fraction $x/(x - 1)$ does not denote a real number if $x = 1$. Likewise, $3/(2x + 5)$ is not a real number if $x = -\frac{5}{2}$. Thus the replacement set for x is the set of all real numbers except 1 and $-\frac{5}{2}$. We indicate the set as $x \neq 1, -\frac{5}{2}$. ■

The **solution set** of an equation is the subset of real number replacements for which the left and right members of the equation are equal. If the solution set of an equation equals its replacement set, the equation is called an **identity.**

Example 2 Solve $6x + 5 = 9x - 15 - 3x + 20$.

Solution

$$6x + 5 = 9x - 15 - 3x + 20$$
$$6x + 5 = (9x - 3x) + (20 - 15)$$
$$6x + 5 = 6x + 5$$

It is sufficient to recognize that, in solving an equation by generating a sequence of equivalent equations, the restrictions on the replacement set of the original equation hold for all the equations in the sequence.

Example 6 Solve $x/(x - 3) - 8 = 3/(x - 3)$.

Solution

$$\begin{aligned} \frac{x}{x-3} - 8 &= \frac{3}{x-3}, \quad x \neq 3 \\ \left(\frac{x}{x-3} - 8\right)(x-3) &= \left(\frac{3}{x-3}\right)(x-3) \\ x - 8x + 24 &= 3 \\ -7x &= -21 \\ x &= 3 \end{aligned}$$

Multiplying both sides of the equation by $x - 3$ produces an equivalent equation only if $x \neq 3$. Hence, the last equation in the sequence should read $x = 3$, $x \neq 3$. We conclude that the solution set of the given equation is the empty set. ■

Linear Equations

The process of generating equivalent equations led us, in examples 4, 5, and 6, to produce a linear equation. An equation that has the form,

$$ax + b = 0, \quad a \neq 0, \tag{1}$$

or that is equivalent to (1), is called a **linear equation.** The solution set of a linear equation has exactly one element.

$$\begin{aligned} ax + b &= 0 \\ x &= -\frac{b}{a} \end{aligned}$$

It will be shown in section 1.5 that a second degree polynomial equation, an equation of the form, $ax^2 + bx + c = 0$, $a \neq 0$, has exactly two solutions. And in chapter 4 it will be shown that a third degree polynomial equation has exactly three solutions, a fourth degree polynomial equation has exactly four solutions, etc.

Mathematical Models

Linear equations often serve as mathematical models of problems, stated in words, that come from almost every area of human endeavor. As an illustration of such a "word problem," consider the following:

> A certain design requires that a 16-inch line segment be divided into two segments. One of the segments is to be three times as long as the other segment. How long must each segment be?

Although the design problem is simple, we employ a relatively lengthy five-step procedure to solve it. Our purpose is to outline a method of solution that is applicable to all word problems. As you become more proficient at solving word problems, you will find yourself shortcutting and refining the procedure.

Step 1 Read the problem carefully and locate the sentence or phrase that tells you what it is you are to find. Prompt yourself with the question, "What do I have to find?"

The length of each segment is what we must find.

Step 2 Call what you have to find, the unknown number or numbers. Assign a variable to represent one of the unknown numbers. If possible, represent the other unknown numbers in terms of the same variable. A sketch is helpful in expressing the unknowns in terms of the variable.

x $3x$

16"

Let x be the length of the shorter segment. Then, $3x$ must be the length of the longer segment.

Step 3 Construct a mathematical model for the problem.

The mathematical model of the design problem is a linear equation. It states that the sum of the lengths of the two segments is 16 inches.

$$x + 3x = 16$$

Step 4 Determine the solution set of the mathematical model. Be aware that the context of the problem may restrict the replacement set for the variable.

We note that x represents the length of a segment. Its replacement set is therefore the set of positive real numbers.

$$4x = 16$$
$$x = 4$$

If $x = 4$, then $3x = 12$.

Step 5 Interpret the model's solutions relative to the problem that produced the model. Sometimes a solution of the model is not a solution of the problem.

The lengths of the segments are 4 inches and 12 inches.

The five step procedure is used to solve the word problems in the examples that follow.

Example 7 A plane flies between cities A and B. The flight from A to B, against a 50 mph headwind, takes 1.5 hours. The flight from B to A, with a 50 mph tailwind, takes 1 hour and 10 minutes. The pilot maintains the same airspeed on both legs of the flight. Determine the plane's airspeed.

Solution

Step 1 "Determine the plane's airspeed" tells us what we must find.

Step 2 Let s represent the plane's airspeed. The sketch below suggests how to construct a mathematical model of the problem.

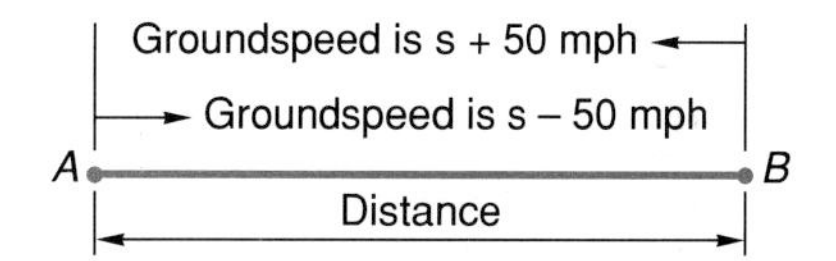

Step 3 We use the relationship between distance, speed, and time to construct a mathematical model.

$$\text{distance} = \text{speed} \cdot \text{time}, \quad d = st$$

The travel time from A to B is $\frac{3}{2}$ hours. Hence, the distance between A and B is $(s - 50)(\frac{3}{2})$. The travel time between B and A is $\frac{7}{6}$ hours. Hence, the distance between A and B is $(s + 50)(\frac{7}{6})$. Our mathematical model is

$$(s - 50)(\tfrac{3}{2}) = (s + 50)(\tfrac{7}{6}).$$

Step 4 Solve the model.
The replacement set for s is the set of nonnegative real numbers. (Why?)

$$\begin{aligned} 6[(s - 50)(\tfrac{3}{2})] &= 6[(s + 50)(\tfrac{7}{6})] \\ 9s - 450 &= 7s + 350 \\ 2s &= 800 \\ s &= 400 \end{aligned}$$

Step 5 Interpret the model's solution relative to the original problem.
The plane's airspeed is 400 miles per hour. ■

Example 8 A 70% acid solution is to be mixed with a 45% acid solution. How many liters of the 70% solution and how many liters of the 45% solution are required to produce 30 liters of a 60% solution?

Solution

Step 1 "How many liters of a 70% solution" and "How many liters of a 45% solution" identify what we are to find.

Step 2 Let x be the number of liters of the 70% acid solution. Then $30 - x$ is the number of liters of the 45% acid solution.

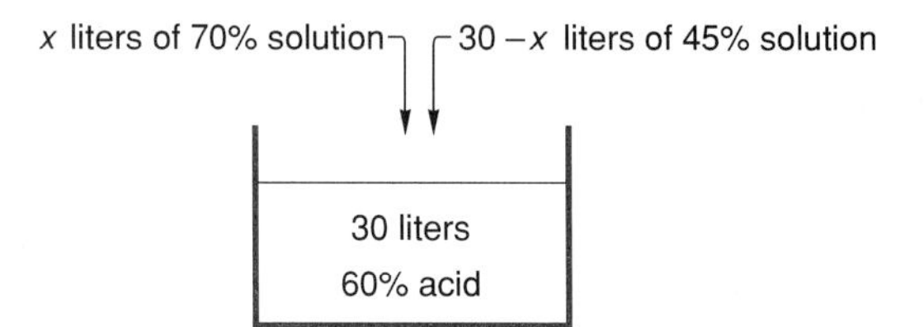

Step 3 The sketch in step 2 suggests how to construct a mathematical model of the problem.
The amount of acid contributed by the 70% solution, $0.7x$, plus the amount of acid contributed by the 45% solution, $0.45(30 - x)$, equals the amount of acid in the 60% solution, $0.6(30)$.

$$0.7x + 0.45(30 - x) = 0.6(30)$$

Step 4

$$\begin{aligned} 100[0.7x + 0.45(30 - x)] &= 100[0.6(30)] \\ 70x - 45x + 1350 &= 1800 \\ 25x &= 450 \\ x &= 18 \\ 30 - x &= 12 \end{aligned}$$

Step 5 Eighteen liters of the 70% solution and 12 liters of the 45% solution are required to produce 30 liters of a 60% acid solution. ■

Example 9 A $\frac{1}{2}$ horsepower pump can fill a tank in 3 hours. A $\frac{3}{4}$ horsepower pump fills the tank in 2 hours. How long does it take to fill the tank if the pumps are operating together?

Solution

Steps 1 and 2 Let x be the number of hours it takes to fill the tank when both pumps are operating together. Since the $\frac{1}{2}$ horsepower pump fills the tank in 3 hours, it fills $\frac{1}{3}$ of a tank per hour. The $\frac{3}{4}$ horsepower pump fills $\frac{1}{2}$ a tank per hour.

Step 3 As it takes x hours to fill the tank, we can write,
x hours ($\frac{1}{3}$ tank/hour + $\frac{1}{2}$ tank/hour) = 1 tank.

Steps 4 and 5

$$x\left(\frac{1}{3} + \frac{1}{2}\right) = 1$$

$$\frac{5x}{6} = 1$$

$$x = \frac{6}{5}$$

The two pumps, working together, take $\frac{6}{5}$ hours to fill the tank. ■

Exercise Set 1.1

(A)

Show that each equation is an identity.

1. $3x - 5 = 7x + 12 - 17 - 4x$
2. $8x + 3x - x + 2 = 12x - 2x - 3 + 5$
3. $(x - 1)(x + 3) + 3 = x^2 + 2x$
4. $6 + x^2 = (2x - 1)(x + 5) + 11 - 9x - x^2$

Solve each equation.

5. $4[x - 5(2x + 3) - 3x] = 6x + 24$
6. $-2(3x - 1) + 4x - 5 = -3[(2x - 1) - (6x + 3)]$
7. $4[y + 3y(y + 1) - 2y^2] = 4y^2 - 3y + 8$
8. $2[2y + y(5y - 4) - 2y^2] = y - 15 + 6y^2$
9. $(w - 1)(w + 4) = w^2$
10. $w(w - 3) = w^2 + 12$
11. $(2x - 3)(5x + 1) = (10x + 6)(x + 4)$
12. $(3x + 2)(4x - 1) = (6x - 5)(2x + 2)$
13. $(y - 4)(y - 3) = (y - 4)(5y - 11)$
14. $y(2y - 5) = (y + 3)y$
15. $(6z + 15)(z - 7) = (2z - 13)(7 - z)$
16. $(2z - 5)(z + 5) = (5 - 2z)(4z + 14)$
17. $\frac{2x + 1}{3} - \frac{5x - 1}{2} = 0$
18. $\frac{8x - 7}{4} = \frac{x + 5}{3}$
19. $\frac{y}{y - 2} - 3 = \frac{2}{y - 2}$
20. $\frac{y + 2}{y - 3} = \frac{5}{y - 3} - 3$
21. $10 - \frac{6}{w + 5} = \frac{2w}{w + 5}$
22. $\frac{3}{7 - 2w} + 2 = \frac{w - 5}{2w - 7}$
23. A 22-inch line segment is to be divided into two segments, one of which is 7 times as long as the other. Find the length of each segment.
24. Are there 3 consecutive even integers whose sum is 92?
25. Are there 3 consecutive even integers whose sum is 144?
26. Are there 3 consecutive integers, two odd and one even, whose sum is 510?
27. A farmer wishes to enclose a flat rectangular region with 120 feet of fencing. The length of the rectangle is to be twice its width. What is the area of the enclosed region?
28. How many gallons of a 40% salt solution must be added to 60 gallons of a 15% salt solution to obtain a 20% salt solution?
29. There are 14 problems on a test, some worth 15 points, some worth 10 points. How many 10 and 15 point problems can there be if the test has a total of 200 points?
30. An investor borrows a total of $80,000 from two different sources. He is able to obtain an 11.8% annual rate on the money borrowed from one source and a 13% rate on the money borrowed from the other source. The total annual interest on the borrowed money is $9,824. How much money is borrowed at each rate?

31. Two families plan to caravan to a camping site. One family's departure is delayed $\frac{1}{2}$ hour. Assume that the cars transporting the families start at the same place and travel the same route. If the first car to leave travels 60 mph, how long does it take the second car, traveling 65 mph, to catch the first car? How far from the starting point will both cars be when the second car catches the first car?

32. A student's grades on his first four exams are 63, 85, 79, and 93. What grade does he need on the next exam to secure an 80 average for his five exams?

33. Each base angle in an isosceles triangle is 15° more than the third angle in the triangle. What is the measure of the third angle?

34. By decreasing the lengths of the sides of a square by 4 inches, the area of the square is reduced by 96 square inches. What is the length of a side of the original square?

35. How many liters of a 40% alcohol solution must be added to 12 liters of a 25% alcohol solution to obtain a 32% alcohol solution?

36. How many ounces of an alloy containing 35% silver must be melted with an alloy containing 55% silver to obtain an alloy with 44% silver?

37. A girl cycled from her home to a bus station at an average speed of 15 mph. Fifteen minutes after arriving at the station she was taken by bus, at an average speed of 50 mph, to her destination. The girl traveled a total of 102 miles. She reached her destination 2 hours and 45 minutes after leaving home. How far did she cycle?

38. Fifty thousand dollars is invested in a certificate of deposit at 8%. How much money must be invested in municipal bonds paying 12% interest so that the total interest earned on the two investments is $10,000?

(B)

39. Two grades of coffee, A and B, are mixed to give a commercial grade C that sells for $1.40 per pound. How many pounds of grade A coffee at $1.20 a pound must be mixed with grade B coffee at $1.80 a pound to obtain 200 pounds of grade C coffee?

40. One copying machine produces 500 copies in 10 minutes; a second copying machine produces 500 copies in 12 minutes. If both machines are employed on a rush job, how long would it take to produce 500 copies?

41. One of two pipes, A or B, is usually used to fill a pool. Using only pipe A, the pool is filled in 4 hours. Using only pipe B, the pool is filled in 6 hours. How long will it take to fill the pool if both pipes are used?

42. How many quarts of a 10-quart mixture that is 20% alcohol must be replaced with pure alcohol to produce a 10-quart mixture that is 30% alcohol?

43. The average speed of a car trip from city A to city B is 55 mph. What must the average speed on the return trip to city A be so that the average speed for the round trip is 60 mph?

44. A jogs a 9-minute mile and B jogs an 11-minute mile. They start jogging together in the same direction on a $\frac{1}{4}$ mile track. If they jog at their respective rates, how many minutes pass after their start before A is again jogging abreast of B?

45. All the conditions in exercise 44 are the same except that A and B jog in opposite directions. How long after they start jogging will A and B pass each other?

46. A and B start jogging together in the same direction on a $\frac{1}{4}$ mile track. A laps B after jogging 3 and $\frac{1}{3}$ laps. Compare the jogging rates of A and B.

47. One pump takes two more hours than a second pump to fill a tank. If both pumps, working together, can fill the tank in 1 hour and 20 minutes, how long does it take each pump, working alone, to fill the tank?

48. A two-pound box of nut mix consists of $\frac{1}{2}$ pound of cashews, 1 pound of pecans, and $\frac{1}{2}$ pound of hazel nuts. The cashews cost twice as much as the hazel nuts. The pecans are $\frac{2}{3}$ the cost of the cashews. The $6.80 price for the 2-pound box of nut mix reflects a 60% markup on the cost of the nuts. What is the per pound cost of each type of nut in the mix?

49. Consider equations $A(x) = B(x)$ and $A(x) + C(x) = B(x) + C(x)$, where $A(x)$, $B(x)$, and $C(x)$ are expressions in x. For example, in the equations $6x - 4 = 2x - 18$ and $(6x - 4) + (4 - 2x) = (2x - 18) + (4 - 2x)$, we might let $A(x) = 6x - 4$, $B(x) = 2x - 18$, and $C(x) = 4 - 2x$. We want to show

that for any $A(x)$, $B(x)$, and $C(x)$, $A(x) = B(x)$ and $A(x) + C(x) = B(x) + C(x)$ are equivalent equations. This can be done by showing that every solution of the first equation is also a solution of the second, and that every solution of the second equation is also a solution of the first. We will then have shown that the solution sets of the two equations are equal. Fill in the missing details in the following proof. Let r be a solution of $A(x) = B(x)$. If x is replaced by r, then $A(x)$, $B(x)$, and $C(x)$ become real numbers that can be denoted as $A(r)$, $B(r)$, and $C(r)$ respectively. Let $A(r) = a$, $B(r) = b$, and $C(r) = c$; then $a = b$ and it follows that $a + c = b + c$. We have shown that every solution of $A(x) = B(x)$ is also a solution of **(a)**. Now let p be a solution of $A(x) + C(x) = B(x) + C(x)$. If $A(p) = m$, $B(p) = n$, and $C(p) = k$, then $m + k = n + k$. It follows that **(b)**, and that p is also a solution of **(c)**. Since every solution of $A(x) = B(x)$ is also a solution of $A(x) + C(x) = B(x) + C(x)$, and vice versa, then the two equations are equivalent.

50. Use exercise 49 as a model to prove that if left and right members of an equation are multiplied by a nonzero expression, then the resulting equation is equivalent to the original equation.

1.2 Exponents

Exponential notation simplifies the writing of a product in which all the factors are the same. We prefer to write 3^5 rather than $3 \cdot 3 \cdot 3 \cdot 3 \cdot 3$. But if we can give meaning to 3^5, it is natural to ask if we can define the symbol 3^{-5}, or if we can define the symbol $3^{4/5}$. Integer and rational number exponents will be defined in this section. Recall that the set of integers is $\{\ldots -3,-2,-1,0,1,2,3 \ldots\}$, and the set of rational numbers consists of numbers of the form p/q, where p and q are integers and $q \neq 0$. The letter J will denote the set of integers, and Q will denote the set of rational numbers.

If b is a real number and n is a natural number, then

$$\overbrace{b \cdot b \cdot b \cdot \quad \cdots \quad \cdot b}^{n \text{ factors}} = b^n.$$

The number b is called the **base**, n is called the **exponent**, and b^n is called the **nth power** of b. The familiar properties of natural number exponents are listed below.

Properties of Natural Exponents

For real numbers b and c, and natural numbers m and n:

I. $b^m \cdot b^n = b^{m+n}$

II. $\dfrac{b^m}{b^n} = b^{m-n}, \quad m > n, \quad b \neq 0$

III. $(b^m)^n = b^{mn}$

IV. $(bc)^n = b^n c^n$

V. $\left(\dfrac{b}{c}\right)^n = \dfrac{b^n}{c^n}, \quad c \neq 0$

Each of the five properties follows directly from the definition of natural number exponents.

$$\text{I.}\quad b^m \cdot b^n = (\overbrace{b \cdot b \cdot \cdots \cdot b}^{m \text{ factors}})(\overbrace{b \cdot b \cdot \cdots \cdot b}^{n \text{ factors}}) = (\overbrace{b \cdot b \cdot b \cdot \cdots \cdot b \cdot b}^{m+n \text{ factors}}) = b^{m+n}.$$

$$\text{III.}\quad (b^m)^n = \overbrace{(\overbrace{b \cdot b \cdot \cdots \cdot b}^{m \text{ factors}}) \cdots (\overbrace{b \cdot b \cdot \cdots \cdot b}^{m \text{ factors}})}^{n \text{ groups of } m \text{ factors}} = (\overbrace{b \cdot b \cdot \cdots \cdot b}^{mn \text{ factors}}) = b^{mn}.$$

The other properties can be similarly verified (see exercise 89).

Integral Exponents

To extend the properties of natural number exponents to all integral exponents, we must give meaning to the symbols b^0 and b^{-n}, in which $b \neq 0$ and n is a natural number.

If Property II of natural number exponents is to apply to all integral exponents, we must be able to write

$$\frac{b^n}{b^n} = b^{n-n} = b^0.$$

But $b^n/b^n = 1$. Thus, we can agree that if $b \neq 0$, then

Definition

$$b^0 = 1.$$

Example 1

a. $8^0 = 1$

b. $(-x)^0 = 1, \quad x \neq 0$

c. $\left(\dfrac{x^2y}{z^3}\right)^0 = 1, \quad x,y,z \neq 0$

If Property I of natural number exponents is to extend to all integral exponents, we must be able to write

$$b^n \cdot b^{-n} = b^{n+(-n)} = b^0 = 1.$$

But $b^n \cdot 1/b^n = 1$. Therefore b^{-n} and $1/b^n$ name the same number. We conclude that for $b \neq 0$, and n a natural number,

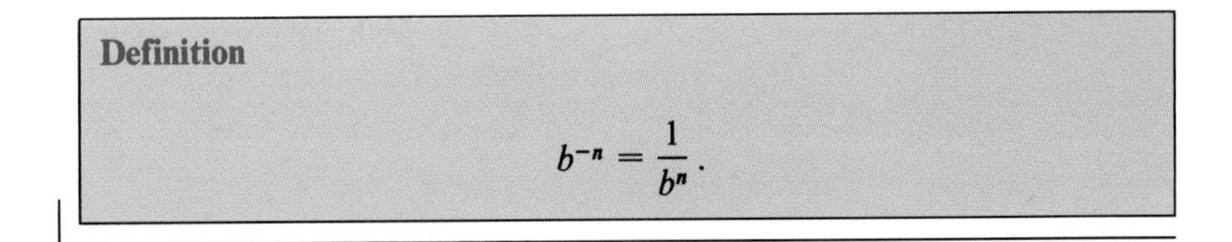

Definition

$$b^{-n} = \frac{1}{b^n}.$$

Example 2

a. $3^{-2} = \dfrac{1}{3^2}$

b. $x^{-1} = \dfrac{1}{x}, \quad x \neq 0$

c. $\left(\dfrac{1}{2}\right)^{-3} = \dfrac{1}{\left(\dfrac{1}{2}\right)^3} = \dfrac{1}{\dfrac{1}{8}} = 8$

d. $(x^4y)^{-2} = \dfrac{1}{(x^4y)^2} = \dfrac{1}{x^8y^2}, \quad x,y \neq 0$

The definition, $b^{-n} = 1/b^n$, enables us to extend the properties of natural number exponents to negative integer exponents. In the extensions of Properties I and III, it is assumed that m and n are positive integers.

$$\text{I.} \quad b^{-m} \cdot b^{-n} = \frac{1}{b^m} \cdot \frac{1}{b^n} = \frac{1}{b^{m+n}} = b^{-(m+n)} = b^{-m-n}$$

$$\text{III.} \quad (b^{-m})^{-n} = \frac{1}{(b^{-m})^n} = \frac{1}{\left(\dfrac{1}{b^m}\right)^n} = \frac{1}{\dfrac{1}{b^{mn}}} = b^{mn} = b^{(-m)(-n)}$$

The remaining properties can be extended in the same way (exercise 90). We can therefore state that

Properties of Integer Exponents

If b and c are nonzero real numbers and m and n are integers, then

I. $b^m \cdot b^n = b^{m+n}$

II. $\dfrac{b^m}{b^n} = b^{m-n}$

III. $(b^m)^n = b^{mn}$

IV. $(bc)^n = b^nc^n$

V. $\left(\dfrac{b}{c}\right)^n = \dfrac{b^n}{c^n}$

Example 3 Write each expression so that each variable appears at most once and all exponents are positive.

a. x^2x^{-4}

b. $\left(\dfrac{x^2}{x^3}\right)^{-5}$

c. $\left(\dfrac{4^0xy^{-2}}{5x^2y^3}\right)^{-2}$

Solution

a. $x^2x^{-4} = x^{2+(-4)} = x^{-2} = \dfrac{1}{x^2}$

b. $\left(\dfrac{x^2}{x^3}\right)^{-5} = (x^{2-3})^{-5} = (x^{-1})^{-5} = x^5$

c.
$$\left(\frac{4^0xy^{-2}}{5x^2y^3}\right)^{-2} = \left[\left(\frac{1}{5}\right)\left(\frac{x}{x^2}\right)\left(\frac{y^{-2}}{y^3}\right)\right]^{-2}$$
$$= \left(\frac{1}{5}\right)^{-2}(x^{-1})^{-2}(y^{-5})^{-2}$$
$$= 25x^2y^{10}$$

■

Exponents in the Form $1/n$

If the symbol $b^{1/2}$ is to be defined so that the properties of integral exponents can be applied to it, then it must follow from Property III that

$$(b^{1/2})^2 = b^{(1/2)(2)} = b.$$

Since the *square* of $b^{1/2}$ is b, then $b^{1/2}$ is a **square root** of b. Similarly, if Property III is to apply to $b^{1/n}$, $n \in N$, then we must have that

$$(b^{1/n})^n = b^{(1/n)(n)} = b,$$

and that $b^{1/n}$ is an $\boldsymbol{n^{th}}$ **root** of b.

Not all real numbers have a real n^{th} root. A negative number has no real number square root because the product of a real with itself must be nonnegative. In fact, no negative real number has an nth root if n is an even number. (Why?) Positive numbers have two real n^{th} roots if n is even. Thus, 2 and -2 are fourth roots of 16. Does the symbol $16^{1/4}$ denote 2 or -2? In order that $b^{1/n}$ be an unambiguous symbol, we insist that it denote the principal n^{th} root of b.

Definition

The principal n^{th} root of b, denoted as $b^{1/n}$ is:

1. The positive n^{th} root of b if b is positive;
2. 0 if b is 0;
3. the negative n^{th} root of b if b is negative and n is odd.

Example 4

a. $25^{1/2} = 5$. The negative square root of 25 is denoted as $-(25^{1/2})$.

b. $81^{1/4} = 3$; $\quad -(81^{1/4}) = -3$.

c. $27^{1/3} = 3$; $\quad (-27)^{1/3} = -3$.

d. $(-81)^{1/4}$ does not define a real number, since there is no real number that, when multiplied by itself four times, yields a negative product.

Exponents in the Form m/n

Having defined $b^{1/n}$ so that it names a unique real number, we can now write the mth power of $b^{1/n}$, where m is an integer, as $(b^{1/n})^m$. The symbol, $(9^{1/2})^3$, denotes the cube of the square root of 9, or 27. Rather than take the root first and then raise to a power, we can reverse the order of the operations. The square root of the cube of 9 is symbolized as $(9^3)^{1/2}$. This number is also 27.

If the properties of integral exponents are to apply to $(b^{1/n})^m$ or to $(b^m)^{1/n}$, we must have that

$$(b^{1/n})^m = b^{(1/n)(m)} = b^{m/n} \quad \text{and} \quad (b^m)^{1/n} = b^{(m)(1/n)} = b^{m/n}. \tag{1}$$

Equations (1) suggest that

Definition

$$b^{m/n} = (b^{1/n})^m = (b^m)^{1/n}, \quad m \in J, \quad n \in N. \tag{2}$$

However, (2) is true for all integers m and natural numbers n only if $b > 0$ (see exercise 91). If we attempt to evaluate $(-9)^{2/4}$ with (2), we have on the one hand $((-9)^{1/4})^2$. This symbol does not define a real number because there is no real number fourth root of -9. On the other hand, $((-9)^2)^{1/4} = (81)^{1/4} = 3$. Thus, we shall do only the rooting and powering, as indicated by the symbols in (2), if $\boldsymbol{b > 0}$. Finally, we note that $b^{-(m/n)}$ is the reciprocal of $b^{m/n}$ because

$$b^{-(m/n)} = b^{-m/n} = (b^{1/n})^{-m} = \frac{1}{(b^{1/n})^m} = \frac{1}{b^{m/n}}.$$

Example 5

a. $8^{2/3} = (8^{1/3})^2 = 2^2 = 4$; or $8^{2/3} = (8^2)^{1/3} = (64)^{1/3} = 4$

b. $32^{-2/5} = (32^{1/5})^{-2} = 2^{-2} = \dfrac{1}{4}$; or

$$32^{-2/5} = (32^{-2})^{1/5} = \left(\frac{1}{1024}\right)^{1/5} = \frac{1}{4}$$

c. $16^{-(3/2)} = \dfrac{1}{16^{3/2}} = \dfrac{1}{4^3} = \dfrac{1}{64}$

Rational Number Exponents

Let us evaluate the product $64^{1/2} \cdot 64^{2/3}$, in which each factor is a power with a rational number exponent.

$$64^{1/2} \cdot 64^{2/3} = 8 \cdot 16 = 128$$

Now, $\frac{1}{2} = \frac{3}{6}$, and $\frac{2}{3} = \frac{4}{6}$. Also, $64^{3/6} = (64^{1/6})^3 = 2^3 = 8$, and $64^{4/6} = 2^4 = 16$. Therefore,

$$64^{3/6} \cdot 64^{4/6} = 128.$$

Lastly, we have that $64^{7/6} = (64^{1/6})^7 = 2^7 = 128$. Thus,

$$64^{1/2} \cdot 64^{2/3} = 64^{3/6} \cdot 64^{4/6} = 64^{7/6} = 128.$$

The calculations suggest that $64^{1/2} = 64^{3/6}$, and that $64^{2/3} = 64^{4/6}$. They further suggest that to multiply powers with rational exponents we must rewrite the rational exponents so that they have the same denominator, and then add the exponents.

It can be shown (exercise 92) that if $m/n = p/q$ and $b > 0$, then

$$b^{m/n} = b^{p/q}. \tag{3}$$

The last result, together with (2), enables us to extend the properties of integral exponents to rational number exponents (exercises 93 and 94).

Properties of Rational Exponents

If m, n, p, and q are integers and b and c are positive real numbers, then

I. $b^{m/n} \cdot b^{p/q} = b^{(m/n) + (p/q)}$

II. $\dfrac{b^{m/n}}{b^{p/q}} = b^{(m/n) - (p/q)}$

III. $(b^{m/n})^{p/q} = b^{mp/nq}$

IV. $(bc)^{m/n} = b^{m/n} \cdot c^{m/n}$

V. $\left(\dfrac{b}{c}\right)^{m/n} = \dfrac{b^{m/n}}{c^{m/n}}$

Example 6

a. $36^{2/4} = 36^{1/2} = 6$

b. $32^{-6/10} = 32^{-3/5} = \dfrac{1}{32^{3/5}} = \dfrac{1}{(32^{1/5})^3} = \dfrac{1}{2^3} = \dfrac{1}{8}$

c. $25^{2/3} \cdot 25^{5/6} = 25^{(4/6 + 5/6)} = 25^{9/6} = 25^{3/2} = 125$

d. $\dfrac{16^{3/4}}{16^{1/2}} = 16^{(3/4 - 2/4)} = 16^{1/4} = 2$

e. $(4^{5/3})^{-3/2} = 4^{(5/3 \cdot -3/2)} = 4^{-5/2} = \dfrac{1}{32}$

f. $(8 \cdot 27)^{2/3} = 8^{2/3} \cdot 27^{2/3} = 4 \cdot 9 = 36$

g. $\left(\frac{81}{16}\right)^{3/4} = \frac{81^{3/4}}{16^{3/4}} = \frac{27}{8}$

Example 7 Write the expression $x^2y^{1/2}z^{-2/3}/x^{3/4}y^{-1/2}z^{-1}$ so that each variable occurs only once and all exponents are positive.

Solution

$$\frac{x^2y^{1/2}z^{-2/3}}{x^{3/4}y^{-1/2}z^{-1}} = \frac{x^2}{x^{3/4}} \cdot \frac{y^{1/2}}{y^{-1/2}} \cdot \frac{z^{-2/3}}{z^{-1}}$$
$$= x^{(8/4 - 3/4)}y^{(1/2 - (-1/2))}z^{(-2/3 - (-3/3))}$$
$$= x^{5/4}z^{1/3}y$$ ■

Example 8 Multiply $(x + a)^{5/3}$ by $(x + a)^{1/4}$.

Solution

$$(x + a)^{5/3}(x + a)^{1/4} = (x + a)^{(5/3 + 1/4)} = (x + a)^{23/12}$$ ■

Radical Notation

Besides $b^{1/n}$, $\sqrt[n]{b}$ is another symbol for the principal n^{th} root of b. The radical, $\sqrt{\ }$, is usually introduced in a basic algebra course to indicate square root. The number n associated with the radical is called the **index** of the radical. The index tells us which principal root is being called for. If no index appears, such as in the symbol $\sqrt{b}$, we seek the principal square root. The expression under the radical is called the **radicand.**

A number of useful properties of radical notation follow from the properties of rational exponents and the following

Definition

$$\sqrt[n]{b} = b^{1/n}. \tag{4}$$

Properties of Radical Notation

If $x \geq 0$, $y > 0$, and $n \varepsilon N$, then

I. $(\sqrt[n]{x})^n = (x^{1/n})^n = x$

II. $\sqrt[n]{xy} = (xy)^{1/n} = x^{1/n}y^{1/n} = \sqrt[n]{x} \cdot \sqrt[n]{y}$

III. $\sqrt[n]{\frac{x}{y}} = \left(\frac{x}{y}\right)^{1/n} = \frac{x^{1/n}}{y^{1/n}} = \frac{\sqrt[n]{x}}{\sqrt[n]{y}}$

Example 9

a. $\sqrt{36} = (36)^{1/2} = 6$

b. $-\sqrt[4]{16} = -(16^{1/4}) = -2$

c. $\sqrt[3]{-27} = (-27)^{1/3} = -3$

d. $\sqrt{49} = \sqrt{7^2} = 7$

e. $\sqrt[3]{27x^6y^9} = \sqrt[3]{(3x^2y^3)^3} = 3x^2y^3$

Example 10 Simplify each radical expression.

a. $\sqrt{98}$

b. $\sqrt[3]{135xy^5}$

c. $\sqrt{\dfrac{18y^5}{x^3}}$

Solution We use Properties I, II, and III to rewrite each radical expression so that its radicand is simpler.

a. $$\begin{aligned}\sqrt{98} &= \sqrt{49 \cdot 2}\\ &= \sqrt{49} \cdot \sqrt{2}\\ &= 7\sqrt{2}\end{aligned}$$

b. $$\begin{aligned}\sqrt[3]{135xy^5} &= \sqrt[3]{(27y^3)(5xy^2)}\\ &= \sqrt[3]{27y^3}\sqrt[3]{5xy^2}\\ &= 3y\sqrt[3]{5xy^2}\end{aligned}$$

c. $$\begin{aligned}\sqrt{\frac{18y^5}{x^3}} &= \frac{\sqrt{(9y^4)(2y)}}{\sqrt{x^2 \cdot x}}\\ &= \frac{\sqrt{9y^4}\sqrt{2y}}{\sqrt{x^2}\sqrt{x}}\\ &= \frac{3y^2\sqrt{2y}}{x\sqrt{x}}\\ &= \left(\frac{3y^2}{x}\right)\left(\sqrt{\frac{2y}{x}}\right)\end{aligned}$$

■

The relationships $b^{m/n} = (b^{1/n})^m = (b^m)^{1/n}$ have their counterparts in radical notation.

$$(\sqrt[n]{b})^m = (b^{1/n})^m = (b^m)^{1/n} = \sqrt[n]{b^m}$$

Thus, if $b \geq 0$,

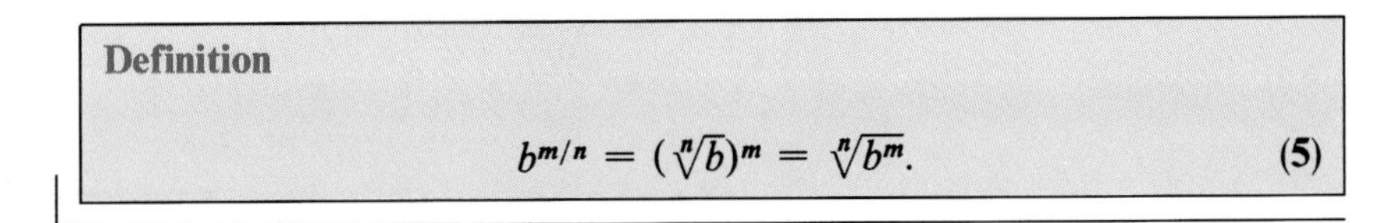

Definition

$$b^{m/n} = (\sqrt[n]{b})^m = \sqrt[n]{b^m}. \tag{5}$$

Example 11

a. $9^{3/2} = (\sqrt{9})^3 = 27;$ or $9^{3/2} = \sqrt{9^3} = \sqrt{729} = 27$

b. $(-8)^{2/3} = (\sqrt[3]{-8})^2(-2)^2 = 4;$
or $(-8)^{2/3} = \sqrt[3]{(-8)^2} = \sqrt[3]{64} = 4$

Example 12 Write each radical expression in exponential form.

a. $\sqrt[5]{(x^2 - 3)^4}$

b. $\sqrt[3]{(x - 2)^2(x + 5)^7}$

Solution

a. From (5), $\sqrt[5]{(x^2 - 3)^4} = (x^2 - 3)^{4/5}$

b. $\sqrt[3]{(x - 2)^2(x + 5)^7} = [(x - 2)^2(x + 5)^7]^{1/3}$
$= (x - 2)^{2/3}(x + 5)^{7/3}$ ■

If numerator and denominator of the expression $2/\sqrt{3}$ are multiplied by $\sqrt{3}$, we obtain the following:

$$\frac{2}{\sqrt{3}} \cdot \frac{\sqrt{3}}{\sqrt{3}} = \frac{2\sqrt{3}}{\sqrt{3} \cdot \sqrt{3}} = \frac{2\sqrt{3}}{3}.$$

The expression $2/\sqrt{3}$, with a radical in its denominator, has been changed to $2\sqrt{3}/3$, an expression whose denominator is free of radicals. We say that the denominator of $2/\sqrt{3}$ has been **rationalized.**

Example 13 Rationalize the denominator of the first three expressions, and rationalize the numerator of the last expression.

a. $\dfrac{6}{\sqrt{x}}$

b. $\dfrac{4}{\sqrt[3]{5}}$

c. $\dfrac{1}{\sqrt{2x}+9}$

d. $\dfrac{\sqrt{7}}{\sqrt{5}}$

Solution Properties I, II, and III, of radical notation, are employed to obtain the desired results.

a. $\dfrac{6}{\sqrt{x}} \cdot \dfrac{\sqrt{x}}{\sqrt{x}} = \dfrac{6\sqrt{x}}{x}$

b. $\dfrac{4}{\sqrt[3]{5}} = \dfrac{4\sqrt[3]{5^2}}{\sqrt[3]{5}\cdot\sqrt[3]{5^2}} = \dfrac{4\sqrt[3]{25}}{\sqrt[3]{5^3}} = \dfrac{4\sqrt[3]{25}}{5}$

c. $\dfrac{1}{\sqrt{2x}+9} = \dfrac{1}{\sqrt{2x}+9}\cdot\dfrac{\sqrt{2x}-9}{\sqrt{2x}-9}$

$= \dfrac{\sqrt{2x}-9}{\sqrt{4x^2}-81}$

$= \dfrac{\sqrt{2x}-9}{2x-81}$

d. $\dfrac{\sqrt{7}}{\sqrt{5}} = \dfrac{\sqrt{7}\sqrt{7}}{\sqrt{5}\sqrt{7}} = \dfrac{7}{\sqrt{35}}$ ■

Exercise Set 1.2

(A)

Write each expression as a single numeral with exponent 1.

1. $(5^{-2} - 4^{-2})^{-2} \cdot 3^2$
2. $(3^{-2} \cdot 2^{-3})^{-1}$
3. $\left(\dfrac{8^3 \cdot 5^{-2}}{2^{-4}}\right)^0$
4. $\left(\dfrac{7^0}{7^{-2}} + 1\right)^{-1}$

Write each expression so that each variable appears once at most, and so that all exponents are positive.

5. x^3x^{-4}
6. $x^{-1}y^{-1}$
7. $\dfrac{1}{x^{-1}y^{-2}}$
8. $\dfrac{x^{-3}}{x^2y^{-2}}$
9. $\left(\dfrac{3x}{2y^3}\right)^{-2}$
10. $\left(\dfrac{3x^{-1}}{2y^{-3}}\right)^2$
11. $(x^3y^{-2})^{-3}$
12. $\dfrac{(x-y)^{-2}}{(x-y)^{-3}}$
13. $\left(\dfrac{2x+y}{3(x-y)^0}\right)^{-2}$
14. $\left(\dfrac{x}{y}\right)^4 y^3x^{-3}$
15. $\left(\dfrac{x^{-1}y^0z^{-3}}{xy^2z^{-2}}\right)^{-1}$
16. $\dfrac{(x-y)^0(y-x)^2}{(y-x)^{-2}}$

Find the principal n^{th} root, if it exists.

17. $16^{1/4}$
18. $(-16)^{1/4}$
19. $(-32)^{1/5}$
20. $(-64)^{1/3}$
21. $729^{1/6}$
22. $(-\frac{216}{343})^{1/3}$

For each power, write an equivalent numeral with exponent 1.

23. $16^{3/4}$
24. $27^{2/3}$
25. $16^{-3/4}$
26. $(-8)^{4/3}$
27. $(-27)^{4/3}$
28. $(-8)^{-2/3}$
29. $(\frac{1}{81})^{-3/4}$
30. $(0.01)^{-5/2}$
31. $(0.0001)^{3/4}$

Write each expression as an equivalent single numeral with exponent 1.

32. $(4^2 + 3^2)^{-3/2}$
33. $2 + 4^{1/2} - 125^{-2/3}$
34. $8^{-2/3} - 64^{-2/3}$
35. $(25^{3/2} \cdot 16^{3/4}) \div 81^{-3/4}$

For each power, write an equivalent numeral with exponent 1.

36. $8^{4/6}$
37. $256^{9/12}$
38. $(-\frac{64}{27})^{6/9}$
39. $32^{-8/10}$
40. $(\frac{1}{8})^{8/6}$
41. $(-0.008)^{-10/15}$

Write each expression so that each variable occurs only once and all exponents are positive. Assume that the replacement set for each variable is the set of positive real numbers.

42. $x^{1/2}x^{3/4}x$ **43.** $3y^{-3/4} \div 6y^{-5/6}$ **44.** $(9x^{-4}y^2)^{1/2}$

45. $(y^{-3/2}y^{5/8}y^{3/16})^{-2/3}$ **46.** $\dfrac{x^{1/3}y^{1/4}z^{1/2}}{x^{-2/3}y^2z^{-3/2}}$

47. $\left(\dfrac{x^{n+1}}{x}\right)^{1/2n}$ **48.** $\left(\dfrac{x^{-n+2}y}{x^2y^{-n+1}}\right)^{-1/n}$

Do the indicated multiplication.

49. $(x+2)^{1/4} \cdot (x+2)^{5/4}$

50. $(2x-5)^{3/2} \cdot (2x-5)^{-1/5}$

51. $(x+y)^{3/4} \cdot (x+y)^{1/12}$

52. $(y+w)^{4/3} \cdot (y+w)^{-3/5}$

53. $(x+y)^{2/3}[(x+y)^{1/3} + (x+y)^{1/2}]$

54. $x^{5/8}(x^{1/4} + x^{3/8}y)$

55. $x^{-1/2}(x^{3/2}y^2 - x^{5/2}y^3)$

56. $(y+z)^{-1/6}[(y+z)^{2/3} - 1]$

Evaluate each radical expression.

57. $\sqrt{2}\sqrt{8}$ **58.** $\dfrac{\sqrt{27}}{\sqrt{3}}$ **59.** $\sqrt[3]{2^3}\,\sqrt[3]{27}$

60. $(\sqrt[3]{-2})^3$ **61.** $\left(\dfrac{\sqrt[3]{-64}}{27}\right)^2$ **62.** $\sqrt{\left(\dfrac{4}{25}\right)^3}$

63. $\sqrt{\dfrac{1}{5^4}}$ **64.** $\sqrt{\dfrac{8}{3}} \cdot \sqrt{\dfrac{3}{4}} \cdot \sqrt{2}$

Simplify each radical expression. Assume all variables represent positive numbers.

65. $\sqrt[3]{16x^4y^3}$ **66.** $\sqrt[4]{32x^5y^6}$

67. $\sqrt{18x^2y^3z^4}$ **68.** $\sqrt[3]{\dfrac{x^3y^2}{w^6}}$

69. $\sqrt[3]{\dfrac{81w^5}{v}} \cdot \sqrt[3]{\dfrac{9v^2}{w^2}}$ **70.** $\sqrt[4]{\dfrac{x^2}{y^3}} \cdot \sqrt[4]{\dfrac{16x^3}{y^5}}$

71. $\sqrt[3]{\dfrac{y^5z^3}{125x^9}}$ **72.** $\sqrt{\dfrac{49w^5}{x^2y^4}}$

73. $\sqrt{\dfrac{8a^2b^4}{c^2}}$ **74.** $\sqrt{192w^4x^5y^3}$

Write each radical expression in exponential form.

75. $\sqrt[4]{(x+2)^3}$ **76.** $\sqrt[6]{(x^2-3)^5}$

77. $\sqrt{(y-1)^3(y+6)^5}$ **78.** $\sqrt[5]{(2y-7)^3(2y+7)^6}$

79. $\sqrt[4]{(w^2+8)^2}$ **80.** $\sqrt[6]{(z+1)^3(z^2+4z+5)^8}$

Rationalize the denominator of each expression. Assume all variables represent positive numbers.

81. $\dfrac{3}{\sqrt{3}}$ **82.** $\dfrac{xy}{\sqrt{xy}}$ **83.** $\dfrac{1}{\sqrt[3]{2}}$

84. $\dfrac{x^2}{\sqrt[3]{x}}$ **85.** $\dfrac{1}{\sqrt{x}-5}$ **86.** $\dfrac{\sqrt{x}}{\sqrt{3x}+2}$

87. $\dfrac{2y}{\sqrt[4]{y^3}}$ **88.** $\dfrac{x^2y}{\sqrt[3]{x^2y}}$

(B)

89. Properties I and III of natural number exponents are verified on page 11. Verify Properties II, IV, and V.

90. Properties I and III of integer exponents are verified on page 13. Verify Properties II, IV, and V.

91. The following is a proof of the statement, "If $b > 0$, then $(b^{1/n})^m = (b^m)^{1/n}$." Fill in the missing details. (b must be positive so that the symbols $b^{1/n}$, $(a^{nm})^{1/n}$, and $(b^m)^{1/n}$ denote real numbers.)

Suppose $a = b^{1/n}$. Then, $a^m =$ (a). If a is the nth root of b, then b is the (b) power of a. Hence, $b = a^n$, and $b^m =$ (c). As a consequence,

$$\begin{aligned}(b^m)^{1/n} &= (a^{nm})^{1/n}\\ &= \underline{(d)}\\ &= a^m.\end{aligned}$$

Now, $(b^m)^{1/n} = a^m = (b^{1/n})^m$.

92. Fill in the details of the following proof of the statement, "If $b > 0$, m, n, p, and q are integers, $m/n = p/q$, and m and n are relatively prime, then $b^{m/n} = b^{p/q}$."

If $m/n = p/q$, there must be a natural number, k, such that $p = km$ and $q =$ (a). Let $b^{1/q} = a$, then

$$b = a^q = a^{kn} = (a^k)^n.$$

Since b is the nth power of a^k, then $b^{1/n} =$ (b). As a result,

$$b^{m/n} = (b^{1/n})^m = (a^k)^m = a^{km}.$$

But $km = p$ and $a = b^{1/q}$. Therefore,

$$a^{km} = a^p = \underline{(c)} = b^{p/q}.$$

It follows that (d) $= a^{km} =$ (e).

(If m and n are not relatively prime, there are relatively prime integers r and s such that $r/s = m/n = p/q$. The above proof can be repeated to show $b^{r/s} = b^{m/n}$, and $b^{r/s} = b^{p/q}$. Thus, $b^{m/n} = b^{p/q}$.)

93. Fill in the missing reasons of the Statement-Reason proof of Property I for rational exponents.

Statement	Reason
$b^{m/n} \cdot b^{p/q} = b^{mq/nq} \cdot b^{pn/nq}$	equation (3)
$= (b^{1/nq})^{mq}(b^{1/nq})^{np}$	(a)
$= (b^{1/nq})^{mq+np}$	(b)
$= b^{(mq+np)/nq}$	equation (2)
$= b^{m/n+p/q}$	simplifying a fraction

94. Argue that if $b = a^{1/n}$ and $c = b^{1/q}$, then $c = a^{1/nq}$.

95. Use the result in exercise 94 to construct a Statement-Reason proof of Property III for rational number exponents.

1.3 Factoring

Factoring helps us solve equations and simplify complex expressions. You will become aware of the great number of applications of factoring as you move through the text. In this section, we review some factoring techniques and some common factored forms.

Recall that an expression of the form ax^n, where a and x are real numbers and n is a whole number, is called a **monomial.** The letter a is called a **numerical coefficient** of the power x^n. A sum of monomial terms is called a **polynomial.** A polynomial of two terms is called a **binomial** and a polynomial of three terms is called a **trinomial.**

Common Monomial Factors

If each term of a polynomial has a common factor, we say the polynomial has a **common monomial factor.** Thus

$$\begin{aligned} 30x^3 - 5x^2 + 5x &= \underline{5x} \cdot 6x^2 - \underline{5x} \cdot x + \underline{5x} \cdot 1 \\ &= 5x(6x^2 - x + 1). \end{aligned}$$

It is possible to find more than one factored form of a polynomial. For example,

$$x(2x + 1),\ 2x\left(x + \frac{1}{2}\right),\quad \text{and}\quad \sqrt{2}x\left(\sqrt{2}x + \frac{1}{\sqrt{2}}\right)$$

are all factorizations of $2x^2 + x$. However, a polynomial with integral coefficients has, except for signs and order of factors, exactly one **completely factored form** in which the terms in each factor have integral coefficients. This form is characterized as follows:

> **Completely Factored Form**
>
> 1. Each factor is a polynomial with integral coefficients.
> 2. No polynomial factor can be further factored into polynomial factors with integral coefficients.

Example 1 Each of the products

$$x(2x + 1), \quad (2x + 1)(x), \quad -x(-2x - 1), \quad (-2x - 1)(-x)$$

is a completely factored form of $2x^2 + x$. The forms differ only in the order in which the factors appear and the signs of the factors.

Example 2 Both $6x^2(4x + 1)$ and $6x(4x^2 + x)$ are factorizations of $24x^3 + 6x^2$. However, $6x^2(4x + 1)$ is the *complete factorization* because $4x + 1$ cannot be factored further, whereas $4x^2 + x$ can be factored further into $x(4x + 1)$.

Trinomials of the Form $ax^2 + bx + c$

The product of the binomial factors $(dx + e)$ and $(fx + g)$ is

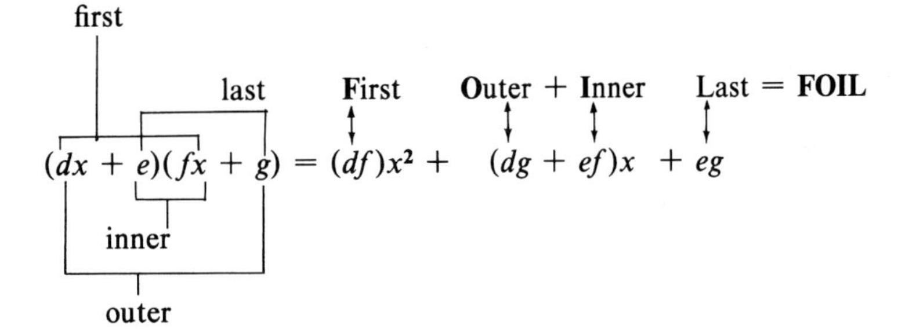

The order of the letters in the word FOIL gives us the order of multiplications for obtaining the product of two binomial factors. The product of $(dx + e)$ and $(fx + g)$ is a trinomial of the form $ax^2 + bx + c$, where $a = df$, $b = ef + dg$, and $c = eg$. Thus, to factor $ax^2 + bx + c$ into the binomial factors $(dx + e)$ and $(fx + g)$, d and f must be integral factors of a, and e and g must be integral factors of c. The next example illustrates how these observations are used to factor a trinomial by the "trial and error" method.

Example 3 Completely factor $6x^2 - 5x + 1$.

Solution We assume there are binomials $(dx + e)$ and $(fx + g)$ such that

$$(dx + e)(fx + g) = 6x^2 - 5x + 1.$$

We know that d and f must be integral factors of 6, and that e and g must be integral factors of 1. Consequently, the possible binomial factor pairs of $6x^2 - 5x + 1$ are

$$\begin{array}{ll} (3x + 1)(2x + 1) & (-3x + 1)(-2x + 1) \\ (3x - 1)(2x - 1) & (-3x - 1)(-2x - 1) \\ (6x + 1)(x + 1) & (-6x + 1)(-x + 1) \\ (6x - 1)(x - 1) & (-6x - 1)(-x - 1). \end{array}$$

Of the eight possibilities, only

$$(3x - 1)(2x - 1) = 6x^2 - 5x + 1$$

and

$$(-3x + 1)(-2x + 1) = 6x^2 - 5x + 1.$$

But $(3x - 1) = -(-3x + 1)$ and $2x - 1 = -(-2x + 1)$ so that, of the two factor pairs that yield the required product, the corresponding factors differ only in sign. ■

Example 3 illustrates the trial and error method of factoring in its most basic form. As you think about the possible factored forms, some possibilities can be eliminated directly, because they do not yield the desired middle term. Thus, factor pairs such as $(3x + 1)(2x + 1)$ and $(6x + 1)(x + 1)$ cannot be seriously considered because they do not produce a negative coefficient for the x term of the product.

Example 4 Completely factor $4x^3 - 10x^2 - 6x$.

Solution First we look for a common monomial factor.

$$4x^3 - 10x^2 - 6x = 2x(2x^2 - 5x - 3)$$

If $2x^2 - 5x - 3$ cannot be further factored, we are finished. However, our trial and error factoring procedure yields

$$2x^2 - 5x - 3 = (2x + 1)(x - 3).$$

Thus

$$4x^3 - 10x^2 - 6x = 2x(2x + 1)(x - 3).$$

■

Some trinomials of the form $ax^2 + bx + c$ cannot be completely factored.

Example 5 Completely factor $6x^2 - x + 1$.

Solution We find that none of the factor pairs obtained with the trial and error method yield the required product.

$$\begin{aligned}(3x + 1)(2x + 1) &= 6x^2 + 5x + 1,\\(3x - 1)(2x - 1) &= 6x^2 - 5x + 1,\\(6x + 1)(x + 1) &= 6x^2 + 7x + 1,\\(6x - 1)(x - 1) &= 6x^2 - 7x + 1\end{aligned}$$

As a result, $6x^2 - x + 1$ cannot be written as a product of binomial factors having only integral coefficients. ■

Example 6 Completely factor $6x^2 + 11xy + 4y^2$.

Solution Testing possible binomial factor pairs, we find that

$$6x^2 + 11xy + 4y^2 = (3x + 4y)(2x + y).$$

■

Difference of Two Squares

It is easy to verify that a difference of squares, $a^2 - b^2$, can be factored as shown in equation (1) below. You need only multiply the right side of (1) to obtain its left side.

Factoring a Difference of Squares

$$a^2 - b^2 = (a - b)(a + b) \tag{1}$$

Example 7 Completely factor $25x^2 - 16$.

Solution If we rewrite $25x^2 - 16$ as $(5x)^2 - (4)^2$, it can be compared to the left side of (1), with $5x$ playing the role of a and 4 playing the role of b. Consequently,

$$25x^2 - 16 = (5x - 4)(5x + 4).$$ ■

Example 8 Completely factor $x^4 - y^4$.

Solution Writing $x^4 - y^4$ as $(x^2)^2 - (y^2)^2$, we have

$$x^4 - y^4 = (x^2 - y^2)(x^2 + y^2).$$

But $x^2 - y^2$ is a difference of squares. Hence,

$$x^4 - y^4 = (x - y)(x + y)(x^2 + y^2).$$ ■

Sums and Differences of Cubes

Cubic polynomials in the form $a^3 + b^3$ and $a^3 - b^3$ are factored as follows:

Factoring a Sum or a Difference of Cubes

$$a^3 + b^3 = (a + b)(a^2 - ab + b^2); \tag{2}$$
$$a^3 - b^3 = (a - b)(a^2 + ab + b^2). \tag{3}$$

Verify (2) and (3) by multiplying the factors on the right side of each equation.

Example 9 Completely factor $8x^3 + 27$.

Solution If we view $8x^3 + 27$ as $(2x)^3 + (3)^3$, where $2x$ plays the role of a and 3 plays the role of b in equation (2), then

$$\begin{aligned} 8x^3 + 27 &= (2x + 3)((2x)^2 - (2x)(3) + (3)^2) \\ &= (2x + 3)(4x^2 - 6x + 9). \end{aligned}$$ ■

Example 10 Completely factor $(x + y)^3 - (x - y)^3$.

Solution Letting $(x + y)$ play the role of a and $(x - y)$ the role of b in equation (3), we write

$$\begin{aligned}(x + y)^3 - (x - y)^3 &= [(x + y) - (x - y)] \\ &\quad [(x + y)^2 + (x + y)(x - y) + (x - y)^2] \\ &= 2y(x^2 + 2xy + y^2 + x^2 - y^2 + x^2 - 2xy + y^2) \\ &= 2y(3x^2 + y^2).\end{aligned}$$

■

Factoring by Grouping

Polynomials that are not binomials or trinomials can sometimes be factored by considering particular groupings of their terms. Often a factorization can be obtained by grouping terms that have common factors.

Example 11 Completely factor $3x^3 - 2x^2 + 6x - 4$.

Solution

a. Grouping the first and third terms together and the second and fourth terms together yields

$$\begin{aligned}(3x^3 + 6x) - (2x^2 + 4) &= 3x(x^2 + 2) - 2(x^2 + 2) \\ &= (3x - 2)(x^2 + 2).\end{aligned}$$

b. Grouping the first two terms together and the last two terms together yields

$$\begin{aligned}(3x^3 - 2x^2) + (6x - 4) &= x^2(3x - 2) + 2(3x - 2) \\ &= (x^2 + 2)(3x - 2).\end{aligned}$$

■

Example 12 Completely factor $4x^2 - 9y^2 - 6y - 1$.

Solution Pairing terms as was done in example 11 does not appear to be promising. However, if we group the last three terms, we obtain

$$\begin{aligned}4x^2 - (9y^2 + 6y + 1) &= 4x^2 - (3y + 1)(3y + 1) \\ &= 4x^2 - (3y + 1)^2 \\ &= (2x)^2 - (3y + 1)^2 \\ &= [2x + (3y + 1)][2x - (3y + 1)] \\ &= (2x + 3y + 1)(2x - 3y - 1).\end{aligned}$$

■

Other Kinds of Factoring

To simplify certain expressions, we may want to factor polynomials with non-integral coefficients, or to find factors that are polynomials with non-integral coefficients. Such factorizations, unlike the complete factorizations that have been discussed throughout most of this section, are not unique.

Example 13

a. $x^2 - 2 = (x - \sqrt{2})(x + \sqrt{2})$

b. $x^3 - \dfrac{1}{8} = x^3 - \left(\dfrac{1}{2}\right)^3 = \left(x - \dfrac{1}{2}\right)\left(x^2 + \dfrac{x}{2} + \dfrac{1}{4}\right)$

Certain expressions, although not polynomials, may be treated as polynomials for the purpose of factoring.

Example 14 Factor $x^{2/3} + 6x^{1/3} + 5$.

Solution View the given expression as having the form $ay^2 + by + c$, in which $x^{1/3}$ plays the role of y. Then apply the trial and error factoring technique to the expression.

$$\begin{aligned} x^{2/3} + 6x^{1/3} + 5 &= (x^{1/3})^2 + 6(x^{1/3}) + 5 \\ &= (x^{1/3} + 5)(x^{1/3} + 1) \end{aligned}$$ ■

Example 15 Factor $y^{3/2} - 27$.

Solution We can view the expression as a difference of cubes.

$$\begin{aligned} y^{3/2} - 27 &= (y^{1/2})^3 - (3)^3 \\ &= (y^{1/2} - 3)(y + 3y^{1/2} + 9) \end{aligned}$$ ■

Example 16 Factor $x^{-2}y^{-3} - x^{-4}y^{-1}$.

Solution There is more than one common monomial factor. If we choose the monomial factor in which the exponents on x and y are the lowest valued exponents that appear on each variable, then the other factor has only positive exponents.

$$\begin{aligned} x^{-2}y^{-3} - x^{-4}y^{-1} &= x^{-4}y^{-3}(x^2) - x^{-4}y^{-3}(y^2) \\ &= x^{-4}y^{-3}(x^2 - y^2) \\ &= x^{-4}y^{-3}(x - y)(x + y) \end{aligned}$$ ■

Exercise Set 1.3

(A)

Completely factor each polynomial.

1. $8x^2 + 24x$
2. $15x^3 - 45x^2 + 30x$
3. $9y^3 + 54y^2 + 81y$
4. $32x^2y - 16xy + 48x^3y^2$
5. $x^2 + 5x + 6$
6. $3y^2 + 5y - 2$
7. $8x^2 + 38x + 35$
8. $2x^2 - 7xy - 4y^2$
9. $11w^2 + 29wz - 12z^2$
10. $9w^2 - 12w + 4$
11. $25 - 40xy + 16x^2y^2$
12. $81 + 36xy + 4x^2y^2$
13. $x^2 - 9$
14. $36y^2 - 1$
15. $25x^2z^2 - 9y^2$
16. $16x^2y^2 - 25w^2$
17. $x^4 - 16$
18. $y^4 - 81$
19. $16a^4 - 81b^4$
20. $2x^4 - 32$

21. $(xy)^2 - 1$
22. $(ab)^2 - 9$
23. $72w^2 + 13w - 15$
24. $-12z^2 - 21z + 33$
25. $ay^2 + 3ay - 5a$
26. $10x^2y - 5xy + 5y$
27. $-6y^2 - 10xy + 56x^2$
28. $-z^2 - 15wz + 54w^2$
29. $9x^2y + 45xy + 36y$
30. $48ax^2y - 80axy^2 - 100ay^3$
31. $x^3 + 27$
32. $z^3 - 125$
33. $8x^3 - a^3y^3$
34. $b^3x^3 + 8y^3$
35. $(x + 2)^3 - 8$
36. $(a + b)^3 + (a - b)^3$
37. $x^6 + 27$
38. $a^3y^6 - 64$

Factor by grouping.

39. $xy + xb + ay + ab$
40. $6ax + 3ay + 2bx + by$
41. $bx^2 + x + b^2x + b$
42. $x^3 - x^2 + xy - y$
43. $x^3 + x^2 - x - 1$
44. $y^3 + y^2 - 4y - 4$
45. $8x^2 - 10x^3 - 4 + 5x$
46. $3x^3 - 2x^2 + 6x - 4$

(B)

Factor each polynomial so that the coefficients in each factor are members of the indicated set. If such a factorization is not possible, state that the polynomial is not factorable. Recall that the letters J, Q, and R denote the sets of Integers, Rationals, and Reals, respectively. (See example 13.)

47. $y^2 + y + \frac{1}{4}, Q$
48. $\frac{x^2}{9} - \frac{x}{3} - 2, Q$
49. $3y^2 + 5y + 1, J$
50. $x^2 - 3, J$
51. $x^2 - 3, R$
52. $x^2y - 6y, R$
53. $y^3 + 2, R$
54. $(xy)^3 - 5, R$

Factor each expression. (See examples 14 and 15.)

55. $w + 2w^{1/2} + 1$
56. $3x^{2/3} - 5x^{1/3} - 2$
57. $4y^{4/5} + 13y^{2/5} + 3$
58. $2x^{4/5} + 5(xy)^{2/5} - 3y^{4/5}$
59. $x^{3/2} - 1$
60. $8y^{3/2} - 27x^{3/2}$
61. $(\sqrt{x})^3 - 8$
62. $(\sqrt{x} - 1)^3 - 1$

Factor each expression. (See example 16.)

63. $x^{-3}y^{-4} - x^{-5}y^{-2}$
64. $7x^{-1} - 7xy^{-2}$
65. $x^{-5}y^{-1} + x^{-2}y^{-4}$
66. $4x^{-1}y^{-1} - 4x^{-4}y^{-2}$
67. $x^{-3}y^{-1} + 3x^{-4} + 2x^{-5}y$
68. $4y^{-4} + 10x^{-1}y^{-3} - 6x^{-2}y^{-2}$

Completely factor each polynomial. Assume n is a natural number.

69. $y^{2n} - 100$
70. $ax^{4n} - ay^{4n}$
71. $4x^2y^n + 5xy^{n+1} + y^{n+2}$
72. $12x^{n+2} + 8x^{n+1}y - 4x^ny^2$
73. $2y^{4n} - 15y^{2n} - 27$
74. $20w^{3n} + 63w^{2n} + 36w^n$

Factor each polynomial by the grouping method. (See example 12.)

75. $a^2 + 2a + 1 - x^2$
76. $b^2 - y^2 + 2y - 1$
77. $4x^2 - 9y^2 - 6y - 1$
78. $16x^2 - y^2 + 2yz - z^2$
79. $4w^2 + 12w + 9 - y^2 + 4y - 4$
80. $y^4 + 2y^2 - z^4 - 4 - 4z^2 + 1$

81. Factor $(3x - 5)^{2n}(y - 1)^3 + (3x - 5)^{2n-1}(y - 1)^4$, $n \in N$.
82. Factor $(x + 2)(y + 3)^{n+1} - (x + 2)^2(y + 3)^{n+2}$, $n \in N$.

1.4 Complex Numbers

Methods for solving equations of the form $ax^2 + bx + c = 0$ are taken up in the next section. Many such equations have no solutions if the replacement set for x is limited to R. As an example, $x^2 + 1 = 0$ has no real number solutions because the square of any real number is non-negative. If x is a real number, the expression $x^2 + 1$ must be greater than or equal to 1. However, $x^2 + 1 = 0$ can be solved if we invent a number whose square is -1. That is exactly what sixteenth century mathematicians did to solve polynomial equations that had no real number solutions. They called their invented numbers *imaginary numbers.* There seemed to be no application for these numbers other than solving some intractable equations. Just as the invention of negative numbers in an earlier time, however, led mathematicians to a larger set of usable numbers, the invention of imaginary numbers led to the eventual construction of the set of complex numbers, *C.* The set *C* contains the real numbers *R* as a subset. A brief introduction to the set of complex numbers follows.

We shall call $\sqrt{-1}$ the **imaginary unit,** denote it as ***i***, and assign to i the property

$$i \cdot i = i^2 = -1.$$

If the property of radicals, $\sqrt{ab} = \sqrt{a}\sqrt{b}$, is applied to $\sqrt{-b}$ where $b > 0$, we have a means of finding the principal square root of a negative number.

$$\sqrt{-b} = \sqrt{b(-1)} = \sqrt{b}\sqrt{-1} = \sqrt{b}i$$

Thus, $\sqrt{-4} = \sqrt{4}\sqrt{-1} = 2i$, and $\sqrt{-27} = \sqrt{27}\sqrt{-1} = 3\sqrt{3}i$.

You must be careful in using the radical property when both a and b are negative.

$$\sqrt{-4}\sqrt{-9} = (2i)(3i) = 6i^2 = -6$$
$$\sqrt{-4}\sqrt{-9} \neq \sqrt{(-4)(-9)} = \sqrt{36} = 6$$

Numbers such as $2i$, $3\sqrt{3}i$, or bi, where $b \in R$, are called **pure imaginary numbers.** With pure imaginary numbers we can find the principal square root of a negative real number. We can also combine a real number with a pure imaginary number to form a complex number.

Let us call a symbol of the form

$$z = a + bi, \quad a,b \in R, \quad \text{and} \quad i^2 = -1$$

a **complex number.** We call a the **real part** and b the **imaginary part** of the complex number z. Complex numbers can be added, subtracted, multiplied, and divided in a way that is consistent with those arithmetic operations in the set of real numbers.

1. Two complex numbers are equal if and only if their real parts are equal and their imaginary parts are equal.

Equality of Complex Numbers

$a + bi = c + di$ if and only if $a = c$ and $b = d$.

2. The sum or difference of two complex numbers, $z_1 \pm z_2$, is *the complex number* whose real part is the sum or difference of the real parts of z_1 and z_2 and whose imaginary part is the sum or difference of the imaginary parts of z_1 and z_2.

Addition and Subtraction of Complex Numbers

$$(a + bi) + (c + di) = (a + c) + (b + d)i$$
$$(a + bi) + (c + di) = (a - c) + (b - d)i$$

3. The product of two complex numbers, $z_1 \cdot z_2$, is *the complex number* obtained by treating z_1 and z_2 as binomials and performing binomial multiplication.

Multiplication of Complex Numbers

$$\begin{aligned}(a + bi)(c + di) &= ac + adi + bci + bdi^2 \\ &= (ac - bd) + (ad + bc)i\end{aligned}$$

Example 1

a. $2 + xi = y - 4i$ if and only if $2 = y$ and $x = -4$.

b. $(3 + 2i) + (5 - 3i) = (3 + 5) + (2 + (-3))i = 8 - i$

c. $(3 + 2i) - (5 - 3i) = (3 - 5) + (2 - (-3))i = -2 + 5i$

d. $$\begin{aligned}(3 + 2i) \cdot (5 - 3i) &= 3 \cdot 5 + 3(-3i) + (2i)5 + (2i)(-3i) \\ &= 15 + i - 6i^2 \\ &= 21 + i\end{aligned}$$

The set of real numbers is a subset of the set of complex numbers, as any real number can be written in the form $a + 0i$. When the arithmetic operations in C are applied to real numbers, the results are real numbers.

$$5 + 7 = (5 + 0i) + (7 + 0i) = 12 + 0i = 12$$
$$5 \cdot 7 = (5 + 0i)(7 + 0i) = 35 + 0i = 35$$

Conjugates and Quotients in C

If two complex numbers differ only in that their imaginary parts are opposite in sign, then each is called the **complex conjugate,** or simply the **conjugate,** of the other. Thus, $2 + 6i$ and $2 - 6i$ are conjugates of each other, as are $a + bi$ and $a - bi$.

Conjugate complex numbers furnish a method of expressing the quotient of two complex numbers as a complex number. Consider the quotient $(a + bi)/(c + di)$, where c and d are not both zero. Multiplying numerator and denominator of this quotient by the conjugate of the denominator, we obtain

$$\begin{aligned}\frac{a+bi}{c+di} &= \frac{(a+bi)(c-di)}{(c+di)(c-di)}\\ &= \frac{(ac+bd)+(bc-ad)i}{c^2+d^2}\end{aligned}$$

from which it follows that

Division of Complex Numbers

$$\frac{a+bi}{c+di} = \frac{ac+bd}{c^2+d^2} + \frac{(bc-ad)i}{c^2+d^2}. \qquad (1)$$

We see from (1) that the quotient of two complex numbers is itself a complex number, provided that the divisor is not $0 + 0i$. Also note that the product of a complex number and its conjugate, $(c + di)(c - di)$, is a real number, $c^2 + d^2$.

Example 2 Write $(2 + i)/(3 - 2i)$ as a complex number in the form $a + bi$.

Solution

$$\begin{aligned}\frac{2+i}{3-2i} &= \frac{(2+i)(3+2i)}{(3-2i)(3+2i)}\\ &= \frac{6+7i+2i^2}{9-4i^2}\\ &= \frac{4+7i}{13}\\ &= \frac{4}{13}+\frac{7}{13}i\end{aligned}$$

The result could have been obtained directly from (1) by letting $a = 2$, $b = 1$, $c = 3$, and $d = -2$. ■

Subsets of *C*

Figure 1.1 shows the relationship between C and its major subsets. Specific instances of numbers in some of these subsets is given in example 3.

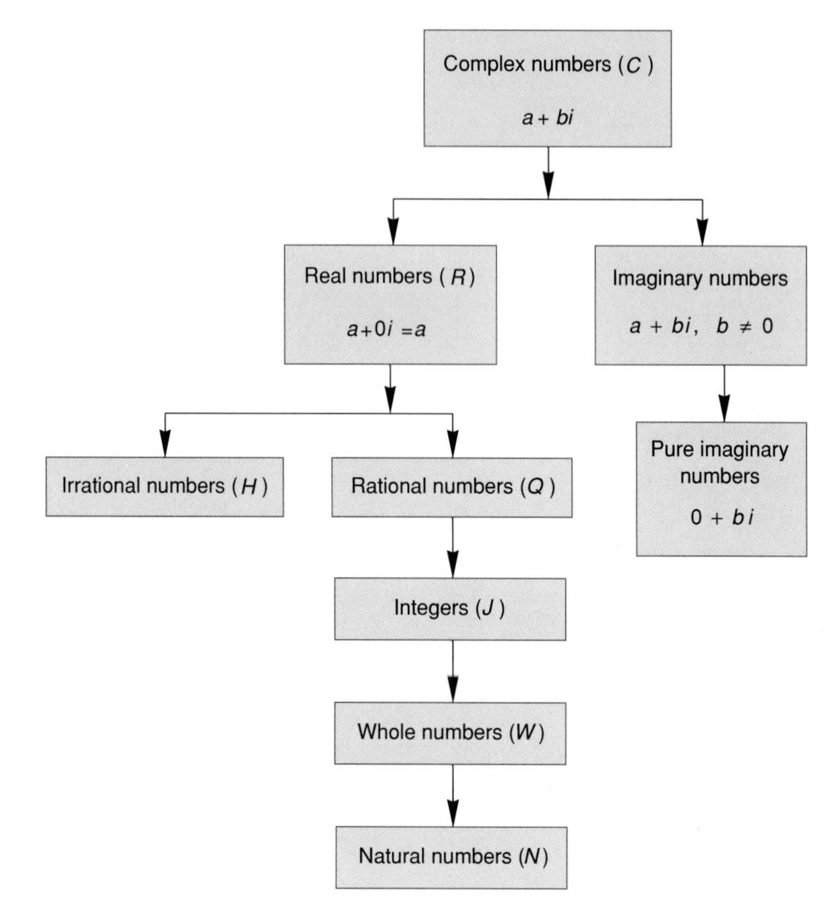

Figure 1.1

Example 3

a. $2 - 3i$ and $5i$ are imaginary numbers.

b. $5i$ and $-i$ are pure imaginary numbers.

c. $5i$, $2 - 3i$, 6, and -7 are complex numbers.

Exercise Set 1.4

(A)

Determine a and b.

1. $a + bi = 3 + 7i$

2. $a + bi = -5 + \frac{1}{4}i$

3. $2a + 4bi = \frac{1}{2} - 12i$

4. $\dfrac{a}{2} + \dfrac{b}{4}i = 6 + 15i$

Write each sum, difference, or product in $a + bi$ form.

5. $(3 + 2i) + (5 - i)$

6. $(\frac{3}{4} - \frac{1}{2}i) + 2i$

7. $(3 + 2i)(5 - i)$

8. $(\frac{3}{4} - \frac{1}{2}i)(2i)$

9. $6i(-2 + 3i)$

10. $5 \cdot 4i$

11. $6i - (-2 + 3i)$

12. $5 - (2 + 4i)$

13. $(5 + 2i) + (-6 + 3i)$ 14. $(1 + i) - (-2 - 2i)$
15. $(5 + i)(3 - 2i)$ 16. $(-7 + i)(6 - i)$

Use the symbol i to rewrite each expression so that the expression has no negative sign under the radical.

17. $\sqrt{-4}$ 18. $\sqrt{-9}$ 19. $\sqrt{-169}$
20. $\sqrt{-324}$ 21. $\sqrt{(-8)(-4)}$ 22. $\sqrt{-8}\sqrt{-4}$
23. $\dfrac{\sqrt{-27}}{\sqrt{-3}}$ 24. $\sqrt{\dfrac{-27}{-3}}$ 25. $\dfrac{\sqrt{-5}}{20}$
26. $\dfrac{1}{\sqrt{-18}}$ 27. $\sqrt{-5}\sqrt{-125}$ 28. $\dfrac{\sqrt{-98}}{14}$

Specify the complex conjugate of each number.

29. $1 + i$ 30. $-3 + 4i$ 31. $-2i$
32. $3i$ 33. $2 - 5i$ 34. $-4 - 2i$
35. $i + 6$ 36. $2i - 3$

Express each quotient in the form $a + bi$.

37. $\dfrac{1 + 3i}{1 + 5i}$ 38. $\dfrac{7 - 3i}{4 - 2i}$ 39. $\dfrac{8}{i}$
40. $\dfrac{2}{-i}$ 41. $\dfrac{11 + 6i}{-5i}$ 42. $\dfrac{8 - 3i}{-2i}$
43. $\dfrac{1 - \sqrt{3}i}{2 - \sqrt{2}i}$ 44. $\dfrac{\sqrt{6}i}{\sqrt{2} - 4i}$ 45. $\dfrac{1}{\sqrt{-18}}$
46. $\dfrac{14}{\sqrt{-98}}$ 47. $\dfrac{1 + \sqrt{2}}{2 - \sqrt{-3}}$ 48. $\dfrac{4 - \sqrt{-9}}{3 + \sqrt{-25}}$

(B)

49. Given that $i^2 = -1$, then $i^3 = i^2 \cdot i = (-1)i = -i$, and $i^4 = i^2 \cdot i^2 = (-1)(-1) = 1$. Determine i^5, i^6, i^7, and i^8. Do you observe a pattern? What is it?
50. Given the following computations,

$$i^{-1} = \frac{1}{i} = \frac{i}{i^2} = -i, \text{ and}$$
$$i^{-2} = \frac{1}{i^2} = \frac{1}{-1} = -1,$$

determine i^{-3}, i^{-4}, i^{-5}, i^{-6}, i^{-7}, and i^{-8}. Do you see a pattern? What is it?

Perform the indicated multiplication. Then simplify the product to $a + bi$ form.

51. $(4 - 2i)^2$ 52. $(5 + 3i)^3$
53. $(4 - i)^3$ 54. $(1 + 2i)^4$

Let $z = a + bi$ and $\bar{z} = a - bi$.

55. Show that $\overline{z_1 + z_2} = \bar{z}_1 + \bar{z}_2$.
56. Show that $\overline{z_1 \cdot z_2} = \bar{z}_1 \cdot \bar{z}_2$.
57. Show that $\overline{(z_1/z_2)} = \bar{z}_1/\bar{z}_2$.

1.5 Quadratic Equations

The general quadratic equation has the standard form

$$ax^2 + bx + c = 0, \quad a \neq 0. \tag{1}$$

The left member of (1) is a second degree, or quadratic, polynomial in which ax^2 is called the **second degree term,** bx is called the **linear term,** and c is called the **constant term.** If the quadratic polynomial can be factored as $(dx + e)(fx + g)$, (1) can be written in the form

$$(dx + e)(fx + g) = 0.$$

The solution set for (1) is then obtained from the equations

$$dx + e = 0 \quad \text{and} \quad fx + g = 0.$$

Example 1 Solve $3x^2 + 5x - 12 = 0$.

Solution $3x^2 + 5x - 12 = (3x - 4)(x + 3)$
Hence, $(3x - 4)(x + 3) = 0$. It follows that

$$3x - 4 = 0 \quad \text{or} \quad x + 3 = 0,$$

and

$$x = \tfrac{4}{3} \quad \text{or} \quad x = -3.$$

The solution set is $\{-3, \frac{4}{3}\}$. ■

Example 2 Solve $2x^2 - 12 = x^2 + 6x - 21$.

Solution

$$2x^2 - 12 = x^2 + 6x - 21$$
$$x^2 - 6x + 9 = 0$$
$$(x - 3)(x - 3) = 0$$

Since the linear factors are the same, the equation has only one solution,

$$x - 3 = 0 \quad \text{or} \quad x = 3.$$

We say *3 is a solution, or root, of multiplicity two.* ■

Example 3 Solve $x^2 = 5$.

Solution

$$x^2 - 5 = 0$$
$$(x - \sqrt{5})(x - \sqrt{5}) = 0$$

The solution set is $\{-\sqrt{5}, \sqrt{5}\}$. ■

An alternate solution for $x^2 - 5 = 0$ is obtained directly by finding a number whose square is 5. That number can be either $-\sqrt{5}$ or $\sqrt{5}$, so we write the solution set as $x = \pm\sqrt{5}$.

Example 4 Solve $(x - 3)^2 = 8$.

Solution

$$(x - 3)^2 = 8$$
$$x - 3 = \pm\sqrt{8} = \pm 2\sqrt{2}$$
$$x = 3 \pm 2\sqrt{2}$$ ■

Completing the Square

If the left side of (1) cannot be easily factored, then (1) can be solved by a general method called *completing the square.* An equation that is equivalent to (1) is constructed so that its left side is a *perfect square trinomial.* In example 2, $x^2 - 6x + 9$ is a perfect square trinomial because its binomial factors are equal.

$$x^2 - 6x + 9 = (x - 3)(x - 3) = (x - 3)^2$$

Similarly, $x^2 - 8x + 16$, $x^2 + 10x + 25$, and $x^2 + 2ax + a^2$, are perfect square trinomials because,

$$x^2 - 8x + 16 = (x - 4)(x - 4) = (x - 4)^2,$$

$$x^2 + 10x + 25 = (x + 5)(x + 5) = (x + 5)^2,$$

and

$$x^2 + 2ax + a^2 = (x + a)(x + a) = (x + a)^2.$$

All perfect square trinomials that have 1 as the coefficient of their second degree terms are represented by the trinomial, $x^2 + 2ax + a^2$. In all such trinomials, *the constant term is one half the square of the coefficient of the linear term,* $a^2 = (2a/2)^2$.

Example 5 Solve $x^2 + 4x + 1 = 0$.

Solution The left side of the equation cannot be easily factored. We attempt a solution by completing the square. After 1 is subtracted from both sides of the equation, the quantity 2^2 is added to both sides of the equation. Note that 2^2 is the square of one half the coefficient of the linear term, $4x$. The expression $x^2 + 4x$ becomes a perfect square trinomial after 4 is added to it. But adding 4 to the left side of the equation requires that we add 4 to the right side to produce an equation that is equivalent to the original equation.

$$\begin{aligned} x^2 + 4x &= -1 \\ x^2 + 4x + 2^2 &= -1 + 2^2 \\ (x + 2)^2 &= 3 \\ x + 2 &= \pm\sqrt{3} \\ x &= -2 \pm \sqrt{3} \end{aligned}$$

■

Example 6 Solve $3x^2 + 9x - 4 = 0$.

Solution Unable to factor, we complete the square. The first step in the process is to make the coefficient of the second degree term, 1. Left and right members of the equation are divided by 3, and $-\frac{4}{3}$ is transposed to the right side of the equation. Then the square is completed on the left side of the equation.

$$\begin{aligned} 3x^2 + 9x - 4 &= 0 \\ x^2 + 3x &= \frac{4}{3} \\ x^2 + 3x + \left(\frac{3}{2}\right)^2 &= \frac{4}{3} + \left(\frac{3}{2}\right)^2 \\ \left(x + \frac{3}{2}\right)^2 &= \frac{43}{12} \\ x + \frac{3}{2} &= \pm\sqrt{\frac{43}{12}} \\ x &= -\frac{3}{2} \pm \sqrt{\frac{43}{12}} \\ x &= -\frac{9}{6} \pm \sqrt{\frac{129}{36}} \\ x &= \frac{-9 \pm \sqrt{129}}{6} \end{aligned}$$

■

The Quadratic Formula

If the method of solution in example 6 is applied to the general quadratic equation, (1), the quadratic formula, (2), is the result.

$$ax^2 + bx + c = 0$$

$$x^2 + \frac{bx}{a} = -\frac{c}{a}$$

$$x^2 + \frac{b}{a}x + \left(\frac{b}{2a}\right)^2 = -\frac{c}{a} + \left(\frac{b}{2a}\right)^2$$

$$\left(x + \frac{b}{2a}\right)^2 = \frac{b^2}{4a^2} - \frac{4ac}{4a^2}$$

$$x + \frac{b}{2a} = \pm\sqrt{\frac{b^2 - 4ac}{4a^2}}$$

$$x = -\frac{b}{2a} \pm \frac{\sqrt{b^2 - 4ac}}{2a}$$

■

The Quadratic Formula

$$x = \frac{-b \pm \sqrt{b^2 - 4ac}}{2a} \qquad (2)$$

The quadratic formula results from the application of the method of completing the square to the general quadratic equation. This means that we need not solve a quadratic equation by completing the square. It is simpler to use the quadratic formula. Solutions of the equations in examples 5 and 6 can be obtained directly by using the quadratic formula.

For $x^2 + 4x + 1 = 0$,

$$x = \frac{-4 \pm \sqrt{16 - 4(1)(1)}}{2(1)} = \frac{-4 \pm \sqrt{12}}{2} = -2 \pm \sqrt{3}.$$

For $3x^2 + 9x - 4 = 0$,

$$x = \frac{-9 \pm \sqrt{81 - 4(3)(-4)}}{2(3)} = \frac{-9 \pm \sqrt{129}}{6}.$$

Discriminant and Roots

The quadratic formula tells us that a quadratic equation has two roots.

Roots of the General Quadratic Equation

$$r_1 = \frac{-b - \sqrt{b^2 - 4ac}}{2a} \text{ and } r_2 = \frac{-b + \sqrt{b^2 - 4ac}}{2a} \qquad (3)$$

Yet, the equation in example 2 has only one root. Further, if the replacement set for x, in (1), is restricted to R, it is possible to exhibit quadratic equations that have no roots. However, if the replacement set for x is C, and if we consider equations such as the one in example 2 as having roots of multiplicity two, then it is possible to show that all quadratic equations have *exactly* two roots (see exercises 67–69). This means that if we know the roots of a quadratic equation, we can construct the equation.

Example 7 Write a quadratic equation in standard form, and with integral coefficients, that has $\{\frac{2}{3}, -5\}$ as its solution set.

Solution The equation $(x - \frac{2}{3})(x + 5) = 0$ has $\frac{2}{3}$ and -5 as its roots. If the factors on the left side are multiplied together we obtain a standard form quadratic equation,

$$x^2 + \frac{13x}{3} - \frac{10}{3} = 0.$$

The last equation has the equivalent form,

$$3x^2 + 13x - 10 = 0.$$

■

The solution of a quadratic equation yields two real roots, as in examples 5 and 6, a root of multiplicity two, as in example 2, or two imaginary roots, as in example 8 below. The nature of the roots depends upon the quantity $b^2 - 4ac$, which is called the **discriminant** of the quadratic formula. Which category of roots an equation has depends upon whether its discriminant is less than, equal to, or greater than zero.

$b^2 - 4ac$	Nature of Roots	Example
> 0	two real and unequal roots	5, 6
$= 0$	one real root of multiplicity two	2
< 0	two imaginary roots that are complex conjugates of each other	8

Example 8 Solve $2x^2 - x + 4 = 0$, $x \in C$.

Solution Using the quadratic formula with $a = 2$, $b = -1$, and $c = 4$, we write

$$x = \frac{-(-1) \pm \sqrt{(-1)^2 - 4(2)(4)}}{2(2)}$$

$$x = \frac{1 \pm \sqrt{-31}}{4}$$

$$x = \frac{1 \pm \sqrt{31}i}{4}.$$

The discriminant is negative, and the roots are a pair of imaginary numbers that are complex conjugates. ■

Applications

In the examples that follow, word problems are solved by the five-step procedure that was outlined in section 1.1. The steps in the procedure are listed here for easy reference.

Procedure for Solving Word Problems

1. Read the problem carefully and locate the sentence or phrase that tells you what it is you are to find.
2. Call what you have to find the unknown number or numbers. Assign a variable to represent one of the unknown numbers. If possible, represent the other unknown numbers in terms of the same variable. A sketch is helpful in expressing the unknowns in terms of the variable.
3. Construct a mathematical model for the problem.
4. Construct the solution set for the mathematical model. Be aware that the context of the problem may restrict the replacement set for the variable.
5. Interpret the model's solutions relative to the problem that produced the model. Sometimes a solution of the model is not a solution of the problem.

Example 9 A homeowner wishes to build a pool on a 40-foot by 100-foot portion of his property. He likes the shape of his neighbor's pool, which is a 20-foot by 60-foot rectangle. Can the homeowner build a pool that has the same shape as his neighbor's pool, but with twice the area of that pool, on the alloted portion of his property?

Solution

Step 1 "What are the dimensions of the pool" tells us what we are to find.

Step 2 Let x be the width of the pool. Then $3x$ must be its length, so that the ratio of length to width is three to one.

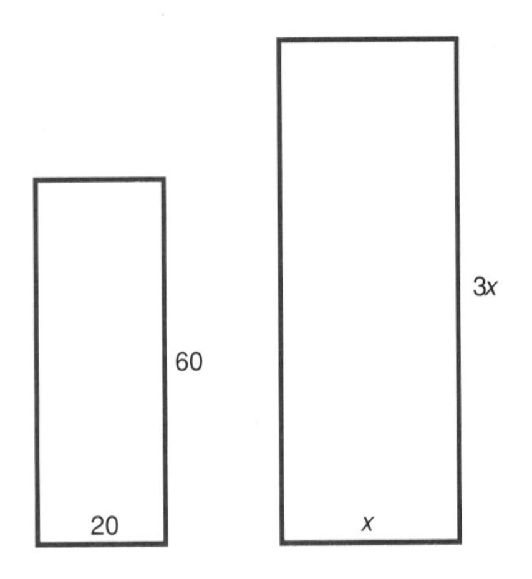

Step 3 The area of the pool must be twice that of the neighbor's pool.

$$x \cdot 3x = 2(20 \cdot 60)$$
$$3x^2 = 2400$$

Step 4 Since x represents the width of the pool, it must be a positive number.

$$x^2 = 800$$
$$x = \sqrt{800}$$
$$x = 20\sqrt{2}$$

Step 5 The width and length of the pool will be $20\sqrt{2}$ feet and $60\sqrt{2}$ feet, respectively. Rounded to the nearest tenth of a foot, the dimensions are 28.3 feet wide by 84.8 feet long. The 40-foot by 100-foot space allotted to the pool is sufficient. ■

Example 10 One car travels 180 miles in one hour's less time than it takes a second car to travel 200 miles. If the speed of the first car is 10 mph greater than the speed of the second car, what is the speed of each car?

Solution

Steps 1 and 2 We wish to determine the speed of each car. Let t be the time it takes the second car to travel 200 miles; then $t - 1$ is the time it takes the first car to travel 180 miles.

Step 3 A relation between speed, distance, and time is given by the equation $s = d/t$. We use the relation to show that the speed of the first car is 10 mph greater than that of the second car.

$$\frac{180}{t-1} = \frac{200}{t} + 10$$

Steps 4 and 5

$$t(t-1)\frac{180}{t-1} = \left(\frac{200}{t} + 10\right)(t)(t-1)$$
$$180t = 200(t-1) + 10(t^2 - t)$$
$$0 = 10t^2 + 10t - 200$$
$$0 = t^2 + t - 20$$
$$0 = (t+5)(t-4)$$

The solutions of the mathematical model are 4 and -5, but t represents hours traveled. It must be a positive number. Hence, the second car traveled 200 miles in 4 hours at a speed of 50 mph. The faster car traveled 180 miles in 3 hours at a speed of 60 mph. ■

Example 11 An airline sets $400 as the ticket price for its charter flight if a minimum of 150 people sign up for the flight. The airline agrees to reduce every passenger's ticket price $1 for each passenger in excess of the 150 minimum. The airline collects $75,625 in ticket revenue. How many people sign up for the flight, and what is the price of a ticket?

Solution

Steps 1 and 2 We have to find the number of people that sign up for the flight and the price of a ticket for the charter flight. More than 150 people sign up for the flight because the ticket revenue from 150 people is $150 \cdot \$400 = \$60{,}000$, which is less than the actual ticket revenue. Let x be the number of tickets purchased in excess of 150.

Step 3 The ticket revenue is the product of the number of tickets purchased and the price per ticket.

$$75{,}625 = (150 + x)(400 - x)$$

Steps 4 and 5

$$75{,}625 = 60{,}000 + 250x - x^2, \; x \in J, \; x > 0$$
$$x^2 - 250x + 15{,}625 = 0$$
$$x = \frac{250 \pm \sqrt{250^2 - 4(1)(15{,}625)}}{2}$$
$$x = \frac{250 \pm \sqrt{62{,}500 - 62{,}500}}{2}$$
$$x = 125$$

The number of tickets sold is $150 + 125 = 275$. The price per ticket is $\$400 - \$125 = \$275$. ■

Exercise Set 1.5

The replacement set for x is C.

(A)

Write each equation in standard form; then factor to find its solution.

1. $x^2 + 3x + 2 = 0$

2. $x^2 - 4x - 5 = 0$

3. $2x^2 = 5 - 3x$

4. $8x^2 - 24 = -4x$

5. $x(2x - 3) = 4(x + 1)$

6. $2x(3x - 5) = x + 10$

7. $x^2 + 1 = 0$

8. $9x^2 + 25 = 0$

Solve each equation by completing the square.

9. $x^2 + 4x - 9 = 0$

10. $x^2 - 6x + 2 = 0$

11. $3x^2 + 6x - 4 = 0$

12. $2x^2 - 8x - 5 = 0$

13. $3x^2 - x - 6 = 0$

14. $4x^2 + 5x - 12 = 0$

Solve each equation using the quadratic formula.

15. $x^2 + 3x + 2 = 0$

16. $8x^2 - 24 = -4x$

17. $x^2 + 4x - 9 = 0$

18. $2x^2 - 8x - 5 = 0$

19. $9x^2 + 25 = 0$

20. $4x^2 + 5x - 12 = 0$

Solve each equation.

21. $x^2 - 6x = 15$

22. $x^2 + 5 = 6x$

23. $5x^2 + 14x + 10 = 0$

24. $3x^2 - 6x - 8 = 0$

25. $2x^2 - 6x + 1 = 0$

26. $4x^2 + x - 2 = 0$

27. $x^2 + 3x + 4 = 0$

28. $x^2 - 3x + 6 = 0$

29. $(3x + 4)^2 = 25$

30. $(2x - 1)^2 = 64$

31. $(x + 2)^2 = 12$

32. $(4x + 5)^2 = 18$

33. $\dfrac{4}{3} + \dfrac{x}{x + 1} = \dfrac{x}{x - 1}$

34. $\dfrac{x}{x + 3} + \dfrac{2x}{3x + 1} = 0$

35. $\dfrac{x + 6}{x^2 - 2x} + 4 = \dfrac{2x}{x - 2}$

36. $\dfrac{4x}{x + 1} - \dfrac{2x}{x + 2} = 1$

In standard form and with integral coefficients, write a quadratic equation that has the indicated solution set.

37. $\{2,1\}$

38. $\{-3,4\}$

39. $\{5,5\}$

40. $\{-1,-1\}$

41. $\{2 + i, 2 - i\}$

42. $\{1 + 3i, 1 - 3i\}$

(B)

43. Determine k so that $kx^2 - 3x + 5 = 0$ has a solution of multiplicity 2.

44. Determine k so that $x^2 - kx + 9 = 0$ has a solution of multiplicity 2.

45. Without solving, determine the nature of the roots of $4x^2 + 2x - 5 = 0$.

46. Without solving, determine the nature of the roots of $x^2 + x + 2 = 0$.

Solve each equation for x.

47. $(x - 1)(x + 2)(x - 3) = 0$

48. $(2x - 1)(3x + 5)(x + 2) = 0$

49. $(x + 1)(x^2 + 3x + 2) = 0$

50. $(x - 2)(2x^2 + 9x + 4) = 0$

51. $x^4 - 13x^2 + 36 = 0$

52. $4x^4 - 5x^2 + 1 = 0$

53. The sum of a number and three times its reciprocal is 4. Find the number.

54. Two planes flying at approximately the same altitude pass each other at noon. One plane is flying due north at 180 mph. The other plane is flying due east at 150 mph. How far apart are the planes at 12:40 P.M.?

55. The sides of a rectangular lot are to be in the ratio of 5:2. The area of the lot is to be 16,000 square feet. What will the dimensions of the lot be?

56. A 25-meter by 50-meter pool is surrounded by a concrete walk of uniform width. The area of the walk is 486 square meters. What is the width of the walk?

57. A commercial jet flew west to east 3,000 miles, and then back to its takeoff point. The 6,000-mile flight took 11 hours. The wind was from the west toward the east at 50 mph. What was the plane's airspeed?

58. It takes 2 hours longer to fill a pool using only water pipe B than it would take if only water pipe A were used. The pool is filled in 3 hours if both pipes are used. How long does it take to fill the pool if only pipe A is used; if only pipe B is used?

59. Cities A and B are 460 miles apart. Car 1 leaves A for B at 8:00 A.M. Car 2 leaves B for A at 9:30 A.M. traveling 20 mph faster than car 1. If the two cars meet 210 miles from B, how fast is each car traveling?

60. By increasing his usual average speed 10 mph, a trucker finds that he can cut his route time on a certain 240-mile run by 2 hours. Find the trucker's usual average speed.

61. Based on past experience, a local theater manager knows that she can fill the theater's 250-seat auditorium for a good production if the ticket price is \$10 or less. The manager also knows that each \$1 increase in ticket price beyond the \$10 results in 10 fewer playgoers per performance. What ticket price should the manager set so that daily revenue for the run of the play is \$3,060?

62. The local ski tour entrepreneur knows that if he sells his 3-day ski package for \$280, he can fill a 60-seat bus. For each \$20 increase in the package, he loses 3 customers. At what price should he set the package so that the revenue from each trip is \$17,340?

63. Brand X copying machine produces 1,200 copies of a document in 3 minutes less time than brand Y copier can produce 1,200 copies. If brand X copier produces 5 copies more per minute than brand Y copier, find the rate at which each machine produces copies.

64. Two integers differ by 13. The sum of their squares is 697. Find the integers.

65. Are there two numbers whose sum is 30 and whose product is 250? If you say yes, find the numbers. If you say no, justify your response.

66. A 40-inch long piece of wire is cut into two pieces. Each piece is shaped into a square. Find the length of each piece of wire if the sum of the areas of the squares is 58 square inches.

Exercises 67–69 show that a quadratic equation cannot have more than two roots in the set of complex numbers. Since the quadratic formula always yields two roots, if we include roots of multiplicity two, we must conclude that a quadratic equation has exactly two roots in C.

67. The roots of the quadratic equation $ax^2 + bx + c = 0$ are

$$r_1 = \frac{-b - \sqrt{b^2 - 4ac}}{2a} \quad \text{and}$$
$$r_2 = \frac{-b + \sqrt{b^2 - 4ac}}{2a}.$$

Show that $r_1 + r_2 = -\dfrac{b}{a}$ and that $r_1 \cdot r_2 = \dfrac{c}{a}$.

68. Show that a quadratic equation, $ax^2 + bx + c = 0$, can be written in the form $(x - r_1)(x - r_2) = 0$, where r_1 and r_2 are defined in exercise 67.

69. Assume that r_3 is a root of a quadratic equation, $ax^2 + bx + c = 0$, and that it is not equal to either r_1 or r_2. The roots r_1 and r_2 are defined in exercise 67. Show that the assumption that the quadratic equation has more than two roots leads to a contradiction and is therefore false. (Hint: Use the factored form of the quadratic equation that is given in exercise 68.)

1.6 Inequalities

Statements such as

$$8 > 2, \quad -3 < 1, \quad \text{and} \quad 5x + 8 \geq 3x - 16$$

are inequalities. The first two inequalities, 8 is *greater* than 2, and -3 is *less* than 1, are statements of fact. The last inequality becomes a true or false statement, depending upon which real number replaces x. Any real number replacement for x that produces a true statement is a solution of the inequality. The set of all its solutions constitutes the solution set of the inequality.

Generating Equivalent Inequalities

It is possible to solve $5x + 8 \geq 3x - 16$ as equations are solved in section 1.1. A sequence of equivalent inequalities are generated until one is obtained for which the solution set is obvious.

$$5x + 8 \geq 3x - 16$$
$$5x + 8 - (8 + 3x) \geq 3x - 16 - (8 + 3x)$$
$$2x \geq -24$$
$$\frac{2x}{2} \geq -\frac{24}{2}$$
$$x \geq -12$$

The last inequality in the sequence tells us that the solution set is the set of all real numbers greater than or equal to -12.

As with equations, an equivalent inequality is generated from an inequality when the same expression is added to, or subtracted from, both members of the inequality (see exercise 49). Thus, $5x + 8 \geq 3x - 16$ and $2x \geq -24$ are equivalent inequalities because the second inequality is obtained from the first when $8 + 3x$ is subtracted from both members of the first inequality. An equivalent inequality is also generated if both members of a given inequality are multiplied, or divided, by the same expression (see exercise 50). Inequalities $2x \geq -24$ and $x \geq -12$ are equivalent. The second is obtained from the first when both members of the first are divided by 2.

In the process of generating equivalent inequalities, you must be concerned with the possibility of a change in "sense" from less than to greater than, or vice versa. Adding (subtracting) the same expression to (from) both members of an inequality does not change its sense. Nor does multiplying (dividing) both members of an inequality by a positive number change its sense. *An inequality changes sense **only** when both its members are multiplied (divided) by a negative number.*

Example 1

a.
$$-8 < 15$$
$$-8 + 4 < 15 + 4$$
$$-4 < 19$$

b.
$$-8 < 15$$
$$-8 - 4 < 15 - 4$$
$$-12 < 11$$

c.
$$-8 < 15$$
$$-8(4) < 15(4)$$
$$-32 < 60$$

d.
$$-8 < 15$$
$$-8(-4) > 15(-4)$$
$$32 > -60$$

Before going on to solve some inequalities, we summarize the above discussion.

Summary

Some inequalities can be solved by generating a sequence of equivalent inequalities until an inequality is obtained for which the solution set is obvious.

Equivalent inequalities are generated from a given inequality by

1. adding (subtracting) the same expression to (from) left and right members of the given inequality,

or by

2. multiplying (dividing) left and right members of the inequality by the same non-zero expression.

The "sense" of an inequality changes *only* if both its members are multiplied (divided) by an expression that represents a negative number.

Example 2 Solve the inequality $6x - 9 \geq 2x + 3$ and graph its solution set.

Solution A sequence of equivalent inequalities is generated by

a. adding 9 to left and right members,

b. subtracting $2x$ from left and right members, and

c. dividing left and right members by 4.

$$6x - 9 \geq 2x + 3$$
$$6x \geq 2x + 12$$
$$4x \geq 12$$
$$x \geq 3$$

The solution set, $\{x | x \geq 3\}$, is the set of all real numbers greater than or equal to 3. The solid circle at 3 in the graph of the solution set (figure 1.2), indicates that 3 is in the solution set.

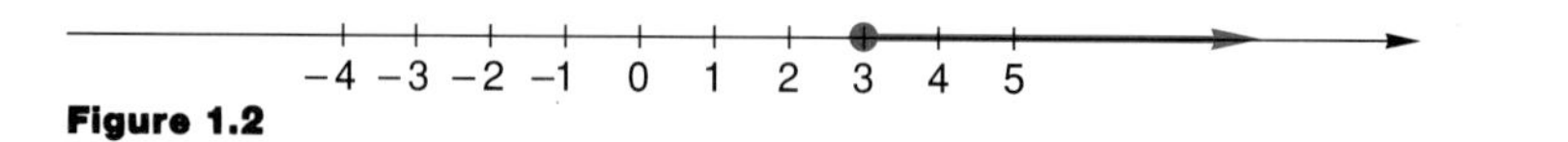

Figure 1.2

■

Example 3 Solve the inequality $-x + 7 > 4x + 27$ and graph its solution set.

Solution The inequality is solved two ways. In the solution on the left, both members of the inequality are divided by -5, changing the sense of the inequality from $>$ to $<$.

$$\begin{aligned} -x + 7 &> 4x + 27 \\ -5x &> 20 \\ x &< -4 \end{aligned} \qquad \begin{aligned} -x + 7 &> 4x + 27 \\ -20 &> 5x \\ -4 &> x \end{aligned}$$

The solution set is $\{x|x < -4\}$. The open circle at -4 in figure 1.3 indicates that -4 is not included in the solution set.

−6 −5 −4 −3 −2 −1 0 1 2 3 4

Figure 1.3

■

A chain, or string, of inequalities such as

$$-3 < \frac{x-2}{2} < \frac{5}{2},$$

is often encountered in mathematics. The chain is read either as "-3 is less than $(x - 2)/2$, which is less than $\frac{5}{2}$," or as "$(x - 2)/2$ is between -3 and $\frac{5}{2}$." It can be viewed as two distinct inequalities,

$$\begin{aligned} -3 &< \frac{x-2}{2} && \text{and} && \frac{x-2}{2} < \frac{5}{2}. \\ 2(-3) &< \left(\frac{x-2}{2}\right)2 && \text{and} && 2\left(\frac{x-2}{2}\right) < \left(\frac{5}{2}\right)2 \\ -6 &< x - 2 && \text{and} && x - 2 < 5 \\ -4 &< x && \text{and} && x < 7 \end{aligned}$$

The solution set of the chain of inequalities includes all the real numbers that are greater than -4 and less than 7 (figure 1.4). It is the **intersection** of the sets, $\{x|x > -4\}$ and $\{x|x < 7\}$, in that it includes all the real numbers that the two sets have in common.

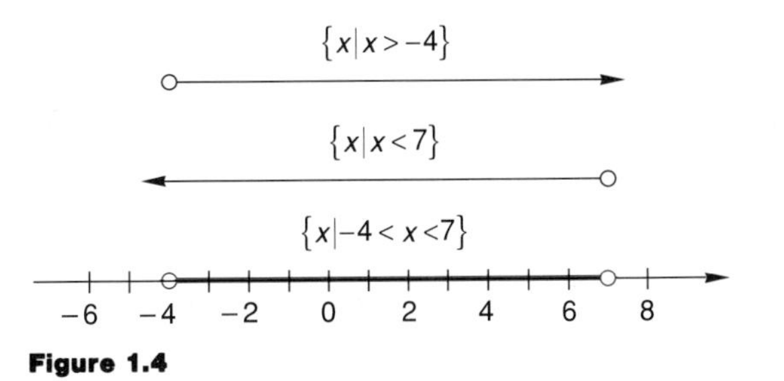

Figure 1.4

Definition

The intersection of sets A and B, denoted as $\mathbf{A} \cap \mathbf{B}$, is the set of elements that are in A *and* in B.

It is often possible to solve a chain of inequalities without breaking the chain into two distinct inequalities. To solve

$$-3 < \frac{x-2}{2} < \frac{5}{2},$$

we multiply each member in the chain by 2 and then add 2 to each member.

$$-6 < x - 2 < 5$$
$$-4 < x < 7$$

Example 4 Solve the chain of inequalities $-\frac{2}{3} \leq (5 - 2x)/6 < 1$, and graph the solution set.

Solution The solution is begun by multiplying each member of the chain by 6, which is the LCD.

$$-4 \leq 5 - 2x < 6$$
$$-9 \leq -2x < 1$$
$$\tfrac{9}{2} \geq x > -\tfrac{1}{2}$$

Note that the inequalities change sense as a result of the division by -2. The solution set includes all the real numbers greater than $-\frac{1}{2}$ and less than or equal to $\frac{9}{2}$.

$$\{x|x > -\tfrac{1}{2}\} \cap \{x|x \leq \tfrac{9}{2}\} = \{x|\ -\tfrac{1}{2} < x \leq \tfrac{9}{2}\}$$

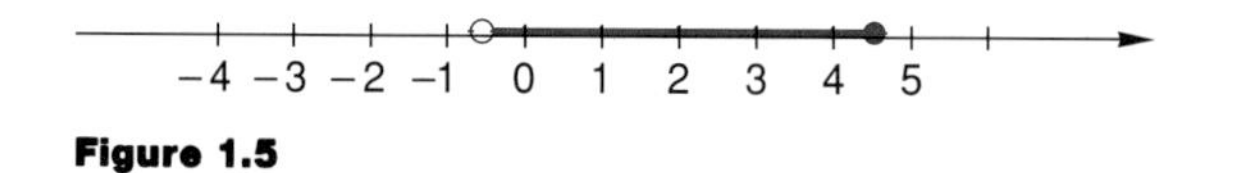

Figure 1.5

■

Intervals

Let a and b be two points on the number line so that a is less than b. The set of points between a and b, but not including a or b, is called an **open interval.** Such an interval is denoted as (a,b). The set of points that includes a and b as well as all the points between a and b, is a **closed interval,** and it is denoted as $[a,b]$. A **half open interval** includes one endpoint and all the points between the endpoints. It is denoted as $[a,b)$ or as $(a,b]$.

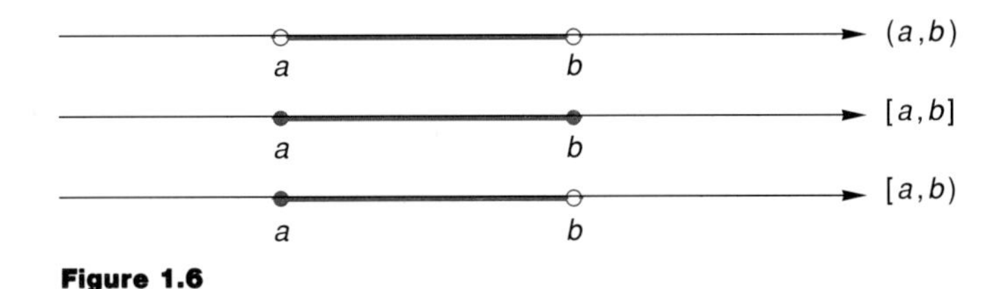

Figure 1.6

With interval notation, the solution sets in examples 2, 3, and 4 are designated as $[3,\infty)$, $(-\infty,-4)$, and $(-\frac{1}{2},\frac{9}{2}]$ respectively. The first interval, $[3,\infty)$, contains 3 and all the points to its right. In other words, the interval extends to the right without end. This state of affairs is indicated with the symbol for infinity, ∞. The symbol $-\infty$, in the second interval, indicates that the interval extends to the left without end.

Sign Graphs

If an inequality can be written in equivalent form so that one side is zero, then it can be solved by the method of sign graphs.

Example 5 Solve $x^2 - 3x + 2 > 0$.

Solution The left number of the inequality is factored.

$$(x - 2)(x - 1) > 0$$

A number line (figure 1.7) is constructed, and the points 1 and 2 are located on it. Point 1 separates the number line into point sets to the right of, or to the left of 1, for which the factor $x - 1$ is respectively, positive or negative. The number line is similarly separated by 2, relative to the factor $x - 2$. Plus and minus signs indicate the intervals over which each of the factors is positive or negative. The product, $(x - 2)(x - 1)$, is positive over intervals where its factors have the *same sign.* As indicated in the sign graph of figure 1.7, those intervals are $(-\infty, 1)$ and $(2,\infty)$. The solution set, which includes both intervals, is said to be the **union** of the intervals. It is designated as $(-\infty,1) \cup (2,\infty)$, where the symbol $\cup$ denotes Union.

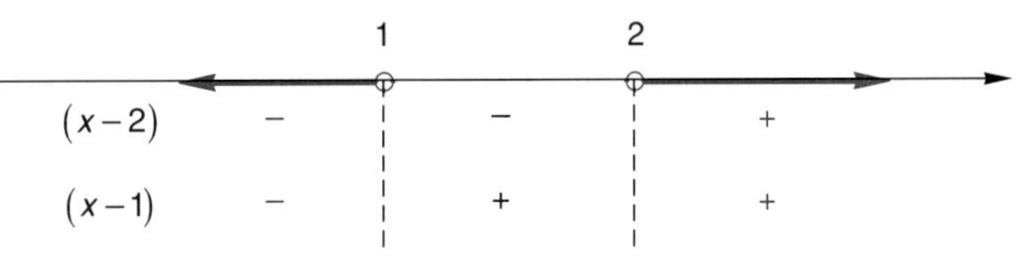

Figure 1.7 ■

> **Definition**
>
> The union of sets A and B, denoted as **A ∪ B,** is the set of elements that belong either to A or to B.

The left side of the inequality in example 5 is easily factored. Such is not the case in the next example.

Example 6 Solve $x^2 - 2x - 1 < 0$.

Solution It was shown in section 1.5 that a quadratic equation can be written in the form $(x - r_1)(x - r_2) = 0$, if its roots r_1 and r_2 are known. Thus, we solve the quadratic equation $x^2 - 2x - 1 = 0$, to factor the left side of the given inequality. The roots of the equation are,

$$r_1 = \frac{2 - \sqrt{4 + 4}}{2} \quad \text{and} \quad r_2 = \frac{2 + \sqrt{4 + 4}}{2}$$
$$= 1 - \sqrt{2} \qquad\qquad = 1 + \sqrt{2}.$$

Therefore, the inequality can be written as,

$$[x - (1 - \sqrt{2})][x - (1 + \sqrt{2}] < 0.$$

A sign graph is constructed with $1 - \sqrt{2}$ and $1 + \sqrt{2}$ as separation points (figure 1.8). The product of factors is negative in those intervals where the factors are of opposite sign. The sign graph shows the solution set to be the interval $(1 - \sqrt{2}, 1 + \sqrt{2})$.

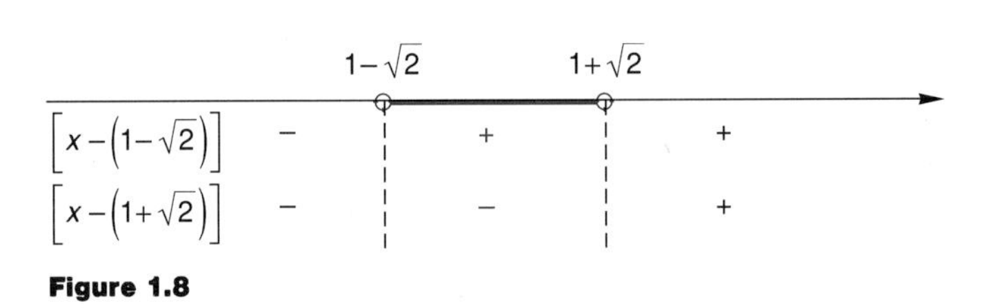

Figure 1.8

■

Example 7 Solve $x/(2 - x) \geq 2x/(3x + 2)$.

Solution The expression $2x/(3x + 2)$ is subtracted from both members of the inequality to make the right side zero.

$$\frac{x}{2-x}-\frac{2x}{3x+2}\geq 0$$
$$\frac{x(3x+2)-2x(2-x)}{(2-x)(3x+2)}\geq 0$$
$$\frac{5x^2-2x}{(2-x)(3x+2)}\geq 0$$
$$\frac{x(5x-2)}{(2-x)(3x+2)}\geq 0$$

A sign graph is then constructed for the factors x, $5x - 2$, $2 - x$, and $3x + 2$ (figure 1.9). It clearly shows the intervals where the numerator and denominator of the left member of the inequality have the same sign. A fraction is positive if its numerator and denominator have the same sign. It is zero only if the numerator is zero. Hence, the solution set is $(-\frac{2}{3},0] \cup [\frac{2}{5},2)$. ■

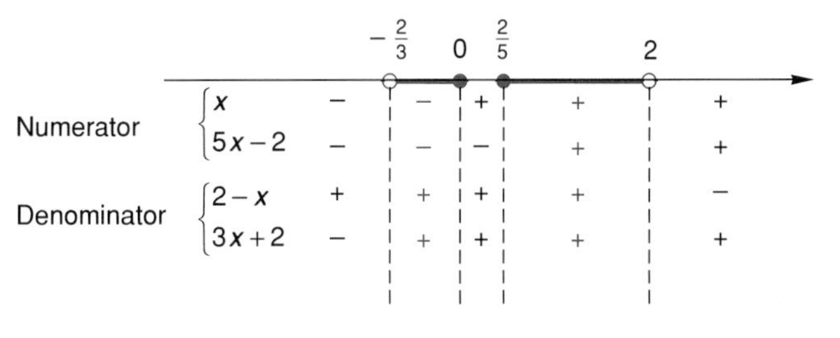

Figure 1.9

Example 8 A manufacturing firm produces bookcases that sell for \$120 each. The dollar cost for producing x bookcases per day is given by the equation $C = x^2 + 55x + 900$. How many bookcases must be manufactured daily in order for the firm to realize a profit?

Solution We use the five-step procedure for solving word problems.

Step 1 Determine the daily production of bookcases that yield a profit.

Step 2 Let x be the number of bookcases manufactured per day.

Step 3 Profit, P, is the difference between revenue and cost. The daily revenue is $120x$. Therefore, the profit is

$$P = 120x - (x^2 + 55x + 900).$$

Step 4 A profit is realized if the difference between revenue and cost is positive.

$$120x - (x^2 + 55x + 900) > 0$$
$$-x^2 + 65x - 900 > 0$$
$$x^2 - 65x + 900 < 0$$
$$(x - 20)(x - 45) < 0$$

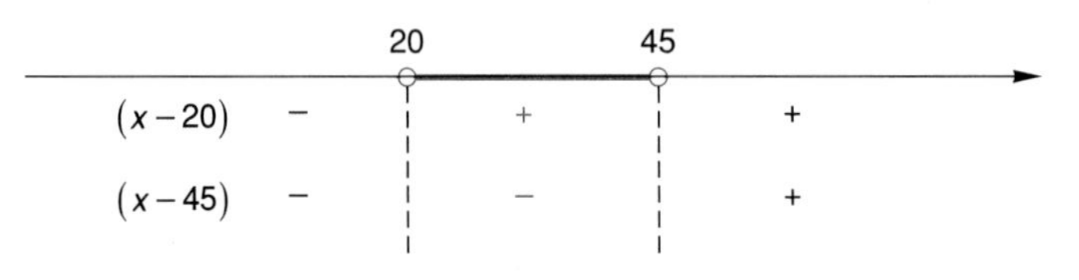

Figure 1.10

Step 5 The manufacturing firm must produce more than 20 but fewer than 45 bookcases daily to realize a profit. ■

Exercise Set 1.6

(A)

In Exercises 1–34, solve each inequality. Denote the solution set in both set builder notation and interval notation.

1. $2x - 3 < 7$
2. $8x + 5 > 21$
3. $-(8 + x) - 3 \geq 1 - 4x$
4. $\dfrac{3x + 2}{4} - 3 \leq \dfrac{2(2x - 1)}{3} + \dfrac{5}{6}$
5. $\dfrac{x + 2}{3} \geq x - 4(x - 5)$
6. $\dfrac{2x - 3}{2} - \dfrac{3(4x - 1)}{5} \leq 4$
7. $-4 \leq 2x \leq 12$
8. $4 \leq 6 - x \leq 9$
9. $-1 \leq 5 - 4x \leq 10$
10. $6 < 2(4 - 3x) < 15$
11. $14 > \dfrac{x}{3} + 10 > 2$
12. $\dfrac{18}{5} \geq \dfrac{3x - 5}{2} \geq -\dfrac{3}{5}$
13. $(x - 1)(x + 3) \leq 0$
14. $(x - 2)(x - 5) \geq 0$
15. $6x^2 - 15x + 6 > 0$
16. $x^2 - 7x + 10 < 0$
17. $x^2 + 2x - 2 < 0$
18. $x^2 - 3x - 1 > 0$
19. $\dfrac{2}{x - 1} < \dfrac{5}{x + 2}$
20. $\dfrac{1}{(3x + 1)} < \dfrac{3}{x - 4}$
21. $\dfrac{1}{2x} > \dfrac{-3}{x + 5}$
22. $\dfrac{3}{x} > \dfrac{2}{4x - 9}$
23. $\dfrac{5}{3x - 2} > 4$
24. $\dfrac{-2}{5x + 3} < -2$

(B)

25. $3x^2 + 4x - 2 > 0$
26. $2x^2 - 8x - 7 < 0$
27. $x(x - 1)(x + 3) < 0$
28. $(x + 1)(x + 5)(4 - 3x) > 0$
29. $x^2 + 2x + 2 > 0$
30. $x^2 - 4x + 5 < 0$
31. $\dfrac{3x}{x - 2} \leq \dfrac{-x}{x + 4}$
32. $\dfrac{x}{2x + 1} < \dfrac{3x}{x + 1}$
33. $\dfrac{x - 1}{x + 3} > \dfrac{x + 2}{x - 2}$
34. $\dfrac{4x + 1}{6x - 5} < \dfrac{-3}{3x - 4}$
35. How many cubic centimeters of a 50cc, 30% acid solution, must be replaced with pure acid to bring the acid strength up to a percentage between 45% and 50%?

36. Within what range should a Celsius temperature thermostat be set so that the temperature in the room is maintained between 68° and 77° Fahrenheit? (Hint: The relationship between Fahrenheit and Celsius temperatures is given by the equation $F = 9C/5 + 32$.)

37. A student's test scores in a certain course, prior to her last test, were 75, 92, 68, and 76. A perfect score is 100. To receive a B grade in the course, she must have a test score average greater than or equal to 80 and less than 90. To receive an A grade, the student must have a test score average of 90 or greater. What is the lowest score the student can earn on her last test to bring her test score average into the B range? Can the student score high enough to earn an A?

38. A bowler rolled scores of 142, 136, 154, and 158 in four successive games. What score must she roll in her next game so that her five game average will be between 150 and 165?

39. A certain brand of unleaded gasoline contains 3% of component G-90. The same brand of super-unleaded gasoline contains 12% of the component. A car owner determines that he obtains the best performance from his car engine, over the longest period of time, if he fills his 15-gallon tank with a mixture of unleaded and super-unleaded gasoline that has between 7% and 9% of component G-90. If the car owner wishes to obtain best performance at minimum cost, with what mixture of unleaded and super-unleaded gasoline should he fill the tank?

40. The Easy-Lease car agency rents its medium-sized cars on one of two plans.

Plan A: $170 per week and unlimited mileage.

Plan B: $90 per week and 15 cents per mile.

What range of weekly mileage makes Plan B the cheaper rental plan?

41. Why is the interval containing all the points to the right of 2 denoted as $(2,\infty)$ rather than $(2,\infty]$?

42. A manufacturing firm produces a non-clogging shower head that sells for $80. The production cost for producing x shower heads per day is given by the formula $C = x^2 + 38x + 400$. What must the daily production be for the firm to realize a profit?

43. Pipe A fills a pool in 6 hours. A second pipe is to be installed so that the two pipes, operating together, can fill the pool in less than 2 hours. At what rate must the second pipe, operating alone, fill the pool?

44. Show that the sum of any positive real number and its reciprocal is greater than or equal to 2.

45. Show that if $x + y > 0$, then $x + y \geq 4xy/(x + y)$. [Hint: Start with $(x - y)^2 \geq 0$].

46. Show that if x and y are positive real numbers, then $\sqrt{xy} \leq (x + y)/2$.

47. What is wrong with the following solution for exercise 19?

$$\frac{2}{x-1} < \frac{5}{x+2}$$
$$(x-1)(x+2)\left(\frac{2}{x-1}\right) < \left(\frac{5}{x+2}\right)(x-1)(x+2)$$
$$2(x+2) < 5(x-1)$$
$$2x + 4 < 5x - 5$$
$$9 < 3x$$
$$3 < x$$

48. What is wrong with the following solution for exercise 21?

$$\frac{1}{2x} > -\frac{3}{x+5}$$
$$(2x)(x+5)\frac{1}{2}x > \left(\frac{-3}{x+5}\right)(2x)(x+5)$$
$$x + 5 > -6x$$
$$7x > -5$$
$$x > -\frac{5}{7}$$

49. Consider the inequalities $A(x) < B(x)$ and $A(x) + C(x) < B(x) + C(x)$, where $A(x)$, $B(x)$, and $C(x)$ are expressions in x. For example, $A(x)$ can be an expression such as $5x - 8$. Symbols such as $A(2)$ or $A(-3)$ are the values of the expression when x is replaced by 2 or -3. If $A(x) = 5x - 8$, then $A(2) = 2$ and $A(-3) = -23$. We want to show that $A(x) < B(x)$ and $A(x) + C(x) < B(x) + C(x)$ are equivalent inequalities. This can be done by showing that any solution of the first inequality is also a solution of the second, and that any solution of the second inequality is also a solution of the first. We will then have shown that the solution sets of the two inequalities are equal, and consequently, that the inequalities are equivalent. Fill in the missing details of the proof.

Let r be a solution of $A(x) < B(x)$. Also, let $A(r) = a$, $B(r) = b$, and $C(r) = c$. Because $A(x) < B(x)$ when $x = r$, a __(a)__ b. As a result, $a - b < 0$.

Now, $(a + c) - (b + c) = (a - b) + (c - c) = a - b$ (b) 0. Therefore, $a + c < b + c$ or $A(r) + C(r) <$ (c). Thus, r is a solution of $A(x) + C(x) < B(x) + C(x)$. Let p be a solution of $A(x) + C(x) < B(x) + C(x)$. Also, let $A(p) = m$, $B(p) = n$, and $C(p) = k$. It follows that $n + k <$ (d), and that $(m + k) - (n + k) = m - n$ (e) 0. As a result $A(p) -$ (f) < 0, or $A(p) < B(p)$. Thus, p is a solution of (g).

Since any solution of $A(x) < B(x)$ is also a solution of $A(x) + C(x) < B(x) + C(x)$, and vice versa, the inequalities have the same solution set. They are therefore, (h) inequalities.

50. Fill in the missing details in the proof of the statement,

If $C(x) < 0$, then $A(x) < B(x)$ and $A(x) \cdot C(x) > B(x) \cdot C(x)$ are equivalent inequalities.

Let r be a solution of $A(x) < B(x)$. Also, let $A(r) = a$, $B(r) = b$, and $C(r) = c$. Because r is a solution of $A(x) < B(x)$, $a < b$. Multiplication of both members of an inequality by a (a) number changes the (b) of the inequality. Therefore, $a \cdot c >$ (c), and so (d) is a solution of $A(x) \cdot C(x) > B(x) \cdot C(x)$.

Let p be a solution of $A(x) \cdot C(x) > B(x) \cdot C(x)$. Also let $A(p) = m$, $B(p) = n$, and $C(p) = k < 0$. It follows that $m \cdot k > n \cdot k$. If both sides of this inequality are divided by k, the inequality m (e) n results. But then p is a (f) of $A(x) < B(x)$.

It has been shown that any solution of (g) is also a solution of $A(x) \cdot C(x) > B(x) \cdot C(x)$, and vice versa. Therefore, the solution sets of the two inequalities are (h), and the inequalities are equivalent.

1.7 Equations Solvable as Linear or Quadratic Equations

Squaring both sides of $x = 1$ produces the equation $x^2 = 1$. The solution set of $x^2 = 1$, $\{-1,1\}$ contains the solution set of $x = 1$ as a subset. It is true in general (exercise 46) that if both sides of an equation are raised to the same natural number power, *the resulting equation's solution set contains the solution set of the original equation.* This fact can be used to solve equations that contain radicals or fractional exponents.

Example 1 Solve $x = \sqrt{2x + 3}$.

Solution Square both sides of the equation to remove the radical.

$$x^2 = 2x + 3$$
$$x^2 - 2x - 3 = 0$$
$$(x - 3)(x + 1) = 0$$

Equation $x^2 - 2x - 3 = 0$ is obtained by squaring both sides of the original equation. Consequently, its solution set $\{-1,3\}$ contains the solution set of $x = \sqrt{2x + 3}$ as a subset. We test both numbers to see if they are solutions of $x = \sqrt{2x + 3}$.

$$-1 \neq \sqrt{2(-1) + 3} = \sqrt{1} = 1$$
$$3 = \sqrt{2(3) + 3} = \sqrt{9} = 3$$

Only 3 satisfies the given equation. The solution set is $\{3\}$. ■

Example 2 Solve $(x + 2)^{1/3} = 4$.

Solution A simple linear equation can be generated from the given equation if both its members are raised to the third power.

$$\begin{aligned} [(x + 2)^{1/3}]^3 &= 4^3 \\ x + 2 &= 64 \\ x &= 62 \end{aligned}$$

The solution set of $x + 2 = 64$ contains the solution set of $(x + 2)^{1/3} = 4$ as a subset. (Why?) Therefore, we must test 62 in the original equation.

$$(62 + 2)^{1/3} = 64^{1/3} = 4$$

The solution set is $\{62\}$. ■

Example 3 Solve $\sqrt{x + 7} + \sqrt{x + 4} = 3$.

Solution Writing the given equation as

$$\sqrt{x + 7} = 3 - \sqrt{x + 4},$$

and then squaring, gives us

$$\begin{aligned} x + 7 &= 9 - 6\sqrt{x + 4} + x + 4 \\ -6 &= -6\sqrt{x + 4} \\ 1 &= \sqrt{x + 4}. \end{aligned}$$

Squaring again yields

$$\begin{aligned} 1 &= x + 4 \\ -3 &= x. \end{aligned}$$

We test -3 in the original equation.

$$\sqrt{-3 + 7} + \sqrt{-3 + 4} = 2 + 1 = 3$$

The solution set is $\{-3\}$. ■

The radical expressions in example 3 are separated before squaring to simplify the algebraic manipulations required to obtain a solution. Here is what happens if the radical expressions are not placed on opposite sides of the equation.

$$\begin{aligned} (\sqrt{x + 7} + \sqrt{x + 4})^2 &= 3^2 \\ x + 7 + 2\sqrt{x + 7}\sqrt{x + 4} + x + 4 &= 9 \\ 2\sqrt{(x + 7)(x + 4)} &= -2 - 2x \\ \sqrt{x^2 + 11x + 28} &= -1 - x \\ x^2 + 11x + 28 &= 1 + 2x + x^2 \\ 9x &= -27 \\ x &= -3 \end{aligned}$$

An equation may have no solutions, even though the equation that is generated from it by squaring has solutions.

Example 4 Solve $\sqrt{x-3} + \sqrt{x+2} = 1$.

Solution

$$\begin{aligned}\sqrt{x-3} &= 1 - \sqrt{x+2}\\ x - 3 &= 1 - 2\sqrt{x+2} + x + 2\\ -6 &= -2\sqrt{x+2}\\ 3 &= \sqrt{x+2}\\ 9 &= x + 2\\ 7 &= x\end{aligned}$$

Testing 7 in the original equation yields

$$\sqrt{7-3} + \sqrt{7+2} = 2 + 3 \neq 1.$$

Hence, the solution set of the given equation is $\emptyset$, the empty set. ■

Equations in Quadratic Form

Some equations can be made to look like a quadratic equation with a proper substitution. Such equations are said to have quadratic form.

Example 5 Put each equation in quadratic form.

a. $x^{1/3} - 3x^{1/6} + 2 = 0$

b. $5x^4 + 15x^2 - 20 = 0$

c. $12x^{-2} - 23x^{-1} = -5$

Solution

a. Let $y = x^{1/6}$, then $x^{1/3} = (x^{1/6})^2 = y^2$.

$$y^2 - 3y + 2 = 0$$

b. Let $z = x^2$, then $x^4 = (x^2)^2 = z^2$.

$$5z^2 + 15z - 20 = 0$$

c. Let $w = x^{-1}$, then $x^{-2} = (x^{-1})^2 = w^2$.

$$12w^2 - 23w = -5$$ ■

Example 6 Solve $x^{1/3} - 3x^{1/6} + 2 = 0$.

Solution The equation is transformed into $y^2 - 3y + 2 = 0$ (example 5a). The quadratic equation is solved, and its roots then serve to form equations involving x.

$$\begin{aligned} y^2 - 3y + 2 &= 0 \\ (y-1)(y-2) &= 0 \\ y = 1 \quad &\text{or} \quad y = 2 \end{aligned}$$

But, $y = x^{1/6}$. Therefore,

$$\begin{aligned} x^{1/6} = 1 \quad &\text{or} \quad x^{1/6} = 2 \\ x = 1 \quad &\text{or} \quad x = 2^6 \\ x = 1 \quad &\text{or} \quad x = 64. \end{aligned}$$

The solution set is $\{1,64\}$. ■

Example 7 Solve $5x^4 + 15x^2 - 20 = 0$, $x \in C$.

Solution With the substitution $z = x^2$ (example 5b), the equation becomes $5z^2 + 15z - 20 = 0$.

$$\begin{aligned} z^2 + 3z - 4 &= 0 \\ (z+4)(z-1) &= 0 \\ z = 1 \quad &\text{or} \quad z = -4 \\ x^2 = 1 \quad &\text{or} \quad x^2 = -4 \\ x = \pm 1 \quad &\text{or} \quad x = \pm 2i \end{aligned}$$

The solution set is $\{-1,1,-2i,2i\}$. ■

Example 8 Solve $2(x + 1/x)^2 - 7(x + 1/x) + 5 = 0$, $x \in C$.

Solution With the substitution $y = x + 1/x$, the equation becomes $2y^2 - 7y + 5 = 0$.

$$\begin{aligned} 2y^2 - 7y + 5 &= 0 \\ (2y-5)(y-1) &= 0 \\ 2y = 5 \quad &\text{or} \quad y = 1 \\ 2x + \frac{2}{x} = 5 \quad &\text{or} \quad x + \frac{1}{x} = 1 \\ 2x^2 + 2 = 5x \quad &\text{or} \quad x^2 + 1 = x \\ 2x^2 - 5x + 2 = 0 \quad &\text{or} \quad x^2 - x + 1 = 0 \\ (2x-1)(x-2) = 0 \quad &\text{or} \quad x = \frac{1 \pm \sqrt{1-4}}{2} \end{aligned}$$

The solution set is $\{\frac{1}{2}, 2, (1 - \sqrt{3}i)/2, (1 + \sqrt{3}i)/2\}$. ■

Exercise Set 1.7

(A)

Solve each equation. Assume that each radical expression denotes a real number.

1. $\sqrt{3x+4}=5$
2. $3=\sqrt{x+2}$
3. $(5+2x)^{1/2}=x+1$
4. $(3x-5)^{1/2}=x-1$
5. $\sqrt[3]{2x-1}=-3$
6. $(x+2)^{1/3}=4$
7. $\sqrt{x-5}-\sqrt{x}=1$
8. $\sqrt{x-3}+\sqrt{x+2}=1$
9. $\sqrt{x+2}-\sqrt{x}=1$
10. $3\sqrt{x}-\sqrt{2}=\sqrt{3x+2}$
11. $\sqrt{2x+5}+\sqrt{4x+3}=3$
12. $\sqrt{x+5}-\sqrt{x-3}=2$
13. $x^{2/3}-4x^{1/3}+3=0$
14. $x^{1/3}-3x^{1/6}+2=0$
15. $x^{1/2}-2x^{1/4}-15=0$
16. $x^{-2}-9x^{-1}-10=0$
17. $12x^{-2}-23x^{-1}+5=0$
18. $6x^{-4}-13x^{-2}=5$
19. $9x^4-37x^2+4=0$
20. $6x^4-5x^2+1=0$
21. $x^6-8x^3+7=0$
22. $2x^6-5x^3+2=0$

(B)

23. $6x^4+x^2-40=0,\ x \varepsilon C$
24. $5y^4+20y^2+15=0,\ x \varepsilon C$
25. $\sqrt{6x-2}-\sqrt{2x}=-\sqrt{2x-1}$
26. $\sqrt{2x+9}-\sqrt{x+1}=\sqrt{x+4}$
27. $\sqrt{3\sqrt{x-1}}=\sqrt{x-1}$
28. $\dfrac{(4-x)}{\sqrt{x^2-8x+32}}=\dfrac{3}{5}$
29. $2\sqrt{x}-\sqrt{4x-3}=\dfrac{1}{\sqrt{4x-3}}$
30. $\dfrac{1}{\sqrt{x-3}}-\sqrt{\dfrac{(2x-5)}{(x-3)}}=2$
31. $x+2+3\sqrt{x+2}-4=0$
32. $\left(\dfrac{x^2+3}{x}\right)^2-\dfrac{6(x^2+3)}{x}+8=0,\ x \varepsilon C$
33. $(x+1)^{2/3}-2(x+1)^{1/3}-3=0$
34. $(2x-3)^{1/3}-(2x-3)^{1/6}-2=0$
35. $\sqrt{2x+1}+\sqrt{6x+1}-\sqrt{17x-4}=0$
36. $\sqrt{x-1}+\sqrt{x+3}-\sqrt{2x-5}=0$
37. $\sqrt{10x+9}=\sqrt{x}+\sqrt{6x+1}$
38. $\sqrt{x-2}-\sqrt{x+2}+\sqrt{x+1}=0$

In Exercises 39–42, solve for the specified symbol.

39. $t=\sqrt{\dfrac{2s}{g}},\ s$
40. $t=\sqrt{\dfrac{2s}{g}},\ g$
41. $f=\dfrac{1}{2\pi\sqrt{LC}},\ C$
42. $v=2\pi n\sqrt{r^2-x^2},\ r$

43. A stone is dropped from a cliff into a stream. The sound of the splash is heard 6 seconds after the stone is released. How high is the cliff above the stream? Assume the velocity of sound is 1060 feet per second, and use the formula $h=16t^2$ to determine the distance h, that the rock falls in t seconds.

44. Let $A(x)=B(x)$ be an equation in which $A(x)$ and $B(x)$ are expressions in x. For example, $A(x)$ can be $\sqrt{3x-5}$. A symbol such as $A(3)$ denotes the real number that results when 3 replaces x in $\sqrt{3x-5}$. Thus, $A(3)=2$. We want to show that if n is a natural number, then the solution set of $A(x)=B(x)$ is a subset of the solution set of $[A(x)]^n=[B(x)]^n$. Fill in the details in the following argument. Let r be a solution of $A(x)=B(x)$. Then, $A(r)=$ <u>(a)</u> and it follows that $[A(r)]^n=$ <u>(b)</u>. Therefore, r is also a solution of <u>(c)</u> = <u>(d)</u>. It has been shown that any solution of <u>(e)</u> = <u>(f)</u> is also a solution of $[A(x)]^n=[B(x)]^n$. Consequently, the solution set of $A(x)=B(x)$ is a subset of the solution set of $[A(x)]^n=[B(x)]^n$.

1.8 Absolute Value in Equations and Inequalities

We are often concerned only with the magnitude of a number and not its sign. To compute gas mileage between two cities, we need know only the distance between the cities. It is not necessary to know the direction of travel. We may be interested only in the magnitude of the error in the diameter of a manufactured part. Whether the diameter is larger or smaller than the design diameter may be of no consequence.

If it is only the magnitude, and not the sign, of a number that is to be denoted, we can use absolute value notation. The magnitude of y is given as $|y|$. It is convenient to think of the absolute value of a number as its distance from the origin. Thus, the numbers that satisfy $|y| = 6$ are depicted in figure 1.11 as -6 and 6. Each number is 6 units from the origin.

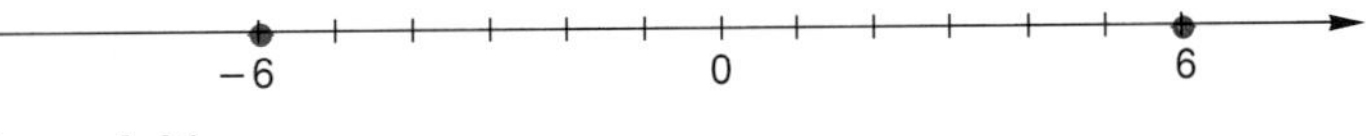

Figure 1.11

Absolute value notation is necessary, but it is very difficult to manipulate algebraically. Here is a definition that gives us the magnitude of a number without using absolute value notation.

$$|y| = \begin{cases} y \text{ if } y \geq 0 \\ -y \text{ if } y < 0 \end{cases} \tag{1}$$

Example 1 Use (1) to write $|6|$, $|-6|$, and $|4x - 12|$ without absolute value notation.

Solution

$$|6| = 6$$

$$|-6| = -(-6) = 6$$

$$|4x - 12| = \begin{cases} 4x - 12 \text{ if } \quad 4x - 12 \geq 0 \quad \text{or if} \quad x \geq 3 \\ -(4x - 12) \text{ if } \quad 4x - 12 < 0 \quad \text{or if} \quad x < 3 \end{cases}$$

■

Absolute Value Properties

Some properties of absolute value notation and arguments verifying those properties are given next.

Some Absolute Value Properties

If a and b are real numbers, then

1. $|a|^2 = a^2$
2. $|ab| = |a||b|$
3. $\left|\dfrac{a}{b}\right| = \dfrac{|a|}{|b|}$
4. $|a + b| \leq |a| + |b|$

1. Either $a \geq 0$ or $a < 0$. If $a \geq 0$, then by (1), $|a| = a$ and therefore, $|a|^2 = a^2$. If $a < 0$, then by (1), $|a| = -a$, and it follows that $|a|^2 = (-a)^2 = a^2$.
2. From part 1, $|ab|^2 = (ab)^2$. But $(ab)^2 = a^2b^2 = |a|^2\,|b|^2$. Hence, $|ab|^2 = |a|^2|b|^2$ and so $|ab| = |a||b|$.
3. From part 1, $|a/b|^2 = (a/b)^2 = |a|^2/|b|^2$. Hence, $|a/b| = |a|/|b|$.
4. From part 1,

$$\begin{aligned} |a + b|^2 &= (a + b)^2 \\ &= a^2 + 2ab + b^2 \\ &= |a|^2 + 2ab + |b|^2. \end{aligned}$$

Now, $ab \leq |ab| = |a||b|$. Therefore,

$$|a|^2 + 2ab + |b|^2 \leq |a|^2 + 2\,|a|\,|b| + |b|^2.$$

It follows that

$$|a + b|^2 \leq |a|^2 + 2\,|a||b| + |b|^2 = (|a| + |b|)^2,$$

and we conclude that $|a + b| \leq |a| + |b|$.

The first three properties are easily remembered as

1. The square of the absolute value of a number equals the square of the number.

$$|x - 3|^2 = (x - 3)^2$$

2. The absolute value of a product is the product of the absolute values.

$$|3(-7)| = |-21| = 21, \text{ and } |3||-7| = 3 \cdot 7 = 21$$
$$|(x + 2)(x - 3)| = |x + 2||x - 3|$$

3. The absolute value of a quotient is the quotient of the absolute values.

$$\left|\frac{3}{-7}\right| = -\left(\frac{3}{-7}\right) = \frac{3}{7}, \text{ and } \frac{|3|}{|-7|} = \frac{3}{7}$$

$$\left|\frac{x+2}{x-3}\right| = \frac{|x+2|}{|x-3|}$$

The fourth property is one of the most important inequalities in mathematics. It is called the Triangle Inequality because, in a more advanced context, it can be related to the lengths of the sides of a triangle. Note that the equality holds only if a and b are of the same sign.

$$|-5-7| = |-12| = 12 = |-5| + |-7|$$
$$|-5+7| = |2| = 2 < |-5| + |7| = 12$$

Solving $|y| = k$, $|y| < k$, and $|y| > k$

Equation $|y| = 6$ can be read as "Y is a number whose distance from the origin is 6 units." Figure 1.11 depicts the solutions, -6 and 6, of the equation.

The inequality $|y| < 6$ can be read as "Y is a number whose distance from the origin is less than 6 units." That is, y must be a number that is between -6 and 6. Figure 1.12 depicts the solution set, $(-6,6)$, of $|y| < 6$.

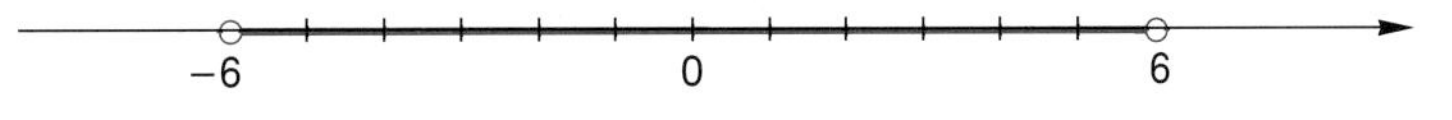

Figure 1.12

If y is a number that satisfies the inequality $|y| > 6$, then it must be a number whose distance from the origin is greater than 6 units. Such numbers must be less than -6 or greater than 6. Therefore, the solution set of $|y| > 6$ is $(-\infty,-6) \cup (6,\infty)$ (figure 1.13).

-6 0 6

Figure 1.13

The discussion above that led to the solution sets for $|y| = 6$, $|y| < 6$, and $|y| > 6$, would be the same for any positive number k. As a result we can say

> **Rewriting Absolute Value Statements**
>
> $|y| = k$ is equivalent to $y = k$ or $y = -k$,
> $|y| < k$ is equivalent to $-k < y < k$, and
> $|y| > k$ is equivalent to $y < -k$ or $y > k$.

Example 2 Solve **a.** $|x - 4| = 7$, **b.** $|x - 4| < 7$, **c.** $|x - 4| > 7$.

Solution Let $y = x - 4$.

a. $|y| = 7$ is equivalent to

$$\begin{aligned} y &= 7 \quad \text{or} \quad & y &= -7 \\ x - 4 &= 7 \quad \text{or} \quad & x - 4 &= -7 \\ x &= 11 \quad \text{or} \quad & x &= -3. \end{aligned}$$

The solution set is $\{-3,11\}$ (figure 1.14).

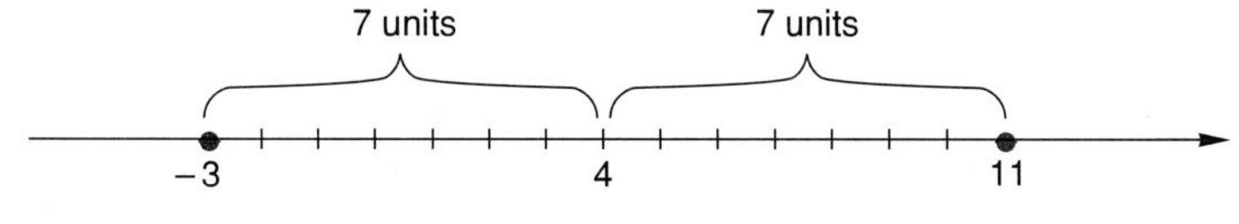

Figure 1.14

b. $|y| < 7$ is equivalent to

$$\begin{aligned} -7 &< y < 7 \\ -7 &< x - 4 < 7 \\ -3 &< x < 11. \end{aligned}$$

The solution set is $(-3,11)$ (figure 1.15).

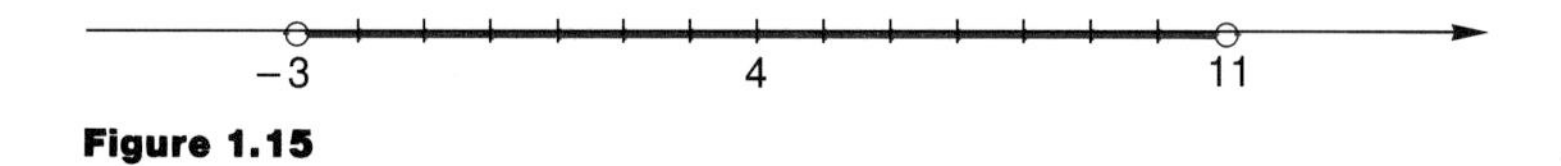

Figure 1.15

c. $|y| > 7$ is equivalent to

$$\begin{aligned} y &< -7 \quad \text{or} \quad & y &> 7 \\ x - 4 &< -7 \quad \text{or} \quad & x - 4 &> 7 \\ x &< -3 \quad \text{or} \quad & x &> 11. \end{aligned}$$

The solution set is $(-\infty,-3) \cup (11,\infty)$ (figure 1.16).

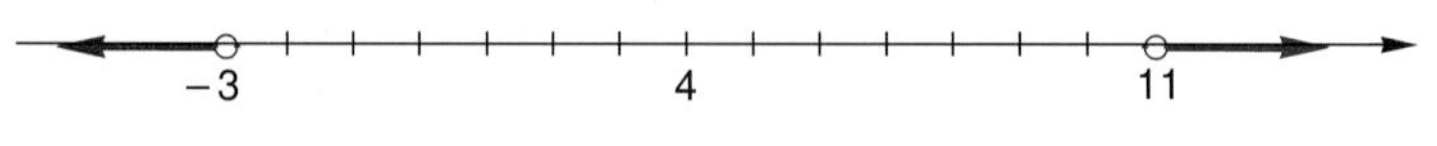

Figure 1.16 ■

Example 3 Solve $|3x^2 + 9x - 8| = 4$.

Solution The equation has the form $|y| = 4$ where $y = 3x^2 + 9x - 8$. Because $|y| = 4$ is equivalent to $y = 4$ or $y = -4$, we may write

$$\begin{aligned} 3x^2 + 9x - 8 &= 4 \quad \text{or} \quad 3x^2 + 9x - 8 = -4 \\ 3x^2 + 9x - 12 &= 0 \quad \text{or} \quad 3x^2 + 9x - 4 = 0 \\ x^2 + 3x - 4 &= 0 \quad \text{or} \quad x = \frac{-9 \pm \sqrt{81 - 4(3)(-4)}}{6} \\ (x + 4)(x - 1) &= 0 \quad \text{or} \quad x = \frac{-9 \pm \sqrt{129}}{6}. \end{aligned}$$

The solution set is $\{-4, (-9 - \sqrt{129})/6, (-9 + \sqrt{129})/6, 1\}$. ■

Example 4 Solve $|(x - 1)/(4 - x)| \geq 3$.

Solution The inequality has the form $|y| \geq 3$, where $y = (x - 1)/(4 - x)$. But $|y| \geq 3$ is equivalent to $y \leq -3$ or $y \geq 3$ (figure 1.17). We therefore write

$$\begin{aligned} \frac{x-1}{4-x} &\leq -3 \quad \text{or} \quad \frac{x-1}{4-x} \geq 3 \\ \frac{x-1}{4-x} + 3 &\leq 0 \quad \text{or} \quad \frac{x-1}{4-x} - 3 \geq 0 \\ \frac{x - 1 + 3(4-x)}{4-x} &\leq 0 \quad \text{or} \quad \frac{x - 1 - 3(4-x)}{4-x} \geq 0 \\ \frac{-2x + 11}{4-x} &\leq 0 \quad \text{or} \quad \frac{4x - 13}{4-x} \geq 0. \end{aligned}$$

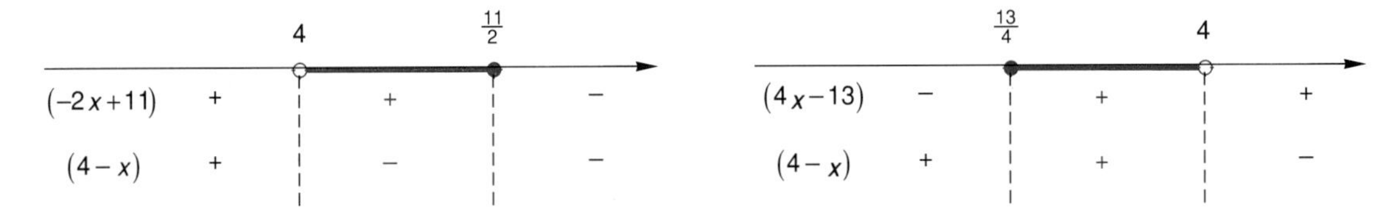

Figure 1.17

The fraction $(-2x + 11)/(4 - x)$ is negative in the interval $(4, \frac{11}{2}]$, where its numerator and denominator have opposite signs. The fraction is 0 at $\frac{11}{2}$.

The fraction $(4x - 13)/(4 - x)$ is positive in the interval $[\frac{13}{4}, 4)$, where its numerator and denominator have the same sign. The fraction is 0 at $\frac{13}{4}$.

The solution set is $[\frac{13}{4}, 4) \cup (4, \frac{11}{2}]$.

Example 5 Use absolute value notation to rewrite each statement.

a. $-2 < x < 8$

b. $x \leq -3$ or $x \geq 1$

Solution

a. x is any real number in the interval $(-2,8)$. The midpoint of the interval is 3, and any point in the interval is less than 5 units from 3 (figure 1.18). This fact can be written with absolute value notation as $|x - 3| < 5$.

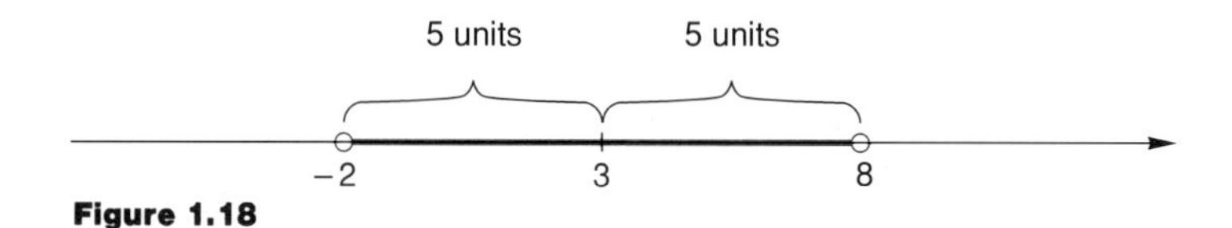

Figure 1.18

b. The midpoint of the interval $[-3,1]$ is -1. Hence, the numbers that belong to $\{x|x \leq -3 \text{ or } x \geq 1\}$ must be 2 or more units from -1 (figure 1.19). This fact is written, with absolute value notation, as $|x - (-1)| \geq 2$ or as $|x + 1| \geq 2$.

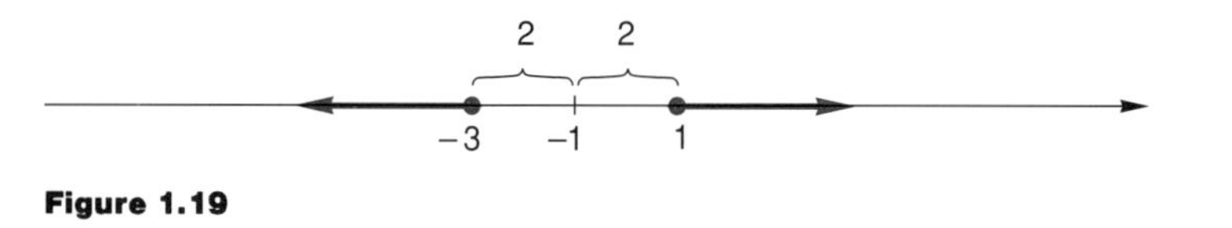

Figure 1.19

■

Exercise Set 1.8

(A)

Each of the first eight exercises has three related parts, $|y| = k$, $|y| < k$, and $|y| > k$. Solve each part and graph its solution set.

1. $|x + 2| = 6$, $|x + 2| < 6$, $|x + 2| > 6$
2. $|x - 5| = 2$, $|x - 5| < 2$, $|x - 5| > 2$
3. $|3x - 2| = 9$, $|3x - 2| < 9$, $|3x - 2| > 9$
4. $|5x + 4| = 12$, $|5x + 4| < 12$, $|5x + 4| > 12$
5. $|8 - 3x| = 7$, $|8 - 3x| < 7$, $|8 - 3x| > 7$
6. $|11 - 4x| = 3$, $|11 - 4x| < 3$, $|11 - 4x| > 3$
7. $\left|\frac{4 - 3x}{2}\right| = \frac{2}{3}$, $\left|\frac{4 - 3x}{2}\right| < \frac{2}{3}$, $\left|\frac{4 - 3x}{2}\right| > \frac{2}{3}$
8. $\left|\frac{x + 6}{3}\right| = 5$, $\left|\frac{x + 6}{3}\right| < 5$, $\left|\frac{x + 6}{3}\right| > 5$

Solve each equation or inequality.

9. $|x^2 - 5x + 2| = 3$
10. $|2x^2 - 6x + 1| = 3$
11. $|5x + 3| \leq 9$
12. $|6x - 4| \leq 5$
13. $|4 - 9x| \geq 2$
14. $|6 - 3x| \geq 1$
15. $|3x - 8| < 13$
16. $|2x + 7| > 11$
17. $|5x^2 + 15x - 2| = 8$
18. $|3x^2 + 12x - 4| = 11$

19. $\left|\frac{1-2x}{3}\right| < 7$

20. $\left|\frac{2x+4}{5}\right| < 2$

21. $\left|\frac{3x+8}{2}\right| > 10$

22. $\left|\frac{x}{2} - \frac{5}{6}\right| > 4$

23. $\left|\frac{11+4x}{5}\right| \leq 11$

24. $\left|\frac{3x}{4} + \frac{6x}{5}\right| \geq \frac{39}{10}$

Use absolute value notation to rewrite each statement.

25. $1 < x < 9$

26. $-12 < x < -2$

27. $x \leq -8$ or $x \geq -2$

28. $x \leq -5$ or $x \geq 15$

29. $-8 < x < 3$

30. $\frac{3}{5} < x < \frac{8}{3}$

31. $x \leq 0.58$ or $x \geq 4.32$

32. $x \leq -\frac{15}{4}$ or $x \geq \frac{5}{6}$

(B)

Solve each equation or inequality.

33. $|x^2 + 2x + 5| = 2$

34. $|2x^2 - x + 3| = 1$

35. $\left|\frac{x-2}{x+3}\right| = 5$

36. $\left|\frac{2x-1}{x-4}\right| = 9$

37. $\left|\frac{x+5}{2x-1}\right| \leq 3$

38. $\left|\frac{3x-2}{x+6}\right| \leq 8$

39. $\left|\frac{2-5x}{2x+3}\right| \geq 6$

40. $\left|\frac{4-2x}{x+1}\right| \geq 3$

41. $\left|\frac{3x+4}{5x-12}\right| < 4$

42. $\left|\frac{x-6}{2-3x}\right| > 5$

43. $|x+3||x-1| \leq 2$

44. $|x-5||x+3| \geq 4$

45. $\frac{|x-7|}{|x+3|} < 1$

46. $\frac{|2x+5|}{|4-x|} \geq 7$

47. $|2x+3||x+1| > 3$

48. $|5x-2||x+2| < 6$

49. $|2x+1| = |x+3|$

50. $|5-2x| = |x+7|$

51. $|x-2| < |x+5|$

52. $|x+1| > |2x-3|$

53. $\left|\frac{2}{x-1}\right| < \frac{5}{x+3}$

54. $\left|\frac{4}{5-2x}\right| > \frac{1}{3x+1}$

55. $|x^2 - 5| \leq 0.1$

56. $|x^2 + 2x - 15| \leq 0.01$

57. Find all the points on the number line that are 3 times as far from -2 as they are from 6.

58. Find all the points on the number line that are 2 times as far from 1 as they are from 9.

In exercises 59–61, a, b, and c are assumed to be real numbers.

59. Show that $|a - b| \geq |a| - |b|$. (Hint: Let $a = a - b + b$. Then use the triangle inequality.)

60. Use the triangle inequality to show that

$$|a + b + c| \leq |a| + |b| + |c|.$$

61. Argue the truth of the statement $|a - b| = |b - a|$.

62. Given that $|x - 9| < 3 \cdot 10^{-4}$ and that $|y - 6| < 5 \cdot 10^{-4}$, use the triangle inequality to show that $|x + y - 15| < 10^{-3}$.

1.9 Absolute Value and Error

Suppose that in the expression $8x + 6$, x is a measured length. Because all measurement is subject to instrument and human error, there is likely to be an error in x and therefore an error in $8x + 6$. Let us assume that x is to be 2 and that $8x + 6$ is to be 22 with an error that is not greater than ± 0.05 units. What maximum error can be tolerated in the measurement x so that the magnitude of the error in $8x + 6$ is less than or equal to 0.05? The maximum tolerable error in x can be determined by making the absolute value of the difference between $8x + 6$ and 22 less than or equal to 0.05 units.

$$\begin{aligned} |(8x + 6) - 22| &\le 0.05 \\ |8x - 16| &\le 0.05 \\ |8(x - 2)| &\le 0.05 \\ |8||x - 2| &\le 0.05 \\ |x - 2| &\le \frac{0.05}{8} = 0.00625 \end{aligned}$$

Now we have that

$$\begin{aligned} -0.00625 \le x - 2 &\le 0.00625 \\ 1.99375 \le \quad x \quad &\le 2.00625. \end{aligned}$$

We see that the magnitude of the error in the measurement x must be less than or equal to 0.00625 to guarantee that $8x + 6$ differs from 22 by no more than ± 0.05 units. If x is a number between 1.994 and 2.006, then

$$21.952 \le 8x + 6 \le 22.048.$$

Now consider a more difficult problem. A 30 by 20 plate is to be fabricated. The maximum allowable error in the area is ± 0.5 square units. What is the maximum allowable error in the measurement of a side of the plate? We let x be the length of the larger side of the plate. Then $x - 10$ is the length of the smaller side. Since the actual area of the plate can differ from 600 square units by no more than ± 0.5 square units we write

$$\begin{aligned} |x(x - 10) - 600| &\le 0.5 \\ |x^2 - 10x - 600| &\le 0.5 \\ |(x - 30)(x + 20)| &\le 0.5 \\ |x - 30||x + 20| &\le 0.5 \\ |x - 30| &\le \frac{0.5}{|x + 20|}. \end{aligned} \tag{1}$$

The allowable area error limits the size of the error in x. For example, the error in x cannot be as large as ± 1 unit. If it is ± 1 unit, the resulting plate areas are

$$\begin{aligned} 31 \cdot 21 &= 651 \text{ square units} \\ 31 \cdot 19 &= 589 \text{ square units} \\ 29 \cdot 21 &= 609 \text{ square units} \\ 29 \cdot 19 &= 551 \text{ square units.} \end{aligned}$$

Each of the above areas differs from 600 square units by more than ± 0.5 square units. It is certain then that x must be a number between 29 and 31. As a result, $x + 20$ in (1) must be a number between 49 and 51.

$$\frac{0.5}{51} \leq \frac{0.5}{|x + 20|} \leq \frac{0.5}{49}$$

Now, if $|x - 30| \leq \frac{0.5}{51}$, where $29 \leq x \leq 31$, then

$$|x - 30| \leq \frac{0.5}{|x + 20|}. \tag{2}$$

From (2) we get

$$|x - 30||x + 20| \leq 0.5$$
$$|(x - 30)(x + 20)| \leq 0.5$$
$$|x^2 - 10x - 600| \leq 0.5$$
$$|x(x - 10) - 600| \leq 0.5.$$

Thus, if

$$|x - 30| \leq \frac{0.5}{51} = 0.0098,$$

then

$$-0.0098 \leq x - 30 \leq 0.0098$$
$$29.9902 \leq x \leq 30.0098,$$

and the area error is less than ± 0.5 square units.

$$30.0098 \cdot 20.0098 = 600.490 \text{ square units}$$
$$30.0098 \cdot 19.9902 = 559.902 \text{ square units}$$
$$29.9902 \cdot 20.0098 = 600.098 \text{ square units}$$
$$29.992 \cdot 19.9902 = 599.510 \text{ square units}$$

In summary, the above area calculations show that if the magnitude of the error in the measurement of a side is less than or equal to $0.5/51$ units, then the magnitude of the area error is less than 0.5 square units.

Exercise Set 1.9

(A)

1. If $6x + 3$ is to differ from 27 by less than 0.5 units, how much can x differ from 4?
2. If $8x - 1$ is to differ from 39 by less than 1 unit, how much can x differ from 5?
3. If the magnitude of the error in the quantity $2x + 6$ is to be less than or equal to ± 0.05 units, how large an error can be tolerated in the measurement of x, which is to be 8 units?
4. If the magnitude of the error in the quantity $12 - 4x$ is to be less than or equal to ± 0.5 units, how large an error can be tolerated in the measurement of x, which is to be 2 units?
5. Show that if $|(3x + 7) - 19| < 1$, then $|x - 4| < \frac{1}{3}$, and $\frac{11}{3} < x < \frac{13}{3}$.
6. Show that if $|(5x - 8) - 22| \leq 0.5$, then $5.9 \leq x \leq 6.1$.

(B)

7. A 50 by 30 rectangular plate is to be fabricated. If the allowable error in the area is ± 1 square unit, what is an allowable error in the measurement of a side of the plate?

8. If $x^2 + 3x - 4$ is to differ from 6 by less than 0.5 units, how much can x differ from 2?

9. If $8/(x - 3)$ is to differ from 4 by less than 0.05 units, how much can x differ from 5?

10. Indicate some interval around $x = 8$ so that if x is any number in that interval, then $x^2 + x - 6$ will differ from 66 by less than 1 unit.

Chapter 1 Review

Major Points for Review

Section 1.1

Set builder notation

Replacement and solution sets of an equation

Null set, subset

Solving an equation by generating a sequence of equivalent equations

The general form of a linear equation

Mathematical models—word problems

Section 1.2

Properties of integral and rational exponents

The principal n^{th} root

Interpreting the symbol $b^{m/n}$

Radical notation

Section 1.3

Factoring trinomials

Factoring a difference of squares and a sum or difference of cubes

Factoring by grouping

Factoring expressions that are not polynomials

Section 1.4

Add, subtract, multiple, and divide complex numbers

The relationship between imaginary numbers and the sets R, H, Q, J, W, and N, as subsets of C, the set of complex numbers

Section 1.5 Solving a quadratic equation by factoring, by completing the square, or by using the quadratic formula

The discriminant, as it relates to the nature of the roots of a quadratic equation

Section 1.6 Solving an inequality by generating a sequence of equivalent inequalities

Changing the sense of an inequality

Solving inequalities with the aid of sign graphs

Interval notation

Set union and set intersection

Section 1.7 Solving equations containing radical expressions

Solving equations in quadratic form

Section 1.8 The absolute value of a real number, and properties of absolute value notation

Solving equations and inequalities that contain absolute value expressions

Review Exercises

Use set builder notation to specify each set.

1. The set of integers between -2 and 6
2. The set of natural number multiples of 7

Write each sum or product in $a + bi$ form.

3. $(8 - 2i) + (-5 + 6i)$
4. $(-23 - 11i) - (18 + 7i)$
5. $(5 + 3i)(-15 + i)$
6. $(2i)(8i + 19)$

Completely factor each expression.

7. $2x^2 + 7x + 6$
8. $x^2 - 8x - 20$
9. $y^3 + 27$
10. $16y^2 - x^2$
11. $x^5 - 3x^4 + x^3$
12. $2y^4 + 3y^3 - 9y^2$
13. $x^2 + xy + xz + yz$
14. $x^3 + 3x^2 + 3x + 1$

Write each expression as a single numeral without exponents.

15. $-8^{2/3}$
16. $25^{3/2}$
17. $81^{3/4}$
18. $-32^{-4/5}$
19. $16^{-3/4} + 64^{-4/3}$
20. $\dfrac{243^{9/15}}{(0.001)^{2/6}}$
21. $[6^{-2} + (49^{-1})^{3/2}]1024^{6/10}$
22. $\dfrac{25^{-1/2}\,(\frac{1}{125})^{2/3}500}{216^{4/6}}$

Completely factor each expression.

23. $y^4 - 25$
24. $(5x - y)^2 - 16$
25. $x^2 - y^2 + 4y - 4$
26. $9x^3y^3 + 12x^2y - 15xy^2$
27. $5x^{4n} - 26x^{2n} + 5$
28. $27x^{6n} + y^{12n}$
29. $10y - 23y^{1/2} - 5$
30. $12x^{4/3} - 18x^{2/3} + 6$

Write the complex conjugate of each number.

31. $4 - i$
32. $7 + 3i$
33. $5i$
34. $-8 + 11i$

Express each quotient in the form $a + bi$.

35. $\dfrac{2 + 5i}{7 - i}$

36. $\dfrac{5 - 2i}{6i}$

37. $\dfrac{4i}{8 - 6i}$

38. $\dfrac{11}{3i}$

Write each expression so that each variable appears once at most and all exponents are positive.

39. $\left(\dfrac{5x^2y^5z}{3xy^{-3}z^4}\right)^{-2}$

40. $\left(\dfrac{x^{-3}y^{-2}z^2}{x^3y^{-5}z}\right)^4$

41. $\left(\dfrac{x^{2/3}y^{1/3}}{(xy)^{-1/2}}\right)^{-1/3}$

42. $\left[\left(\dfrac{x^{3n}y^{2n}}{x^ny^{3n}}\right)^{1/6}\right]^{12/n}$

Factor each expression.

43. $27y^3 - 12y^2 + 72y - 32$

44. $-9y^2 + 25 + 12xy - 4x^2$

45. $x^2y^{-2} + x^{-1}y$

46. $2x^{-2}y^{-3} + 5x^{-3}y^{-2} - 3x^{-4}y^{-1}$

Solve for x, $x \varepsilon R$.

47. $8[x + 3(2x - 5) - 6] = 5x - 12$

48. $-4[(3x - 2) - (8 - x)] = 6x + 4(5 - 2x)$

49. $(x - 2)(3x + 1) = (5x - 7)(x - 2)$

50. $(4x - 1)(7x + 4) = (4x - 1)(x + 12)$

51. $\dfrac{5}{4 - 3x} - 1 = \dfrac{x + 3}{3x - 4}$

52. $\dfrac{x^2 + 2}{x^2 - 9} = \dfrac{2}{x - 3} + \dfrac{x}{x + 3}$

53. $x^2 + x - 6 = 0$

54. $x^2 + 10x + 16 = 0$

55. $6x^2 - 19x + 10 = 0$

56. $20x^2 + 17x + 3 = 0$

57. $x^2 + 6x - 2 = 0$

58. $3x^2 - 4x - 5 = 0$

59. $(2x + 5)^2 = 9$

60. $(3x - 4)^2 = (x + 6)^2$

61. $\dfrac{2x - 1}{x^2 - 4x} - 3 = \dfrac{6}{x}$

62. $\dfrac{2}{5} - \dfrac{x}{x + 3} = \dfrac{x}{x - 3}$

63. $8x + 5 < 12$

64. $3x - 11 > -2$

65. $-3 \le 5 - 2x \le 8$

66. $12 \ge 3x + 2 \ge 1$

67. $2 \le \dfrac{4x - 9}{3} \le 14$

68. $-18 \le 8 - \dfrac{5x}{4} \le -5$

69. $\dfrac{3}{x - 2} < \dfrac{4}{x + 1}$

70. $\dfrac{7}{x} > \dfrac{-2}{3x - 4}$

71. $2x^2 + 11x + 5 \le 0$

72. $3x^2 - 14x + 8 \le 0$

73. $x^2 + 8x + 3 > 0$

74. $5x^2 - 2x - 1 < 0$

75. $\dfrac{3x}{4 - x} \ge \dfrac{2x}{x + 2}$

76. $\dfrac{3x + 2}{2x + 1} \le \dfrac{-4}{5x - 6}$

77. $\sqrt{x + 32} - \sqrt{x} = 4$

78. $2\sqrt{x} - 3 = \sqrt{2x + 5}$

79. $x^{2/3} + 3x^{1/3} - 10 = 0$

80. $3x^{-2} - 22x^{-1} - 16 = 0$

81. $10x^{-2/3} - 13x^{-1/3} + 15 = 0$

82. $9x^4 - 229x^2 + 100 = 0$

83. $\sqrt{x} - \sqrt{5x - 2} = \dfrac{4}{\sqrt{5x - 2}}$

84. $\sqrt{2x - 1} = \sqrt{2x} - \sqrt{6x - 2}$

85. $|6x + 11| \le 7$

86. $|3x - 10| \ge 8$

87. $|x^2 + 3x - 1| = 5$

88. $|2x^2 - 8x + 3| = 7$

89. $\left|\dfrac{12x - 5}{4}\right| > 8$

90. $\left|\dfrac{18 - 3x}{5}\right| < 10$

91. $\left|\dfrac{3 - 2x}{x + 6}\right| \ge 2$

92. $\left|\dfrac{x+4}{3x+8}\right| \leq 13$

93. $|5x+6| = |8x-3|$

94. $|7-3x| = |x+2|$

95. $|3x+2| < |2x+9|$

96. $|6x+11| > |4x-7|$

Solve each equation by the method of completing the square, $x \varepsilon C$.

97. $x^2 + 6x + 13 = 0$

98. $x^2 + 8x + 25 = 0$

99. $2x^2 - 3x - 6 = 0$

100. $3x^2 + x - 2 = 0$

Without solving the equation, determine the nature of its roots.

101. $5x^2 + 3x + 4 = 0$

102. $2x^2 - 6x - 7 = 0$

Write a quadratic equation in standard form that has the indicated solution set.

103. $\{-6,1\}$

104. $\{3,8\}$

105. $\{1+i, 1-i\}$

106. $\{-4,-2\}$

107. Determine k so that the equation $kx^2 + 5x - 4 = 0$ has one root of multiplicity 2.

108. Determine k so that the equation $x^2 - kx + 9 = 0$ has imaginary roots.

109. Can the sum of 3 consecutive integers be 286?

110. The perimeter of a rectangular lot is 480 feet. If the length of the lot is one and one half times its width, what are the dimensions of the lot?

111. Walnuts selling at $1.10 per pound and almonds selling at $1.60 per pound are combined to form 140 pounds of a nut mix that sells for $1.45 a pound. How many pounds of each type of nut are in the mix?

112. Machine A can fabricate a gidget in 4 minutes. Machine B fabricates the same gidget in 5 minutes. How long would it take the two machines, working together, to fabricate 60 gidgets?

113. Car A leaves Los Angeles at 10:00 A.M. traveling towards San Francisco on route 5 at 65 mph. Car B leaves San Francisco at 10:00 A.M. and travels towards Los Angeles on route 5. If the distance between Los Angeles and San Francisco on route 5 is 360 miles, and if the cars pass each other 2 hours and 40 minutes after beginning their trips, determine the average speed of car B.

114. A 50 by 25 meter pool is surrounded by a concrete walk of uniform width. Together, the area of the pool and walk is 1.75 times the area of the pool. What is the width of the walk?

115. A canoeist rowed upstream 10 miles and then back to the starting point in 2 hours. If the stream current was 3 mph, what was the canoeist's rowing speed in still water?

116. A tank is filled through pipe B one hour faster than it is filled through pipe A. It takes 2.5 hours to fill the tank if both pipes are used simultaneously. How long does it take each pipe, operating alone, to fill the tank?

117. Based on past sales records, a car dealership knows it can sell 1,000 units of its "Young Executive" model at $10,000 each. It also knows that each $500 decrease in price will result in 100 more units being sold. What selling price for the "Young Executive" model will yield $11,250,000 in revenue for the dealership?

118. The percentage of antifreeze in an automobile's 9-liter radiator has to be increased from its current 20% to a percentage between 30% and 40%. How much of the solution in the radiator must be drained and replaced with pure antifreeze to achieve this goal?

119. The length of a rectangle is 10 units more than its width. Letting w be the width, determine the values of w for which the ratio of the perimeter of the rectangle to its area is less than 0.5.

120. A rock is dropped down a dry well. The sound of the rock hitting bottom is heard 2.5 seconds after the rock is released. Assume that the velocity of sound is 1,040 feet per second. Determine the depth of the well if the distance that the rock falls in t seconds is given by the formula $h = 16t^2$.

121. If the magnitude of the error in the quantity $10x - 4$ is to be less than $|0.05|$ units, how large an error can be tolerated in the measurement of x, which is to be 3 units?

122. If $x^2 - 2x + 5$ is to differ from 13 by less than $|0.5|$ units, how much can x differ from 4?

Chapter 1 Test 1

1. Write each expression as an equivalent numeral having no exponent.
   a. $-25^{10/6}$
   b. $\dfrac{25^{-1/2} \cdot 81^{3/4}}{216^{2/3}}$
2. Let $z_1 = 8 + 5i$, and $z_2 = -3 + 7i$. Find $z_1 - z_2$ and z_1/z_2 in $a + bi$ form.
3. Factor
   a. $x^2 + 3x - 10$ **b.** $27x^3 - 8$
4. Factor $6xw + 3wy + 2yz + 4xz$.

Solve for x, $x \in R$.

5. $6x^2 - 2x - 1 = 0$
6. $12x - 4 < -3x + 16$
7. $x^{2/3} - 5x^{1/3} - 6 = 0$
8. $(x - 6)(2x + 5) \geq 0$
9. $|4x - 9| \leq 7$
10. $\sqrt{x + 16} - \sqrt{x} = 2$
11. Find three consecutive odd integers that sum to 375.

Chapter 1 Test 2

1. Write
   a. $(9^{3/2} \cdot 27^{2/3})/256^{3/4}$ as an equivalent numeral having no exponent.
   b. $[(x^{-2}y^{5/3}z^{1/4})/(x^0y^{-7/3}z^{3/4})]^{-1}$ so that each variable occurs only once and all exponents are positive.
2. If z_1 and z_2 are complex conjugates, show that $z_1 \cdot z_2$ is a real number.
3. Factor
   a. $30x^3 - 48x^2 - 9x$ **b.** $x^{3/4} + 64$
4. Factor $9x^2 - 4y^2 - 12y - 9$.

Solve for x, $x \in R$.

5. $2x^2 + 8x + 3 = 0$
6. $\sqrt{2x + 5} + \sqrt{x - 4} = 1$
7. $12x^2 - 37x + 21 \geq 0$
8. $|7x + 12| \leq 9$
9. $\dfrac{2x}{1 - x} < -\dfrac{x}{3x + 5}$
10. Determine k so that the equation $2x^2 + kx - 8 = 0$ has one root of multiplicity 2.
11. A plane made a 4,000 mile roundtrip in 8 hours. It encountered a 30 mph headwind in one direction, but it was aided by a 30 mph tailwind when flying in the opposite direction. What was the plane's airspeed?

Chapter 2

Functions and Graphing

The notion of a function is one of the central concepts in mathematics. In this chapter we define a function. We show how to construct functions, and how, with graphing, it is possible to realize a function in a concrete way. We show that a function is the mathematical model for cause and effect relationships in the real world.

2.1 Functions

A student was determined to pass his math course with minimum effort. Was it possible, he wondered, to earn a B with only twenty minutes of study a night? He kept a record of daily study time in the week prior to each test and the corresponding test scores, and he graphed the results (figure 2.1). It became depressingly obvious to the student that to earn a B he would have to study about two hours a night. Earning an A appeared to be impossible.

Time (T) (hours per day)	Grade (G)
$\frac{1}{3}$	1.0 (D)
$\frac{3}{4}$	1.6 (C−)
1	2.0 (C)
$\frac{4}{3}$	2.2 (C+)
$\frac{3}{2}$	2.5 (C+)
$\frac{7}{4}$	2.8 (B−)
2	3.0 (B)
$\frac{7}{3}$	3.1 (B+)
$\frac{5}{2}$	3.1 (B+)

Figure 2.1

Relations

The table created by the student established a set of ordered pairs $\{(\frac{1}{3},1.0), (\frac{3}{4},1.6), \ldots, (\frac{5}{2},3.1)\}$ between the numbers in the sets $T = \{\frac{1}{3},\frac{3}{4}, \ldots, \frac{5}{2}\}$ and $G = \{1.0,1.6, \ldots, 3.1\}$. *Any set of ordered pairs defines a relation.*

Definition

A *relation F,* between the members of two sets X and Y, is any subset of $\{(x,y) | x \varepsilon X \text{ and } y \varepsilon Y\}$. The *domain* of F is the set of all first components x such that $(x,y) \varepsilon F$. The *range* of F is the set of all second components y such that $(x,y) \varepsilon F$.

Example 1 Construct the sets of ordered pairs that determine the relations F and G between the numbers in the sets $X = \{1,3,12\}$ and $Y = \{-5,-2,3,4\}$. Indicate the domain and range of each relation.

a. F is the relation "y is a divisor of x."

b. H is the relation "y is greater than x."

Solution

a. There is no element in Y that is a divisor of 1 in X. However, 3 in Y is a divisor of 3 in X, and -2, 3, and 4 in Y are divisors of 12 in X. As a result $F = \{(3,3), (12,-2), (12,3), (12,4)\}$. The domain and range of F are $\{3,12\}$ and $\{-2,3,4\}$ respectively.

b. Three and 4 in Y are greater than 1 in X, and 4 in Y is greater than 3 in X. Consequently, $H = \{(1,3), (1,4), (3,4)\}$. The domain and range of H are $\{1,3\}$ and $\{3,4\}$, respectively. ■

Functions

Of particular importance in mathematics are those relations in which no first component is paired with more than one second component.

Definition

A function is a relation in which *each element in the domain is paired with exactly one element in the range.* The variable representing the domain elements is called the *independent* variable, and the variable representing the range elements is called the *dependent* variable.

The relation constructed by the student, between time studied (T), and grade earned (G), is a function. Associated with each value of the independent variable T is exactly one value of the dependent variable G. The relation F in example 1 is not a function because the number 12 in its domain is paired with -2, 3, and 4 in its range (see figure 2.2). In F, x is the independent variable, and y is the dependent variable. The relation H in example 1 is not a function because 1 in its domain is paired with 3 and 4 in its range (see figure 2.2).

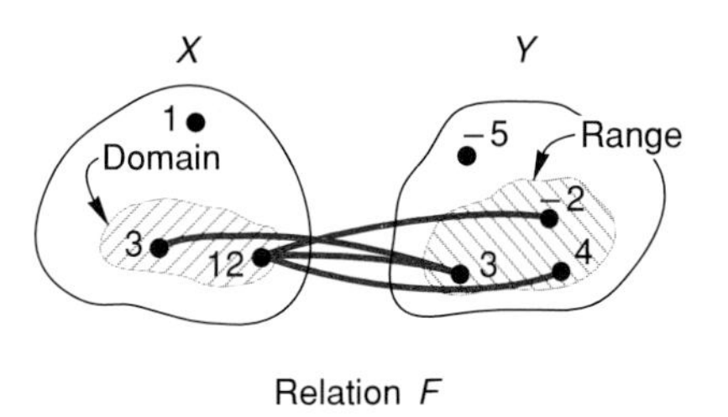

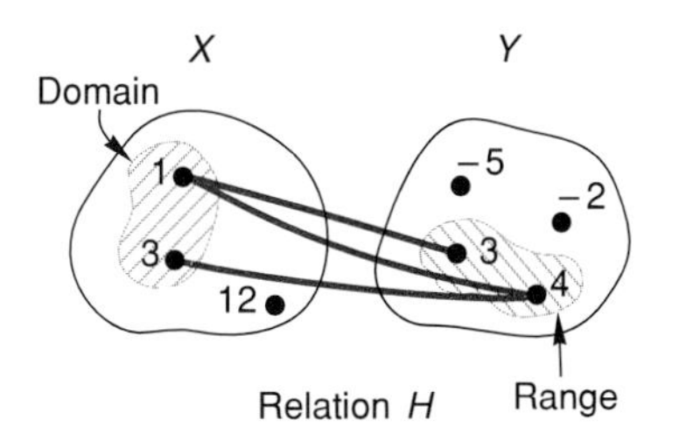

Figure 2.2

Functions Represented by Equations

It is natural to collect the data of an experiment involving two variables as a set of ordered pairs. Our student did this with his Time studied-Grade earned experiment. The collection of data is a first step that the experimenter hopes will lead to a formulation relating the variables in the experiment.

Suppose that the student took his results to his math instructor and asked her to produce a formula relating time and grade. She might have produced the equation

$$G = \frac{-1 + \sqrt{1 + 24T}}{2}. \quad \textbf{(1)}$$

The student would have accepted the formulation because, as the table below shows, the equation closely predicts the grades he earned.

Time (hours)	Grade Earned	Grade by Formula
$\frac{1}{3}$	1.0	1.00
$\frac{3}{4}$	1.6	1.68
1	2.0	2.00
$\frac{4}{3}$	2.2	2.37
$\frac{3}{2}$	2.5	2.54
$\frac{7}{4}$	2.8	2.78
2	3.0	3.00
$\frac{7}{3}$	3.1	3.27
$\frac{5}{2}$	3.1	3.41

Equation (1) defines a functional relationship between the variables T and G as long as

$$1 + 24T \geq 0 \quad \text{or} \quad T \geq -\tfrac{1}{24}.$$

That is, to each real number T, where $T \geq -\frac{1}{24}$, equation (1) associates exactly one real number G. Because the equation produces an infinite set of ordered pairs (T,G), we cannot display the function it defines as a set of ordered pairs. We can write $\{(T,G) | G = (-1 + \sqrt{1 + 24T})/2\}$ and call this set the function. However, for the sake of simplicity we drop the set builder notation and say "the function $G = (-1 + \sqrt{1 + 24T})/2$." As G is solved in terms of T, we call T the *independent variable* and G the *dependent variable* of the equation.

Equation (1) inherently restricts the domain of the function it defines because G cannot be a real number unless $T \geq -\frac{1}{24}$. It is not uncommon that the domain of a function is restricted to a set smaller than R by its defining equation. Example 2c contains such an equation. It is also not uncommon to find the domain of a function being restricted by the context in which the function is constructed. Equation (1) comes out of the student's attempt to relate grade earned (G) to time studied (T). The data of the experiment (figure 2.1) limit T to the interval $[\frac{1}{3}, \frac{5}{2}]$.

Example 2 Argue that each equation defines a function. Find the domain and range of the function.

a. $y = 3x + 2$

b. $y = x^2 - 6x + 4$

c. $y = \dfrac{x - 1}{x + 1}$

Solution

a. Any real number replacement for x in $y = 3x + 2$ determines *exactly one real y-value.* Hence, the equation defines a function whose domain is R. The range of the function is found by solving its defining equation for x in terms of y, and then testing the resulting equation to see if there are any real values of y for which x is not a real number. Noting that $x = (y - 2)/3$, and that $(y - 2)/3$ is a real number for any real number replacement of y, we conclude that the range of the function is R.

b. Any real number replacement for x in $y = x^2 - 6x + 4$ determines *exactly one real y-value.* Hence, the equation defines a function whose domain is R. Applying the quadratic formula to

$$x^2 - 6x + 4 - y = 0$$

with $a = 1$, $b = -6$, and $c = 4 - y$, we get

$$x = \frac{6 \pm \sqrt{36 - 4(4 - y)}}{2} = 3 \pm \sqrt{5 + y}.$$

The expression $3 \pm \sqrt{5 + y}$ is a real number if $5 + y \geq 0$. Thus, the range of the function is $[-5,\infty)$.

c. Any real number replacement for x except -1, in $y = (x - 1)/(x + 1)$ determines exactly one real y-value. If $x = -1$, y is not defined. Hence, the equation defines a function whose domain is the set of all real numbers except -1. Now,

$$\begin{aligned} y &= \frac{x - 1}{x + 1} \\ xy + y &= x - 1 \\ xy - x &= -1 - y \\ x(y - 1) &= -1 - y \\ x &= \frac{y + 1}{1 - y}. \end{aligned}$$

The expression $(y + 1)/(1 - y)$ denotes a real number if $y \neq 1$. Therefore, the range of the function consists of all real numbers except 1. ■

Function Notation

Rather than repeat its defining equation each time we wish to refer to a particular function, it is simpler to name the function with a single letter and refer to that letter. The letters most often used for naming functions are f, g, and h. Suppose we name $y = x^3 + 2x - 8$, f. Then the symbol $f(x)$, read "f of x," denotes the y-value, or the function value, or the range value that is associated with a particular x-value in the domain of f. For f, $f(x)$ and y are interchangeable symbols. The number obtained when x is replaced by 5 in the equation defining f, is named $f(5)$ (read "f of 5"). It is the range value associated with the domain value 5.

$$f(5) = 5^3 + 2(5) - 8 = 127$$

Example 3 Functions f and g are defined by the equations $f(x) = x^2 - 7x + 4$ and $g(x) = 5x - 3$ respectively. Find

a. $f(-2)$
b. $g(-4)$
c. $g(0)$
d. $f(-x)$
e. $f(2x - 1)$
f. $g(6 - x)$

Solution

a. $f(-2) = (-2)^2 - 7(-2) + 4 = 22$
b. $g(-4) = 5(-4) - 3 = -23$
c. $g(0) = 5(0) - 3 = -3$
d. $f(-x) = (-x)^2 - 7(-x) + 4 = x^2 + 7x + 4$
e. $f(2x - 1) = (2x - 1)^2 - 7(2x - 1) + 4 = 4x^2 - 18x + 12$
f. $g(6 - x) = 5(6 - x) - 3 = 27 - 5x$ ■

Graphs

A graph is a picture of a relation. It is the set of points whose coordinates are the ordered pairs (x,y) that determine the relation.

Example 4 Construct a graph of the relation F in example 1a.

Solution Since $F = \{(3,3), (12,-2), (12,3), (12,4)\}$, its graph is the set of points $P_1(3,3)$, $P_2(12,-2)$, $P_3(12,3)$, and $P_4(12,4)$ shown in figure 2.3. The *x-coordinate,* or **abscissa,** of P_1 is 3. The *y-coordinate,* or **ordinate,** of P_1 is also 3. The x and y coordinates of P_2 are 12 and -2 respectively. The abscissa and ordinate of P_3 are 12 and 3 respectively.

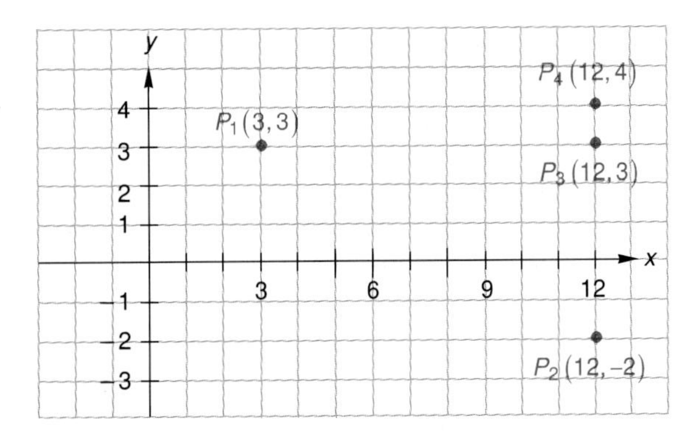

Figure 2.3

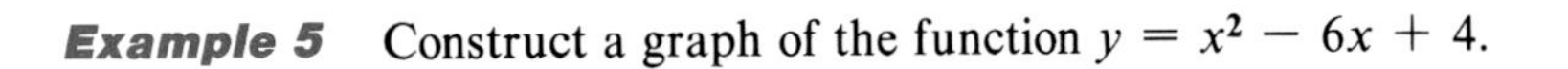

Example 5 Construct a graph of the function $y = x^2 - 6x + 4$.

Solution Because $x^2 - 6x + 4$ is a second degree polynomial, the function is called a polynomial function. It can be shown with the aid of calculus that polynomial functions are **continuous** functions. A graph of a continuous function is an unbroken curve or an unbroken straight line. It is usually constructed by plotting enough points to get a sense of the shape of the graph and then connecting the points. The graph of $y = x^2 - 6x + 4$, in figure 2.4, is a continuous curve passed through five points. The arrowheads on the curve indicate that the graph continues to rise at both ends.

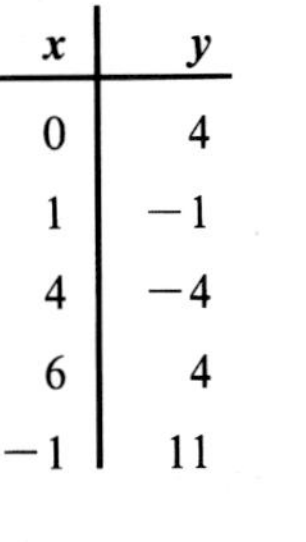

x	y
0	4
1	−1
4	−4
6	4
−1	11

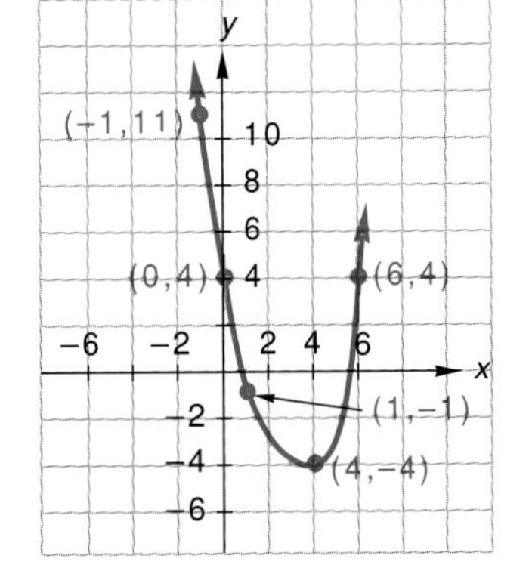

Figure 2.4

There is an easy way to determine if a graph depicts a function. We simply imagine a vertical line moving across the plane in the direction of increasing x-values. If, for every fixed position of the vertical line, it crosses the graph in *at most one point,* then the graph depicts a function. The **vertical line test** tells us that the first graph in figure 2.5 depicts a function, and that the second graph in figure 2.5 does not depict a function. (Why?) What does the vertical line test tell you about the graphs in figures 2.3 and 2.4?

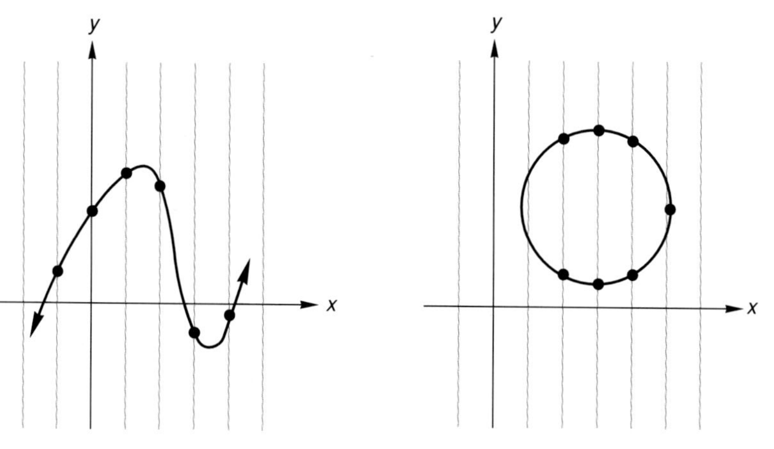

Figure 2.5

Increasing and Decreasing Functions

Let A be a subset of the domain of some function f. We say

Definition

1. f is increasing on A if whenever x_1 and x_2 are in A and $x_2 > x_1$, then $f(x_2) > f(x_1)$.
2. f is decreasing on A if whenever x_1 and x_2 are in A and $x_2 > x_1$, then $f(x_2) < f(x_1)$.
3. f is constant on A if whenever x_1 and x_2 are in A and $x_1 \neq x_2$, then $f(x_1) = f(x_2)$.

The function f (figure 2.6) is increasing over the intervals $(-2,-1)$ and $(2,3)$ because as x increases, the functional values increase. The function is decreasing on the intervals $(-3,-2)$ and $(3,6)$, as functional values decrease with increasing x. For each x in the interval $(-1,2)$, $f(x) = 2$. We say that f is constant on $(-1,2)$.

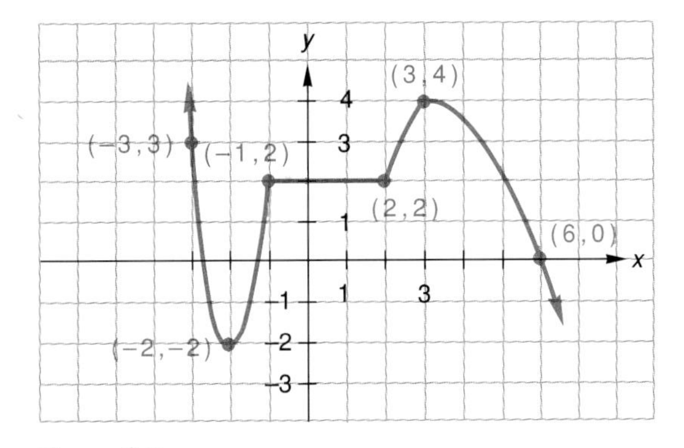

Figure 2.6

Constructing Functions

Equations that define functions do not come to us out of the void. They are often constructed to serve as mathematical models of some aspect of the real world.

Example 6 The volume of a rectangular box having a square base is 10 cubic feet. Express the surface area of the box as a function of the length of a side of its base. Determine the domain of the function.

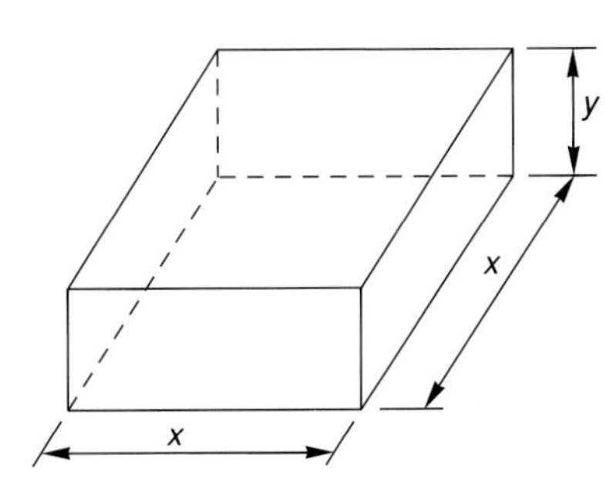

Figure 2.7

Solution Let A be the surface area of the box. From figure 2.7 we get

$$A = 2x^2 + 4xy.$$

The volume of the box is the product of its length, width, and height.

$$10 = x^2y$$

Solving the last equation for y and substituting into the area equation gives us the surface area as a function of x, the length of a side of the base.

$$A = 2x^2 + \frac{40}{x}$$

The length of a side of the box must be a positive number. Therefore the domain of the area function is $(0,\infty)$. ■

Exercise Set 2.1

(A)

Indicate if the relation is, or is not, a function, and give a reason for your answer.

1. $\{(1,2), (3,4), (5,6), (7,8), (9,10)\}$
2. $\{(3,-2), (1,5), (4,-2), (8,5)\}$
3. $\{(-2,1), (-8,4), (3,3), (-8,6)\}$
4. $\{(-5,3), (-2,4), (-2,6), (-2,9)\}$
5. $\{(x,y) | x = 1, 2 < y < 9, y \varepsilon N\}$
6. $\{(x,y) | y = 3, -4 \leq x \leq 5, x \varepsilon J\}$

Specify the domain and range of each function.

7. $y = 4x + 6$
8. $y = -2x - 5$
9. $y = x^2 + x - 1$
10. $y = 3x^2 - 2x + 5$
11. $y = 2x^2 - 5x + 3$
12. $y = 4x^2 + 6x + 9$
13. $y = \sqrt{x + 2}$
14. $y = \sqrt{x^2 - 4}$
15. $y = \sqrt{x^2 + 2x}$
16. $y = \sqrt{9 - x^2}$
17. $y = \dfrac{2x + 5}{x - 2}$
18. $y = \dfrac{5}{x + 4}$
19. $y = \dfrac{5x}{x^2 - 5x}$
20. $y = \dfrac{x^2 - 2x - 3}{x + 1}$

21. Let $f(x) = 4x + 6$. Find $f(0)$, $f(2)$, $f(-x)$, and $f(2x - 1)$.
22. Let $g(x) = 3x^2 - 2x + 5$. Find $g(0)$, $g(7)$, $g(-x)$, and $g(3 - x)$.
23. Let $h(x) = \sqrt{x + 2}$. Find $h(3)$, $h(0)$, $h(-x)$, and $h(13 - 3x)$.
24. Let $q(x) = \sqrt{x^2 - 4}$. Find $q(2)$, $q(-7)$, $q(5)$, $q(-x)$, and $q(3x/2)$.
25. Let $p(x) = \sqrt{x^2 + 2x}$. Find $p(0)$, $p(-2)$, $p(-x)$, $p(3x + 2)$, and $p(x + a)$.
26. Let $F(x) = \sqrt{9 - x^2}$. Find $F(-3)$, $F(5)$, $F(-4x)$, $F(3 - 2x)$, and $F(x - b)$.
27. Let $G(x) = (2x + 5)/(x - 2)$. Find $G(0)$, $G(-1)$, $G(5x - 2)$, and $G(1/x)$.
28. Let $H(x) = 5/(x + 4)$. Find $H(-5)$, $H(-2x)$, $H(6 - 2x)$, and $H(-2/x)$.

Use the vertical line test to determine if the graph depicts a function.

29. **30.**

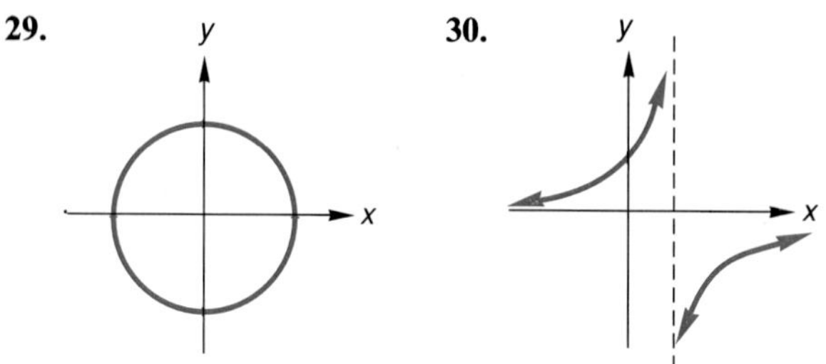

31. **32.**

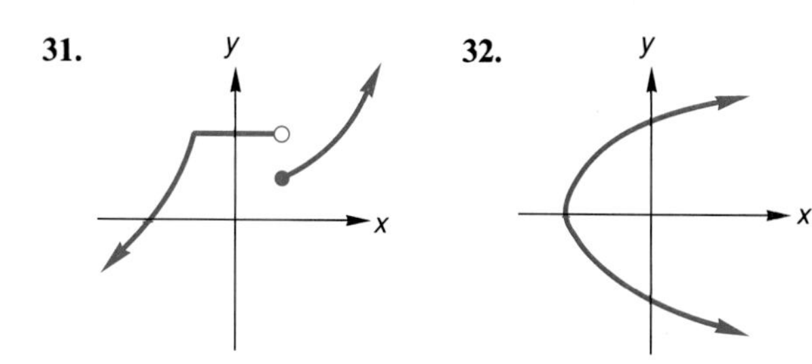

33. **34.**

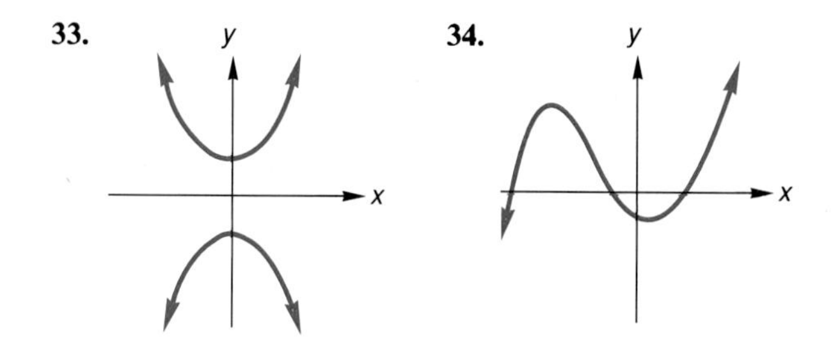

Determine the intervals over which each function is increasing, decreasing, or constant.

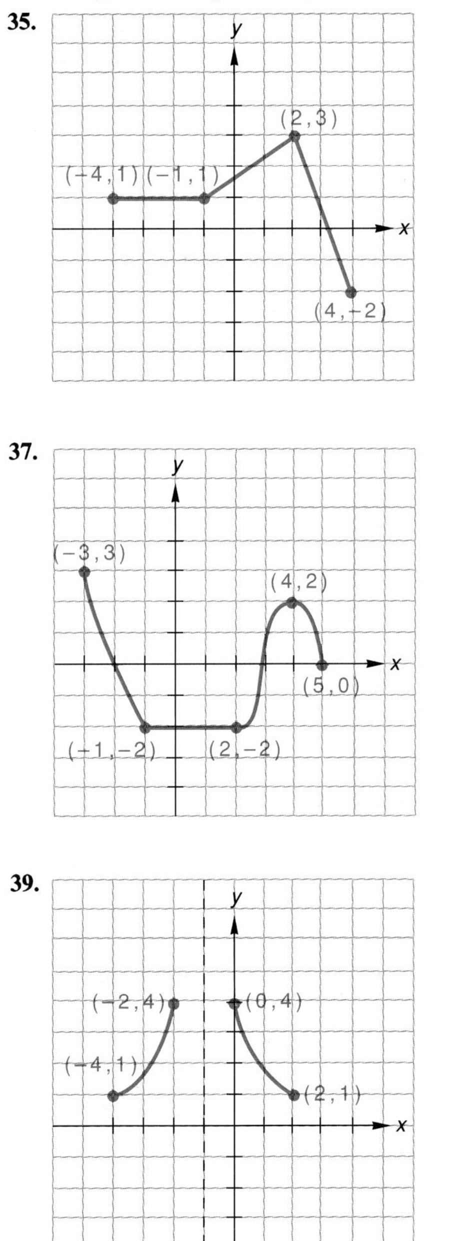

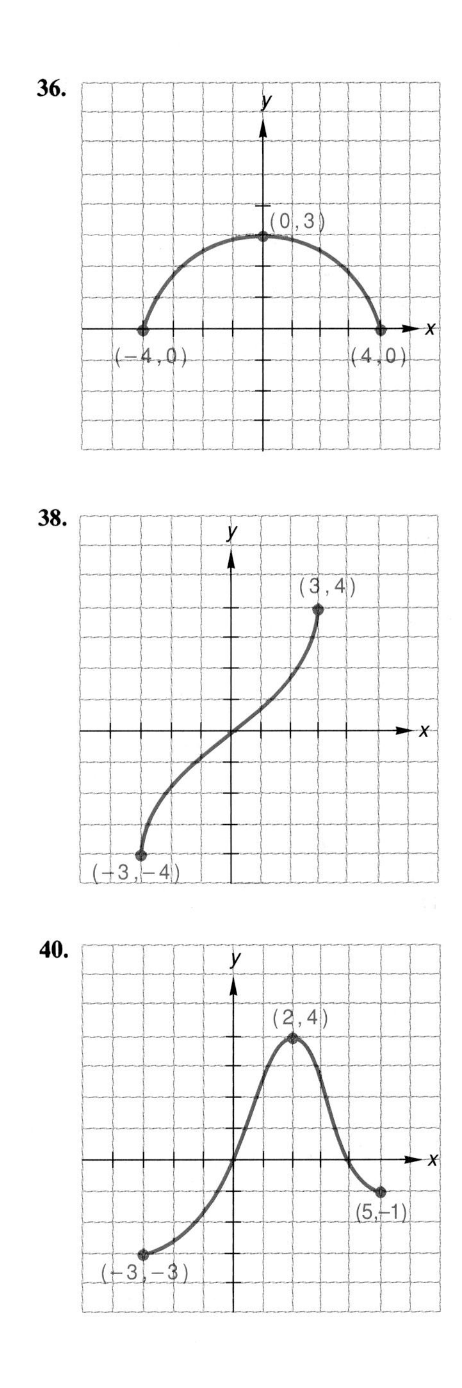

41. Select a stock on the New York Stock Exchange. Maintain a five-day record of a stock's closing price. Graph the ordered pairs (day, closing price). Is the relation between closing price and day, numbered 1 through 5, a function?

42. Using the stock chosen for exercise 41, maintain a five-day record of closing price versus the number of shares traded. Graph the ordered pairs (thousands of shares traded, closing price). Is the relation between closing price and thousands of shares traded a function?

43. Maintain a record of the number of encounters you have with other people in any one day. If your sense of the encounter is favorable, label the encounter 1. If an encounter is unfavorable, label it 0. Graph the ordered pairs (number of the encounter, 0 or 1). Does the graph depict a function?

44. Why must the low point of the graph of $y = x^2 - 6x + 4$ be $(3, -5)$?

Graph each function using the point plotting technique of example 5.

45. Exercise 7 **46.** Exercise 8

47. Exercise 9 **48.** Exercise 10

49. Exercise 13 **50.** Exercise 14

51. Exercise 15 **52.** Exercise 16

53. Exercise 17 **54.** Exercise 18

(B)

55. An 8-inch wide flat piece of sheet metal is made into a rain gutter by bending its ends up 90°. (See the figure). Express the area of the gutter's cross section as a function of x. What is the domain of the function?

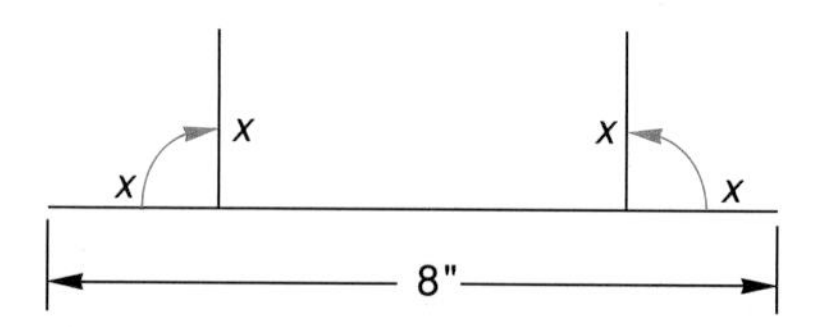

56. The area of a rectangular lot is 6,000 square feet. Express the length of the lot's perimeter as a function of the length of the lot's frontage. What is the domain of the function?

57. A piece of wire of length L is cut into two pieces, one of which is x inches long. The piece of length x is shaped into a circle. The other piece is shaped into an equilateral triangle. Express the total area of the two shapes as a function of x. What is the domain of the function?

58. A rectangle is inscribed in a circle of radius r. Let x be the length of one side of the rectangle. Express the area of the rectangle as a function of x. What is the domain of the function?

59. A rectangular poster is to contain a rectangular layout of 160 square inches. The layout will have 4-inch margins top and bottom, and 2-inch margins on the sides. Let x be the horizontal length of the layout. Express the total area of the poster as a function of x. What is the domain of the function?

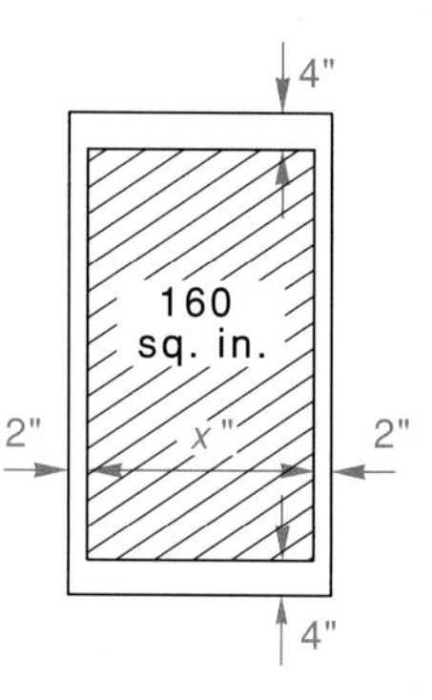

60. The volume of a right circular cylinder of radius r is 15 cubic feet. Express the surface area of the cylinder, including top and bottom, as a function of r. What is the domain of the function? (The volume of a right circular cylinder is $\pi r^2 h$.)

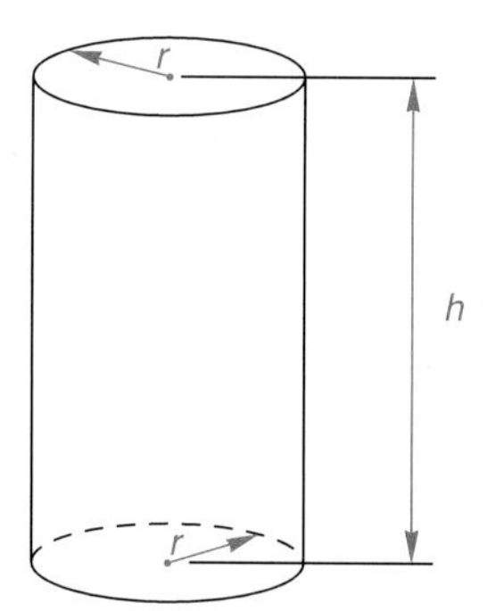

61. The expression for surface area in example 6 is $A = 2x^2 + 40/x$. The domain of the area function is the set of positive real numbers. You can use your calculator to check that as x assumes values closer and closer to zero, or values that become greater and greater, the area becomes greater and greater. For some x between the extremes of zero and a large positive number, the surface area must be minimized. Graph the area function and estimate the value of x that minimizes the surface area.

2.2 Linear Functions

Imagine a bug crawling from P to Q along the line determined by $P(x_1,y_1)$ and $Q(x_2,y_2)$ (figure 2.8). It will rise $y_2 - y_1$ units and move to the right $x_2 - x_1$ units. The ratio $(y_2 - y_1)/(x_2 - x_1)$ is a measure of the tilt of the line. It is called the **slope** of the line, and it is designated with the letter ***m.*** Because $(y_2 - y_1)/(x_2 - x_1) = (y_1 - y_2)/(x_1 - x_2)$, we can write

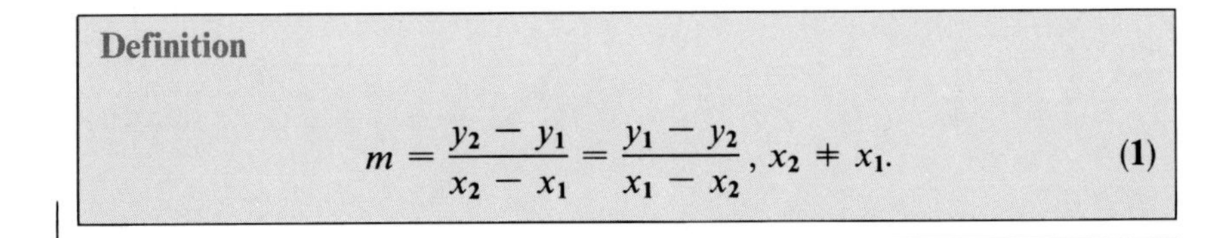

Definition

$$m = \frac{y_2 - y_1}{x_2 - x_1} = \frac{y_1 - y_2}{x_1 - x_2},\ x_2 \neq x_1. \tag{1}$$

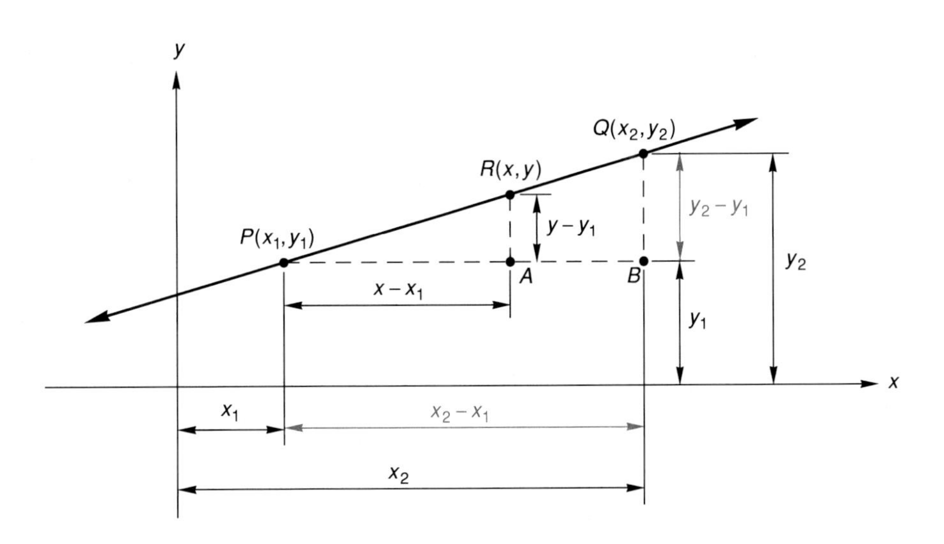

Figure 2.8

Think of the computation for slope as the evaluation of the fraction $\Delta y/\Delta x$, where Δy and Δx are the changes that occur in the y and x directions when moving from one point to another on the line. You can imagine moving from a first point to a second point, *always subtracting the coordinates of the first point from the corresponding coordinates of the second point.*

Example 1 Compute the slope of the line containing $(-5,4)$ and $(2,-3)$.

Solution Imagine moving from $(-5,4)$ to $(2,-3)$ (figure 2.9) and use (1).

$$m = \frac{-3 - 4}{2 - (-5)} = -1$$

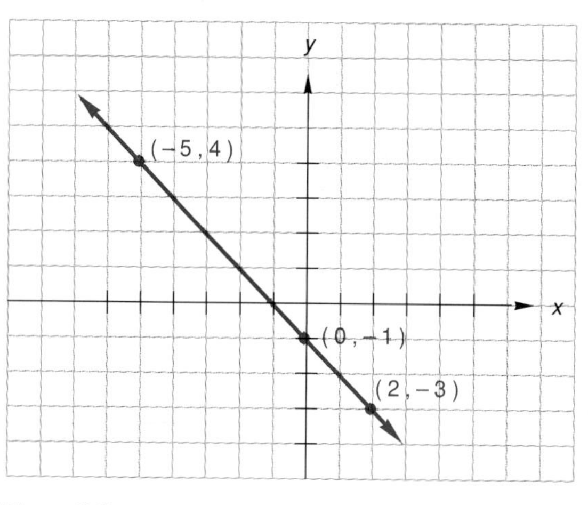

Figure 2.9

■

Any two points on a line can be used to determine its slope. Let $R(x,y)$ be a point on the line determined by $P(x_1,y_1)$ and $Q(x_2,y_2)$ (figure 2.8). Triangles RPA and QPB are similar (Why?) so that their corresponding sides are proportional. We can write

$$\frac{\overline{AR}}{\overline{PA}} = \frac{\overline{BQ}}{\overline{PB}}$$

or

$$\frac{y - y_1}{x - x_1} = \frac{y_2 - y_1}{x_2 - x_1}.$$

But $(y_2 - y_1)/(x_2 - x_1) = m$. Therefore, points R and P determine the same slope as do points Q and P.

The fact that any two distinct points on a line determine its slope enables us to write an equation for the line. In figure 2.8, $P(x_1,y_1)$ and $Q(x_2,y_2)$ determine a straight line whose slope is $m = (y_2 - y_1)/(x_2 - x_1)$. Let $R(x,y)$ be any other point on the line. Then $(y - y_1)/(x - x_1) = m$ or

Point-Slope Form

$$y - y_1 = m(x - x_1). \tag{2}$$

Equation (2) is called the **point-slope** form of the equation of a straight line. It determines the y-coordinate of any point on the line if the x-coordinate of the point is known.

Example 2 Write a point-slope form equation for the line determined by $(-5,4)$ and $(2,-3)$.

Solution In example 1, the slope of the line is computed to be -1. If we let $(-5,4)$ play the role (x_1,y_1) in (2), then the point-slope equation is

$$y - 4 = -1(x - (-5))$$
$$y - 4 = -(x + 5).$$

If $(2,-3)$ plays the role of (x_1,y_1), then the equation is

$$y + 3 = -(x - 2).$$

Both equations define the same line. They are equivalent equations because each can be rewritten in the form $y = -x - 1$. ■

Slope-Intercept Form

The slope of the line in example 2 is -1. It is also the coefficient of x in the equation $y = -x - 1$. The constant, -1, is called the ***y*-intercept** of the line. It is the y-coordinate of the point of intersection $(0,-1)$ between the line and the y-axis (figure 2.9). Equation $y = -x - 1$ is a specific instance of the *slope-intercept* form of the equation of a straight line. The general slope-intercept form is

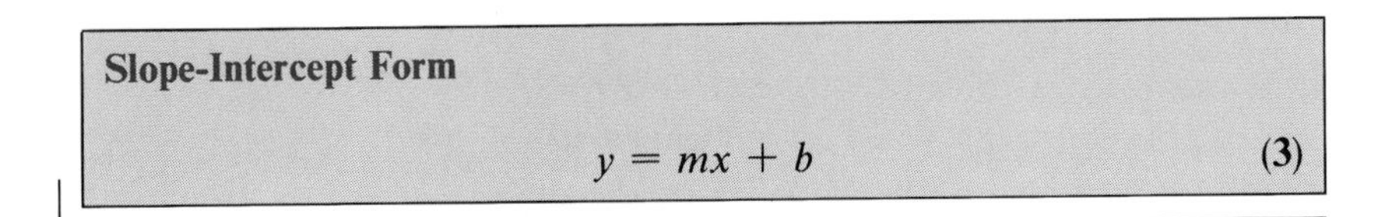
Slope-Intercept Form

$$y = mx + b \tag{3}$$

where m is the slope, and b is the y-intercept of the line. Note that (3) is obtained from (2) if $(x_1,y_1) = (0,b)$.

Example 3 Write the slope-intercept form of the equation of the line that contains $(3,-2)$ and has slope 4.

Solution The equation has the form $y = 4x + b$. Substituting 3 for x and -2 for y gives us $-2 = 4(3) + b$, and $b = -14$. Hence,

$$y = 4x - 14.$$

■

Example 4 Graph the line $y - 4 = 2(x + 1)$.

Solution The equation is in point-slope form (equation (2)) so that $(-1,4)$ is a point on the line. Another point on the line is gotten by assigning a value to x and solving for y. Let $x = 2$, then

$$y - 4 = 2(2 + 1) \quad \text{and} \quad y = 10.$$

The points $(-1,4)$ and $(2,10)$ are plotted, and a straight line is passed through them (figure 2.10).

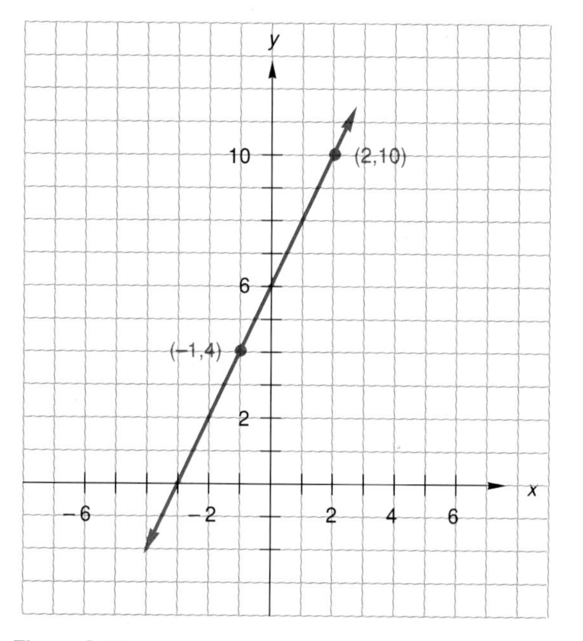

Figure 2.10

■

Example 5 Graph $y = -3x + 2$.

Solution The equation is in slope-intercept form, with slope -3 and y-intercept 2. Point (0,2) is on the line. (Why?) A second point is located so that the ratio $\Delta y/\Delta x$, between the points, is -3. One such point is $(2,-4)$ because $\Delta y = -6$ and $\Delta x = 2$ (figure 2.11).

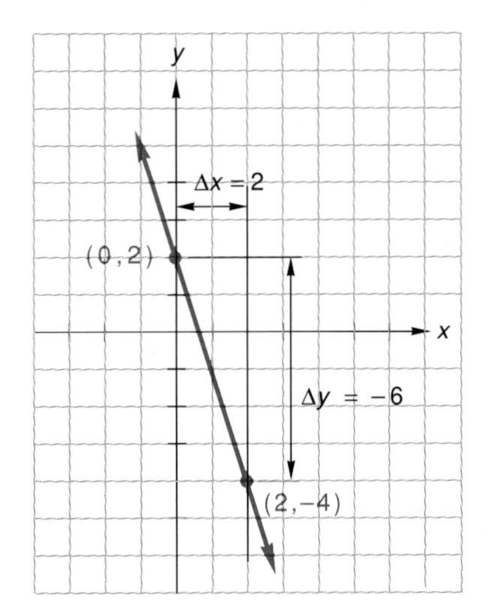

Figure 2.11 ■

General Form

A third common form of an equation of a straight line is the *general form*

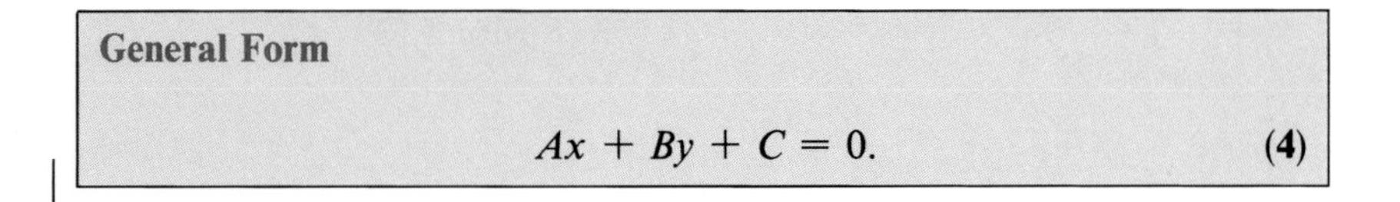
General Form

$$Ax + By + C = 0. \tag{4}$$

Example 6

a. Write $y - 2 = \frac{2}{3}(x - 5)$ in general form.

b. Write $6x - 3y + 5 = 0$ in slope-intercept form.

Solution

a.
$$3(y - 2) = 2(x - 5)$$
$$-2x + 3y + 4 = 0$$

b.
$$-3y = -6x - 5$$
$$y = 2x + \tfrac{5}{3}$$

■

Lines Parallel to the Coordinate Axes

If $A = 0$ and $B \neq 0$, (4) becomes $0x + By + C = 0$. Since the coefficient of x is 0, $y = -C/B$ no matter what value x assumes. Thus, all the points on the line $y = -C/B$ are $|-C/B|$ units from the x-axis. The line is parallel to and $|-C/B|$ units from the x-axis. As an example, either of the equivalent equations, $0x + 2y - 8 = 0$ or $y = 4$, define a straight line that is parallel to, and 4 units above, the x-axis. In general,

Horizontal and Vertical Lines

$y = c$ is a horizontal line that is $|c|$ units from the x-axis. The slope of the line is 0.
$x = k$ is a vertical line that is $|k|$ units from the y-axis. The slope of the line is *undefined.*

Example 7

a. Write the equations of the horizontal and vertical lines that contain (5,3).

b. Graph the lines.

c. Show why the slope of the horizontal line is 0, and why the slope of the vertical line is undefined.

Solution

a. The y-coordinate of every point on a horizontal line is the same. Hence, the horizontal line is $y = 3$ (see figure 2.12). A similar argument gives us the vertical line $x = 5$.

b. Figure 2.12.

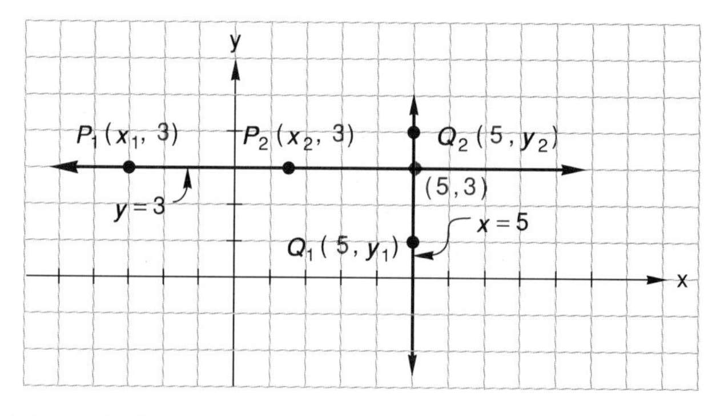

Figure 2.12

c. The slope of the horizontal line is

$$m = \frac{3 - 3}{x_2 - x_1} = 0.$$

The slope of the vertical line

$$m = \frac{y_2 - y_1}{5 - 5},$$

is not defined because we are attempting to divide by zero. ■

Parallel and Perpendicular Lines

Knowing that lines l_1 and l_2 are parallel (figure 2.13), it can be shown that triangles ABC and PQR are congruent (exercise 71). As a result $\overline{BC}/\overline{AC} = \overline{QR}/\overline{PR}$, so that the *slopes of nonvertical parallel lines are equal.* Conversely, if the slopes of l_1 and l_2 are equal, the lines are parallel (exercise 72).

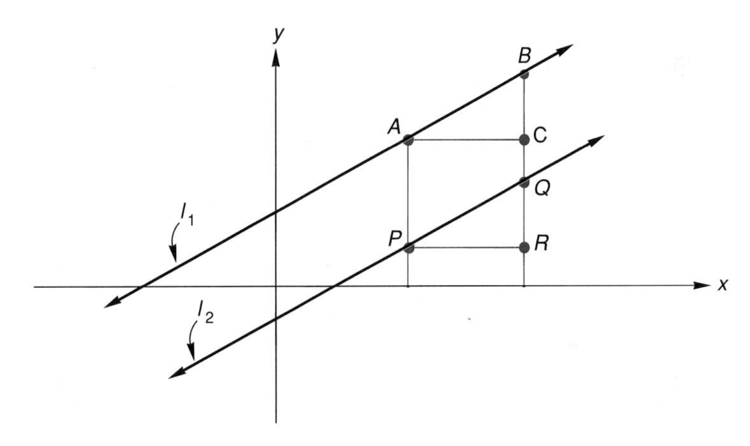

Figure 2.13

Example 8 Write an equation of the line containing $(-2,5)$ that is parallel to $3x + y - 6 = 0$.

Solution The line must have the same slope as $3x + y - 6 = 0$. That slope is -3, as seen when $3x + y - 6 = 0$ is written in slope-intercept form, $y = -3x + 6$. Point $(-2,5)$ and slope -3 are substituted into the point-slope equation (2), to give us $y - 5 = -3(x + 2)$. ■

Assume l_1 and l_2 (figure 2.14) are perpendicular and have slopes m_1 and m_2. Then $m_1 = h/a$ and $m_2 = -h/b$. (Why?) Triangles ABD and BCD are similar because $\measuredangle 1 = \measuredangle 2$ and $\measuredangle 3 = \measuredangle 4$. (Why?) Since corresponding sides of similar triangles are proportional, we may write $h/a = b/h$. But b/h is the negative reciprocal of $-h/b$. Therefore h/a is the negative reciprocal of $-h/b$. We have shown that if two lines are perpendicular, their slopes are negative reciprocals. In exercise 73 you are asked to show the converse, "If the slopes of two lines are negative reciprocals, the lines are perpendicular."

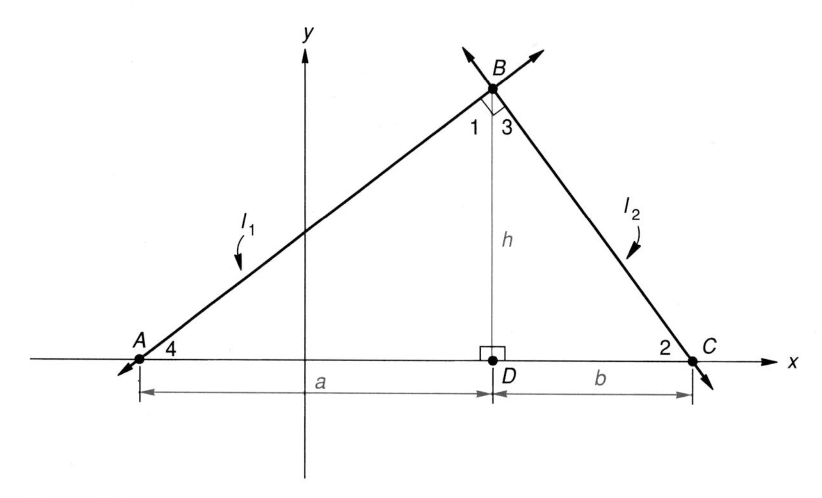

Figure 2.14

> **Parallel and Perpendicular Lines**
>
> If m_1 and m_2 are the slopes of non-vertical lines l_1 and l_2, then
>
> **1.** l_1 is parallel to l_2 if and only if $m_1 = m_2$.
> **2.** l_1 is perpendicular to l_2 if and only if $m_1 = -1/m_2$.

Example 9 Write an equation of the line that contains (0,0) and is perpendicular to $y = -x/2 + 9$.

Solution The perpendicular line must have slope 2, which is the negative reciprocal of $-\frac{1}{2}$. With slope 2 and (0,0) substituted into the point-slope equation, we get $y - 0 = 2(x - 0)$ or $y = 2x$. ■

Example 10 Which pair of lines are perpendicular?

a. $y = \dfrac{2x}{3} + 6, \quad 2y + 3x - 4 = 0$

b. $5x - y + 8 = 0, \quad x - 5y + 2 = 0$

Solution

a. The lines are perpendicular because their slopes, $\frac{2}{3}$ and $-\frac{3}{2}$, are negative reciprocals.

b. The lines are not perpendicular. Their slopes, 5 and $\frac{1}{5}$, are not negative reciprocals. ■

Distance and Midpoint Formulas

Let $P(x_1,y_1)$ and $Q(x_2,y_2)$ be any two points in the plane (figure 2.15). A line through P, parallel to the x-axis ($y = y_1$), intersects a line through Q, that is parallel to the y-axis ($x = x_2$) in the point $T(x_2,y_1)$. $\overline{PQ}$ is the hypotenuse of a right triangle whose sides have lengths $|x_2 - x_1|$ and $|y_2 - y_1|$. Hence,

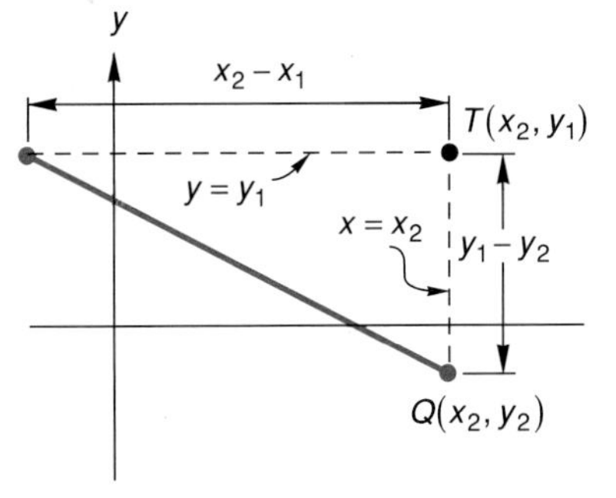

Figure 2.15

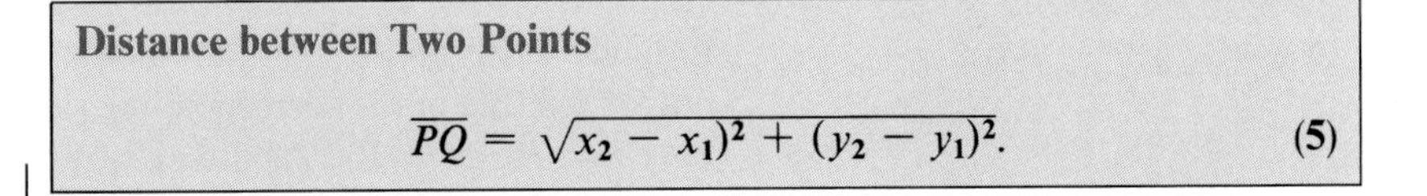

Distance between Two Points

$$\overline{PQ} = \sqrt{(x_2 - x_1)^2 + (y_2 - y_1)^2}. \tag{5}$$

Equation (5) gives the distance between any two points, $P(x_1,y_1)$ and $Q(x_2,y_2)$, in the plane. If P and Q determine a vertical line, (5) becomes $\overline{PQ} = |y_2 - y_1|$. (Why?) If P and Q determine a horizontal line, (5) becomes $\overline{PQ} = |x_2 - x_1|$. (Why?)

Example 11

a. The distance between $(-2,6)$ and $(8,-1)$ is

$$\sqrt{[8 - (-2)]^2 + [-1 - 6]^2} = \sqrt{149}.$$

b. The distance between $(-2,6)$ and $(-2,-4)$ is

$$\sqrt{[-2 - (-2)]^2 + [-4 - 6]^2} = \sqrt{100} = 10 = |-4 - 6|.$$

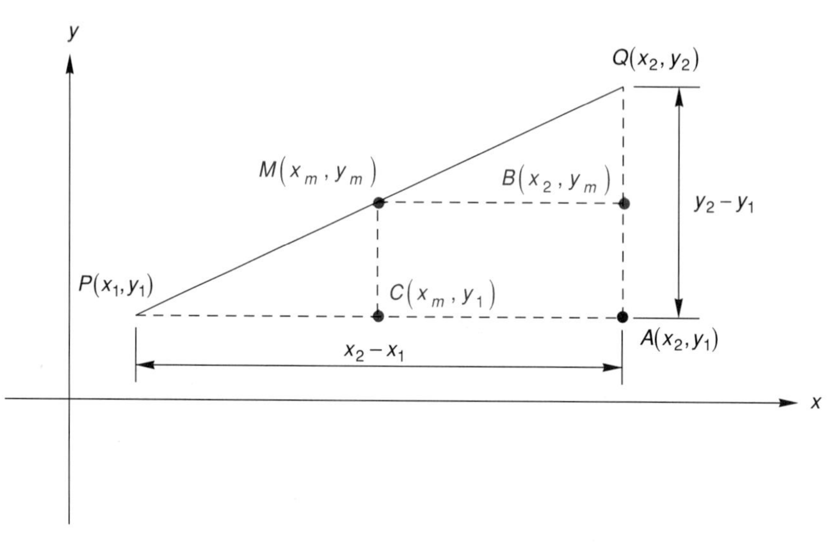

Figure 2.16

In figure 2.16, $M(x_m,y_m)$ is the midpoint of segment $\overline{PQ}$. It follows from the proportionality of the corresponding sides of similar triangles that because $\overline{PM} = \frac{1}{2}\,\overline{PQ}$, $\overline{PC} = \frac{1}{2}\,\overline{PA}$. As a result, the x-coordinate of C is

$$x_m = x_1 + \frac{1}{2}\,\overline{PA} = x_1 + \frac{x_2 - x_1}{2} = \frac{x_1 + x_2}{2}.$$

Similarly, $\overline{AB} = \frac{1}{2}\,\overline{AQ}$ so that $y_m = (y_1 + y_2)/2$. Thus, the coordinates of the midpoint of a segment with endpoints $P(x_1,y_1)$ and $Q(x_2,y_2)$ are

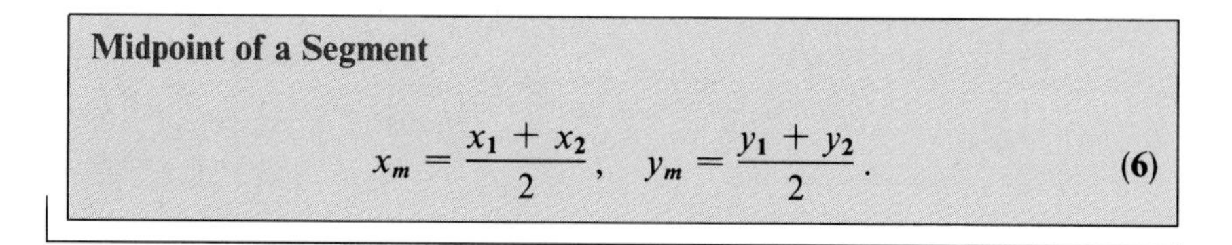

Midpoint of a Segment

$$x_m = \frac{x_1 + x_2}{2}, \quad y_m = \frac{y_1 + y_2}{2}. \tag{6}$$

Equations (6) are applicable to vertical and horizontal line segments. (Why?)

Example 12

a. The coordinates of the midpoint of the segment with endpoints $(2,-4)$ and $(9,8)$ are

$$x_m = \frac{9 + 2}{2} = \frac{11}{2}, \quad y_m = \frac{8 + (-4)}{2} = 2.$$

b. The midpoint of the segment determined by $(2,-4)$ and $(-5,-4)$ has coordinates

$$x_m = \frac{2 + (-5)}{2} = -\frac{3}{2}, \quad y_m = \frac{-4 + (-4)}{2} = -4.$$

Example 13 Write an equation for the perpendicular bisector of the segment in example 12a.

Solution We want an equation of the line that contains the midpoint $(\frac{11}{2},2)$ of the segment, and that is perpendicular to the segment. The slope of the segment is

$$\frac{8 - (-4)}{9 - 2} = \frac{12}{7}.$$

Therefore, an equation of the perpendicular bisector is

$$y - 2 = -\tfrac{7}{12}(x - \tfrac{11}{2}).$$ ■

Some Graphs

There are functions whose graphs are made up of straight line segments that require some analysis before their graphs can be constructed. As an example, let us graph $y = |x - 2|$. We could plot some points and then try to connect them. But when the function being graphed is unfamiliar, there is always the question of how to connect the points. The point-connecting dilemma for the graph of $y = |x - 2|$ is avoided if we recall that

$$|x - 2| = \begin{cases} x - 2, & x - 2 \geq 0 \quad \text{or} \quad x \geq 2 \\ -(x - 2), & x - 2 < 0 \quad \text{or} \quad x < 2. \end{cases}$$

Thus, if $x \geq 2$, then $y = x - 2$. If $x < 2$, then $y = -(x - 2)$. The graph (figure 2.17) is constructed from parts of two straight lines.

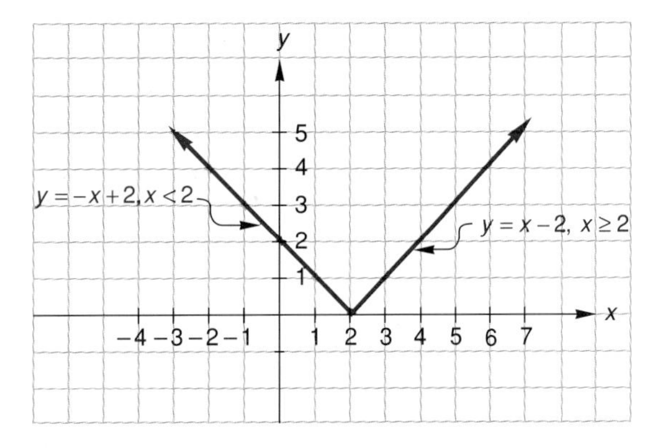

Figure 2.17

The "greatest integer" function, $y = [x]$, associates with each real number x, the greatest integer *less than or equal to* x.

$$[2.83] = [2 + 0.83] = 2$$
$$[-2.83] = [-3 + 0.17] = -3$$

If we picture x as a point on the number line, then the greatest integer function assigns to x the closest integer to the left of x (figure 2.18).

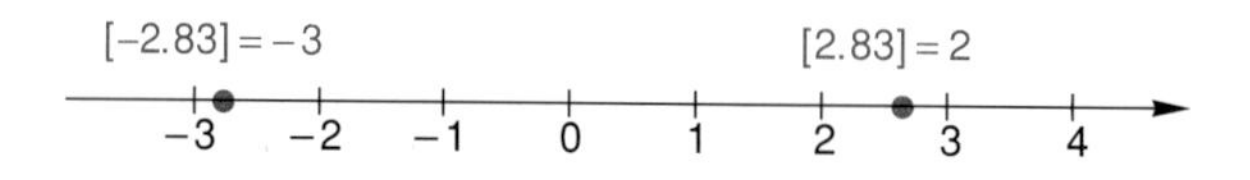

Figure 2.18

Example 14 Graph $y = x + [x]$, $-2 \le x \le 2$.

Solution By definition, $[-2] = -2$, and if $-2 < x < -1$, then $[x] = -2$. Likewise, $[-1] = -1$, and $[x] = -1$ if $-1 < x < 0$. Since it is possible to continue this way for each of the intervals $[0,1)$ and $[1,2)$, we can rewrite the given function as a collection of linear functions.

$$\begin{aligned} y &= x - 2, & -2 \le x < -1 \\ y &= x - 1, & -1 \le x < 0 \\ y &= x, & 0 \le x < 1 \\ y &= x + 1, & 1 \le x < 2 \\ y &= 4, & x = 2 \end{aligned}$$

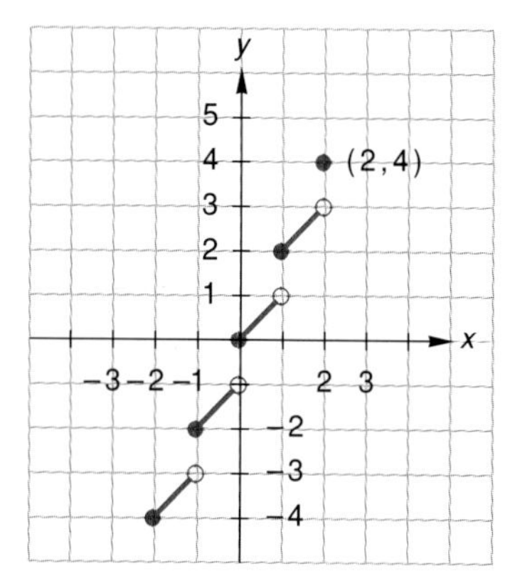

Figure 2.19

The graph of the function (figure 2.19) consists of four line segments, each having slope 1, and the point (2,4). The function values associated with each integer in the domain are the y-coordinates of the points depicted as solid circles. ■

Example 14 suggests that a function may be defined with two or more equations.

Example 15 A graph of $f(x)$ (shown in figure 2.20)

$$f(x) = \begin{cases} 2x - 3, & 0 \le x < 3 \\ -3x + 12, & 3 \le x \le 5 \end{cases}$$

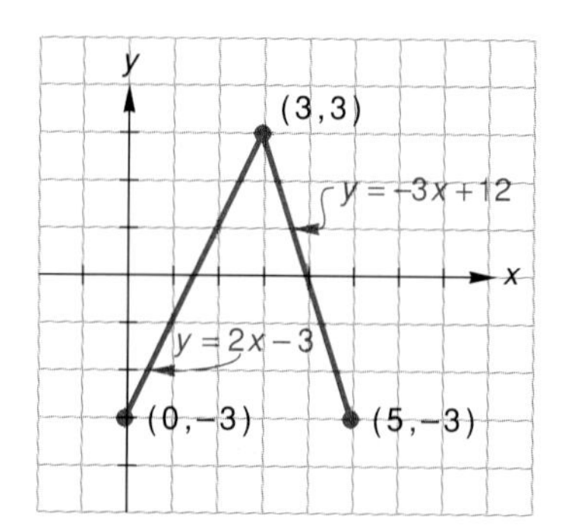

Figure 2.20

Exercise Set 2.2

(A)

Write a point-slope form of the equation of the line containing the given pair of points. Convert the point-slope equation into the slope-intercept equation. Then write the general equation of the line.

1. $(1,3), (2,5)$ **2.** $(-1,4), (6,2)$

3. $(-3,2), (7,-4)$ **4.** $(0,5), (2,0)$

5. $(0,0), (-2,5)$ **6.** $(6,3), (0,0)$

7. $(5,4), (8,-4)$ **8.** $(-2,-3), (7,-3)$

Write an equation of the horizontal or vertical line, as stipulated by H or V, that contains the given point.

9. $H, (5,2)$ **10.** $H, (-3,4)$

11. $V, (6,1)$ **12.** $V, (2,2)$

13. $H, (0,0)$ **14.** $H, (8,-2)$

15. $V, (0,0)$ **16.** $V, (-4,0)$

Write an equation of the line whose graph is shown.

17.

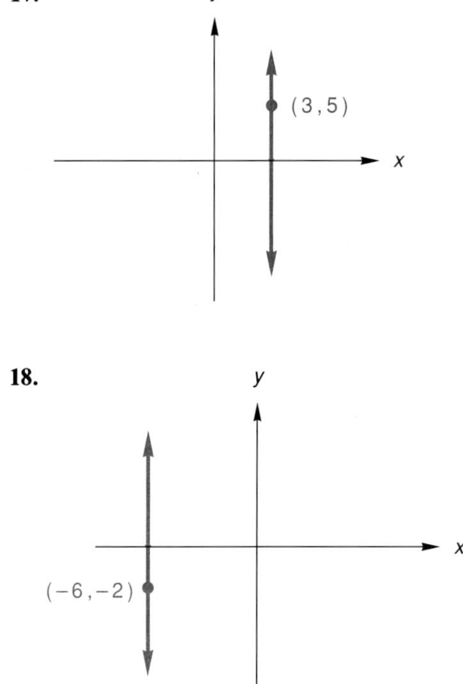

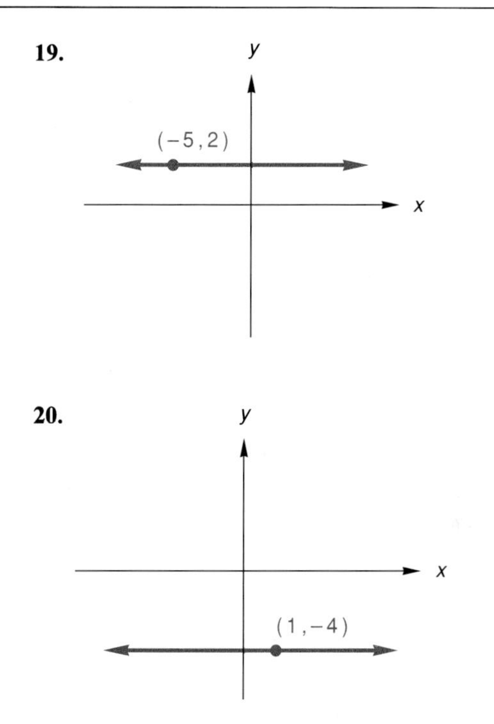

Write an equation of the line containing the given point that is, (a) parallel to, and (b) perpendicular to, the given line.

21. $(0,0)$, $2x + y - 4 = 0$

22. $(3,-2)$, $x - 3y - 8 = 0$

23. $(-5,-2)$, $y = 5x - 4$

24. $(0,6)$, $3y = 9x + 7$

25. $(4,3)$, $-3x + 2y = 9$

26. $(1,-2)$, $6x - 9y = 5$

27. $(7,0)$, $3y = -2x + 9$

28. $(-11,8)$, $-15y = 6x + 20$

Determine if the pair of lines are parallel, perpendicular, or neither parallel nor perpendicular.

29. $x + 2y + 5 = 0,\ 2x - y + 8 = 0$

30. $y = \frac{3x}{4} + 9,\ 3x + 4y + 2 = 0$

31. $5y + 11x = 8,\ y = \frac{5x}{11} - 6$

32. $2x + 7y + 3 = 0,\ y = -\frac{2x}{7}$

33. $3x + 8y - 4 = 0,\ 9x + 2y = 5$

34. $y = x,\ 5x - 5y - 3 = 0$

35. $y = 3x,\ 4x + 12y - 2 = 0$

36. $x - y = 8,\ 6x + 6y + 17 = 0$

Find (a) the distance between the points, (b) the midpoint of the segment determined by the points, and (c) an equation of the perpendicular bisector of the segment.

37. (3,5), (1,9)

38. (−2,8), (−9,4)

39. (−5,−3), (6,4)

40. (4,−2), (−6,8)

41. (0,0), (5,0)

42. (1,9), (1,−4)

43. Show that the triangle determined by (2,4), (3,6), and (6,2) is a right triangle.

44. Show that the triangle determined by (0,5), (0,0), and (3,4) is an isosceles triangle.

Graph each function.

45. $y = 3x - 5$

46. $y = -2x + 4$

47. $2x + 4y - 9 = 0$

48. $8x - 3y + 12 = 0$

49. $y = 5x$

50. $y = x + 7$

51. $x - 2y + 8 = 0$

52. $3x + 2y + 6 = 0$

53. $y = |x|$

54. $y = |x + 2|$

(B)

55. $y = [x]$

56. $y = [x + 1]$

57. $y = \begin{cases} -1, & x < 0 \\ 1, & x \geq 0 \end{cases}$

58. $y = \begin{cases} 4, & x \leq 0 \\ 2, & x > 0 \end{cases}$

59. $y = |x| + 2$

60. $y = -|x| + 2$

61. $y = [3x]$

62. $y = [x] - x$

63. $y = \begin{cases} x - 3, & x < 2 \\ -x + 1, & x \geq 2 \end{cases}$

64. $y = \begin{cases} -2x + 5, & x \leq 1 \\ x + 6, & x > 1 \end{cases}$

65. The set of ordered pairs {(−2,−1), (−1,1), (0,2), (1,6), (2,7)} represents data from an experiment relating a variable y to a variable x. The experimenter assumes that the relationship is linear. Plot the data and "fit" a straight line to the points. Determine the slope of the line and write its equation. How closely does the equation predict the y-values in the data?

66. Repeat exercise 65 for variables s and t, where s represents distance in feet, and t represents time in seconds. The set of data points (s,t) is {(0,0), (1,9), (2,21), (3,30), (4,38)}. Interpret the meaning of the slope in this linear relationship.

67. Assume the baggage charge on international air carriers is 0 for the first 50 pounds and \$3 per pound for each pound greater than 50. Write a linear baggage cost function. Interpret the meaning of the slope in this equation.

68. The total cost (C) for producing x units of a certain product includes a tooling up cost (B) as well as a per unit cost (A). Write C as a linear function of x. What meaning can be ascribed to the slope?

69. A business is allowed to claim as a loss the depreciation of a piece of equipment over its useful life. If a \$100,000 computer has a useful life of ten years, after which it is worthless, at what linear rate does the computer lose value? Write a linear depreciation equation that relates value (V) to time (t). What meaning can be ascribed to the slope?

70. Assume that the cost for mailing a letter weighing an ounce or less is c cents, and that the cost for mailing a letter weighing more than 1 ounce and less than or equal to 2 ounces is $2c$ cents, etc. Construct a graph of the "postage stamp" function that relates mailing cost to letter weight.

71. Prove that triangles ABC and PQR in figure 2.13 are congruent.

72. Prove that if the slopes of two lines are equal, then the lines are parallel.

73. Prove that if the slopes of two lines are negative reciprocals of each other, then the lines are perpendicular. (Hint: Use figure 2.14. You cannot, however, assume a right angle at B. Starting with

$$m_1 = -\frac{1}{m_2} \quad \text{or} \quad \frac{h}{a} = -\frac{1}{-\frac{h}{b}},$$

show that the sides of triangle ABC satisfy the Pythagorean Theorem.)

74. Given the points $(-5,1)$ and $(6,4)$, find points $(2,y)$, $(x,-1)$, and $(5,y)$ that are equidistant from the given points. Show that the three points found are on a straight line that is the perpendicular bisector of the segment determined by $(-5,1)$ and $(6,4)$.

75. Let (x_1,y_1), (x_2,y_2) and (x_3,y_3) be the vertices of a triangle. Prove that the segment determined by the midpoints of two sides of the triangle is parallel to the third side and half the length of the third side.

76. Let (x_1,y_1), (x_2,y_2), (x_3,y_3), and (x_4,y_4) be the vertices of a quadrilateral. Prove that the figure obtained when the midpoints of the sides of the quadrilateral are joined is a parallelogram.

77. The vertices of a rhombus are $(0,0)$, $(5,0)$, $(3,4)$, and $(8,4)$. Prove that the figure obtained by joining the midpoints of the sides of the rhombus is a rectangle.

78. Determine an equation for line L if the area enclosed between the line and the coordinate axes is 26 square units.

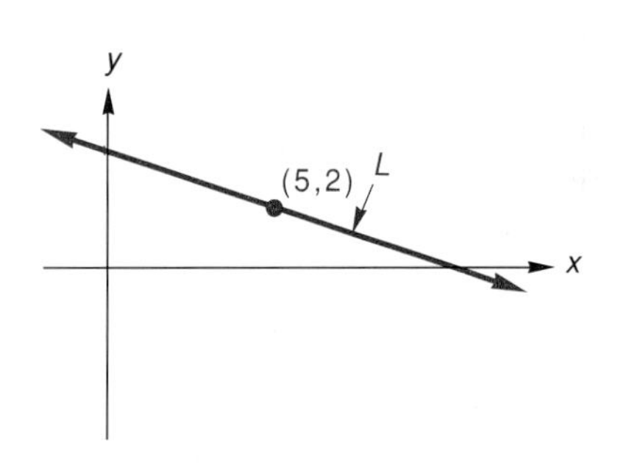

2.3 Quadratic Functions

An equation of the form

$$y = ax^2 + bx + c, \quad a \neq 0, \tag{1}$$

defines a **quadratic function.** If y is set equal to zero, then (1) becomes the general quadratic equation that is discussed in section 1.5. Unlike the graph of a linear function that can be generated from two points, the graph of a quadratic function is a curve that requires at least three points to construct. It is shown in chapter 3 that if three ordered pairs in a quadratic function are known, its equation can be determined.

Example 1 Construct a graph of $y = x^2 - 2x - 3$.

Solution The graph of a polynomial function is a continuous curve. Thus, all we need do is plot a sufficient number of points and connect them.

x	y
0	-3
1	-4
3	0
-1	0
-2	5

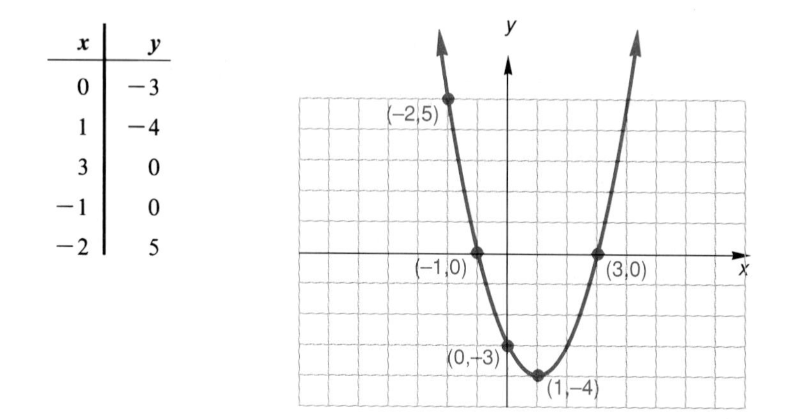

Figure 2.21 ■

The graph of a quadratic function is called a **parabola.** In figure 2.21, the parabola opens upward, and it has a low point. A parabola that opens downward has a high point. The high or low point is called the **vertex** of the graph. Knowing the vertex of a parabola enables us to refine the construction of its graph, as we would know the graph's true low or high point. As it is now, we cannot be sure that $(1,-4)$ is the low point of the parabola of example 1.

Finding the Vertex of a Parabola

Let us complete the square on the right side of equation (1).

$$\begin{aligned} y &= a\left(x^2 + \frac{bx}{a}\right) + c \\ &= a\left(x^2 + \frac{bx}{a} + \frac{b^2}{4a^2}\right) + c - \frac{b^2}{4a} \\ &= a\left(x + \frac{b}{2a}\right)^2 + \frac{4ac - b^2}{4a} \\ &= a\left(x + \frac{b}{2a}\right)^2 + K \qquad \textbf{(2)} \end{aligned}$$

The constant $(4ac - b^2)/4a$ is denoted by K. Also, $(x + b/2a)^2 > 0$ if $x \neq -b/2a$. (Why?) If $a > 0$ and $x \neq -b/2a$, then $y > K$. If $x = -b/2a$, then $y = K$. Thus, if $a > 0$, the low point of a graph of $y = ax^2 + bx + c$ occurs at $\boldsymbol{x = -b/2a}$ and the graph opens upward. A parallel argument shows that if $a < 0$, then a graph of (1) opens downward and has a high point at $x = -b/2a$.

> **Vertex of a Parabola**
>
> A graph of $f(x) = ax^2 + bx + c$ is called a **parabola.** The parabola opens upward if $a > 0$. It opens downward if $a < 0$. The coordinates of the parabola's vertex are $(-b/2a, f(-b/2a))$.

Example 2 Find the vertex of the parabola in example 1.

Solution In the function $y = x^2 - 2x - 3$, $a = 1$ and $b = -2$. Because $a > 0$, the graph of the function must open upward (see figure 2.21), and its vertex must be a low point. The x-coordinate of the low point is

$$\begin{aligned} x &= \frac{-b}{2a} \\ &= -\frac{(-2)}{2(1)} \\ &= 1. \end{aligned}$$

The y-coordinate of the low point is $(1)^2 - 2(1) - 3 = -4$. ■

Example 3 Graph the function $s = -2t^2 + 11t - 5$.

Solution The given equation has the form of (1) with $a = -2$, $b = 11$, and $c = -5$. Its graph is a parabola that opens downward ($a < v$). The coordinates of the vertex are,

$$\begin{aligned} t &= \frac{-11}{2(-2)} = \frac{11}{4} \\ s &= -2\left(\frac{11}{4}\right)^2 + 11\left(\frac{11}{4}\right) - 5 \\ &= -\frac{121}{8} + \frac{242}{8} - \frac{40}{8} \\ &= \frac{81}{8}. \end{aligned}$$

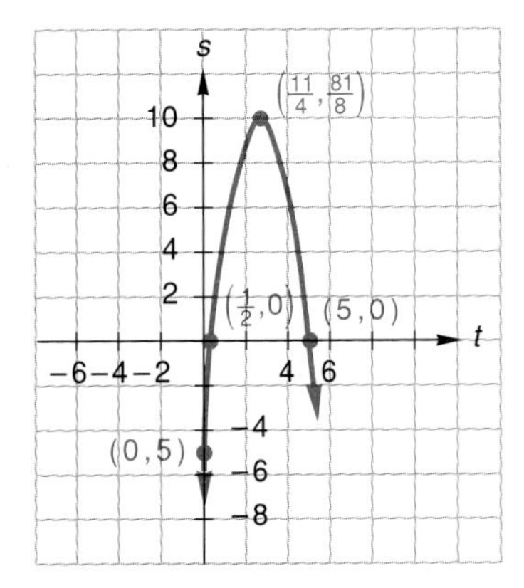

Figure 2.22

The graph's intercepts with the s and t axes (figure 2.22) are found by first setting $t = 0$, and then setting $s = 0$.

$$\begin{aligned} s &= -2(0)^2 + 11(0) - 5 \\ &= -5 \end{aligned}$$

$$\begin{aligned} 0 &= -2t^2 + 11t - 5 \\ 0 &= (-2t + 1)(t - 5) \\ t &= \tfrac{1}{2} \quad \text{or} \quad t = 5 \end{aligned}$$

■

Max-Min Problems

There is a class of problems in which a quantity is to be maximized or minimized. The mathematical model of many such problems is a quadratic function. As a result, the solution of the problem requires that we find the coordinates of the vertex of a parabola.

Example 4 A ski tour entrepreneur knows that if he sells his three-day ski package for \$280, he can fill a sixty-seat bus. He also knows from past experience that he will lose three customers for each \$20 increase in the package. At what price should he set the package to maximize his revenue?

Solution Let x be the number of \$20 increases. Then the price of the ski package is $280 + 20x$ and the number of purchasers of the package is $60 - 3x$. The resulting revenue for the entrepreneur is

$$\begin{aligned} R &= (60 - 3x)(280 + 20x) \\ &= -60x^2 + 360x + 16{,}800. \end{aligned}$$

The revenue equation represents a quadratic function in which $a = -60$ and $b = 360$. Since a is negative, the graph of the revenue function opens downward, and the x-coordinate of its highpoint is

$$\begin{aligned} x &= \frac{-360}{2(-60)} \\ &= 3. \end{aligned}$$

Thus, a ski package price of \$340 yields a maximum revenue of (51)(\$340) = \$17,340. ■

Example 5 Find the rectangle of maximum area that can be inscribed in triangle ABC.

Solution The area of an inscribed rectangle is $A = wh$. But w and h are related, as the triangles ABC and DEC are similar. (Why?) We can write

$$\frac{w}{30} = \frac{15 - h}{15},$$

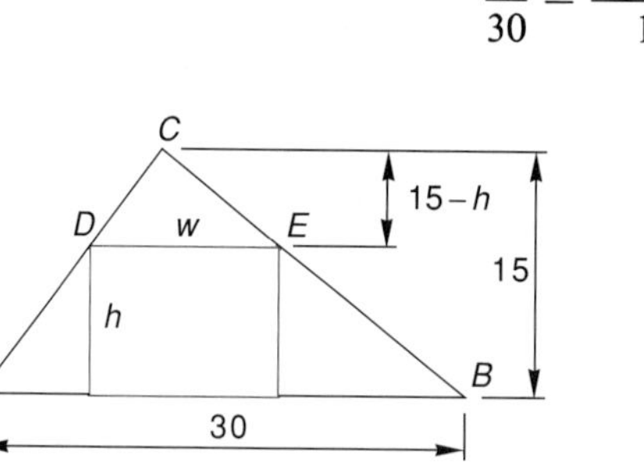

because the lengths of the sides and altitudes of similar triangles are proportional. Now

$$w = 30 - 2h \quad \text{and} \quad A = (30 - 2h)h = -2h^2 + 30h.$$

Area is expressed as a quadratic function of h in which $a = -2$ and $b = 30$. Consequently, a graph of the area function opens downward and the h-coordinate of its high point is

$$\begin{aligned} h &= \frac{-30}{2(-2)} \\ &= 7.5. \end{aligned}$$

It follows that $w = 15$. The area of this maximum area rectangle is 112.5 square units. ■

The Difference Quotient

Let P be a function that gives the population of a certain country when the year, x, is known (figure 2.23). Let Δx represent a change in x, then ΔP, which is the corresponding change in P, is equal to $P(x + \Delta x) - P(x)$. The ratio of the two changes,

$$\frac{\Delta P}{\Delta x} = \frac{P(x + \Delta x) - P(x)}{\Delta x},$$

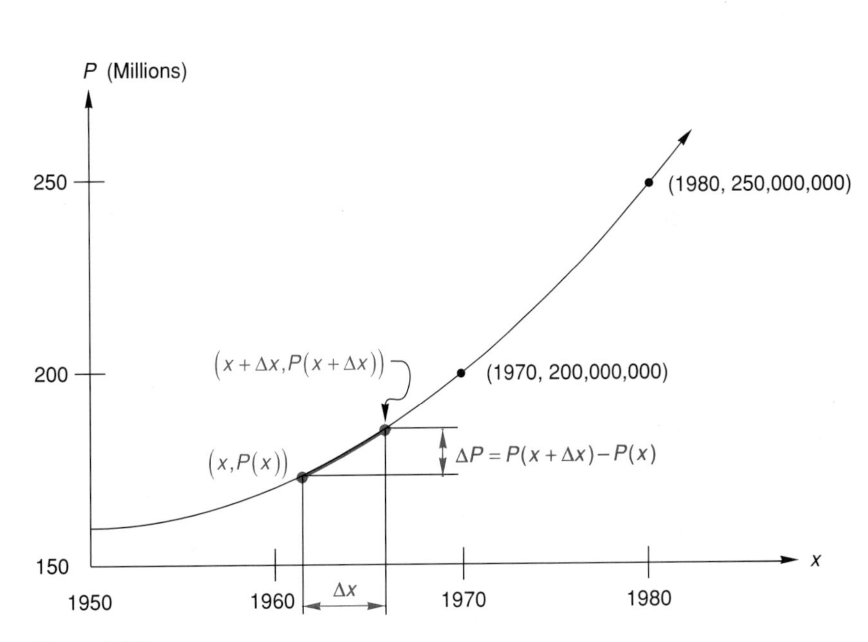

Figure 2.23

is called the *difference quotient.* It is the **slope** of the line segment between the points $(x,P(x))$ and $(x + \Delta x, P(x + \Delta x))$. If $P(1970) = 200{,}000{,}000$ and $P(1980) = 250{,}000{,}000$, then the difference quotient

$$\frac{P(1970 + 10) - P(1970)}{10} = \frac{50{,}000{,}000}{10} = 5{,}000{,}000,$$

represents the *average rate of change* of population per year from 1970 to 1980. In this instance $x = 1970$ and $\Delta x = 10$.

Example 6 Let h represent a change in x for the function $f(x) = x^2 + 5$. Find the difference quotient between x and $x + h$.

Solution

$$\begin{aligned}\frac{f(x+h) - f(x)}{h} &= \frac{(x+h)^2 + 5 - (x^2 + 5)}{h} \\ &= \frac{x^2 + 2hx + h^2 + 5 - x^2 - 5}{h} \\ &= \frac{2hx + h^2}{h} \\ &= 2x + h\end{aligned}$$

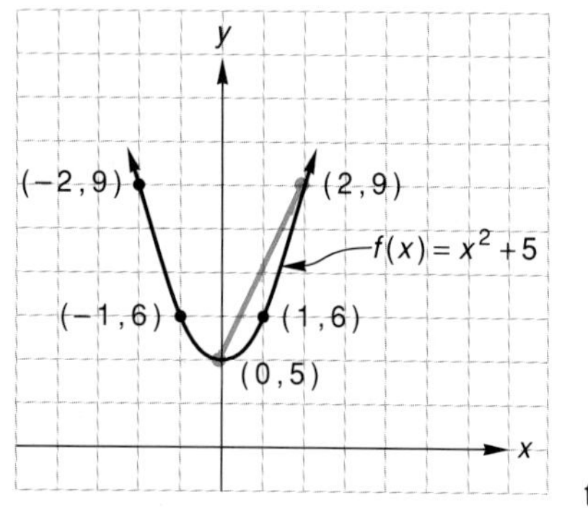

Figure 2.24

The result of example 6 says that the slope of the chord drawn between the points $(x,f(x))$ and $(x + h, f(x + h))$ is $2x + h$. If $x = 0$ and $h = 2$, then the slope of the chord between (0,5) and (2,9) is 2 (see figure 2.24).

Example 7 Let $C = -0.1x^2 + 90x + 800,\ 0 \le x \le 600$ be the cost (in hundreds of dollars) of producing x units of a certain product. Find

a. a general expression for the average rate of change in cost.

b. the average rate of change in cost between 100 and 150 units.

c. the average rate of change in cost between 100 and 101 units.

Solution

a. If $x > 0$, then x and $x + \Delta x$ are different levels of production. An expression for the average rate of change in cost over a change of Δx units of production is

$$\begin{aligned}&\frac{C(x + \Delta x) - C(x)}{\Delta x} \\ &= \frac{-0.1\,(x + \Delta x)^2 + 90\,(x + \Delta x) + 800 - (-0.1x^2 + 90x + 800)}{\Delta x} \\ &= \frac{-0.2x\,(\Delta x) - 0.1\,(\Delta x)^2 + 90\,\Delta x}{\Delta x} \\ &= -0.2x - 0.1\Delta x + 90.\end{aligned}$$

b. Applying the result in part a to $x = 100$ and $\Delta x = 50$, we obtain an average rate of change in cost between 100 and 150 units of \$6,500.

c. Applying the result in part a to $x = 100$ and $\Delta x = 1$, we obtain \$6,990 as the average rate of change in cost between 100 and 101 units of production. ■

The result in part c of example 7 is called the *marginal cost*. It is the cost of producing one more unit at a level of production of 100 units.

Exercise Set 2.3

(A)

Graph each quadratic function. Label the vertex of each parabola.

1. $y = x^2$

2. $y = -2x^2$

3. $y = x^2 + 3x$

4. $y = x^2 - 2x$

5. $y = -3x^2 + 4x - 2$

6. $y = 2x^2 - 5x + 1$

7. $y = (x - 4)^2$

8. $y = (x + 4)^2$

9. $y = -(x + 2)^2$

10. $y = 6x^2 + 4x + 3$

Given f, construct the difference quotient $[f(x + \Delta x) - f(x)]/\Delta x$ and simplify it if possible.

11. $f(x) = 3x + 5$

12. $f(x) = -2x$

13. $f(x) = -12$

14. $f(x) = 5$

15. $f(x) = x^2$

16. $f(x) = -2x^2$

17. $f(x) = 6x^2 + 4x + 3$

18. $f(x) = -3x^2 + 4x - 2$

19. $f(x) = 1/x$

20. $f(x) = x^3$

In exercises 21–28, given g, construct the difference quotient $[g(x + h) - g(x)]/h$ and simplify it if possible.

21. $g(x) = 6x - 2$

22. $g(x) = (x + 4)^2$

23. $g(x) = -(x + 2)^2$

24. $g(x) = \dfrac{5}{x + 3}$

25. $g(x) = -\dfrac{2}{x + 1}$

26. $g(x) = \sqrt{x}$

27. $g(x) = \sqrt{3x + 7}$

28. $g(x) = \dfrac{1}{x^2}$

(B)

29. Based on past experience, a local theater manager knows that she can fill the theater's 250-seat auditorium for a good production if the ticket price for the production is \$10 or less. The manager also knows that each \$1 increase in ticket price beyond \$10 results in 10 fewer playgoers per performance. What ticket price should the manager set to maximize daily revenue?

30. Based on past sales records, the management of a car dealership knows it can sell 1,000 units of its "Young Executive" model at \$10,000 each, and that each \$500 dollar decrease in price will result in 100 more units being sold. What price should management set on its "Young Executive" model to maximize revenue from sales of the model?

31. Express the dimensions of the rectangle of maximum area, in terms of b and h, that can be inscribed in triangle ABC.

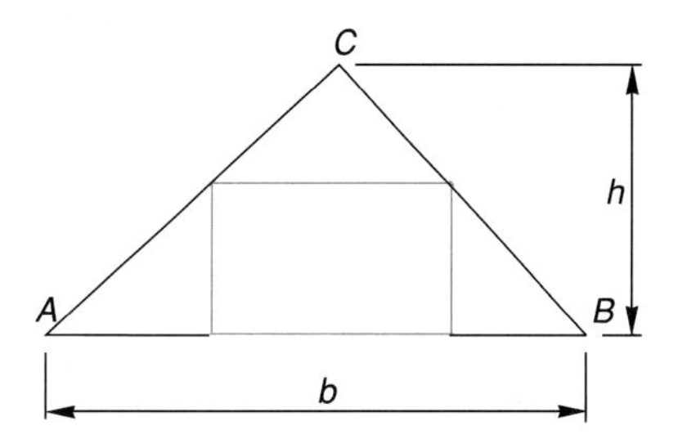

32. Find two positive numbers whose sum is 612 and whose product is the maximum possible for all such pairs of numbers.

33. Determine the dimensions of the rectangle of maximum area that has a perimeter of 420 feet.

34. The perimeter of a rectangle is 420 feet. Determine the dimensions of the rectangle that will minimize the length of its diagonal.

35. The cost of producing x units of a product is given by the formula

$$C = 4.20x^3 - 126.00x^2 + 22.65x.$$

How many units should be produced to minimize the average cost per unit, which is expressed as $C(x)/x$?

36. The height to which a ball rises when thrown from a height of 5 feet with a velocity of 128 feet per second is given by the equation

$$h = -16t^2 + 128t + 5,$$

where h represents the height in feet and t represents the time in seconds. Find the maximum height to which the ball rises. Also, write the difference quotient $\Delta h/\Delta t$ and determine its physical meaning.

37. The population growth of a certain animal population is given by the formula $P = 5t^2 + 20t + 2{,}000$ where t is time in years. Write an expression for the average rate of change of population. Then determine the average rate of change of population between $t = 2$ and $t = 7$.

38. A manufacturer's cost function is given as $C(x) = 0.05x^2 + 20x + 800$. Write an expression for the average rate of change in cost. Find the marginal cost at a level of 50 units of production.

39. Find the marginal average cost of the function in exercise 35 when $x = 25$.

40. A rectangular-shaped region is to be fenced as shown with 1,000 feet of fencing. Determine the dimensions of the rectangle so that the total area enclosed is maximized.

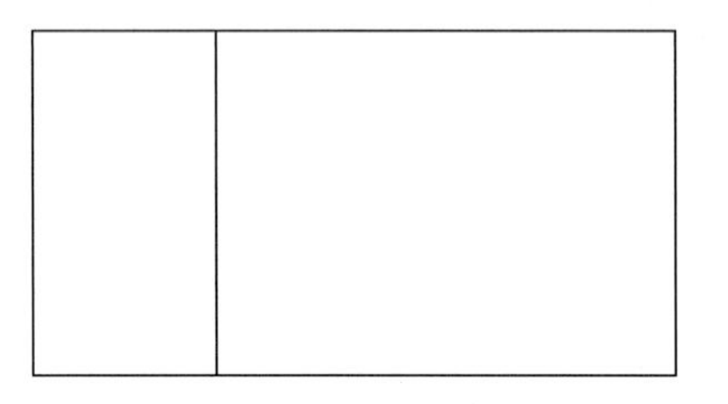

41. A flat piece of sheet metal 10 inches wide is to be bent into a U-shaped rain gutter. Determine the dimensions of the gutter of greatest capacity.

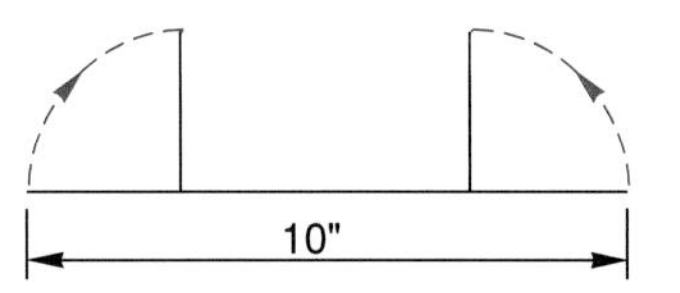

42. Find the shortest distance between (0,4) and the graph of $y = x^2$, $x \geq 0$.

43. Prove that for all x, $1/(3x^2 - 4x + 5) \leq \frac{3}{11}$.

44. Referring to figure 2.23, determine whether the rate of population increase is greater in 1960 or 1970. Support your answer.

2.4 Graphing Techniques

A plane graph provides us with a picture of the relationship between two variables that is defined by a function. It enables us to see visual features of the relationship. Seeing a function is what graphing is all about and why the ability to graph quickly and accurately is essential in the study and application of mathematics.

When it is possible we want to move away from laborious point plotting as the only means of constructing a graph. We would prefer to use properties of the function to construct its graph. If we know that the graph of a function is symmetric to the y-axis, then we need only duplicate on one side of the

y-axis the points that are plotted on the other side of the axis. Knowing the domain and range of a function enables us to limit a graph of the function to a certain region of the plane. Our intent then is to extract properties of a function from its equation and to utilize those properties together with a minimum of point plotting to sketch an accurate graph of the function.

Symmetry

Two points are symmetric to each other with respect to a line if they are equidistant from the line, and if the segment they determine is perpendicular to the line. P_1 and P_2 (figure 2.25) are symmetric to each other with respect to the y-axis, and Q_1 and Q_2 are symmetric to each other with respect to the x-axis. Two points are symmetric to each other with respect to a point if they are equidistant from the point and if the three points lie in a straight line. Thus, T_1 and T_2 are symmetric to each other with respect to the origin.

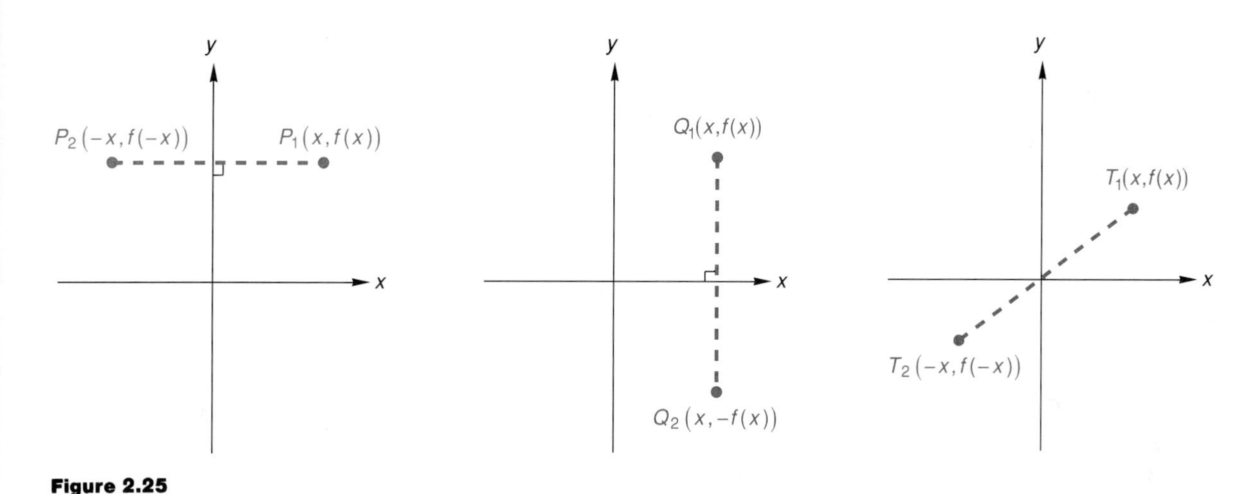

Figure 2.25

From figure 2.25, we see that a graph of a function f is symmetric to the y-axis if to each point $P_1(x,f(x))$ in the graph, there corresponds another point $P_2(-x,f(-x))$ in the graph, so that $f(-x) = f(x)$. Hence, a graph of f is symmetric to the y-axis if

$$f(-x) = f(x). \tag{1}$$

The function $y = |x|$ (figure 2.26) is symmetric to the y-axis because

$$f(-x) = |-x| = |x| = f(x).$$

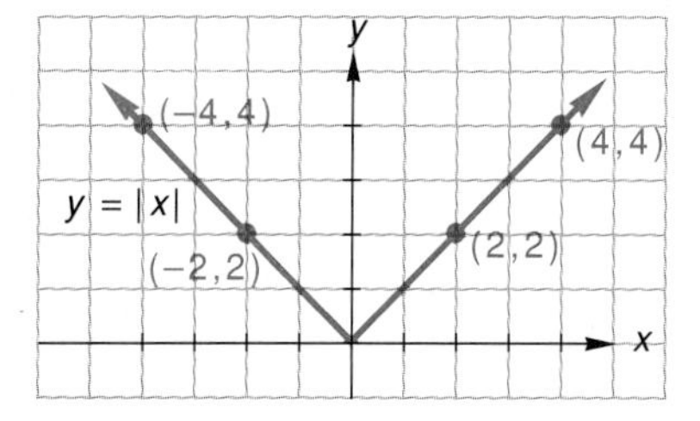

Figure 2.26

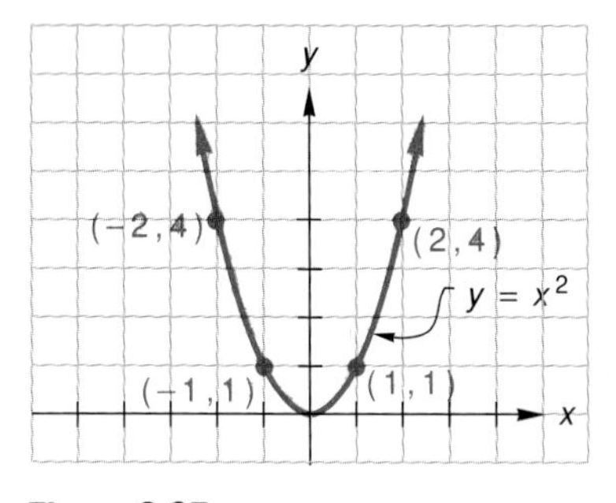

Figure 2.27

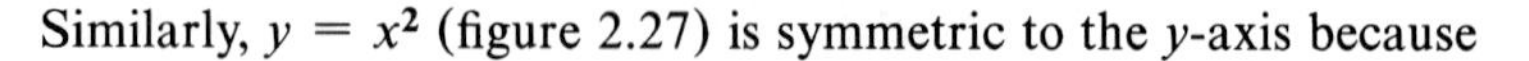
Similarly, $y = x^2$ (figure 2.27) is symmetric to the y-axis because

$$f(-x) = (-x)^2 = x^2 = f(x).$$

The function $y = x^2 - x$, is not symmetric to the y-axis because

$$f(-x) = (-x)^2 - (-x) = x^2 + x \neq f(x).$$

From figure 2.25, we see that a graph of f is symmetric to the origin if to each point $T_1(x,f(x))$, there corresponds another point $T_2(-x,f(-x))$, so that $f(-x) = -f(x)$. In other words, symmetry with respect to the origin is assured if

$$\boldsymbol{f(-x) = -f(x).} \qquad \textbf{(2)}$$

The function $y = x^3 + x$ is symmetric to the origin because

$$f(-x) = (-x)^3 + (-x) = -x^3 - x = -(x^3 + x) = -f(x).$$

But, $y = x^4 - 3x$ is not symmetric to the origin because

$$f(-x) = (-x)^4 - 3(-x) = x^4 + 3x \neq -x^4 + 3x = -f(x).$$

Symmetry Tests

A graph of f is symmetric to the y-axis if $\boldsymbol{f(-x) = f(x)}$.
A graph of f is symmetric to the origin if $\boldsymbol{f(-x) = -f(x)}$.

A graph of a function cannot be symmetric to the x-axis. (Why?) Consequently, if a graph of f is symmetric to the y-axis, it cannot be symmetric to the origin (see exercise 49). Also, if a graph of f is symmetric to the origin, it cannot be symmetric to the y-axis (see exercise 50).

Intercepts

The points of intersection between a graph and the coordinate axes are called the **intercepts** of the graph.

Example 1 Find the x and y intercepts of $y = x^2 - x - 6$.

Solution The y-coordinate of a point of intersection of the graph and the x-axis is 0. The x-coordinate of the point of intersection is called the ***x*-intercept.** Therefore, the x-intercepts of the graph are obtained by setting y to 0 and solving the resulting equation.

$$0 = (x + 2)(x - 3)$$

The x-intercepts are -2 and 3 (see figure 2.28).

The y-intercept is determined by setting x to 0 and solving the resulting equation.

$$y = 0^2 - 0 - 6$$

The y-intercept is -6. ■

Domain and Range

Restrictions on the domain and range of a function are also restrictions on the graph of the function. The domain of $y = \sqrt{x-1}$ is $x \geq 1$. A graph of this function must be to the right of, or on, the line $x = 1$. The graph must also be on or above the x-axis. (Why?)

Example 2 Sketch a graph of $y = x^2 - x - 6$.

Solution We find the function's intercepts, its domain, and if possible, its range. We determine if the function is symmetric to the y-axis or the origin. In addition, we plot as many points as are necessary to produce an accurate graph.

x-intercepts: 3 and -2, from example 1

y-intercept: -6

Domain: y is defined for *all real number replacements* for x. The domain is R.

Range: The range is often found by solving for x in terms of y.

$$x^2 - x - (6 + y) = 0$$
$$x = \frac{1 \pm \sqrt{1 + 24 + 4y}}{2}$$

We see that x is a real number if $4y + 25 \geq 0$, or if $y \geq -\frac{25}{4}$.

Because the function is quadratic, its range can also be determined by finding the low point of its graph.

$$x = -\frac{b}{2a} = \frac{-(-1)}{2(1)} = \frac{1}{2}$$
$$y = \left(\frac{1}{2}\right)^2 - \frac{1}{2} - 6 = -\frac{25}{4}$$

Since the low point of the graph is $(\frac{1}{2}, -\frac{25}{4})$, the range of the function must be $\{y \mid y \geq -\frac{25}{4}\}$.

Symmetries: The graph is not symmetric to the y-axis because

$$f(-x) = (-x)^2 - (-x) - 6 = x^2 + x - 6 \neq f(x).$$

The graph is not symmetric to the origin because

$$f(-x) = x^2 + x - 6 \neq -x^2 + x + 6 = -f(x).$$

x	y
-3	6
4	6

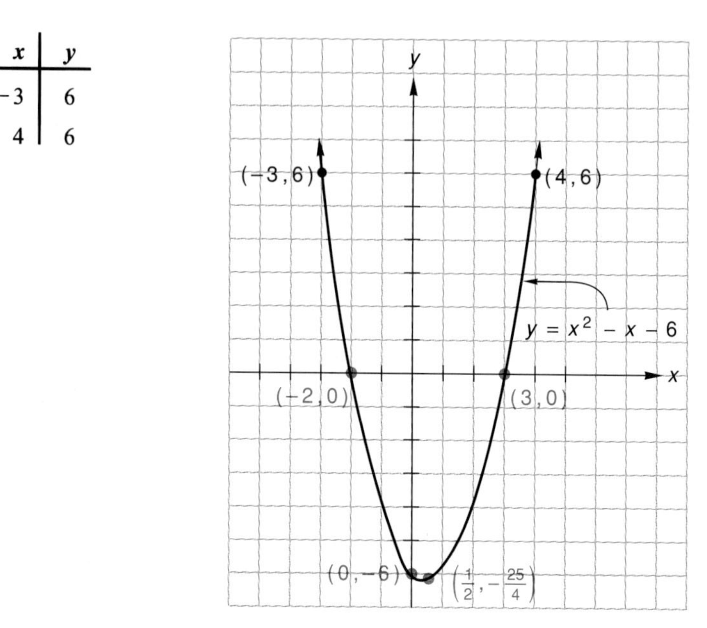

Figure 2.28

■

Example 3 Graph $y = x^4 - 4x^2$.

Solution

The y-intercept is 0.

The x-intercepts, as determined from

$$0 = x^4 - 4x^2 = x^2(x^2 - 4) = x^2(x - 2)(x + 2),$$

are -2, 0, and 2.

Domain: R

Range:

$$0 = x^4 - 4x^2 - y. \text{ Let } z = x^2.$$

$$0 = z^2 - 4z - y$$

$$x^2 = z = \frac{4 \pm \sqrt{16 + 4y}}{2} = 2 \pm \sqrt{4 + y}$$

The range is $[-4,\infty)$, as x is a real number only if $y \geq -4$. When $y = -4$, $x = \pm\sqrt{2}$. Hence, the lowest points in the graph are $(-\sqrt{2},-4)$ and $(\sqrt{2},-4)$.

The graph is symmetric to the y-axis because

$$f(-x) = (-x)^4 - 4(-x)^2 = x^4 - 4x^2 = f(x).$$

The graph is not symmetric to the origin because

$$f(-x) = x^4 - 4x^2 \neq -x^4 + 4x^2 = -f(x) \text{ (figure 2.29).}$$

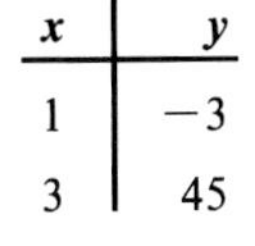

x	y
1	-3
3	45

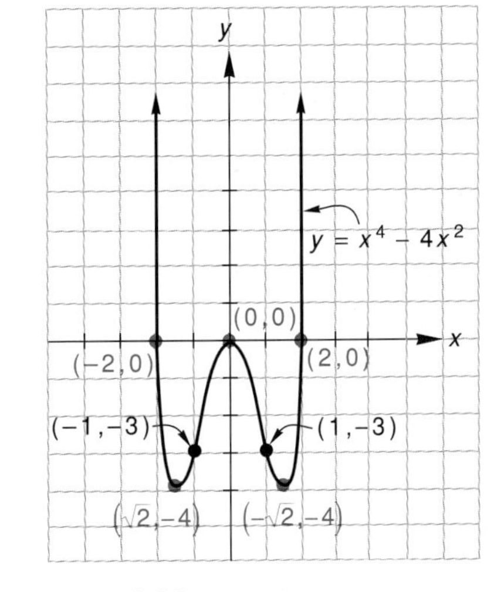

Figure 2.29

New Graphs from Old: Reflections

We have shown that a graph of $y = f(x)$ is symmetric to the y-axis if

$$f(-x) = f(x).$$

The test assured us that each point (x,y) in the graph had a mirror image $(-x,y)$ with respect to the y-axis. It follows that the graphs of $y = f(x)$ and $y = f(-x)$ are *reflections in the y-axis*. To each (x,y) in $y = f(x)$ there corresponds a point $(-x,y)$ in $y = f(-x)$ (see figure 2.30).

Graphs of $y = f(x)$ and $y = -f(x)$ are reflections in the x-axis. To each point (x,y) in $y = f(x)$ there corresponds its reflection in the x-axis $(x,-y)$, in $y = -f(x)$ (figure 2.30).

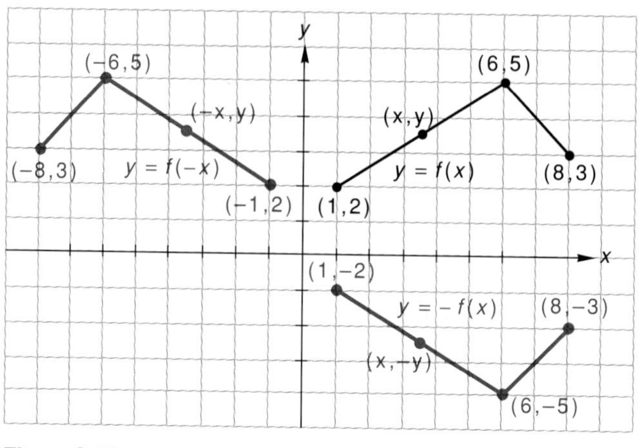

Figure 2.30

Example 4 Graph $f(x) = x^2 - x$ and $f(-x) = x^2 + x$ on the same coordinate axes.

Solution (See figure 2.31.)

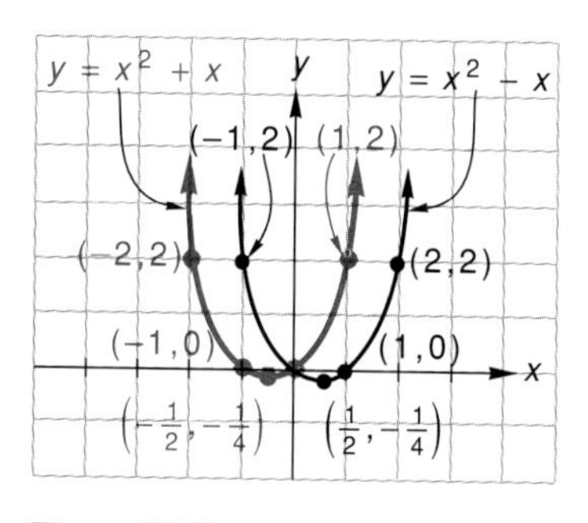

Figure 2.31

	$f(x) = x^2 - x$	$f(-x) = x^2 + x$
x-intercepts	$0 = x(x - 1); 0,1$	$0 = x(x + 1); 0, -1$
y-intercepts	0	0
Domain	R	R
Range	$y \geq -\frac{1}{4}$	$y \geq -\frac{1}{4}$
Symmetry	none	none

■

Example 5 Graph $f(x) = x^3 - 2$, $x \geq 0$ and $-f(x) = -(x^3 - 2)$, $x \geq 0$ on the same coordinate axes.

Solution (See figure 2.32.)

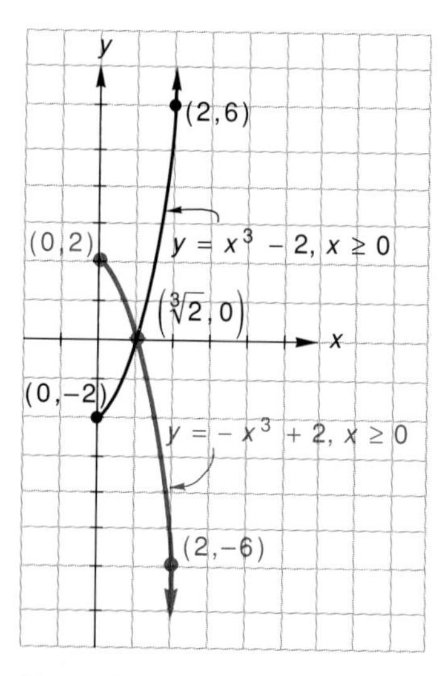

Figure 2.32

	$f(x) = x^3 - 2$	$-f(x) = -x^3 + 2$
x-intercepts	$2^{1/3}$	$2^{1/3}$
y-intercepts	-2	2
Domain	R^+	R^+
Range	$x = (y + 2)^{1/3}$ and $x \geq 0$ thus $y \geq -2$	$x = (2 - y)^{1/3}$ and $x \geq 0$ thus $y \leq 2$
Symmetry	none	none

■

Tests for Graphs that are Reflections

Graphs of $y = f(x)$ and $y = f(-x)$ are reflections in the y-axis.
Graphs of $y = f(x)$ and $y = -f(x)$ are reflections in the x-axis.

Translations

For each x-value, the y-values obtained from the equations $y = f(x)$ and $y = f(x) + C$ differ by C units. Hence, the graph of $y = f(x) + C$ is *translated C units vertically* with respect to the graph of $y = f(x)$.

Example 6 The graphs of $y = |x|$ and $y = |x| - 2$ (figure 2.33) are on the same coordinate axes.

The graph of $y_2 = f(x - c)$ is translated c units horizontally relative to the graph of $y_1 = f(x)$. If $x = a$ then $y_1 = f(a)$. However, $y_2 = f(a)$ only if $x - c = a$, or $x = a + c$. Thus, the y-values of the two functions are equal if their x-values differ by c units. If $c > 0$, the graph of $y_2 = f(x - c)$ is translated c units to the right with respect to the graph of $y_1 = f(x)$. The translation is to the left if $c < 0$.

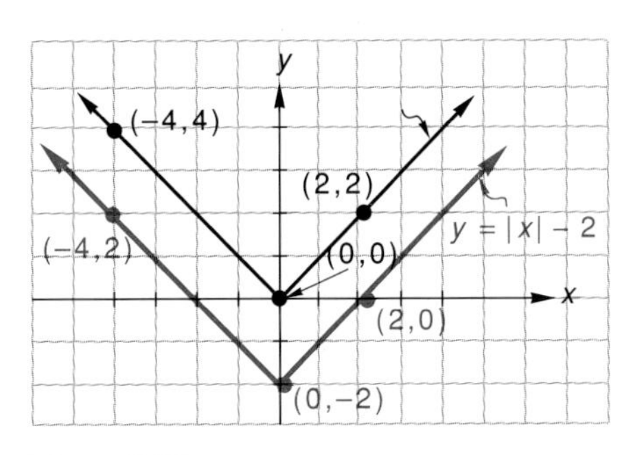

Figure 2.33

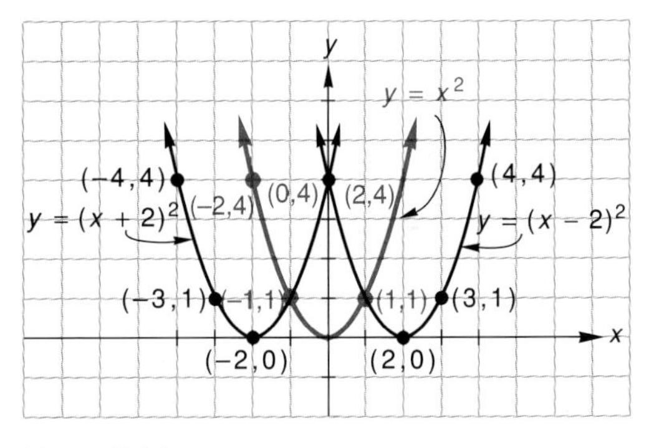

Figure 2.34

Example 7 Graph $y = (x - 2)^2$ and $y = (x + 2)^2$ on the same coordinate axes.

Solution The graph of $y = (x - 2)^2$ is the graph of $y = x^2$ translated 2 units to the right ($c > 0$). The graph of $y = (x + 2)^2 = (x - (-2))^2$ is the graph of $y = x^2$ translated two units to the left ($c < 0$) (figure 2.34). ■

Translations

The graph of $y = f(x) + C$ is translated vertically with respect to the graph of $y = f(x)$; up if $C > 0$ and down if $C < 0$.
The graph of $y = f(x - c)$ is translated horizontally with respect to the graph of $y = f(x)$; to the right if $c > 0$ and to the left if $c < 0$.

Stretchings and Contractions

The graph of $y = cf(x)$, $c > 1$, is a stretching away from the x-axis, of the graph of $y = f(x)$. This is so because for each x in the domain of f, $|cf(x)| > |f(x)|$. If $0 < c < 1$, the graph of $y = cf(x)$ is a contraction toward the x-axis, of the graph of $y = f(x)$ (figure 2.35).

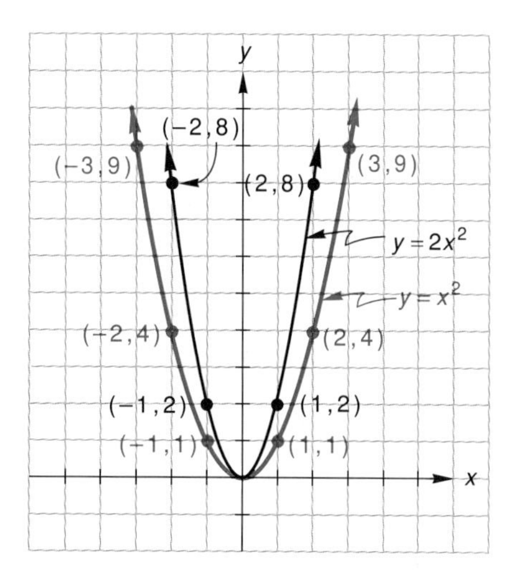

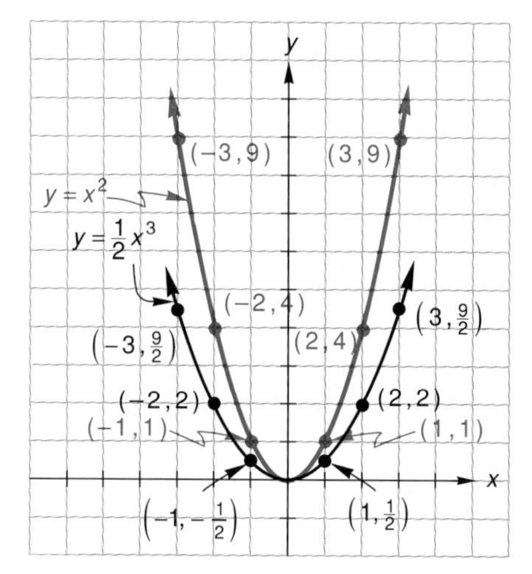

Figure 2.35

The graph of $y = f(cx)$, $c > 1$, is a contraction toward the y-axis, of the graph of $y = f(x)$. As an example, the same y-values are produced in $y = f(x)$ and $y = f(2x)$ when the x-values in the second equation are half the x-values in the first equation (figure 2.36). If $0 < c < 1$, then the graph of $y = f(cx)$ is a stretching away from the y-axis, of the graph of $y = f(x)$.

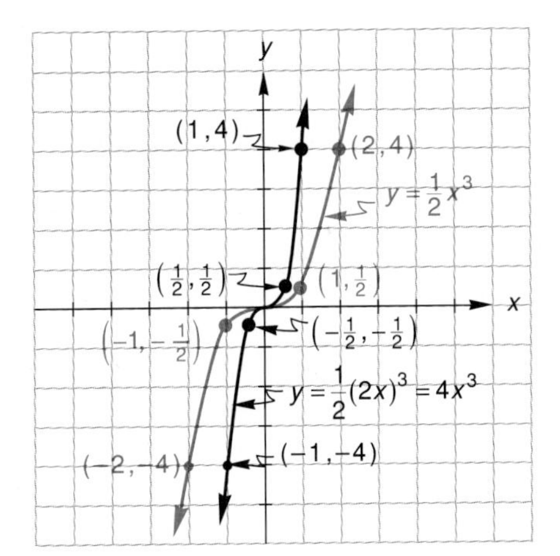

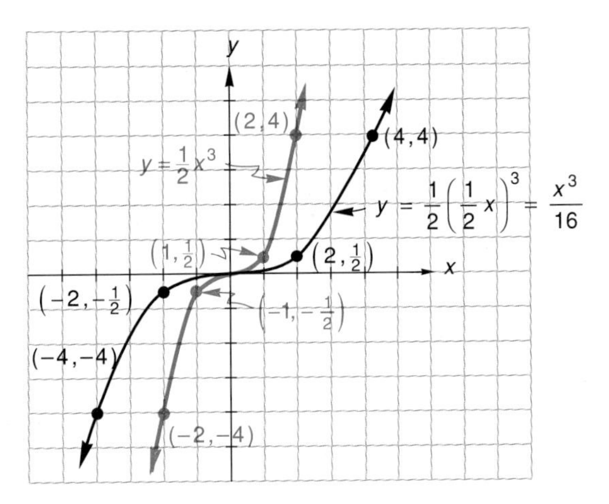

Figure 2.36

> **Stretchings and Contractions**
>
> The graph of $y = cf(x)$ is a stretching away from, or a contraction towards, the x-axis of the graph of $y = f(x)$, as $c > 1$ or $0 < c < 1$. The graph of $y = f(cx)$ is a contraction towards, or a stretching away from, the y-axis of the graph of $y = f(x)$, as $c > 1$ or $0 < c < 1$.

Exercise Set 2.4

(A)

Find a point symmetric to the given point with respect to the (a) y-axis, (b) x-axis, and the (c) origin.

1. (2,5) **2.** (−4,7) **3.** (−1,−5) **4.** (3,−6)
5. (−6,8) **6.** (4,0) **7.** (0,−2) **8.** (0,0)

Sketch a graph of each function after first determining (a) the x and y intercepts of the graph, (b) the domain and, if possible, the range of the function, (c) if the graph is symmetric to the y-axis or to the origin.

9. $y = x^2 - 1$ **10.** $y = x^2 + 4$
11. $y = x^2 - 2x - 3$ **12.** $y = 2x^2 - 6x$
13. $y = \sqrt{x}$ **14.** $y = x$
15. $y = \sqrt{x^3}$ **16.** $y = \sqrt{x^2}$
17. $y = \sqrt{9 - x^2}$ **18.** $y = \sqrt{x^2 + 1}$
19. $y = x^3 - 16x$ **20.** $y = x^4 - 4x^2$

(a) Graph $f(x)$. Then sketch graphs of its reflections, $f(-x)$ and $-f(x)$, on the same coordinate system.
(b) Construct a second graph of $f(x)$. Then sketch graphs of its translations, $f(x - c)$ and $f(x) - c$, on the same coordinate system.

21. $f(x) = 2x - 4$, $c = 1$
22. $f(x) = -3x - 2$, $c = -1$
23. $f(x) = x^2 - 1$, $c = -4$
24. $f(x) = 2x^2 - 6x$, $c = 3$
25. $f(x) = \sqrt{9 - x^2}$, $c = 2$
26. $f(x) = \sqrt{x^2 + 1}$, $c = -2$
27. $f(x) = x^3 + 16x$, $c = -5$
28. $f(x) = x^4 + 4x^2$, $c = 6$

29. Use their relationship to the graph of $y = |x|$, to sketch graphs of $y = |x + 3|$, $y = |x - 3|$, $y = |x| + 3$, and $y = -|x + 3|$ on the same set of coordinate axes.

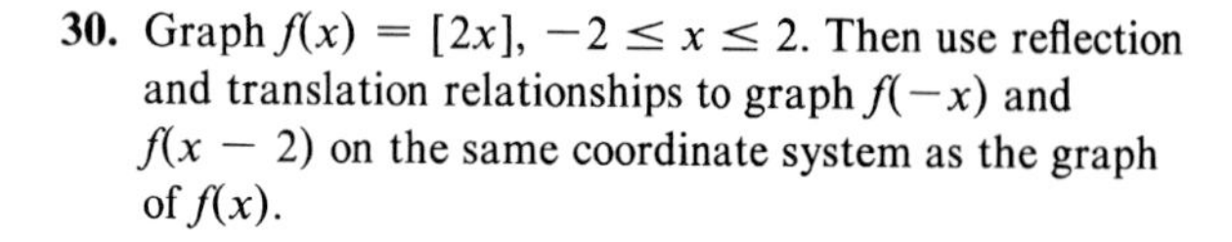

30. Graph $f(x) = [2x]$, $-2 \le x \le 2$. Then use reflection and translation relationships to graph $f(-x)$ and $f(x - 2)$ on the same coordinate system as the graph of $f(x)$.

(a) Graph $f(x)$. Then, on the same coordinate system, sketch graphs of $y = cf(x)$ and $y = f(cx)$. (b) Graph on a second coordinate system $y = -cf(x)$ and $y = f(-cx)$.

31. $f(x) = x^2 - 1$, $c = 2$
32. $f(x) = x^2 + 4$, $c = \frac{1}{2}$
33. $f(x) = x^2 - 2x - 3$, $c = \frac{1}{3}$
34. $f(x) = 2x^2 - 6x$, $c = 3$
35. $f(x) = x^3 - 16x$, $c = 4$
36. $f(x) = x^4 - 4x^2$, $c = \frac{1}{4}$

(B)

Indicate what reflections, translations, stretchings, or contractions of a simpler function are required to obtain the graph of the given function. Do not graph the function.

37. $y = 3(x + 1)^2$
38. $y = -[2(x - 1)]^2$
39. $y = -\frac{1}{2}|x - 1| + 5$
40. $y = 2|x + 3| - 1$
41. $y = \left(-\frac{x}{2}\right)^3$
42. $y = \frac{1}{4}(x + 2)^3$

43. Sketch the graph of $y = 3x^2 - 12x + 12$ by the method used in examples 2 and 3. Then observe that $3x^2 - 12x + 12 = 3(x - 2)^2$ so that the desired graph could be constructed by translating the graph of $y = x^2$ two units to the right and then stretching by a factor of 3.

44. Graph $y = -2(x - 1)^2 + 3$ by (a) translating a graph of $y = x^2$ one unit to the right, and (b) reflecting that graph about the x-axis and stretching it by a factor of 2, and (c) translating the resulting graph upwards three units. Then graph the function directly by noting that $-2(x - 1)^2 + 3 = -2x^2 + 4x + 1$.

45. Graph the function in exercise 11 after completing the square.

46. Graph the function in exercise 12 after completing the square.

47. Let $f(x) = x + 1$. Let g be the function obtained by translating f upwards 1 unit and then reflecting with respect to the x-axis. Let h be the function obtained from f by reflecting f with respect to the x-axis and then translating upwards 1 unit. Write equations for g and h. Does $g = h$?

48. Write equations for g and h in exercise 47 if each function is formed by translating 1 unit to the right instead of 1 unit upwards. Does $g = h$?

49. Draw a sketch similar to figure 2.25 to prove that if a graph of a function is symmetric to the y-axis, it cannot be symmetric to the origin. (Hint: Assume the graph is symmetric to both the y-axis and the origin. Show that the assumption leads to the conclusion that the graph is also symmetric to the x-axis. This conclusion contradicts the fact that you started with the graph of a function.)

50. Prove that if a graph of a function is symmetric to the origin, it cannot be symmetric to the y-axis. (Hint: See exercise 49.)

Construct the graph of each function from the graph of f (figure 2.37).

51. $y = f(x - 1)$
52. $y = -2f(x - 1)$
53. $y = f(2x)$
54. $y = f(-2x)$
55. $y = 3f(x + 2) - 4$
56. $y = \frac{1}{3}f(x - 3) + 1$
57. $y = f(6 - 3x)$
58. $y = f(5 - x) - 2$

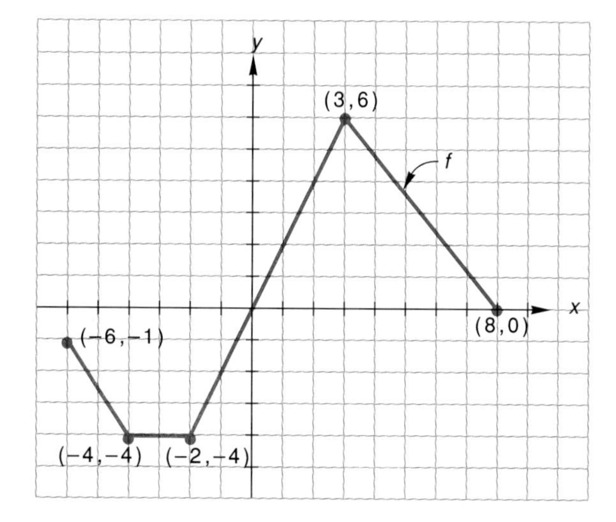

Figure 2.37

2.5 Algebra of Functions

Is the sum or difference of two functions a function? Is the product or quotient of two functions a function? If the answer to either question is yes, then we have some simple ways of constructing new functions from old.

Let f and g be any two functions *whose domains intersect*. Let $s(x) = f(x) + g(x)$ for each x in the common domain of f and g. Then s must be a function because associated with each x in its domain is exactly one real number, the sum $f(x) + g(x)$. We call s the sum of f and g and denote it as $f + g$.

Example 1 Let $f(x) = x^2$, $x \geq 0$ and $g(x) = x + 1$. Then

$$(f + g)(x) = f(x) + g(x) = x^2 + x + 1.$$

The domain of $f + g$ is the intersection of the domains of f and g. It is the set of nonnegative real numbers (figure 2.38).

Example 2 Let $f(x) = x^2$, $x \geq 1$ and $g(x) = x + 1$, $x \leq 0$. The domains of f and g do not intersect. Therefore, $f + g$ is not defined (figure 2.38).

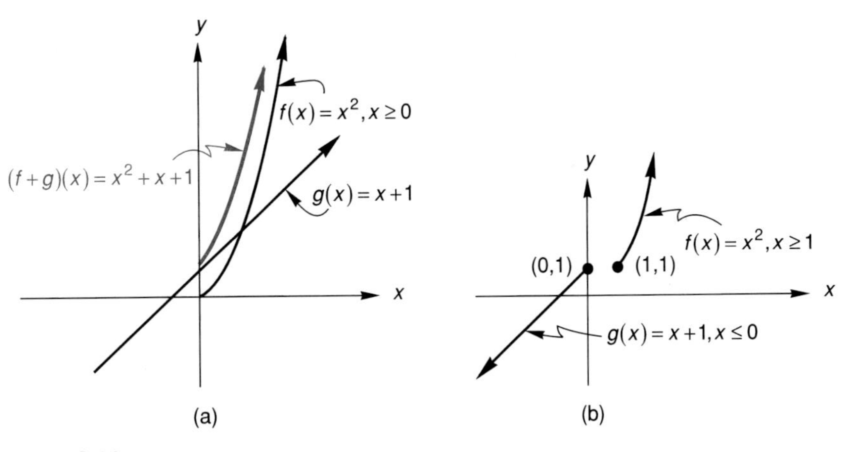

Figure 2.38

The difference and product and quotient functions are defined below. The difference and product functions are defined over the intersection of the domains of f and g. The domain of the quotient function is further restricted since it cannot contain values of x for which $g(x) = 0$.

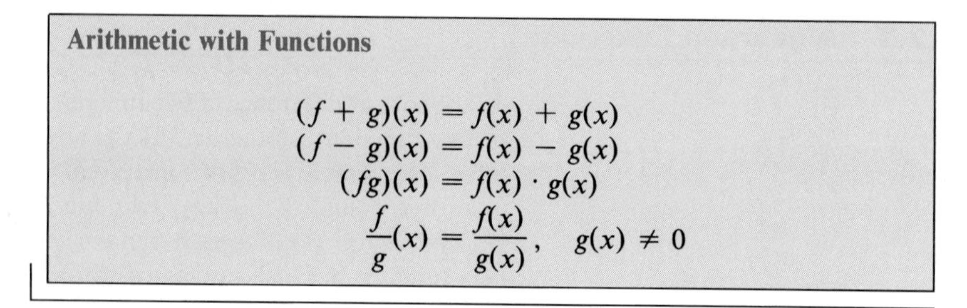

Arithmetic with Functions

$$(f+g)(x) = f(x) + g(x)$$
$$(f-g)(x) = f(x) - g(x)$$
$$(fg)(x) = f(x) \cdot g(x)$$
$$\frac{f}{g}(x) = \frac{f(x)}{g(x)}, \quad g(x) \neq 0$$

Example 3 Let $f(x) = \sqrt{x+3}$ and $g(x) = x/(x-1)$. Determine equations and domains for the difference and product and quotient functions.

Solution The domains of f and g are $\{x|x \geq -3\}$ and $\{x|x \neq 1\}$ respectively.

$$(f-g)(x) = \sqrt{x+3} - \frac{x}{x-1}, \quad \{x|x \geq -3 \text{ and } x \neq 1\}$$

$$(fg)(x) = \sqrt{x+3} \cdot \frac{x}{x-1} = \frac{x\sqrt{x+3}}{x-1}, \quad \{x|x \geq -3 \text{ and } x \neq 1\}$$

$$\frac{f}{g}(x) = \frac{\sqrt{x+3}}{\dfrac{x}{(x-1)}} = \frac{(x-1)\sqrt{x+3}}{x}, \quad \{x|x \geq -3 \text{ and } x \neq 0,1\}$$

■

Composition of Functions

New functions can be formed from old by ways other than addition, multiplication, subtraction, and division. One such method is called composition. Let

$$f(x) = \sqrt{x} \quad \text{and} \quad g(x) = x + 3.$$

For $x = 1$, $g(1) = 4$ and $f(4) = 2$. We took a number $(x = 1)$ in the domain of g, found its associated function value (4), and then evaluated f at 4. A schematic of the process is shown below.

$$1 \rightarrow 4 \rightarrow f(4) = 2$$
$$x = 1 \rightarrow g(1) = 4 \rightarrow f(g(1)) = 2$$
$$x \rightarrow g(x) \rightarrow f(g(x))$$

The two step process, depicted in the schematic, is a function that assigns to an x in the domain of g a number $f(g(x))$ in the range of f. The domain of the function is the subset of the domain of g for which $f(g(x))$ is a real number. We call the function **the composition of f and g** and we denote it as $f \circ g$.

Definition

The composition of functions f and g is defined by the equation

$$(f \circ g)(x) = f(g(x)).$$

The domain of $f \circ g$ is the set of all x in the domain of g for which $g(x)$ is in the domain of f.

Example 4 $f(x) = \sqrt{x}$, $g(x) = x + 3$. Find an equation for $f \circ g$ and determine its domain. Sketch graphs of f, g, and $f \circ g$ in the same coordinate system.

Solution

$$\begin{aligned}(f \circ g)(x) &= f(g(x)) \\ &= \sqrt{g(x)} \\ &= \sqrt{x + 3}\end{aligned}$$

$g(x)$ is in the domain of f if $x + 3 \geq 0$, or if $x \geq -3$. Thus, the domain of $f \circ g$ is $[-3,\infty)$ (figure 2.39).

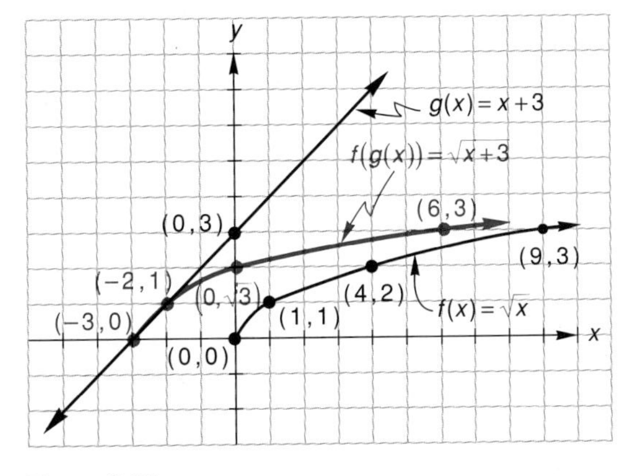

Figure 2.39

■

Example 5 $f(x) = \sqrt{x}$, $g(x) = x + 3$. Find an equation for $g \circ f$ and determine its domain. Graph f, g, and $g \circ f$ in the same coordinate system.

Solution

$$\begin{aligned}(g \circ f)(x) &= g(f(x)) \\ &= f(x) + 3 \\ &= \sqrt{x} + 3\end{aligned}$$

$f(x)$ is in the domain of g if $\sqrt{x}$ is a real number. Hence, the domain of $g \circ f$ is $[0,\infty)$ (figure 2.40).

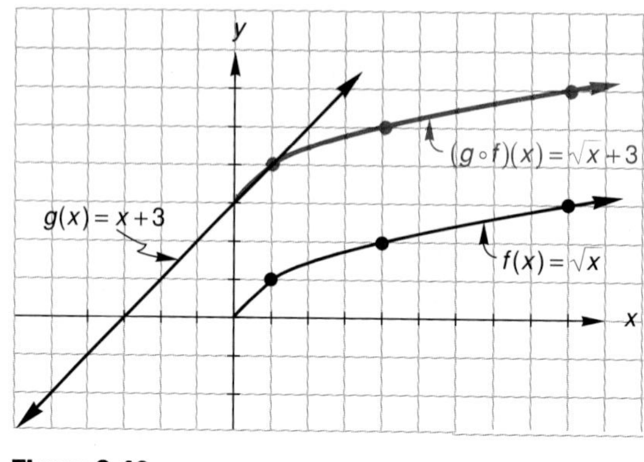

Figure 2.40

We see, in examples 4 and 5, that $f \circ g \neq g \circ f$. *The composition of two functions is not commutative.*

Example 6 $f(x) = x^4$, $g(x) = x^2 - 5$. Find

a. $(f \circ g)(x)$
b. $(f \circ g)(2)$
c. $(g \circ f)(x)$
d. $(g \circ f)(2)$

Solution

a. $(f \circ g)(x) = f(g(x))$
$= (x^2 - 5)^4$

b. $(f \circ g)(2) = (2^2 - 5)^4$
$= 1$

c. $(g \circ f)(x) = g(f(x))$
$= (x^4)^2 - 5$
$= x^8 - 5$

d. $(g \circ f)(2) = 2^8 - 5$
$= 251$

There is more than one way to write a function as a composition of two functions, as the next example illustrates.

Example 7 Write $h(x) = \dfrac{5}{x} - 3$ as a composition of two functions.

Solution

Let $g(x) = 5/x$ and $f(x) = x - 3$, then $f(g(x)) = \dfrac{5}{x} - 3$;

or Let $g(x) = 1/x$ and $f(x) = 5x - 3$, then $f(g(x)) = \dfrac{5}{x} - 3$. ■

The symbol $f(g(x))$ is designed to suggest *a function of a function,* a very useful notion. Consider an expanding balloon whose radius is increasing at the rate of $\frac{1}{2}$ inch per minute. The volume of the balloon, $V = \frac{4}{3}\pi r^3$, is a function of the radius r. The radius is a function of time, $r = t/2$. Let

$$f(r) = \frac{4}{3}\pi r^3 \quad \text{and} \quad g(t) = \frac{t}{2},$$

then

$$V = f(r) = f(g(t))$$

or

$$V = \frac{4}{3}\pi r^3 = \frac{4}{3}\pi\left(\frac{t}{2}\right)^3 = \frac{\pi}{6}t^3.$$

Hence, volume, a function of t through r, can be thought of as a function of a function.

Inverse Functions

In section 2.1 we defined a relation F as a set of ordered pairs. If the components of each ordered pair are interchanged, another relation G is obtained. *The relation G is called the inverse of F.*

Example 8 The inverse of $F = \{(1,2), (3,4), (5,6)\}$
is $G = \{(2,1), (4,3), (6,5)\}$.

Being a relation, each function has an inverse. But its inverse need not be a function. The functions

$$F = \{(1,2), (3,4), (5,6)\} \quad \text{and} \quad f = \{(2,3), (4,1), (6,3)\}$$

have as their respective inverses

$$G = \{(2,1), (4,3), (6,5)\} \quad \text{and} \quad g = \{(3,2), (1,4), (3,6)\}.$$

G is a function but g is not a function because associated with 3 in its domain are 2 and 6 in its range.

Each number in the domain of F is paired with exactly one number in its range, and each number in its range is paired with exactly one number in its domain.

Definition

A function in which there is a one-to-one correspondence between the numbers in its domain and range is called a one-to-one function.

Let f be a one-to-one function. If the components in its ordered pairs are interchanged, a second function is obtained, which is called the inverse of f. The inverse function is usually denoted as f^{-1}. Be cautioned that $f^{-1} \neq 1/f$ (see exercises 29–34). Since f^{-1} is obtained from f by interchanging the ordered pairs of f, the domain and range of f^{-1} are, respectively, the range and domain of f.

Example 9 $f = \{(-4,3), (-1,5), (0,1), (2,-1), (3,6)\}$ is one-to-one. It has the inverse function,

$$f^{-1} = \{(3,-4), (5,-1), (1,0), (-1,2), (6,3)\}.$$

The domain of f^{-1}, $\{3,5,1,-1,6\}$, is the range of f. The range of f^{-1}, $\{-4,-1,0,2,3\}$, is the domain of f.

A function can be seen to be one-to-one with the **horizontal line test.** If any horizontal line intersects a graph of the function in more than one point, the function is not one-to-one. A function that is not one-to-one must have at least one range element that is associated with two or more domain elements (figure 2.41). Such a function cannot have an inverse function.

Example 10 The horizontal line test applied to the graphs of $y = x^2$ and $y = x^3$ (figure 2.41) shows that $y = x^3$ has an inverse function while $y = x^2$ does not.

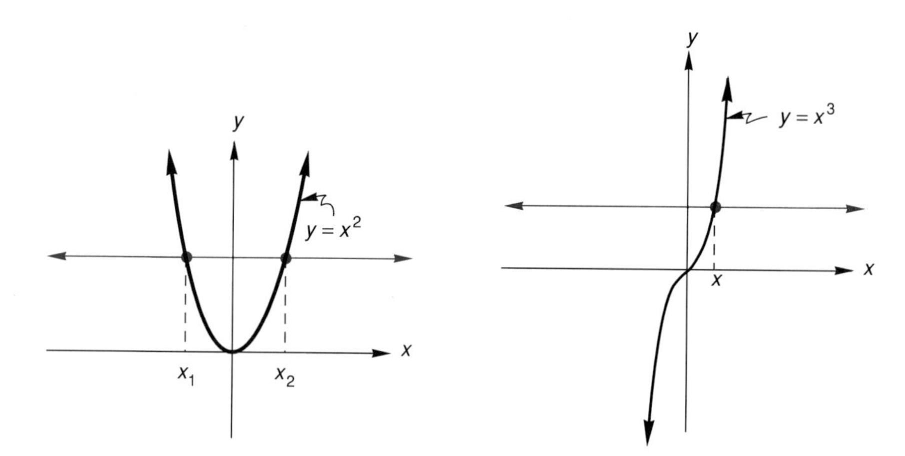

Figure 2.41

Inverses and Composition

Let $f = \{(1,3), (2,5), (4,9)\}$, then $f^{-1} = \{(3,1), (5,2), (9,4)\}$. Note that

because	$f(1) = 3$	and	$f^{-1}(3) = 1,$
then	$f(f^{-1}(3)) = 3$	and	$f^{-1}(f(1)) = 1;$
because	$f(2) = 5$	and	$f^{-1}(5) = 2,$
then	$f(f^{-1}(5)) = 5$	and	$f^{-1}(f(2)) = 2;$
because	$f(4) = 9$	and	$f^{-1}(9) = 4,$
then	$f(f^{-1}(9)) = 9$	and	$f^{-1}(f(4)) = 4.$

Inverse Function Property

If (a,b) is an ordered pair in f, then (b,a) is an ordered pair in f^{-1} and

$$f(f^{-1}(b)) = b \quad \text{and} \quad f^{-1}(f(a)) = a. \tag{1}$$

Equations (1) provide us with a way to construct an inverse function, as well as a means for showing that two functions are inverses of each other.

Example 11 Construct the inverse function for each of the given functions. Verify that the constructed function is the inverse.

a. $f(x) = 3x + 5$

b. $g(x) = x^3$

Solution

a. Replace x in the given equation by $f^{-1}(x)$.

$$f(f^{-1}(x)) = 3\,f^{-1}(x) + 5$$
$$x = 3\,f^{-1}(x) + 5 \quad [f(f^{-1}(x)) = x \text{ by } (1)]$$
$$\frac{x-5}{3} = f^{-1}(x)$$

We use (1) to verify that $f^{-1}(x) = (x - 5)/3$ is the inverse of $f(x) = 3x + 5$.

$$f(f^{-1}(x)) = 3\left(\frac{x-5}{3}\right) + 5 = x$$
$$f^{-1}(f(x)) = \frac{(3x+5)-5}{3} = x$$

b.

$$\begin{aligned} g(g^{-1}(x)) &= (g^{-1}(x))^3 \\ x &= (g^{-1}(x))^3 \\ x^{1/3} &= g^{-1}(x) \end{aligned}$$

Now use (1) to verify that $g^{-1}(x) = x^{1/3}$ is the inverse of $g(x) = x^3$.

$$\begin{aligned} g(g^{-1}(x)) &= (x^{1/3})^3 = x \\ g^{-1}(g(x)) &= (x^3)^{1/3} = x \end{aligned}$$

■

The method for constructing inverse functions that is exhibited in example 11 can be formalized as follows:

Constructing an Inverse Function

To construct f^{-1} given $y = f(x)$:

1. Interchange the roles of x and y.
2. Solve the resulting equation for y.
3. Verify the results by showing that

$$f(f^{-1}(x)) = x \quad \text{and} \quad f^{-1}(f(x)) = x.$$

$$\begin{aligned} f(x) = y &= 3x + 5 \\ x &= 3y + 5 && \text{(interchange } x \text{ and } y\text{)} \\ \frac{x-5}{3} &= y = f^{-1}(x) && \text{(solve for } y\text{)} \\ f(f^{-1}(x)) &= 3\frac{(x-5)}{3} + 5 = x && \text{(verify results)} \\ f^{-1}(f(x)) &= \frac{(3x+5)-5}{3} = x \end{aligned}$$

Graphing Inverse Functions

Graphs of $f(x) = x^3$ and $f^{-1}(x) = x^{1/3}$ are shown in figure 2.42.

In figure 2.42 a point $P(x,y)$ in the graph of $f(x) = x^3$ has a corresponding point $Q(y,x)$ in the graph of $f^{-1}(x) = x^{1/3}$. A consequence of this interchange of components is that *the graphs of a function and its inverse*

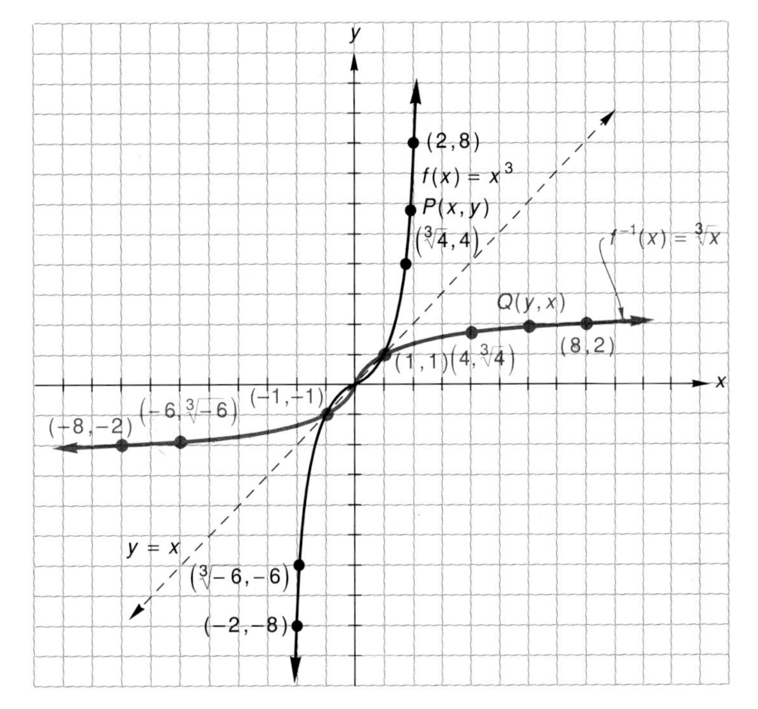

Figure 2.42

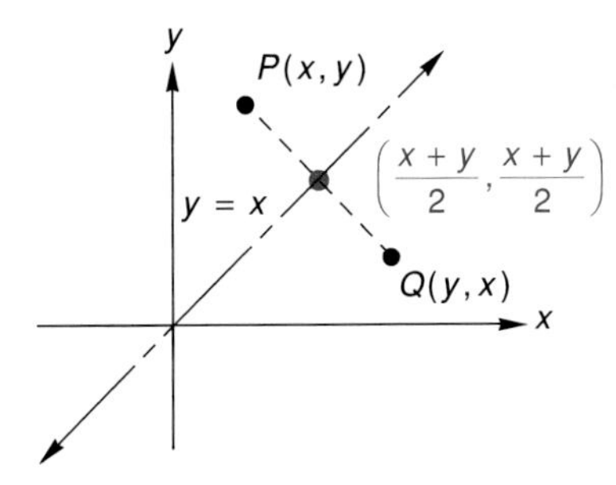

Figure 2.43

function are reflections of each other with respect to the line $y = x$. To see that this must be true, note that the segment through $P(x,y)$ and $Q(y,x)$ (figure 2.43) has slope

$$\frac{y - x}{x - y} = -1$$

while line $y = x$ has slope 1. The segment and line are perpendicular. Also note that the midpoint of $\overline{PQ}$,

$$\left(\frac{x + y}{2}, \frac{y + x}{2}\right),$$

has equal first and second coordinates and must therefore be a point on $y = x$. Since P and Q are equidistant from $y = x$, and $\overline{PQ}$ is perpendicular to $y = x$, it follows that P and Q are reflections of each other with respect to the line $y = x$. But if $P(x,y)$ is a point in the graph of f, then $Q(y,x)$ must be a point in the graph of f^{-1}. As a result, the graphs of f and f^{-1} are reflections of each other with respect to the line $y = x$.

Example 12 Graph the function $f(x)$ and its inverse function on the same coordinate axes.

$$f(x) = \begin{cases} 2x + 4, & -3 \leq x < -1 \\ (x + 1)^2 + 2, & -1 \leq x < 2 \\ x/2 + 10, & 2 \leq x \leq 4 \end{cases}$$

Solution A graph of f is constructed along with the line $y = x$. Then a graph of f^{-1} can be sketched as the reflection of the graph of f with respect to the line $y = x$ (figure 2.44).

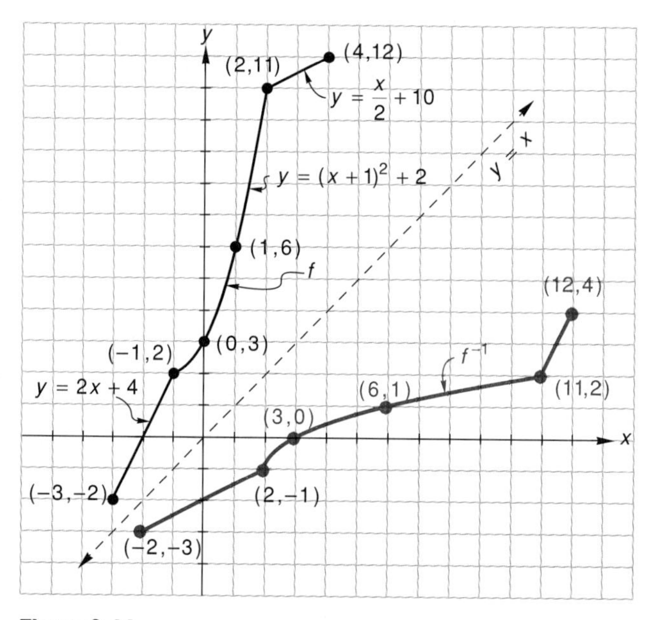

Figure 2.44

■

Exercise Set 2.5

(A)

For each pair of functions f and g, find $f + g$, $f - g$, fg, and f/g, and indicate their domains.

1. $f(x) = x - 3$, $g(x) = 2x + 1$
2. $f(x) = -2x - 5$, $g(x) = -x + 2$
3. $f(x) = x^2 - 1$, $g(x) = x^2 + 4x + 3$
4. $f(x) = x^3 + 8$, $g(x) = x^2 - 2x - 8$
5. $f(x) = \sqrt{x}$, $g(x) = |x|$
6. $f(x) = \sqrt{x^2 - 4}$, $g(x) = \sqrt{9 - x^2}$
7. $f(x) = \sqrt{16 - x^2}$, $g(x) = \sqrt{x^2 - 1}$
8. $f(x) = \sqrt{-3 - x}$, $g(x) = 2x$, $x \geq 2$
9. $f(x) = \begin{cases} 1, x < 0 \\ x + 1, x > 1 \end{cases}$ $g(x) = \begin{cases} x, x \leq 1 \\ x^2, x > 1 \end{cases}$
10. $f(x) = \begin{cases} 3x - 4, x \leq 2 \\ -x, x > 2 \end{cases}$ $g(x) = \begin{cases} x^2 + 1, x < -1 \\ 2x + 4, x \geq 1 \end{cases}$

For each pair of functions f and g, find the composition $f \circ g$ and its domain.

11. $f(x) = x - 3$, $g(x) = 2x + 1$
12. $f(x) = -2x - 5$, $g(x) = -x + 2$
13. $f(x) = \sqrt{x}$, $g(x) = |x|$
14. $f(x) = \sqrt{x - 5}$, $g(x) = x^2 - 4$
15. $f(x) = x^5$, $g(x) = 3x^2 - 2x + 1$
16. $f(x) = x^4$, $g(x) = x^3 + 5$
17. $f(x) = \sqrt{16 - x^2}$, $g(x) = \sqrt{x^2 - 1}$
18. $f(x) = \sqrt{x^2 - 4}$, $g(x) = \sqrt{9 - x^2}$

The composition of two functions is not commutative. Show that $f \circ g \neq g \circ f$.

19. $f(x) = x - 3$, $g(x) = 2x + 1$
20. $f(x) = x^2 - 1$, $g(x) = x^2 + 4x + 3$
21. $f(x) = \sqrt{x}$, $g(x) = |x|$
22. $f(x) = \sqrt{16 - x^2}$, $g(x) = \sqrt{x^2 - 1}$

Consider the given equation as defining a composition $(f \circ g)$ of two functions f and g. Determine f and g. (More than one answer may be possible.)

23. $y = (x - 3)^2$
24. $y = x^2 - 2x + 1$
25. $y = \sqrt[3]{2x}$
26. $y = \sqrt{x^2 - 1}$
27. $y = |x^3|$
28. $y = |(x - 3)^5|$

Each of the following functions has an inverse function. (a) Find the inverse function. (b) Verify that the function found in part a is the inverse (see example 11). (c) Determine the domain and range for both f and f^{-1} and how the domains and ranges of f and f^{-1} are related. (d) Write an equation for $1/f$. Is the equation for $1/f$ the same as the equation for f^{-1}?

29. $f(x) = 2x$, $x \geq 2$
30. $f(x) = 3x + 1$, $x \leq 3$
31. $f(x) = x^3 + 1$, $x \geq -1$
32. $f(x) = x^2 - 4$, $x \leq 0$
33. $f(x) = \sqrt{2x - 1}$, $x \geq 5$
34. $f(x) = \sqrt{x}$, $x \geq 1$

(B)

35. The radius of a circular oil slick is increasing at the rate of one foot every ten minutes. Express the area of the oil slick at any time, in terms of t (time in minutes).
36. The volume of a cylindrical water tank is $\pi r^2 h$, where r is the radius and h is the height of the water tank. Water is being pumped from the tank at a constant rate so that the height of water in the tank drops one foot per hour. The tank is 15 feet high, and it has a radius of 10 feet. It is full of water when the pumping begins. Express the volume of water in the tank as a function of time. How much water is in the tank 6 hours and 20 minutes after pumping begins?

37. The base of a fifteen-foot ladder is five feet from a wall that is perpendicular to the ground on which the ladder is based. The top of the ladder, which rests against the wall, begins to slide down the wall as the base of the ladder slides away from the wall at the rate of 0.1 feet per second. Express the height of the top of the ladder as a function of time. What is the height of the top of the ladder 25 seconds after its base begins to slide away from the wall?

38. The volume of a sphere is given as $V = \frac{4}{3}\pi r^3$. Its surface area is given as $S = 4\pi r^2$. Express V as a function of S.

Graph the function and use the horizontal line test to determine if it has an inverse function. If there is an inverse function, write its equation. Sketch a graph of the inverse function in the same coordinate system as the given function, as a reflection in the line $y = x$.

39. $y = 2x + 1$

40. $y = -x - 3$

41. $y = x^2, x \geq 0$

42. $y = x^4, x \leq 0$

43. $y = \sqrt{9 - x^2}$

44. $y = \sqrt{4 - x^2}$

45. $y = 2x^3 - 1$

46. $y = \sqrt{1 - x^2}, x \geq 0$

47. $y = \dfrac{1}{x + 3}, x \geq -2$

48. $y = \dfrac{5}{2x - 6}, x \leq 1$

49. $y = [2x], 0 \leq x \leq 3$

50. $y = -3|x| + 2$

51. $y = \begin{cases} 2x, & -2 \leq x < 1 \\ x^2 + 1, & 1 \leq x < 3 \\ x/3 + 9, & 3 \leq x \leq 6 \end{cases}$

52. $y = \begin{cases} -x^2 - 6x - 4, & -3 \leq x < -1 \\ -x, & -1 \leq x < 1 \\ -x^2, & 1 \leq x \leq 3 \end{cases}$

Let $f(x) = 2x$, $g(x) = x^2 - 1$, and $h(x) = x/5$. Write an equation for

53. $[(f + g) \circ h](x)$.

54. $[f \circ (g + h)](x)$.

55. $[(fg) \circ h](x)$.

56. $[f \circ (gh)](x)$.

57. $\left(\left(\dfrac{h}{g}\right) \circ f\right)(x)$.

58. $\left(f \circ \left(\dfrac{h}{g}\right)\right)(x)$.

59. $(f \circ g \circ h)(x)$.

60. $(g \circ h \circ f)(x)$.

61. A function is called strictly increasing if for x_1 and x_2 in its domain, $f(x_2) > f(x_1)$ when $x_2 > x_1$.
    - **a.** Give an example of a strictly increasing linear function.
    - **b.** Prove that a strictly increasing function is a one-to-one function. (Hint: You must show that for each y in the range of the function there is associated exactly one x in its domain. Assume this is not true and reach a contradiction.)

62. Define a strictly decreasing function and then repeat parts a and b in exercise 61.

63. The graph of an "even" function is symmetric to the y-axis. Thus, if f is an even function,
$$f(-x) = f(x).$$
    - **a.** Show that the sum of two even functions is even.
    - **b.** Show that the product of two even functions is even.

64. The graph of an "odd" function is symmetric to the origin. Thus, if g is an odd function,
$$g(-x) = -g(x).$$
    - **a.** Show that the sum of two odd functions is odd.
    - **b.** Show that the product of two odd functions is even.
    - **c.** Show that the product of an odd and an even function is odd.

65. Show that if f and g have inverse functions, then $(f \circ g)^{-1}(x) = g^{-1}(f^{-1}(x))$. (Hint: $f^{-1}(f(g(x))) = g(x)$. Now continue.)

66. Verify the result in exercise 65 with the functions, $f(x) = x^3$ and $g(x) = 2x - 1$.
    - **a.** Show that $f(g(x)) = (2x - 1)^3$.
    - **b.** Show that $(f \circ g)^{-1}(x) = (\sqrt[3]{x} + 1)/2$.
    - **c.** Find $f^{-1}(x)$ and $g^{-1}(x)$ and then show that $g^{-1}(f^{-1}(x)) = (\sqrt[3]{x} + 1)/2$.

67. Repeat exercise 66 with $f(x) = x^2 - 5, x \geq 0$, and $g(x) = 3x + 2$.

Chapter 2 Review

Major Points for Review

Section 2.1 Definition of a function

The domain and range of a function

Independent and dependent variables in the equation defining a function

Function notation

Increasing, decreasing, and constant functions

Vertical line test

Section 2.2 Slope of a line

The point-slope form, slope-intercept form, and general form of the equation of a straight line

Equations of horizontal and vertical lines

The relationship between the slopes of two parallel lines or two perpendicular lines

Distance between two points in the plane

Midpoint of a line segment

Graphs of absolute value functions and bracket functions

Section 2.3 Graphs of quadratic functions

Locating the vertex of a parabola

Quadratic functions as models for max-min problems

Interpreting the difference quotient

Section 2.4 Symmetry tests

How to find the intercepts and domain and range of a graph

Reflections of a graph with respect to the x and y axes

Vertical and horizontal translations of a graph

Stretchings and contractions of graphs with respect to the x and y axes

Section 2.5 Sum, difference, product, and quotient functions

Composition of two functions

Inverse functions

Horizontal line test

How to construct f^{-1} given an equation defining f

How to graph f^{-1} given a graph of f

Review Exercises

Which relation is a function? If the relation is a function, find its inverse function, if it exists.

1. $\{(5,2), (2,6), (6,4), (4,3)\}$
2. $\{(-7,8), (-2,6), (0,4), (1,2)\}$
3. $\{(8,5), (2,5), (2,9), (3,8)\}$
4. $\{(-1,4), (0,4), (3,7), (8,7)\}$

Given two points, (a) find the distance between the points, (b) find the midpoint of the line segment determined by the points, and (c) write equations of the line determined by the points in point-slope form, slope-intercept form, and general form.

5. $(-3,4), (2,-6)$
6. $(7,-3), (-14,6)$
7. $(-8,-5), (-1,12)$
8. $(8,6), (-5,2)$

9. Write an equation of the horizontal line through $(4,-3)$.
10. Write an equation of the vertical line through $(7,12)$.

Given f, construct the difference quotient, $[f(x + h) - f(x)]/h$, and simplify it if possible.

11. $f(x) = 2x^2 - 6x + 5$
12. $f(x) = x^3 + 2x$

Graph each quadratic function. Show the vertex of the parabola in your graph.

13. $y = x^2 + 6x + 9$
14. $y = 3x^2 - 9x + 10$
15. $y = -x^2 - x + 5$
16. $y = -2x^2 + 8x + 3$

Graph each function after determining (a) the x and y intercepts of the graph, (b) the domain and, if possible, the range of the function, and (c) if the graph is symmetric to the y-axis or the origin.

17. $y = -3x^4 + x^2$
18. $y = |2x - 5| - 3$
19. $y = \sqrt{16 - 4x^2}$
20. $y = -x^3/3$

For each pair of functions, f and g, write an equation for, and find the domain of, the indicated function: (a) $f + g$, (b) $f - g$, (c) fg, (d) f/g, (e) $f \circ g$, (f) $g \circ f$.

21. $f(x) = 2x - 5$, $g(x) = 8x + 1$
22. $f(x) = \sqrt{x^2 - 2}$, $g(x) = 3x + 4$
23. $f(x) = \sqrt{4 - x^2}$, $g(x) = x - 2$
24. $f(x) = x^2 + 1$, $g(x) = (x - 4)^2$

Write an equation of the line that contains the given point and is (a) parallel to the given line, and (b) perpendicular to the given line.

25. $(0,0)$, $y = 3x + 5$
26. $(4,-2)$, $y = -6x - 8$
27. $(-3,-7)$, $y = -2x + 9$
28. $(7,2)$, $y = 8x - 1$

Construct the graph of each function from the graph of f.

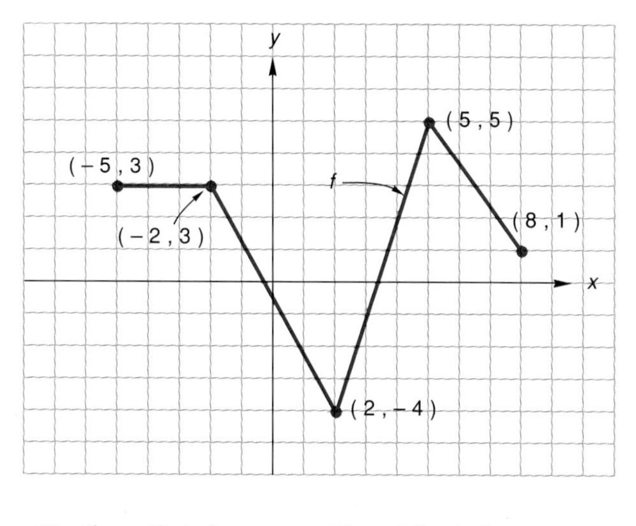

29. $f(x - 2) + 2$
30. $-2f(x + 1) - 3$
31. $f(3x)$
32. $f(-3x)$
33. $3f(x) - 5$
34. $f(4 + 2x)$

If the function has an inverse function, find it. Then graph the function and its inverse on the same coordinate system.

35. $y = 4x - 3$
36. $y = \sqrt{4 - x^2}$
37. $y = \dfrac{-2}{x - 1}, x \geq 2$
38. $y = x^2 - 2x + 3$

39. A city bus company carries 10,000 riders daily at a fare of 75 cents. On the basis of field surveys the management of the company knows that each 10-cent fare increase will result in 500 fewer daily riders. What fare should management set to maximize the bus company's daily revenue?

40. A certain farm allots \$1,000 for fencing a rectangular area that is to abut a highway. Because the fencing on the highway side must be attractive, it costs \$4 per lineal foot. The other three sides of the area are fenced at \$2 per foot. What are the dimensions of the rectangle that maximize its area?

41. An electric circuit develops power according to the formula $W = EI - RI^2$ where W is watts, E is voltage, I is current, and R is resistance. What is the maximum power generated in a 220-volt circuit that has 4 ohms of resistance?

42. Find the shortest distance from (6,0) to the graph of $y = \sqrt{x}$.

43. An object moves along a straight line. Its position with respect to the origin is given by the formula $s = t^2 - 6t + 7$ where s is distance from the origin in feet and t is time in seconds. What is the change in position between $t = 1$ and $t = 4$? Write a general expression for the average change in position with respect to time.

44. A manufacturer's cost function is given as $C(x) = 0.04x^2 + 80x + 1{,}200$. Write a general expression for the average change in cost. Find the marginal cost at a level of 100 units of production.

Indicate what reflections, translations, stretchings, or contractions are required to obtain the graph of

45. $y = -3(x + 1)^2 + 5$ from the graph of $y = x^2$.

46. $y = (2(x - 1))^2 - 2$ from the graph of $y = x^2$.

47. $y = 4|x - 3| + 2$ from the graph of $y = |x|$.

48. $y = -\frac{1}{4}(2x + 1)^3$ from the graph of $y = x^3$.

Chapter 2 Test 1

1. Specify the domain and range of the function $y = \sqrt{9 - x^2}$.

2. Write an equation of the line containing the points $(-6,8)$ and $(4,1)$.

3. Write an equation of the line containing the point (5,0) that is parallel to the line $2x + 3y - 8 = 0$.

4. Graph $y = x^2 - 6x + 4$. Label the vertex of the graph.

5. Graph $y = 2|x + 3| - 4$.

6. Graph $y = x^3 - 4x$. Show intercepts and indicate if the graph is symmetric to the y-axis or the origin.

7. Given f, graph $f(2x) - 3$.

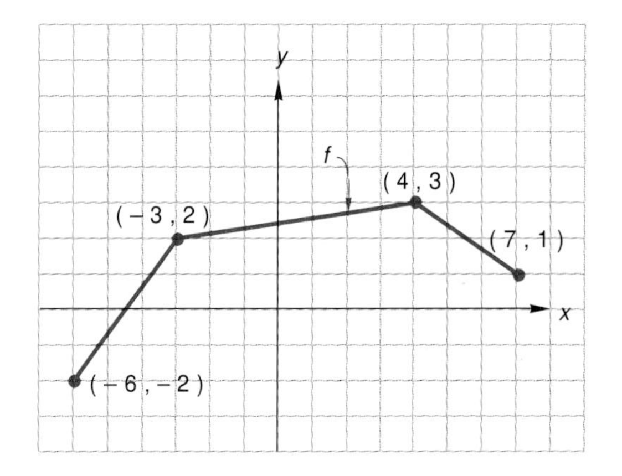

8. Given $f(x) = 2/(x + 4)$, $-2 \leq x \leq 4$, find f^{-1}. Graph f and f^{-1} on the same coordinate axes.

9. Two hundred feet of fencing is used to enclose three sides of a rectangular piece of land that is bounded on the fourth side by a river. What are the dimensions of the rectangle of greatest area?

10. The relationship between Celsius temperature and Fahrenheit temperature is linear. A temperature of 32° Fahrenheit is equivalent to 0° Celsius, and a temperature of 212° Fahrenheit is equivalent to 100° Celsius. Write an equation relating Fahrenheit and Celsius temperatures.

Chapter 2 Test 2

1. Specify the domain and the range of the function $y = 2/(x^2 - 3x)$.

2. Write an equation of the perpendicular bisector of the segment determined by the points (2,7) and $(4,-3)$.

3. Graph $y = -x^2 - 10x - 22$. Label the graph's intercepts and vertex.

4. Graph $y = [x + 1]$, $-2 \leq x \leq 2$.

5. Graph $y = \sqrt{8 - 2x^2}$. Show intercepts and indicate any symmetries.

6. Find and simplify the difference quotient

$$\frac{(f(x + h) - f(x))}{h}$$

for $f(x) = 3/x$. Evaluate the difference quotient for $x = 1$ and $h = 2$ and give a geometric interpretation to the number obtained.

7. What reflections, translations, stretchings, or contractions would you impose on the graph of $y = x^2$ to obtain the graph of $y = -3(x + 2)^2 - 1$? Do not construct a graph.

8. Find f^{-1} given that

$$f = \begin{cases} x^2 - 2x + 5, & -2 \leq x < 1 \\ -3x + 7, & 1 \leq x \leq 4. \end{cases}$$

Graph f and f^{-1} on the same coordinate axes.

9. A piece of machinery is purchased for \$75,000. Its value, V, is depreciated linearly over fifteen years at which time the piece of machinery is considered worthless. Express V as a linear function of time, t.

10. A certain television set can be manufactured for \$180. With p the selling price of the set, market research shows that $5{,}000 - 4p$ is the number of sets that can be sold per month. At what price should the sets be sold to maximize the manufacturer's monthly profit?

Chapter 3

Systems of Linear Equations

Many problems encountered in the physical and social sciences lead to mathematical models involving more than one linear equation. Such models are called systems of linear equations.

3.1 2 × 2 Linear Systems

A pair of linear equations,

$$\begin{aligned} a_1x + b_1y &= c_1 \\ a_2x + b_2y &= c_2, \end{aligned} \quad (1)$$

is called a **2 by 2 (2 × 2) system of linear equations.** System (1) is said to be written in **standard form.** Also, a_1 and b_1 are not both zero and a_2 and b_2 are not both zero. *The solution set of (1) is the set of all ordered pairs that are solutions of both equations.* Thus, in the system

$$\begin{aligned} x - 4y &= -8 \\ 3x - 2y &= 6, \end{aligned}$$

(0,2) is a solution of the first equation, but it is not a solution of the second equation. Therefore, (0,2) is not in the solution set of the system. On the other hand, (4,3) is a solution of both equations and so must be an element in the solution set of the system.

Since a graph of each equation in (1) is a straight line, the system consists of

1. two lines that intersect in exactly one point, or
2. two lines that are parallel, or
3. two lines that are coincident.

The three possibilities are shown in figure 3.1.

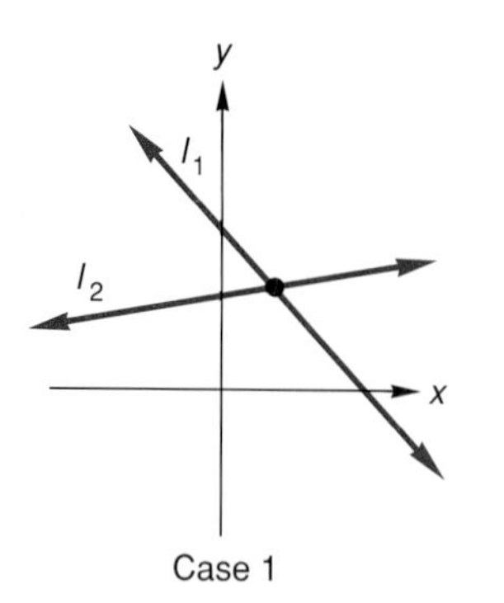

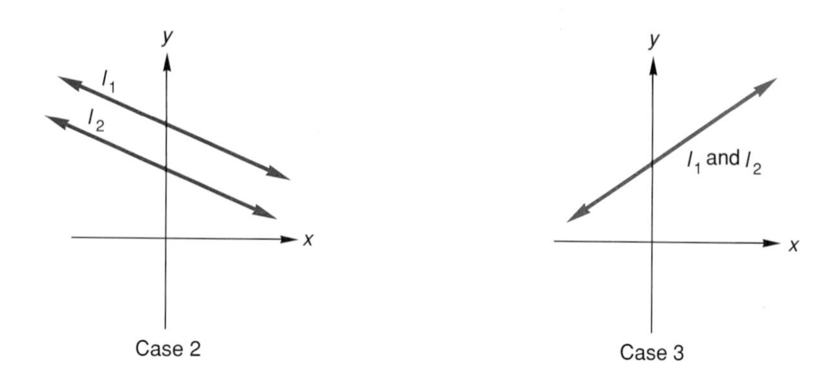

Figure 3.1

It is clear from figure 3.1 that system (1) has solutions only when its lines intersect, and that the solutions are ordered pairs of numbers that are the coordinates of the points of intersection. We do not want to seek solutions where none exist. That is, before attempting to solve (1), we should be assured that it has solutions. If (1) has solutions, we say the system is **consistent.** Otherwise we say the system is **inconsistent.**

Test for Consistency

Let l_1 and l_2 name the lines that are the respective graphs of the equations in (1). If in those equations, $b_1 \neq 0$ and $b_2 \neq 0$, then system (1) can be rewritten as

$$y = \left(\frac{-a_1}{b_1}\right)x + \frac{c_1}{b_1}$$
$$y = \left(\frac{-a_2}{b_2}\right)x + \frac{c_2}{b_2}. \qquad \textbf{(2)}$$

Since the equations in (2) are in slope-intercept form, the slopes of l_1 and l_2 are $-a_1/b_1$ and $-a_2/b_2$. The y-intercepts of the lines are c_1/b_1 and c_2/b_2. If the lines are parallel and distinct, as in case 2 of figure 3.1, their slopes are equal, but their y-intercepts are unequal.

If $\quad \dfrac{-a_1}{b_1} = \dfrac{-a_2}{b_2} \quad$ and $\quad \dfrac{c_1}{b_1} \neq \dfrac{c_2}{b_2}$, then

$$\left(\frac{-a_1}{b_1}\right)\left(\frac{b_1}{-a_2}\right) = \left(\frac{-a_2}{b_2}\right)\left(\frac{b_1}{-a_2}\right) \quad \text{and} \quad \left(\frac{c_1}{b_1}\right)\left(\frac{b_1}{c_2}\right) \neq \left(\frac{c_2}{b_2}\right)\left(\frac{b_1}{c_2}\right)$$

or $\quad \boldsymbol{\dfrac{a_1}{a_2} = \dfrac{b_1}{b_2}} \quad$ **and** $\quad \boldsymbol{\dfrac{b_1}{b_2} \neq \dfrac{c_1}{c_2}}. \qquad$ **(3)**

Hence, system (1) is **inconsistent** and its solution set is empty if (3) holds.

If the slopes of lines l_1 and l_2 are equal, and the y-intercepts of the lines are equal, then the lines coincide (case 3, figure 3.1). Thus, if

$$\frac{a_1}{a_2} = \frac{b_1}{b_2} \quad \text{and} \quad \frac{b_1}{b_2} = \frac{c_1}{c_2},$$

system (1) is **consistent** and its solution set is the infinite set of ordered pairs that are the coordinates of the points of l_1 or l_2.

If the slopes of lines l_1 and l_2 are not equal, then the lines intersect (case 1, figure 3.1). Thus, if

$$\frac{a_1}{a_2} \neq \frac{b_1}{b_2},$$

system (1) is **consistent,** and its solution set is the ordered pair that are the coordinates of the point of intersection of lines l_1 and l_2.

Consistency Tests

A pair of linear equations in standard form

$$a_1x + b_1y = c_1$$
$$a_2x + b_2y = c_2$$

1. has exactly one solution if $a_1/a_2 \neq b_1/b_2$. **(4)**
2. has an infinite number of solutions if $a_1/a_2 = b_1/b_2 = c_1/c_2$. **(5)**
3. has no solutions if $a_1/a_2 = b_1/b_2 \neq c_1/c_2$. **(6)**

Example 1 Test each of the systems for consistency, and state the expected number of solutions.

a. $2x + y = 6$
$4x + 2y = 9$

b. $2x + y = 6$
$4x + 2y = 12$

c. $2x + y = 6$
$4x + y = 12$

Solution

a. $a_1/a_2 = \frac{2}{4}$, $b_1/b_2 = \frac{1}{2}$, $c_1/c_2 = \frac{6}{9}$, and $\frac{2}{4} = \frac{1}{2} \neq \frac{6}{9}$. Hence, according to (6), the system is inconsistent and has no solutions.

b. $a_1/a_2 = \frac{2}{4}$, $b_1/b_2 = \frac{1}{2}$, $c_1/c_2 = \frac{6}{12}$, and $\frac{2}{4} = \frac{1}{2} = \frac{6}{12}$. Hence, according to (5), the system is consistent and has an infinite number of solutions. The solution set is $\{(x,y)|2x + y = 6\}$.

c. $a_1/a_2 = \frac{2}{4}$, $b_1/b_2 = \frac{1}{1}$, and $\frac{2}{4} \neq \frac{1}{1}$. Hence, according to (4), the system is consistent and has exactly one solution. ■

It should be apparent that the consistency tests are a simple and useful tool in helping to determine the solution set of a 2×2 linear system. It is only in the case where the lines in the system intersect in exactly one point that we need go beyond the consistency tests to find the solution of the system. We do that next.

The Point of Intersection of Nonparallel Lines

The family of parallel lines in figure 3.2 is represented by the single equation $y = 3x + b$. The letter b may be regarded as a "varying constant" in that a particular member of the family is determined when b is assigned a specific value. *A varying constant such as* b *is called a* **parameter.**

In figure 3.3, all lines containing the point (2,5), except the vertical line, are represented by the equation $y - 5 = m(x - 2)$ in which the slope, m, is a parameter. This family of lines is called a **pencil of lines.**

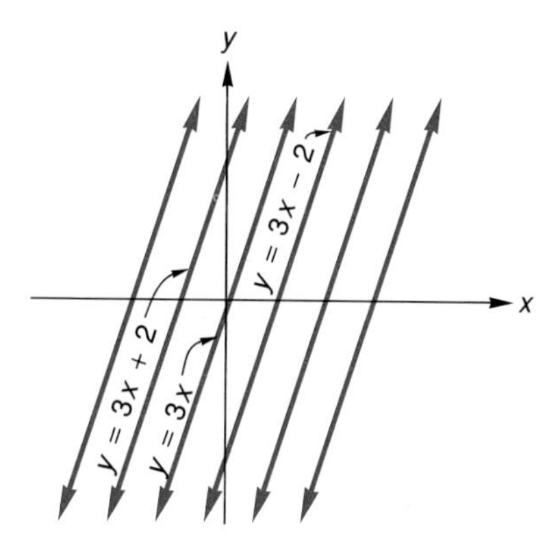

Figure 3.2

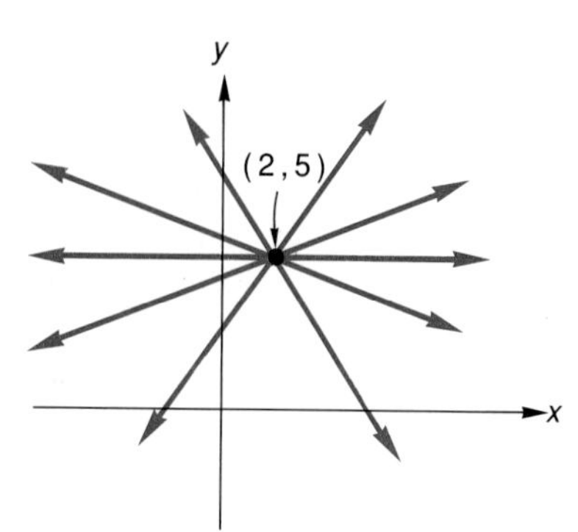

Figure 3.3

It is possible to write a single equation representing a pencil of lines in which the vertical line is included. Given two nonparallel lines, say

$$l_1\colon 3x - y + 4 = 0 \quad \text{and} \quad l_2\colon x + 2y - 6 = 0$$

the equation,

$$s(3x - y + 4) + t(x + 2y - 6) = 0, \tag{7}$$

in which s and t are parameters, represents the pencil of lines containing the point of intersection of l_1 and l_2. As long as s and t are not both zero, (7) is a linear equation in x and y. Now if (x_0,y_0) is the point of intersection of l_1 and l_2, then

$$3x_0 - y_0 + 4 = 0 \quad \text{and} \quad x_0 + 2y_0 - 6 = 0.$$

It follows that

$$s(3x_0 - y_0 + 4) + t(x_0 + 2y_0 - 6) = 0.$$

Therefore, as long as s and t are not both zero, (x_0,y_0) is a solution of (7). It remains to show that an equation for any line containing (x_0,y_0) can be obtained from (7) by assigning appropriate values to s and t (see exercises 29–34).

Example 2 Find the point of intersection of lines $l_1\colon 2x + y - 9 = 0$ and $l_2\colon -3x + 2y - 4 = 0$.

Solution The pencil of lines containing the point of intersection of l_1 and l_2 is given by

$$s(2x + y - 9) + t(-3x + 2y - 4) = 0. \tag{8}$$

Letting $s = -2$ and $t = 1$, equation (8) becomes

$$\begin{aligned} -4x - 2y + 18 - 3x + 2y - 4 &= 0 \\ -7x + 14 &= 0 \\ x &= 2. \end{aligned}$$

The vertical member of the pencil of lines defined by (8) is $x = 2$ (see figure 3.4). Therefore the x-coordinate of the point of intersection of l_1 and l_2 is 2. If we let $s = 3$ and $t = 2$, then (8) becomes $y = 5$, which is the equation of the horizontal member of the pencil of lines. Also, $y = 5$ is the y-coordinate of the point of intersection.

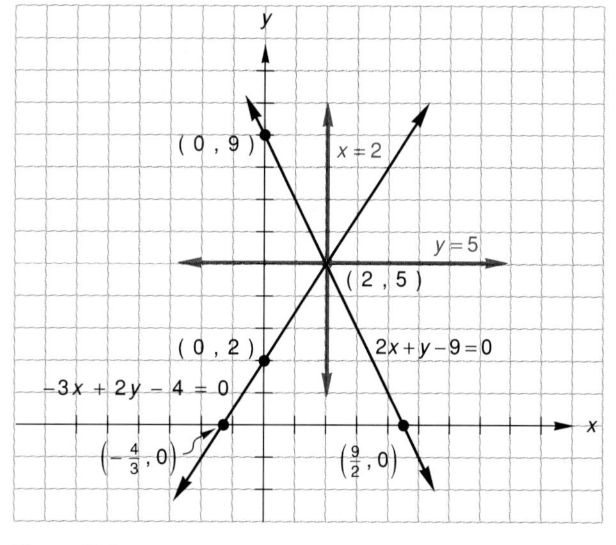

Figure 3.4

■

The values chosen for s and t in example 2 enabled us to eliminate either y or x from (8) and thereby produce either the vertical or horizontal line in the pencil of lines. Since all the lines in the pencil contain the point of intersection of l_1 and l_2, the vertical line gives us the x-coordinate of the point of intersection. The y-coordinate of the point of intersection is given by the horizontal member of the pencil of lines (see figure 3.4).

Linear Combinations

The technique for finding the point of intersection of two nonparallel lines, exhibited in example 2, is called the "Method of Linear Combinations." With slight modification it is the method most often used to solve a system of two linear equations.

Example 3 Solve the system

$$\begin{aligned} 2x &= 9 - y \\ -3x + 2y - 4 &= 0 \end{aligned}$$

Solution First, rewrite the system in standard form.

$$\begin{aligned} 2x + y &= 9 \\ -3x + 2y &= 4 \end{aligned}$$

Then apply the consistency tests. Because

$$\frac{a_1}{a_2} = \frac{2}{-3} \neq \frac{1}{2} = \frac{b_1}{b_2},$$

the system has exactly one solution. If we multiply the first equation by -2, the second equation by 1, and add the resulting equations, we get

$$\begin{array}{rcll} 2x + y &=& 9 & \quad (-2 \\ -3x + 2y &=& 4 & \quad (1 \\ \hline -7x &=& -14 & \\ x &=& 2. & \end{array}$$

Now we substitute 2 for x in either of the equations of the system to solve for y.

$$\begin{aligned} 2(2) + y &= 9 \\ y &= 5 \end{aligned}$$

The solution set of the system is $\{(2,5)\}$. ■

The system in example 3 could also have been solved by multiplying the first equation by 3, the second equation by 2, and adding the resulting equations.

$$\begin{array}{rcll} 2x + y &=& 9 & \quad (3 \\ -3x + 2y &=& 4 & \quad (2 \\ \hline 7y &=& 35 & \\ y &=& 5 & \end{array}$$

Substituting 5 for y in the second equation yields

$$\begin{aligned} -3x + 2(5) &= 4 \\ x &= 2. \end{aligned}$$

Applications

Finding the shortest distance between a point and a line has any number of physical applications; the shortest pipe link or cable link, the shortest distance to the nearest highway, etc. Let us find the distance between the point (4,1) and the line $2x - 3y = -12$ (see figure 3.5).

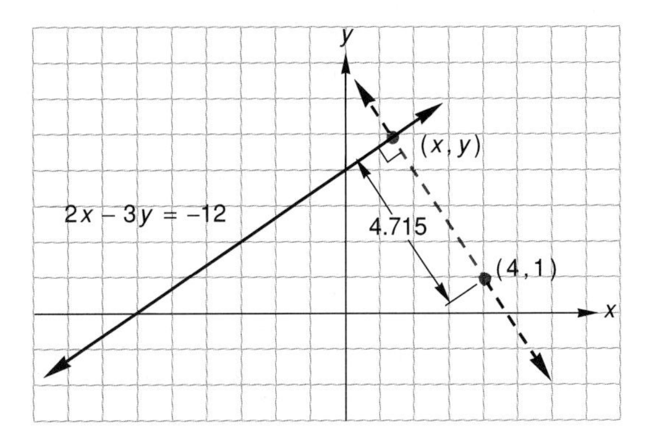

Figure 3.5

The distance sought is the length of the line segment from (4,1) that is perpendicular to the given line. To find that length we

1. write an equation of the line through (4,1) that is perpendicular to the line $2x - 3y = -12$,
2. determine the point of intersection, (x_1,y_1), of the perpendicular lines, and
3. apply the distance formula to the points (x_1,y_1) and (4,1).

The slope-intercept form of the given line is $y = \frac{2}{3}x + 4$. Hence, the slope of a perpendicular line is $-\frac{3}{2}$, and an equation of the line through (4,1) that is perpendicular to the given line is

$$y - 1 = -\tfrac{3}{2}(x - 4) \quad \text{or} \quad 3x + 2y = 14.$$

The point of intersection of the lines is found by solving the 2 × 2 linear system

$$\begin{aligned} 2x - 3y &= -12 \\ 3x + 2y &= 14. \end{aligned}$$

We leave the solving to you. The point of intersection is $(\frac{18}{13},\frac{64}{13})$. Finally, we are able to compute the shortest distance from (4,1) to the line $2x - 3y = -12$.

$$\sqrt{\left(\frac{18}{13} - 4\right)^2 + \left(\frac{64}{13} - 1\right)^2} = \frac{17\sqrt{13}}{13} \approx 4.715$$

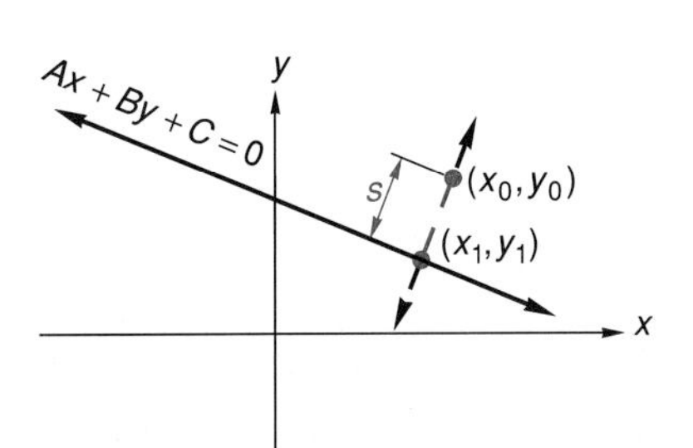

Figure 3.6

The method for finding the distance from (4,1) to line $2x - 3y = -12$ can be generalized (see exercise 57). If (x_0,y_0) is a point that is not on line $Ax + By + C = 0$ (figure 3.6), then distance s from the point to the line is given by the equation

$$s = \frac{|Ax_0 + By_0 + C|}{\sqrt{A^2 + B^2}}. \tag{9}$$

With (9), the distance from (4,1) to $2x - 3y = -12$ is

$$s = \frac{|2 \cdot 4 + (-3) \cdot 1 + 12|}{\sqrt{2^2 + (-3)^2}} = 4.715.$$

Note that in order to use (9), the equation of the line must be written in the form $Ax + By + C = 0$.

Example 4 A diet is designed to provide a minimum weekly requirement of 50 units of carbohydrates and 15 units of protein. One pound of food A provides 8 units of carbohydrates and 2 units of protein. One pound of food B provides 3 units of protein and 2 units of carbohydrates. Both foods must be included in the diet. How many pounds of each food are necessary to satisfy the diet's minimum requirements?

Solution The five-step procedure for solving word problems (section 1.1) is applied to the problem.

We wish to determine the number of pounds of foods A and B that are required to meet the diet's minimum requirements.

Let x be the number of pounds of food A.

Let y be the number of pounds of food B.

A mathematical model of the problem is given by the following system of linear equations. The first equation equates the carbohydrate contributions of foods A and B to the minimum weekly requirement of 50. The protein contributions of foods A and B to the minimum weekly requirement is given by the second equation.

$$\begin{aligned} 8x + 2y &= 50 \qquad (3 \\ 2x + 3y &= 15 \qquad (-2 \\ \hline 20x \qquad &= 120 \\ x &= 6 \\ y &= 1 \end{aligned}$$

Six pounds of food A and one pound of food B are necessary to meet the diet's weekly minimum requirements in carbohydrates and protein. ■

Exercise Set 3.1

(A)

Find the solution of each system after testing each system for consistency.

1. $x + y = 8$, $x - y = 0$
2. $2x + y = 3$, $x - 3y = 5$
3. $x - 2y = 11$, $2x - 4y = 22$
4. $2x + 3y = -14$, $x + 3y = 13$
5. $-5x + 3y = 8$, $2x + 2y = 7$
6. $4x + y = 3$, $7x - 2y = 9$
7. $11x + 14y = 1$, $3x - 5y = -2$
8. $5x - 8y = 10$, $2x + 3y = 16$
9. $3x + 2y = 12$, $5x - 3y = 8$
10. $8x - 5y = -4$, $-3x + 2y = 2$
11. $4x - 6y = 10$, $8x - 12y = -1$
12. $9x + 3y = -7$, $2x + 11y = 18$

13. $\dfrac{x}{3} - \dfrac{2y}{5} = -1$
$x + \dfrac{6y}{5} = -3$

14. $v - \dfrac{w}{3} = -4$
$\dfrac{v}{2} - \dfrac{w}{6} = -\dfrac{2}{5}$

15. $0.75x - 3y = 8$
$x - 4y = 24$

16. $0.14p - 0.05t = 0.13$
$0.04p - 0.11t = 0.4$

17. $0.26r + 1.24s = 2.38$
$0.78r + 0.81s = 3.12$

18. $0.7x + 1.5y = 0.8$
$2.3x + 3.5y = 3.2$

19. $\dfrac{3v}{5} + \dfrac{7w}{4} = 2$
$-\dfrac{8v}{5} + \dfrac{3w}{4} = -\dfrac{1}{2}$

20. $\dfrac{3p}{8} - \dfrac{5q}{3} = -2$
$\dfrac{5p}{8} + \dfrac{2q}{3} = \dfrac{7}{4}$

View each system as a 2 × 2 linear system in the variables $1/x$ and $1/y$, and solve the system.

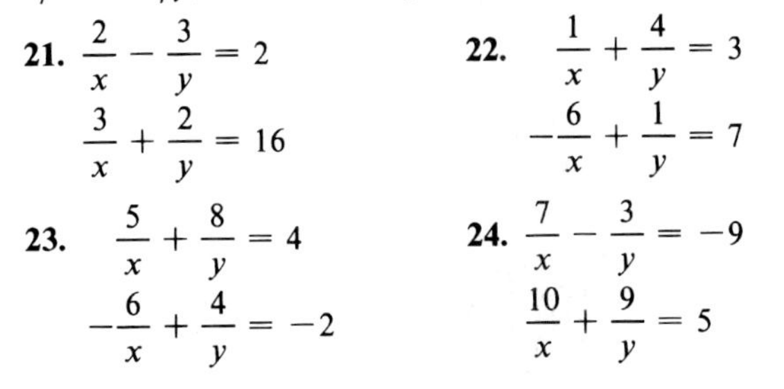

21. $\dfrac{2}{x} - \dfrac{3}{y} = 2$
$\dfrac{3}{x} + \dfrac{2}{y} = 16$

22. $\dfrac{1}{x} + \dfrac{4}{y} = 3$
$-\dfrac{6}{x} + \dfrac{1}{y} = 7$

23. $\dfrac{5}{x} + \dfrac{8}{y} = 4$
$-\dfrac{6}{x} + \dfrac{4}{y} = -2$

24. $\dfrac{7}{x} - \dfrac{3}{y} = -9$
$\dfrac{10}{x} + \dfrac{9}{y} = 5$

25. Write an equation that represents the family of lines with slope 2.

26. Write an equation that represents the family of lines perpendicular to the line $y = 2x$.

27. Write an equation that represents the family of lines containing (5,1).

28. Write an equation that represents a family of parallel lines that has line $2x + 3y = 8$ as a specific member.

Given the equation of a line and a point not on the line, find the distance from the point to the line. Use the method described on page 137. Then use (9) to verify your result.

29. $x - 5y = 8$, (3,2)

30. $4x + y = -2$, (1,0)

31. $9x - 12y = -8$, (−3,4)

32. $7x - 4y = 12$, (−1,−4)

33. $6x - 14y = 11$, (2,−1)

34. $10x + 3y = 9$, (3,−5)

(B)

Two linear equations are given. Form the linear combination $s(a_1x + b_1y + c_1) + t(a_2x + b_2y + c_2) = 0$ that determines the pencil of lines through the point of intersection of the given lines. Then (a) compute values of s and t that produce the vertical line in the pencil of lines, (b) compute values of s and t that produce the horizontal line in the pencil of lines, (c) compute values of s and t that produce the member of the pencil of lines that has the indicated slope. (Hint: The slope of $ax + by + c = 0$ is $-a/b$.)

There is more than one correct answer for each of the parts a, b, and c.

35. $3x + 2y - 4 = 0$, $x - 5y + 8 = 0$, $m = 2$

36. $-x + 6y - 10 = 0$, $4x + y + 2 = 0$, $m = -3$

37. $5x + 2y + 7 = 0$, $8x - 5y - 4 = 0$, $m = -\frac{1}{4}$

38. $9x - 12y + 8 = 0$, $2x - 7y + 13 = 0$, $m = \frac{3}{5}$

39. $10x + 3y - 9 = 0$, $6x - 14y - 11 = 0$, $m = -4$

40. $7x - 4y - 12 = 0$, $14x + 8y + 1 = 0$, $m = \frac{3}{2}$

41. The perimeter of a rectangle is 108 feet. Its width is 6 feet less than twice its length. Find the dimensions of the rectangle.

42. The measure of one of the acute angles of a right triangle is 3° less than twice the measure of the other acute angle. Find the measure of each acute angle.

43. Part of \$18,000 is invested at 12% and the remainder at 9%. If the annual income from both investments is \$1,836, determine the amount of each investment.

44. An investor earns 8% interest annually on \$20,000 by investing in municipal bonds that yield 7% annually and in second mortgages that yield 10% annually. How much of the \$20,000 is invested in municipal bonds and how much in second mortgages?

45. The line $ax + by + 6 = 0$ contains the points (−5,2) and (4,−1). Determine a and b.

46. A small theater is scaled for \$12 and \$9 seats. If on a certain night the receipts from 420 tickets was \$4,752, how many tickets of each kind were sold?

47. One alloy is $\frac{1}{5}$ gold and another is $\frac{1}{2}$ gold. The two alloys are mixed to obtain 60 ounces of an alloy that is $\frac{3}{10}$ gold. How many ounces of each alloy go into the mixture?

48. One brine solution is 20% salt. A second brine solution is 40% salt. How many gallons of each solution are necessary to produce 50 gallons of brine that is 28% salt?

49. A west to east cross-country flight of 2,400 miles takes four and one-half hours. The return flight takes five hours. The difference in flight time is entirely the result of the constant velocity west-to-east wind. Find the average airspeed of the plane and the wind velocity.

50. A freight train travels from city A to city B at an average speed of 45 miles per hour. A passenger train leaves city A two and one-half hours later, bound for city B. If the average speed of the passenger train is 70 miles per hour, when does the passenger train pass the freight train? How far are the trains from city A when the passenger train passes the freight train?

51. Classic economic theory proclaims that consumer demand for a product will decrease as the price of the product increases. However, the manufacturer's supply of the product will increase with increased price. That price p, at which the supply of a product equals its demand, is called the market equilibrium price. Assume that a certain product has the following supply and demand functions:

$$S = \frac{3p}{2} + 1{,}000, \quad D = 1{,}225 - 3p.$$

Find the market equilibrium price of the product and its supply and demand at that price.

52. Based on previous sales records the marketing department of a car radio manufacturer determines that the supply and demand for their top model radio are related to price by the following equations:

$$S = p + 80, \quad D = -2p + 740.$$

What is the equilibrium price for the radios, and how many radios are expected to be sold at this price?

53. A contractor builds two different home models. Model A requires $120,000 and 150 man-days of labor to build. Model B requires $150,000 and 180 man-days of labor to build. If the contractor has $15,480,000 and 19,080 man-days of labor available for building, how many homes of each type can he build so as to use all his capital and labor?

54. A certain diet is designed to provide a minimum weekly requirement of 45 units of carbohydrates and 17 units of protein. One pound of food A provides 12 units of carbohydrates and 4 units of protein. One pound of food B provides 2 units of protein and 2 units of carbohydrates. If the foods can be purchased only in $\frac{1}{2}$ pound increments and both foods must be included in the diet, how many pounds of each food are necessary to meet the minimum weekly diet requirements?

55. Find the equation, $y = mx + b$, for the line determined by $(-6,8)$ and $(2,5)$.

56. Write an equation representing the family of lines in which the y-intercept of each member is always twice its x-intercept.

57. Find the distance from (x_0,y_0) to $Ax + By + C = 0$ (see figure 3.6).

a. Find the point of intersection (x_1,y_1) between $Ax + By + C = 0$ and the line through (x_0,y_0) that is perpendicular to $Ax + By + C = 0$. Show that

$$x_1 = \frac{B^2x_0 - ABy_0 - AC}{A^2 + B^2} \quad \text{and}$$

$$y_1 = \frac{A^2y_0 - ABx_0 - BC}{A^2 + B^2}.$$

b. Use the distance formula, $s = \sqrt{(x_1 - x_0)^2 + (y_1 - y_0)^2}$ to show that

$$s^2 = \frac{A^2(Ax_0 + By_0 + C)^2}{(A^2 + B^2)^2} + \frac{B^2(Ax_0 + By_0 + C)^2}{(A^2 + B^2)^2}.$$

Equation (9) (page 137) follows from the last equation.

3.2 3 × 3 Systems

A linear equation in 3 variables such as

$$3x + 2y + z = 6$$

is the equation of a plane. Let T be the name of the plane defined by $3x + 2y + z = 6$. Solutions of the equation are ordered triples of real numbers, (x,y,z), that are the coordinates of points in T. Thus (2,0,0), (0,3,0), and (0,0,6) are all solutions of the equation as well as coordinates of points in T.

In figure 3.7, three mutually perpendicular lines, the x, y, and z axes, form a rectangular coordinate system in three-dimensional space. Point P, with (2,3,6) as its x, y, and z coordinates, is plotted in the coordinate system. Pairs of coordinate axes determine the x-y, x-z, and y-z coordinate planes. Where in two dimensions, the x and y axes divide the plane into four quadrants, the coordinate axes in three-dimensional space divide the space into eight octants. Only the first octant portion of plane T is shown in figure 3.8 with the lines of intersection between the plane and the coordinate planes.

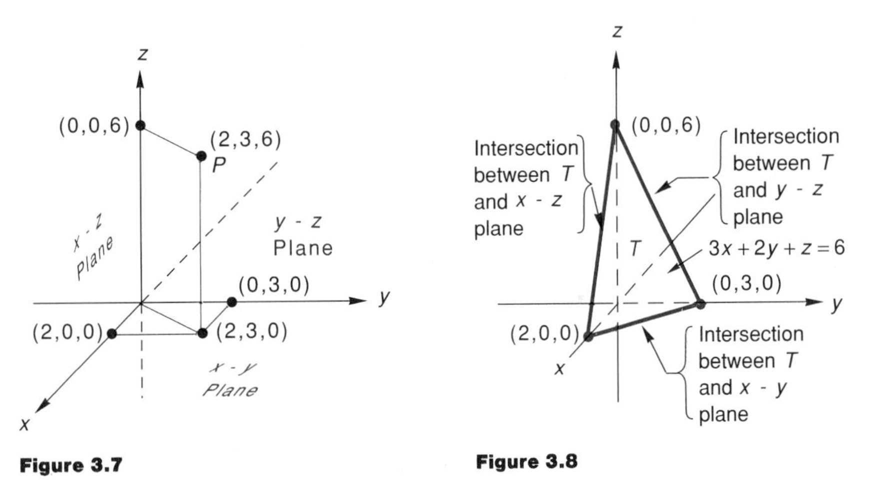

Figure 3.7

Figure 3.8

3 × 3 Linear Systems

A set of 3 linear equations in 3 variables is called a 3 × 3 linear system. As each equation in the system determines a plane, the solution set of the system is the set of ordered triples that determines the points in the common intersection of the planes. Some of the ways three planes can be related to each other are shown in figure 3.9.

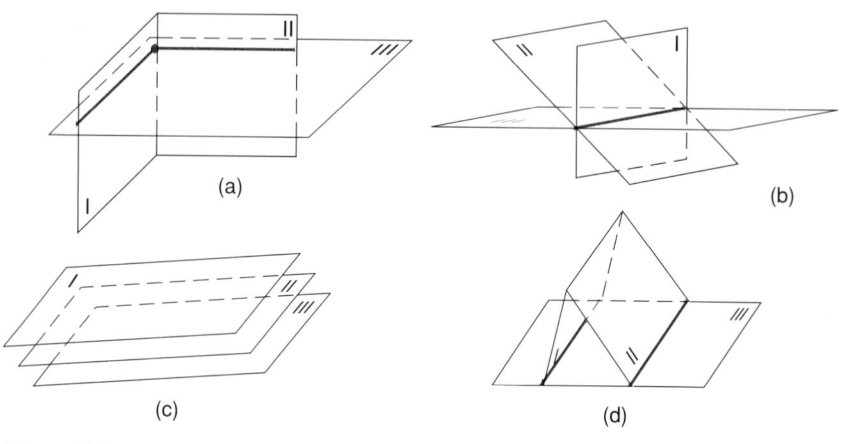

Figure 3.9

The systems represented in parts (a) and (b) of figure 3.9 are **consistent** in that the three planes in each part have a common intersection. The planes in parts (c) and (d) have no common intersections, so the systems they represent are **inconsistent.** There are no simple consistency tests for 3×3 systems. We shall proceed with each system as if it has solutions. In the process of finding solutions we shall discover if the system is consistent or inconsistent.

Linear Combinations

The method of linear combinations used in section 3.1 to solve 2×2 systems can be extended to 3×3 systems.

Example 1 Solve the system

$$\begin{aligned} 3x + 2y - z &= 7 \\ x + y + z &= 5 \\ -2x + y + 2z &= 2. \end{aligned}$$

Solution

a. $3x + 2y - z = 7$ (1

b. $x + y + z = 5$ (1 (−2

c. $-2x + y + 2z = 2$ (1

Multiply the first and second equations by 1 and add the resulting equations to get

d. $4x + 3y = 12.$

Now multiply the second equation by -2 and the third equation by 1 and add the resulting equations to get

e. $-4x - y = -8.$

We have reduced our original system of three equations in three variables to a system of two equations in two variables.

$$\textbf{d.}\quad 4x + 3y = 12 \qquad (1$$

$$\textbf{e.}\quad -4x - y = -8 \qquad (1$$

Multiply each of these equations by 1 and add the resulting equations to obtain

$$2y = 4$$
$$y = 2.$$

The system of two equations in two variables is reduced to one equation in one variable. That equation is solved, and the value of y that is obtained is substituted into either **d** or **e** to find x.

$$\textbf{e.}\quad -4x - (2) = -8$$
$$x = \tfrac{3}{2}.$$

Now $\frac{3}{2}$ is substituted for x, and 2 is substituted for y in one of the equations, **a**, **b**, or **c**, to find z

$$\textbf{a.}\quad 3(\tfrac{3}{2}) + 2(2) - z = 7$$
$$z = \tfrac{3}{2}.$$

The solution set of the system is $\{(\frac{3}{2}, 2, \frac{3}{2})\}$. ■

Our basic strategy for solving a 3 × 3 system is to reduce the system to a 2 × 2 system by forming two linear combinations of the form

$$s(a_1x + b_1y + c_1z + d_1) + t(a_2x + b_2y + c_2z + d_2) = 0. \qquad \textbf{(1)}$$

In both linear combinations s and t are chosen so as to eliminate the same variable. In example 1 equation **d** is a linear combination of equations **a** and **b** in which $s = 1$, $t = 1$, and z is eliminated. Equation **e** is a linear combination of **b** and **c**, in which $s = -2$, $t = 1$, and z is eliminated.

Equation (1) represents a family of planes, each of which contains the line of intersection between the planes

$$a_1x + b_1y + c_1z + d_1 = 0 \quad \text{and} \quad a_2x + b_2y + c_2z + d_2 = 0.$$

Figure 3.9b depicts such a family. By choosing s and t so as to eliminate z we choose the member of the family that is parallel to the z-axes and perpendicular to the x-y plane. Thus equation **d** is an equation of a vertical plane. It is also an equation of the line of intersection between that vertical plane and the x-y plane (see figure 3.10). Similarly equation **e** is an equation of a vertical plane as well as an equation of the line of intersection between the vertical plane and the x-y plane.

The point common to the planes defined by equations **a–c** must be the point where the lines of intersection between planes **a** and **b** and planes **b** and **c** meet (see figure 3.9a). But plane **d** must contain the line of intersection between planes **a** and **b.** After all, equation **d** is a linear combination of equations **a** and **b.** Similarly plane **e** must contain the line of intersection between planes **b** and **c.** The common point of intersection between planes **a, b,** and **c** must therefore be a point on the line of intersection between vertical planes **d** and **e.** That line of intersection is a vertical line (see figure 3.10) and so the x and y coordinates for all points on the line are the same. They are the numbers obtained as the solution of the 2×2 system

d. $4x + 3y = 12$

e. $-4x - y = -8$

where $x = \frac{3}{2}$ and $y = 2$.

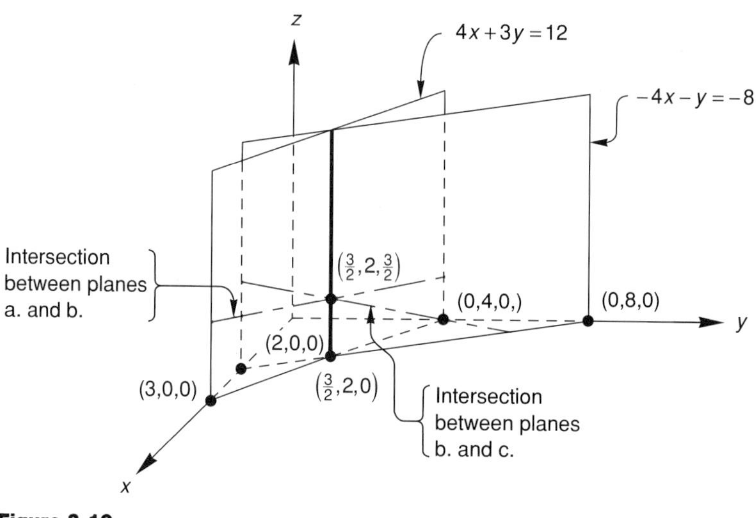

Figure 3.10

Now that we have justified solving a 3×3 linear system by the method of linear combinations, let us solve another 3×3 system without all the explanation and detail.

Example 2 Solve the system

$$\begin{aligned} 4x - y + 3z &= 11 \\ 7x + 7y + z &= 2 \\ 2x + y - 5z &= -9. \end{aligned}$$

Solution

$$
\begin{array}{llllll}
\textbf{a.} & 4x - y + 3z = 11 & (7 & (1 \\
\textbf{b.} & 7x + 7y + z = 2 & (1 & \\
\textbf{c.} & \underline{2x + y - 5z = -9} & & (1 \\
\textbf{d.} & 35x + 22z = 79 & (1 & \\
\textbf{e.} & \underline{6x - 2z = 2} & (11 & \\
 & 101x = 101 & & \\
 & \underline{ x = 1} & & \\
\textbf{e.} & 6(1) - 2z = 2 & & \\
 & z = 2 & & \\
\textbf{c.} & 2(1) + y - 5(2) = -9 & & \\
 & y = -1 & &
\end{array}
$$

check:

$$
\begin{array}{ll}
\textbf{a.} & 4(1) - (-1) + 3(2) = 11 \\
\textbf{b.} & 7(1) + 7(-1) + 2 = 2
\end{array}
$$

The check is not extended to **c** because that equation was used to determine that $y = -1$. The solution set is $\{(1,-1,2)\}$. ■

Note that z is eliminated in equations **d** and **e** of example 1. In example 2, y is eliminated from those equations. It does not matter which variable is eliminated. The same solution set will be obtained whether you choose to eliminate x or y or z. Try redoing example 2 by eliminating x from equations **d** and **e.**

Applications

Let us return to our student who tried to discover, by experimentation, the optimum study time to maximize his math grade (see figure 2.1 on page 71). When the student brought his data to his instructor, she produced the equation

$$G = \frac{-1 + \sqrt{1 + 24T}}{2}$$

that related grade earned, G, to time studied, T. The instructor observed that a graph of the data points had the shape of a quadratic function. She attempted to find a function of the form

$$T = aG^2 + bG + c, \tag{2}$$

whose graph would contain or pass close to the student's data points. She chose three data points, $(\frac{1}{3},1)$, $(1,2)$, and $(2,3)$ to be in the function. Substituting the coordinates of the data points into (2) she constructed a 3×3 linear system that was used to find a, b, and c.

$$\begin{aligned} \tfrac{1}{3} &= a(1)^2 + b(1) + c \\ 1 &= a(2)^2 + b(2) + c \\ 2 &= a(3)^2 + b(3) + c \end{aligned}$$

$$\begin{array}{rcll} \tfrac{1}{3} &=& a + b + c & (-1 \\ 1 &=& 4a + 2b + c & (1 \qquad (-1 \\ 2 &=& 9a + 3b + c & \qquad\quad\; (1 \end{array}$$

$$\begin{array}{rcll} \tfrac{2}{3} &=& 3a + b & (-1 \\ 1 &=& 5a + b & (1 \end{array}$$

$$\begin{aligned} \tfrac{1}{3} &= 2a \\ \hline \tfrac{1}{6} &= a \end{aligned}$$

$$\begin{aligned} 1 &= 5(\tfrac{1}{6}) + b \\ \hline \tfrac{1}{6} &= b \end{aligned}$$

$$\begin{aligned} 1 &= 4(\tfrac{1}{6}) + 2(\tfrac{1}{6}) + c \\ \hline 0 &= c \end{aligned}$$

Equation (2) becomes $T = G^2/6 + G/6$. Solving this equation for G in terms of T yields the equation that the instructor produced for the student.

$$O = G^2 + G - 6T$$

$$G = \frac{-1 \pm \sqrt{1 + 24T}}{2}$$

Exercise Set 3.2

(A)

Solve each 3×3 linear system.

1. $\begin{aligned} x + 2y + z &= 3 \\ 3x + y + z &= 2 \\ 2x - 5y - z &= 12 \end{aligned}$

2. $\begin{aligned} x + 2y - z &= -3 \\ 2x - y + z &= -5 \\ 3x + 2y - 2z &= -3 \end{aligned}$

3. $\begin{aligned} 2x - 2y + 3z &= 1 \\ x - 3y - 2z &= -9 \\ x + y + z &= 6 \end{aligned}$

4. $\begin{aligned} 2x + y + z &= 0 \\ 6x - 2y + 3z &= 1 \\ x + 2y - 7z &= 1 \end{aligned}$

5. $\begin{aligned} 2x + y + z &= 2 \\ 3x - y + z &= 2 \\ 7x - 5y + 3z &= 0 \end{aligned}$

6. $\begin{aligned} 5x - y - z &= -1 \\ 2x + 3y - z &= -4 \\ x - 7y + z &= -7 \end{aligned}$

7. $\begin{aligned} 3x + y + z &= 6 \\ x + y - z &= 0 \\ x - y + 2z &= 5 \end{aligned}$

8. $\begin{aligned} x + 2y - z &= 3 \\ 3x + y + z &= 2 \\ 2x - 5y - z &= 12 \end{aligned}$

9. $\begin{aligned} x + y + 2z &= 0 \\ y + z &= 2 \\ x - z &= 4 \end{aligned}$

10. $\begin{aligned} x + y + z &= 1 \\ 2x - 3y - z &= 0 \\ - y + z &= 1 \end{aligned}$

11. $\begin{aligned} 2r + s + t &= 2 \\ 3r - s + t &= 2 \\ 7r - 5s - 3t &= 0 \end{aligned}$

12. $\begin{aligned} w + 2u - v &= -1 \\ w - 3u + v &= 2 \\ 2w + u + 2v &= 6 \end{aligned}$

13. $\begin{aligned} 5r - 2s - 5t &= 1 \\ 3r + s - t &= 10 \\ 8r - s - 6t &= -3 \end{aligned}$

14. $\begin{aligned} 2w - u + v &= 4 \\ 3w - 2u + 2v &= 3 \\ w - u + 3v &= 2 \end{aligned}$

15. $\begin{aligned} x + y + 2z &= 4 \\ x - 5y + z &= 5 \\ 3x - 4y + 7z &= 24 \end{aligned}$

16. $\begin{aligned} 5x - 4y + 5z &= 6 \\ 6x + y - 2z &= 4 \\ 4x - 9y + 12z &= 5 \end{aligned}$

17. $\begin{aligned} p + q + 2r &= 0 \\ q + r &= 2 \\ p - r &= 4 \end{aligned}$

18. $\begin{aligned} 5r + t &= 8 \\ s - 3t &= -5 \\ 2r + 5s &= 12 \end{aligned}$

19. $\begin{aligned} 4w + 3u - 2v &= -11 \\ w - u + 6v &= 4 \\ 2u - 5v &= -8 \end{aligned}$

20. $\begin{aligned} 11x + 5y &= 12 \\ 2x - 6z &= -16 \\ 8x + 3y + 2z &= 4 \end{aligned}$

(B)

21. The sum of the measures of the angles of a triangle is 180°. The measure of the first angle is three times the measure of the second angle. The third angle measures 20° more than the sum of the measures of the first two angles. Find the measures of the three angles.

22. A collection of nickels, dimes, and quarters is worth \$5.55. The number of nickels in the collection is one less than three times the number of quarters. There are nine more dimes in the collection than the number of nickels and quarters combined. Find the number of nickels, dimes, and quarters in the collection.

23. The sum of three numbers is 131. The first number is five less than three times the second number. The third number is five more than the sum of the first two numbers. Find the three numbers.

24. There are 44 coins in a collection of pennies, nickels, and dimes. There are 3 more pennies than nickels and 4 more nickels than dimes. How many pennies, nickels, and dimes are in the collection?

25. A nut mix is to consist of cashews, peanuts, and almonds. The per pound selling prices of cashews, almonds, and peanuts are \$2.20, \$1.70, and \$0.90, respectively. A ten-pound batch of the nut mix sells for \$1.50 per pound. The weight of peanuts in the mix is twice that of the cashews. How many pounds of each type of nut are in the mix?

26. Each pound of food A contains 10 units of protein and 10 units of fat. Each pound of food B contains 20 units of protein and 5 units of fat, and each pound of food C contains 12 units of protein and 8 units of fat. Foods A, B, and C are mixed to produce 30 pounds of a food that contains 14 units of protein and 7.5 units of fat per pound. How many pounds of each food are in the mix?

27. A pool has three inlet pipes A, B, and C. The pool is filled in 6 hours if pipe A alone is open. If pipes B and C are open and A is closed, the pool fills in 5 hours. If pipes A and C are open and B is closed, the pool fills in 4 hours. How long does it take to fill the pool if all three pipes are open?

28. Three different candy bars A, B, and C are made from the same ingredients; chocolate, nougat, and hazel nuts. The ingredient mix for each candy bar is given (in ounces) in the following table.

Candy	Chocolate	Nougat	Nuts
A	2.0	1.5	0.5
B	1.5	2.0	0.5
C	1.0	2.0	1.0

If 24.75 pounds of chocolate, 26.5 pounds of nougat, and 8.75 pounds of nuts are available, how many bars of each type of candy can be made so that the ingredients are completely exhausted?

29. A cargo airline has three types of cargo planes. Their specifications are listed in the table.

Type	Cargo Capacity	Crew Size	Fuel Required/ 1,000 mi.
A	50 tons	5	260 gal.
B	40 tons	4	220 gal.
C	65 tons	7	300 gal.

If the airline ships 1,105 tons of cargo 1,000 miles using 116 crew members and 5,400 gallons of fuel, how many of each type of cargo plane does it use?

30. A manufacturer produces three models, P, Q, and R, of a product. Three machines, A, B, and C, are used to manufacture each model. The machine time (hours) needed for the manufacture of each model and the available machine time (hours) per week for each type of machine are given in the table.

Machine	Model P	Model Q	Model R	Available Time/Week
A	6	8	5	194
B	5	5	6	165
C	8	4	3	162

If the manufacturer utilizes all of the machine time available, how many of each type of model can be produced?

The graph of a quadratic function $y = ax^2 + bx + c$ contains the given three points. Determine the equation of the function.

31. $(0,2), (3,0), (5,-2)$

32. $(-6,4), (-2,-1), (3,3)$

33. $(-4,-3), (0,0), (4,-3)$

34. $(1,2), (3,5), (6,11)$

Experimental data are collected in tabular form. It is assumed, after plotting the data points, that a quadratic function can be found whose graph will pass close to the given data points. Find such a function and graph it on the same coordinate axes that locate the given data points. There is no one correct answer.

35.

x	y
-1	2
2	1
4	-1
7	-2
10	2

36.

x	y
-2	8
1	7
4	9
6	11
8	14

37.

x	y
0	-2
3	1
6	3
8	2
10	0

38.

x	y
2	4
3	7
4	10
6	5
9	1

39. Equations **d** and **e** in example 2 define planes that are both parallel to the (a) axis and perpendicular to the (b) plane. Fill in blanks (a) and (b).

3.3 Elementary Transformations and Matrices

Decision making in industry and government is very often supported by mathematical models that are linear systems involving 100 or more linear equations. It is virtually impossible to solve such systems without the aid of powerful computers. Although we will not construct computer solutions here, we will develop a method of solution that is easily programmed for a computer.

Elementary Transformations

Each of the following changes imposed on a system of linear equations is called an *elementary transformation.*

1. Two equations are interchanged.
2. Each member of an equation is multiplied by a nonzero real number.
3. Each member of an equation is multiplied by a nonzero real number and added to the corresponding member of another equation.

Each elementary transformation changes the form, but not the solution set, of a linear system. Certainly interchanging the position of two equations in the system will not alter the solution set of the system. Nor will multiplying each member of an equation by a nonzero real number change the solution set of the system (see exercise 39). The third elementary transformation is nothing more than a linear combination. It too changes only the form of the system but not the solution set. To see this consider systems I and II below. The second equation in system II is gotten by multiplying the first equation in system I by -2 and adding the resulting equation to the second equation of system I. The second system is generated from the first with elementary transformation 3. Note that the solution sets for the two systems are identical.

I.

$$\begin{array}{rcl|ll}
x + 2y - z & = & -3 & (1 & \\
2x - y + z & = & 5 & (1 & (2 \\
\underline{3x + 2y - 2z} & = & \underline{-3} & & (1 \\
3x + y & = & 2 & & \\
\underline{7x} & = & \underline{7} & & \\
x & = & 1 & & \\
y & = & -1 & & \\
z & = & 2 & &
\end{array}$$

II.

$$\begin{array}{rcl|l}
x + 2y - z & = & -3 & (-3 \\
-5y + 3z & = & 11 & \\
\underline{3x + 2y - 2z} & = & \underline{-3} & (1 \\
-5y + 3z & = & 11 & (1 \\
\underline{-4y + z} & = & \underline{6} & (-3 \\
7y & = & -7 & \\
y & = & -1 & \\
z & = & 2 & \\
x & = & 1 &
\end{array}$$

Now we shall use elementary transformations to convert a 3×3 linear system into an equivalent 3×3 linear system in which the solution set is "obvious."

Example 1 Use elementary transformations to solve the system of linear equations.

$$A\begin{cases} \textbf{1.}\ 2x - y + z = 5 \\ \textbf{2.}\ x + 2y - z = -3 \\ \textbf{3.}\ 3x + 2y - 2z = -3 \end{cases}$$

Solution

Objective: to obtain a first equation in which the term involving x has the coefficient 1.

$$B\begin{cases} \textbf{1.}\ x + 2y - z = -3 & \text{(Interchange } A\text{–1 and } A\text{–2)} \\ \textbf{2.}\ 2x - y + z = 5 \\ \textbf{3.}\ 3x + 2y - 2z = -3 \end{cases}$$

Objective: to eliminate the terms involving x in B–2 and B–3.

$$C\begin{cases} \textbf{1.}\ x + 2y - z = -3 \\ \textbf{2.}\ 0 - 5y + 3z = 11 & \text{(Multiply } B\text{–1 by } -2 \text{ and add to } B\text{–2)} \\ \textbf{3.}\ 0 - 4y + z = 6 & \text{(Multiply } B\text{–1 by } -3 \text{ and add to } B\text{–3)} \end{cases}$$

Objective: to obtain a coefficient of 1 for the y term in C–2.

$$D\begin{cases}\textbf{1.}\ x + 2y - z = -3 \\ \textbf{2.}\ 0 + y - \dfrac{3z}{5} = -\dfrac{11}{5} & \left(\text{Multiply } C\text{–2 by } -\dfrac{1}{5}\right) \\ \textbf{3.}\ 0 - 4y + z = 6\end{cases}$$

Objective: to eliminate the y term in D–3.

$$E\begin{cases}\textbf{1.}\ x + 2y - z = -3 \\ \textbf{2.}\ 0 + y - \dfrac{3z}{5} = -\dfrac{11}{5} \\ \textbf{3.}\ 0 + 0 - \dfrac{7z}{5} = -\dfrac{14}{5} & (\text{Multiply } D\text{–2 by 4 and add to } D\text{–3})\end{cases}$$

From E–3 it follows that $z = 2$. Substituting 2 for z in E–2, we have

$$0 + y - \tfrac{6}{5} = -\tfrac{11}{5}, \quad \text{or} \quad y = -1.$$

Substituting -1 for y and 2 for z in E–1, we have

$$x - 2 - 2 = -3, \quad \text{or} \quad x = 1,$$

and the solution set of system E as well as systems A, B, C, and D is $\{(1,-1,2)\}$. ■

Because of the *triangular form* of system E in example 1, the solution set of the system can be thought of as "obvious." The value of z is immediately obtainable from the third equation. Then by substituting up into equations 2 and 1, the values of y and x are obtained.

Matrices

The process of solving linear systems with elementary transformations can be speeded up if the method is somewhat formalized with the introduction of a symbol called a matrix.

Consider the 3×3 system A and the symbols labeled M and G.

$$A\begin{cases}\textbf{1.}\ a_1x + b_1y + c_1z = d_1 \\ \textbf{2.}\ a_2x + b_2y + c_2z = d_2 \\ \textbf{3.}\ a_3x + b_3y + c_3z = d_3,\end{cases}$$

$$M: \begin{bmatrix} a_1 & b_1 & c_1 \\ a_2 & b_2 & c_2 \\ a_3 & b_3 & c_3 \end{bmatrix} \qquad G: \begin{bmatrix} a_1 & b_1 & c_1 & d_1 \\ a_2 & b_2 & c_2 & d_2 \\ a_3 & b_3 & c_3 & d_3 \end{bmatrix}.$$

Rectangular arrays of numbers such as M and G are called **matrices.** The horizontal listings of entries are called **rows.** The vertical listings are called **columns.** A matrix with m rows and n columns is called an $\boldsymbol{m \times n}$ matrix. The 3×3 matrix M, in which the entries are the coefficients of the variables of system A, is called the **coefficient matrix** of the system. The 3×4 matrix G, is obtained from matrix M by inserting into M a last column in which the entries are the constant terms d_1, d_2, and d_3 of the equations of system A. Matrix G is called the **augmented matrix** of the system.

Example 2 For each augmented matrix, write the corresponding linear system. Use x, y, and z for the variables.

a. $\begin{bmatrix} 1 & 2 & 3 \\ 4 & 5 & 6 \end{bmatrix}$

b. $\begin{bmatrix} 2 & 0 & -3 & 5 \\ 1 & 2 & 1 & -2 \\ 0 & 1 & 3 & 0 \end{bmatrix}$

Solution

a. $\begin{aligned} x + 2y &= 3 \\ 4x + 5y &= 6 \end{aligned}$

b. $\begin{aligned} 2x \qquad - 3z &= 5 \\ x + 2y + z &= -2 \\ y + 3z &= 0 \end{aligned}$

■

Corresponding to each of the three elementary transformations on linear systems of equations, there is an associated operation on matrices called a **row operation.** Referring to matrix A

$$A\colon \begin{bmatrix} a_1 & b_1 & c_1 \\ a_2 & b_2 & c_2 \\ a_3 & b_3 & c_3 \end{bmatrix},$$

we describe each of the three row operations, together with an example of the resulting matrix.

1. Interchange any two rows of A:

$$\begin{bmatrix} a_1 & b_1 & c_1 \\ a_3 & b_3 & c_3 \\ a_2 & b_2 & c_2 \end{bmatrix}$$

2. Multiply each entry of any row of A by a nonzero real number k:

$$\begin{bmatrix} a_1 & b_1 & c_1 \\ a_2 & b_2 & c_2 \\ ka_3 & kb_3 & kc_3 \end{bmatrix}$$

3. To each entry of a row of A add the same nonzero multiple of the corresponding entry of any other row:

$$\begin{bmatrix} a_1 + ka_2 & b_1 + kb_2 & c_1 + kc_2 \\ a_2 & b_2 & c_2 \\ a_3 & b_3 & c_3 \end{bmatrix}$$

Example 3 Apply a row operation to the following matrix so that the first nonzero entry of row two shall be 1.

$$\begin{bmatrix} 1 & -2 & 1 & 0 \\ 0 & -3 & -2 & -4 \\ 0 & -5 & 1 & -11 \end{bmatrix}$$

Solution The first nonzero entry in row two is -3. Hence, we multiply each entry of row two by $-\frac{1}{3}$ (row operation two) to get

$$\begin{bmatrix} 1 & -2 & 1 & 0 \\ 0 & 1 & \frac{2}{3} & \frac{4}{3} \\ 0 & -5 & 1 & -11 \end{bmatrix}.$$

■

The application of row operations to the augmented matrix of a given system enables us to generate a sequence of matrices, each of which is the augmented matrix of a linear system that is equivalent to the given system.

Echelon Form

Consider the rows of matrix G

$$G: \begin{bmatrix} 1 & 3 & 0 & -2 \\ 0 & 1 & \frac{4}{3} & \frac{11}{3} \\ 0 & 0 & 1 & 3 \end{bmatrix},$$

and observe that:

1. The first nonzero entry (called the *leading coefficient*) of each row is 1.
2. If there is a leading coefficient in a row, then all the entries in the column below that leading coefficient are zeros.
3. The leading coefficient in any row is to the right of the leading coefficient in any row above it.
4. The matrix has at least one nonzero entry.

Matrices that satisfy these four requirements are said to be in **echelon form.**

Example 4 Specify which of the following matrices is in echelon form. If a matrix is not in echelon form, explain why.

a. $\begin{bmatrix} 1 & 0 & 1 & 1 \\ 0 & 1 & 0 & 1 \\ 0 & 0 & 0 & 0 \end{bmatrix}$

b. $\begin{bmatrix} 1 & 2 & 3 & 4 \\ 0 & 1 & 2 & 3 \\ 0 & 1 & 3 & 5 \end{bmatrix}$

c. $\begin{bmatrix} 1 & 2 & 3 & 4 \\ 0 & 1 & 0 & 2 \\ 0 & 0 & 2 & 5 \end{bmatrix}$

Solution

a. The matrix is in echelon form.

b. The matrix is not in echelon form because the leading coefficient in row three is not to the right of the leading coefficient in row two.

c. The matrix is not in echelon form because the leading coefficient in row three is not 1. ■

A matrix can have more than one echelon form. Thus, the echelon matrix of example 4a can be transformed into the echelon matrix

$$\begin{bmatrix} 1 & 2 & 1 & 3 \\ 0 & 1 & 0 & 1 \\ 0 & 0 & 0 & 0 \end{bmatrix}$$

by adding to each entry in the first row twice the corresponding entry in the second row. Although the two echelon forms differ, the linear systems represented by the matrices are equivalent because one system is obtained from the other with an elementary row transformation.

Having an augmented matrix of a linear system in echelon form is equivalent to having the system in the triangular form of system E in example 1. Thus, an $n \times n$ system of linear equations, whether n is 3 or 100, can be solved with matrices. Furthermore, the row operations on matrices are easily programmed for a computer so that a large system of linear equations can be solved quickly with a computer.

Example 5 Use matrices to solve the following linear system.

$$A: \begin{cases} \textbf{1.} \quad x - 2y + z = 0 \\ \textbf{2.} \quad -2x + y - 4z = -4 \\ \textbf{3.} \quad -4x + 3y - 3z = -11 \end{cases}$$

Solution The augmented matrix G of system A is

$$G: \begin{bmatrix} 1 & -2 & 1 & 0 \\ -2 & 1 & -4 & -4 \\ -4 & 3 & -3 & -11 \end{bmatrix}.$$

Applying row operations to obtain an echelon form of G, we have:

$$B: \begin{bmatrix} 1 & -2 & 1 & 0 \\ 0 & -3 & -2 & -4 \\ 0 & -5 & 1 & -11 \end{bmatrix} \quad \begin{matrix} \text{(Twice } G\text{–1 added to } G\text{–2)} \\ \text{(Four times } G\text{–1 added to } G\text{–3)} \end{matrix}$$

$$C: \begin{bmatrix} 1 & -2 & 1 & 0 \\ 0 & 1 & \frac{2}{3} & \frac{4}{3} \\ 0 & -5 & 1 & -11 \end{bmatrix} \quad \text{(Multiply } B\text{–2 by } -\tfrac{1}{3}\text{)}$$

$$D: \begin{bmatrix} 1 & -2 & 1 & 0 \\ 0 & 1 & \frac{2}{3} & \frac{4}{3} \\ 0 & 0 & \frac{13}{3} & -\frac{13}{3} \end{bmatrix} \quad \text{(Five times } C\text{–2 added to } C\text{–3)}$$

$$E: \begin{bmatrix} 1 & -2 & 1 & 0 \\ 0 & 1 & \frac{2}{3} & \frac{4}{3} \\ 0 & 0 & 1 & -1 \end{bmatrix} \quad \text{(Multiply } D\text{–3 by } \tfrac{3}{13}\text{)}$$

From matrix E, we have the system

$$E: \begin{cases} \textbf{1.}\ x - 2y + z = 0 \\ \textbf{2.}\ y + \dfrac{2z}{3} = \dfrac{4}{3} \\ \textbf{3.}\ z = -1 \end{cases}$$

Substituting -1 for z in E–2, we obtain $y = 2$. Substituting -1 for z and 2 for y in equation E–1, we obtain $x = 5$. Because system E is equivalent to system A, we have that the solution set of system A is $\{(5,2,-1)\}$. ■

Consistency

A simple test allows us to determine, before attempting a solution, whether a 2×2 linear system is consistent. No such simple test exists for 3×3 or larger systems. We must attempt a solution and then decide consistency from the structure of the final matrix obtained in the solution process.

Example 6 Use matrices to solve the following linear system.

$$A: \begin{cases} \textbf{1.}\ \ x - 2y + 5z = -3 \\ \textbf{2.}\ \ x - 5y - 14z = 11 \\ \textbf{3.}\ 2x - 7y - 9z = 5 \end{cases}$$

Solution The augmented matrix G of system A is

$$G: \begin{bmatrix} 1 & -2 & 5 & -3 \\ 1 & -5 & -14 & 11 \\ 2 & -7 & -9 & 5 \end{bmatrix}.$$

Applying row operations to obtain an echelon form of G, gives us

$$B: \begin{bmatrix} 1 & -2 & 5 & -3 \\ 0 & -3 & -19 & 14 \\ 0 & -3 & -19 & 11 \end{bmatrix} \quad \begin{matrix} \\ (-1 \text{ times } G\text{–}1 \text{ added to } G\text{–}2) \\ (-2 \text{ times } G\text{–}1 \text{ added to } G\text{–}3) \end{matrix}$$

$$C: \begin{bmatrix} 1 & -2 & 5 & -3 \\ 0 & -3 & -19 & 14 \\ 0 & 0 & 0 & -3 \end{bmatrix} \quad \begin{matrix} \\ \\ (-1 \text{ times } B\text{–}2 \text{ added to } B\text{–}3) \end{matrix}$$

Matrix C corresponds to the linear system

$$C: \begin{cases} \textbf{1.} \;\; x - 2y + 5z = -3 \\ \textbf{2.} \;\; 0x - 3y - 19z = 14 \\ \textbf{3.} \;\; 0x + 0y + 0z = -3, \end{cases}$$

which is equivalent to system A. The solution set of system C must be the set of all ordered triples that are solutions of each of the three equations. But equation C–3 has no solutions because there is no ordered triple (x,y,z) such that $0x + 0y + 0z = -3$. Hence system C, and therefore system A, have no solutions. System A is *inconsistent*. ■

Example 6 illustrates that, if in the attempted solution of a linear system by matrix methods, a matrix is obtained that has *a row in which the numbers in every column except the last are zeros, then the system is inconsistent.*

It is possible for a linear system to produce a matrix in which *all* the numbers in *one or more* rows are zero. We consider such systems in the next section.

Exercise Set 3.3

(A)

Use elementary row transformations to solve each system.

1. $\begin{aligned} 3x + 2y + z &= 7 \\ x + y - 3z &= 3 \\ -2x + y - z &= 0 \end{aligned}$

2. $\begin{aligned} x - y - z &= 2 \\ x + 2y + z &= 7 \\ 3x - y + 2z &= 12 \end{aligned}$

3. $\begin{aligned} 3x + 2y + 4z &= 5 \\ 2x + y + z &= 1 \\ x - 2y - 3z &= 1 \end{aligned}$

4. $\begin{aligned} 2x + y + z &= 1 \\ x + 2y + z &= 0 \\ x + y + 2z &= 0 \end{aligned}$

5. $\begin{aligned} x + y &= 2 \\ y + z &= 1 \\ x - 2z &= -2 \end{aligned}$

6. $\begin{aligned} 2x - 3y + z &= 5 \\ x + 2y - 3z &= -15 \\ x - 4y + 2z &= 12 \end{aligned}$

Write an augmented matrix for the linear system in the indicated exercise of exercise set 3.2. Reduce the augmented matrix to echelon form and solve the system.

7. exercise 1	**8.** exercise 2	**9.** exercise 3
10. exercise 4	**11.** exercise 5	**12.** exercise 6
13. exercise 7	**14.** exercise 8	**15.** exercise 9
16. exercise 10	**17.** exercise 11	**18.** exercise 12
19. exercise 13	**20.** exercise 14	**21.** exercise 15
22. exercise 16	**23.** exercise 17	**24.** exercise 18
25. exercise 19	**26.** exercise 20	

Solve each system by matrix methods. If the system is inconsistent, say so.

27.
$$\begin{aligned} x + y - 3z &= 3 \\ 3x + 2y + z &= 7 \\ -2x + y - z &= 0 \end{aligned}$$

28.
$$\begin{aligned} x + y + z &= 1 \\ x - y + 2z &= 7 \\ x + y - z &= -3 \end{aligned}$$

29.
$$\begin{aligned} x + 2y + 4z &= 13 \\ 3x + 5y + 6z &= 2 \\ 2x + 3y + 2z &= 7 \end{aligned}$$

30.
$$\begin{aligned} 3x + 4y + 2z &= 0 \\ x + y + z &= 2 \\ 4x + 5y + 3z &= 7 \end{aligned}$$

31.
$$\begin{aligned} x + y \quad &= 2 \\ y + z &= 1 \\ x \quad - 2z &= -2 \end{aligned}$$

32.
$$\begin{aligned} 3x \quad - z &= 16 \\ x - y + 5z &= 0 \\ 4y + 8z &= -5 \end{aligned}$$

33.
$$\begin{aligned} x + 3y + 2z &= 5 \\ x + 5y - 4z &= 6 \\ 2x + 8y - 2z &= -2 \end{aligned}$$

34.
$$\begin{aligned} x - y + 2z &= 2 \\ y + 3z &= -1 \\ x + 2y + 11z &= -3 \end{aligned}$$

(B)

35.
$$\begin{aligned} x + y \quad &= 2 \\ y + z \quad &= 0 \\ z + w &= 0 \\ x \quad + z \quad &= -2 \end{aligned}$$

36.
$$\begin{aligned} x + y + z + w &= 1 \\ 2x - y + 3z - w &= 2 \\ -x + 4y - 2z + 5w &= 0 \\ 3x + 2y + z + 3w &= 6 \end{aligned}$$

37.
$$\begin{aligned} 3x - 4y - z + 2w &= 3 \\ x + y + 2z + 3w &= 1 \\ -2x + 3y - z + w &= 0 \\ -2x - y - 9z + w &= 5 \end{aligned}$$

38.
$$\begin{aligned} y + z + w &= 0 \\ x \quad + 2z - w &= 4 \\ 2x \quad - z + 2w &= 3 \\ -2x \quad + 2z - w &= -2 \end{aligned}$$

39. Show that the equation obtained by multiplying $ax + by + cz = d$ by a nonzero constant k has the same solution set as $ax + by + cz = d$. (Hint: Let (x_0,y_0,z_0) be a solution of $ax + by + cz = d$. Show that it is also a solution of $k(ax + by + cd) = kd$. Then show that if (x_1,y_1,z_1) is a solution of $k(ax + by + cz) = kd$, it must also be a solution of $ax + by + cz = d$.)

3.4 *M* × *N* Linear Systems

We have so far considered only systems in which the number of equations is the same as the number of variables. But systems in which the number of equations is different from the number of variables can also be solved by the matrix methods introduced in the last section.

In the examples that follow, particular row operations are no longer specified. It will be up to you to determine which row operations generated a matrix from its predecessor matrix. Arrows are used to show a sequence of matrices ending with the matrix from which the solution of the system is determined.

Example 1 Use matrices to solve the following system.

$$\begin{aligned} x + 2y - 5z &= 4 \\ -3x - 7y + z &= 0 \end{aligned}$$

Solution Starting with the augmented matrix of the system, we obtain:

$$\begin{bmatrix} 1 & 2 & -5 & 4 \\ -3 & -7 & 1 & 0 \end{bmatrix} \to \begin{bmatrix} 1 & 2 & -5 & 4 \\ 0 & -1 & -14 & 12 \end{bmatrix} \to \begin{bmatrix} 1 & 2 & -5 & 4 \\ 0 & 1 & 14 & -12 \end{bmatrix}.$$

The third matrix, in echelon form, corresponds to the linear system

$$\begin{aligned} x + 2y - 5z &= 4 \\ y + 14z &= -12. \end{aligned}$$

From the second equation $y = -14z - 12$. Substituting $-14z - 12$ for y in the first equation yields

$$x + 2(-14z - 12) - 5z = 4, \quad \text{or} \quad x = 33z + 28.$$

The solution set of the system is

$$\{(x,y,z) \mid x = 33z + 28, \quad y = -14z - 12, \quad z \in R\},$$

which is an infinite set. A particular solution in the set is gotten by assigning some real value to z and then computing x and y. If $z = 0$, then $(28,-12,0)$ is a solution of the system. If $z = -2$, then $(-38,16,-2)$ is a solution of the system, etc. ■

Example 2 Use matrices to solve the system

$$\begin{aligned} 3w + 2x - y + z &= 8 \\ w + x - 4y - 2z &= -6 \\ 3w + x + 10y + 8z &= 3. \end{aligned}$$

Solution

$$\begin{bmatrix} 3 & 2 & -1 & 1 & 8 \\ 1 & 1 & -4 & -2 & -6 \\ 3 & 1 & 10 & 8 & 3 \end{bmatrix} \to \begin{bmatrix} 1 & 1 & -4 & -2 & -6 \\ 3 & 2 & -1 & 1 & 8 \\ 3 & 1 & 10 & 8 & 3 \end{bmatrix} \to$$

$$\begin{bmatrix} 1 & 1 & -4 & -2 & -6 \\ 0 & -1 & 11 & 7 & 26 \\ 0 & -2 & 22 & 14 & 21 \end{bmatrix} \to \begin{bmatrix} 1 & 1 & -4 & -2 & -6 \\ 0 & 1 & -11 & -7 & -26 \\ 0 & -2 & 22 & 14 & 21 \end{bmatrix} \to$$

$$\begin{bmatrix} 1 & 1 & -4 & -2 & -6 \\ 0 & 1 & -11 & -7 & -26 \\ 0 & 0 & 0 & 0 & -31 \end{bmatrix} \to \begin{bmatrix} 1 & 1 & -4 & -2 & -6 \\ 0 & 1 & -11 & -7 & -26 \\ 0 & 0 & 0 & 0 & 1 \end{bmatrix}$$

The last row of the final matrix corresponds to the equation

$$0w + 0x + 0y + 0z = 1,$$

which has no solutions. Therefore, the solution set of the given system is the null set, and the system is inconsistent. ■

Example 3 Use matrices to solve the system

$$\begin{aligned} x - y + z &= 0 \\ 4x + y - 2z &= 4 \\ 3x - 5y - 6z &= -13 \\ 8x \phantom{{}-5y} - 13z &= -5. \end{aligned}$$

Solution

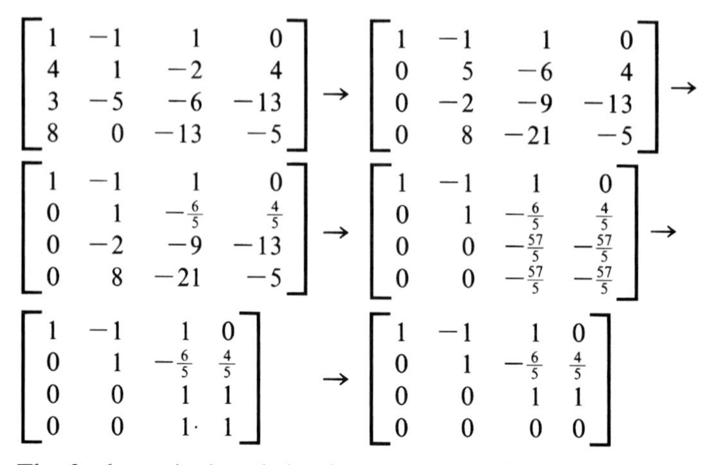

The final matrix, in echelon form, corresponds to the linear system

$$\begin{aligned} x - y + z &= 0 \\ y - \frac{6z}{5} &= \frac{4}{5} \\ z &= 1, \end{aligned}$$

which has the solution set $\{(1,2,1)\}$. ■

Although we cannot pursue the matter in this course, the row of zeros in the echelon matrix of example 3 is significant. That elementary row operations produced a fourth row having a zero in each of its columns tells us that the fourth equation in the given system of equations is a linear combination of the other three equations. The fourth equation gives us no more information than is already available in the other three equations.

The three examples above were selected as specific instances of what is true in general about the solution set of an $m \times n$ linear system. Either the system is inconsistent and its solution set is empty, or if the system is consistent, its solution set contains either one solution or an infinite number of solutions. Which type of solution set is obtained for a given linear system is easily determined from the final matrix.

1. If the final matrix has a row in which each column position except the last is occupied by a 0, and the last column position is occupied by a number that is not 0, then the linear system is **inconsistent.** (The final matrix in example 2 is such a matrix.)
2. If the final matrix has a coefficient matrix that has at least one non-zero number in each row, and if the coefficient matrix has as many rows as it has columns, then **the system is consistent and has exactly one solution.** (Example 3 exhibits such a matrix.)

3. If the final matrix has a coefficient matrix that has at least one non-zero number in each row, and if the coefficient matrix has fewer rows than it has columns, then **the system is consistent and its solution set is infinite.** (Example 1 exhibits such a matrix.)

Current flow in an electric circuit is often determined with the aid of a system of linear equations. Consider the schematic diagram of an electric circuit that is shown in figure 3.11. A battery, which is a voltage source, is symbolized by a pair of parallel line segments (||) of unequal length. The longer segment is considered the positive side and the shorter segment the negative side of the battery. Current flows through a conductor in the circuit from the positive to the negative side of the battery. The saw tooth (—/\/\/—) symbols in the circuit are resistors. Given the voltage (in volts) of each battery in the circuit, and the resistance (in ohms) of each resistor, we wish to find the currents (in amps) through each conductor in the circuit. To find currents I_1, I_2, and I_3 we employ Kirchhoff's* Laws.

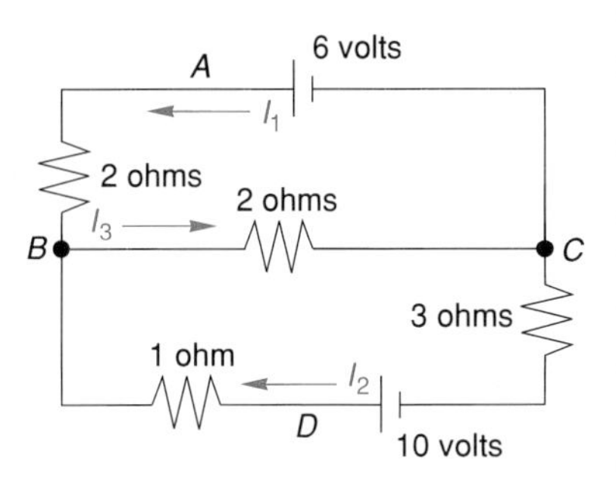

Figure 3.11

1. The sum of the currents flowing into a junction must equal the sum of the currents flowing out of the junction.
2. The algebraic sum of the voltage drops in any direction around a circuit branch is equal to the voltage impressed in that branch.

In figure 3.11, points B and C, where two or more conductors join, are called junctions of the circuit. Paths such as ABC, DBC, or $ABDC$ are called branches of the circuit. The product $\boldsymbol{IR}$, where I is the current through a resistor and R is the resistance of the resistor, is called the voltage drop across the resistor.

We begin the determination of currents I_1, I_2, and I_3 by assuming directions for each current. The direction of currents I_1 and I_2 must be, as shown in the circuit, from the positive to the negative side of the battery. However, the direction of current I_3 in conductor BC is up for grabs. We guess that the current moves from B to C. If our guess is wrong, the computed value of I_3 will be negative. Having assigned directions for the currents, we apply Kirchhoff's First Law.

at B: $I_1 + I_2 = I_3$
at C: $I_3 = I_1 + I_2$

The first law yields one distinct equation

1. $I_1 + I_2 - I_3 = 0.$

*German physicist Gustav Kirchhoff, (1824–1887).

We seek two more equations involving the currents to have a system of three equations in three variables. Examples 1, 2, and 3 and the discussion following those examples tell us that *exactly one solution* is obtained *if the number of equations is the same as the number of variables,* and that an infinite number of solutions is obtained if there are more variables than equations. Certainly we do not expect a circuit with fixed voltages and resistances to have more than one current flowing through a conductor. Therefore, we apply Kirchhoff's Second Law to two of the three branches, *BCA, BCD,* or *ABDC,* of the circuit.

branch *BCA:* 2. $2I_1 + 2I_3 = 6$

branch *BCD:* 3. $3I_2 + I_2 + 2I_3 = 10$

We note in passing that Kirchhoff's Second Law, applied to branch *ABDC,* yields equation $2I_1 - 4I_2 = -4$, which is a linear combination of equations 2 and 3. In other words, the equation obtained from branch *ABDC* does not give us any new information. Thus, to obtain two equations from Kirchhoff's Second Law, we choose any two of the three branches of the circuit. Now we have three linear equations in three variables and can proceed with the determination of the circuit's currents.

$$\begin{array}{lrcl}
\textbf{1.} & I_1 + I_2 - I_3 &=& 0 \qquad (-4 \\
\textbf{2.} & 2I_1 \qquad + 2I_3 &=& 6 \\
\textbf{3.} & 4I_2 + 2I_3 &=& 10 \qquad (1 \\
\hline
& 2I_1 \qquad + 2I_3 &=& 6 \qquad (2 \\
& -4I_1 \qquad + 6I_3 &=& 10 \qquad (1 \\
\hline
& 10I_3 &=& 22 \\
& I_3 &=& 2.2 \text{ amps} \\
& 2I_1 \quad + 2(2.2) &=& 6 \\
& I_1 &=& 0.8 \text{ amps} \\
& 0.8 + I_2 - 2.2 &=& 0 \\
& I_2 &=& 1.4 \text{ amps}
\end{array}$$

Our second application of linear systems is in the analysis of traffic flow through city streets. Vehicle flow, in vehicles per hour, entering and leaving the downtown portions of some of the main thoroughfares of a city is shown in figure 3.12. The streets are all one-way as indicated by the arrowheads.

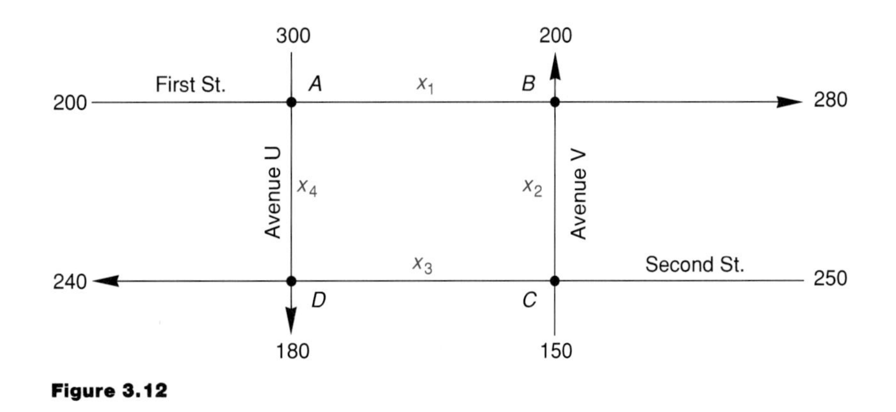

Figure 3.12

We wish to determine traffic flow on First and Second Streets between Avenues U and V as well as traffic flow on Avenues U and V between First and Second Streets. Labeling those traffic flows as x_1, x_2, x_3, and x_4 and noting that *traffic into a junction must equal traffic out of the junction* we can write the following system of linear equations.

$$\begin{array}{llcl}
\text{at } A\text{:} & 200 + 300 = x_1 + x_4 & \quad & x_1 \phantom{{}+x_2+x_3} + x_4 = 500 \\
\text{at } B\text{:} & x_1 + x_2 = 200 + 280 & & x_1 + x_2 \phantom{{}+x_3+x_4} = 480 \\
\text{at } C\text{:} & 250 + 150 = x_2 + x_3 & & x_2 + x_3 \phantom{{}+x_4} = 400 \\
\text{at } D\text{:} & x_3 + x_4 = 180 + 240 & & x_3 + x_4 = 420
\end{array}$$

Converting the linear system to an augmented matrix, and then using matrix methods to solve the system yields

$$\begin{bmatrix} 1 & 0 & 0 & 1 & 500 \\ 1 & 1 & 0 & 0 & 480 \\ 0 & 1 & 1 & 0 & 400 \\ 0 & 0 & 1 & 1 & 420 \end{bmatrix} \to \begin{bmatrix} 1 & 0 & 0 & 1 & 500 \\ 0 & 1 & 0 & -1 & -20 \\ 0 & 1 & 1 & 0 & 400 \\ 0 & 0 & 1 & 1 & 420 \end{bmatrix} \to$$

$$\begin{bmatrix} 1 & 0 & 0 & 1 & 500 \\ 0 & 1 & 0 & -1 & -20 \\ 0 & 0 & 1 & 1 & 420 \\ 0 & 0 & 1 & 1 & 420 \end{bmatrix} \to \begin{bmatrix} 1 & 0 & 0 & 1 & 500 \\ 0 & 1 & 0 & -1 & -20 \\ 0 & 0 & 1 & 1 & 420 \\ 0 & 0 & 0 & 0 & 0 \end{bmatrix}$$

$$\begin{aligned} x_1 \phantom{{}+x_2+x_3} + x_4 &= 500 \\ x_2 \phantom{{}+x_3} - x_4 &= -20 \\ x_3 + x_4 &= 420. \end{aligned}$$

Since the final matrix converts to three equations in four variables, the linear system has an infinite number of solutions. We solve x_1, x_2, and x_3 in terms of x_4.

$$\begin{aligned} x_1 &= 500 - x_4 \\ x_2 &= -20 + x_4 \\ x_3 &= 420 - x_4 \end{aligned}$$

A negative traffic flow would indicate traffic moving in the wrong direction along a one-way street. Thus, in order for x_1, x_2, x_3, and x_4 to be nonnegative, x_4 must be a number between 20 and 420. The maximum traffic flows that can be expected along First and Second Streets, between Avenues U and V, are 480 and 400 vehicles per hour, respectively. The minimum flows to be expected along those streets are 80 and 0 vehicles per hour respectively.

If street repair is planned for Avenue V between First and Second Streets so that only 100 vehicles per hour ($x_2 = 100$) could be accommodated along that portion of Avenue V, then $x_4 = 120$, $x_1 = 380$ and $x_3 = 300$.

Exercise Set 3.4

(A)

Use matrices to solve each system. If the system has more than one solution, give three specific solutions.

1. $\begin{aligned} x + 3y - z &= 5 \\ 2x - 2y + 4z &= -1 \end{aligned}$

2. $\begin{aligned} 3x - y - 3z &= 0 \\ x - 2y + z &= 2 \end{aligned}$

3. $\begin{aligned} 2x - y + z &= -4 \\ -6x + 3y - 3z &= 1 \end{aligned}$

4. $\begin{aligned} 6x + 8y + 4z &= 2 \\ 3x + 4y + 2z &= 1 \end{aligned}$

5. $\begin{aligned} w + 2x + 3y - 9z &= 5 \\ -w - x + 2y + 2z &= -3 \\ 4w - 3x + 2y - 4z &= -2 \end{aligned}$

6. $\begin{aligned} w + 2x + 3y - z &= 8 \\ -w - x + 2y + 5z &= 3 \\ 4w - 3x + 2y - 3z &= 1 \end{aligned}$

7. $\begin{aligned} w - x - 6y + 2z &= -1 \\ 2w - 2x - 15y + 12z &= -6 \\ 3w - 3x - 12y - 10z &= 5 \end{aligned}$

8. $\begin{aligned} 3w + 7x - 5y + 5z &= 16 \\ 2w + x - 18y - 4z &= 7 \\ w + 2x - 3y + z &= 5 \end{aligned}$

9. $\begin{aligned} w - 2x + z &= 0 \\ 2w + 3y &= 0 \\ x - y &= 0 \end{aligned}$

10. $\begin{aligned} -2w + 3y + 5z &= 0 \\ w + 2x - z &= 0 \\ x - 4y + z &= 0 \end{aligned}$

11. $\begin{aligned} 3x - y - 4z &= 5 \\ x - 2z &= 0 \\ -x + 4y &= 0 \end{aligned}$

12. $\begin{aligned} x + y - z &= 1 \\ 2x + 2y + 3z &= 4 \\ -x + 7y - 2z &= 8 \end{aligned}$

13. $\begin{aligned} w + 2x - y + 2z &= 0 \\ 5w - 3x - 4y + 3z &= 2 \end{aligned}$

14. $\begin{aligned} w + 4x + 4y - 5z &= 0 \\ 5w - x - y - 4z &= 6 \end{aligned}$

(B)

Use Kirchhoff's Laws to find the currents in the conductors of each circuit.

15.

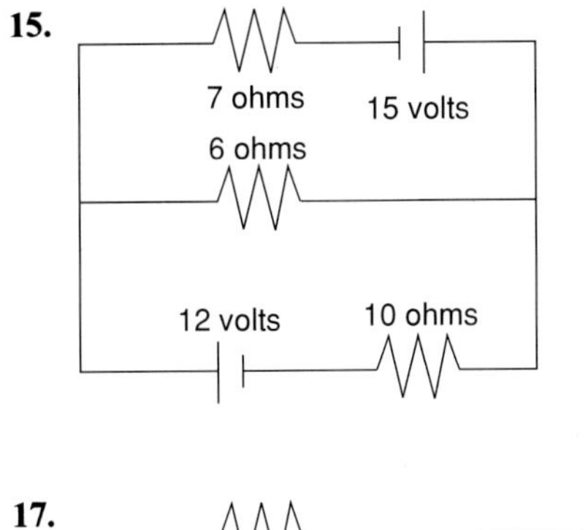

16.

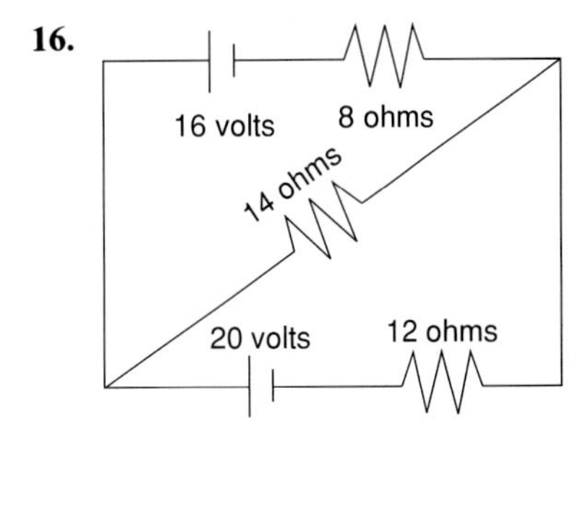

17.

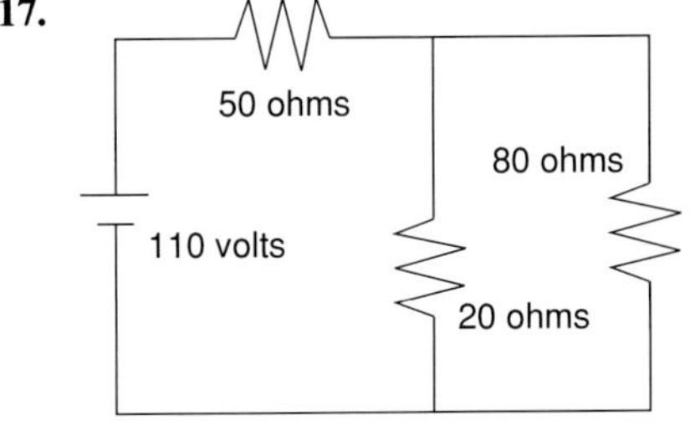

18.

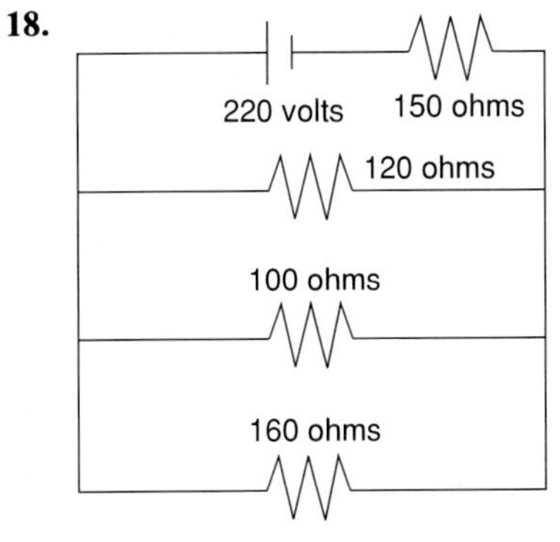

19. What is the minimum traffic flow that can be expected along *CD?* What is the maximum traffic flow that can be expected along *CB?*

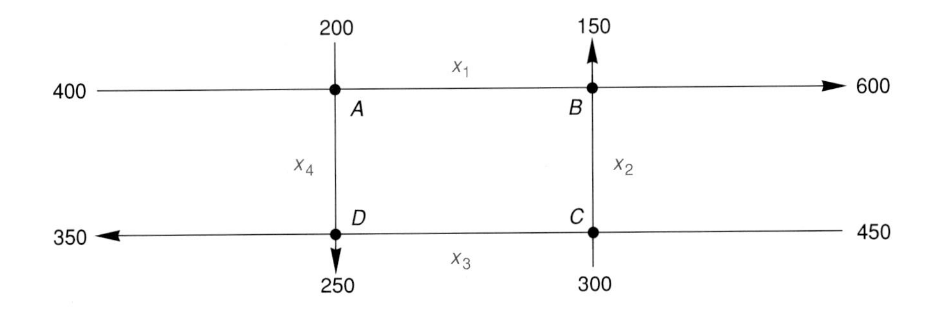

20. Find the minimum and maximum values of traffic flow that can be expected along DA.

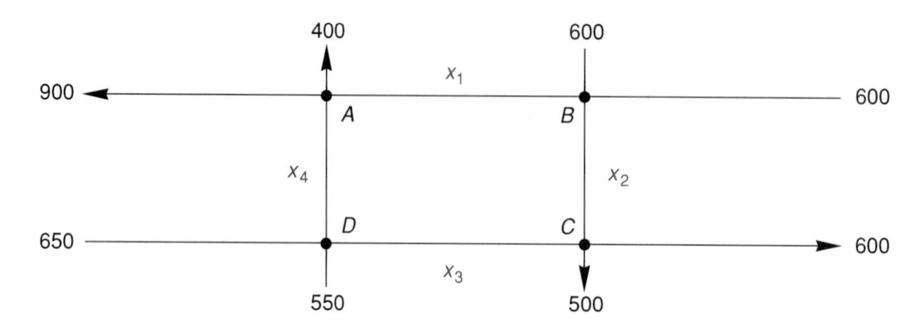

21. Determine one possible traffic flow pattern for the following system.

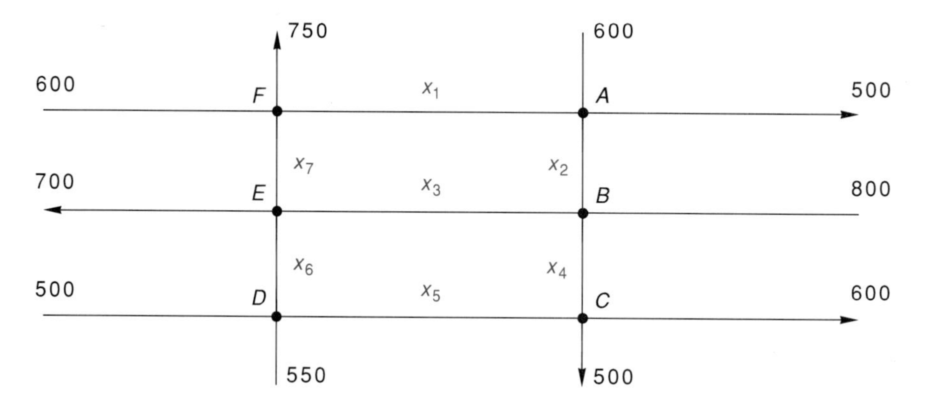

22. Give a geometric interpretation to the solution set of example 1.

23. A collection of nickels, dimes, and quarters is worth \$3.65. The collection has eight more dimes than quarters. How many nickels, dimes, and quarters are in the collection?

24. Four commercially packaged foods, A, B, C, and D, require different amounts of ingredients U, V, and W, as shown in the following table.

	A	B	C	D
U	0.2	0.6	0.4	0.1
V	0.2	0.2	0.3	0.3
W	0.4	0.1	0.1	0.2

A tabular value represents pounds of ingredient per pound of packaged food. There is 170 pounds of U, 70 pounds of V, and 50 pounds of W available. Determine the number of *whole* pounds of each packaged food that can be produced so that only a fractional part of a pound of each ingredient remains.

25. Find a relationship between a, b, and c that guarantees that the system of equations will have a solution.

$$\begin{aligned} 2x - 3y &= a \\ 4x + 2y &= b \\ 3x + y &= c \end{aligned}$$

26. For what value of c does the system of equations have exactly one solution? Determine that solution.

$$\begin{aligned} x + y - 3z &= 5 \\ 3x + 2y \qquad &= 4 \\ 4y + 6z &= -1 \\ x - y + cz &= -2 \end{aligned}$$

Chapter 3 Review

Major Points for Review

Section 3.1 Consistency tests for 2×2 linear systems

Solving a 2×2 system by the method of linear combinations

2×2 systems as mathematical models

Section 3.2 Solving a 3×3 system by the method of linear combinations

3×3 systems as mathematical models

Section 3.3 Using elementary transformations to solve a system of linear equations

The coefficient matrix of a linear system

The augmented matrix of a linear system

Using elementary row operations to transform an augmented matrix into an echelon matrix

Section 3.4 Using matrix methods to solve $n \times n$ and $m \times n$ linear systems

The structure of the matrix of a linear system that is inconsistent

The structure of the matrix of a linear system that is consistent and has exactly one solution

The structure of the matrix of a linear system that is consistent and has an infinite number of solutions

Review Exercises

Use the method of linear combinations to solve each system.

1. $\begin{aligned} 3x - 2y &= 8 \\ 2x + 3y &= 14 \end{aligned}$

2. $\begin{aligned} x - 4y &= 30 \\ 3x + 5y &= 29 \end{aligned}$

3. $\begin{aligned} 3x - 5y &= 8 \\ 9x - 15y &= -4 \end{aligned}$

4. $\begin{aligned} 3x + 2y &= 6 \\ 6x - y &= -6 \end{aligned}$

5. $\begin{aligned} 1.2x + 0.3y &= 2 \\ 4x - 6y &= 0 \end{aligned}$

6. $\begin{aligned} 2.4x + 1.8y &= 5.2 \\ 0.6x + 0.45y &= 1.3 \end{aligned}$

7. $\begin{aligned} x + y + z &= 6 \\ x - 2y + 3z &= 6 \\ 3x + 6y - 3z &= 6 \end{aligned}$

8. $\begin{aligned} x - y - z &= 2 \\ x + y + z &= 0 \\ x + 5y - 3z &= 4 \end{aligned}$

9. $\begin{aligned} 2x - y + 4z &= 5 \\ 7x + 7y - 10z &= 4 \\ x + 3y - 6z &= -2 \end{aligned}$

10. $\begin{aligned} x - y + z &= 2 \\ x - 5z &= 4 \\ x - y + 11z &= -6 \end{aligned}$

11. $\begin{aligned} 4x + 3y + 6z &= 22 \\ 6x - 5y - 10z &= 14 \\ 3x - 2y - 4z &= 8 \end{aligned}$

12. $\begin{aligned} -x + 2y + 3z &= 5 \\ x + 3y - z &= -3 \\ 3x + 7y + 6z &= 0 \end{aligned}$

Use matrix methods to solve each system.

13. $\begin{aligned} x - 3y + 6z &= 10 \\ 4x + 2y - z &= 9 \\ -x - y + 2z &= -5 \end{aligned}$

14. $\begin{aligned} x + 8y - 11z &= 18 \\ 3x - 5y + z &= 0 \\ 2y + 7z &= 10 \end{aligned}$

15. $\begin{aligned} x + 2y - 7z &= -4 \\ 4x + 5y - 13z &= 5 \\ 2x + y + z &= 13 \end{aligned}$

16. $\begin{aligned} 5x \qquad - z &= 0 \\ -8y + 3z &= -7 \\ 2x + 6y \qquad &= -2 \end{aligned}$

17. $\begin{aligned} 3x + 6y - 3z &= 13 \\ 2x + 6y + 3z &= 11 \end{aligned}$

18. $\begin{aligned} 10x - 5y + 8z &= 20 \\ x + 6y - z &= 8 \end{aligned}$

19. $\begin{aligned} w + 2x - 5y - z &= 0 \\ 3w \qquad + 2y + 7z &= 6 \\ -5x \qquad - 3z &= 1 \\ -6w + x + y - 4z &= 9 \end{aligned}$

20. $\begin{aligned} 7w - x - y \qquad &= -2 \\ -2w + 2x + 11y - z &= -10 \\ w + 4x + 2y + 6z &= 14 \\ -3w - x \qquad + 2z &= 5 \end{aligned}$

21. $\begin{aligned} w - x + y - z &= 0 \\ 4w + 2x - 2y \qquad &= 3 \\ 3x - y + 5z &= 7 \end{aligned}$

22. $\begin{aligned} 2w \qquad + 2z &= 10 \\ w + 5x \qquad &= -1 \\ 3x + y - 8z &= 16 \end{aligned}$

23. $\begin{aligned} x - y - z &= 0 \\ x + y + z &= 6 \\ x - 2y + z &= 0 \\ 2x + 3y - 6z &= 6 \end{aligned}$

24. $\begin{aligned} x + 3y + 4z &= 10 \\ -3x + y + 2z &= 6 \\ 5x - 7y + 10z &= 0 \\ -6y + 12z &= -4 \end{aligned}$

Find the distance from the point to the line.

25. $3x + 6y = -4$, $(-5,-2)$

26. $y = 5x + 1$, $(-1,8)$

The given three points are in a graph of the quadratic function $y = ax^2 + bx + c$. Find the function.

27. $(6,-5)$, $(8,-3)$, $(12,-6)$

28. $(1,9)$, $(5,4)$, $(9,7)$

29. The perimeter of an isosceles triangle is 54 centimeters. The base of the triangle is 5 centimeters less than the length of one of the equal sides. Find the lengths of the sides of the triangle.

30. An investor earns $4,660 annually from an investment of $36,000. Part of the $36,000 is invested in a high-risk real estate venture that yields 16%. The remainder of the $36,000 is in a mutual fund yielding 11%. How much of the investor's money is in real estate?

31. The intercept form of the equation of a straight line is $x/a + y/b = 1$, where $(a,0)$ and $(0,b)$ are the points of intersection of the line and the x and y axes, respectively. Write the intercept form of the equation of the line containing the points $(3,2)$ and $(-4,6)$.

32. How many liters of a 20% acid solution should be mixed with a 36% acid solution to produce 18 liters of a 25% acid solution?

33. Brand A cameras sell for $34.80. Brand B cameras sell for $39.45. One day's total sales in both cameras is $1,946.70. If 53 cameras were sold, how many were of brand A?

34. A man canoes six miles upstream in 45 minutes. He canoes the six miles back to his starting point in 24 minutes. Find the man's canoeing rate in still water. Also find the rate of the current.

35. The sum of three numbers is 513. The sum of the second two numbers is 33 less than the first number. Twice the second number is 18 more than the third number. Find the three numbers.

36. An oil tank has three inlet pipes, A, B, and C. The tank is filled in 2 hours if all three pipes are open. If A is closed and B and C open, the tank is filled in 4 hours. If B is closed and A and C are open, the tank is filled in 3 hours. How long does it take to fill the tank if only pipe A is open? How long does it take to fill the tank if only B is open?

37. An industrial landscaping firm determines that the phosphate-nitrogen mix that best promotes grass growth is 5% phosphate and 13% nitrogen. The firm's stock of commercial fertilizers contain brands A, B, and C. Brand A contains 3% phosphate and 15% nitrogen. Brand B contains 4% phosphate and 16% nitrogen. Brand C contains 7% phosphate and 10% nitrogen. How many tons of each brand of fertilizer should be mixed together to produce 30 tons of fertilizer that contains 5% phosphate and 13% nitrogen?

38. A manufacturer produces three models, A, B, and C, of a product. Two processes, I and II, are needed to produce each model. The process time needed for the manufacture of each model and the available process time per week are given in the following table.

Process	A	B	C	Available Time per Week
I	5	4	3	180 hours
II	8	5	5	250 hours

How many of each model can be produced, using much of the available processing time as is possible?

39. A pasta mix consists of three kinds of pasta, A, B, and C. The per pound selling prices of A, B, and C are \$1.10, \$1.40, and \$1.90, respectively. Twenty-five pounds of the mix is made to sell for \$1.50 per pound. The weight of pasta A in the mix is $\frac{3}{4}$ that of pasta B. How many pounds of each pasta is in the mix?

40. A collection of nickels, dimes, and quarters is worth \$4.80. There are 40 coins in the collection. How many coins of each type are in the collection?

Find the current in each conductor in the circuit.

41.

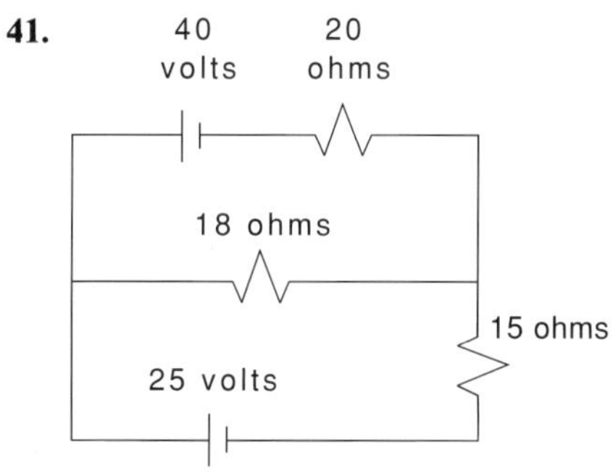

42.

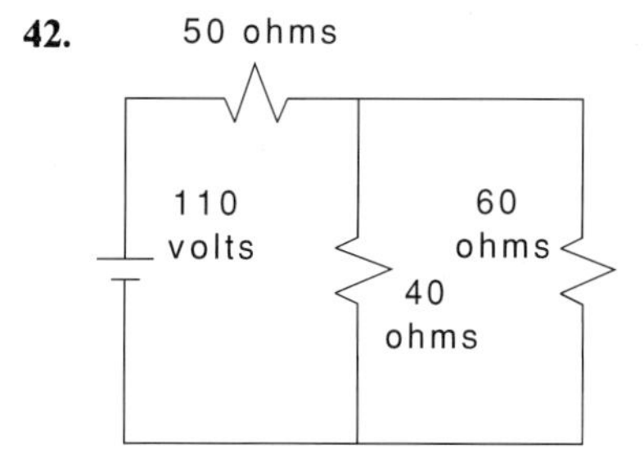

43. What is the minimum and maximum traffic flow that can be expected along *AD*?

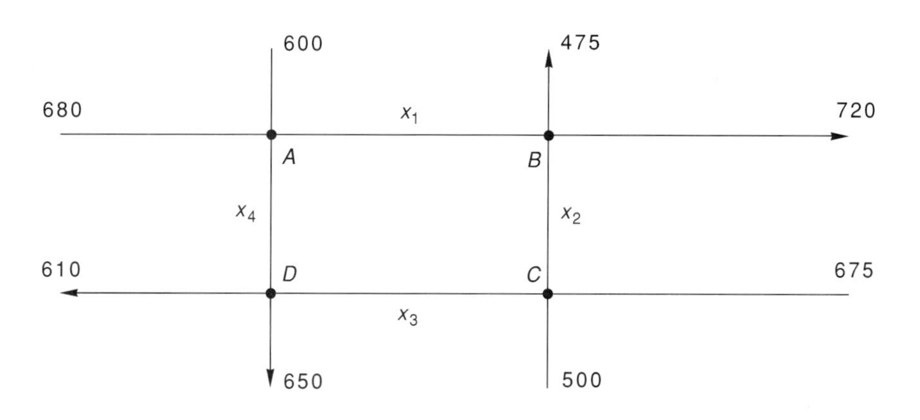

44. Find a relationship between a, b, and c that will permit the system to have a solution.

$$\begin{aligned} x + 2y &= a \\ 5x - y &= b \\ 2x + 3y &= c \end{aligned}$$

45. For what value of c does the system have a solution?

$$\begin{aligned} x + 2y - z &= -3 \\ 3x + y + 2z &= 2 \\ y + z &= 5 \\ x + cz &= 4 \end{aligned}$$

Chapter 3 Test 1

Solve each system with the method of linear combinations.

1. $\begin{aligned} 5x + 8y &= 12 \\ 3x - 5y &= -2 \end{aligned}$

2. $\begin{aligned} x + 2y - z &= -2 \\ 4x - 3y + 2z &= 9 \\ 2x + y + 6z &= 15 \end{aligned}$

3. $\begin{aligned} -3x + 5y + 2z &= 11 \\ 5x - y - z &= 1 \\ 9x + 3y + 8z &= 28 \end{aligned}$

Use matrices to solve each system.

4. $\begin{aligned} x + 5y + 4z &= 18 \\ -2x - y + 3z &= 6 \\ 3x + 2y - z &= 11 \end{aligned}$

5. $\begin{aligned} 2x - y + z &= 4 \\ 3x + 2y + 2z &= 10 \\ x - 4y &= 2 \end{aligned}$

6. $\begin{aligned} 3x - 8y + 5z &= 12 \\ 10x + 6y - 2z &= 21 \end{aligned}$

7. The perimeter of a rectangle is 210 feet. Its length is 2 feet less than three times its width. Find the dimensions of the rectangle.

8. One alloy is 25% silver. Another alloy is 40% silver. The two alloys are mixed to produce 45 ounces of an alloy that is 34% silver. How many ounces of each alloy are in the mix?

9. A collection of nickels, dimes, and quarters is worth \$4.20. The combined number of nickels and quarters is 15 less than the number of dimes. There are 2 more quarters than nickels. How many coins of each type are in the collection?

10. Find the quadratic function, $y = ax^2 + bx + c$, whose graph contains the points $(-4,-1)$, $(2,8)$, and $(6,-2)$.

Chapter 3 Test 2

Use linear combinations to solve each system.

1. $\begin{aligned} 7x - 4y &= 18 \\ 6x + 9y &= 33 \end{aligned}$

2. $\begin{aligned} -2x + 4y + 3z &= 13 \\ 5x - 3y + 2z &= 9 \\ 4x - 2y + 5z &= 15 \end{aligned}$

3. $\begin{aligned} 8x - 3y + 5z &= 13 \\ 6x + 2y - 3z &= 8 \\ 5x + 4y - 9z &= -11 \end{aligned}$

Use matrices to solve each system.

4. $\begin{aligned} 6x + 5y - 2z &= 11 \\ -3x + 2y + 7z &= -4 \\ 9x + 3y + 8z &= 28 \end{aligned}$

5. $\begin{aligned} w + 2x - y + 3z &= 14 \\ 3w - x - 2z &= -2 \\ 5x + 4y - 4z &= 12 \\ 7w + 6y &= 15 \end{aligned}$

6. $\begin{aligned} w + y - 2z &= 5 \\ x + 2y + 4z &= 8 \\ w - 2x + z &= 3 \end{aligned}$

7. A commercial airliner flies 2,600 miles across the country, from west to east, in 5 hours. The prevailing west to east wind decreases the airliner's flight time. The return flight takes 5.75 hours. What is the wind velocity and the average airspeed of the plane?

8. A food mix of 40 pounds is obtained from foods A, B, and C. Each pound of food A contains 8 units of protein and 12 units of carbohydrates. Each pound of food B contains 10 units of protein and 6 units of carbohydrates, and each pound of food C contains 6 units of protein and 10 units of carbohydrates. Each pound of the food mix contains 9 units of protein and 8 units of carbohydrates. How many pounds of each food are in the mix?

9. There are 21 coins in a collection of nickels, dimes, and quarters. The collection has twice as many dimes as nickels. What are the possible dollar values of the coin collection?

10. What value(s) can c assume so that the following system is consistent?

$$\begin{aligned} x - y + z &= 6 \\ 2x + 3z &= 12 \\ 2y - z &= -2 \\ 3x + y &= c \end{aligned}$$

Chapter 4

Polynomial Functions

In chapter 2 we addressed ideas such as the definition of a function, how to determine properties of a function from its equation, and how to graph a function with a minimum of point plotting. The same ideas are again addressed in this chapter, but for a specific class of functions called polynomial functions. We will also construct a compact theory for the solution of polynomial equations and introduce a numerical method for approximating irrational roots of polynomial equations.

4.1 The Arithmetic Operations on Polynomials

An equation such as

$$f(x) = a_n x^n + a_{n-1} x^{n-1} + \cdots + a_1 x + a_0, \quad a_n \neq 0, \tag{1}$$

in which n is a natural number and $a_0, a_1, \ldots, a_n$ are real numbers, defines a **polynomial function.** The right member of (1) is an ***n*th degree polynomial** that is written in **standard form.** The terms of the polynomial are ordered, left to right, from highest degree to lowest degree. We call a_n the **leading coefficient** and a_0 the **constant term** of the polynomial. The fourth degree polynomial,

$$5x^4 - 6x^2 + 2x + 1,$$

has 5 as its leading coefficient and 1 as its constant term. A fourth degree polynomial function is defined by the equation

$$f(x) = 5x^4 - 6x^2 + 2x + 1.$$

The values of x for which $f(x) = 0$ are the **solutions,** or **roots,** of the polynomial equation

$$0 = 5x^4 - 6x^2 + 2x + 1.$$

A number of interesting results concerning the roots of polynomial equations are stated as theorems in the first three sections of this chapter. Taken together, the theorems belong to a compact body of mathematics known as "The Theory of Equations." Much of the theory was developed between the fifteenth and nineteenth centuries.

Polynomial functions are called the "simplest" functions in mathematics because they are the easiest to evaluate and the easiest to manipulate algebraically.

The Arithmetic of Polynomials

Recall that polynomials are added (subtracted) by adding (subtracting) *like terms*. If $f(x) = 4x^4 + 3x^2 - x - 8$ and $g(x) = 7x^3 + 2x^2 + 5x - 1$, then

$$\begin{aligned} f(x) + g(x) &= (4x^4 + 3x^2 - x - 8) + (7x^3 + 2x^2 + 5x - 1) \\ &= (4 + 0)x^4 + (0 + 7)x^3 + (3 + 2)x^2 \\ &\quad + (-1 + 5)x + (-8 + (-1)) \\ &= 4x^4 + 7x^3 + 5x^2 + 4x - 9, \end{aligned}$$

and

$$\begin{aligned} f(x) - g(x) &= (4 - 0)x^4 + (0 - 7)x^3 + (3 - 2)x^2 \\ &\quad + (-1 - 5)x + (-8 - (-1)) \\ &= 4x^4 - 7x^3 + x^2 - 6x - 7. \end{aligned}$$

The product of two polynomials is obtained by multiplying each term of one polynomial by every term of the other polynomial and then combining like terms.

$$\begin{aligned} (2x + 3)(x^2 - 5x + 2) &= 2x^3 - 10x^2 + 4x + 3x^2 - 15x + 6 \\ &= 2x^3 - 7x^2 - 11x + 6. \end{aligned}$$

If each of the factor polynomials is of degree 2 or greater, you may wish to find their product as follows:

1. Write each polynomial in standard form.
2. Multiply the highest degree terms of the factor polynomials together to find the highest degree term of the product.
3. Find the coefficients of the next highest degree term down to the constant term, in order.

$$\begin{aligned} &(3x^4 - 5x^2 + 4x)(x^3 - 2x^2 - 3x + 3) \\ &= 3x^7 - 6x^6 + (-9 - 5)x^5 + (9 + 10 + 4)x^4 + (15 - 8)x^3 \\ &\quad + (-15 - 12)x^2 + 12x \\ &= 3x^7 - 6x^6 - 14x^5 + 23x^4 + 7x^3 - 27x^2 + 12x \end{aligned}$$

The last polynomial illustrates what is always the case, that the degree of the product polynomial is the *sum of the degrees of the factor polynomials.*

It is shown above, that the sum, difference, or product of two polynomials is a polynomial. However, division of one polynomial by another need not yield a polynomial.

Example 1 Use the long division process to divide $x^3 + 3x^2 - 7x + 2$ by $x^2 + 2x + 1$.

Solution

$$\begin{array}{r|l} & x + 1 \\ \hline x^2 + 2x + 1 & x^3 + 3x^2 - 7x + 2 \\ & \underline{x^3 + 2x^2 + x} \\ & \quad x^2 - 8x + 2 \\ & \quad \underline{x^2 + 2x + 1} \\ & \qquad -10x + 1 \end{array}$$

Hence, $x^3 + 3x^2 - 7x + 2 = (x^2 + 2x + 1)(x + 1) + (-10x + 1)$. ■

In example 1, $x^2 + 2x + 1$ is the **divisor,** $x^3 + 3x^2 - 7x + 2$ is the **dividend,** $x + 1$ is the **quotient,** and $-10x + 1$ is the **remainder.** The long division process is carried forward until the remainder is either zero or a polynomial of lower degree than the divisor polynomial. *If the remainder is zero, then the quotient and the divisor are factors of the dividend.* Example 1 gives us a specific instance of the "Division Algorithm for Polynomials," which is stated next without proof.

Theorem 4.1 (Division Algorithm for Polynomials) If $M(x)$ and $N(x)$ are polynomials and $N(x) \neq 0$, then there are *unique* polynomials $Q(x)$ and $R(x)$, such that

$$M(x) = N(x) \cdot Q(x) + R(x).$$

Either $R(x) = 0$ or the degree of $R(x)$ is less than the degree of $N(x)$.

Synthetic Division

Long division can be replaced by a simpler process called synthetic division if the divisor is a first degree polynomial of the form $x - c$, where c is a rational number.[1] We shall employ long division and then synthetic division to divide $2x^3 - 5x^2 + 3x + 6$ by $x - 3$.

$$\begin{array}{r|l} & 2x^2 + x + 6 \\ \hline x - 3 & 2x^3 - 5x^2 + 3x + 6 \\ & \underline{2x^3 - 6x^2} \\ & \qquad x^2 + 3x \\ & \qquad \underline{x^2 - 3x} \\ & \qquad\qquad 6x + 6 \\ & \qquad\qquad \underline{6x - 18} \\ & \qquad\qquad\qquad 24 \end{array}$$

1 Synthetic division can be performed if c is a complex number. We will not do synthetic division with complex numbers in this text.

Thus, $2x^3 - 5x^2 + 3x + 6 = (x - 3)(2x^2 + x + 6) + 24$. Since $24 = 24x^\circ$, the remainder can be viewed as a zero degree polynomial that is of lower degree than the divisor polynomial, $x - 3$.

The synthetic division algorithm proceeds as follows:

1. List the coefficients of the dividend in the order of descending powers of x, entering 0 for any missing power of x.

$$2 \quad -5 \quad 3 \quad 6$$

2. Prefix this listing by the value of x that makes the divisor zero. In this case $x = 3$ is the prefix.

$$3\rfloor \; 2 \quad -5 \quad 3 \quad 6$$

3. Bring down the leading coefficient in the listing, multiply it by the prefix, and add the product to the next coefficient in the list.

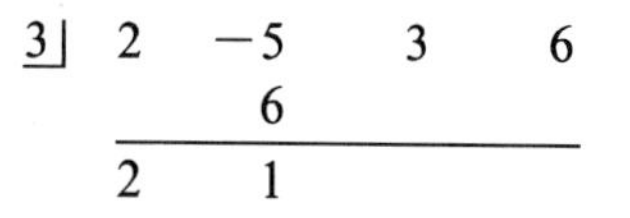

3⌋	2	−5	3	6
		6		
	2	1		

4. Multiply by the prefix the sum obtained in step 3 and add the product to the next coefficient. Repeat this step until all the coefficients in the listing are used.

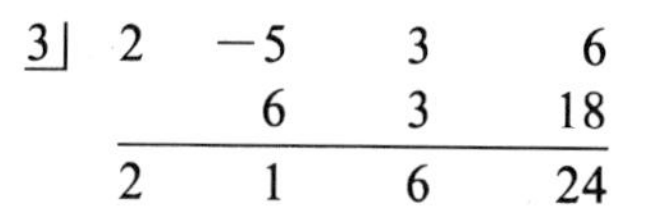

3⌋	2	−5	3	6
		6	3	18
	2	1	6	24

5. All the entries in the third row except the last are the coefficients of the quotient polynomial, arranged in descending powers. The last entry in the third row is the remainder.

$$Q(x) = 2x^2 + x + 6, \quad R(x) = 24$$

Example 2 Use synthetic division to divide $3x^4 - 2x^2 + 6x - 5$ by $x + 2$.

Solution

−2⌋	3	0	−2	6	−5
		−6	12	−20	28
	3	−6	10	−14	23

Hence, $3x^4 - 2x^2 + 6x - 5 = (x + 2)(3x^3 - 6x^2 + 10x - 14) + 23$. ■

To see why synthetic division works, let us go back to the problem that was used to introduce the algorithm.

Long Division **A**

$$\begin{array}{r} 2x^2 + x + 6 \\ x - 3 \overline{) \, \textcircled{2}x^3 - 5x^2 + 3x + 6} \\ \underline{2x^3 - 6x^2 } \\ \textcircled{1}x^2 + 3x + 6 \\ \underline{x^2 + 3x } \\ \textcircled{6}x + 6 \\ \underline{6x - 18} \\ \textcircled{24} \end{array}$$

Synthetic Division **B**

3	2	−5	3	6
		6	3	18
	2	1	6	24

The term by term construction of the quotient in **A** is designed to eliminate, at each step in the long division process, the highest degree term in the remainder polynomial. The remainder polynomials in the first three steps are $2x^3 - 5x^2 + 3x + 6$, $x^2 + 3x + 6$, and $6x + 6$. The leading coefficient of each of these polynomials (circled in **A**) appear as the first three numbers in the third line of **B.**

Method **B** synthesizes method **A.** The variable is unnecessary. The long division process is carried forward by the leading coefficients of the remainder polynomials. The first term in the third line of **B** has to be 2 because the leading coefficient of the divisor is 1. If the divisor was $2x - 3$, the leading coefficient of the quotient would be 1. Since method **B** is applicable only if the divisor has the form $x - c$, the first number in the third line will always be the leading coefficient of the dividend (the first remainder polynomial). In **A** the leading coefficient of the second remainder polynomial, 1, is obtained by multiplying 2 by *−3* and *subtracting* the result from −5. In **B,** that number is obtained by multiplying 2 by *3* and *adding* the result to −5. How is the leading coefficient 6, of the third remainder polynomial, obtained in **A?** How is the corresponding number obtained in **B?** How is the remainder 24 computed in **A** and in **B?**

Example 3 Show that $x + 1$ is a factor of $x^5 + 1$.

Solution If the remainder is 0 when $x^5 + 1$ is divided by $x + 1$, then $x + 1$ is a factor of $x^5 + 1$.

−1	1	0	0	0	0	1
		−1	1	−1	1	−1
	1	−1	1	−1	1	0

$$x^5 + 1 = (x + 1)(x^4 - x^3 + x^2 - x + 1)$$

■

Exercise Set 4.1

(A)

Let $P(x)$ be the first polynomial and $Q(x)$ be the second polynomial. Find (a) $P(x) + Q(x)$, (b) $P(x) - Q(x)$, and (c) $P(x) \cdot Q(x)$.

1. $x - 6, 3x - 8$
2. $5x + 2, -6x + 7$
3. $2x^2 + 4x - 3, 6x + 1$
4. $2x + 2, -x^2 + 4x - 1$
5. $-8x + 3, -5x^2 + 3x - 9$
6. $10x^2 + 4, 5x + 7$
7. $x^2 + 4, 3x^2 + 2x - 1$
8. $2x^2 - 1, 6x^2 - 4x + 12$
9. $7x^2 - 2x - 3, -x^2 - 5x + 8$
10. $-3x^2 + 10x + 12, 6x^2 - 4x + 12$
11. $x^3 + 7x^2 - 5x + 4, x^2 - 2x + 5$
12. $5x^2 + x - 9, 3x^3 - 6x^2 - 10x + 15$
13. $4x^3 - 5x + 8, 10x^3 - 8x^2 + 6$
14. $x^3 + 9x^2 + 3x - 8, x^3 - 2x^2 + x + 4$
15. $2x^3 - 5x^2 + 8x + 3, 12x^3 + x^2 + 5x - 10$
16. $8x^3 - 1, 15x^3 - 11x^2 + 4x + 3$

Let $M(x)$ be the first polynomial and $N(x)$ be the second polynomial. (a) Use long division in exercises 17–24 and synthetic division in exercises 25–30 to divide $M(x)$ by $N(x)$. (b) For exercises 17–30, write $M(x)$ as $N(x) \cdot Q(x) + R(x)$, where $Q(x)$ is the quotient and $R(x)$ is the remainder.

17. $6x^2 - x - 2, 2x + 1$
18. $5x^2 + 4x - 9, x - 3$
19. $8x^3 - 1, 2x - 1$
20. $x^3 + 27, x + 3$
21. $8x^4 + 12x^3 - 6x^2 + 8x + 3, 4x^2 + 6x - 1$
22. $3x^4 - 7x^3 - 10x^2 + x - 4, x^2 + 5x$
23. $6x^4 - 2x^2 + 5, 3x^2 + 4x + 3$
24. $12x^4 + 8x^3 + 4x^2 + 7x - 3, 2x^2 - 13x - 8$
25. $x^3 + 5x^2 - 7x + 2, x + 3$
26. $4x^3 - 6x^2 + 11x + 10, x - 2$
27. $6x^3 - 10x, x - 1$
28. $5x^3 + x^2 + 9, x + 5$
29. $8x^4 - 32x^3 + 27x - 108, x - 4$
30. $-x^4 + 6x^3 - 18x - 35, x - 5$

(B)

31. Find the coefficients of x^{n+m}, x^{n+m-1}, and x^{n+m-2} in the product
$$(a_nx^n + a_{n-1}x^{n-1} + \cdots + a_0) \cdot$$
$$(b_mx^m + b_{m-1}x^{m-1} + \cdots + b_0).$$
32. How must $M(x) = 4x^3 - 3x^2 + 6x - 10$ and $N(x) = 2x - 6$ be altered so that $M(x)$ can be divided by $N(x)$ synthetically? Perform the synthetic division.
33. Repeat exercise 32 for $M(x) = 3x^3 - 6x^2 - 8x + 4$ and $N(x) = 2x - 1$.
34. Show that $x - 1$ is a factor of $x^6 - 1$.
35. Show that $x - 1$ is a factor of $x^n - 1, n \in N$.
36. Show that $x + 1$ is a factor of $x^n + 1, n \in N$, only if n is odd.

4.2 Zeros of Polynomial Functions I

Values of x for which $F(x) = 0$ in the equation

$$F(x) = a_nx^n + a_{n-1}x^{n-1} + \cdots + a_1x + a_0,\ a_n \neq 0 \tag{1}$$

are called zeros of the polynomial function defined by (1). These x-values are solutions, or roots, of the nth degree polynomial equation

$$0 = a_nx^n + a_{n-1}x^{n-1} + \cdots + a_1x + a_0. \tag{2}$$

Use synthetic division, as in example 3, to write $F(x)$ in the form

$$F(x) = a_n(x - r_1)(x - r_2) \cdots (x - r_n).$$

You may assume that each polynomial function has integer zeros that are not less than -5 or greater than 5.

17. $F(x) = x^3 - x^2 - 4x + 4$

18. $F(x) = x^3 + 6x^2 + 11x + 6$

19. $F(x) = x^3 + 3x^2 - 4$

20. $F(x) = x^3 - 11x^2 + 39x - 45$

21. $F(x) = x^3 + 9x^2 + 27x + 27$

22. $F(x) = x^3 - 15x^2 + 75x - 125$

23. $F(x) = x^4 + 2x^3 - 11x^2 - 12x + 36$

24. $F(x) = x^4 + 2x^3 - 7x^2 - 8x + 12$

25. $F(x) = x^4 + 11x^3 + 33x^2 + 5x - 50$

26. $F(x) = x^4 + 4x^3 - 13x^2 - 64x - 48$

(B)

Use synthetic division to show that each equation has a root between the given pair of integers.

27. $2x^3 + 7x^2 + 2x - 6 = 0; -2, -1$

28. $3x^3 - 4x^2 + 2x + 4 = 0; -1, 0$

29. $6x^4 - 11x^3 + 15x^2 - 22x + 6 = 0; 1, 2$

30. $x^5 - 8x^3 - 4x^2 + 3x - 2 = 0; 3, 4$

31. Prove: If $x - c$ is a factor of a polynomial function $F(x)$, then $F(c) = 0$.

32. Prove: If each coefficient of

$$F(x) = a_nx^n + a_{n-1}x^{n-1} + a_{n-2}x^{n-2} + \cdots + a_1x + a_0$$

is positive, then equation $F(x) = 0$ has no positive real solutions.

4.3 Zeros of Polynomial Functions II

From theorem 4.4 we have that an nth degree polynomial function,

$$F(x) = a_nx^n + a_{n-1}x^{n-1} + \cdots + a_o, a_n \neq 0$$

can be expressed as a product of n linear factors.

$$F(x) = a_n(x - r_1)(x - r_2) \cdots (x - r_n) \tag{1}$$

Because $F(r_1) = F(r_2) = \ldots = F(r_n) = 0$, the equation $F(x) = 0$ has roots $r_1, r_2, \ldots, r_n$. The roots need not be distinct. If a root occurs k times, it is called **a root of multiplicity k.**

Example 1 $F(x) = 4x^3 - 4x^2 - 32x + 48 = 4(x - 2)(x - 2)(x + 3)$. The roots of $F(x) = 0$ are 2 and -3, with 2 a root of multiplicity 2.

Since an nth degree polynomial function can be written in the form of (1), it must have at least n zeros, $r_1, r_2, \ldots, r_n$, if a zero of multiplicity k is counted k times. Can an nth degree polynomial function have more than n zeros? The answer is no. It can be shown that, aside from the order in which the factors appear, (1) is the only way to express an nth degree polynomial function as a product of linear factors.

Theorem 4.5 An nth degree polynomial function has exactly n zeros if zeros of multiplicity k are counted k times.

Proof Let F be an nth degree polynomial function. By theorem 4.4, F can be represented as

$$F(x) = a_n(x - r_1)(x - r_2) \cdots (x - r_n).$$

Assume that there is a number r different from $r_1, r_2, \ldots,$ and r_n such that $F(r) = 0$. Then

$$0 = F(r) = a_n(r - r_1)(r - r_2) \cdots (r - r_n) \neq 0.$$

The assumption that r is a zero of F leads to a contradiction. Therefore the only zeros of F are $r_1, r_2, \ldots, r_n$.

Example 2 Write a polynomial equation whose roots are -3, 4, and the double root $\frac{1}{2}$.

Solution Let $F(x) = 0$ in (1) and replace r_1, r_2, r_3, and r_4 by -3, 4, $\frac{1}{2}$, and $\frac{1}{2}$ respectively.

$$(x + 3)(x - \tfrac{1}{2})(x - \tfrac{1}{2})(x - 4) = 0$$

Multiplying the factors together gives us

$$x^4 - 2x^3 - \tfrac{43}{4}x^2 + \tfrac{47}{4}x - 3 = 0.$$

Fractional coefficients can be eliminated if both sides of the last equation are multiplied by 4.

$$4x^4 - 8x^3 - 43x^2 + 47x - 12 = 0$$ ■

Among the zeros of a polynomial function F there may be complex numbers. It turns out that the complex zeros must come in complex conjugate pairs. If $a + bi$ is a zero of F, so is $a - bi$. Equation $x^2 - 4x + 5 = 0$ illustrates this fact. Verify that the roots of the equation are $2 + i$ and $2 - i$.

Theorem 4.6 The complex zeros of a polynomial function F, of degree $n > 1$, occur in complex conjugate pairs.

Proof Assume $a + bi$, $b \neq 0$, is a zero of F. Then $[x - (a + bi)]$ is a factor of F. Construct the polynomial

$$B(x) = [x - (a + bi)][x - (a - bi)] = x^2 - 2ax + a^2 + b^2.$$

If it is shown that B is a factor of F, then $[x - (a - bi)]$ would also be a factor of F (exercise 29). It would follow that $a - bi$ is a zero of F. By the Division Algorithm (theorem 4.1),

$$F(x) = B(x) \cdot Q(x) + R(x),$$

where R must be of lower degree than B. Let $R(x) = cx + d$. Now,

$$\begin{aligned} F(a + bi) &= [(a + bi) - (a + bi)][(a + bi) - (a - bi)] \\ &\quad Q(a + bi) + c(a + bi) + d \\ 0 &= 0[2bi]Q(a + bi) + (ca + d) + cbi \\ 0 + 0i &= (ca + d) + cbi. \end{aligned}$$

Hence, $0 = ca + d$ and $0 = cb$. But $b \neq 0$. Therefore $c = 0$, and it must follow that $d = 0$. Consequently $R(x) = cx + d = 0$. With the remainder zero, B is a factor of F.

Finding Rational Zeros

Not one of the theorems 4.1 through 4.6 tells us how to find the n zeros of an nth degree polynomial function. Only theorem 4.6 says anything about the nature of those zeros—that complex zeros come in complex conjugate pairs. We do not as yet know how many of the zeros are real and how many are complex, how many are positive and how many are negative. Because of theorem 4.6 we can say that a third degree polynomial can have three real zeros, or one real and two complex zeros. (Why?) A fourth degree polynomial can have four real zeros, or two real and two complex zeros, or four complex zeros, etc. But knowing how many of each type of zero that a polynomial function can have isn't nearly as satisfying as finding the zeros.

Theorem 4.7 (Rational Roots Theorem) Let the coefficients of $F(x) = a_n x^n + a_{n-1}x^{n-1} + \cdots + a_0$, $a_n \neq 0$, be integers. If p/q is a rational number zero of F, and if p and q have no factors in common, then p is a factor of a_0 and q is a factor of a_n.

Proof Because p/q is a zero of F we may write

$$a_n\left(\frac{p}{q}\right)^n + a_{n-1}\left(\frac{p}{q}\right)^{n-1} + \cdots + a_1\left(\frac{p}{q}\right) + a_0 = 0.$$

Multiplying both sides of the equation by q^n yields

$$a_n p^n + a_{n-1}p^{n-1}q + \cdots + a_1 pq^{n-1} + a_0 q^n = 0$$

or

$$a_n p^n + a_{n-1}p^{n-1}q + \cdots + a_1 pq^{n-1} = -a_0 q^n. \tag{2}$$

Factoring out p on the left side gives us

$$p(a_n p^{n-1} + a_{n-1}p^{n-2}q + \cdots + a_1 q^{n-1}) = -a_0 q^n. \tag{3}$$

The left side of (3) has p as a factor. Hence, p must be a factor of $-a_0 q^n$ also. But p has no factor in common with q. It therefore cannot be a factor of q^n. We must conclude that p is a factor of a_0.

Writing (2) in the form

$$-a_n p^n = a_{n-1} p^{n-1} q + \cdots + a_1 p q^{n-1} + a_0 q^n,$$

and factoring out q on the right, produces

$$-a_n p^n = (a_{n-1} p^{n-1} + \cdots + a_1 p q^{n-2} + a_0 q^{n-1}) q.$$

The right side has q as a factor so that q must also be a factor of $-a_n p^n$. Since p and q have no factors in common, q cannot be a factor of p^n. Therefore, q must be a factor of a_n.

The last theorem opens the door. If a polynomial function has rational zeros, they will be of the form p/q, where p is a factor of the constant term, and q is a factor of the leading coefficient. We need only construct all the possible rational zeros, and then use synthetic division to find the actual rational zeros.

Example 3 Find the rational zeros of $F(x) = 4x^3 + 8x^2 - x - 2$.

Solution If F has rational zeros, they must be of the form p/q, where p is a factor of -2 and q is a factor 4.

$$p \in \{\pm 1, \pm 2\}, \quad q \in \{\pm 1, \pm 2, \pm 4\}, \quad \text{and} \quad \frac{p}{q} \in \left\{\pm\frac{1}{4}, \pm\frac{1}{2}, \pm 1, \pm 2\right\}$$

	4	8	−1	−2
$\frac{1}{4}$	4	9	$\frac{5}{4}$	$-\frac{27}{16}$
$\frac{1}{2}$	4	10	4	0

As $\frac{1}{2}$ is a zero of F, we may write

$$4x^3 + 8x^2 - x - 2 = (x - \tfrac{1}{2})(4x^2 + 10x + 4).$$

The other zeros of F can be found by solving $4x^2 + 10x + 4 = 0$. They are $-\frac{1}{2}$ and -2. The zeros of F are -2, $-\frac{1}{2}$, and $\frac{1}{2}$. ■

Example 4 Find the rational zeros of $G(x) = x^3 - x^2/6 - 11x/9 - \frac{4}{9}$.

Solution Theorem 4.7 is applicable to polynomials with *integer* coefficients. Hence, we write

$$18G(x) = 18x^3 - 3x^2 - 22x - 8$$

and note that the zeros of $18G$ are also the zeros of G.

$$p \in \{\pm 1, \pm 2, \pm 4, \pm 8\}, \quad q \in \{\pm 1, \pm 2, \pm 3, \pm 6, \pm 9, \pm 18\}, \quad \text{and}$$

$$\tfrac{p}{q} \in \{\pm\tfrac{1}{18}, \pm\tfrac{1}{9}, \pm\tfrac{1}{6}, \pm\tfrac{2}{9}, \pm\tfrac{1}{3}, \pm\tfrac{4}{9}, \pm\tfrac{1}{2}, \pm\tfrac{2}{3}, \pm\tfrac{8}{9}, \pm 1, \pm\tfrac{4}{3}, \pm 2, \pm\tfrac{8}{3}, \pm 4, \pm 8\}$$

Rather than test each of the 30 possible rational zeros of G, we proceed as follows:

	18	-3	-22	-8
1	18	15	-7	-15
2	18	33	44	80

Division by $x - 1$ and $x - 2$ yield remainders of opposite sign. Therefore, G has a zero between 1 and 2. We test $\frac{4}{3}$, the only possible rational root between 1 and 2.

	18	-3	-22	-8
1	18	15	-7	-15
2	18	33	44	80
$\frac{4}{3}$	18	21	6	0

Had we not found that $G(x)$ changes sign between 1 and 2, we would have tested -1 and -2 to see if either number is a zero, or if $G(x)$ changes sign between -1 and -2. Now we have that

$$18\,G(x) = (x - \tfrac{4}{3})(18x^2 + 21x + 6).$$

The remaining zeros of G, $-\frac{2}{3}$ and $-\frac{1}{2}$, are obtained by solving $18x^2 + 21x + 6 = 0$. ■

Shortening the Search for Rational Zeros

The next three theorems greatly reduce the computation required to find the zeros of a polynomial.

Theorem 4.8 (Upper Bound Theorem) If synthetic division is used to divide a polynomial function F by $x - c$, where $c > 0$, and a line of *non-negative coefficients* is obtained, then F can have no zero greater than c.

Proof $F(x) = (x - c)Q(x) + R(x)$, by theorem 4.1. Also, the coefficients of Q and R are, by hypothesis, non-negative. If p is any number greater than c, then $p - c > 0$, and $Q(p) > 0$ because the coefficients of Q are non-negative. R is the last non-negative number in the line of coefficients obtained when dividing F synthetically by $x - c$. As a result, $F(p) = (p - c)Q(p) + R > 0$ and so p cannot be a zero of F.

The conclusion of theorem 4.8 is nicely illustrated by the synthetic division performed in example 4.

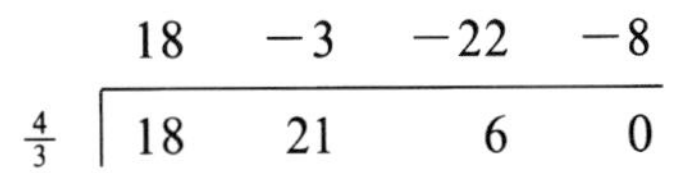

	18	-3	-22	-8
$\frac{4}{3}$	18	21	6	0

The division gives us $F(x) = (x - \frac{4}{3})(18x^2 + 21x + 6) + 0$. If p is any number greater than $\frac{4}{3}$, then $F(p) = (p - \frac{4}{3})(18p^2 + 21p + 6) > 0$. Hence, any number p that is greater than $\frac{4}{3}$ cannot be a zero of F. Because of theorem 4.8, the search for positive rational zeros of F stops at $\frac{4}{3}$. But what of the negative rational zeros? How do we limit that search?

Theorem 4.9 (Lower Bound Theorem) If a polynomial function F is divided synthetically by $x - c$, $c < 0$, and a line of coefficients having alternating signs is obtained, then F can have no zero less than c.

Example 5 Find a lower bound for the negative zeros of $2x^3 + 4x^2 - 2x - 5$.

Solution

	2	4	−2	−5
−1	2	2	−4	−1
−2	2	0	−2	−1
−3	2	−2	4	−17

The line of coefficients associated with -3 alternate in sign. Hence, -3 is a lower bound for the zeros of F. ■

Finally, we turn to a result obtained by Rene Descartes, a seventeenth century mathematician and philosopher, that enables us to determine the possible number of positive, negative, and imaginary zeros of a polynomial F. The determination is based on the number of changes in sign of $F(x)$ and $F(-x)$. Reading $4x^3 + 5x^2 - 23x - 6$ from highest to lowest degree terms, there is only one change of sign. It is between $+5x^2$ and $-23x$. However, $F(-x) = -4x^3 + 5x^2 + 23x - 6$ has two changes of sign. One change is between $-4x^3$ and $+5x^2$, and the other is between $+23x$ and -6.

Theorem 4.10 (Descartes' Rule of Signs) The number of positive zeros of a polynomial function F is either equal to the number of sign changes in $F(x)$, or it is less than the number of sign changes by an even number. The number of negative zeros of F is either equal to the number of sign changes in $F(-x)$, or it is less than the number of sign changes by an even number.

Recall that the graph of $F(-x)$ is a reflection, with respect to the y-axis, of the graph of $F(x)$. Thus, the negative zeros of $F(x)$ are the positive zeros of $F(-x)$. It is for this reason that the sign test method of theorem 4.10 is exactly the same for counting positive and negative zeros.

The next example is designed as an application of the theorems of this section to finding the rational zeros of an nth degree polynomial function. The procedure followed in the example is outlined below.

Procedure	**Theorems Used**
1. Determine the number of, and the nature of, the zeros.	4.5 An nth degree polynomial has n zeros. 4.6 Complex zeros occur in complex conjugate pairs. 4.10 Descartes' Rule of Signs.
2. List the possible rational zeros.	4.7 With p and q relatively prime, p/q can be a zero only if p is a factor of the constant term and q is a factor of the leading coefficient.
3. Use synthetic division to test possible rational roots, p/q. Do not test rational numbers that are greater than an upper bound, or less than a lower bound, for the zeros of the polynomial.	4.8 A positive rational number that produces a line of non-negative numbers in the synthetic division process is an upper bound for the zeros. 4.9 A negative rational number that in the synthetic division process produces a line of numbers that alternate in sign is a lower bound for the zeros.

Example 6 Find the zeros of $F(x) = 2x^4 - 5x^3 + 11x^2 - 20x + 12$.

Solution $F(x)$ has four sign changes. According to Descartes' Rule of Signs, F has 4, 2, or 0 positive zeros.

$$\begin{aligned} F(-x) &= 2(-x)^4 - 5(-x)^3 + 11(-x)^2 - 20(-x) + 12 \\ &= 2x^4 + 5x^3 + 11x^2 + 20x + 12 \end{aligned}$$

Since $F(-x)$ has no sign changes, F has no negative zeros. Because F is a fourth degree polynomial, it must have four zeros. The possible nature of those zeros is shown in the table.

Positive Zeros	Negative Zeros	Complex Zeros
4	0	0
2	0	2
0	0	4

If F has rational zeros, they must be of the form p/q where $q \in \{1,2\}$ and $p \in \{1,2,3,4,6,12\}$. Hence, $p/q \in \{\frac{1}{2}, 1, \frac{3}{2}, 2, 3, 4, 6, 12\}$.

	2	−5	11	−20	12	
1	2	−3	8	−12	0	← 1 is a zero
1	2	−1	7	−5		← 1 is not a double zero
						← zero between 1 and 2
2	2	1	10	8		← 2 is an upper bound
$\frac{3}{2}$	2	0	8	0		← $\frac{3}{2}$ is a zero

At this point we have

$$F(x) = (x - 1)(x - \tfrac{3}{2})(2x^2 + 8).$$

Solving $2x^2 + 8 = 0$, we find the remaining zeros to be $-2i$ and $2i$. ■

Exercise Set 4.3

(A)

Write a polynomial equation that has the given roots.

1. $-5, -3, 2$
2. $-4, 0, 1$
3. $-1, -1, 2, 2$
4. $1, 3, 3, 3$
5. $-2, -2, 1 + i, 1 - i$
6. $2, 5, -3i, 3i$

Use example 6 as a guide in finding the zeros of each polynomial.

7. $x^3 - x^2 - 4x + 4$
8. $x^3 - 7x + 6$
9. $x^3 + 6x^2 + 11x + 6$
10. $x^3 - 9x^2 + 23x - 15$
11. $4x^3 - x^2 - 2x + \frac{3}{4}$
12. $\frac{2x^3}{3} - x^2 - \frac{8x}{3} - 11$
13. $3x^3 + 2x^2 + 12x + 8$
14. $3x^3 + 8x^2 - x - 20$
15. $x^4 + 2x^3 - 7x^2 - 8x + 12$
16. $x^4 + 5x^3 - 10x^2 - 20x + 24$
17. $3x^4 - 11x^3 + 9x^2 + 13x - 10$
18. $x^4 + x^3 - 5x^2 - 15x - 18$
19. $x^3 + \frac{7x^2}{12} - \frac{7x}{24} - \frac{1}{8}$
20. $x^3 - \frac{x^2}{6} - \frac{11x}{9} - \frac{4}{9}$
21. $6x^4 + 13x^3 + 24x^2 - 8$
22. $4x^4 + 20x^3 + 97x^2 + 84x + 20$
23. $3x^5 - x^4 - 23x^3 + 25x^2 - 4$
24. $2x^5 - 6x^4 - x^2 + 4x + 1$

(B)

Show that each equation has no rational roots.

25. $2x^4 - 3x^3 - 8x^2 - 5x - 3 = 0$
26. $2x^4 + 2x^3 + 9x^2 - x - 5 = 0$

27. Show that $\sqrt{3}$ is not a rational number. (Hint: Consider the equation $x^2 = 3$.)

28. Show that $\sqrt[3]{5}$ is not a rational number.

29. The proof of theorem 4.6 depends upon the truth of the statement, "If A, B, and F are polynomial functions, and if B is a factor of F and A is a factor of B, then A is a factor of F." Prove this statement.

30. In the proof of theorem 4.7 we say, "If p and q have no factors in common, then p cannot be a factor of q^n." Prove this statement. (Hint: If two integers have no factors in common, their prime factorizations have no common factors. Consider the prime factorization of q^n.)

31. A proof of the Lower Bound theorem that is similar to the proof of the Upper Bound theorem can be constructed as follows: The division algorithm permits us to write

$$F(x) = (x - c)Q(x) + R.$$

a. Assume that the leading coefficient of $F(x)$ is positive. Then the leading coefficient of $Q(x)$ must be positive. (Why?)

b. Let $Q(x) = b_n x^n + b_{n-1}x^{n-1} + \cdots + b_1 x + b_0$ and consider two cases: n is even and n is odd. By hypothesis, the coefficients of $Q(x)$ alternate in sign. Hence, if n is even, $b_0 > 0$ and $R < 0$. If n is odd, $b_0 < 0$ and $R > 0$. Why?

c. If p is a real number less than c, then

$$F(p) = (p - c)Q(p) + R,$$

where $p - c < 0$. If n is even, $Q(p) > 0$, $R < 0$, and it follows that $F(p) < 0$. Why?

d. If n is odd, $Q(p) < 0$, $R > 0$, and it follows that $F(p) > 0$. Why?

e. We conclude that any real number p that is less than c cannot be a zero of $F(x)$. Why?

f. Essentially the same proof holds in the case that the leading coefficient of $F(x)$ is negative?

4.4 Graphing Polynomials

A polynomial function

$$F(x) = a_n x^n + a_{n-1}x^{n-1} + \cdots + a_1 x + a_0,\ a_n \neq 0$$

may be graphed quickly and accurately by focusing on its zeros and on its functional values for large positive and negative values of x. Suppose that we want to graph $F(x) = x^2 + 3x - 10$. Because

$$x^2 + 3x - 10 = (x + 5)(x - 2),$$

the zeros of F, or the x-intercepts of the graph, are 2 and -5. The y-intercept is -10. A sign graph reveals that $F(x) > 0$ if $x < -5$ or $x > 2$, and that $F(x) < 0$ if $-5 < x < 2$.

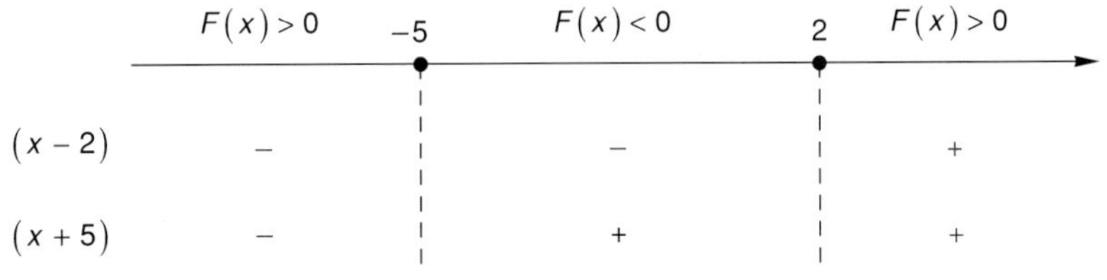

	$F(x)>0$	-5	$F(x)<0$	2	$F(x)>0$
$(x-2)$	$-$		$-$		$+$
$(x+5)$	$-$		$+$		$+$

To analyze how the polynomial behaves for large positive and negative x we write

$$x^2 + 3x - 10 = x^2\left(1 + \frac{3}{x} - \frac{10}{x^2}\right).$$

As $|x|$ assumes greater and greater values, $3/x$ and $-10/x^2$ assume values closer and closer to zero (see the table). Consequently, if $|x|$ is a large number, $F(x) \approx x^2$.

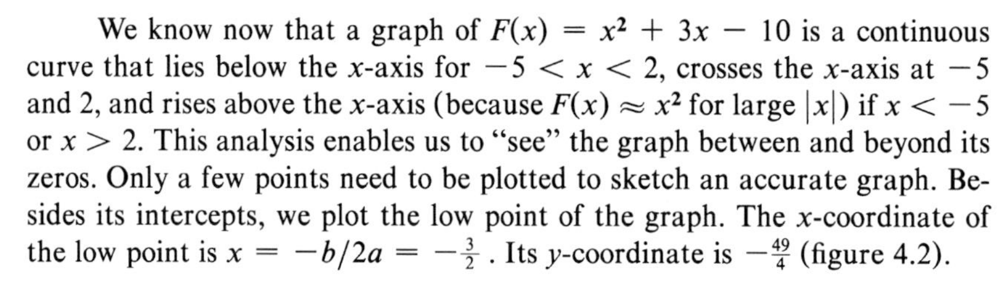

x	$3/x$	$-10/x^2$	x^2	$F(x) = x^2(1 + 3/x - 10/x^2)$
10	3/10	−1/10	100	120
100	3/100	−1/1,000	10,000	10,290
1,000	3/1,000	−1/100,000	1,000,000	1,002,990
−100	−3/100	−1/1,000	10,000	9,690
−1,000	−3/1,000	−1/100,000	1,000,000	996,990

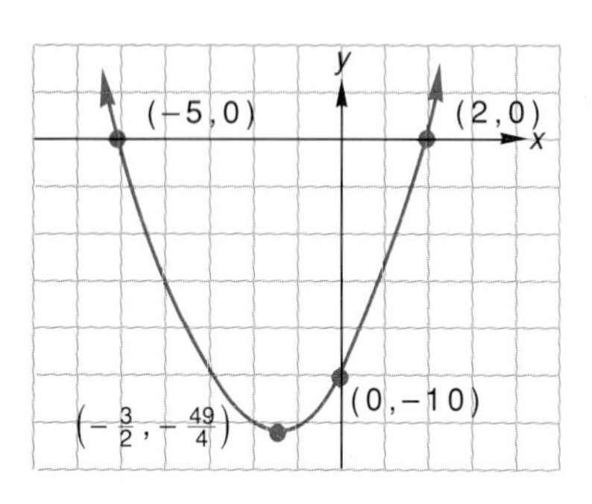

Figure 4.2

We know now that a graph of $F(x) = x^2 + 3x - 10$ is a continuous curve that lies below the x-axis for $-5 < x < 2$, crosses the x-axis at -5 and 2, and rises above the x-axis (because $F(x) \approx x^2$ for large $|x|$) if $x < -5$ or $x > 2$. This analysis enables us to "see" the graph between and beyond its zeros. Only a few points need to be plotted to sketch an accurate graph. Besides its intercepts, we plot the low point of the graph. The x-coordinate of the low point is $x = -b/2a = -\frac{3}{2}$. Its y-coordinate is $-\frac{49}{4}$ (figure 4.2).

Example 1 Graph **a.** $y = x^3$ and **b.** $y = x^5$.

Solution

a. The x and y intercepts are both 0. The domain of the function is R, and because $\sqrt[3]{y} = x$, its range is R. Also,

$$f(-x) = (-x)^3 = -(x^3) = -f(x),$$

so that the graph is symmetric to the origin (figure 4.3a).

b. Verify that the discussion for part a applies to part b (figure 4.3b). ■

Note that the graphs of $y = x^3$ and $y = x^5$ have essentially the same shape. In fact, the whole family of graphs, $y = x^n$ where n is a positive odd integer, have essentially the same shape. Graph $y = x^7$ to verify this statement.

Graphs of $y = x^n$, where n is a positive even integer, all have essentially the same shape. Each graph is symmetric to the y-axis. (Why?) Each graph contains the origin and is tangent to the x-axis (figure 4.4).

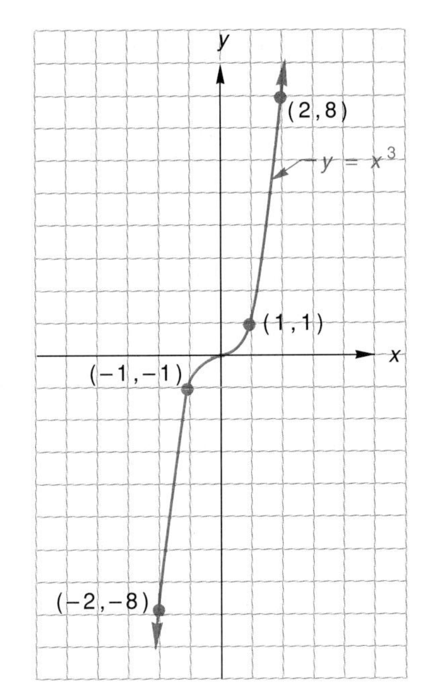

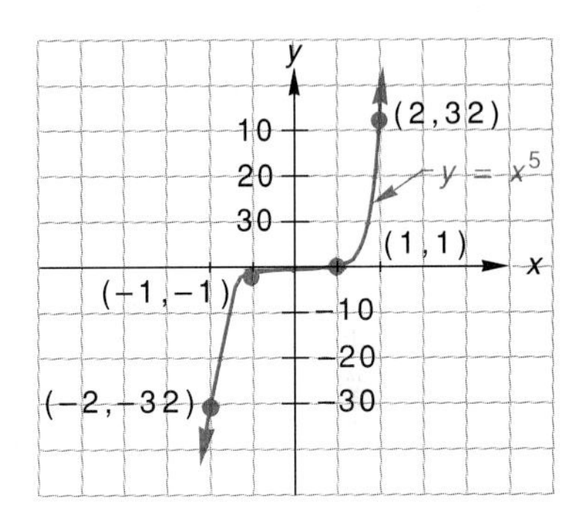

Figure 4.3

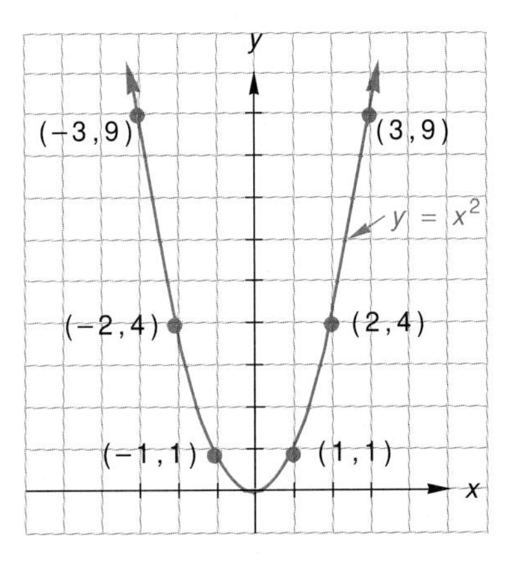

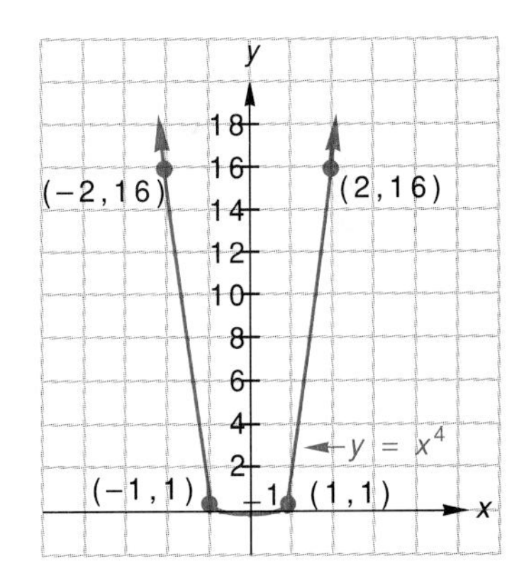

Figure 4.4

Example 2 Graph $y = x^3 + 2x^2 - 5x - 6$.

Solution The y-intercept is -6. The x-intercepts are the roots of the equation $0 = x^3 + 2x^2 - 5x - 6$. By Descartes' Rule of Signs there are one positive root and two negative roots, or one positive root and two imaginary roots.

$$p \,\varepsilon \{\pm 1, \pm 2, \pm 3, \pm 6\}, \quad q \,\varepsilon \{\pm 1\}, \quad \frac{p}{q} \,\varepsilon \{\pm 1, \pm 2, \pm 3, \pm 6\}$$

$$\begin{array}{c|rrrr} & 1 & 2 & -5 & -6 \\ \hline 1 & 1 & 3 & -2 & -8 \\ 2 & 1 & 4 & 3 & 0 \end{array}$$

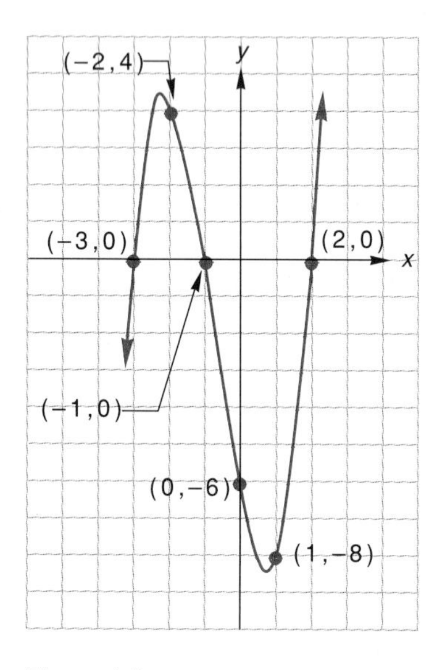

Figure 4.5

Now,
$$\begin{aligned} x^3 + 2x^2 - 5x - 6 &= (x - 2)(x^2 + 4x + 3) \\ &= (x - 2)(x + 3)(x + 1). \end{aligned}$$

The x-intercepts are -3, -1, and 2. Synthetic division gives us $(1,-8)$ as a point on the graph. The portion of the graph close to the origin is obtained by passing a continuous curve through the intercepts, the point $(1,-8)$ and the calculated point $(-2,4)$ (figure 4.5). The portion of the graph remote from the origin is constructed by noting that because

$$x^3 + 2x^2 - 5x - 6 = x^3\left(1 + \frac{2}{x} - \frac{5}{x^2} - \frac{6}{x^3}\right),$$

$y \approx x^3$ for large $|x|$. The graph rises on the right and falls on the left (see example 1). ■

The highest degree term of a polynomial function determines the behavior of its graph for large $|x|$. This is the case in example 2, and it is true in general. An nth degree polynomial $F(x)$ can be written in the form

$$x^n\left(a_n + \frac{a_{n-1}}{x} + \frac{a_{n-2}}{x^2} + \cdots + \frac{a_0}{x^n}\right).$$

As $|x|$ assumes greater and greater values, the terms that have powers of x in their denominators tend to become smaller and smaller and approach zero. Consequently, $F(x) \approx a_n x^n$ for large $|x|$. The table below summarizes the possible behaviors of polynomial graphs for large $|x|$.

a_n	n	Behavior of Graph for Large $\lvert x \rvert$	Example
positive	odd	rises on right, falls on left	1, 2
positive	even	rises on right and left	3
negative	odd	rises on left, falls on right	4
negative	even	falls on left and right	

Example 3 Graph $y = (x + 1)(x + 2)^2(x - 1)^3$.

Solution The graph has x-intercepts at -1, -2, and 1. The y-intercept is -4. (Why?) For large $|x|$ the graph behaves as the graph of $y = x^6$, rising on the right and left (figure 4.6). For values of x close to 1

$$y \approx (1 + 1)(1 + 2)^2(x - 1)^3 = 18(x - 1)^3.$$

In the neighborhood of $x = 1$, a root of multiplicity three, the graph has the appearance of the cubic polynomial $y = 18(x - 1)^3$. That is, near $x = 1$ it behaves as the graph of $y = x^3$ does at the origin (see figure 4.3). For values of x close to -2

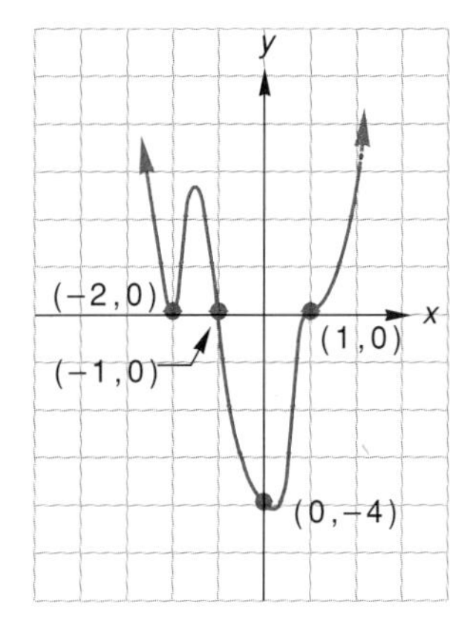

Figure 4.6

$$y \approx (-2 + 1)(x + 2)^2(-2 - 1)^3 = 27(x + 2)^2.$$

Near $x = -2$, which is a root of multiplicity two, the graph behaves as if it is a graph of the second degree polynomial $y = 27(x + 2)^2$. We see that $y = 0$ at -2 and that $y > 0$ near -2. Thus, the graph opens upward from -2. It is tangent to the x-axis at $x = -2$ just as the graphs of $y = x^2$ and $y = x^4$ are tangent to the x-axis at $x = 0$ (figure 4.4). For values of x close to -1

$$y \approx (x + 1)(-1 + 2)^2(-1 - 1)^3 = -8(x + 1).$$

The graph appears to be the straight line $y = -8(x + 1)$, as it passes through the x-axis at -1. Having analyzed the behavior of the polynomial near its x-intercepts and remote from the origin, we can, with the aid of a few plotted points, draw an accurate graph. ■

Example 4 Graph $y = -x^2(x - 3)^5(x - 1)^4$.

Solution The x-intercepts are 0, 1, and 3. The y-intercept is 0. Because the leading term of the polynomial is $-x^{11}$ (why?), the graph rises on the left and falls on the right. For values of x near 0

$$y \approx -x^2(-3)^5(-1)^4 = 243x^2.$$

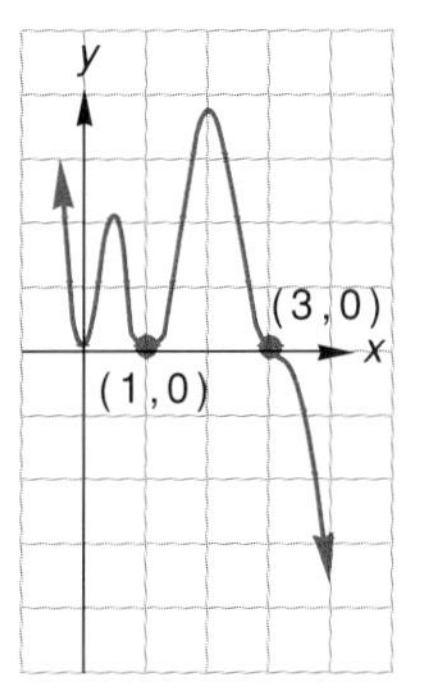

Figure 4.7

The graph behaves like a graph of $y = 243x^2$ near the origin. It is tangent to the x-axis at the origin, and it is positive on either side of the origin. For values of x near 1

$$y \approx -(1)^2(1 - 3)^5(x - 1)^4 = 32(x - 1)^4.$$

The graph behaves like a graph of $y = 32(x - 1)^4$ near $x = 1$. It is zero at $x = 1$ and positive on either side of 1. Thus, the graph is tangent to the x-axis at 1, and it rises from the x-axis on either side of 1 (figure 4.7). Finally, for x close to 3

$$y \approx -(3)^2(x - 3)^5(3 - 1)^4 = -144(x - 3)^5.$$

Near $x = 3$, the graph behaves like a graph of $y = -144(x - 3)^5$. It crosses the x-axis at 3 because functional values are positive to left of 3 and negative to the right of 3. In addition, the shape of the graph near 3 is similar to the shape of a graph of $y = -x^5$ near the origin (see figure 4.3). ■

Examples 3 and 4 suggest the following generalization.

Behavior of Polynomial Graph at the *X*-Axis

A polynomial graph is tangent to the x-axis at zeros with even multiplicities. The graph crosses the x-axis at zeros with odd multiplicities. At a zero with odd multiplicity greater than 1, the graph has the appearance of $y = x^3$ at $x = 0$.

Example 5 (an application) A 492 cubic foot rectangular container is to be fabricated from 400 square feet of sheet material. The base of the container is to be a square. Determine the dimensions of the container. Is it possible to fabricate a square base container from the given amount of sheet material whose volume is greater than 492 cubic feet?

Solution Let the height and width of the container be h and x respectively. Then the volume and surface area of the container can be expressed as $V = x^2h$ and $A = 2x^2 + 4xh$. Since $V = 492$ and $A = 400$, we can write

$$400 = 2x^2 + 4xh \quad \text{and} \quad 492 = x^2h$$

$$h = \frac{400 - 2x^2}{4x} \quad \text{and} \quad 492 = x^2\left(\frac{400 - 2x^2}{4x}\right)$$

$$492 = 100x - \frac{x^3}{2} \tag{1}$$

$$0 = x^3 - 200x + 984$$

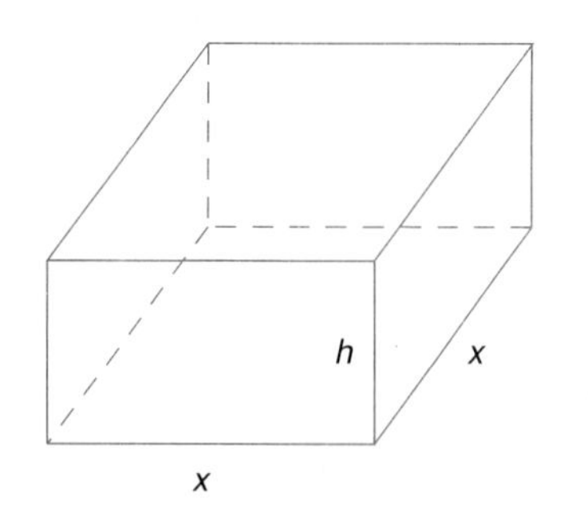

Rational roots of the cubic equation must be factors of 984. (Why?) We need not list all the possible rational factors because the 400 square foot surface area limits the values x can assume. Certainly, x must be positive and less than 20 feet. (Why?) The positive factors of 984 that are less than 20 are 2, 4, 6, 8, and 12.

	1	0	−200	984
2	1	2	−198	588
4	1	4	−184	248
6	1	6	−164	0

$$x^3 - 200x + 984 = (x - 6)(x^2 + 6x - 164)$$

The roots of $x^2 + 6x - 984 = 0$, correct to two decimal places, are -16.15 and 10.15. Since $h = (400 - 2x^2)/4x$, the possible dimensions of the container are $6 \times 6 \times 13.67$ or $10.15 \times 10.15 \times 4.78$.

To determine if it is possible to fabricate a container that has a volume greater than 492 cubic feet, we replace 492 in equation (1) on page 192 with $V(x)$ and graph the resulting volume function.

$$V(x) = 100x - \frac{x^3}{2}, \quad x \geq 0$$

We see, in the graph below, that a maximum volume of about 540 cubic feet is obtained if x is about 8 feet. The graph also clearly shows that a volume of 492 is obtained for two different x values.

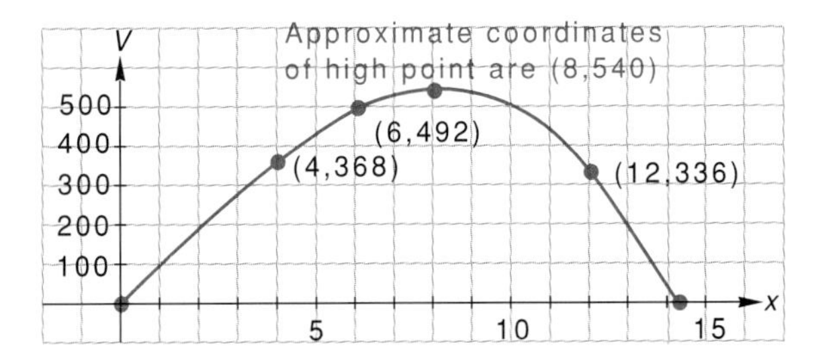

■

Exercise Set 4.4

Graph each equation after (a) determining the y-intercept of the graph, (b) determining the x-intercepts of the graph by finding, where possible, the zeros of the polynomial function, (c) determining if the graph is symmetric to the y-axis or to the origin. If the equation defines a second degree polynomial function, locate the high or low point of its graph.

1. $y = x^2 - x - 6$
2. $y = 2x^2 + 9x + 10$
3. $y = 4x^2 + 6x - 1$
4. $y = -3x^2 + 2x + 5$
5. $y = -(x + 3)^2$
6. $y = 6x^2 + 4x + 3$
7. $y = 2x^3 - x$
8. $y = x^3 - 4x^2 - x + 4$
9. $y = (x + 3)^2(x - 1)$
10. $y = (x - 5)^2(x + 2)$
11. $y = (x - 2)^3$
12. $y = (x + 4)^3$
13. $y = 2x^3 - 3x^2 - 12x + 13$
14. $y = x^3 - 4x^2 - \frac{x}{2} + 2$
15. $y = x^4 + 3x^2 - 6$
16. $y = 2x^4 - x^2 - 5$
17. $y = 3x^4 - 11x^3 + 9x^2 + 13x - 10$
18. $y = x^4 + x^3 - 5x^2 - 15x - 18$
19. $y = (x + 3)^2(x - 1)^2$
20. $y = (x - 2)^3(x + 5)$
21. $y = x^3(x + 2)^2$
22. $y = (x + 1)^4(x - 6)$
23. $y = x^2(x + 4)(x + 1)(x - 3)$
24. $y = x^5 - 2x^3 + x$

In exercises 25–28, determine the nature of the zeros of the polynomial function whose graph and degree are given.

25.

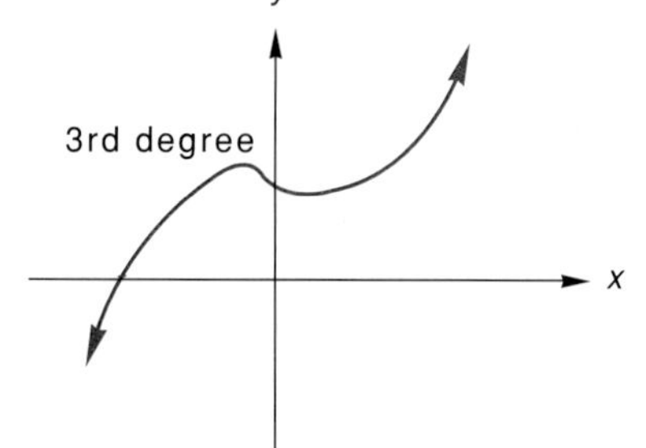

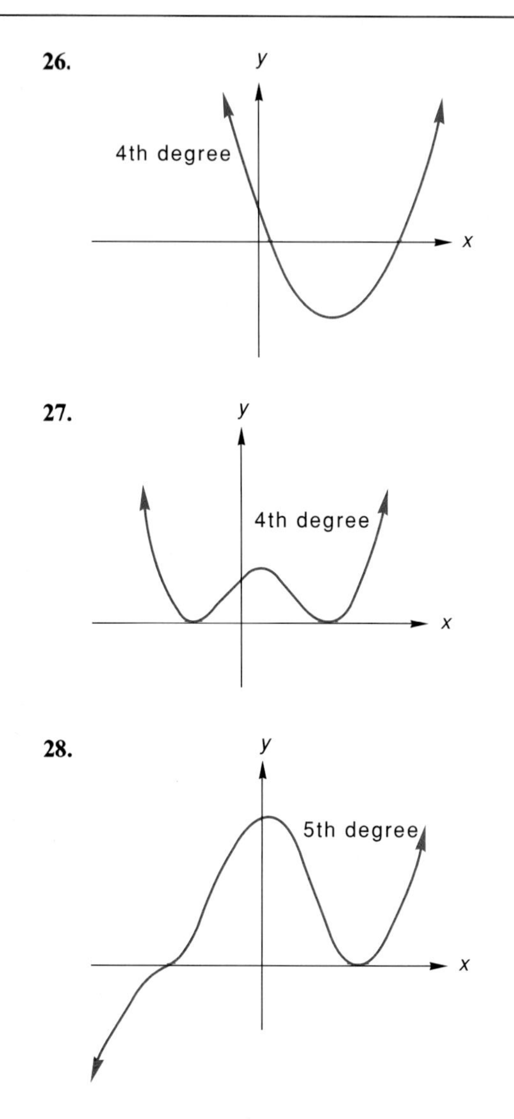

In exercises 29–32, write an equation for the polynomial function of indicated degree whose graph is shown.

29.

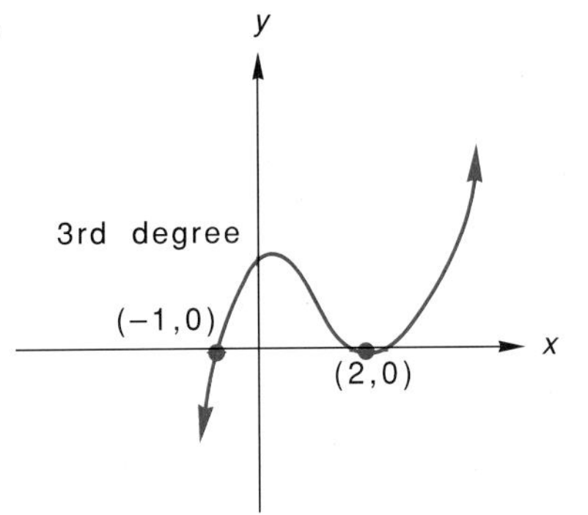

30. Use the same graph as exercise 29, but with a fifth degree polynomial.

31.

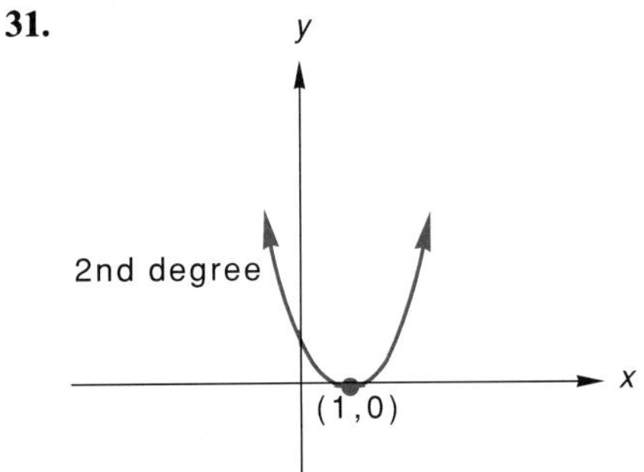

32. Use the same graph as exercise 31, but with a fourth degree polynomial.

33. Given $f(x) = (x - 1)^2(x + 1)$, find c so that the only real zero of $f(x) + c$ is -2.

34. Given $f(x) = (x + 3)(x)(x - 2)$, find h so that the real zeros of $f(x + h)$ are -1, 2, and 4.

35. Construct a third degree polynomial function $F(x)$, that has three real zeros. Show that the zeros of $F(-x)$ are the negatives of the zeros of $F(x)$.

36. Find the second degree polynomial function $y = a_2x^2 + a_1x + a_0$ that contains the ordered pairs $(-1,18)$, $(2,15)$, and $(3,30)$.

37. Find the third degree polynomial $y = a_3x^3 + a_2x^2 + a_1x + a_0$ that contains the ordered pairs $(-4,-50)$, $(-2,-18)$, $(-1,-14)$, and $(2,22)$.

38. Exercise 36 suggests that three ordered pairs determine a second degree polynomial function. Can you find three ordered pairs that do not determine a second degree polynomial function?

39. A rectangular box whose volume is 200 cubic inches is to be constructed from a 15″ × 18″ piece of cardboard. Squares of equal area are cut from the corners of the cardboard. The cardboard is then bent up along the dashed lines (see figure). What is the length of the side of the square that is cut from each corner? Express the volume of the box as a function of the length of a side of a corner square. Graph the function and estimate the side length that produces the maximum volume.

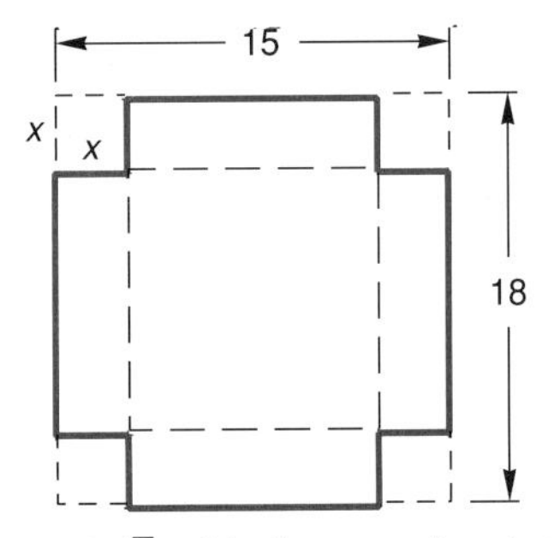

40. A $4\sqrt{5}$ cubic foot container is to be constructed so that its depth will be twice its width. A diagonal of the box is to be three feet more than the width. Approximate to the nearest tenth of a foot the dimensions of the box. (Hint: Use synthetic division.)

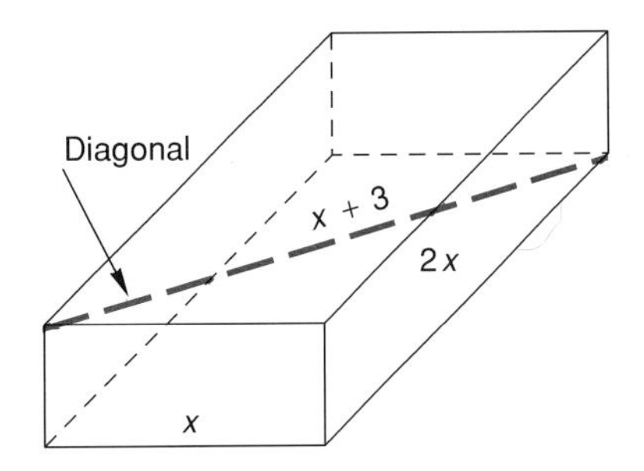

41. A gas storage tank is to be fabricated in the shape of a right circular cylinder with hemispheres attached at each end. The total length of the tank is to be 20′. Express the volume of the tank as a function of the cylinder's radius (r). Determine r and h if the volume of the tank is to be 576π cubic feet.

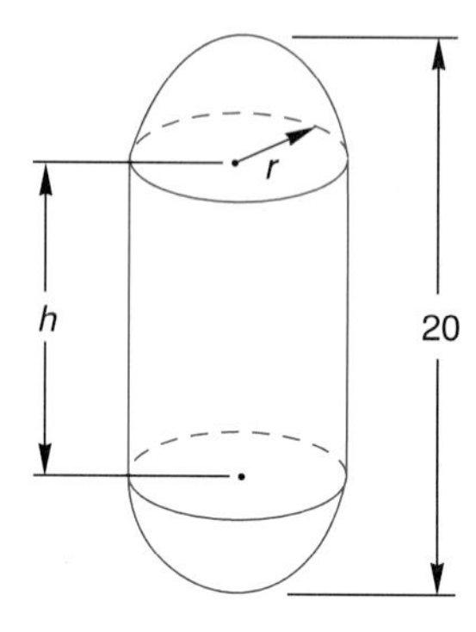

42. Express the volume of a storage shack as a function of x. Determine x if the shack is to be 833 cubic feet.

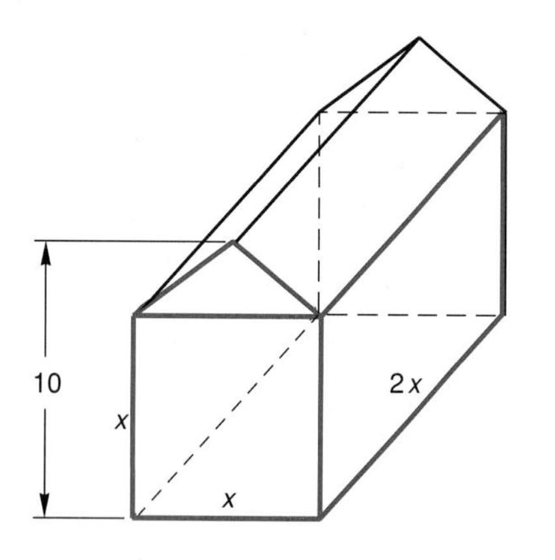

43. The projected population of a certain range animal, over the next 20 years, is given by the function

$P(t) = -0.03t^5 + 10t^3 - 5t^2 + 400t + 16{,}000,$
$0 \le t \le 20,$

where t is time in years. Graph the function. Estimate the time interval(s) over which the population is increasing.

44. A particle moves back and forth along a straight line according to the following formulation

$$s = t(t - 2)(t - 5),$$

where s is positioned with respect to some starting point, when time (t) is 0. Graph the function. Approximate the time intervals when the particle is moving away from, or toward, the starting point.

45. A cylindrical container of 45π cubic feet has radius r and height h. The container is to be fabricated from 48π square feet of sheet material. The container has a top and bottom. Determine r and h. Express the volume of the cylinder in terms of r. Graph the volume function, and from it estimate the radius that produces maximum volume. Estimate maximum volume.

Let $F(x) = a_2x^2 + a_1x + a_0$ and $G(x) = b_2x^2 + b_1x + b_0$ each contain the non-collinear ordered pairs (x_1,y_1), (x_2,y_2), and (x_3,y_3). Then we can write the following equations,

$$\left.\begin{aligned} F(x_1) - G(x_1) &= 0 \\ &= (a_2 - b_2)x_1^2 + (a_1 - b_1)x_1 + (a_0 - b_0) \\ F(x_2) - G(x_2) &= 0 \\ &= (a_2 - b_2)x_2^2 + (a_1 - b_1)x_2 + (a_0 - b_0) \\ F(x_3) - G(x_3) &= 0 \\ &= (a_3 - b_3)x_3^2 + (a_1 - b_1)x_3 + (a_0 - b_0). \end{aligned}\right\} \text{(I)}$$

Equations (I) imply that the second degree polynomial

$$F(x) - G(x) = (a_2 - b_2)x^2 + (a_1 - b_1)x + (a_0 - b_0),$$

has three zeros. But it is impossible for a second degree polynomial to have three zeros unless it is identically zero. Hence, $a_2 = b_2$, $a_1 = b_1$, and $a_0 = b_0$. We conclude that three non-collinear, ordered pairs determine a unique second degree polynomial.

46. Explain the construction of equations (I), and why those equations imply that $F(x) - G(x)$ has three zeros.

47. Why is it impossible for a second degree polynomial to have three zeros? Referring to the discussion that precedes exercise 46, why must $F(x) - G(x)$ be identically zero?

4.5 Approximating Irrational Zeros

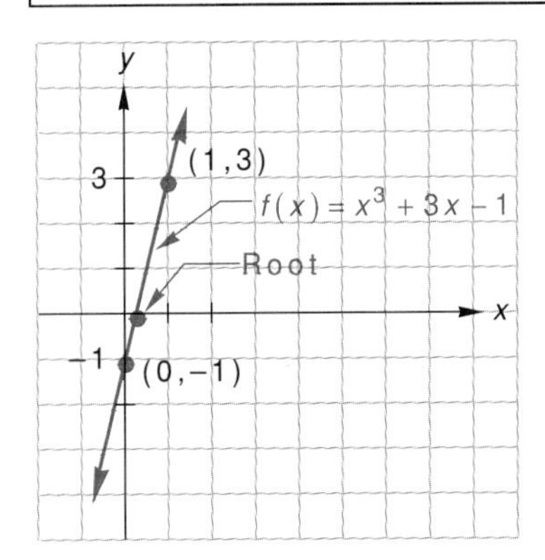

Figure 4.8

Suppose we want to find the roots of $x^3 + 3x - 1 = 0$. Let $F(x) = x^3 + 3x - 1$, then $F(-x) = -x^3 - 3x - 1$. Descartes' Rule of Signs tells us that $F(x)$ has one positive zero and no negative zeros. If the positive root of the equation is a rational number, it must be 1. (Why?) Because $F(0) = -1$ and $F(1) = 3$, synthetic division reveals that the positive root is not 1, but an irrational number between 0 and 1 (figure 4.8).

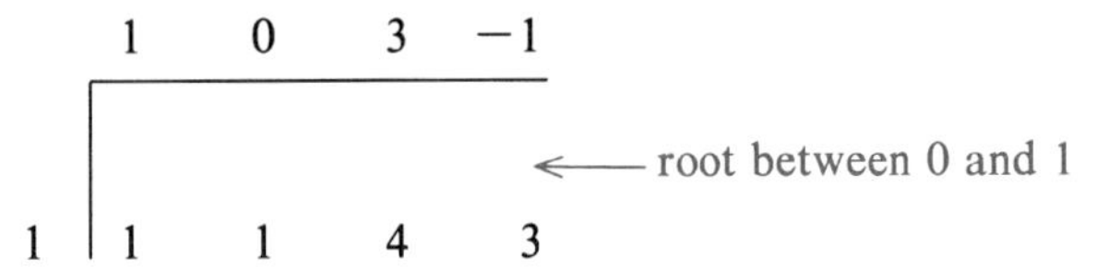

	1	0	3	−1	
1	1	1	4	3	← root between 0 and 1

The synthetic division is continued (below) with 0.5 being tested as a possible root. It produces a remainder of 0.625. The irrational root is now located between 0 and 0.5. We assume that the root is closer to 0.5 than it is to 0 because 0.625 is closer to 0 than −1 is. We next test 0.3, which produces a remainder of −0.0073. The root is now captured between 0.3 and 0.5. Our next estimate of the root should be closer to 0.3 than it is to 0.5. (Why?) We try 0.33. By continuing the synthetic division ad nauseum we are able to obtain as good an approximation of the irrational root as is desired. In the synthetic division tableau below, we show the root captured between 0.322 and 0.325. Thus, 0.32 is an approximation of the root that is correct to two decimal places.

	1	0	3	−1	
0.5	1	0.5000	3.2500	0.6250	
0.3	1	0.3000	3.0900	−0.0073	
0.33	1	0.3300	3.1089	0.0259	
0.32	1	0.3200	3.1024	−0.0072	root
0.325	1	0.3250	3.1056	0.0093	
0.322	1	0.3220	3.1037	−0.0006	

Iteration

There are many numerical methods for finding irrational roots of equations. We consider one such method that, with the aid of a calculator, can produce very good approximations quickly. The method is called iteration.

If $x^3 + 3x - 1 = 0$ is written in the equivalent form

$$x = \frac{1 - x^3}{3},$$

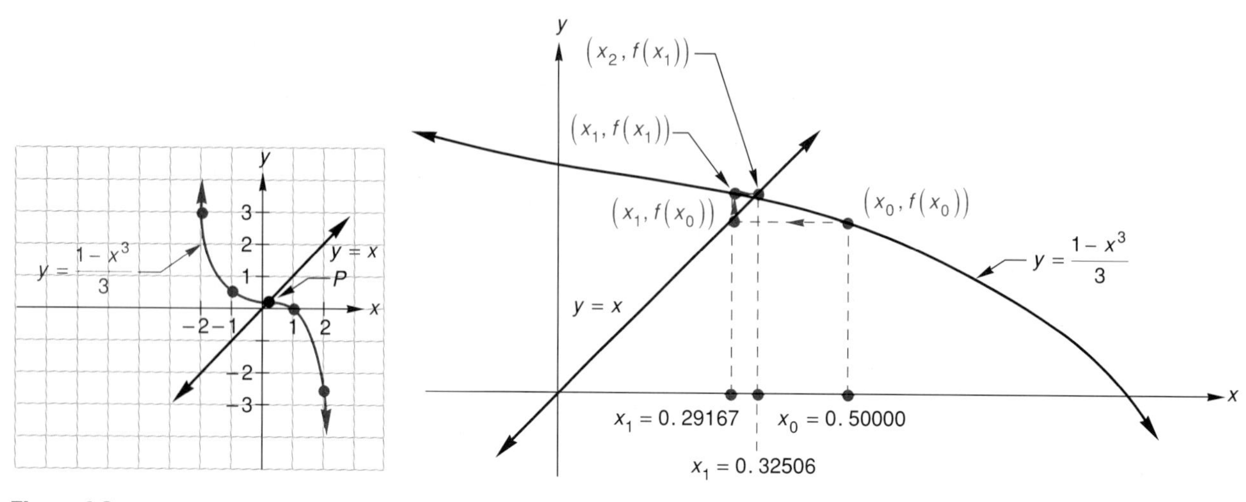

Figure 4.9

Figure 4.10

then the value of x that solves the second equation must also solve the first equation. The solution of $x = (1 - x^3)/3$ is the x-coordinate of the point of intersection (P) of the graphs of $y = x$ and $y = (1 - x^3)/3$ (figure 4.9).

It is possible to approximate the x-coordinate of P as follows: (see figure 4.10)

Let $x_0 = 0.5$ be a first approximation. Then

$$x_1 = f(x_0) = \frac{1 - 0.5^3}{3} = 0.29167$$

$$x_2 = f(x_1) = \frac{1 - 0.29167^3}{3} = 0.32506$$

$$x_3 = f(x_2) = \frac{1 - 0.32506^3}{3} = 0.32188$$

$$x_4 = f(x_3) = \frac{1 - 0.32188^3}{3} = 0.32221$$

$$x_5 = f(x_4) = \frac{1 - 0.32221^3}{3} = 0.32222.$$

Since the fourth and fifth iterates (x_4 and x_5) differ by less than five units in the fifth decimal place, we may conclude that 0.3222 approximates the irrational root to four decimal places.

Iteration is a method in which the ith approximation (x_i) produces the next approximation $x_{i+1} = f(x_i)$. The method works if x_{i+1} is a better approximation of the root than is x_i. This is not always the case (see exercise 18). When using iteration you should look for convergence of the iterates to some fixed value. Such a convergence occurs in the above illustration, where x_0, x_1, x_2, etc. converge to a number that is approximated by 0.3222. If the iterates diverge, as they do in exercise 18, abandon the process.

Example 1 Use iteration to approximate a real root of $x^3 + 3x^2 + 8x + 2 = 0$ to three decimal places.

Solution As an initial approximation, we attempt to locate a root of the equation between two integers. Descarte's Rule of Signs indicates no positive zeros and one or three negative zeros. We test -1 as a possible rational root.

$$\begin{array}{r|rrrr} & 1 & 3 & 8 & 2 \\ \hline -1 & 1 & 2 & 6 & -4 \end{array} \quad \leftarrow \left\{ \begin{array}{l} \text{irrational root} \\ \text{between 0 and } -1 \end{array} \right.$$

Write the equation in the form $x = f(x)$.

$$x = \frac{-x^3 - 3x^2 - 2}{8}$$

As a first approximation, let $x_0 = -0.5$ and then use iteration.

$$x_1 = f(x_0) = -\frac{(-0.5)^3 - 3(-0.5)^2 - 2}{8} = -0.3281$$

$$x^2 = f(x_1) = -\frac{(-0.3281)^3 - 3(-0.3281)^2 - 2}{8} = -0.2860$$

$$x_3 = f(x_2) = -\frac{(-0.2860)^3 - 3(-0.2860)^2 - 2}{8} = -0.2777$$

$$x_4 = f(x_3) = -\frac{(-0.2777)^3 - 3(-0.2777)^2 - 2}{8} = -0.2762$$

$$x_5 = f(x_4) = -\frac{(-0.2762)^3 - 3(-0.2762)^2 - 2}{8} = -0.2760$$

Approximating a root to three decimal places means that the error in the fourth decimal place can be no larger than five units. The error between the root and its approximation must be less than 0.0005. If we wanted to approximate the root to two decimal places, the error in the approximation would have had to be less than 0.005, or five units in the third decimal place. Since the root between 0 and -1 is to be approximated to three decimal places, we stop the iteration process when the absolute value of the difference between two successive iterates is less than 0.0005 units.

$$|x_4 - x_5| = |-0.2762 - (-0.2760)| = 0.0002 < 0.0005$$

We conclude that -0.276 is an approximation, correct to three decimal places, of the root of the equation between 0 and -1. ■

Evaluating Polynomials with a Calculator

There are more efficient ways to evaluate a polynomial than the brute force approach that is used in example 1 to evaluate $(-x^3 - 3x^2 - 2)/8$. The programmable capability of a moderately priced calculator enables one to easily evaluate a polynomial for many different values of the variable. It is also possible to evaluate a polynomial without using the programmable feature of your calculator and without having to store partial results as we did in example 1. We simply write the polynomial in such a form that its evaluation involves only a sequence of basic key strokes. If the polynomial in example 1 is rewritten as

$$\frac{((x + 3)x)x + 2}{8},$$

it can be evaluated for $x = -0.5$ as follows.

[.5] [±] [STOR] [+] [3] [=] [×] [RCL] [×]
[RCL] [+] [2] [=] [÷] [8] [±] [=] -0.328125

The keystrokes, STOR and RCL, refer to your calculator's ability to store and to recall data.

Example 2 $F(x) = 2x^4 - x^3 + 4x^2 + 3x - 5$. Find $F(3)$ and $F(1.8)$.

Solution Rewrite the polynomial as $F(x) = (((2x - 1)x + 4)x + 3)x - 5$.

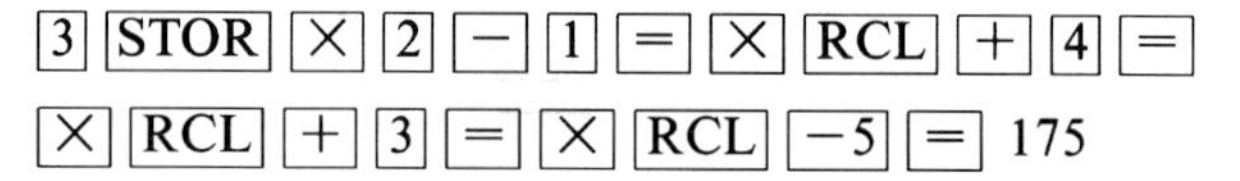

With the exception that 3 is replaced by 1.8, the same keystrokes permit us to evaluate $F(1.8)$. $F(1.8) = 28.5232$. ■

Exercise Set 4.5

(A)

Approximate the indicated root of each equation to two decimal places. Use synthetic division.

1. $x^3 - 3x + 1 = 0$; between 0 and 1
2. $x^3 - 6x^2 + 2x + 4 = 0$; between 1 and 2
3. $x^3 + x^2 + x + 2 = 0$; between -2 and -1
4. $x^3 - 2x^2 - 2x + 2 = 0$; between 2 and 3
5. $x^4 - 8x^2 + 2 = 0$; between -3 and -2
6. $x^4 - x^3 - 4x^2 - 5x + 5 = 0$; between 2 and 3

Use the method of example 2 to evaluate each polynomial for the indicated values.

7. $x^3 - 3x + 1$; $x = 6$, $x = -2.7$
8. $x^3 - 2x^2 - 2x + 2$; $x = -4$, $x = 3.6$
9. $x^4 - 8x^2 + 2$; $x = 7$, $x = -4.3$
10. $x^4 - x^3 - 4x^2 - 5x + 5$; $x = -3$, $x = 5.8$
11. $x^4 - 2x^3 - 10x + 18$; $x = 8$, $x = 9.1$
12. $3x^4 - 7x^3 + 2x^2 - 6x + 9$; $x = -9$; $x = 6.7$

(B)

Approximate the indicated root of each equation to two deimal places. Use iteration.

13. $x^3 - 3x + 1 = 0$; between 0 and 1

14. $x^3 - 2x^2 + 12x + 8 = 0$; between -1 and 0

15. $x^4 - x^2 + 9x + 3 = 0$; between -1 and 0

16. $x^4 - 2x^3 - 10x + 18 = 0$; between 1 and 2

17. Figure 4.10 illustrates the iteration process used to find a root of $x^3 + 3x - 1 = 0$. In the figure, the iterates oscillate around the root. No such oscillation occurs in example 1, where iteration is used to find a root of $x^3 + 3x^2 + 8x + 2 = 0$. Explain the non-oscillatory behavior of the iterates in example 1 by constructing a graph similar to that in figure 4.10.

18. Iteration, as described in this section, may not work. Using figure 4.11, in which x_0 is an initial guess, show that iteration yields a sequence of numbers x_1, x_2, etc., that diverge from the root.

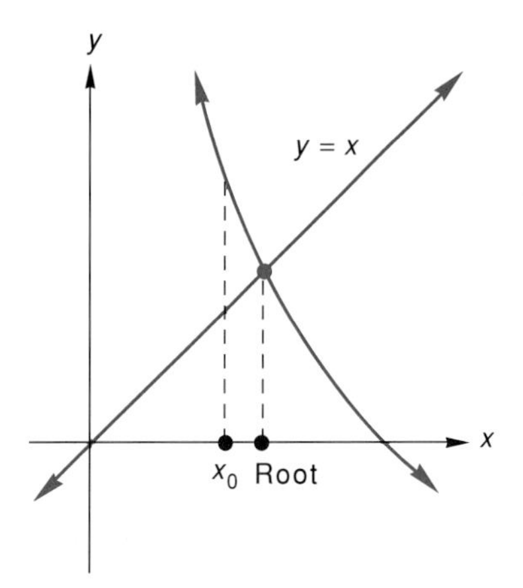

Figure 4.11

Chapter 4 Review

Major Points for Review

Section 4.1 The form of an nth degree polynomial

The leading coefficient and constant term of a polynomial

Adding and multiplying polynomials

Long division and synthetic division of polynomials

The division algorithm of polynomials

Section 4.2 Zeros of a polynomial function or roots of a polynomial equation

The Fundamental Theorem of algebra

The Remainder Theorem

The use of synthetic division to locate the roots of a polynomial equation

Section 4.3 An nth degree polynomial function has exactly n zeros in C

The imaginary zeros of a polynomial function occur in complex conjugate pairs

The rational zeros of a polynomial function are of the form p/q where p is a factor of the constant term and q is a factor of the leading coefficient

Using synthetic division to find the rational roots of a polynomial equation

The Upper and Lower Bound theorems

Descartes' Rule of Signs

Section 4.4 Graphing polynomial functions

The appearance of the graph of a polynomial function for large $|x|$ and for values of x that are near the zeros of the function

Section 4.5 Using synthetic division to approximate the irrational roots of a polynomial equation

Approximating the irrational roots of a polynomial equation by iteration

Review Exercises

Let $P(x)$ be the first polynomial and $Q(x)$ be the second polynomial. Find (a) $P(x) + Q(x)$, (b) $P(x) - Q(x)$, and (c) $P(x) \cdot Q(x)$.

1. $2x^2 - 5x, -3x^2 + 6x + 8$
2. $-4x^2 - 2x + 9, 7x^2 + 8x - 11$
3. $3x^3 + 5x^2 - 10x + 12, 6x^3 - 2x^2 - 5x + 9$
4. $9x^3 + 3x - 4, 10x^3 + 6x^2 - 8x - 7$
5. $x^4 + 15x^2 - 4, -7x^3 - 9x^2 + 10$
6. $-11x^3 + 6x^2 - 4x, -5x^2 - 3x + 9$

Divide the first polynomial by the second. Express the first polynomial as a product of the second polynomial and the quotient, plus the remainder.

7. $5x^3 - 8x^2 + 6x - 9, x^2 + 3x$
8. $6x^3 + 9x^2 - 4x + 14, 2x^2 + x - 3$
9. $20x^4 - 5x^3 + 9x - 4, 4x^2 + 5x - 7$
10. $x^4 + 2x, -x^3 + 1$

Use synthetic division to enable you to write the first polynomial as a product of the second polynomial and the quotient, plus the remainder.

11. $5x^3 + 7x^2 - 11x + 4, x + 2$
12. $-10x^3 + 12x^2 - 6x - 1, x - 8$
13. $4x^3 - 15x^2 + 10x - 3, x - 3$
14. $-2x^3 - 7x^2 + 36x + 36, x + 6$
15. $x^4 + 2x - 9, x - 5$
16. $8x^4 + x^2 + 2, x + 1$

Use synthetic division to evaluate $F(c)$. Then evaluate $F(c)$ directly.

17. $F(x) = -5x^3 + 6x^2 + 6x - 8, c = -2$
18. $F(x) = x^4 - x^2 + 9x - 2, c = 3$
19. $F(x) = 2x^5 - 3x^3 + 5x^2 - 12, c = 2$
20. $F(x) = x^4 + 4x^3 - 7, c = -3$

Use synthetic division to show that the second polynomial is a factor of the first polynomial.

21. $3x^3 + 3x^2 - 2x - 4, x - 1$
22. $4x^3 - 5x^2 + 10x + 12, x + \frac{3}{4}$
23. $6x^4 + 5x^3 - 14x^2 + 3x + 1, x - \frac{1}{2}$
24. $x^4 + 3x^3 + 6x^2 + 13x - 15, x + 3$

Use synthetic division to write $F(x)$ in the form

$$F(x) = a_n(x - r_1)(x - r_2) \cdots (x - r_n).$$

25. $F(x) = 2x^3 - 5x^2 - x + 6$
26. $F(x) = -10x^3 + 41x^2 - 2x - 8$
27. $F(x) = 4x^4 + 16x^3 + 17x^2 + 4x + 4$
28. $F(x) = 36x^4 - 12x^3 - 23x^2 + 4x + 4$

Find the rational roots of each equation. Use Descartes' Rule of Signs and the Upper and Lower Bound theorems to limit your search for roots.

29. $3x^3 + 16x^2 + 15x - 18 = 0$
30. $2x^3 - 17x^2 + 20x + 75 = 0$
31. $x^4 - 2x^3 - 7x^2 - 2x - 8 = 0$

32. $x^4 - 18x^2 + 81 = 0$

33. $-2x^5 + 8x^3 + 7x = 0$

34. $3x^5 + 20x^4 + 28x^3 + 16x^2 + 64x - 256 = 0$

Convert the polynomial equation to a polynomial function and graph the function. Show the x and y intercepts, if they are rational numbers. Also, state if the graph is symmetric with respect to the y-axis or the origin.

35. Exercise 29 **36.** Exercise 30

37. Exercise 31 **38.** Exercise 32

39. Exercise 33 **40.** Exercise 34

Use synthetic division to approximate the indicated root of each equation to two decimal places.

41. $x^3 - 5x + 3 = 0$, root between 0 and 1

42. $-x^4 + 4x^3 - 2x + 12 = 0$, root between -1 and -2

Use synthetic division to divide $A(x)$ by $B(x)$.

43. $A(x) = 6x^3 - 9x + 12$, $B(x) = 3x + 9$

44. $A(x) = 4x^3 + 14x^2 + 10x - 6$, $B(x) = 2x - 4$

45. Show that $x + 1$ is a factor of $x^n - 1$ only if n is even.

46. Show that $2x - 5$ is a factor of $8x^3 - 14x^2 - 69x + 135$.

47. Prove that the equation $2x^4 + 6x^3 - 7x^2 + 9x - 5 = 0$ has no rational roots.

48. Prove that $\sqrt{7}$ is not a rational number.

49. Prove that $\sqrt[3]{9}$ is not a rational number.

50. A rectangular container is to be fabricated from 320 square feet of sheet material. The depth of the container's base is to be three times its width. Express the volume of the container as a function of the base width. Graph the function and estimate the volume of the container of maximum volume that can be fabricated from 320 square feet of sheet material. Estimate the dimensions of the maximum volume container. Also determine the dimensions of the container if its volume is to be 336 cubic feet.

Determine the nature of the zeros of the polynomial function whose graph and degree are given.

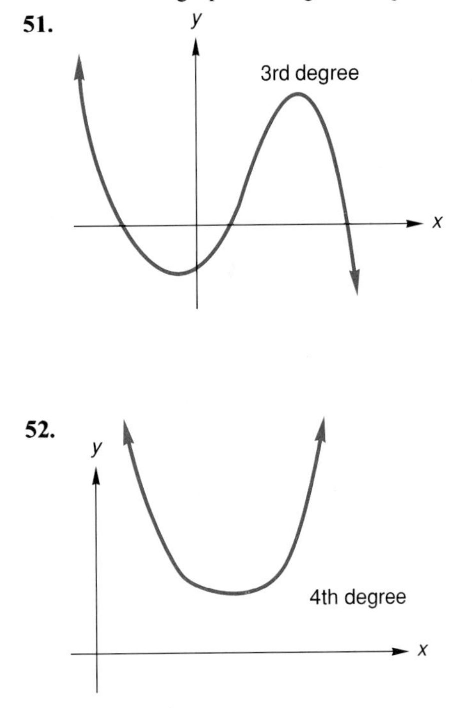

53. A graph of a polynomial function and the degree of the function are given below. Write a possible defining equation for the function.

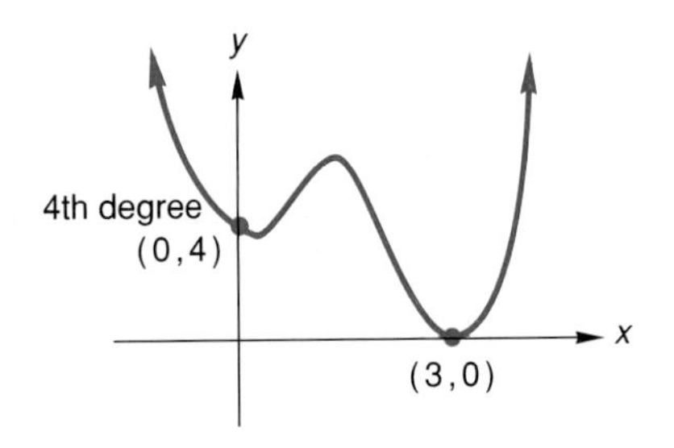

Graph each polynomial function.

54. $y = (x + 1)^3$

55. $y = x^4 + 2x^3 - 11x^2 - 12x + 36$

56. $y = 2x^4 - 6x^2 + 1$

57. $y = 2x^3(x - 1)^2$

58. $y = x(x + 3)^2(x + 5)(x + 2)$

59. $y = (x - 3)^3(x + 2)^2$

Use iteration to approximate the indicated root to two decimal places.

60. $x^3 + 2x - 2 = 0$, root between 0 and 1

61. $x^4 + 3x^3 - 8x^2 + 6x - 4 = 0$, root between 1 and 2

62. Find the third degree polynomial function, $y = a_3x^3 + a_2x^2 + a_1x + a_0$, that contains the ordered pairs, $(-2,-17)$, $(1,-14)$, $(2,-9)$, and $(3,18)$.

Chapter 4 Test 1

1. Find the sum and product of the polynomials, $-5x^3 + 4x^2 + 9x - 2$ and $2x^3 + 6x - 7$.

2. $A(x) = 2x^4 - 10x^3 + 5x^2 + 8x - 1$ and $B(x) = x + 3$. Find the quotient, $Q(x)$, and the remainder, $R(x)$, when $A(x)$ is divided by $B(x)$. Express $A(x)$ in terms of $B(x)$, $Q(x)$, and $R(x)$.

3. Use synthetic division to evaluate $F(c)$ if $F(x) = 5x^3 + 6x^2 - 3x + 9$ and $c = -3$.

Find the rational roots of each equation.

4. $x^3 - 5x^2 - 8x + 12 = 0$

5. $x^4 + 6x^3 + 13x^2 + 24x + 36 = 0$

Graph each polynomial function. Show the y-intercept. Label x-intercepts if they are rational numbers. State if the graph has symmetry with respect to the y-axis or the origin.

6. $y = x^3 - 5x^2 - 8x + 12$

7. $y = (x + 2)^2(x - 4)^3$

8. Prove that the equation $3x^3 + 7x^2 - 6x + 5 = 0$ has no rational roots.

Chapter 4 Test 2

Find the rational roots of each equation.

1. $12x^3 - 47x^2 - 10x + 24 = 0$

2. $x^4 - 10x^3 + 26x^2 - 10x + 25 = 0$

Graph each polynomial function. Show the y-intercept. Label x-intercepts if they are rational numbers. State if the graph has symmetry with respect to the y-axis or the origin.

3. $y = 12x^3 - 47x^2 - 10x + 24$

4. $y = x^4 - 10x^3 + 26x^2 - 10x + 25$

5. $y = 2x^4 - x^2 + 5$

6. Prove that $\sqrt[3]{4}$ is not a rational number.

7. The graph of a third degree polynomial function is given below. Determine the nature of the zeros of the function.

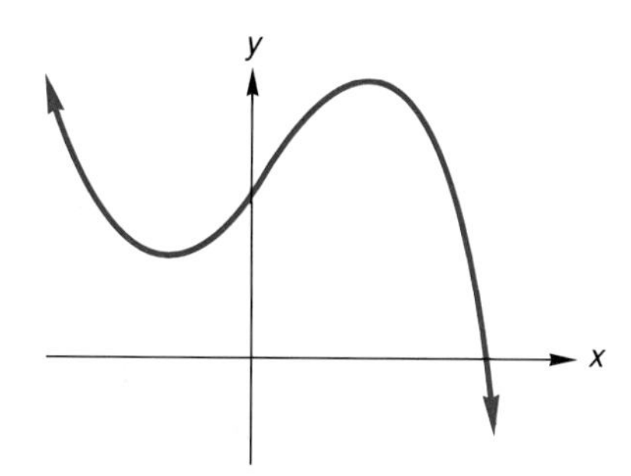

8. The graph of a fourth degree polynomial function is given below. Write a defining equation for the function.

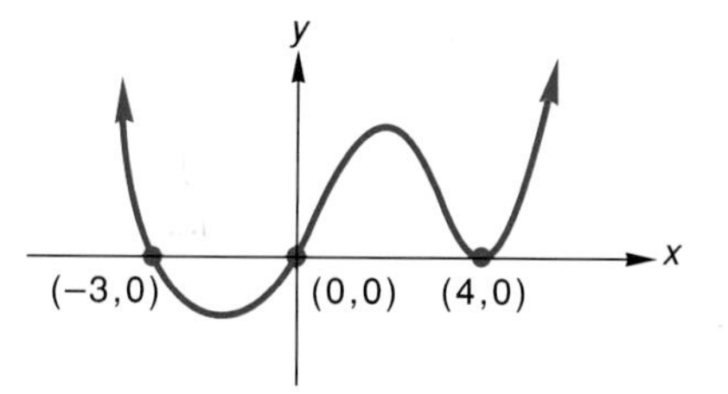

Chapter 5

Rational Functions

A rational function is a ratio of polynomial functions. But unlike polynomial functions, whose graphs are continuous, graphs of rational functions may have breaks in them. Such graphs are said to be **discontinuous.** A graph of a rational function is also likely to have vertical and horizontal asymptotes. In this chapter, we learn how to locate the asymptotes of a graph, and how to use the asymptotes to construct the graph.

5.1 Graphing Rational Functions I

We call $R(x) = P(x)/Q(x)$, where $P(x)$ and $Q(x)$ are polynomials, a **rational function.** $R(x)$ denotes a real number if $Q(x) \neq 0$. Being a polynomial, Q has a finite set of zeros. Hence, the domain of a rational function consists of all real numbers except the finite set of zeros of the denominator polynomial.

Suppose

$$R(x) = \frac{(x-3)(x-2)}{(x+1)(x-2)} \quad \text{and} \quad S(x) = \frac{x-3}{x+1}.$$

The domain of S includes all real numbers except -1. If we further restrict the domain of S so that it does not include 2, then $R(x) = S(x)$. Otherwise, $R(x) \neq S(x)$. Remember that *a requirement for the equality of two functions is that their domains be equal.*

Example 1 Determine the domain of each rational function. Where possible, simplify the fraction that defines the function.

a. $R(x) = \dfrac{x+1}{x^2-1}$

b. $S(x) = \dfrac{x^2 + 5x - 6}{x^2 + 2x - 3}$

c. $T(x) = \dfrac{3x + 5}{2x - 1}$

Solution

a. $R(x) = (x + 1)/[(x - 1)(x + 1)]$. The domain of R consists of all real except -1 and 1.

$$R(x) = \frac{1}{x - 1}, \quad x \neq -1,1$$

b. $S(x) = [(x - 1)(x + 6)]/[(x - 1)(x + 3)]$. The domain of S consists of all real numbers except -3 and 1.

$$S(x) = \frac{x + 6}{x + 3}, \quad x \neq -3,1$$

c. The domain of T is the set of all real numbers except $\frac{1}{2}$. Since numerator and denominator polynomials have no common factor, the fraction cannot be simplified. ■

Vertical Asymptotes

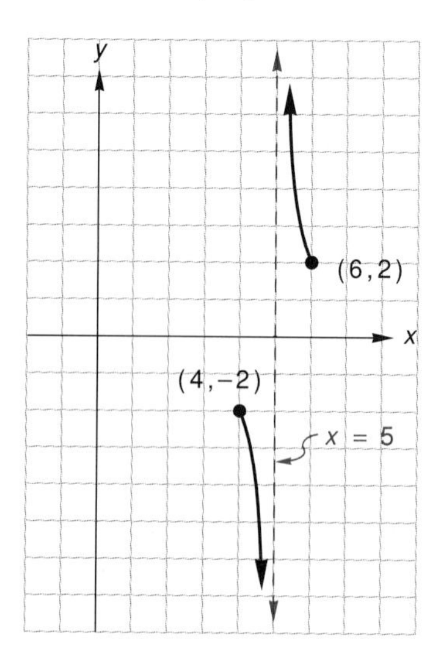

Figure 5.1

The behavior of a graph of $f(x) = 2/(x - 5)$ near $x = 5$ is shown in figure 5.1.

x	$\lvert x - 5 \rvert$	$f(x)$
6.0	1	2
5.1	0.1	20
5.01	0.01	200
5.001	0.001	2,000
4.0	1	−2
4.9	0.1	−20
4.99	0.01	−200
4.999	0.001	−2,000

Apparently $|f(x)|$ can be made greater and greater, without bound, by making $|x - 5|$ smaller and smaller. A graph of f is said to *approach the line $x = 5$ asymptotically.* The graph *cannot* cross the line because 5 is not in the domain of f. We call the line $x = 5$ a **vertical asymptote** of the graph.

In the above discussion, it is possible to substitute the phrase "$|f(x)|$ tends to infinity (∞)" for the phrase "$|f(x)|$ becomes greater and greater without bound." We could also say "$|x - 5|$ tends to 0" rather than "$|x - 5|$ becomes smaller and smaller."

> **Definition**
>
> A graph of a rational function f has a vertical asymptote at $x = c$ if $|f(x)|$ tends to ∞ as $|x - c|$ tends to 0.

Example 2 Show that a graph of $g(x) = 2x^2/[(x - 1)(x + 3)]$ has vertical asymptotes at $x = 1$ and $x = -3$.

Solution 1 and -3 are not in the domain of g. Therefore, a graph of g cannot cross either of the lines $x = 1$ or $x = -3$. To claim that $x = 1$ and $x = -3$ are vertical asymptotes of the graph we must show that $|g(x)|$ tends to ∞ as $|x - 1|$ and $|x - (-3)|$ tend to 0.

x	$\|x - 1\|$	$g(x)$	x	$\|x + 3\|$	$g(x)$
0	1	0	−4	1	−6.4
0.9	0.1	−4.154	−3.1	0.1	−46.878
0.99	0.01	−49.128	−3.01	0.01	−451.875
0.999	0.001	−499.125	−3.001	0.001	−4,501.875
2	1	1.6	−2	1	2.667
1.1	0.1	5.902	−2.9	0.1	43.128
1.01	0.01	50.878	−2.99	0.01	448.125
1.001	0.001	500.875	−2.999	0.001	4,498.125

We often make the mistake of assuming that a rational function, $R'(x) = P(x)/Q(x)$, has vertical asymptotes at those values of x for which $Q(x) = 0$. *The assumption is correct if $P(x)$ and $Q(x)$ have no factors in common.* ■

Example 3 Show that a graph of $h(x) = (x^2 - 1)/(x - 1)$ does not have a vertical asymptote at $x = 1$. Graph h.

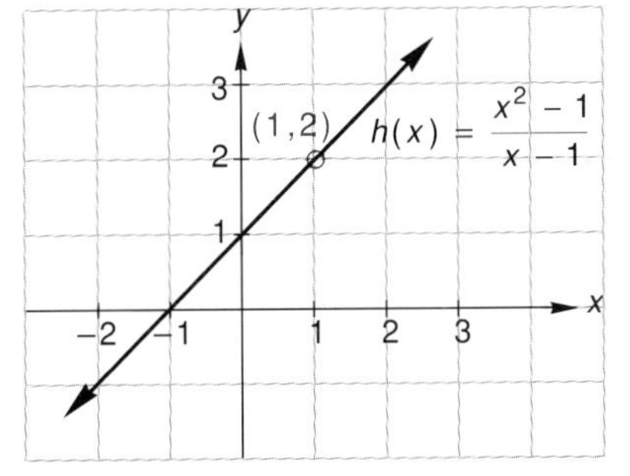

Figure 5.2

Solution Since

$$\frac{x^2 - 1}{x - 1} = \frac{(x - 1)(x + 1)}{x - 1}$$
$$= x + 1, \quad x \neq 1,$$

then

$$h(x) = x + 1, \quad x \neq 1$$

$|h(x)|$ does not tend to ∞ as $|x - 1|$ tends to 0. In fact, as x assumes values closer and closer to 1, $h(x)$ assumes values closer and closer to 2. The point (1,2) is not in the graph of $h(x) = (x^2 - 1)/(x - 1)$ because 1 is not in the domain of h. We say that the graph has a "hole" at $x = 1$ (figure 5.2). ■

Vertical Asymptotes and "Holes"

A graph of a rational function $R(x) = P(x)/Q(x)$ has

a. a vertical asymptote at an x-value that is a zero of Q but is not a zero of P.

b. a "hole" at an x-value that is a zero of both P and Q, providing that it is not a zero of lower multiplicity of P than it is of Q.

Example 4

a. The equation

$$y = \frac{(x-1)^2(x+1)}{x-1}$$

is equivalent to

$$y = (x-1)(x+1), \quad x \neq 1.$$

Therefore its graph has no vertical asymptote at $x = 1$, but it does have a hole at $x = 1$.

b. The equation

$$y = \frac{(x-1)(x+1)}{(x-1)^2}$$

is equivalent to

$$y = \frac{x+1}{x-1}.$$

Its graph has a vertical asymptote at $x = 1$.

Horizontal Asymptotes

If we evaluate $2/x$ for large $|x|$, we see that as $|x|$ tends to ∞, $2/x$ tends to 0.

x	10	100	1,000	10,000	−10	−100	−1,000	−10,000
$2/x$	0.2	0.02	0.002	0.0002	−0.2	−0.02	−0.002	−0.0002

It is also true that for any constant c and natural number n, $|c/x|$ tends to 0 as $|x|$ tends to ∞. This fact is useful in determining the horizontal asymptote of a graph of a rational function. As an illustration, we can write

$$f(x) = \frac{2}{x-5} = \frac{\frac{2}{x}}{\frac{x-5}{x}} = \frac{\frac{2}{x}}{1-\frac{5}{x}}, \quad x \neq 0.$$

As $|x|$ tends to ∞,

$$f(x) = \frac{\frac{2}{x}}{1-\frac{5}{x}} \approx \frac{0}{1-0} = 0.$$

Because $f(x)$ tends to 0 as $|x|$ tends to ∞, we call $y = 0$ a **horizontal asymptote** of a graph of f.

A portion of the graph of $f(x) = 2/(x - 5)$, near its vertical asymptote, is shown in figure 5.1. The entire graph of f is drawn in figure 5.3.

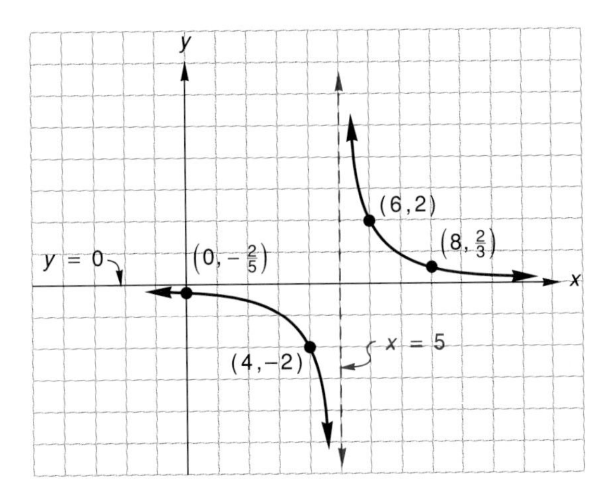

Figure 5.3

Definition

$y = k$ is a horizontal asymptote of a graph of f if, when $|x|$ tends to ∞, $f(x)$ tends to k.

The vertical and horizontal asymptotes of a rational function provide a frame for a rough sketch of its graph (figure 5.3). It is therefore useful to be able to find horizontal asymptotes as easily as it is to find vertical asymptotes.

Example 5 Find the horizontal asymptote for each rational function.

a. $f(x) = \dfrac{3x^2 - 7x + 2}{x^3 + 5x^2}$

b. $g(x) = \dfrac{4x^2 + 8x - 6}{x^2 - 2x - 3}$

c. $h(x) = \dfrac{x^3 + 2}{x^2 - 4}$

Solution Each rational function is rewritten in an equivalent form in which *numerator and denominator are divided by the highest power of x in the denominator.*

a. $$f(x) = \frac{\dfrac{3x^2 - 7x + 2}{x^3}}{\dfrac{x^3 + 5x^2}{x^3}} = \frac{\dfrac{3}{x} - \dfrac{7}{x^2} + \dfrac{2}{x^3}}{1 + \dfrac{5}{x}}$$

As $|x|$ tends to ∞, $f(x) \approx (0 - 0 + 0)/(1 + 0) = 0$. Thus, the horizontal asymptote of a graph of f is $y = 0$.

b. $$g(x) = \frac{\dfrac{4x^2 + 8x - 6}{x^2}}{\dfrac{x^2 - 2x - 3}{x^2}} = \frac{4 + \dfrac{8}{x} - \dfrac{6}{x^2}}{1 - \dfrac{2}{x} - \dfrac{3}{x^2}}$$

As $|x|$ tends to ∞, $g(x) \approx (4 + 0 - 0)/(1 - 0 - 0) = 4$. Thus, the horizontal asymptote is $y = 4$.

c. $$h(x) = \frac{\dfrac{x^3 + 2}{x^2}}{\dfrac{x^2 - 4}{x^2}} = \frac{x + \dfrac{2}{x^2}}{1 - \dfrac{4}{x^2}}$$

As $|x|$ tends to ∞, $h(x) \approx (x + 0)/(1 - 0) = x$. Because $h(x)$ does not tend to a constant, a graph of h has no horizontal asymptote. ■

Finding Horizontal Asymptotes of Rational Function Graphs

Given a rational function

$$F(x) = \frac{a_n x^n + a_{n-1}x^{n-1} + \cdots + a_0}{b_m x^m + b_{m-1}x^{m-1} \cdots b_0},$$

I. F has the horizontal asymptote $y = 0$ if $n < m$. (example 5a)
II. F has the horizontal asymptote $y = a_n/b_m$ if $n = m$. (example 5b)
III. F has no horizontal asymptote if $n > m$. (example 5c)

Graphs

Rough graphs of rational functions can be drawn quickly after the graph's asymptotes are in place. We proceed as follows:

1. Find the vertical and horizontal asymptotes.
2. Determine the behavior of the graph near the vertical asymptotes.
3. Plot x and y intercepts.
4. Plot other points if needed and then sketch the graph.

Example 6 Graph $f(x) = (3x^2 - 7x + 2)/(x^3 + 5x^2)$.

Solution $f(x) = [(3x - 1)(x - 2)]/[x^2(x + 5)]$. Numerator and denominator have no factors in common, and the denominator is zero at $x = 0$ and $x = -5$. Therefore f has vertical asymptotes at $x = 0$ and $x = -5$. The horizontal asymptote is $y = 0$ because the numerator is a polynomial of lower degree than the denominator.

Near $x = 0$: $f(x) \approx [(3 \cdot 0 - 1)(0 - 2)]/[x^2(0 + 5)] = 2/5x^2$. We see that $f(x) > 0$ if $x < 0$, or if $x > 0$.

Near $x = -5$: $f(x) \approx [(-16)(-7)]/[25(x + 5)] = 112/[25(x + 5)]$. $f(x) > 0$ if $x > -5$, and $f(x) < 0$ if $x < -5$.

Set $f(x) = 0$ to find x-intercepts, $\frac{1}{3}$ and 2. There is no y-intercept. (Why?) Plot enough points, $(-6, -\frac{38}{9})$, $(-2, \frac{7}{3})$, $(1, -\frac{1}{3})$, $(4, \frac{11}{72})$, to sketch an accurate graph. ■

The graph in figure 5.4 crosses the horizontal asymptote $y = 0$ at $x = \frac{1}{3}$ and $x = 2$. This is not unusual. A horizontal asymptote dictates the behavior of a graph for large $|x|$. It says nothing about the behavior of the graph near the origin.

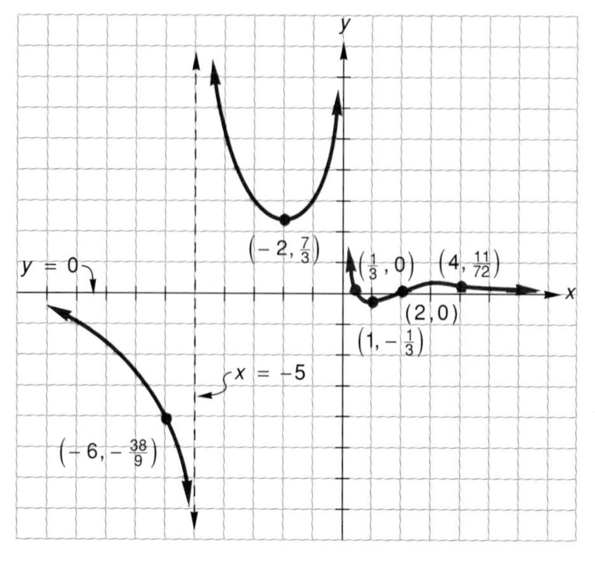

Figure 5.4

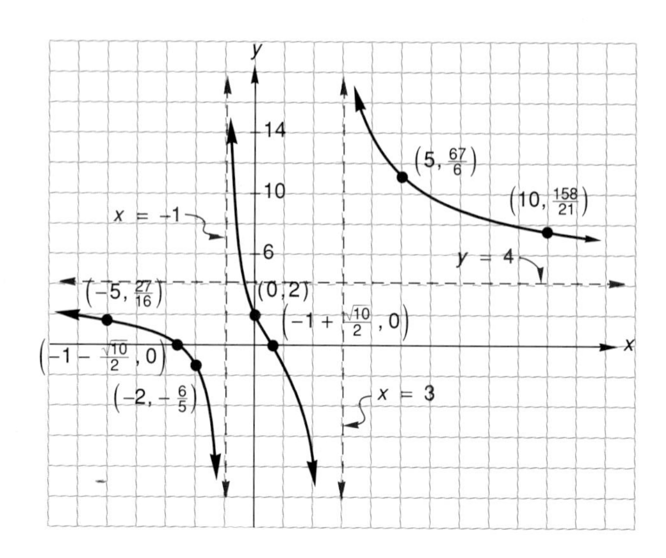

Figure 5.5

Example 7 Graph $g(x) = (4x^2 + 8x - 6)/(x^2 - 2x - 3)$.

Solution $g(x) = (4x^2 + 8x - 6)/(x - 3)(x + 1)$. The vertical asymptotes, $x = 3$ and $x = -1$, and the horizontal asymptote $y = 4$ (see example 5b) are put in place. We then determine the behavior of the graph near $x = 3$ and $x = -1$.

Near $x = 3$: $g(x) \approx 54/[(x - 3)4] = 27/[2(x - 3)]$. Thus, $g(x) > 0$ if $x > 3$, and $g(x) < 0$ if $x < 3$.

Near $x = -1$: $g(x) \approx -10/[-4(x + 1)] = 5/[2(x + 1)]$. Thus, $g(x) > 0$ if $x > -1$, and $g(x) < 0$ if $x < -1$.

Finally, we plot the x and y intercepts and a few other points and then sketch the graph (figure 5.5). ■

Example 8 (an application)

a. A 12π cubic inch cylindrical container is to be fabricated from 20π square inches of sheet material. Determine the dimensions of cylinder.

b. Estimate the dimensions of the 12π cubic inch cylindrical container that requires the least amount of sheet material for fabrication. Estimate that least amount of sheet material.

Solution

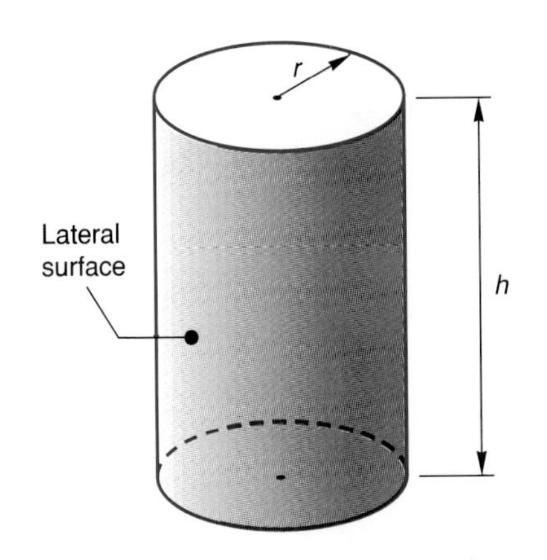

a. The volume and surface area of a cylinder of radius r and height h are given by the formulas $V = \pi r^2 h$ and $A = 2\pi rh + 2\pi r^2$. The terms of the area formula represent the area of the lateral surface ($2\pi rh$) and the combined area of top and bottom ($2\pi r^2$). We solve the volume formula, with $V = 12\pi$, for h. The result is substituted into the area formula to obtain area as a rational function of the radius.

$$h = \frac{12\pi}{\pi r^2} = \frac{12}{r^2}$$

$$A = 2\pi r \frac{12}{r^2} + 2\pi r^2$$

$$A = \frac{24\pi}{r} + 2\pi r^2, \quad r > 0 \qquad \textbf{(1)}$$

In (1), 20π is substituted for A. The resulting equation is solved for r with the methods of "Theory of Equations" from chapter 4.

$$20\pi = \frac{24\pi}{r} + 2\pi r^2$$

$$10 = \frac{12}{r} + r^2$$

$$0 = r^3 - 10r + 12$$

$$r = 2, \quad r = -1 + \sqrt{7}, \quad r = -1 - \sqrt{7}$$

As the radius must be positive, the negative root is rejected. The dimensions of the container are either

$$r = 2,\ h = \frac{V}{\pi r^2} = \frac{12\pi}{4\pi} = 3$$

or

$$r = -1 + \sqrt{7},\ h = \frac{12\pi}{\pi(8 - 2\sqrt{7})}$$

$$r \approx 1.646,\ h \approx 4.431.$$

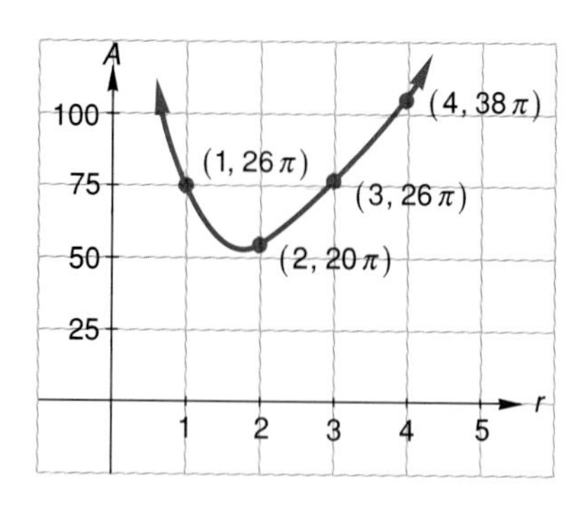

b. To estimate the dimensions that require the least amount of sheet material for fabrication, we graph (1) and locate the lowest point in the graph. The approximated coordinates of the low point are $r \approx 1.8$ and $A \approx 62$. With $r \approx 1.8$, $h \approx 3.7$.

The asymptotic behavior of the graph at $r = 0$ suggests that as r tends to 0, the surface area of the cylinder tends to infinity. Why do you think this is so? ■

r	A
1	26π
2	20π
3	26π
4	38π

Exercise Set 5.1

(A)

Determine the domain of each rational function. Where possible, simplify the fraction.

1. $R(x) = \dfrac{x+1}{x^2-1}$

2. $R(x) = \dfrac{2x-3}{4x^2-9}$

3. $S(x) = \dfrac{x^2-x-6}{x^2+4x+4}$

4. $S(x) = \dfrac{x^3-8}{x^2-7x+10}$

5. $T(x) = \dfrac{3x^2-x-10}{6x^2+10x}$

6. $T(x) = \dfrac{x^2}{(3-x)^2}$

Graph each rational function using examples 6 and 7 as a guide. Show vertical and horizontal asymptotes and x and y intercepts in your graphs.

7. $y = \dfrac{1}{x}$

8. $y = \dfrac{5}{x-3}$

9. $y = \dfrac{2}{(x+2)^2}$

10. $y = \dfrac{3}{x^2}$

11. $y = \dfrac{x+4}{x+3}$

12. $y = \dfrac{2x-1}{3x+6}$

13. $y = \dfrac{3x+6}{x^2-1}$

14. $y = \dfrac{5}{2x^2+5x-3}$

15. $y = \dfrac{3x-2}{4x^2+2}$

16. $y = \dfrac{6x^2+1}{x^2+5}$

17. $y = \dfrac{2x-3}{4x^2-9}$

18. $y = \dfrac{3x^2-x-10}{6x^2+10x}$

19. $y = \dfrac{4x^2}{(2x+3)^2}$

20. $y = \dfrac{x^3}{(x-2)^2}$

21. $y = \dfrac{(x+2)(x-3)}{(2x-5)(x+1)^2}$

22. $y = \dfrac{x-6}{(x+3)^2(x-3)^3}$

23. $y = \dfrac{x^2+5x+6}{x+2}$

24. $y = \dfrac{9x^2-16}{3x+4}$

25. $y = \dfrac{(x+1)^2}{x^2+3x+2}$

26. $y = \dfrac{8x^2-14x-15}{2x^2-3x-5}$

27. $y = \dfrac{2x^2+1}{x^4+x^3+4x^2+4x}$

28. $y = \dfrac{x(x-3)^2}{x^3+5x^2+2x-8}$

(B)

29. Translate and/or reflect the graph of $f(x) = 1/x$ to obtain graphs of

a. $f(x-2)$

b. $-f(x)$

c. $f(-x)$

d. $f(x+1)+2$.

Write a defining equation for each of the functions in **a–d.**

30. Repeat exercise 29 for $g(x) = (x+4)/(x+3)$.

31. The volume of a rectangular container is to be 144 cubic inches. The base of the container is square.
    a. Express the surface area of the container as a rational function of the width of the base.
    b. Assume that the surface area of the container is to be 168 square inches. Find the dimensions of the container.
    c. From a graph of the function in **a**, estimate the dimensions of the 144 cubic inch container that requires the least amount of sheet material for its fabrication. Estimate the minimum surface area.
    d. What is the implication of the vertical asymptote in the graph?
32. The volume of a cylindrical container is to be 60π cubic centimeters.
    a. Express the surface area of the cylinder as a rational function of its radius.
    b. Assume that the surface area of the cylinder is to be 62π square centimeters. Find the dimensions of the cylinder.
    c. From a graph of the function in **a**, estimate the dimensions of the 60π cc container that requires the least amount of sheet material for its fabrication. Estimate the minimum surface area.
    d. What is the implication of the vertical asymptote in the graph?
33. The effectiveness (E) of a painkilling drug is given by the equation $E = 2t/(t^2 + 10)$, where t is time in hours. From a graph of the equation, estimate when, after taking it, the drug is most effective. What is the implication of the horizontal asymptote in the graph?
34. The annual cost (C) (in units of 100,000) to a company for the removal of a certain pollutant from its smokestack emissions is given by the formula $C = 4.8p/(100 - p)$. The percentage of pollutant removed is denoted by p. Graph the rational function. What is the domain of the function? What is the implication of the vertical asymptote in the graph?

A graph of $f(x) = 2/(x - 1)$ has the horizontal asymptote $y = 0$. This means that for large $|x|$, $|f(x) - 0|$ can be made very small (see figure). If

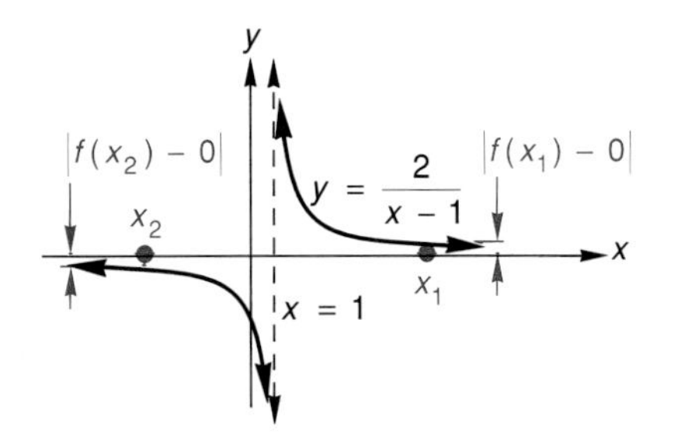

we want to find an integer c so that if $|x| > c$, then $|f(x) - 0| < 0.01$, we can proceed as follows. We write

$$\left|\frac{2}{x-1} - 0\right| < 0.01 \quad (1)$$

$$\frac{|x-1|}{2} > 100 \quad (2)$$

$$|x - 1| > 200. \quad (3)$$

But (3) can be written in the form

$$x - 1 < -200 \quad \text{or} \quad x - 1 > 200. \quad (4)$$

Therefore, $$x < -199 \quad \text{or} \quad x > 201. \quad (5)$$

If $|x| > 201$, then $x < -201$ or $x > 201$, and (5) is satisfied. It follows that if $|x| > 201$, then (1) is true. Thus $c = 201$ or any integer greater than 201.

35. Let $f(x) = 5/(x + 4)$. Find an integer c so that if $|x| > c$, then $|f(x) - 0| < 0.01$.
36. Let $g(x) = (3x + 1)/(2x - 1)$. Find an integer c so that if $|x| > c$, then $|g(x) - \frac{3}{2}| < 0.01$.
37. Let $h(x) = (-3x + 2)/(x + 5)$. Find an integer c so that if $|x| > c$, then $|h(x) - (-3)| < 0.005$.

38. The graphs of the rational functions in examples 6 and 7 cross their horizontal asymptotes. However, a graph of a rational function cannot cross its vertical asymptotes. Why?

The functions defined in exercises 39–42 are not rational. Nevertheless, a graph of each function has asymptotes. Sketch a graph of each function and show the graph's asymptotes.

39. $y = \dfrac{x}{\sqrt{x^2 - 1}}$
40. $y = \dfrac{3x}{\sqrt{x^2 + 1}}$
41. $y = \dfrac{x + 1}{\sqrt{x^2 - 6x + 9}}$
42. $y = \dfrac{\sqrt{x - 1} - 1}{\sqrt{x - 2}}$

5.2 Graphing Rational Functions II Oblique Asymptotes

The rational function $h(x) = (x^3 + 2)/(x^2 - 4)$ does not have a horizontal asymptote because $h(x) \approx x$ for large $|x|$ (see example 5c, in section 5.1). As $|x|$ tends to ∞, $h(x)$ assumes values closer and closer to x. A graph of h does not "settle down" to some horizontal line, but it tends to approach closer and closer to the oblique line $y = x$. We call that line an **oblique asymptote** of the graph of h.

Example 1 Graph $h(x) = (x^3 + 2)/(x^2 - 4)$.

Solution Because $h(x) = (x^3 + 2)/[(x - 2)(x + 2)]$, a graph of h has vertical asymptotes $x = 2$ and $x = -2$. We have already shown that the graph has the oblique asymptote $y = x$.

> Near $x = -2$: $h(x) \approx -6/[-4(x + 2)] = 3/[2(x + 2)]$. If $x < -2$, $h(x) < 0$. If $x > -2$, $h(x) > 0$.
>
> Near $x = 2$: $h(x) \approx 10/[4(x - 2)] = 5/[2(x - 2)]$. If $x < 2$, $h(x) < 0$. If $x > 2$, $h(x) > 0$.

We now know how the graph behaves near its vertical asymptotes. We also know that for large $|x|$, the graph draws close to the line $y = x$. All that remains to be done before sketching the graph is to locate the x and y intercepts and to plot a few points (figure 5.6).

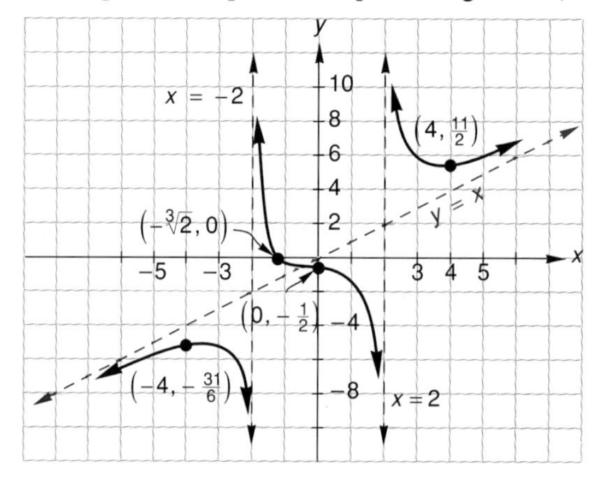

Figure 5.6

■

Requirements for an Oblique Asymptote

If $R(x) = P(x)/Q(x)$ is a rational function in which P and Q have no factors in common, and *the degree of P is one greater than the degree of Q,* then a graph of R has an oblique asymptote.

Example 2 Show that each rational function has an oblique asymptote.

a. $f(x) = \dfrac{5x^2 - 3x + 4}{x + 1}$

b. $g(x) = \dfrac{3x^4 - 2x^3 + 5x^2 - x + 4}{x^3 + x^2 + 2x - 1}$

Solution

a. Divide $5x^2 - 3x + 4$ by $x + 1$.

$$\begin{array}{r|rrr} & 5 & -3 & 4 \\ \hline -1 & 5 & -8 & 12 \end{array}$$

$$5x^2 - 3x + 4 = (x + 1)(5x - 8) + 12, \quad \text{or}$$
$$\frac{5x^2 - 3x + 4}{x + 1} = (5x - 8) + \frac{12}{x + 1}.$$

For large $|x|$, $12/(x + 1) \approx 0$. Therefore, if $|x|$ is large,

$$f(x) \approx 5x - 8.$$

A graph of f has the oblique asymptote $y = 5x - 8$.

b. Divide $3x^4 - 2x^3 + 5x^2 - x + 4$ by $x^3 + x^2 + 2x - 1$.

$$\begin{array}{r|l} & 3x - 5 \\ \hline x^3 + x^2 + 2x - 1 & 3x^4 - 2x^3 + 5x^2 - x + 4 \\ & 3x^4 + 3x^3 + 6x^2 - 3x \\ \hline & -5x^3 - x^2 + 2x + 4 \\ & -5x^3 - 5x^2 - 10x + 5 \\ \hline & 4x^2 + 12x - 1 \end{array}$$

$$\frac{3x^4 - 2x^3 + 5x^2 - x + 4}{x^3 + x^2 + 2x - 1} = (3x - 5) + \frac{4x^2 + 12x - 1}{x^3 + x^2 + 2x - 1}$$

For large $|x|$,

$$\frac{4x^2 + 12x - 1}{x^3 + x^2 + 2x - 1} = \frac{\dfrac{4x^2 + 12x - 1}{x^3}}{\dfrac{x^3 + x^2 + 2x - 1}{x^3}}$$
$$= \frac{\dfrac{4}{x} + \dfrac{12}{x^2} - \dfrac{1}{x^3}}{1 + \dfrac{1}{x} + \dfrac{2}{x^2} - \dfrac{1}{x^3}}$$
$$\approx \frac{0 + 0 - 0}{1 + 0 + 0 - 0}$$
$$= 0.$$

Thus, if $|x|$ is large, $g(x) \approx 3x - 5$. A graph of g has the oblique asymptote $y = 3x - 5$. ■

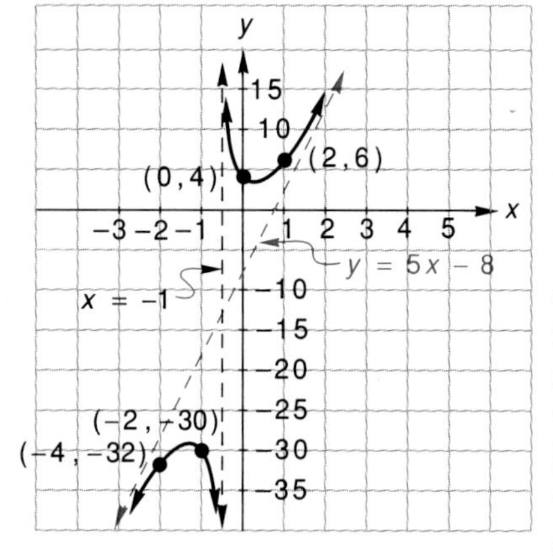

Figure 5.7

Example 3 Graph $f(x) = (5x^2 - 3x + 4)/(x + 1)$.

Solution A graph has the vertical asymptote $x = -1$ and the oblique asymptote $y = 5x - 8$ (example 2a). In addition, the numerator of the fraction defining f is always positive. Equation $5x^2 - 3x + 4 = 0$ has no real roots so that a graph of $y = 5x^2 - 3x + 4$ must lie above the x-axis. (Why?) Since the numerator of f is positive for all real numbers, $f(x) > 0$ if $x > -1$. Also, $f(x) < 0$ if $x < -1$. The behavior of the graph near its vertical asymptote is now known. We also know that for large $|x|$, the graph draws close to its oblique asymptote $y = 5x - 8$. We plot a few points and sketch the graph (figure 5.7). ■

Exercise Set 5.2

(A)

Use examples 1, 2, and 3 as a guide for graphing each rational function. Label the x and y intercepts and the vertical and oblique asymptotes on your graph.

1. $y = \dfrac{x^2 - 4x + 9}{x - 4}$

2. $y = \dfrac{x^3 - 4x}{x^2 - 1}$

3. $y = \dfrac{(x + 3)(x - 2)}{x + 1}$

4. $y = \dfrac{(x - 5)(x + 2)}{x - 3}$

5. $y = \dfrac{x(x^2 \quad 1)}{x^2 - 4}$

6. $y = \dfrac{(x + 3)^3}{x^2}$

7. $y = \dfrac{x^3 - 8}{x^2 - 4}$

8. $y = \dfrac{x^3 + 1}{x^3 + 3x + 2}$

(B)

9. $y = \dfrac{x^3 + 3x^2 - 6x - 8}{2x^2 - 9x - 5}$

10. $y = \dfrac{x^3 + x^2 - 17x + 15}{3x^2 + 14x + 16}$

Show that a graph of each function has a second degree polynomial as an asymptote.

11. $f(x) = \dfrac{x^3 - 1}{x - 2}$

12. $f(x) = \dfrac{x^4 - 2x^3 + x^2 + 6x - 9}{x^2 + 3x - 2}$

5.3 Partial Fractions

It is not very difficult to find the sum of two rational expressions.

$$\frac{3}{x+2} + \frac{7}{x-1} = \frac{10x+11}{(x+2)(x-1)}$$

A reversal of the addition process, in which we decompose the fraction on the right side of the equation to obtain the sum of fractions on the left side, is more challenging. We call $3/(x+2) + 7/(x-1)$ a **partial fraction decomposition** of $(10x+11)/[(x+2)(x-1)]$.

Assume that it is possible to find values for constants A and B so that

$$\frac{10x+11}{(x+2)(x-1)} = \frac{A}{x+2} + \frac{B}{x-1}.$$

The assumption is reasonable because the sum of the fractions

$$\frac{A(x-1)+B(x+2)}{(x+2)(x-1)},$$

has the same denominator as the left member of the equation. The fraction sum also has a linear polynomial in the numerator. Thus, we write

$$\frac{10x+11}{(x+2)(x-1)} = \frac{A(x-1)+B(x+2)}{(x+2)(x-1)},$$

and

$$10x + 11 = A(x-1) + B(x+2). \tag{1}$$

It can be shown that if two nth degree polynomials are equal, their corresponding coefficients must be equal (see exercise 21). Hence, from (1)

$$10x + 11 = (A + B)x + (-A + 2B)$$

$$\begin{aligned} 10 &= A + B \\ 11 &= -A + 2B \\ \hline 21 &= 3B \\ 7 &= B \\ 3 &= A. \end{aligned}$$

It follows that $(10x+11)/[(x+2)(x-1)] = 3/(x+2) + 7/(x-1)$.

There is another way of determining coefficients A and B. We alternately let x equal 1 and -2 in $10x + 11 = A(x-1) + B(x+2)$, to obtain

$$10(1) + 11 = A(1-1) + B(1+2)$$

and

$$10(-2) + 11 = A(-2-1) + B(-2+2).$$

The two equations yield $B = 7$ and $A = 3$.

Example 1 Find a partial fraction decomposition of

$$\frac{3x^2 - 5x + 12}{x^3 + 2x^2 - 7x + 4}.$$

Solution The zeros of the denominator polynomial are found by the methods detailed in chapter 4.

$$\frac{3x^2 - 5x + 12}{x^3 + 2x^2 - 7x + 4} = \frac{3x^2 - 5x + 12}{(x-1)^2(x+4)}$$
$$= \frac{A}{(x-1)^2} + \frac{B}{(x-1)} + \frac{C}{(x+4)},$$

where $A/(x-1)^2 + B/(x-1) + C/(x+4)$ is the assumed form of the partial fraction decomposition.

$$\frac{3x^2 - 5x + 12}{(x-1)^2(x+4)} = \frac{A(x+4) + B(x-1)(x+4) + C(x-1)^2}{(x-1)^2(x+4)}$$

The fractions are equal and their denominators are equal. Therefore, their numerators must be equal.

$$3x^2 - 5x + 12 = A(x+4) + B(x-1)(x+4) + C(x-1)^2 \quad (2)$$

Let $x = 1$, then $10 = 5A$ and $A = 2$. Let $x = -4$, then $80 = 25C$ and $C = \frac{16}{5}$. Equating second degree coefficients in (2) yields $3 = B + C$. This equation together with $C = \frac{16}{5}$, gives us $B = -\frac{1}{5}$. We conclude that

$$\frac{3x^2 - 5x + 12}{(x-1)^2(x+4)} = \frac{2}{(x-1)^2} - \frac{\frac{1}{5}}{x-1} + \frac{\frac{16}{5}}{x+4}.$$

■

On a second reading of example 1, it might occur to you that a possible partial fraction decomposition of $(3x^2 - 5x + 12)/[(x-1)^2(x+4)]$ is

$$\frac{D}{(x-1)^2} + \frac{E}{x+4}. \quad (3)$$

After all, the fraction sum

$$\frac{D(x+4) + E(x-1)^2}{(x-1)^2(x+4)},$$

has the same denominator as the given fraction. It also has a second degree polynomial numerator, as does the given fraction. If we assume the partial fraction decomposition given in (3), then

$$\begin{aligned}\frac{3x^2 - 5x + 12}{(x-1)^2(x+4)} &= \frac{D}{(x-1)^2} + \frac{E}{x+4} \\ &= \frac{D(x+4) + E(x-1)^2}{(x-1)^2(x+4)} \\ &= \frac{Ex^2 + (D-2E)x + (4D+E)}{(x-1)^2(x+4)}.\end{aligned}$$

As a result,

$$\begin{aligned}3x^2 - 5x + 12 &= Ex^2 + (\mathrm{D} - 2E)x + (4D + E), \quad \text{and} \\ 3 &= E \\ -5 &= D - 2E \\ 12 &= 4D + E.\end{aligned}$$

This set of three equations in two unknowns has no solution. Hence, (3) is not a viable partial fraction decomposition of the given rational expression. We consider still another possible decomposition of this fraction in exercise 19.

Example 2 Find a partial fraction decomposition of $(6x - 8)/(x^3 - 2x + 4)$.

Solution

$$\begin{aligned}\frac{6x-8}{x^3 - 2x + 4} &= \frac{6x-8}{(x+2)(x-(1+i))(x-(1-i))} \\ &= \frac{6x-8}{(x+2)(x^2-2x+2)}\end{aligned}$$

We assume a partial fraction decomposition of the form

$$\frac{A}{x+2} + \frac{Bx+C}{x^2-2x+2}$$

rather than

$$\frac{D}{x+2} + \frac{E}{x^2-2x+2},$$

even though each sum has the same denominator, $(x + 2)(x^2 - 2x + 2)$.

$$\begin{aligned}\frac{6x - 8}{(x + 2)(x^2 - 2x + 2)} &= \frac{A}{x + 2} + \frac{Bx + C}{x^2 - 2x + 2}\\ &= \frac{A(x^2 - 2x + 2) + (Bx + C)(x + 2)}{(x + 2)(x^2 - 2x + 2)}\\ &= \frac{(A + B)x^2 + (-2A + 2B + C)x + (2A + 2C)}{(x + 2)(x^2 - 2x + 2)}\end{aligned}$$

$$\begin{aligned}0 &= A + B\\ 6 &= -2A + 2B + C\\ -8 &= 2A + 2C\end{aligned}$$

Solving the system of equations gives us $A = -2$, $B = 2$, and $C = -2$. ■

If, in example 2, we assume a different decomposition, namely

$$\begin{aligned}\frac{6x - 8}{(x + 2)(x^2 - 2x + 2)} &= \frac{D}{x + 2} + \frac{E}{x^2 - 2x + 2}\\ &= \frac{Dx^2 + (-2D + E)x + (2D + 2E)}{(x + 2)(x^2 - 2x + 2)},\end{aligned}$$

then

$$\begin{aligned}0 &= D\\ 6 &= -2D + E\\ -8 &= 2D + 2E.\end{aligned}$$

This system of equations has no solution. You are asked to consider another possible decomposition of the fraction in example 2 in exercise 18.

Although we have discussed only two examples, the rational functions decomposed in those examples are representative. A real polynomial (one having only real coefficients) can be factored into some combination of linear and irreducible quadratic factors (see exercise 20). Hence, any rational expression decomposition must involve a sum of fractions whose denominators contain only first or second degree polynomial factors. It is possible to show that each rational expression has a unique partial fraction decomposition that is determined by the linear and quadratic factors of its denominator. But the development is long and cumbersome, and unnecessary for our purposes here. The partial fraction decomposition forms that come out of the formal theory can be obtained from the trial and error approach of examples 1 and 2.

Exercise Set 5.3

(A)

In Exercises 1–16, decompose each rational expression into partial fractions.

1. $\dfrac{5x - 3}{(2x - 5)(3x + 1)}$

2. $\dfrac{x + 12}{(2x - 1)(x + 2)}$

3. $\dfrac{3x + 1}{x^2 - x - 2}$

4. $\dfrac{2x - 3}{x^2 + 5x + 6}$

5. $\dfrac{3x + 5}{(x + 1)^2}$

6. $\dfrac{4x - 1}{(x - 2)^2}$

7. $\dfrac{x^2 - 5x + 2}{x^2(x - 1)}$

8. $\dfrac{6x^2 - 1}{(x + 3)^2(x + 5)}$

9. $\dfrac{5x^2 + 6x}{(x - 2)(x^2 + 4)}$

10. $\dfrac{2x^2 + x - 3}{(x + 4)(x^2 + 9)}$

11. $\dfrac{2x - 3}{(x + 1)(x^2 - 6x + 13)}$

12. $\dfrac{x + 7}{(x - 3)(x^2 - 2x + 17)}$

13. $\dfrac{x}{(x + 1)(x + 2)(x + 3)}$

14. $\dfrac{2x^2 + 10x + 2}{(x + 1)(x - 2)(x + 3)}$

(B)

15. $\dfrac{3x^3 + x - 2}{(x + 2)^2(x - 1)^2}$

16. $\dfrac{2x^2 + 5x + 1}{(x + 1)^2(x - 5)^2}$

17. In example 1, we found a partial fraction decomposition of
$$\frac{3x^2 - 5x + 12}{(x - 1)^2(x + 4)}$$
to have the form
$$\frac{A}{(x - 1)^2} + \frac{B}{x - 1} + \frac{C}{x + 4}.$$
Another possible decomposition form,
$$\frac{D}{(x - 1)^2} + \frac{E}{x + 4},$$
was not viable. Is it possible to decompose the rational expression in the form
$$\frac{Fx + G}{(x - 1)^2} + \frac{H}{x + 4}?$$
Support your answer.

18. In example 2, we found a decomposition of
$$\frac{6x - 8}{(x + 2)(x^2 - 2x + 2)}$$
to have the form
$$\frac{A}{x + 2} + \frac{Bx + C}{x^2 - 2x + 2}.$$
A second possible decomposition, in the form $D/(x + 2) + E/(x^2 - 2x + 2)$, did not work. Do you think that you can decompose the given rational expression in the form $A/(x + 2) + Bx/(x^2 - 2x + 2)$? Support your answer.

19. Assume that $P(x) = a_nx^n + a_{n-1}x^{n-1} + \cdots + a_1x + a_0$ and $Q(x) = b_nx^n + b_{n-1}x^{n-1} + \cdots + b_1x + b_0$ are equal for all x. It must follow that $a_i = b_i$, $i = 0, 1, 2 \ldots, n$. To see this, subtract Q from P.
$$P(x) - Q(x) = (a_n - b_n)x^n + (a_{n-1} - b_{n-1})x^{n-1} + \cdots + (a_1 - b_1) + (a_0 - b_0).$$
Assume that $a_i - b_i$ is not equal to 0 for all i. Then, reading from the left, let $a_j - b_j$ be the first nonzero coefficient. $P - Q$ is therefore a polynomial of degree j, $1 \le j \le n$, (**a.** Why can't $j = 0$?) that has more than j roots. (**b.** Why?) This is a contradiction. (**c.** Why?) Hence, $a_j - b_j = 0$ for all j.

20. Knowing that a real nth degree polynomial having exacty n zeros in C can be factored as
$$a_n(x - r_1)(x - r_2) \cdots (x - r_n),$$
and that a complex root must have a complex conjugate partner, argue that any real polynomial can only be factored into real linear factors and real irreducible quadratic factors.

Chapter 5 Review

Major Points for Review

Section 5.1 Definition of a rational function

Domain of a rational function

When a graph of a rational function has a vertical asymptote

Behavior of a graph of a rational function near its vertical asymptote

Rules for finding horizontal asymptotes of graphs of rational functions

Behavior, for large $|x|$, of a graph that has a horizontal asymptote

Graphing rational functions

Section 5.2 When a graph of a rational function has an oblique asymptote

Constructing rational function graphs that have oblique asymptotes

Section 5.3 Determining the partial fraction decomposition of a rational expression

Review Exercises

Graph each rational function. Label horizontal, vertical, and oblique asymptotes.

1. $y = \dfrac{-1}{x + 2}$

2. $y = \dfrac{3}{x - 5}$

3. $y = \dfrac{3x - 6}{x + 1}$

4. $y = \dfrac{x + 4}{x}$

5. $y = \dfrac{2}{x^2 + 2}$

6. $y = \dfrac{x - 3}{x^2 - 4}$

7. $y = \dfrac{x(x - 5)}{(2x + 3)(x - 2)^2}$

8. $y = \dfrac{x + 3}{x^2(x - 3)^3}$

9. $y = \dfrac{2x + 5}{(x - 1)(x + 4)^3}$

10. $y = \dfrac{2x^2}{3x^2 + 10x - 8}$

11. $y = \dfrac{(x - 1)^2}{x + 1}$

12. $y = \dfrac{3x^2 + 16x - 12}{x - 5}$

13. $y = \dfrac{x^3 - 2x^2 - x + 2}{2x^2 + 3x - 9}$

14. $y = \dfrac{4x^3 - 4x^2 - 39x - 36}{3x^2 - 5x - 8}$

Find a partial fraction decomposition for each rational expression.

15. $\dfrac{8x + 5}{(x - 2)(3x + 4)}$

16. $\dfrac{x^2 + 3x - 5}{(x - 1)(2x - 3)(x + 2)}$

17. $\dfrac{2x^2 + x + 3}{(x - 2)^2(x - 6)}$

18. $\dfrac{x^2 + 6}{(x - 3)^2(x + 1)}$

19. $\dfrac{x^3 + 3x^2 - 6x - 8}{(x^2 + 2)(x^2 - 2x + 3)}$

20. $\dfrac{5x^2 + 3x - 2}{(2x + 1)^2(x^2 + x + 1)}$

Graph each function. Label any asymptotes.

21. $y = \frac{2x}{\sqrt{x^2 - 9}}$ **22.** $y = -\frac{3}{\sqrt{x + 2}}$

Translate and/or reflect the graph of each function to obtain the graphs of (a) $f(x + 3)$, (b) $-f(x)$, (c) $f(-x)$, (d) $f(x - 1) - 2$.

23. $f(x) = -\frac{2}{x^2}, x > 0$ **24.** $f(x) = \frac{x - 1}{x + 2}$

25. The volume of a rectangular container whose base width is twice its depth, is to be 120 cubic inches.

a. Express the surface area of the container as a rational function of the depth of the base.

b. If the surface area of the container is to be 172 square inches, what must the container's dimensions be?

c. Graph the function in **a.** Use the graph to estimate the dimensions of the 120 cubic inch container that requires the least amount of sheet material for fabrication.

d. What is the significance of the vertical asymptote in the graph?

26. The population of a certain range animal living within a national park is given by $P = (t^3 + 10)/(t^2 + 1)$, where t is time in years. The formula was designed to predict population over a 100-year period beginning ten years ago.

a. What was the park population of this particular range animal ten years ago?

b. A graph of the population function has an asymptote if the domain of t is $t \geq 0$. What is the implication of the asymptote?

Chapter 5 Test 1

Graph each rational function. Label all asymptotes.

1. $y = \frac{5}{x - 2}$ **2.** $y = \frac{2x - 5}{x + 3}$

3. $y = \frac{x^2 - 5x + 6}{(x + 1)^2(x - 4)}$ **4.** $y = \frac{x^2 + x - 6}{x - 1}$

Find a partial fraction decomposition of each fraction.

5. $\frac{8x + 3}{(2x - 1)(x + 4)}$ **6.** $\frac{3x^2 - 4x - 1}{(x - 2)^2(x + 5)}$

Chapter 5 Test 2

Graph each rational function. Label all asymptotes.

1. $y = \frac{2x}{(x - 1)^2}$ **2.** $y = \frac{x + 2}{(x - 3)^3}$

3. $y = \frac{x^3 - 3x^2 - x + 3}{(x - 5)^2(2x + 7)}$ **4.** $y = \frac{8x^3 - 125}{2x^2 - 9}$

Find a partial fraction decomposition for each fraction.

5. $\frac{3x - 1}{(x + 1)^2(x - 1)}$ **6.** $\frac{5x^2 - 6x - 4}{(x - 2)^2(x^2 - 2x + 4)}$

Chapter 6

Exponential and Logarithmic Functions

Growth and decay phenomena abound in nature. Mathematical models are constructed to aid us in the study of these phenomena. We investigate how human and animal populations grow and decline, how radioactive materials decay, how a current builds up in an electric circuit, and much more. Models designed to study growth and decay are often constructed with exponential and logarithmic functions.

6.1 Exponential Functions

An experimenter collects data on the growth of a certain bacteria population.

Time	2:00 P.M.	3:00 P.M.	4:00 P.M.	5:00 P.M.	6:00 P.M.
Population	100	205	390	790	1,620

She wishes to construct a population prediction formula based on the assumption that the bacteria population will double every hour as long as its environment remains undisturbed. The experimenter reasons that her formula must take into account

1. the hourly doubling of the population,
2. the starting time,
3. and the starting population.

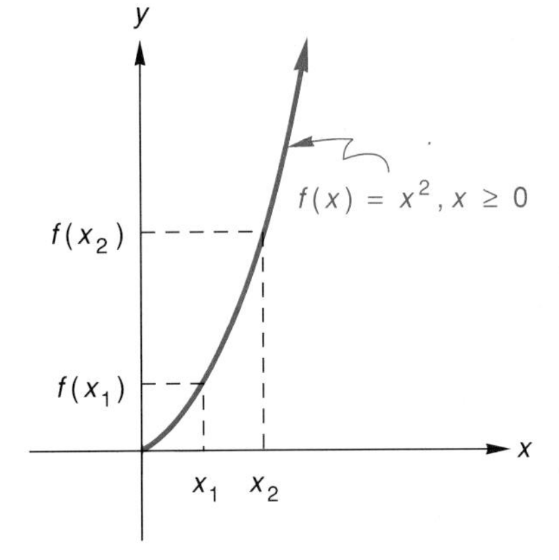

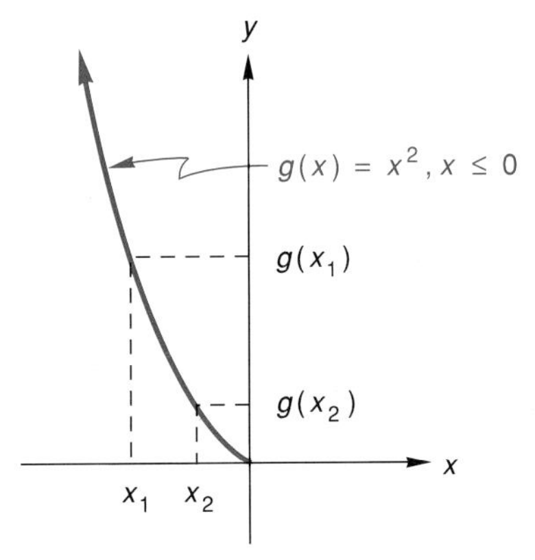

Figure 6.1

She writes

$$P = 100 \cdot 2^t, \tag{1}$$

where P is population at any time, t is the number of hours after 2:00 P.M., and 100 is the bacteria population at 2:00 P.M. She then checks her formula against the observed data.

2:00 P.M., $t = 0$	$P = 100 \cdot 2^0 = 100$
3:00 P.M., $t = 1$	$P = 100 \cdot 2^1 = 200$
4:00 P.M., $t = 2$	$P = 100 \cdot 2^2 = 400$
5:00 P.M., $t = 3$	$P = 100 \cdot 2^3 = 800$
6:00 P.M., $t = 4$	$P = 100 \cdot 2^4 = 1,600$

Since (1) determines populations that closely approximate the observed data, the experimenter accepts the equation as a viable population prediction formula.

Our experimenter may be delighted with her result, but for us (1) poses a quandry. The replacement set for t is the set of nonnegative real numbers, and we have not yet defined 2^t when t is irrational. We do this next, making use of the properties of rational exponents (section 1.2) as well as the properties of increasing and decreasing functions (section 2.1).

The defining properties of increasing and decreasing functions are displayed in figure 6.1. For the increasing function f, if $x_2 > x_1$, then $f(x_2) > f(x_1)$. For the decreasing function g, if $x_2 > x_1$, then $g(x_2) < g(x_1)$.

Now it is not difficult to see that, over the domain $\{x | -4 \leq x \leq 4, x \in J\}$, $f(x) = 2^x$ is an increasing function and $g(x) = (\frac{1}{2})^x$ is a decreasing function (figure 6.2).

Figure 6.2 suggests the following generalization (see exercises 53 and 54).

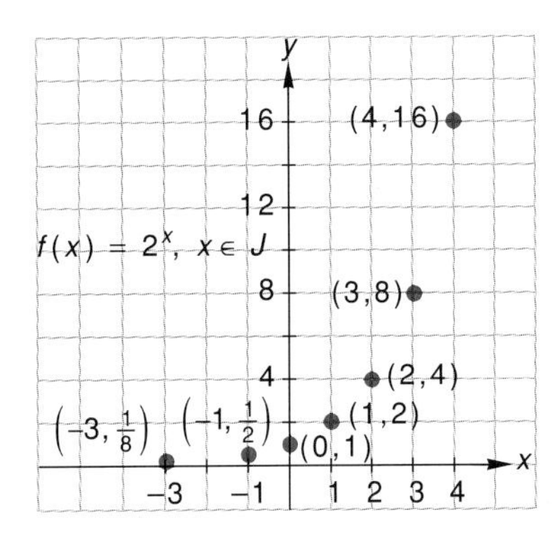

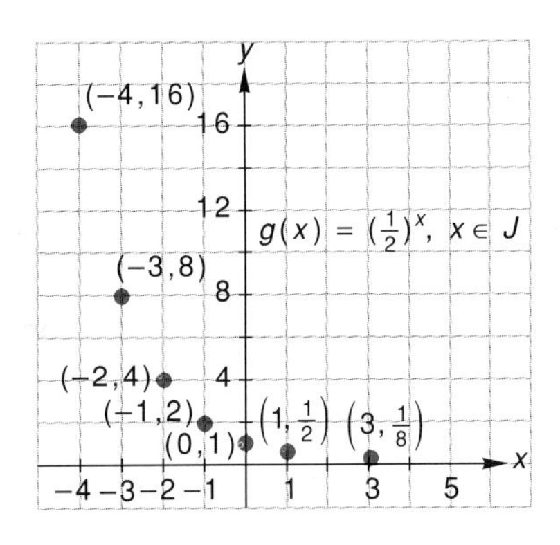

Figure 6.2

Definition

If $f(x) = b^x$ and x is a rational number, then

1. f is increasing if $b > 1$.
2. f is decreasing if $0 < b < 1$.

Defining b^x when x Is Irrational

We use the fact that 2^x is an increasing function when x is rational to argue the existence of the number 2^x when x is irrational. In the argument we let $x = \sqrt{3}$ ($\sqrt{3} \approx 1.73205$), although the discussion that follows can be duplicated for any expression b^x, where b is a positive number different from 1 and x is any irrational number.

Consider the inequalities (2), (3), (4), (5), . . . , each of which follows from the fact that if x is rational, 2^x is an increasing function.

$$2^{1.7} < 2^{1.8} \tag{2}$$
$$2^{1.73} < 2^{1.74} \tag{3}$$
$$2^{1.732} < 2^{1.733} \tag{4}$$
$$2^{1.7320} < 2^{1.7321} \tag{5}$$

Think of the numbers in each inequality as the coordinates of the endpoints of an interval on the number line (figure 6.3). Let us refer to the interval by the number of its associated inequality. Then, because 2^x is an increasing function, interval (2) contains interval (3), (3) contains (4), etc. This infinite set of intervals, called a nested set of intervals, "squeezes" down to a point. There must be a real number associated with this point because of the one-to-one correspondence between the set of real numbers and the points in the number line. We let $2^{\sqrt{3}}$ name this number.

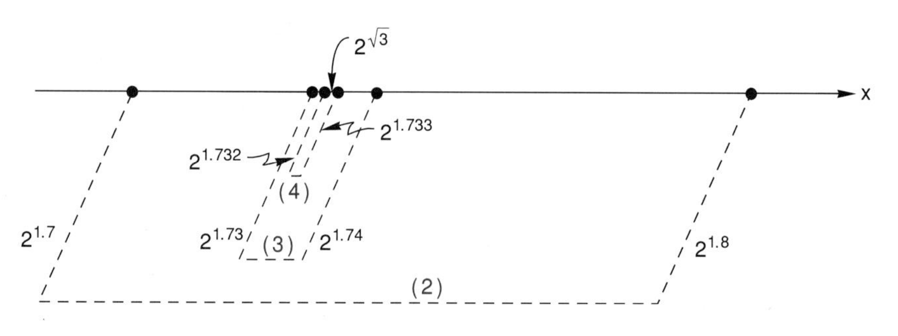

Figure 6.3

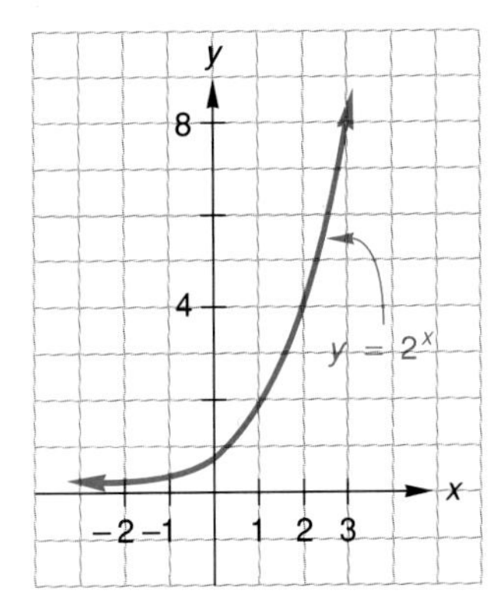

Figure 6.4

Now 2^x is defined for all real x. Hence, $y = 2^x$, $x \in R$ defines a function that is called the exponential function with base 2 (figure 6.4). The general exponential function is defined as follows:

> **Definition**
>
> For each positive number b, $b \neq 1$,
>
> $$y = b^x, \quad x \in R$$
>
> is called the exponential function with base b.

Example 1 Graph $y = 4^x$, $y = (\frac{1}{4})^x$, and $y = 4^{x-2}$.

Solution If $f(x) = 4^x$, then $(\frac{1}{4})^x = (4^{-1})^x = 4^{-x} = f(-x)$. Hence, graphs of $y = 4^x$ and $y = (\frac{1}{4})^x$ are reflections of each other with respect to the y-axis. Also, $4^{x-2} = f(x - 2)$, so that a graph of $y = 4^{x-2}$ can be gotten by shifting a graph of $y = 4^x$ two units to the right.

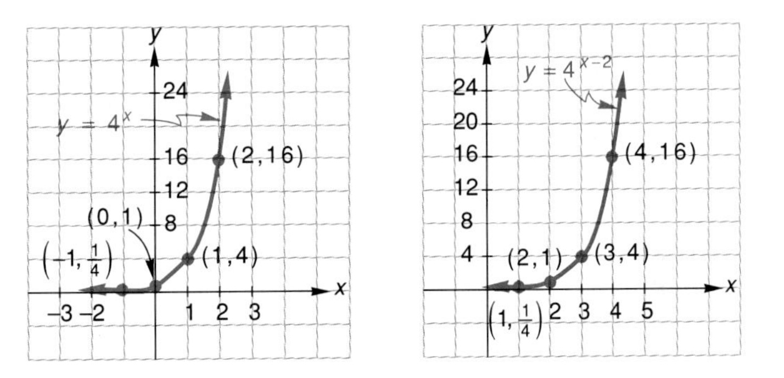

Figure 6.5

■

Example 2 Solve $(x \cdot 4^x - 7 \cdot 4^x)/(x^2 + 5) = 0$.

Solution A fraction is zero only if its numerator is zero.

$$x \cdot 4^x - 7 \cdot 4^x = 0$$
$$4^x(x - 7) = 0$$

If $b > 0$, then $b^x > 0$ for all x (see figures 6.4 and 6.5). Since $4^x > 0$ for all x, the left member of the last equation is zero only if

$$x - 7 = 0$$

or

$$x = 7.$$

■

Example 3 Write an exponential function that predicts the population at any time of a certain range animal whose population is observed to triple every five years. Assume a starting population of 50.

Solution The following table, which depicts population growth over five-year periods, suggests the form of the required exponential function.

t = time in years	0	5	10	15
P = population	50	$150 = 50 \cdot 3$	$450 = 150 \cdot 3 = 50 \cdot 3^2$	$1{,}350 = 450 \cdot 3 = 50 \cdot 3^3$

$$P = 50 \cdot 3^{t/5}$$

■

Properties of Irrational Exponents

The properties of rational exponents in section 1.2 can be extended to irrational exponents. For example, you might be willing to accept that

$$5^{\sqrt{2}} \cdot 5^{\sqrt{3}} = 5^{\sqrt{2}+\sqrt{3}},$$

because

$$5^{3.1} = 5^{1.4} \cdot 5^{1.7} < 5^{\sqrt{2}} \cdot 5^{\sqrt{3}} < 5^{1.5} \cdot 5^{1.8} = 5^{3.3}$$
$$5^{3.14} = 5^{1.41} \cdot 5^{1.73} < 5^{\sqrt{2}} \cdot 5^{\sqrt{3}} < 5^{1.42} \cdot 5^{1.74} = 5^{3.16}$$
$$5^{3.146} = 5^{1.414} \cdot 5^{1.732} < 5^{\sqrt{2}} \cdot 5^{\sqrt{3}} < 5^{1.415} \cdot 5^{1.733} = 5^{3.148}.$$

The intervals $(5^{3.1}, 5^{3.3})$, $(5^{3.14}, 5^{3.16})$, $(5^{3.146}, 5^{3.148})$, . . . , etc., squeeze down to one number that we call $5^{\sqrt{2}+\sqrt{3}}$.

Arguments such as that above can be formalized with the aid of calculus to show that all the properties that hold for rational exponents hold also for irrational exponents. Thus,

1. $\dfrac{5^{3\sqrt{2}}}{5^{\sqrt{2}}} = 5^{2\sqrt{2}}$
2. $(3 \cdot 5)^{\sqrt{3}} = 3^{\sqrt{3}} \cdot 5^{\sqrt{3}}$
3. $\left(\dfrac{2}{x}\right)^{\sqrt{10}} = \dfrac{2^{\sqrt{10}}}{x^{\sqrt{10}}}$
4. $(x^{\sqrt{5}})(x)(x^{3\sqrt{5}}) = x^{1+4\sqrt{5}}$
5. $(x^{-\sqrt{2}})^{4\sqrt{2}} = x^{-8}$.

Exercise Set 6.1

(A)

Specify whether the depicted function is increasing, decreasing, or neither increasing nor decreasing.

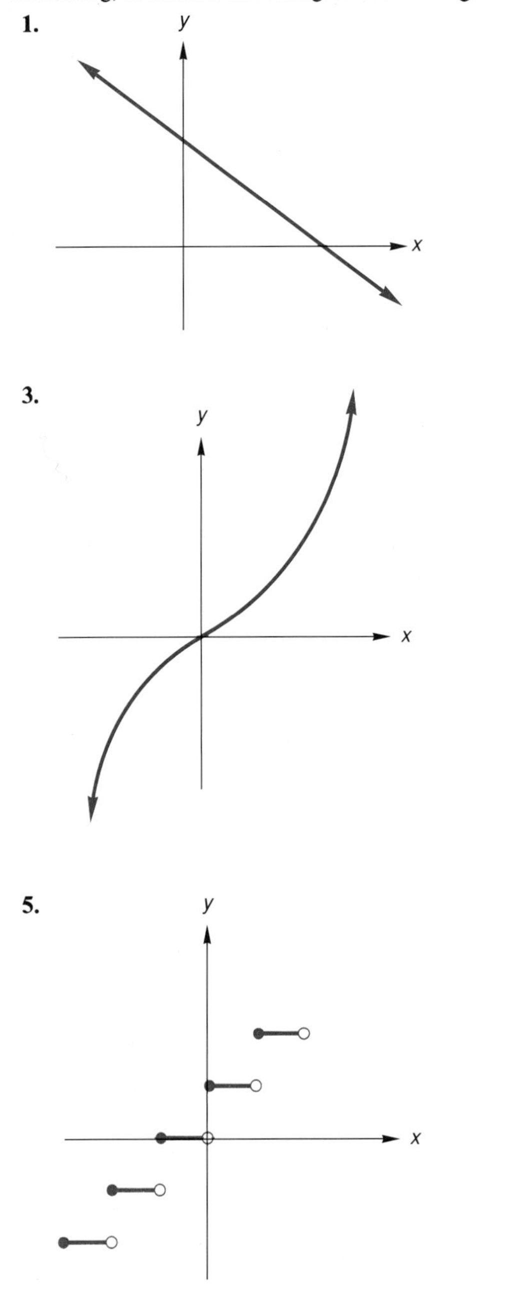

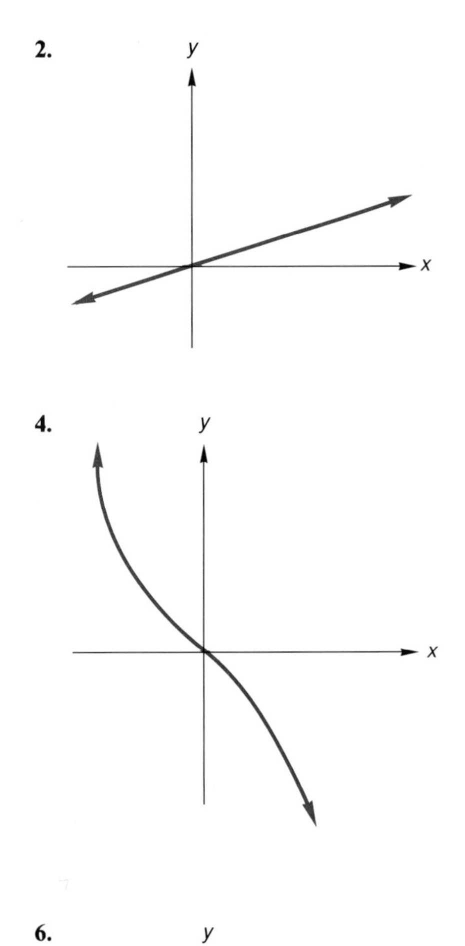

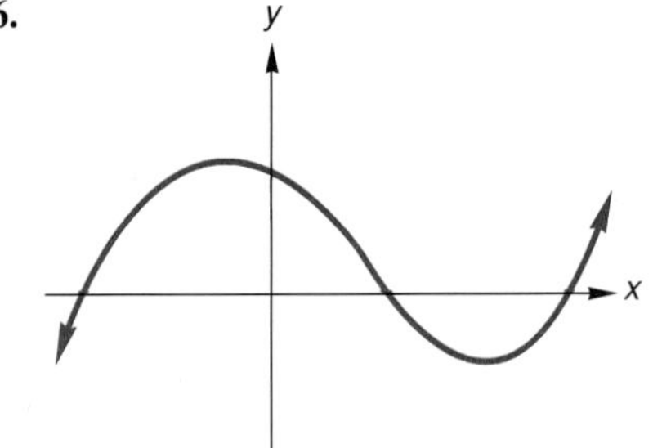

Indicate the portions of the domain over which each depicted function is increasing or decreasing.

7.

8.

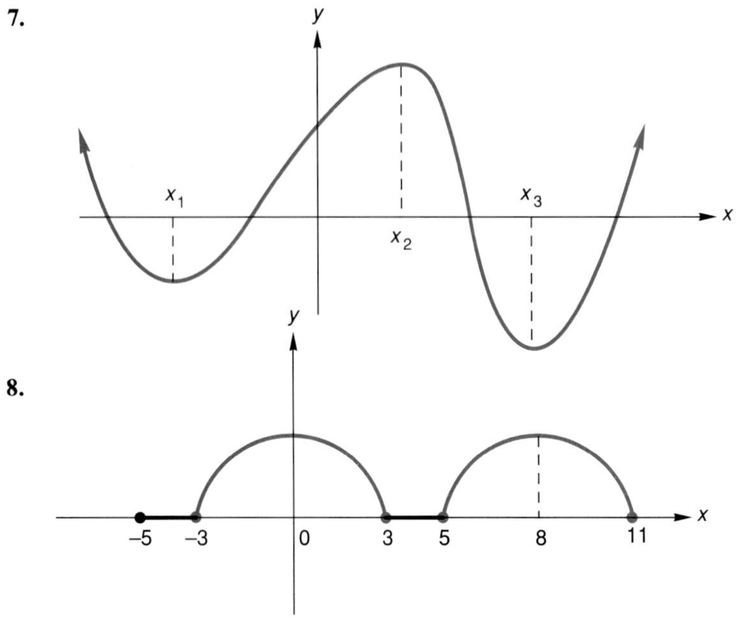

Graph each pair of functions on the same coordinate axes.

9. $y = 3^x,\ y = (\frac{1}{3})^x$

10. $y = 5^x,\ y = (\frac{1}{5})^x$

11. $y = 2^x,\ y = 6^x$

12. $y = (\frac{1}{6})^x,\ y = (\frac{1}{10})^x$

13. $y = |3^x|,\ y = 3^{|x|}$

14. $y = 3^{-x},\ y = -(3^x)$

15. The number of bacteria in a certain culture is given by the formula $P = 1{,}000 \cdot 2^{t/3}$, where t is time in hours after the instant the culture population becomes 1,000. Find P when

a. $t = 6$,

b. $t = 15$,

c. $t = 20$.

16. The size of a certain animal population is predicted by the formula $P = 200 \cdot (\frac{1}{2})^{t/10}$, where t is the time in years after the current year. Find P when

a. $t = 10$,

b. $t = 20$,

c. $t = 35$.

17. The highway department in a certain city predicts the future population of that city from the formula $P = 20{,}000 \cdot 10^{0.0132t}$, where t is the difference in years between the prediction year and the current year. Find P if

a. $t = 10$,

b. $t = 20$,

c. $t = 30$.

18. The pressure p, and volume v, of a gas are related by the formula $pv^{1.7} = 200$. Find the pressure if the volume is 8 cubic feet.

Simplify each expression. Assume that the properties of rational exponents extend to irrational exponents.

19. $3^{\sqrt{2}} \cdot 3^{\sqrt{3}}$

20. $x^{\sqrt{2}+1} \cdot x^{1-\sqrt{2}}$

21. $\dfrac{5^{\sqrt{7}+3}}{5^{\sqrt{5}+2}}$

22. $\dfrac{y^{\sqrt{12}}}{y^{-\sqrt{3}}}$

23. $[(x+1)^{\sqrt{2}}]^{\sqrt{5}}$

24. $[4^{\sqrt{3}+1}]^2$

Rewrite each expression as distinct powers in x and y.

25. $(x^2y)^{\sqrt{3}-2}$

26. $(xy^3)^{\sqrt{2}+1}$

27. $\left(\dfrac{x}{y}\right)^{\pi}$

28. $\left(\dfrac{x+1}{y-3}\right)^{\sqrt{5}+\sqrt{2}}$

(B)

Sketch a graph for each function given that $f(x) = 2^x$.

29. $f(x + 3)$

30. $f(x - 1)$

31. $f(-x)$

32. $-f(x)$

33. $f(x) + 2$

34. $f(x) - 1$

35. $2f(x)$

36. $\frac{1}{2}f(x)$

37. $3f(x + 1) - 2$

38. $-3f(x + 2) + 4$

39. A certain bacteria population doubles every four hours. Write an equation that gives the bacteria population as a function of t, where t is time in hours. Assume that the initial population is 500.

40. A certain range animal population triples every ten years. Write an equation that gives the animal population as a function of t, where t is time in years. Assume an initial population of 200 animals.

41. A certain fish population is observed over a number of years in a controlled environment. The initial population is 1,000. Subsequent population counts are taken at the end of each year for five years. Write an equation relating fish population P, to time t, where t is in years. Check your equation against the data in the table.

Time (years)	1	2	3	4	5
Fish Population	660	450	312	201	130

42. A piece of cardboard, $\frac{1}{4}$ inch thick, is cut in half, and the two halves are placed one on top of the other. The $\frac{1}{2}$ inch thickness of cardboard is cut in half, and again the two halves are placed one on top of the other. The cutting and stacking continues until a column of cardboard is formed that reaches from the earth to the moon, 240,000 miles away. How many cuts are necessary to reach the moon?

Solve each equation by the iteration method discussed in section 4.5. Find solutions correct to the indicated number of decimal places.

43. $x + 2^{x-1} = 2.5$, 2 decimal places

44. $200x - 250 = 10^x$, 3 decimal places

Given a function f, graph f and f^{-1} (inverse of f) on the same set of coordinate axes.

45. $f(x) = 2^x$

46. $f(x) = (\frac{1}{3})^x$

Given functions f and g, graph functions $f \circ g$ and $g \circ f$.

47. $f(x) = 2^x$, $g(x) = x^2$

48. $f(x) = 2^x$, $g(x) = 2 - x$

Solve for x.

49. $x^2 \cdot 2^x - 6x \cdot 2^x + 5 \cdot 2^x = 0$

50. $x^2 \cdot 3^x - 9 \cdot 3^x = 0$

51. $\dfrac{x \cdot 5^x + 2 \cdot 5^x}{x^2 + 3} = 0$

52. $\dfrac{x \cdot 8^x - 12 \cdot 8^x}{4 - x} = 8^x$

53. Fill in the missing details in the proof of the following statement.

If $f(x) = b^x$ and x is a rational number, then f is an increasing function if $b > 1$.

Let p and q be positive integers, then $b^{p/q}$ is a real number that is greater than, equal to, or less than 1. Assume $b^{p/q} = 1$. Then $b^p = 1^q =$ (a). But b^p (b) 1 because $b > 1$. Hence, the assumption that $b^{p/q} = 1$ is false. Now assume that $b^{p/q} < 1$. It follows that b^p (c) $1^q = 1$. But $b^p > 1$, so that the assumption $b^{p/q} < 1$ is false. We are left with the only other possibility, that $b^{p/q} > 1$.

Now let x_2 and x_1 be rational numbers that satisfy the relationship $x_2 > x_1$. Since $x_2 - x_1$ is a (d) rational number, it can be written as p/q, where p and q are positive (e). Thus,

$$\frac{b^{x_2}}{b^{x_1}} = b^{x_2 - x_1} = \text{(f)} > 1,$$

and so $b^{x_2} > b^{x_1}$. Hence, we must conclude that b^x is an increasing function.

54. Using the proof in exercise 53 as a guide, prove that if $f(x) = b^x$ and x is a rational number, then f is a decreasing function when $0 < b < 1$.

6.2 The Natural Base *e*

Population Growth

Many different models of population growth are used to predict the size of future human and animal populations. Most of the models have an exponential form in which the base is the irrational number e ($e \approx 2.718282$). Why this is so is taken up next.

One may wish to consider a variety of environmental, political, and economic factors in constructing a growth model for a human or an animal population. It is almost impossible to determine the exact effect of any of these factors on growth. One growth factor, however, is almost self-evident and therefore easily acceptable. Large populations will generally produce more offspring than small populations. As a result, the larger a population is, the faster it grows. This phenomenon is illustrated by the graph in figure 6.6, where population (P) is a function of time (t), and P_0 is some initial population.

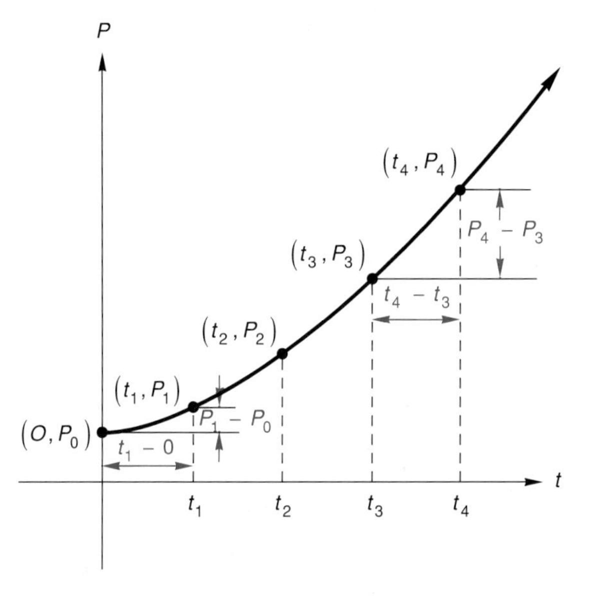

Figure 6.6

We see that equal time intervals, $(t_4 - t_3) = (t_1 - 0)$, do not produce equal population changes, $(P_4 - P_3) > (P_1 - P_0)$. In fact, as population grows with time, the **rate of growth,** which is *the change in population divided by the change in time,* increases.

$$\frac{P_4 - P_3}{t_4 - t_3} > \frac{P_3 - P_2}{t_3 - t_2}$$

Thus, the graph in figure 6.6 depicts a simple but reasonable model of population growth. Our task is to discover a function, $P = f(t)$, whose graph has the form shown in figure 6.6.

Let us momentarily focus on the small interval of time from 0 to t_1. Knowing the population P_0 at time 0, can we predict P_1? We are assuming that the rate of change in population with respect to time, $(P_1 - P_0)/(t_1 - 0)$, which we have called the rate of growth, is some fixed proportion of the known population (remember that the number of offspring depends upon the size of the population). Hence, we write

$$\frac{P_1 - P_0}{t_1 - 0} = kP_0, \tag{1}$$

where k is called a constant of proportionality. You might reasonably guess that k is a number between 0 and 1. In any case, we will be able to evaluate k from population data.

Letting $t_1 - 0 = \Delta t$, and solving (1) for P_1, we get

$$P_1 = (k\Delta t + 1)P_0. \tag{2}$$

Using the predicted value of P_1 from (2), we repeat the above process. We let $t_2 - t_1 = \Delta t$, and solve

$$\frac{P_2 - P_1}{t_2 - t_1} = kP_1$$

to obtain a predicted value for P_2.

$$P_2 = (k\Delta t + 1)P_1 = (k\Delta t + 1)^2 P_0$$

If we continue this process over n equal time intervals,

$$(t_1 - 0 = t_2 - t_1 = t_3 - t_2 = \cdots = t_n - t_{n-1} = t),$$

we obtain a prediction for P_n.

$$P_n = (k\Delta t + 1)^n P_0 \tag{3}$$

If n is 5, then (3) generates 6 points over the time interval t (see figure 6.7). If $n = 10$, then (3) generates 11 points over time interval t. Note that when $n = 10$, Δt is one half the Δt obtained when $n = 5$. As n becomes larger and larger, Δt becomes smaller and smaller and (3) generates more and more points over time interval t.

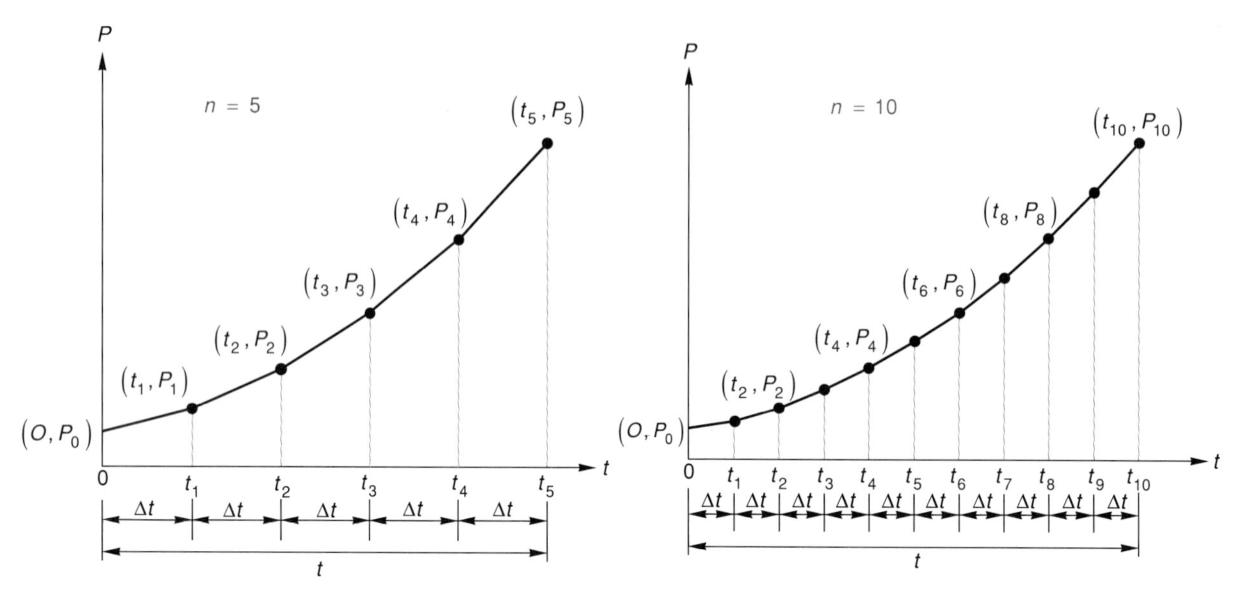

Figure 6.7

We rewrite (3) as

$$\begin{aligned} P_n &= [(1 + k\Delta t)^{n/k\Delta t}]^{k\Delta t}\, P_0 \\ &= [(1 + k\Delta t)^{1/k\Delta t}]^{nk\Delta t}\, P_0 \\ &= [(1 + k\Delta t)^{1/k\Delta t}]^{kt}\, P_0, \end{aligned} \tag{4}$$

where t is the time after n time intervals of length Δt (see figure 6.7). Now let us compute $(1 + k\Delta t)^{1/k\Delta t}$ for values of Δt that become progressively smaller.

Δt	$\frac{1}{k}$	$\frac{1}{10k}$	$\frac{1}{100k}$	$\frac{1}{1{,}000k}$	$\frac{1}{10{,}000k}$
$(1 + k\Delta t)^{1/k\Delta t}$	2	2.5937	2.7048	2.7169	2.7181

The table indicates that with smaller and smaller Δt, the value of the expression $(1 + k\Delta t)^{1/k\Delta t}$ approaches closer and closer to the number e. Thus, as Δt approaches zero, populations predicted by equation (3) differ very little from populations predicted by

$$P = P_0 e^{kt}. \tag{5}$$

Equation (5) is obtained from (4) by substituting e for $(1 + k\Delta t)^{1/k\Delta t}$.

Example 1 A certain small town has a population of 10,000 in 1960. Predict the town's population in 1970 and 1990, if $k = 0.041$.

Solution With 1960 as the base year, $t = 10$ in 1970, and $t = 30$ in 1990.

$$P(10) = 10{,}000\, e^{0.041 \cdot 10} = 15{,}068.$$
$$P(30) = 10{,}000\, e^{0.041 \cdot 30} = 34{,}212.$$
■

Radioactive Decay

Many natural phenomena can be modeled by equations such as (5). It is known that the rate of decay of a radioactive substance depends upon the mass of the substance. Initially, decay will be relatively rapid. But as time passes, and there is less and less mass due to decay, the rate of decay slows down. The process is modeled by

$$M = M_0 e^{kt}, \tag{6}$$

where M is the mass of the substance at any time, M_0 is the initial mass, k is a negative constant that determines the rate of decay of the substance, and t is time.

Example 2 Given that $k = -0.00041$ for radium, determine how much radium remains after 1,000 years if M_0 is 10 grams.

Solution Substitution in (6) gives us

$$M = 10\, e^{(-0.00041)(1{,}000)}$$
$$= 6.64 \text{ grams.}$$
■

Compound Interest

A variation of equation (5) shows up again as a model for the continuous compounding of interest. Assume that you deposit one dollar in a bank that offers an annual interest rate of r percent. Due to interest alone, your account will grow with time. How much money will be in the account after n years?

Let A_1 be the amount in your account after one year, A_2 the amount after two years, and A_n the amount after n years. A_1 must equal your original dollar plus the interest, r cents, on that dollar

$$A_1 = 1 + r.$$

A_2 must be the amount in your account at the beginning of the second year, $1 + r$, plus the interest on that amount.

$$A_2 = (1 + r) + (1 + r)r = (1 + r)(1 + r) = (1 + r)^2$$
$$A_3 = (1 + r)^2 + (1 + r)^2 r = (1 + r)^2(1 + r) = (1 + r)^3$$
$$A_4 = (1 + r)^4 \text{ (Why?)}$$
$$\vdots$$
$$A_n = (1 + r)^n \tag{7}$$

Earning interest on interest is what is meant by compound interest. The computations for $A_1, A_2, \ldots A_n$ show the compounding of interest at r percent on an *annual basis*. Equation (7) must be altered if compounding occurs more frequently, such as quarterly, or monthly, or daily.

If a bank compounds its annual interest rate of r percent t times a year, the interest rate per compounding period is r/t percent. The number of compounding periods in n years is tn. Thus, each dollar in a savings account becomes

$$A_n = \left(1 + \frac{r}{t}\right)^{tn}$$

dollars after n years. A savings account having A_0 dollars initially, becomes

$$A_n = A_0\left(1 + \frac{r}{t}\right)^{tn} \tag{8}$$

dollars after n years.

Example 3 A bank offers savers 6% annual interest on savings accounts having a minimum of $1,000. How large does a $1,000 account become after ten years if

a. interest is compounded annually?
b. interest is compounded monthly?
c. interest is compounded daily?

Solution

a. $A_{10} = 1{,}000(1 + 0.06)^{10} = \$1{,}790.85$
b. $A_{10} = 1{,}000(1 + 0.06/12)^{12 \cdot 10} = \$1{,}819.40$
c. $A_{10} = 1{,}000(1 + 0.06/365)^{365 \cdot 10} = \$1{,}822.03$ ■

The results in example 3 suggest that with the same interest rate and initial investment, the investment that experiences the greatest frequency of compounding will earn the most interest. This being the case, why don't banks compound interest hourly, or by the minute? Why don't banks give their customers the ultimate in compounding, continuous compounding?

Equation (8) tells us how large an investment of A_0 dollars becomes after n years at r percent annual interest, if interest is compounded t times annually. To realize what continuous compounding would do to A_0 dollars we need only let t become larger and larger and larger. We write

$$A_n = A_0\left(1 + \frac{r}{t}\right)^{tn} = A_0\left[\left(1 + \frac{r}{t}\right)^{t/r}\right]^{rn},$$

and allow t to assume values r, $10r$, $100r$, $1{,}000r$, etc.

t	r	$10r$	$100r$	$1{,}000r$	$10{,}000r$	$1{,}000{,}000r$
$\left(1+\frac{r}{t}\right)^{t/r}$	2	2.5937	2.7048	2.7169	2.71815	2.7182818

The expression $(1 + r/t)^{t/r}$ approaches closer and closer to e. With calculus we can show that under continuous compounding (8) becomes

$$A_n = A_0 e^{rn}. \tag{9}$$

Example 4 Determine how large a \$1,000 investment becomes after 10 years of continuous compounding at 6% annual interest.

Solution

$$A_n = 1{,}000e^{(0.06)10} = \$1{,}822.12$$ ■

Exercise Set 6.2

(A)

Given a country's current population P_0, and its rate of growth k, determine its population P, in (a) 10 years, (b) 20 years, (c) 50 years, using the formula $P = P_0e^{kt}$.

1. $P_0 = 210$ million, $k = 0.01$

2. $P_0 = 80$ million, $k = 0.04$

3. $P_0 = 3$ million, $k = 0.02$

4. $P_0 = 1$ billion, $k = 0.02$

Let M_0 be the current mass, in grams, of a radioactive substance. Find the number of grams of the substance (M) that remain after (a) 100 years, (b) 1,000 years, (c) 5,000 years. Use the formula $M = M_0e^{kt}$, where k is the rate of decay of the substance.

5. $M_0 = 100g$, $k = -0.0693$

6. $M_0 = 100g$, $k = -0.00693$

7. $M_0 = 100g$, $k = -0.000347$

8. $M_0 = 100g$, $k = -0.000121$

Use the compound interest formula $A = A_0(1 + r/t)^{tn}$ to compute the amount (A) of an initial investment (A_0) after n years. The annual interest rate is r, and t is the number of compounding periods per year.

9. $A_0 = \$1{,}000$, $r = 0.08$, $t = 4$, $n = 20$

10. $A_0 = \$1{,}000$, $r = 0.08$, $t = 12$, $n = 20$

11. $A_0 = \$1{,}000$, $r = 0.12$, $t = 12$, $n = 20$

12. $A_0 = \$1{,}000$, $r = 0.08$, $t = 365$, $n = 10$

13. How large will a \$1,000 investment become after ten years of continuous compounding at 8% annual interest?

14. A heated object will cool in air at a rate proportional to the difference in temperature between the object and the surrounding air. If the object's temperature at a given instant is 200°, and the surrounding air is 70°, then the object's temperature (T) at some later time (t) will be $T = 130e^{kt} + 70$, where t is in minutes. In the equation, the proportionality constant k is a material property of the object. Assuming that $k = -0.062$, find T when

a. $t = 10$ minutes,

b. $t = 30$ minutes,

c. $t = 1$ hour.

15. The current in an inductive electrical circuit is given by the formula

$$i = \frac{E\left(1 - e - \frac{Rt}{L}\right)}{R},$$

where E is voltage, R is resistance, t is time (in seconds), L is inductance, and i is current. If $E = 100$ volts, $R = 50$ ohms, and $L = 0.5$ henries, what is the current in amperes when

a. $t = 0.001$ seconds,

b. $t = 10$ seconds,

c. $t = 1$ minute?

16. Consumer demand for a particular item depends upon the price of the item. It is generally true that the higher the price, the less the demand. A demand (D)-price (P) function, $D = 2{,}000{,}000e^{-0.0011P}$, has been determined for a certain color TV set. What is D when

a. $P = \$500$,

b. $P = \$800$,

c. $P = \$1{,}000$?

17. Given $y = (1 + x)^{1/x}$. Evaluate y for x-values, 1, 0.1, 0.01, 0.001, 0.0001, and 0.000001. What number do the y-values approach closer and closer to as the x-values approach zero?

18. Given $y = (1 + 1/x)^x$. Evaluate y for x-values 1, 10, 100, 1,000, 10,000, and 100,000. What number do the y-values approach closer and closer to as the x-values tend to infinity?

Let $f(x) = e^x$. Graph

19. $f(-x)$

20. $-f(x)$

21. $f(2x)$

22. $2f(x)$

23. $f(x - 2)$

24. $f(x) - 2$

25. $f(x) + f(-x)$

26. $f(x) - f(-x)$

27. Graph $y = 5xe^{-x}$, $x \geq 0$.

28. Graph $y = 5xe^x$, $x \geq 0$.

29. Use the formula and values in exercise 15 to graph current against time. Assume $t \geq 0$. What fixed current does i approach closer and closer to as time increases?

30. Let $f(x) = e^x$. Graph f and f^{-1} (inverse of f) on the same coordinate axes.

31. Graph $f \circ g$ and $g \circ f$, where $f(x) = e^x$ and $g(x) = x^2$.

32. A country's current population is 50 million. Its population grows 3% annually. What will the country's population be in

a. 10 years,

b. 50 years?

Solve for x. (Hint: See example 2, section 6.1.)

33. $x^2e^{2x} - xe^{2x} - 20e^{2x} = 0$

34. $32e^{x-1} - 2x^2e^{x-1} = 0$

35. $\dfrac{xe^{3x+2} + 4e^{3x+2}}{5x^2 + 1} = 0$

36. $\dfrac{3xe^x - 2e^x}{x + 1} = e^x$

Find the simple annual interest rate that would yield the same interest as

37. 8% compounded monthly.

38. 8% compounded daily.

39. An investment yields 10% compounded monthly. How much must be invested now to realize \$10,000 in three years? The amount invested now is called the present value of \$10,000.

40. What is the present value of \$30,000, twenty years from now, if money can be invested at 9% compounded monthly? (See exercise 39.)

Use iteration to solve for x to three decimal places. (See section 4.5.)

41. $xe^x - 1 = 0$

42. $3x + e^{-2x} = 5$

43. Graph $f(t) = (6 - 2t)e^{-0.1t}$; f is called a damping function, or an exponential decay function. Refer to your graph to comment on why you think f is called a damping function.

44. The function $g(t) = 60(1 - e^{-0.25t})$ is considered a mathematical model of the learning process. Graph the function and comment on why you think it is appropriate to view g as a model for learning. A graph of g is called a learning curve.

45. The function $h(t) = 54{,}000/(1 + 2e^{-0.14t})$ can be considered a model for inhibited population growth. That is, population increase is dependent not only upon the size of the existing population, but also upon limited resources such as food and water. Graph h and comment on why you think h is an appropriate model for inhibited population growth. A graph of h is called a logistic curve.

46. Graph $G(x) = e^{-x^2}$. G is associated with the curve that students refer to when they ask an instructor, "Do you grade on a curve?" Why do you think that a graph of G is an appropriate model for the distribution of course grades (the number of A's, B's, etc.) in a particular math course?

6.3 Logarithmic Functions

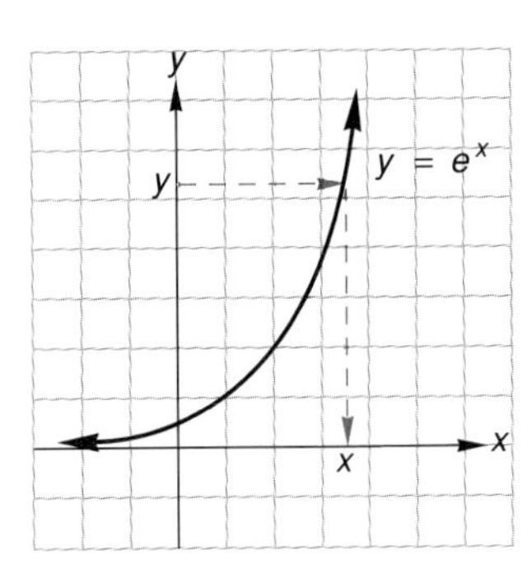

Figure 6.8

Equation (1) is called an exponential equation because it contains a variable, in this case k, as an exponent. Such equations can be solved with the aid of logarithms as we shall see.

$$1.5 = e^{10k} \tag{1}$$

The graph of $y = e^x$, in figure 6.8, shows that e^x is an increasing function. Increasing and decreasing functions have inverse functions (section 2.5). Each y in the range of e^x is associated with exactly one x in its domain. Thus, if we view (1) as having the form $y = e^x$, where $y = 1.5$ and $x = 10k$, we simply have to use the inverse function of e^x to find the x associated with $y = 1.5$. That is what a calculator does when 1.5 is entered and the [ln] key is then depressed.

It seems that we should be able to do the calculator's work by solving the equation $y = e^x$ for x, when $y = 1.5$. Unfortunately, no equation of the form $y = b^x$, $b > 0$, $b \neq 1$, can be solved for x in terms of y using only algebraic methods. To find x in terms of y, we use the inverse of b^x (see figure 6.9). We call the unique x that is associated with a particular y, in $y = b^x$, the **logarithm to the base b of y,** and we write

$$x = \log_b y. \tag{2}$$

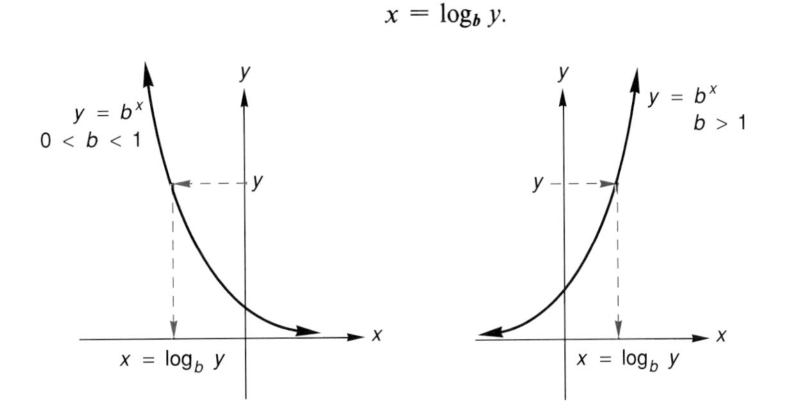

Figure 6.9

Now we have that

$$x = \log_b y \text{ if and only if } y = b^x. \qquad (3)$$

Equations (2) and (3) define the logarithm function, which is the inverse of $y = b^x$. However, the equations are in the form $x = f(y)$, rather than the common form $y = f(x)$. Therefore, we write (3) as

> **Definition**
>
> $$y = \log_b x \quad \text{if and only if} \quad x = b^y. \qquad \textbf{(4)}$$

If $b = 10$, then $\log_b x$ is called the **common log** of x. Common logarithms are usually written **log x,** without base notation. If $b = e$, then $\log_e x$ is called the **natural log** of x, and it is usually written as **ln x.**

Evaluating $\log_b x$

Equations (4) define the same function, just as the equations, $y = x + 7$ and $x = y - 7$, define the same function. Hence equations (4) can be used interchangeably, as shown in example 1, to evaluate some logarithms. Figure 6.10 illustrates how simple it is to go from log form to exponential form or vice versa.

Figure 6.10

Example 1 Evaluate each logarithm.

a. $\log_5 625$

b. $\log 1{,}000$

c. $\log_{1/2} 4$

Solution

a. If $y = \log_5 625$, then $625 = 5^y$, and $y = 4$. Hence, $\log_5 625 = 4$.

b. If $y = \log 1{,}000$, then $1{,}000 = 10^y$, and $y = 3$. Hence, $\log 1{,}000 = 3$.

c. If $y = \log_{1/2} 4$, then $4 = (\frac{1}{2})^y$, and $y = -2$. Hence, $\log_{1/2} 4 = -2$. ■

Equations (4) enable us to evaluate some logarithms, but they cannot be used as a general tool for evaluating all logarithms. For example, if $y = \log 500$, then $500 = 10^y$. We know that $2 < y < 3$ because $10^2 < 500 < 10^3$, but we cannot find the exact value of y by converting from $y = \log 500$ to $500 = 10^y$. Again we are stuck with the fact that algebraic techniques alone are insufficient to solve exponential equations.

If we want to evaluate log 500, we use logarithm tables or calculators. The tables in appendix 1 are constructed by estimation methods developed in calculus that are based on the idea of infinite series. The internal electronics of calculators makes use of these methods so that when a number, call it x, is inserted into the calculator and the [ln] key is depressed, $\log_e x$ is displayed. If the [log] key is depressed, then $\log_{10} x$ is displayed. If your calculator has no e^x key, you can take advantage of the inverse relationship between the logarithm and exponential functions to evaluate e^x. The [INV] and then the [ln] keys are depressed and e^x is displayed.

Example 2 Use a calculator to evaluate

a. $\log 500$

b. $\ln 500$

c. $e^{6.2146}$

d. $10^{2.6989}$

Solution

a. [500] [log] → 2.69897

b. [500] [ln] → 6.21461

c. [6.2146] [INV] [ln] → 499.99595

d. [2.6989] [INV] [log] → 499.91941 ■

Properties of Logarithmic Functions

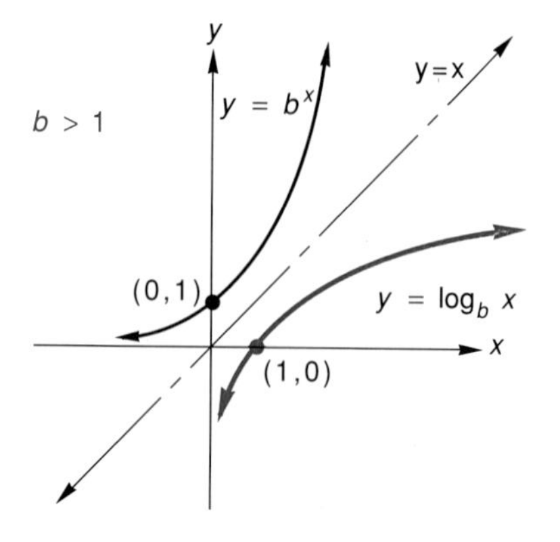

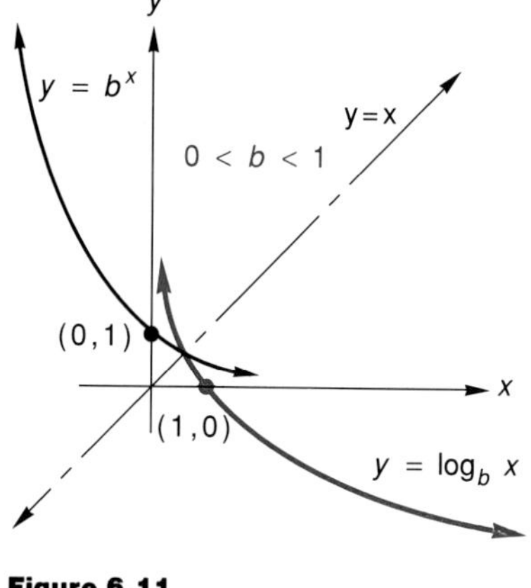

Figure 6.11

Certain properties of logarithmic functions are apparent from the graphs in figure 6.11, in which the inverse relationship between $y = b^x$ and $y = \log_b x$ is shown.

> **Properties of the Logarithmic Functions**
>
> 1. $\log_b x$ is an increasing function if $b > 1$.
> 2. $\log_b x$ is a decreasing function if $0 < b < 1$.
> 3. The domain of $\log_b x$ is the set of positive real numbers, and its range is R.

The next set of properties of the log function come directly from the inverse relationship between b^x and $\log_b x$.

> **More Properties of the Logarithmic Function**
>
> If b is a positive number different than 1, p is a real number, and x, x_1, and x_2 are positive real numbers, then
>
> I. $\log_b x_1x_2 = \log_b x_1 + \log_b x_2$
>
> II. $\log_b \dfrac{x_1}{x_2} = \log_b x_1 - \log_b x_2$
>
> III. $\log_b x^p = p \log_b x$
>
> IV. $\log_b b = 1$
>
> V. $\log_b 1 = 0$
>
> VI. $\log_b x = \dfrac{\log x}{\log b}$
>
> VII. $b^{\log_b x} = x, \quad \log_b b^x = x.$

Proof of I: Let $m = \log_b x_1$ and $n = \log_b x_2$, then $x_1 = b^m$ and $x_2 = b^n$. Also, $x_1x_2 = b^mb^n = b^{m+n}$. Now use (4) to convert the exponential form, $x_1x_2 = b^{m+n}$, to logarithmic form.

$$\log_b x_1x_2 = m + n = \log_b x_1 + \log_b x_2$$

Proof of III: Let $y = \log_b x$, then $x = b^y$. Also, $x^p = (b^y)^p = b^{py}$. With (4), the exponential form $x^p = b^{py}$, can be converted to logarithmic form.

$$\log_b x^p = py = p \log_b x$$

Proof of VI: If $y = \log_b x$, then $x = b^y$, and it follows that

$$\log x = \log b^y.$$

With property III, the last equation bcomes

$$\log x = y \log b$$

or

$$\frac{\log x}{\log b} = y = \log_b x.$$

Proof of VII: If we view $b^{\log_b x}$ as a composition of

$$f(x) = b^x \quad \text{and} \quad f^{-1}(x) = \log_b x,$$

then $b^{\log_b x} = f(f^{-1}(x)) = x$ (see section 2.5).

Similarly, $\log_b b^x$ can be viewed as the composition of $f^{-1}(x) = \log_b x$ and $f(x) = b^x$.

$$\log_b b^x = f^{-1}(f(x)) = x$$

Properties III and IV could also be used to show that

$$\log_b b^x = x \log_b b = x.$$

Example 3

a. $\log_2(4 \cdot 2) = \log_2 4 + \log_2 2 = 2 + 1 = 3$ (Property I)

b. $\log_5 \left(\frac{625}{5}\right) = \log_5 625 - \log_5 5 = 4 - 1 = 3$ (Property II)

c. $\log 10^2 = 2 \log 10 = 2 \cdot 1 = 2$ (Property III)

d. $\log_2 2 = \log_5 5 = \log 10 = \ln e = 1$ (Property IV)

e. $\log_2 1 = \log_5 1 = \log 1 = \ln 1 = 0$ (Property V)

f. $\ln 500 = \dfrac{\log 500}{\log e} = \dfrac{2.69897}{0.43429} = 6.21467$ (Property VI)

g. $4^{3 \log_4(x-2)} = 4^{\log_4(x-2)^3} = (x-2)^3$,
$\log_7 7^{(x^2+3)} = x^2 + 3$ (Property VII)

Example 4 Write $\log_b(\sqrt{x}y^3w)/z^2$ as an expression involving only logarithms of the individual variables.

Solution

$$\begin{aligned}
\log_b\left(\frac{\sqrt{x}y^3w}{z^2}\right) &= \log_b \sqrt{x}y^3w - \log_b z^2 && \text{(Property II)}\\
&= \log_b \sqrt{x} + \log_b y^3 + \log_b w - \log_b z^2 && \text{(Property I)}\\
&= \tfrac{1}{2} \log_b x + 3 \log_b y + \log_b w - 2 \log_b z && \text{(Property III)} \blacksquare
\end{aligned}$$

Example 5 Write $\frac{1}{2}(\log_b x + \log_b y + 3\log_b z)$ as a single logarithm with coefficient 1.

Solution

$$\begin{aligned}\tfrac{1}{2}(\log_b x + \log_b y + 3\log_b z) &= \tfrac{1}{2}(\log_b x + \log_b y + \log_b z^2) && \text{(Property III)}\\ &= \tfrac{1}{2}\log_b xyz^3 && \text{(Property I)}\\ &= \log_b \sqrt{xyz^3} && \text{(Property III)}\end{aligned}$$

■

Exercise Set 6.3

(A)

Evaluate each logarithm by changing to exponential form.

1. $\log_2 16$
2. $\log_5 125$
3. $\log_{16} 2$
4. $\log_8 2$
5. $\log 10{,}000$
6. $\log \frac{1}{100}$
7. $\log_{1/10} \frac{1}{100}$
8. $\log_{1/10} 100$
9. $\log_7 7$
10. $\log_{1/3} \frac{1}{3}$
11. $\log_3 \sqrt[3]{3}$
12. $\log_4 \sqrt[3]{4}$

Solve for x or b by converting the equation to exponential form.

13. $\log_2 x = 5$
14. $\log_3 x = 5$
15. $\log_b 49 = 2$
16. $\log_b 32 = 5$
17. $\log_b 1{,}000 = 3$
18. $\log_{1/10} x = -2$

Locate y between two successive integers. Do not use a calculator.

19. $y = \log_2 90$
20. $y = \log_3 600$
21. $y = \log 1{,}235$
22. $y = \log 0.0046$
23. $y = \log_5 1{,}235$
24. $y = \log_5 0.0046$

Use a calculator and, where necessary, Property VI to find y.

25. Exercise 19
26. Exercise 20
27. Exercise 21
28. Exercise 22
29. Exercise 23
30. Exercise 24

Use properties of the logarithm function to evaluate each logarithm.

31. $\log_3(81 \cdot 243)$
32. $\log_4(64 \cdot 16)$
33. $\log_5(625 \div 3{,}125)$
34. $\log_7(2{,}401 \div 49)$
35. $\log(1{,}000)^{2/5}$
36. $\log_8 8^{25}$

Graph each pair of functions on the same coordinate axes.

37. $y = 2^x$, $y = \log_2 x$
38. $y = \log x$, $y = \log_{1/10} x$
39. $y = \ln x$, $y = \log x$
40. $y = \log_3 x$, $y = \log_9 x$

Use the properties of logarithms to write each logarithm as an expression involving only the logarithms of the individual variables.

41. $\log_b xyz$
42. $\log_b\left(\frac{xy^{2/3}}{z^2}\right)^{3/2}$
43. $\log_b\left(\frac{xy}{z}\right)$
44. $\log_b\left(\frac{x^2}{y^{1/3}z^{1/2}}\right)$
45. $\log_b \sqrt[3]{\frac{x}{y^2}}$
46. $\log_b \sqrt{xz}$

Use the properties of logarithms to write each expression as a single logarithm with coefficient 1.

47. $\log_b x - \log_b y + 3\log_b z^2$
48. $\frac{1}{2}\log_b x - 2\log_b y - \frac{1}{3}\log_b z$
49. $2\log_b \sqrt[3]{x} - \log_b y^3 - 5\log_b z$
50. $\frac{1}{3}(\log_b x - 2\log_b y + 4\log_b z)$
51. $-3[\log_b(x - y) + 2\log_b(y - z) - \log_b(z - x)]$
52. $-\frac{1}{2}[3\log_b(x + y) - 4\log_b(y + z)]$

(B)

Let $f(x) = \ln x$. Graph each function.

53. $f(x - 3)$
54. $-f(x)$
55. $f(x) - 3$
56. $f(-x)$
57. $|f(x)|$
58. $f(x^2)$

59. Show that $\log_9 x = \frac{1}{2} \log_3 x$.

60. Show that $\log_4 x = \frac{1}{2} \log_2 x$.

61. Solve for x: $(\log_b x)(\log_5 b) = 3$.

62. Prove Logarithm Property II.

63. Prove Logarithm Property IV.

64. Prove the general form of Logarithm Property VI, "If neither a nor b is 1, and both a and b are positive, then

$$\log_a x = \frac{\log_b x}{\log_b a}."$$

Each function may be viewed as a composition $f \circ g$. Determine f and g.

65. $\log_b(x^2 - 4)$

66. $\log_b(2x + 3)$

67. $e^{\log_b 2x}$

68. $e^{x^2 - 2x + 5}$

In section 2.5 it is shown that $f(f^{-1}(x)) = x$ and that $f^{-1}(f(x)) = x$. View each of the following expressions as a composition of a function and its inverse, and simplify the expression. (See Logarithm Property VII.)

69. $e^{\ln x^2}$

70. $e^{2 \ln(x - 3)}$

71. $\ln e^{3x}$

72. $\ln e^{7x + 2}$

Without using a calculator, express y in terms of x.

73. $\log x = 0.2380$ and $\log y = 4.2380$

74. $\log x = 2.1573$ and $\log y = 5.1573$

75. $\log x = -0.1296$ and $\log y = 2.8704$

76. $\log x = 3.6053$ and $\log y = -2.3947$

6.4 Logarithmic and Exponential Equations

Equation $\log(80x + 40) = 3$ can be solved by converting it to exponential form.

$$\begin{aligned} 80x + 40 &= 10^3 \\ 80x &= 960 \\ x &= 12 \end{aligned}$$

Before proclaiming $\{12\}$ the solution set, let us make sure that 12 is in the replacement set for x. We know that the domain of the log function is the set of positive real numbers. Hence,

$$80x + 40 > 0$$

and

$$x > -\tfrac{1}{2}.$$

The replacement set for x, $\{x | x > -\frac{1}{2}\}$, contains 12.

Example 1 Solve the equation $\log x + \log(2x - 5) = 1$.

Solution The replacement set for x is $\{x | x > 0 \text{ and } 2x - 5 > 0\}$ or $\{x | x > \frac{5}{2}\}$. Logarithm Property I, a sum of logs is equal to the log of the product, enables us to write the equation as

$$\log[x(2x - 5)] = 1.$$

Converting the last equation to exponential form yields

$$
\begin{aligned}
2x^2 - 5x &= 10 \\
2x^2 - 5x - 10 &= 0 \\
x &= \frac{5 \pm \sqrt{105}}{4}.
\end{aligned}
$$

$(5 - \sqrt{105})/4$ is not in the replacement set for x because it is less than $\frac{5}{2}$. The solution set of the equation is $\{(5 + \sqrt{105})/4\}$. ■

The equations solved above are called **logarithmic equations.** Equations with terms in which the variable appears in an exponential expression are called **exponential equations.** Such equations are often solved with the use of logarithms.

Example 2 Solve the equation $5^x = 8$.

Solution

$$
\begin{aligned}
5^x &= 8, \quad x \in R \\
\log 5^x &= \log 8 \\
x \log 5 &= \log 8 \\
x &= \frac{\log 8}{\log 5} \approx 1.29
\end{aligned}
$$

Note 1: $(\log 8)/(\log 5) \neq \log(\frac{8}{5})$
Note 2: $(\log 8)/(\log 5) \neq \log 8 - \log 5$. Log Property II applies to the log of a quotient. It does not apply to a quotient of logs.
Note 3: Using natural logs, rather than common logs, would produce the same result. ■

Example 3 Solve the equation $14^x = 6^{3x-1}$.

Solution

$$
\begin{aligned}
14^x &= 6^{3x-1}, \quad x \in R \\
\log 14^x &= \log 6^{3x-1} \\
x \log 14 &= (3x - 1)\log 6 \\
\log 6 &= 3x \log 6 - x \log 14 \\
\log 6 &= x(3 \log 6 - \log 14) \\
\log 6 &= x\left[\log\left(\frac{6^3}{14}\right)\right] \\
x &= \frac{\log 6}{\log\left(\frac{6^3}{14}\right)} \approx 0.655
\end{aligned}
$$

■

Some Applications

The sense we have of the loudness of a sound is attributable to the rate at which acoustic energy enters our ears. That rate of energy flow is called the intensity of the sound, and it is measured in "bels." Two sounds differ by one bel if their intensities are in the ratio 10:1. The human ear can discriminate between two sounds if the ratio of their intensities is about $\frac{1}{10}$ of a bel, or a decibel (db). Hence, the decibel is a convenient unit for the measure of the "loudness" of a sound. Now all that is needed to measure the intensity of a sound in decibels is some reference sound. A commonly used reference sound is the "least audible sound." The intensity (I) of a sound in decibels is given by

$$\text{db} = 10 \log\left(\frac{I}{I_0}\right),$$

where I_0 is the intensity of the least audible sound. The difference in intensity between two sounds whose intensities are I_1 and I_2 is

$$\begin{aligned} \text{db}_2 - \text{db}_1 &= 10 \log\left(\frac{I_2}{I_0}\right) - 10 \log\left(\frac{I_1}{I_0}\right) \\ &= 10 \log\left(\frac{\frac{I_2}{I_0}}{\frac{I_1}{I_0}}\right) \\ &= 10 \log\left(\frac{I_2}{I_1}\right). \end{aligned} \qquad (1)$$

Example 4 How much louder is rock music at 125 db than conversation at 65 db?

Solution Using (1), we write

$$\begin{aligned} 125 - 65 &= 10 \log\left(\frac{I_2}{I_1}\right) \\ \frac{60}{10} &= \log\left(\frac{I_2}{I_1}\right) \\ 10^6 &= \frac{I_2}{I_1} \\ 10^6 I_1 &= I_2. \end{aligned}$$

Rock music is one million times as loud as conversation. ■

The radioactive decay model of section 6.2 can be used to date artifacts that are found in archaeological digs. The method compares the carbon-14 content to total carbon content. The ratio of carbon-14 molecules to carbon molecules remains fixed in an organism as long as it lives. When the organism dies, its carbon-14 begins to decay. By comparing the amount of carbon-14 in a sample of a dead organism to the amount of carbon-14 in a similar sample of a living organism of the same kind, it is possible to determine the amount of carbon-14 that has decayed. Knowing the rate of decay, it is then possible to determine when the organism died.

Example 5 A piece of wood taken from an archaeological dig has lost 18% of its carbon-14. Date the piece of wood, knowing that the half-life of carbon-14 is 5,730 years.

Solution Starting with equation (6) section 6.2

$$M = M_0\, e^{kt},$$

we compute k using the fact that half the mass decays in 5,730 years.

$$\begin{aligned} \tfrac{1}{2} M_0 &= M_0\, e^{5{,}730k} \\ \tfrac{1}{2} &= e^{5{,}730k} \\ \ln 0.5 &= 5{,}730k \\ -0.000121 &= k. \end{aligned}$$

Now we find the t associated with $M = 0.82\, M_0$.

$$\begin{aligned} 0.82\, M_0 &= M_0\, e^{-0.000121t} \\ 0.82 &= e^{-0.000121t} \\ \ln 0.82 &= -0.000121t \\ 1{,}640 \text{ years} &= t \end{aligned}$$

■

As a final example, we return to compound interest and the siren call of exponential growth.

Example 6 If $1,000 is deposited today in a 6% account that is compounded daily, how long would it be before the account is worth $1,000,000?

Solution Substituting into equation (8), section 6.2

$$A_n = A_0\left(1 + \frac{r}{t}\right)^{tn},$$

we have

$$\begin{aligned} 1{,}000{,}000 &= 1{,}000\left(1 + \frac{0.06}{365}\right)^{365n} \\ 1{,}000 &= (1.0001644)^{365n} \\ \log 1{,}000 &= 365\, n \log(1.0001644) \\ 3 &= 365\, n\, (0.00007135) \\ 115.2 \text{ years} &= n. \end{aligned}$$

Six generations from now your heirs will look back and praise your foresight.

■

Exercise Set 6.4

(A)

Solve each equation.

1. $\log(x + 3) = 2$
2. $\log(5x - 40) = 3$
3. $\ln(2x - 6) = 3.8$
4. $\ln(x + 5) = 1.2$
5. $\log(x + 2) + \log(x - 1) = 1$
6. $\ln 3x + \ln(4x + 8) = \ln(30 + 6x)$
7. $\ln(2x - 5) - \ln(x + 3) = 3$
8. $\log x^2 - \log(x - 1) = 1$
9. $\log(2x^2 + 23) - \log(2x - 5) = 1 + \log(x + 1)$
10. $\log(3x + 4) - \log(x - 2) = 2 + \log\left(\frac{1}{2x}\right)$
11. $5^{x+2} = 28$
12. $26^{x-5} = 145$
13. $2^{x^2} = 9$
14. $4^{x^2-1} = 98$
15. $e^{3x} = -0.86$
16. $e^{-x} = 8$
17. $e^{x+2} = 2e^{2x}$
18. $3e^x = 4e^{5-x}$
19. $126^{x+2} = 76^{x+1}$
20. $1{,}000^{x^2+2x} = 100^{x^2-2}$

21. How much louder is freeway traffic at 100 db than suburban street traffic at 45 db?
22. An operating garbage disposal is 100 million times as loud as the least audible sound. What decibel level defines the sound intensity of the garbage disposal?
23. An animal bone has lost 25% of its carbon-14. Date the bone, given that the half-life of carbon-14 is 5,730 years.
24. The half-life of a radioactive substance is 4,000 years. How many years are required for $\frac{3}{4}$ of the mass of the substance to decay?
25. How many years of daily compounding are required to double the amount of money in an 8% account?
26. What interest rate is required to double the value of an account in seven years if the account is subject to monthly compounding?
27. A certain town had a population of 8,000 in 1970 and 10,000 in 1980. Use equation (5) section 6.2 to predict the town's population in the year 2000. In what year will the town's population be 20,000?
28. At a given temperature the pressure and volume of a certain gas are given by the formula $pv^k = 300$. Find k if $p = 20$ and $v = 9$.

(B)

29. If a country's population increases at the rate of 1% a year, how long will it be before the country's population doubles?
30. Repeat exercise 29 using a 2% annual rate increase.
31. The current in an inductive electrical circuit is given by the formula
$$i = \frac{E(1 - e^{-RT/L})}{R},$$
where E is voltage, R is resistance, T is time (in seconds), L is inductance, and i is current. If $E = 100$, $R = 50$, and $L = 0.5$, how many seconds does it take for the current in the circuit to become 1.2 amperes?
32. The Richter scale for measuring the magnitude of an earthquake is, like the decibel scale for measuring sound intensity, a logarithmic scale. The magnitude (M) of an earthquake (E) is given by
$$M = \log E - \log E_0,$$
where E_0 is called the "zero earthquake." Thus, the magnitude of any earthquake is really a comparison of the intensity of that quake to some arbitrarily chosen zero earthquake.
   Write a formula that compares the magnitudes M_1 and M_2, of earthquakes E_1 and E_2. (Hint: Read example 4 and the discussion preceding the example.)
33. Use the formula constructed in exercise 32 to determine how much stronger a 7.9 earthquake is than a 6.1 earthquake.
34. Why do you think logarithmic scales are used to measure sound intensity and earthquake magnitude?
35. If inflation is held to 3% per year over the course of a decade, how much is the \$100 that was hidden in the mattress at the beginning of the decade worth at the end of the decade?
36. An accident in a nuclear power plant releases strontium-90 into the air. Readings taken soon after the accident occurs show that within a radius of 20 miles of the plant strontium-90 levels are 5,000 times the maximum safe level. If the half-life of strontium-90 is 28 years, how long will it be before it is safe for humans to return to within 20 miles of the plant?

37. A bacteria population grows at the rate of 10% every 45 minutes. How long will it be before the population triples?

38. A certain range animal population triples over a ten-year period. At what yearly rate is the population increasing?

39. Newton's Law of Cooling states that the rate at which a body cools is proportional to the difference in temperature between the cooling body and the ambient temperature. Hence, the greater the difference in temperature, the faster the cooling. The mathematical model for Newton's law is exactly the same as the models for population growth, radioactive decay, and continuous compound interest.

$$T - T_0 = De^{kt}$$

T is the temperature of the cooling body at any time. T_0 is the ambient temperature. D is the positive difference between the temperature of the body and the ambient temperature when the cooling process begins. Time is denoted by t, and k is a constant associated with the material of the cooling body. Use Newton's law to determine how long it takes for a cup of tea to cool to 100° F from 154° F if it takes six minutes for the tea to cool from 200° F to 154° F. The ambient temperature is 68° F.

40. Assume that equation $f(t) = (6 - 2t)e^{-0.1t}$ represents the up and down motion, with respect to a horizontal datum, that a certain car experiences after hitting a certain bump at a certain speed. If $f(t)$ represents vertical distance in centimeters, and t represents time in seconds, determine t to one decimal place when $f(t)$ is -5 centimeters. (See exercise 43, section 6.2. You may have to use iteration.)

41. The rate, in words per minute, that a student can type increases with each day of typing training according to the formula $g(t) = 60(1 - e^{-0.25t})$. (See exercise 44, section 6.2). If $g(t)$ is the student's current typing rate and t is time in days measured from the first day of training, how many days after the beginning of training is the student typing 50 words per minute?

42. The population of an exclusive east coast village is limited by space and by zoning and building code laws. Assume that the formula for population growth in the village is $h(t) = 54{,}000/(1 + 2e^{-0.14t})$, where h is the village population at any time, and t is time in years. What was the village population three years ago when it was first incorporated? What is the maximum population that the village can accommodate? When will the village have a population of 30,000? (See exercise 45 section 6.2.)

Solve each equation.

43. $\log x^2 = (\log x)^2$

44. $\sqrt{\log x} = \log \sqrt{x}$

45. $\log(\log x) = 1$

46. $\dfrac{e^x + 3e^{-x}}{2} = 2$

Find the root, nearest the given value of x_0, correct to the indicated number of decimal places. Use iteration.

47. $x = \ln x + 2$, $x_0 = 3$, 3 decimal places.

48. $x^4 = 3e^x$, $x_0 = 2.5$, 2 decimal places. (Hint: Rewrite the equation as $x = 3e^x/x^3$.)

49. Exponential and logarithmic functions can be evaluated from their infinite series representations. For example,

$$e^x = 1 + x + \frac{x^2}{2!} + \frac{x^3}{3!} + \cdots + \frac{x^{100}}{100!} + \text{etc.}$$

Hence,

$$e^x \approx 1 + x + \frac{x^2}{2} + \frac{x^3}{6}.$$

Find $|e_p^x - e_c^x|$ for x values 0.5, 1.0, 1.5, and 2.0, where e_p^x is the cubic polynomial approximation of e^x and e_c^x is your calculator value of e^x.

50. Find a second degree polynomial $y = ax^2 + bx + c$ that contains the points $(0,2^0)$, $(1,2^1)$, and $(2,2^2)$. Does the polynomial give good approximations to $y = 2^x$ for x values -0.5, 0.5, 1.5, and 2.5?

Chapter 6 Review

Major Points for Review

Section 6.1 Definition of the exponential function $y = b^x$

Graphs of exponential functions

For what base values an exponential function is increasing (decreasing)

Extension of the properties of rational exponents to irrational exponents

Section 6.2 The exponential function with base e

Formulas for population growth, compound interest, and radioactive decay

Section 6.3 Logarithmic functions

Common and natural logs

The inverse relationship between the log and exponential functions

Graphs of log functions

Properties of log functions

Section 6.4 Solving log and exponential equations

Applications of log and exponential functions

Review Exercises

Graph each pair of functions on the same coordinate axes.

1. $y = 4^x$, $y = (\frac{1}{4})^x$

2. $y = 2^{-x}$, $y = -(2^x)$

Simplify each expression.

3. $6^{\sqrt{5}} \cdot 6^2$

4. $\dfrac{8^{\sqrt{3}+4}}{8^{\sqrt{2}-2}}$

5. $[(x-5)^{\sqrt{3}}]^{\sqrt{6}}$

6. $x^{\sqrt{2}+3} \cdot x^{3-\sqrt{2}}$

Let $f(x) = e^x$.

7. Graph $f(-x)$.

8. Graph $f(x+5)$.

Let $g(x) = \ln x$.

9. Graph $g(2x-3)$.

10. Graph $-3g(x)$.

The population of a certain city is given by P_0 and the rate of growth of that population is given by k. Use the formula $P = P_0e^{kt}$ to determine the city's population in twenty years.

11. $P_0 = 2{,}200{,}000$, $k = 0.03$

12. $P_0 = 430{,}000$, $k = 0.006$

M_0 is the current mass, in grams, of a radioactive substance, and k is the rate of decay of the substance. Use the formula $M = M_0e^{kt}$ to determine how many grams of the substance remain after 100 years.

13. $M_0 = 50g$, $k = -0.72$

14. $M_0 = 10g$, $k = -0.00028$

Use the compound interest formula, $A = A_0(1 + r/t)^{tn}$, to compute the amount of an investment after n years.

15. $A_0 = \$5{,}000$, $r = 0.06$, $t = 365$, $n = 10$

16. $A_0 = \$2{,}000$, $r = 0.08$, $t = 12$, $n = 20$

17. How large will a $3,000 investment become after five years of continuous compounding at 10% annual interest?

18. An animal population increases 5% per year. How much larger will the population be six years from now?

Evaluate each logarithm. Do not use a calculator.

19. $\log_2 8$

20. $\log_{27} 3$

21. $\log_{1/5} 25$

22. $\log_{1/2} \sqrt[5]{4}$

Solve for x or b.

23. $\text{Log}_3\, x = 4$

24. $\log_b 9 = -2$

Graph each pair of functions on the same coordinate axes.

25. $y = \log|x|$, $y = |\log x|$

26. $y = \log_2 x$, $y = \log_8 x$

Write each logarithm as an expression involving only the logarithms of the individual variables.

27. $\log_b \dfrac{w^{1/2}x^3}{y^5}$

28. $\log_b\left(\dfrac{x^{3/4}y}{z^4}\right)^{2/3}$

Write each expression as a single logarithm with coefficient 1.

29. $2 \log_b x - \frac{1}{3} \log_b y + 8 \log_b z$

30. $6[\log_b(w + x) - 3 \log_b(w - y) + \frac{1}{2} \log_b(x + y)]$

Solve each equation.

31. $\log(2x - 3) + \log(x + 2) = 2$

32. $\ln 3x - \ln(2x - 9) = 4$

33. $\ln(x + 4) + \ln(3x + 5) = 3$

34. $\log(x^2 + 2) - \log(4x - 5) = \log(x - 4) - 1$

35. $e^{x^2 + 3x} = 5e^{2x - 1}$

36. $10^{3x^2 - 7} = 10{,}000^{2x + 4}$

37. $3x^2 \cdot 5^x - 10x \cdot 5^x - 8 \cdot 5^x = 0$

38. $\dfrac{x \cdot 2^x - 6 \cdot 2^x}{2 - x} = 2^x$

39. $12e^{x+3} - 3x^2e^{x+3} = 0$

40. $\dfrac{x^2e^{5x-1} - 3e^{5x-1}}{2x^2 + 7} = 0$

41. $\log_2(\log x) = -3$

42. $\dfrac{e^x - 2e^{-x}}{3} = 1$

43. Under ideal conditions a certain animal population will double in eight years. Write an equation that determines the animal population as a function of time (t). Assume the initial population is 1,200.

44. What simple annual interest rate would yield the same interest as 8.6% compounded daily?

45. An investment yields 11.3% compounded monthly. How much must be invested now in order to realize $25,000 in five years?

46. A country's population increases at the rate of 1.5% per year. How long will it be before the population doubles?

47. What annual interest rate, compounded daily, would allow you to double your investment in seven years?

48. The population of a town ten years ago was 12,000. The town's population today is 15,000. What will the population of the town be in twenty years?

49. The half-life of a certain radioactive substance is three days. There is currently 60 grams of the substance. How many days ago did the substance have a mass of 100 grams?

50. A radioactive substance loses 28% of its mass in twelve years. What is the half-life of the substance?

51. One hundred years ago your great-great grandfather invested $100 in an account that drew 3% simple interest annually. You are the sole heir to your great-great grandfather's estate. If you claim his musty old account today, how much money do you get?

52. The time it takes for a new production worker to perform a certain task is given by the equation $g(t) = 12(1 + e^{-0.6t})$, where g is minutes and t is time in days. Graph the function. What is the maximum time needed by the worker to perform the task? How fast will the worker perform the task once she is experienced? When, in the learning process, is the worker performing the task in fifteen minutes?

Graph.

53. $y = x^2e^{-x}$, $x \geq 0$

54. $y = e^x - e^{-x}$

55. $y = \dfrac{12}{1 + 3e^{-0.25t}}$, $t \geq 0$

56. $y = e^x + e^{-x}$

Simplify.

57. $e^{3 \ln (2x - 5)}$

58. $\ln e^{x^2 + 6}$

59. Find the root of $x^2 = \ln(x + 2) + 3$, between 2 and 3, and correct to two decimal places. Use iteration.

Chapter 6 Test 1

1. Graph $y = e^x$ and $y = \ln x$ on the same coordinate axes.
2. Determine the exact value of
   a. $\log_3 81$
   b. $\log_{1/4} 64$
3. Write $\log_b[(y^3z^{1/2})/x^2]$ as an expression involving only the logarithms of the individual variables.

Solve each equation.

4. $\log(3x - 4) + \log(x + 1) = 1$
5. $2^{x^2 - 5} = 64$
6. $12x^2e^{x+2} + 13xe^{x+2} + 3e^{x+2} = 0$
7. Graph $y = \ln|x - 2|$.
8. A bacteria population increases at the rate 3% per hour. How long will it be before the population doubles?
9. What annual rate of interest, compounded monthly, is required to double an investment in five years?
10. The half-life of a radioactive substance is four years. What percentage of the current mass of the substance will remain twenty-five years from now?

Chapter 6 Test 2

1. Determine the exact value of
   a. $\log_4 \sqrt[3]{16}$
   b. $e^{3 \ln 5}$
2. Write $\log_b[(x^2y)^{3/2}/z^5]^{4/5}$ as an expression involving only the logarithms of the individual variables.

Solve each equation.

3. $\log_2(3x - 1) + \log_2(x + 5) = \log_2(2x + 1) + 3$
4. $4^{4x+1} + 4^{2x+1} - 24 = 0$
5. $e^x - 4e^{-x} = 3$

Graph.

6. $y = 3xe^{-x}, x \geq 0$
7. $y = -\ln(2x + 4)$
8. What will be the percentage decrease, in five years, of the value of today's dollar if the inflation rate in each of the next five years is 2.5%?
9. What initial investment in a 9% account will yield $25,000 in ten years, if interest is compounded monthly?
10. A flu epidemic spread through a small community according to the formula
$$F(t) = \frac{15{,}000}{3 + 297e^{-0.28t}},$$
where F is the number of people contracting flu and t is time in days. How large was the group that contracted flu initially? What was the largest number of people contracting the disease? How long was it, after the initial onslaught of the disease, before 3,000 people had the flu?

Chapter 7

Trigonometric Functions

We have been able to construct growth and decay models with exponential and logarithmic functions. Now we turn our attention to a class of functions that can be used to model periodic phenomena.

Any phenomenon that recurs at regular intervals is said to be periodic. The motion of the earth around the sun, the back and forth motion of a piston in a reciprocating engine, and the vibrations of the A string of a violin are all examples of periodic phenomena.

7.1 Periodic Functions

Consider the water pump in figure 7.1. A plate and the pin welded to it are rotated counterclockwise by a crankshaft to which the plate is attached. The pin, constrained by the guidebar attached to the piston shaft, moves up $2r$ units and down $2r$ units with each revolution of the crankshaft. Over each period of time in which the crankshaft makes a revolution, water is sucked into the cylinder on the upstroke and forced out of the cylinder on the downstroke. The circular motion of the pin, *P*, is converted to the linear motion of the piston.

With each revolution of the crankshaft, the piston rises and falls the same distance as the pin does. It is possible to predict the position of the piston at any time if we know the position of the pin. Assume that the crankshaft rotates uniformly at k revolutions per minute, $k \in N$. Imagine a coordinate system through the shaft and a radial line of length r between the shaft and the pin, *P*. At a particular instant in time let θ be the angle between the radial line and the positive x-axis. Let (x,y) be the coordinates of the center of the

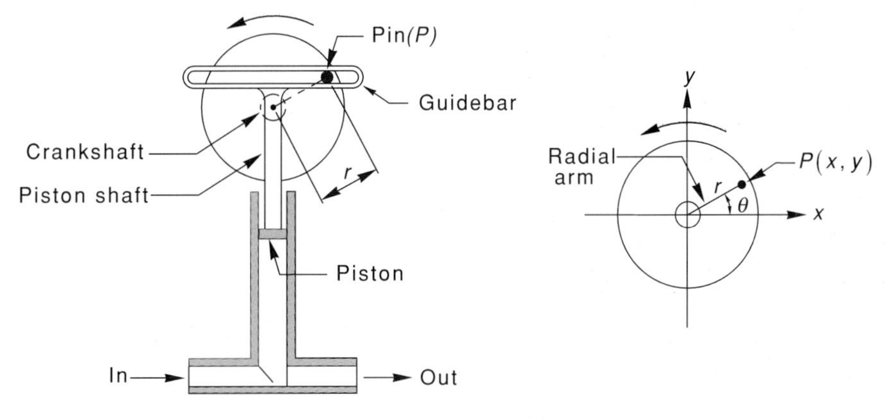

Figure 7.1

pin. To each θ there can be associated exactly one ratio y/r. Thus, the set of ordered pairs $(\theta, y/r)$ constitute a function. We shall call it S. The domain of S is the set of non-negative angles because as the crankshaft rotates through consecutive revolutions, θ assumes all values between 0° and 360°, 360° and 720°, etc. The range of S is $[1, -1]$, as we shall see after determining values of y/r associated with certain θ values.

If $\theta = 0°$ (figure 7.2a) the y-coordinate of the center of the pin P, is 0. Thus,

$$\frac{y}{r} = \frac{0}{r} = 0, \quad \text{and} \quad S(0°) = 0.$$

If $\theta = 30°$ (figure 7.2b) $y/r = \overline{PQ}/r$, where $\overline{PQ}$ is the length of a side of the 30°–60°–90° triangle OPQ. We know from geometry, that if the length of the hypotenuse of a 30°–60°–90° triangle is 2 units, then the length of the side opposite the 30° angle is 1 unit, and the length of the side opposite the 60° angle is $\sqrt{3}$ units (figure 7.3).

As triangles OPQ and ABC are similar, $\overline{PQ}/r = \frac{1}{2}$. When $\theta = 30°$,

$$\frac{y}{r} = \frac{\overline{PQ}}{r} = \frac{1}{2}, \quad \text{and} \quad S(30°) = \frac{1}{2}.$$

If $\theta = 45°$ (figure 7.2c) $y/r = \overline{PQ}/r$, where $\overline{PQ}$ is the length of the side of 45°–45°–90° triangle OPQ. A 45°–45°–90° triangle whose sides are 1 unit

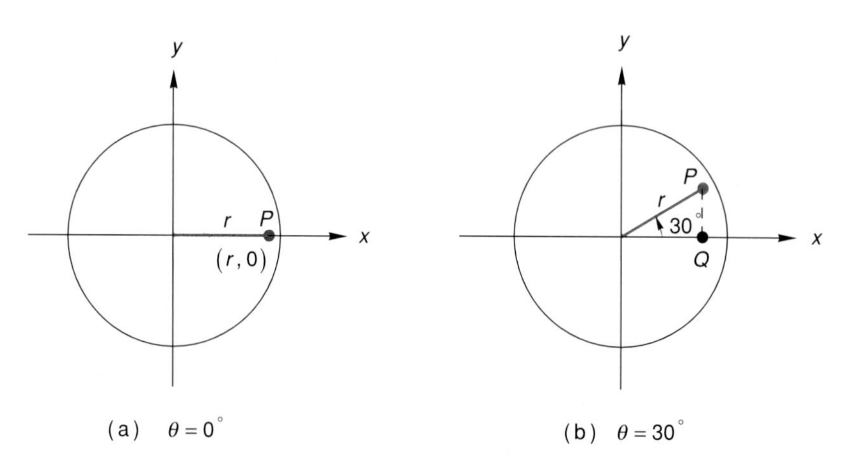

(a) $\theta = 0^\circ$ (b) $\theta = 30^\circ$

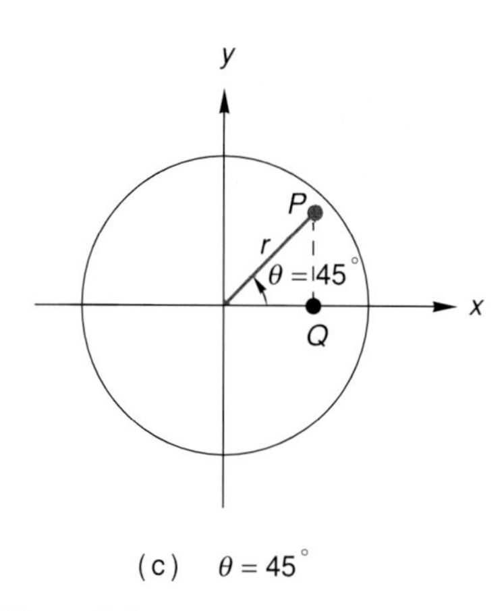

(c) $\theta = 45^\circ$

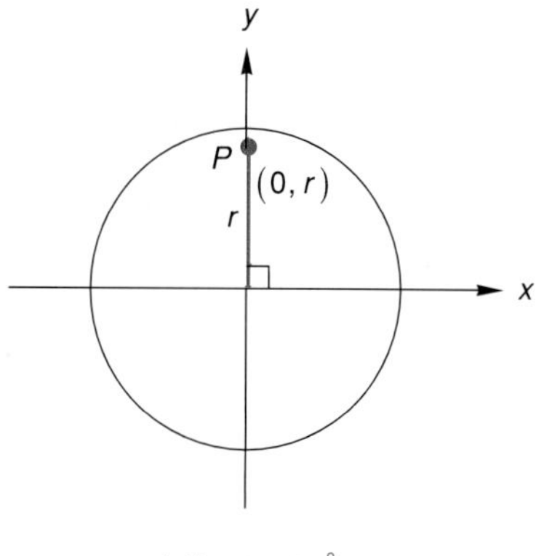

(d) $\theta = 90^\circ$

Figure 7.2

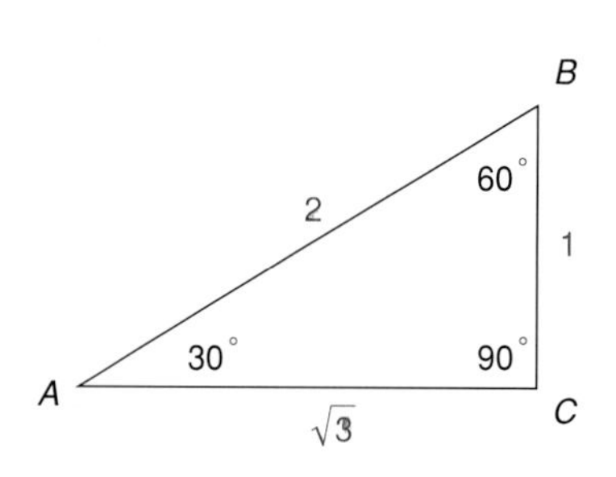

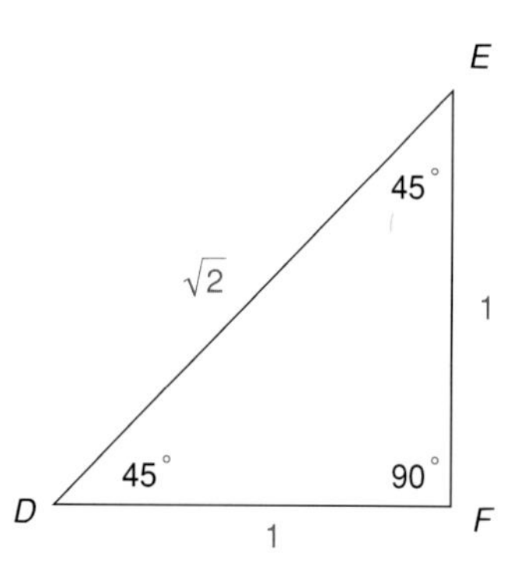

Figure 7.3

in length has a hypotenuse whose length is $\sqrt{2}$ (figure 7.3). Because triangles OPQ and DEF are similar, $\overline{PQ}/r = 1/\sqrt{2}$. Therefore, if $\theta = 45°$,

$$\frac{y}{r} = \frac{\overline{PQ}}{r} = \frac{1}{\sqrt{2}}, \quad \text{and} \quad S(45°) = \frac{1}{\sqrt{2}}.$$

If $\theta = 90°$ (figure 7.2d) $S(90°) = r/r = 1$.

At this point it should not be too difficult to accept that, as the radial arm OP rotates counterclockwise through 90° from its initial position on the x-axis, the y-coordinate of P assumes all real values between 0 and r. Consequently, the set of y/r values in the range of S must include all real numbers between 0 and 1. In the exercises, you are asked to show that

1. as θ increases from 90° to 180°, $S(\theta)$ decreases from 1 to 0.
2. as θ increases from 180° to 270°, $S(\theta)$ decreases from 0 to -1.
3. as θ increases from 270° to 360°, $S(\theta)$ increases from -1 to 0.

We are now able to construct a formula that predicts the position of the pump piston relative to its starting position shown in figure 7.4. The piston is raised y units, the same vertical movement of the pin, as the pin rotates at the rate of k revolutions per minute (rpm). We know that in t minutes $\theta = 360° kt$. As an illustration, if the crankshaft rotates at 60 rpm and $t = \frac{1}{2}$ second, then

$$\theta = (360° \text{ per revolution})(60 \text{ rpm})(\tfrac{1}{120} \text{ minute}) = 180°.$$

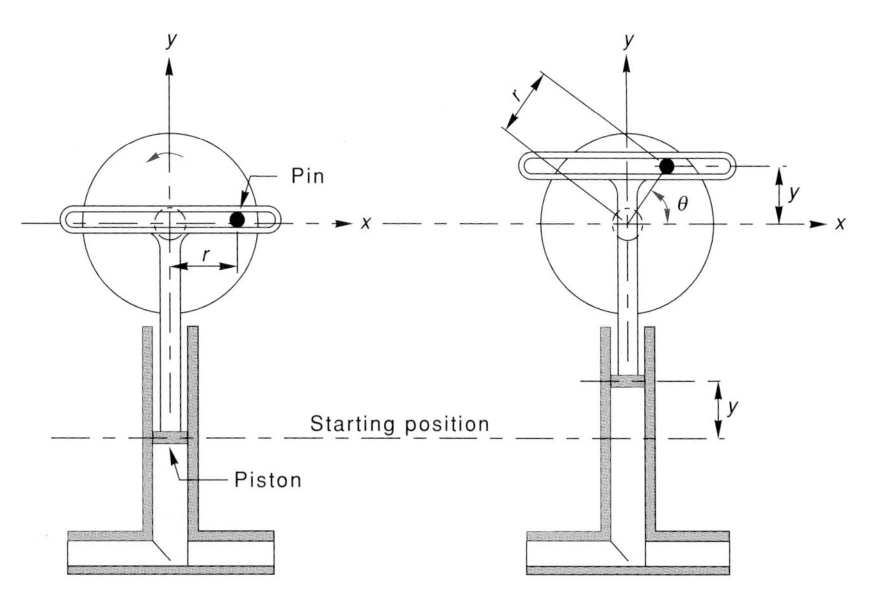

Figure 7.4

We also know that $S(\theta) = y/r$ so that $y = r \cdot S(\theta)$. Therefore, the position of the piston at any time t is given by the equation

$$y = r \cdot S(360° kt). \qquad \textbf{(1)}$$

Example 1 Assume that $r = 4''$, and that the crankshaft rotates 30 rpm. Use (1) to determine the location of the pump piston, relative to its starting position, at time

a. $\frac{1}{6}$ second.
b. $\frac{1}{2}$ second.
c. 1 second.
d. $\frac{13}{6}$ seconds.
e. $\frac{9}{2}$ seconds.
f. 7 seconds.

Solution

a. $\frac{1}{6}$ second $= \frac{1}{360}$ minute.
$y = 4\,S(360° \cdot 30 \cdot \frac{1}{360}) = 4\,S(30°) = 4 \cdot \frac{1}{2} = 2''$

b. $\frac{1}{2}$ second $= \frac{1}{120}$ minute.
$y = 4\,S(360° \cdot 30 \cdot \frac{1}{120}) = 4\,S(90°) = 4 \cdot 1 = 4''$

c. 1 second $= \frac{1}{60}$ minute.
$y = 4\,S(360° \cdot 30 \cdot \frac{1}{60}) = 4\,S(180°) = 4 \cdot 0 = 0''$

d. $\frac{13}{6}$ seconds $= \frac{13}{360}$ minute.
$y = 4\,S(360° \cdot 30 \cdot \frac{13}{360}) = 4\,S(390°)$.
Since the position of the pin is the same when $\theta = 30°$ or $390°$, $S(390°)$ must equal $S(30°)$. As a result, $y = 4 \cdot \frac{1}{2} = 2''$.

e. $\frac{9}{2}$ seconds $= \frac{9}{120}$ minute.
$y = 4\,S(360° \cdot 30 \cdot \frac{9}{120}) = 4\,S(810°) = 4\,S(90°) = 4''$

f. 7 seconds $= \frac{7}{60}$ minute.
$y = 4\,S(360° \cdot 30 \cdot \frac{7}{60}) = 4\,S(1{,}260°) = 4\,S(180°) = 0''$ ■

Periodicity

In example 1

$$S(30°) = S(30° + 360°) = S(390°) = \tfrac{1}{2},$$
$$S(90°) = S(90° + 2 \cdot 360°) = S(810°) = 1, \quad \text{and}$$
$$S(180°) = S(180° + 3 \cdot 360°) = S(1{,}260°) = 0.$$

Because the pattern displayed on page 260 represents specific instances of the general statement

$$S(\theta) = S(\theta + k \cdot 360°), \quad k \,\varepsilon\, W, \tag{2}$$

we call S a periodic function with period 360°. In fact, any function f that satisfies the equation

$$f(x) = f(x + p), \tag{3}$$

where p is the smallest possible number for which (3) is true, is called a **periodic function with period p.**

Example 2 Given $f(x) = |x|, -1 \leq x \leq 1$, extend f so that it is periodic over the set of real numbers.

Solution To extend f so that it is periodic requires that the graph of f (figure 7.5a) be repeated over each two-unit interval to the right of $x = 1$ and to the left of $x = -1$. The graph in figure 7.5b displays that extension. We must translate the vertex of the graph of f to *even integral values* of x.

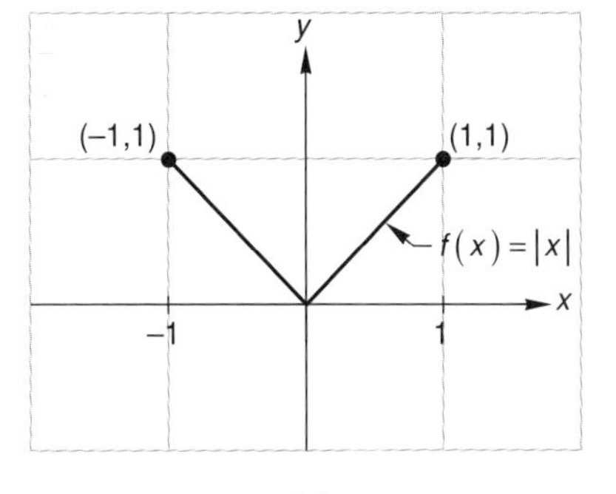

(a)

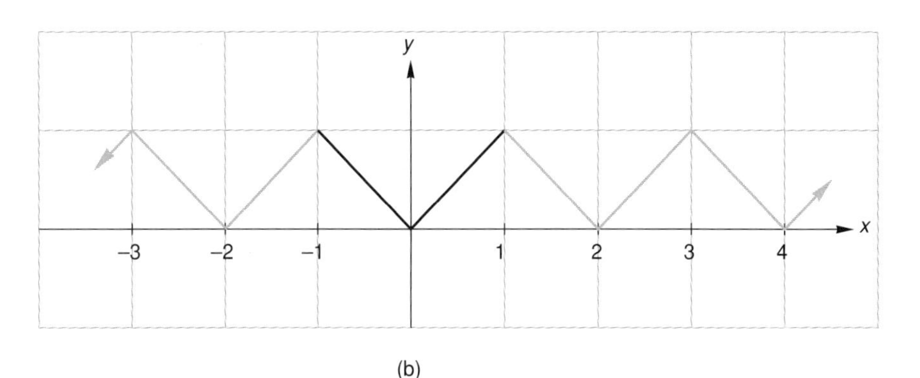

(b)

Figure 7.5

A graph of $f(x-2)$ has its vertex at $x=2$. A graph of $f(x-4)$ has its vertex at $x=4$, and a graph of $f(x-2n)$, $n \in N$, has its vertex at $x=2n$. Similarly, a graph of $f(x+2)$ has its vertex at $x=-2$. A graph of $f(x+4)$ has its vertex at $x=-4$, and a graph of $f(x+2n)$ has its vertex at $x=-2n$. We conclude that the extension of f is

$$f(x) = |x-2n|, \quad 2n-1 \le x \le 2n+1, \quad n \in J.$$

Thus, if $n=5$, the vertex of the extension is at $x=10$ and the graph extends from $2 \cdot 5 - 1 = 9$ to $2 \cdot 5 + 1 = 11$. Where is the vertex of the extension, and what is the interval of the extension if $n=-6$? ■

Example 3 Show that $f(x) = x - [x]$ is periodic with period $p = 1$ by graphing f on the interval $[-2,2]$.

Solution If $-2 \le x < -1$, then $[x] = -2$. Thus, on the interval $[-2,-1)$, $f(x) = x + 2$. If $-1 \le x < 0$, then $[x] = -1$. Thus, on the interval $[-1,0)$, $f(x) = x + 1$. Continuing this way, finding defining equations for f on the intervals $[0,1)$, and $[1,2)$, yields the following definition of f on the interval $[-2,2]$.

$$f(x) = \begin{cases} x+2, & -2 \le x < -1 \\ x+1, & -1 \le x < 0 \\ x, & 0 \le x < 1 \\ x-1, & 1 \le x < 2 \\ 0, & x = 2 \end{cases}$$

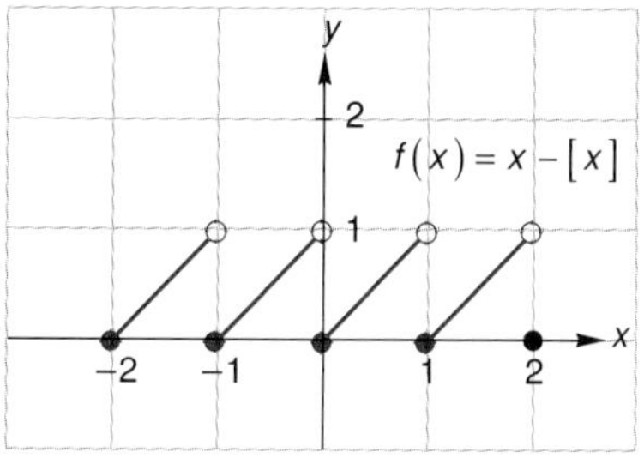

Figure 7.6

The graph of f in figure 7.6, shows that the function is periodic with period 1. ■

The movement of the pump piston (figure 7.1) is periodic. The function S, that describes the piston's motion, is periodic. That function belongs to a class of periodic functions, called the Trigonometric Functions, whose properties will be investigated in this chapter.

Exercise Set 7.1

(A)

Function S is defined on page 257, and it is evaluated for certain θ in this section.

1. Evaluate $S(60°)$.
2. Evaluate $S(\theta)$ for
   a. $\theta = 120°$,
   b. $\theta = 135°$,
   c. $\theta = 150°$,
   d. $\theta = 180°$.
3. Evaluate $S(\theta)$ for
   a. $\theta = 210°$,
   b. $\theta = 225°$,
   c. $\theta = 240°$,
   d. $\theta = 270°$.
4. Evaluate $S(\theta)$ for
   a. $\theta = 300°$,
   b. $\theta = 315°$,
   c. $\theta = 330°$,
   d. $\theta = 360°$.
5. Evaluate $S(\theta)$ for
   a. $\theta = 450°$,
   b. $\theta = 480°$,
   c. $\theta = 495°$,
   d. $\theta = 510°$,
   e. $\theta = 540°$.
6. Evaluate $S(\theta)$ for
   a. $\theta = 630°$,
   b. $\theta = 660°$,
   c. $\theta = 675°$,
   d. $\theta = 690°$,
   e. $\theta = 720°$.

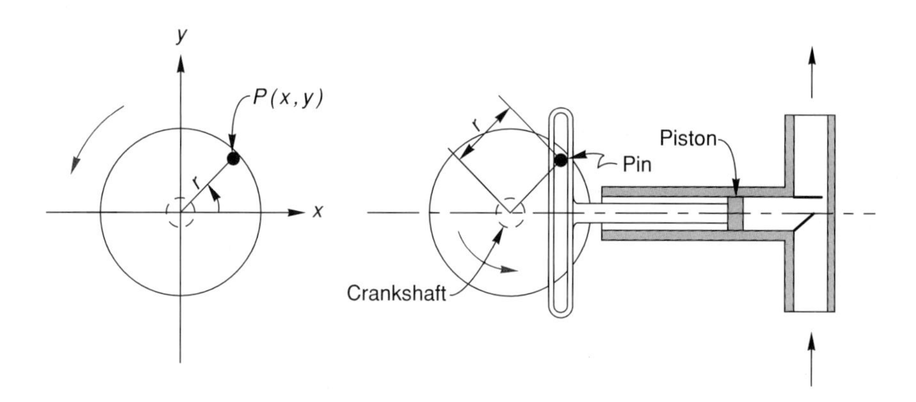

A pump whose axis is horizontal is shown in the figure. Let r be the distance between the centers of the crankshaft and pin, P. Imagine a coordinate system through the center of the crankshaft and let (x,y) be the coordinates of the center of the pin. Define a function C as the set of ordered pairs $(\theta,x/r)$, where θ is the set of positive angles. (See figures 7.2 and 7.3.)

7. Evaluate $C(\theta)$ for
- **a.** $\theta = 0°$,
- **b.** $\theta = 30°$,
- **c.** $\theta = 45°$,
- **d.** $\theta = 60°$,
- **e.** $\theta = 90°$.

8. Evaluate $C(\theta)$ for
- **a.** $\theta = 120°$,
- **b.** $\theta = 135°$,
- **c.** $\theta = 150°$,
- **d.** $\theta = 180°$.

9. Evaluate $C(\theta)$ for
- **a.** $\theta = 210°$,
- **b.** $\theta = 225°$,
- **c.** $\theta = 240°$,
- **d.** $\theta = 270°$.

10. Evaluate $C(\theta)$ for
- **a.** $\theta = 300°$,
- **b.** $\theta = 315°$,
- **c.** $\theta = 330°$,
- **d.** $\theta = 360°$.

11. Evaluate $C(\theta)$ for
- **a.** $\theta = 450°$,
- **b.** $\theta = 480°$,
- **c.** $\theta = 495°$,
- **d.** $\theta = 510°$,
- **e.** $\theta = 540°$.

12. Evaluate $C(\theta)$ for
- **a.** $\theta = 630°$,
- **b.** $\theta = 660°$,
- **c.** $\theta = 675°$,
- **d.** $\theta = 690°$,
- **e.** $\theta = 720°$.

A formula for the position of the pump piston, relative to its starting position, is given by equation (1) of this section. Determine the position of the pump piston for the given r and k, and at the given times.

13. $r = 10$ cm, $k = 60$ rpm
- **a.** $\frac{1}{6}$ second,
- **b.** $\frac{1}{2}$ second,
- **c.** $\frac{13}{8}$ seconds,
- **d.** $\frac{7}{2}$ seconds

14. $r = 15$ cm, $k = 40$ rpm
- **a.** $\frac{1}{8}$ second,
- **b.** 1 second,
- **c.** 3 seconds,
- **d.** $\frac{93}{16}$ seconds

(B)

Extend each function so that it is periodic over the interval $[0,6)$. Graph the periodic function and write an equation for it.

15. $f(x) = x,\ 0 \le x < 1$

16. $f(x) = x/2,\ 0 \le x < 2$

17. $f(x) = x^2,\ 0 \le x < 2$

18. $f(x) = x^3,\ 0 \le x < 1$

Construct a graph of each function. If the function is periodic, determine its period.

19. $f(x) = \begin{cases} 1, & [x] \text{ is even} \\ -1, & [x] \text{ is odd} \end{cases}$

20. $f(x) = \begin{cases} 2, & [x] \text{ is odd} \\ 0, & [x] \text{ is even} \end{cases}$

21. $f(x) = \begin{cases} x, & [x] \text{ is even} \\ -x, & [x] \text{ is odd} \end{cases}$

22. $f(x) = \begin{cases} 2x - 1, & [x] \text{ is even} \\ 1 - 2x, & [x] \text{ is odd} \end{cases}$

23. $f(x) = [x] - x$

24. $[x/2] - (x/2 - 1)$

25. Can a periodic function have an inverse function?

26. Show that if a function is periodic with period p, it is also periodic with period np, where n is a positive integer. (Hint: Use equation (3), $f(x) = f(x + p) = ?$).

27. Show that $f + g$ is periodic with period p if both f and g are periodic with period p. (Hint: $(f + g)(x + p) = f(x + p) + \ldots$).

28. Assume that f and g are periodic with periods 2 and 3 respectively. Is $f + g$ a periodic function? If you say it is, then what is its period? (Hint: $(f + g)(x) = f(x) + g(x) = f(x + 2) + g(x + 3) = ?$ Use result in exercise 26.)

7.2 Radian Measure

As defined in section 7.1, function S cannot be combined with any of the functions defined in chapters 1 through 6 to form new functions. It would make no sense to write the sum $x^2 + S(\theta)$ or the quotient $S(\theta)/\ln x$, because S does not share a common domain with either x^2 or $\ln x$. The domains of the latter two functions are subsets of the real numbers whereas the domain of S is the set of positive angles. We shall be able to form the above sum or quotient after S is redefined in section 7.3, so that its domain is the set of real numbers. A first step in that direction is the introduction of another angle measure.

Circular Arcs and Radian Measure

The length s, of a circular arc, is related to the measure θ*, of the central angle POQ that intercepts it (figure 7.7). The arc length is the same fractional part of the circumference, $2\pi r$, as θ degrees is of 360°.

$$\frac{s}{2\pi r} = \frac{\theta°}{360°} \tag{1}$$

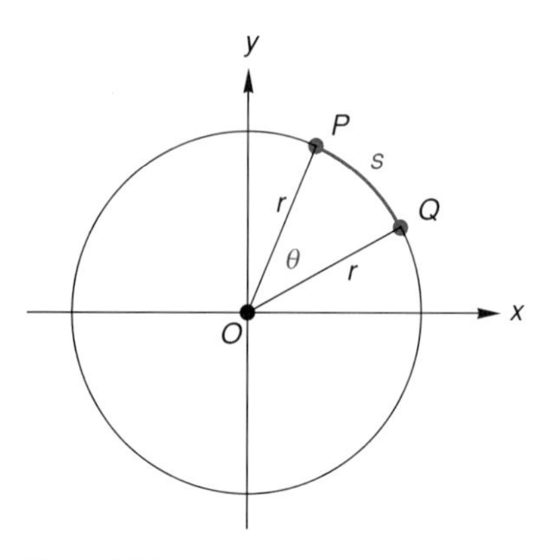

Figure 7.7

A central angle that intercepts an arc whose length is equal to the radius of the circle has a measure of 1 **radian.** If the measure of θ is 1 radian, then (1) becomes

$$\frac{r}{2\pi r} = \frac{1}{2\pi} = \frac{\theta°}{360°}. \tag{2}$$

From (2) it is seen that if a measure of *1 radian* is associated with a central angle θ, then a measure of *2π radians* must be associated with a 360° central angle. It follows that π radians is associated with a 180° central angle, and that $\pi/2$ radians is the measure associated with a central angle of 90°. In general, the proportion

*We will often use the measure of an angle to name the angle. We will say, "the 30° angle," or we will refer to angle θ rather than angle POQ.

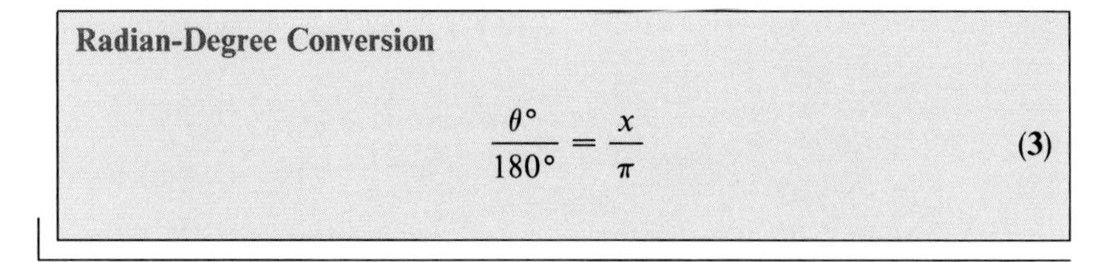

Radian-Degree Conversion

$$\frac{\theta^\circ}{180^\circ} = \frac{x}{\pi} \qquad (3)$$

allows us to convert from degree measure to radian measure and vice versa. In (3) x is the radian measure that is equivalent to θ°. No formal symbol such as the small circle used to denote degrees will be used to denote radians.

Example 1 Covert each degree measure to radian measure.

a. 90°

b. 60°

c. 150°

d. 28°

Solution Solving for x in (3),

a. $\dfrac{90^\circ}{180^\circ} = \dfrac{x}{\pi}$, $x = \dfrac{\pi}{2}$ radians.

b. $\dfrac{60^\circ}{180^\circ} = \dfrac{x}{\pi}$, $x = \dfrac{\pi}{3}$ radians.

c. $\dfrac{150^\circ}{180^\circ} = \dfrac{x}{\pi}$, $x = \dfrac{5\pi}{6}$ radians.

d. $\dfrac{28^\circ}{180^\circ} = \dfrac{x}{\pi}$, $x = \dfrac{7\pi}{45}$ radians. ■

Example 2 Convert each radian measure to degree measure.

a. $\dfrac{\pi}{6}$

b. $\dfrac{\pi}{4}$

c. 1

d. 7

Solution Solving for θ in (3),

a. $\dfrac{\theta°}{180°} = \dfrac{\frac{\pi}{6}}{\pi}$

$$\theta° = 180 \cdot \frac{1}{6}$$

$$\theta° = 30°.$$

b. $\dfrac{\theta°}{180°} = \dfrac{\frac{\pi}{4}}{\pi}$

$$\theta° = 180 \cdot \frac{1}{4}$$

$$\theta° = 45°.$$

c. $\dfrac{\theta°}{180°} = \dfrac{1}{\pi}$

$$\theta° = \frac{180}{\pi} \text{ degrees}$$

$$\theta° \approx 57.3°.$$

d. $\dfrac{\theta°}{180°} = \dfrac{7}{\pi}$

$$\theta° = \frac{1{,}260}{\pi} \text{ degrees}$$

$$\theta° \approx 401.1°.$$

■

With angles measured in radians rather than in degrees, (1) becomes

$$\frac{s}{2\pi r} = \frac{\theta}{2\pi}$$

or

Arc Length

Arc Length

$$s = r\theta. \qquad \textbf{(4)}$$

The length of arc intercepted by a central angle θ is the product of the length of the radius and the *radian measure* (not degree measure) of θ.

Example 3 Find the length of arc intercepted by a central angle with the following measure

a. $\pi/4$
b. $75°$

in a circle with a six-inch radius.

Solution From (4),

a. $s = 6 \cdot \pi/4 = 3\pi/2$ inches.

b. In (4), θ is measured in radians. We must convert 75° to radian measure.

$$\frac{x}{\pi} = \frac{75°}{180°}, \quad x = \frac{75\pi}{180} = \frac{5\pi}{12} \text{ radians}$$

$$s = 6 \cdot \frac{5\pi}{12} = \frac{5\pi}{2} \text{ inches}$$

■

If in (4) r is one unit, then s and θ are *numerically* equal. Thus, in a circle with a one-inch radius, a central angle having measure $\pi/3$ radians will intercept an arc of $\pi/3$ inches.

Quadrant Position of Terminal Side

The central angle POQ in the circle in figure 7.8, is formed from a *fixed* **initial side** $\overline{OQ}$ that is a segment in the positive x-axis, and a **terminal side** $\overline{OP}$ that rotates counterclockwise or clockwise with respect to the initial side. A central angle, so positioned in a circle, is said to be in **standard position.** If the terminal side is rotated counterclockwise, the measure of the angle formed is a *positive* real number. If the terminal side rotates clockwise, then the angle formed has a *negative* measure.

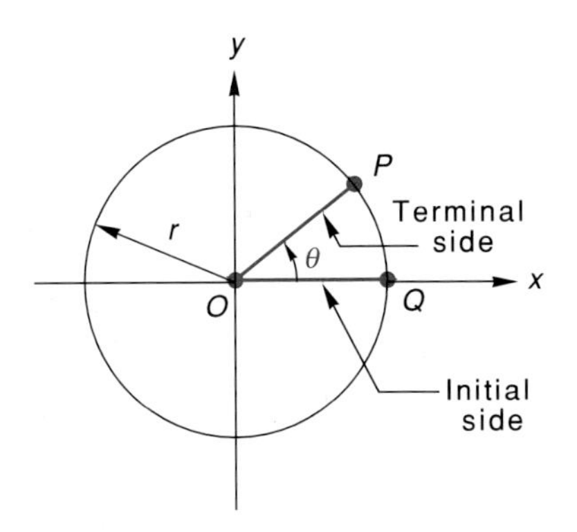

Figure 7.8

Example 4 Sketch each angle in standard position. Note the quadrant position of the terminal side.

a. 3.6 radians

b. 8.3 radians

c. -5.2 radians

Solution Figure 7.9 depicts the positions of the terminal sides of angles $\pm\pi/4$, $\pm\pi/2$, $\pm 3\pi/4$, $\pm\pi$, $\pm 5\pi/4$, $\pm 3\pi/2$, $\pm 7\pi/4$, and $\pm 2\pi$. Each angle is in standard position. Using the figure, it is a simple matter to sketch angles of 3.6, 8.3, and -5.2 radians. We see in figure 7.10 that the terminal sides are in the third, second, and first quadrants.

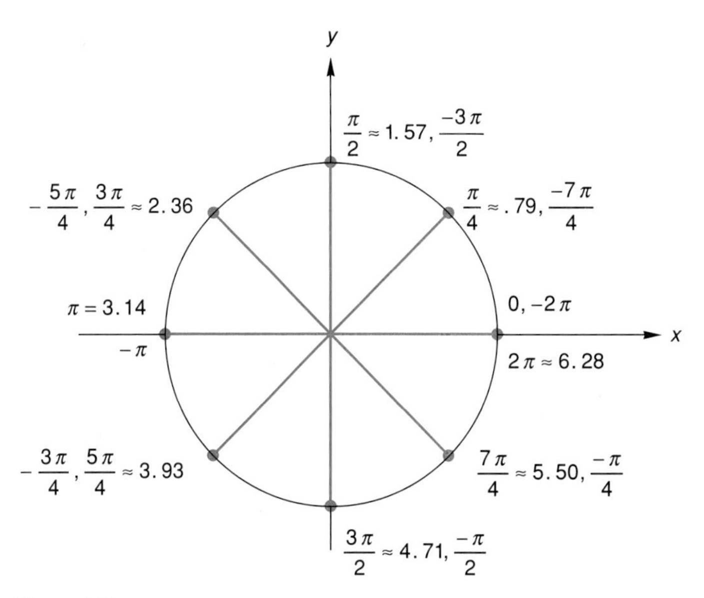

Figure 7.9

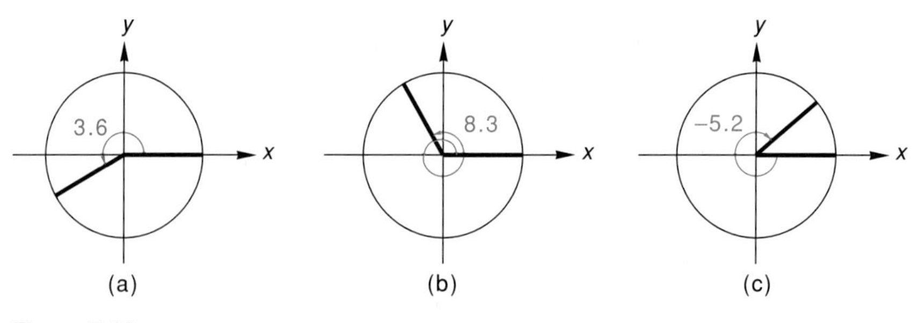

Figure 7.10

■

Exercise Set 7.2

(A)

Convert each degree measure to radian measure.

1. 55° **2.** 75° **3.** 130° **4.** 250°
5. 412° **6.** 508° **7.** 36° **8.** 118°

Convert each radian measure to degree measure.

9. $\frac{3\pi}{4}$ **10.** $\frac{7\pi}{6}$ **11.** $\frac{5\pi}{3}$ **12.** $\frac{17\pi}{6}$
13. 2.0 **14.** 5.0 **15.** 7.8 **16.** 10.3

Find the arc length associated with the given central angle and radius.

17. $\frac{\pi}{3}$, 4 in. **18.** $\frac{\pi}{6}$, 4 cm **19.** 80°, 2 m
20. 70°, 3 ft. **21.** 5.3, 10 cm **22.** 4.7, 10 in.
23. 0.34, 1,000 ft. **24.** 126°, 50 m

In exercises 25–32, sketch each angle in standard position and determine in which quadrant is its terminal side.

25. $\frac{5\pi}{6}$ **26.** $\frac{11\pi}{3}$ **27.** $-\frac{19\pi}{6}$ **28.** $-\frac{5\pi}{4}$
29. -1.3 **30.** -3.9 **31.** 0.8 **32.** 9.1

(B)

33. The region enclosed between the angle POQ and the circular arc PQ of length s, in figure 7.7, is called a sector of the circle. The area of a circular sector is determined from a proportion similar to the proportion that leads to equation (4). Show that the area of a sector of a circle is given by the formula $A = r^2\theta/2$, where r is the radius of the circle and θ is the radian measure of the central angle of the sector.

The central angle of a circular sector and the radius of the circle, in centimeters, are given. Find the area of the sector. Use the formula in exercise 33.

34. $r = 4.2, \theta = 0.82$ **35.** $r = 12.8, \theta = 2.63$
36. $r = 1.73, \theta = 5.53$ **37.** $r = 8.2, \theta = 1.7$

38. A central angle in a circle whose radius is 3.7 cm intercepts an arc of length 7.4 cm. What is the measure of the central angle?

39. A central angle in a circle whose radius is 6.46 in. intercepts an arc of 3.23 in. What is the measure of the central angle?

40. A circular sector in a circle whose radius is 2.63 ft. has an area of 28.4 square ft. What is the measure of the central angle of the sector?

41. The flywheel on a piece of machinery rotates 32 revolutions per minute. Through how many radians does a bolt on the flywheel rotate in one minute?

42. An automobile crankshaft rotates through 3,200 revolutions per minute (rpm). Through how many radians does a pin on the crankshaft turn in one minute?

7.3 The Trigonometric Functions

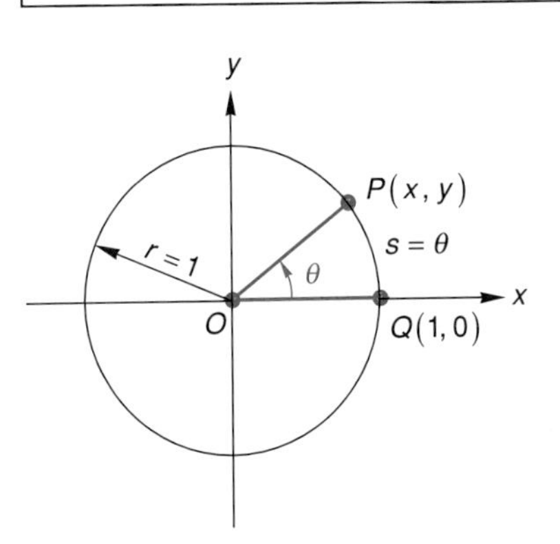

Figure 7.11

An angle of measure θ radians is constructed in a unit circle (figure 7.11). Let (x,y) be the coordinates of the endpoint P, of the terminal side of θ. Each angle θ has exactly one point P associated with it. Thus, it is possible to establish some functional relationships between θ and the coordinates of P. In fact, it is possible to establish exactly six such relationships, and they are called the **Trigonometric Functions.** The six functions are named **sine, cosine, tangent, cosecant, secant,** and **cotangent.** Their defining equations are given below, where the names of the functions are written in the abbreviated forms that are in common use.

Trigonometric Function Definitions

(sine)	$\sin\theta = y$	(cosecant)	$\csc\theta = \dfrac{1}{y}$
(cosine)	$\cos\theta = x$	(secant)	$\sec\theta = \dfrac{1}{x}$
(tangent)	$\tan\theta = \dfrac{y}{x}$	(cotangent)	$\cot\theta = \dfrac{x}{y}$

Because the arc length formula, $s = r\theta$, becomes $s = \theta$ in a unit circle (see figure 7.11), it does not matter if we think of the domain of each trigonometric function as consisting of radian measures or arc lengths. Although either point of view yields a domain that is a subset of R, it is standard practice to read sin 2.8 as, "The sine of 2.8 radians."

Example 1 Evaluate the six trigonometric functions for $\theta = \pi/3$ radians.

Solution The radian measure of θ is equivalent to 60°. Therefore, triangle OPQ (figure 7.12) that is formed by dropping a perpendicular from P onto the x-axis, is a 30°–60°–90° triangle. As the length of $\overline{OP}$ is 1 unit, the length of $\overline{OQ}$, which is the side opposite the 30° angle, must be $\frac{1}{2}$ unit. The length of $\overline{PQ}$, the side opposite the 60° angle is $\sqrt{3}/2$ units. Thus, the x and y coordinates of P are $\frac{1}{2}$ and $\sqrt{3}/2$ respectively, and the trigonometric functional values associated with $\pi/3$ are given below.

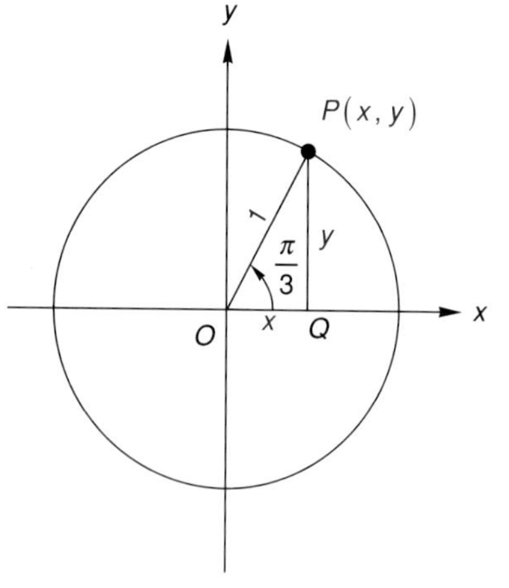

Figure 7.12

$$\sin\frac{\pi}{3} = y = \frac{\sqrt{3}}{2} \qquad \csc\frac{\pi}{3} = \frac{1}{y} = \frac{2}{\sqrt{3}}$$

$$\cos\frac{\pi}{3} = x = \frac{1}{2} \qquad \sec\frac{\pi}{3} = \frac{1}{x} = 2$$

$$\tan\frac{\pi}{3} = \frac{y}{x} = \sqrt{3} \qquad \cot\frac{\pi}{3} = \frac{x}{y} = \frac{1}{\sqrt{3}}$$

■

The method employed in example 1 to find the six trigonometric functional values associated with $\theta = \pi/3$, enables us to evaluate the trigonometric functions for $\theta = 0$, $\theta = \pi/6$, $\theta = \pi/4$, and $\theta = \pi/2$. We use figure 7.13 and note that POQ in figure 7.13b is a 30°–60°–90° triangle and that POQ in figure 7.13c is a 45°–45°–90° triangle. The values of the trigonometric functions for these angles are shown in table 7.1.

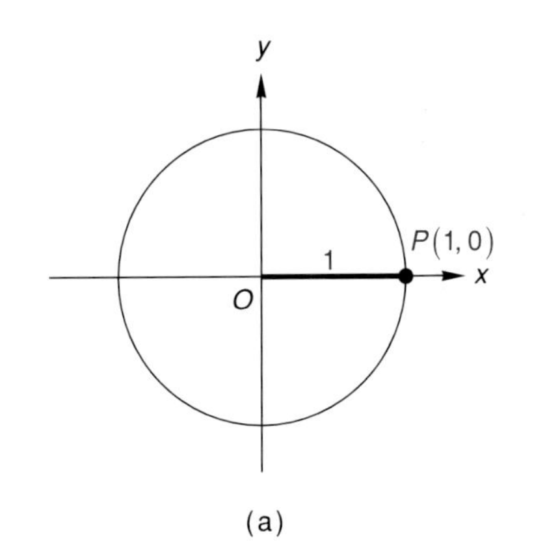

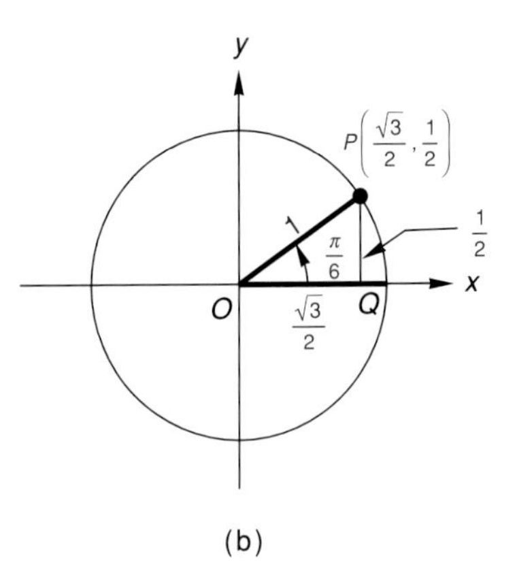

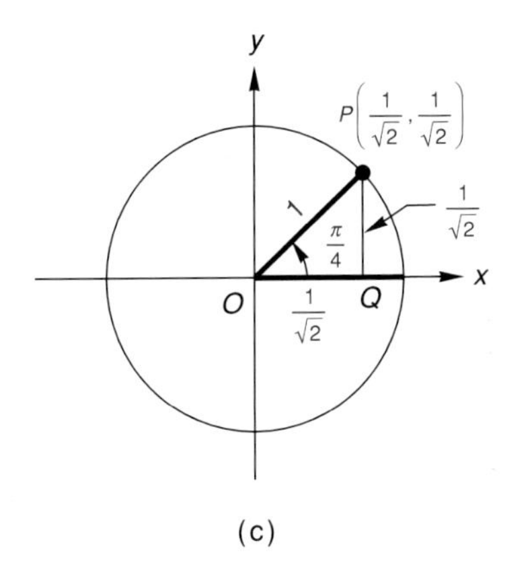

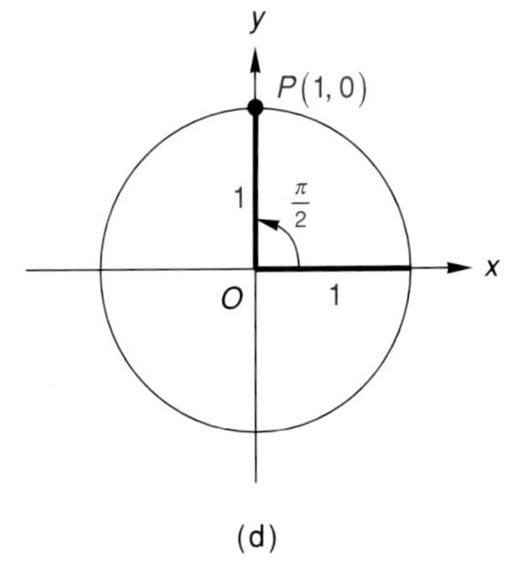

Figure 7.13

Table 7.1

θ	$\sin\theta = y$	$\cos\theta = x$	$\tan\theta = y/x$	$\csc\theta = 1/y$	$\sec\theta = 1/x$	$\cot\theta = x/y$
0	0	1	0	undefined	1	undefined
$\frac{\pi}{6}$	$\frac{1}{2}$	$\frac{\sqrt{3}}{2}$	$\frac{1}{\sqrt{3}}$	2	$\frac{2}{\sqrt{3}}$	$\sqrt{3}$
$\frac{\pi}{4}$	$\frac{1}{\sqrt{2}}$	$\frac{1}{\sqrt{2}}$	1	$\sqrt{2}$	$\sqrt{2}$	1
$\frac{\pi}{3}$	$\frac{\sqrt{3}}{2}$	$\frac{1}{2}$	$\sqrt{3}$	$\frac{2}{\sqrt{3}}$	2	$\frac{1}{\sqrt{3}}$
$\frac{\pi}{2}$	1	0	undefined	1	undefined	0

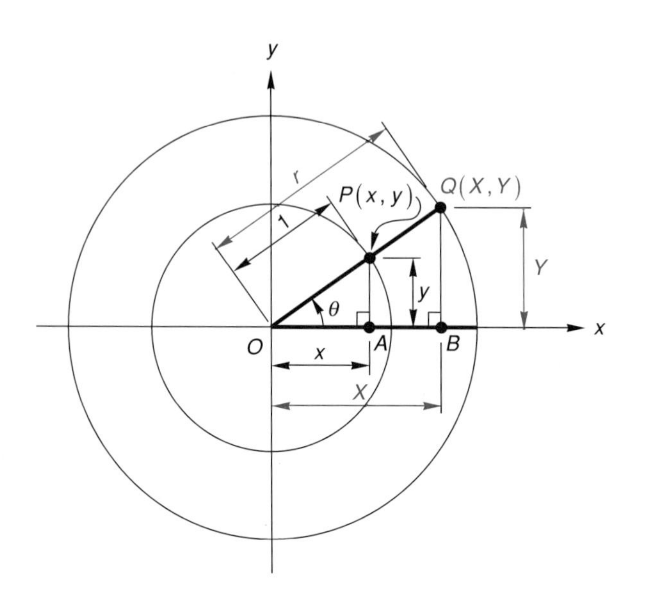

Figure 7.14

The trigonometric functions can be defined in a circle with any radius. Consider the concentric circles with radii 1 and r, that are in figure 7.14. Triangles OPA and OQB are similar so that

$$\frac{y}{1} = \frac{Y}{r}, \quad \frac{x}{1} = \frac{X}{r}, \quad \text{and} \quad \frac{y}{x} = \frac{Y}{X}.$$

Consequently,

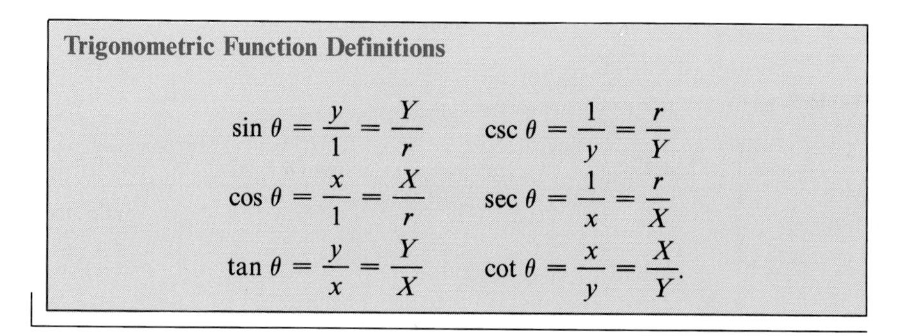

Trigonometric Function Definitions

$$\sin\theta = \frac{y}{1} = \frac{Y}{r} \qquad \csc\theta = \frac{1}{y} = \frac{r}{Y}$$

$$\cos\theta = \frac{x}{1} = \frac{X}{r} \qquad \sec\theta = \frac{1}{x} = \frac{r}{X}$$

$$\tan\theta = \frac{y}{x} = \frac{Y}{X} \qquad \cot\theta = \frac{x}{y} = \frac{X}{Y}.$$

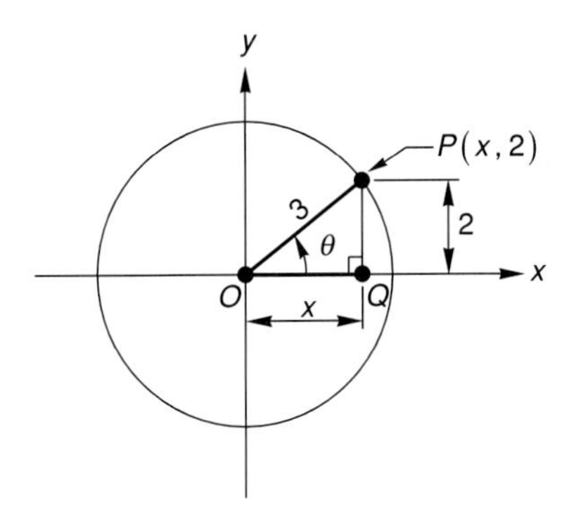

Figure 7.15

Example 2 Evaluate the five other trigonometric functions if $\sin\theta = \frac{2}{3}$ and $0 < \theta < \pi/2$.

Solution Let the terminal side of θ intersect the circumference of a circle whose radius is 3 units, at a point P that has y-coordinate 2 (figure 7.15). Then, $\sin\theta = y/r = \frac{2}{3}$. Now drop a perpendicular from P to the initial side of θ. Call the point at the foot of the perpendicular Q. The length of $\overline{OQ}$ is the x-coordinate of P. By the Pythagorean Theorem,

$$x^2 + 2^2 = 3^2$$
$$x = \sqrt{5}.$$

It follows that

$$\sin\theta = \frac{y}{r} = \frac{2}{3} \qquad \csc\theta = \frac{r}{y} = \frac{3}{2}$$
$$\cos\theta = \frac{x}{r} = \frac{\sqrt{5}}{3} \qquad \sec\theta = \frac{r}{x} = \frac{3}{\sqrt{5}}$$
$$\tan\theta = \frac{y}{x} = \frac{2}{\sqrt{5}} \qquad \cot\theta = \frac{x}{y} = \frac{\sqrt{5}}{2}.$$

■

Evaluating the Trigonometric Functions for $\pi/2 < \theta < 2\pi$

The terminal side of θ is in the second quadrant if $\pi/2 < \theta < \pi$ (figure 7.16). Also the measure of the acute angle between the negative x-axis and the terminal side of θ is $\pi - \theta$. In Figure 7.16 we construct $\pi - \theta$ in the first quadrant and label the endpoint of its terminal side, P. We call the endpoint of the terminal side of θ, Q. Point Q must be a reflection of P with respect to the y-axis.

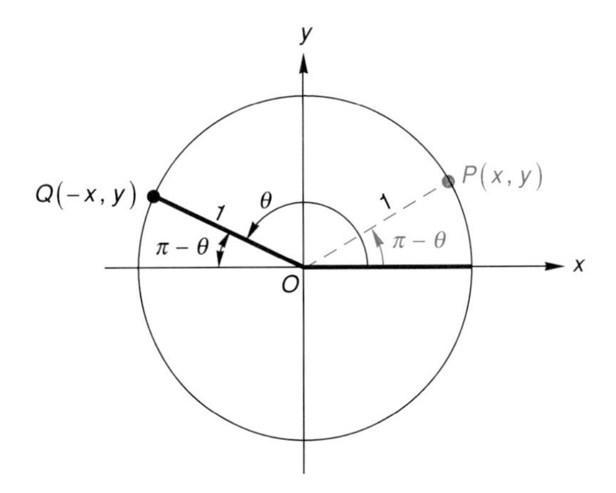

Figure 7.16

(Why? See exercise 51.) Hence, if the coordinates of P are (x,y), then the coordinates of Q are $(-x,y)$. It follows, as is shown below, that each of the six trigonometric functions of θ *can be expressed as a function of the acute angle,* $\pi - \theta$.

$$\sin \theta = y = \sin (\pi - \theta)$$
$$\cos \theta = -x = -\cos (\pi - \theta)$$
$$\tan \theta = \frac{y}{(-x)} = -\frac{y}{x} = -\tan (\pi - \theta)$$
$$\csc \theta = \frac{1}{y} = \csc (\pi - \theta)$$
$$\sec \theta = \frac{1}{(-x)} = -\frac{1}{x} = -\sec (\pi - \theta)$$
$$\cot \theta = \frac{-x}{y} = -\frac{x}{y} = -\cot (\pi - \theta)$$

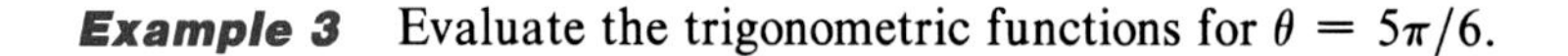

Example 3 Evaluate the trigonometric functions for $\theta = 5\pi/6$.

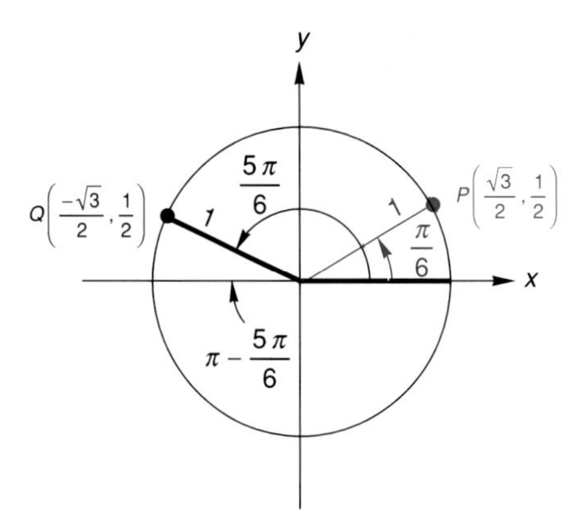

Figure 7.17

Solution As shown in figure 7.17, the terminal side of θ is in the second quadrant. Also, the angle between the terminal side of θ and the negative x-axis is $\pi - 5\pi/6 = \pi/6$. We know from the discussion above that it is possible to express each trigonometric function of θ as a function of $\pi/6$. We need only be aware that negative signs must be attached to the cosine, tangent, secant, and cotangent functions.

$$\sin \frac{5\pi}{6} = \sin\left(\pi - \frac{5\pi}{6}\right) = \sin \frac{\pi}{6} = \frac{1}{2}$$

$$\cos\frac{5\pi}{6} = -\cos\left(\pi - \frac{5\pi}{6}\right) = -\cos\frac{\pi}{6} = -\frac{\sqrt{3}}{2}$$

$$\tan\frac{5\pi}{6} = -\tan\left(\pi - \frac{5\pi}{6}\right) = -\tan\frac{\pi}{6} = -\frac{1}{\sqrt{3}}$$

$$\csc\frac{5\pi}{6} = \csc\left(\pi - \frac{5\pi}{6}\right) = \csc\frac{\pi}{6} = 2$$

$$\sec\frac{5\pi}{6} = -\sec\left(\pi - \frac{5\pi}{6}\right) = -\sec\frac{\pi}{6} = -\frac{2}{\sqrt{3}}$$

$$\cot\frac{5\pi}{6} = -\cot\left(\pi - \frac{5\pi}{6}\right) = -\cot\frac{\pi}{6} = -\sqrt{3}$$

It is possible to construct geometric arguments (see exercises 52 and 53) to show that a trigonometric function of a positive angle θ, whose terminal side is in the third or fourth quadrant, can be expressed as the trigonometric function of an acute angle. In the discussion that follows, that acute angle is called $\boldsymbol{\alpha}$. *It is the angle between the terminal side of θ and the x-axis.* The plus and minus signs associated with the trigonometric functions of positive angles whose terminal sides are in the third or fourth quadrants are given below with the aid of figure 7.18.

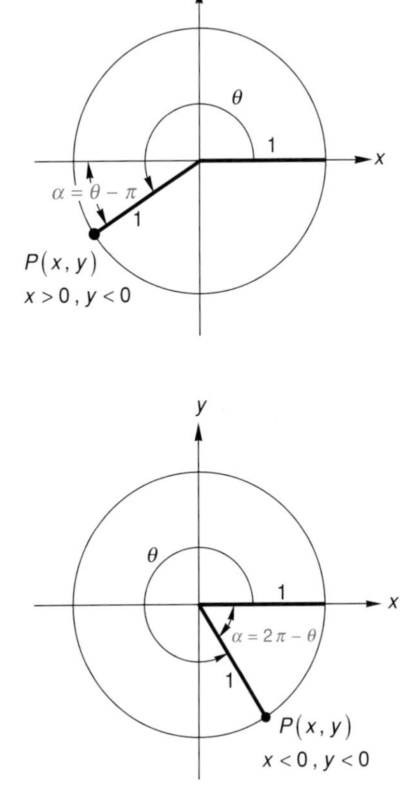

Figure 7.18

Third Quadrant	*Fourth Quadrant*
$\sin\theta < 0$ because $\sin\theta = y$	$\sin\theta < 0$
$\cos\theta < 0$ because $\cos\theta = x$	$\cos\theta > 0$
$\tan\theta > 0$ because $\tan\theta = \dfrac{y}{x}$	$\tan\theta < 0$
$\csc\theta < 0$ because $\csc\theta = \dfrac{1}{y}$	$\csc\theta < 0$
$\sec\theta < 0$ because $\sec\theta = \dfrac{1}{x}$	$\sec\theta > 0$
$\cot\theta > 0$ because $\cot\theta = \dfrac{x}{y}$	$\cot\theta < 0$

We conclude that a trigonometric function of a positive angle θ, whose terminal side lay in the first, second, third, or fourth quadrant, can be expressed as the trigonometric function of *the acute angle formed by the x-axis and the terminal side of θ*. In addition, to insure that the functional values are correct, a plus or a minus sign, as determined by the quadrant position of the terminal side of θ, must be attached to the functional value of the **associated acute angle.** Table 7.2 displays the sign that is to be attached to each trigonometric function for $0 \le \theta \le 2\pi$. In the table, α is the **associated acute angle** formed by the terminal side of θ and the x-axis.

Table 7.2

Terminal Side of θ	α	Sign					
		$\sin\theta$	$\cos\theta$	$\tan\theta$	$\csc\theta$	$\sec\theta$	$\cot\theta$
First Quadrant	$\alpha = \theta$	+	+	+	+	+	+
Second Quadrant	$\alpha = \pi - \theta$	+	−	−	+	−	−
Third Quadrant	$\alpha = \theta - \pi$	−	−	+	−	−	+
Fourth Quadrant	$\alpha = 2\pi - \theta$	−	+	−	−	+	−

Example 4 Evaluate

a. $\cos 2\pi/3$,
b. $\sin 5\pi/4$, and
c. $\cos 11\pi/6$.

Solution Using figure 7.19 and tables 7.1 and 7.2, we have

$$\textbf{a.}\ \cos\frac{2\pi}{3} = -\cos\left(\pi - \frac{2\pi}{3}\right) = -\cos\frac{\pi}{3} = -\frac{1}{2}.$$

$$\textbf{b.}\ \sin\frac{5\pi}{4} = -\sin\left(\frac{5\pi}{4} - \pi\right) = -\sin\frac{\pi}{4} = -\frac{1}{\sqrt{2}}.$$

$$\textbf{c.}\ \cos\frac{11\pi}{6} = \cos\left(2\pi - \frac{11\pi}{6}\right) = \cos\frac{\pi}{6} = \frac{\sqrt{3}}{2}.$$

We note in closing that equation (1) in section 7.1, which determines the position of the pump piston as a function of time, can be written as $y = r\sin(2\pi kt)$.

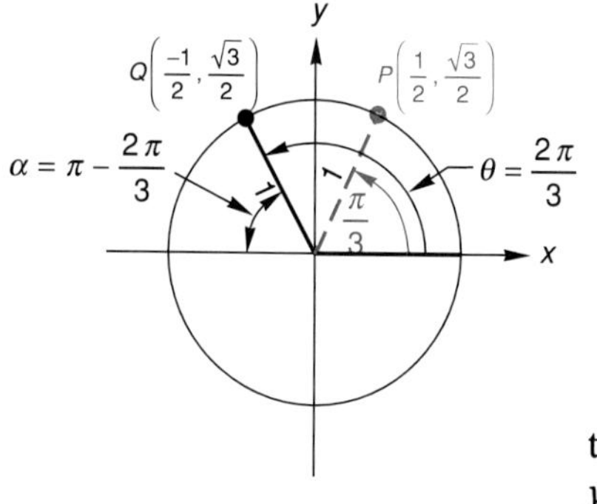

(a)

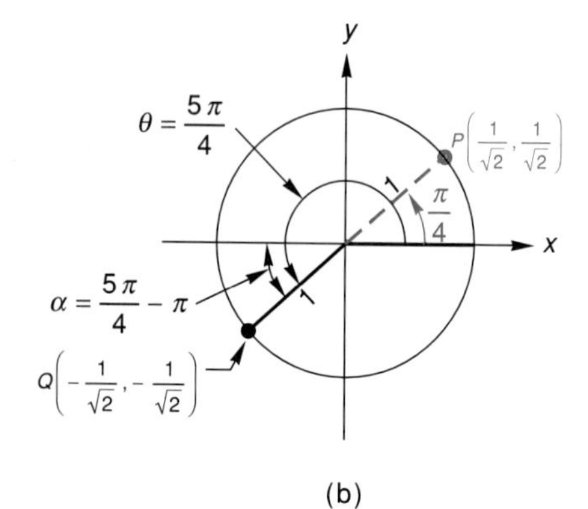

(b)

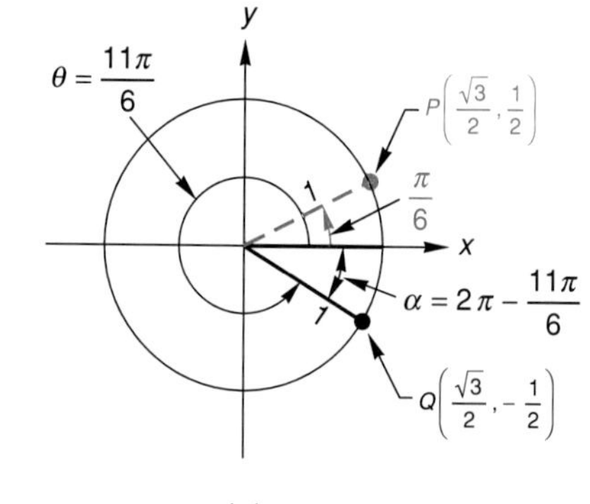

(c)

Figure 7.19

Exercise Set 7.3

(A)

The value of a trigonometric function is given. Assume that $0 < \theta < \pi/2$ and evaluate the other five trigonometric functions.

1. $\sin\theta = \frac{3}{4}$ **2.** $\cos\theta = \frac{1}{8}$
3. $\tan\theta = \frac{1}{2}$ **4.** $\sin\theta = \frac{4}{5}$
5. $\cos\theta = \frac{12}{13}$ **6.** $\tan\theta = \frac{8}{15}$
7. $\csc\theta = \frac{5}{2}$ **8.** $\cos\theta = \frac{1}{5}$
9. $\cot\theta = 3$ **10.** $\sin\theta = \frac{5}{6}$
11. $\sec\theta = \frac{4}{3}$ **12.** $\tan\theta = 4$

Use tables 7.1 and 7.2 to help you determine each functional value exactly. Do not use a calculator.

13. $\sin\frac{3\pi}{4}$ **14.** $\cos\frac{5\pi}{6}$ **15.** $\tan\frac{2\pi}{3}$
16. $\sin\frac{7\pi}{6}$ **17.** $\cos\frac{4\pi}{3}$ **18.** $\tan\frac{5\pi}{4}$
19. $\csc\frac{5\pi}{3}$ **20.** $\cos\frac{7\pi}{4}$ **21.** $\cot\frac{3\pi}{4}$
22. $\sin\frac{11\pi}{6}$ **23.** $\sec\frac{7\pi}{6}$ **24.** $\tan\frac{11\pi}{6}$

Assume that θ is in standard position in a unit circle and that $0 \le \theta \le 2\pi$. What is the quadrant position of the terminal side of θ if:

25. $\sin\theta > 0$ and $\cos\theta < 0$.
26. $\sin\theta < 0$ and $\tan\theta > 0$.
27. $\cos\theta > 0$ and $\tan\theta > 0$.
28. $\csc\theta < 0$ and $\sec\theta > 0$.
29. $\cot\theta > 0$ and $\cos\theta < 0$.
30. $\sec\theta < 0$ and $\sin\theta < 0$.

Assume that θ is in standard position in a unit circle and that $0 \le \theta \le 2\pi$. Determine the radian measure of θ if the endpoint of its terminal side has the coordinates:

31. $\left(\frac{1}{\sqrt{2}}, \frac{1}{\sqrt{2}}\right)$. **32.** $\left(-\frac{\sqrt{3}}{2}, \frac{1}{2}\right)$.
33. $\left(\frac{1}{2}, \frac{\sqrt{3}}{2}\right)$. **34.** $\left(\frac{1}{\sqrt{2}}, -\frac{1}{\sqrt{2}}\right)$.
35. $(0,1)$. **36.** $(-1,0)$.
37. $\left(\frac{\sqrt{3}}{2}, -\frac{1}{2}\right)$. **38.** $(0,-1)$.

The value of a trigonometric function is given. Assume that $\pi/2 < \theta < \pi$ and evaluate the other five trigonometric functions.

39. $\sin\theta = \frac{3}{10}$ **40.** $\cos\theta = -\frac{3}{8}$
41. $\tan\theta = -2$ **42.** $\sin\theta = \frac{1}{4}$
43. $\cos\theta = -\frac{3}{4}$ **44.** $\tan\theta = -\frac{1}{6}$

The value of a trigonometric function is given. Assume that $\pi < \theta < 3\pi/2$ and evaluate the other five trigonometric functions.

45. $\sec\theta = -2$ **46.** $\csc\theta = -\frac{4}{3}$
47. $\cot\theta = \frac{1}{5}$ **48.** $\sec\theta = -\frac{6}{5}$
49. $\csc\theta = -\frac{13}{12}$ **50.** $\cot\theta = \frac{9}{4}$

(B)

51. It is stated on page 273 that points P and Q are reflections of one another with respect to the y-axis. Construct a geometric argument to support this statement. (Hint: congruent triangles.)

Construct a geometric argument to show that

52. if $\pi < \theta < 3\pi/2$ then $\sin\theta = -\sin(\theta - \pi)$, $\cos\theta = -\cos(\theta - \pi)$, and $\tan\theta = \tan(\theta - \pi)$.

53. if $3\pi/2 < \theta < 2\pi$ then $\sin\theta = -\sin(2\pi - \theta)$, $\cos\theta = \cos(2\pi - \theta)$, and $\tan\theta = -\tan(2\pi - \theta)$.

7.4 Some Properties of the Trigonometric Functions

A number of relationships among the trigonometric functions can be constructed directly from the definitions on page 270.

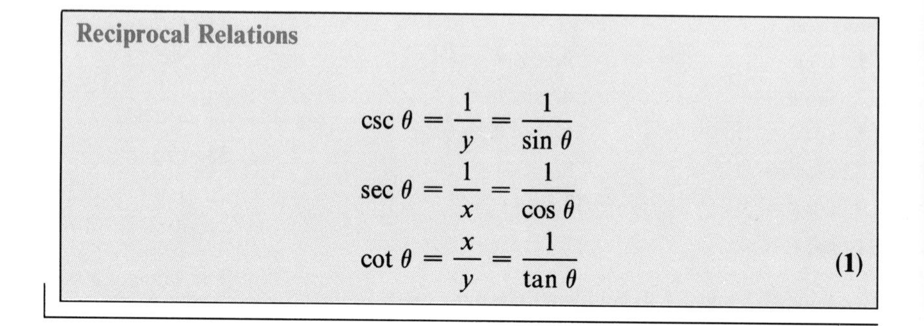

Reciprocal Relations

$$\csc\theta = \frac{1}{y} = \frac{1}{\sin\theta}$$
$$\sec\theta = \frac{1}{x} = \frac{1}{\cos\theta}$$
$$\cot\theta = \frac{x}{y} = \frac{1}{\tan\theta} \qquad \textbf{(1)}$$

Equations (1) define the *reciprocal* relationships among the trig functions. If $\sin\theta = \frac{5}{6}$, then $\csc\theta = \frac{6}{5}$, and if $\cos\theta = \frac{1}{10}$, then $\sec\theta = 10$.

A Tangent Identity

$$\tan\theta = \frac{y}{x} = \frac{\sin\theta}{\cos\theta} \qquad \textbf{(2)}$$

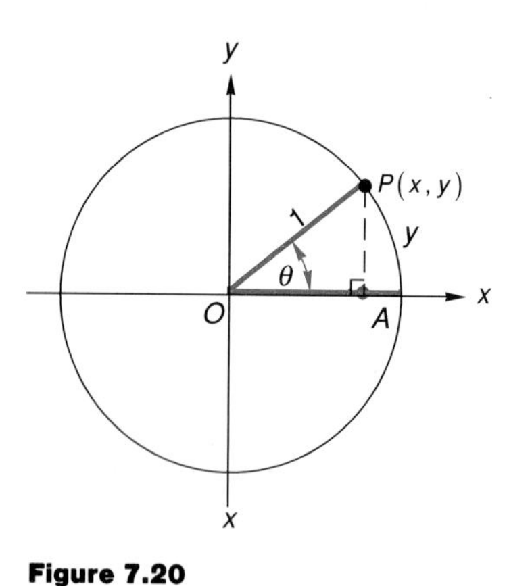

Figure 7.20

If we know $\sin\theta$ and $\cos\theta$, then by (2) we also know $\tan\theta$.

$$\tan\frac{\pi}{6} = \frac{\sin\frac{\pi}{6}}{\cos\frac{\pi}{6}} = \frac{\frac{1}{2}}{\frac{\sqrt{3}}{2}} = \frac{1}{\sqrt{3}}$$

Applying the Pythagorean Theorem to triangle OPA in figure 7.20 gives us $x^2 + y^2 = 1$. But $x = \cos\theta$, and $y = \sin\theta$. Hence, we have

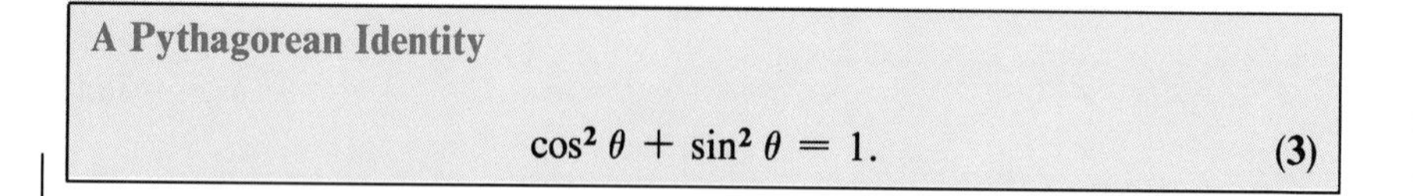

A Pythagorean Identity

$$\cos^2\theta + \sin^2\theta = 1. \qquad \textbf{(3)}$$

Equations (1), (2), and (3) establish the interdependence of the trig functions. Knowing one functional value enables us to find the other five functional values.

Example 1 Given that $\sin\theta = \frac{5}{6}$, and that $0 < \theta < \pi/2$, evaluate $\cos\theta$, $\tan\theta$, $\csc\theta$, $\sec\theta$, and $\cot\theta$.

Solution From (3)

$$\cos^2\theta + \left(\frac{5}{6}\right)^2 = 1$$
$$\cos^2\theta = \frac{11}{36}$$
$$\cos\theta = \pm\frac{\sqrt{11}}{6}.$$

But $\cos\theta$ must be positive because $0 < \theta < \pi/2$. Hence,

$$\cos\theta = \frac{\sqrt{11}}{6}.$$

From (2)

$$\tan\theta = \frac{\sin\theta}{\cos\theta} = \frac{\frac{5}{6}}{\frac{\sqrt{11}}{6}} = \frac{5}{\sqrt{11}}.$$

From (1)

$$\csc\theta = \frac{1}{\sin\theta} = \frac{6}{5},$$
$$\sec\theta = \frac{1}{\cos\theta} = \frac{6}{\sqrt{11}},$$

and

$$\cot\theta = \frac{1}{\tan\theta} = \frac{\sqrt{11}}{5}.$$

■

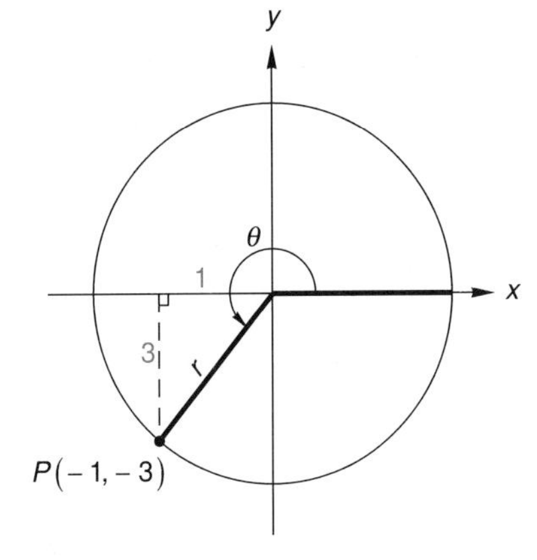

Figure 7.21

Example 2 Given that $\tan\theta = 3$ and $\pi < \theta < 3\pi/2$, evaluate $\sin\theta$, $\cos\theta$, $\csc\theta$, $\sec\theta$, and $\cot\theta$.

Solution Imagine θ to be a central angle in a circle of radius r. In order for θ to satisfy the conditions, $\tan\theta = 3$ and $\pi < \theta < 3\pi/2$ (figure 7.21), we assign the coordinates $(-1, -3)$ to the endpoint P, of the terminal side of the central angle. The radius of the circle is determined from the Pythagorean Theorem.

$$r^2 = 1^2 + 3^2$$
$$r = \sqrt{10}$$

Now,

$$\sin\theta = -\frac{3}{\sqrt{10}}, \qquad \csc\theta = \frac{1}{\sin\theta} = -\frac{\sqrt{10}}{3}$$
$$\cos\theta = -\frac{1}{\sqrt{10}}, \qquad \sec\theta = \frac{1}{\cos\theta} = -\sqrt{10}$$
$$\cot\theta = \frac{1}{\tan\theta} = \frac{1}{3}.$$

Because the solution set equals the replacement set in each of the equations (1), (2), and (3), those equations are *identities*.

More Identities

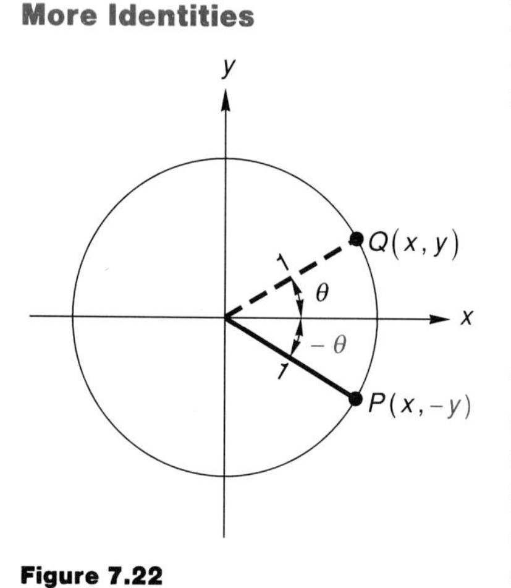

Figure 7.22

Points P and Q (figure 7.22) are reflections of each other with respect to the x-axis. (Why?) That fact and the definitions for the sine, cosine, and tangent functions enable us to write the following equalities.

$$\sin(-\theta) = -y = -\sin\theta$$
$$\cos(-\theta) = x = \cos\theta$$
$$\tan(-\theta) = -\frac{y}{x} = -\tan\theta$$

The relationships established above between functional values of negative angles and functional values of the corresponding positive angles hold for any θ in the interval $[0,2\pi]$. The relationships depend only on the fact that P and Q are reflections of each other with respect to the x-axis. It doesn't matter if the terminal side of θ is in the first, second, third, or fourth quadrants (see exercises 17 and 18). The above relationships are easily extended to the reciprocal trigonometric functions.

$$\csc(-\theta) = \frac{1}{\sin(-\theta)} = \frac{1}{-\sin\theta} = -\csc\theta$$

Hence, we have the following identities.

Function Values for Negative Angles

$$\sin(-\theta) = -\sin\theta, \qquad \csc(-\theta) = -\csc\theta$$
$$\cos(-\theta) = \cos\theta, \qquad \sec(-\theta) = \sec\theta$$
$$\tan(-\theta) = -\tan\theta, \qquad \cot(-\theta) = -\cot\theta \qquad (4)$$

Example 3 Evaluate $\sin\left(-\frac{\pi}{6}\right)$, $\cos\left(-\frac{7\pi}{4}\right)$, and $\tan\left(-\frac{4\pi}{3}\right)$.

Solution

$$\begin{aligned}
\sin\left(-\frac{\pi}{6}\right) &= -\sin\frac{\pi}{6} \\
&= -\frac{1}{2} && \text{(table 7.1)} \\
\cos\left(-\frac{7\pi}{4}\right) &= \cos\frac{7\pi}{4} \\
&= \cos\left(2\pi - \frac{7\pi}{4}\right) && \text{(table 7.2)} \\
&= \cos\left(\frac{\pi}{4}\right) \\
&= \frac{1}{\sqrt{2}} && \text{(table 7.1)} \\
\tan\left(-\frac{4\pi}{3}\right) &= -\tan\frac{4\pi}{3} \\
&= -\tan\left(\frac{4\pi}{3} - \pi\right) && \text{(table 7.2)} \\
&= -\tan\frac{\pi}{3} \\
&= -\sqrt{3} && \text{(table 7.1)}
\end{aligned}$$

Evaluating the Trigonometric Functions over $(-\infty,\infty)$

Taken together, equations (4) and the results summarized in table 7.2 allow us to conclude that if f is a trigonometric function and $-2\pi \leq \theta \leq 2\pi$, then there is an acute angle α such that $f(\theta) = \pm f(\alpha)$.

Example 4 Let f name the trig function and let θ be the angle measure in **a, b,** and **c.** Show that $f(\theta) = \pm f(\alpha)$, where $0 < \alpha < \pi/2$.

a. $\tan\frac{3\pi}{4}$

b. $\cos\frac{5\pi}{3}$

c. $\sin\left(-\frac{7\pi}{6}\right)$

Solution

a. $\tan \frac{3\pi}{4} = -\tan \frac{\pi}{4}$ (table 7.2)

b. $\cos \frac{5\pi}{3} = \cos \frac{\pi}{3}$ (table 7.2)

c. $\sin\left(-\frac{7\pi}{6}\right) = -\sin \frac{7\pi}{6}$ (equation (4))

$= -\left(-\sin \frac{\pi}{6}\right)$ (table 7.2)

$= \sin \frac{\pi}{6}$ ■

We have narrowed the process of evaluating a trig function over the interval $[-2\pi,2\pi]$ to evaluating the function over the smaller interval $[0,\pi/2]$. In fact we can show that if f is a trig function and θ is a real number in the interval $(-\infty,\infty)$, it is still true that there is an acute angle α such that $f(\theta) = \pm f(\alpha)$. The whole matter hinges on our being able to express θ as an integral multiple of 2π plus an angle β, where $0 < \beta < 2\pi$. For example, if $\theta = 23\pi/5$, then $\theta = 2(2\pi) + 3\pi/5$ and $\beta = 3\pi/5$.

Let us evaluate $\sin(\beta + 2\pi)$ and $\cos(\beta + 2\pi)$ in figure 7.23.

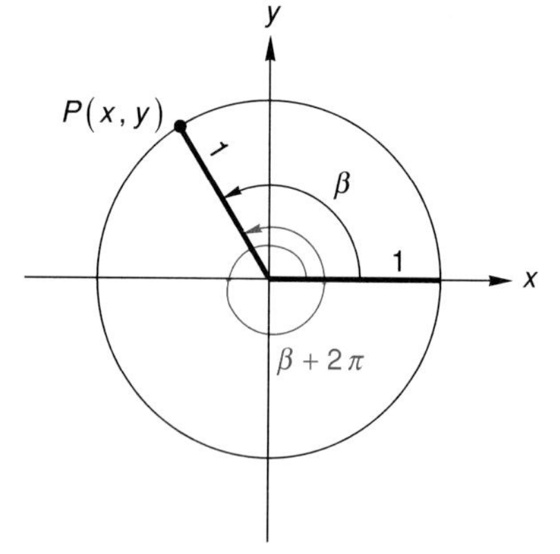

Figure 7.23

$$\sin(\beta + 2\pi) = y = \sin \beta$$
$$\cos(\beta + 2\pi) = x = \cos \beta$$

Observe that the terminal sides of $\beta + 2\pi$ and β are the same. It is also true that any angle $\beta + k \cdot 2\pi$, $k \varepsilon J$, has the same terminal side as β. Consequently we have that

$$\sin(\beta + k \cdot 2\pi) = \sin \beta,$$
$$\cos(\beta + k \cdot 2\pi) = \cos \beta,$$
$$\csc(\beta + k \cdot 2\pi) = \frac{1}{\sin(\beta + k \cdot 2\pi)} = \frac{1}{\sin \beta} = \csc \beta,$$

and
$$\sec(\beta + k \cdot 2\pi) = \frac{1}{\cos(\beta + k \cdot 2\pi)} = \frac{1}{\cos \beta} = \sec \beta.$$

We conclude that the sine and cosine functions and their reciprocal functions are periodic with period 2π.

We have shown that the sine or cosine of an angle greater than 2π, or less than -2π, can be evaluated as the sine or cosine of an angle β, where $0 \le \beta < 2\pi$. And so it is with the tangent and cotangent functions.

$$\tan(\beta + k \cdot 2\pi) = \frac{\sin(\beta + k \cdot 2\pi)}{\cos(\beta + k \cdot 2\pi)} = \frac{\sin \beta}{\cos \beta} = \tan \beta$$

$$\cot(\beta + k \cdot 2\pi) = \frac{1}{\tan(\beta + k \cdot 2\pi)} = \frac{1}{\tan \beta} = \cot \beta$$

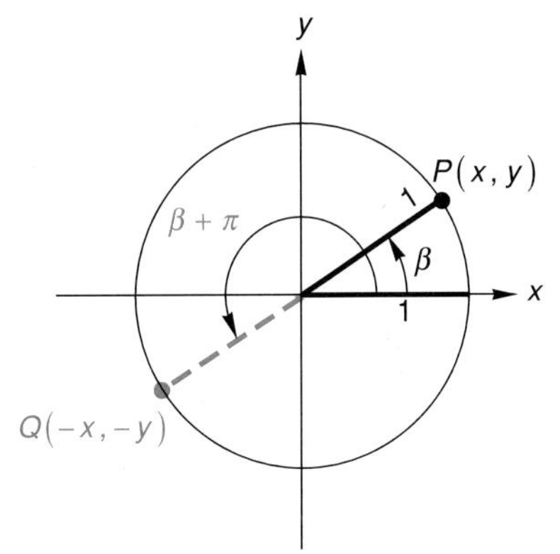

Figure 7.24

However, unlike the sine and cosine and their reciprocal functions, the tangent and cotangent functions are periodic with period π. In figure 7.24, the endpoint Q of the terminal side of $\beta + \pi$ is a reflection through the origin of P, the endpoint of the terminal side of β. This is true whatever the measure of β. Therefore, the coordinates of Q are always opposite in sign from the corresponding coordinates of P. It follows that

$$\tan(\beta + \pi) = \frac{-y}{-x} = \frac{y}{x} = \tan \beta,$$
$$\cot(\beta + \pi) = \frac{1}{\tan(\beta + \pi)} = \frac{1}{\tan \beta} = \cot \beta,$$

and it is true in general that if $k \varepsilon J$, then

$$\tan(\beta + k \cdot \pi) = \tan \beta$$
$$\cot(\beta + k \cdot \pi) = \cot \beta.$$

Thus, the tangent and cotangent functions are periodic with period π.

Example 5 Evaluate

a. $\sin\left(\frac{-13\pi}{2}\right)$

b. $\cos\left(\frac{11\pi}{4}\right)$, and

c. $\tan\left(\frac{17\pi}{6}\right)$.

Solution

a. $$\sin\left(\frac{-13\pi}{2}\right) = \sin\left(\frac{-\pi}{2} - 6\pi\right) = \sin\left(\frac{-\pi}{2}\right) = -\sin\frac{\pi}{2} = -1$$

b. $$\cos\left(\frac{11\pi}{4}\right) = \cos\left(\frac{3\pi}{4} + 2\pi\right) = \cos\frac{3\pi}{4} = -\cos\frac{\pi}{4} = -\frac{1}{\sqrt{2}}$$

$$\text{c. } \tan\left(\frac{17\pi}{6}\right) = \tan\left(\frac{5\pi}{6} + 2\pi\right)$$
$$= \tan\frac{5\pi}{6}$$
$$= -\tan\frac{\pi}{6}$$
$$= -\frac{1}{\sqrt{3}}$$

■

Cofunctions and Tables

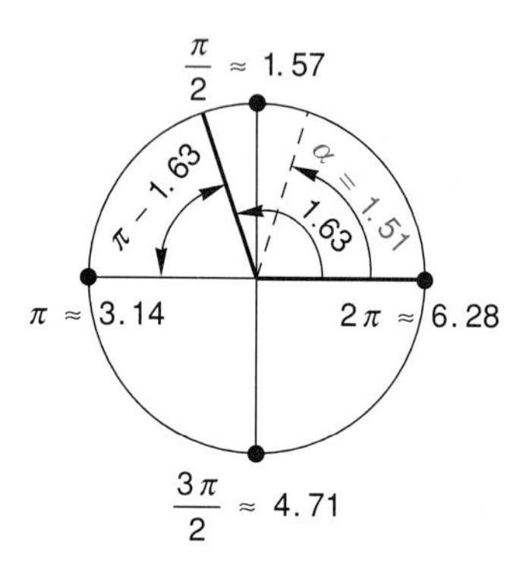

Figure 7.25

Function values have so far been determined from geometric arguments. We shall not continue this way. Calculators are internally programmed to give function values at the touch of a key. For example, with the calculator in radian mode we insert 1.63 and then depress the sine key to obtain

$$\sin 1.63 = \boxed{1.63}\ \boxed{\sin} = 0.99824798.$$

If a calculator is not available, then a trigonometric table (see appendix A) may be used.

It has been shown that if f is a trig function and $|\theta| > \pi/2$, then $f(\theta) = \pm f(\alpha)$, where $0 \le \alpha \le \pi/2$. Hence, the trig table need only contain entries in the interval $[0,\pi/2]$. Because $\pi/2 < 1.63 < \pi$, we cannot find sin 1.63 directly from the trig table. We must enter the table with the associated acute angle, $\pi - 1.63 = 1.51$ (figure 7.25).

$$\sin 1.63 = \sin 1.51 = 0.9981$$

In the trig table, sin 1.51 (read from the bottom headings) equals cos 0.06 (read from the top headings). We have that

$$\sin 1.51 = \cos\left(\frac{\pi}{2} - 1.51\right) = \cos 0.06.$$

In fact it is shown below that the statement

A Complementary Angle Relationship

$$\sin\theta = \cos\left(\frac{\pi}{2} - \theta\right) \tag{5}$$

is true for all θ. Equation (5) tells us that the *sine of an angle is equal to the cosine of the complementary angle.* For this reason sine and cosine are called *complementary functions,* or *co-functions.*

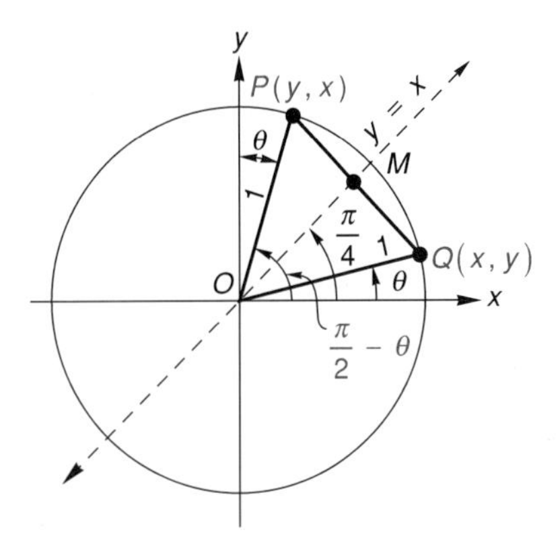

Figure 7.26

Triangle OPQ in figure 7.26 is isosceles. Also,

$$\text{angle } POM = \text{angle } QOM = \frac{\pi}{4} - \theta,$$

making line $y = x$ the angle bisector of angle POQ. The angle bisector of the angle formed by the equal sides of an isosceles triangle bisects, and is perpendicular to, the base of the triangle. Therefore, points P and Q are reflections of each other with respect to $y = x$. (Why?) As a result, if the coordinates of Q are (x,y), then the coordinates of P are (y,x). It follows that

$$\sin\theta = y = \cos\left(\frac{\pi}{2} - \theta\right).$$

The above argument, showing the complementary relationship between sine and cosine, holds whatever the measure of θ (exercise 47). Similar arguments can be developed to show the complementary relationship between tangent and cotangent, and between secant and cosecant (exercises 45 and 46). The trig table in appendix A takes advantage of these complementary relationships in that it has angle entries only in the interval $[0,\pi/4]$.

Example 6 Evaluate sin 2.34, cos 2.34, and tan −2.34 with a calculator and then with the trigonometric table.

Solution Make sure your calculator is in radian mode.
By calculator:

$$\boxed{2.34}\ \boxed{\sin} = 0.7184648$$

$$\boxed{2.34}\ \boxed{\cos} = -0.6955633$$

$$\boxed{2.34}\ \boxed{+/-}\ \boxed{\tan} = 1.032925.$$

By table:

$$\begin{aligned}
\sin 2.34 &= \sin(\pi - 2.34) = \sin(0.80) = \cos\left(\frac{\pi}{2} - 0.80\right) \\
&= \cos(0.77) = 0.7173 \\
\cos 2.34 &= -\cos(\pi - 2.34) = -\cos(0.80) = -\sin\left(\frac{\pi}{2} - 0.80\right) \\
&= -\sin(0.77) = -0.6967 \\
\tan(-2.34) &= -\tan 2.34 = -(-\tan(\pi - 2.34)) = \tan(0.80) \\
&= \cot\left(\frac{\pi}{2} - 0.80\right) = 1.030.
\end{aligned}$$

■

A Summary of the Section's Identities

$$\csc\theta = \frac{1}{\sin\theta}$$
$$\sec\theta = \frac{1}{\cos\theta}$$
$$\cot\theta = \frac{1}{\tan\theta}$$
$$\sin^2\theta + \cos^2\theta = 1$$
$$\sin(-\theta) = -\sin\theta, \quad \csc(-\theta) = -\csc\theta$$
$$\cos(-\theta) = \cos\theta, \quad \sec(-\theta) = \sec\theta$$
$$\tan(-\theta) = -\tan\theta, \quad \cot(-\theta) = -\cot\theta$$
$$\sin(\beta + k\cdot 2\pi) = \sin\beta, \quad \csc(\beta + k\cdot 2\pi) = \csc\beta$$
$$\cos(\beta + k\cdot 2\pi) = \cos\beta, \quad \sec(\beta + k\cdot 2\pi) = \sec\beta$$
$$\tan(\beta + k\cdot \pi) = \tan\beta, \quad \cot(\beta + k\cdot \pi) = \cot\beta$$
$$\sin\theta = \cos\left(\frac{\pi}{2} - \theta\right)$$
$$\tan\theta = \cot\left(\frac{\pi}{2} - \theta\right)$$
$$\sec\theta = \csc\left(\frac{\pi}{2} - \theta\right)$$

Exercise Set 7.4

(A)

Given a trigonometric function value, write the reciprocal function and its value.

1. $\sin\theta = \frac{2}{3}$

2. $\cos\theta = \frac{7}{10}$

3. $\tan\theta = 4$

4. $\cot\theta = \frac{6}{5}$

5. $\sec\theta = 2$

6. $\csc\theta = \frac{11}{8}$

Use equation (3) and the given information to find the indicated function value.

7. $\sin\theta = \frac{2}{3}, \frac{\pi}{2} < \theta < \pi, \quad \cos\theta = ?$

8. $\cos\theta = \frac{9}{10}, \frac{3\pi}{2} < \theta < 2\pi, \quad \sin\theta = ?$

9. $\cos\theta = -\frac{1}{8}, \pi < \theta < \frac{3\pi}{2}, \quad \sin\theta = ?$

10. $\sin\theta = -\frac{2}{5}, \pi < \theta < \frac{3\pi}{2}, \quad \cos\theta = ?$

Given one trigonometric function value, find the other five trigonometric function values.

11. $\sin\theta = \frac{1}{4}, 0 < \theta < \frac{\pi}{2}$

12. $\sin\theta = -\frac{4}{5}, \pi < \theta < \frac{3\pi}{2}$

13. $\cos\theta = \frac{2}{3}, \frac{3\pi}{2} < \theta < 2\pi$

14. $\cos\theta = -\frac{2}{7}, \frac{\pi}{2} < \theta < \pi$

15. $\tan\theta = -2, \frac{\pi}{2} < \theta < \pi$

16. $\tan\theta = \frac{3}{10}, \pi < \theta < \frac{3\pi}{2}$

Construct a geometric argument similar to the one given on page 280 to show that $\sin(-\theta) = -\sin\theta$, $\cos(-\theta) = \cos\theta$, and $\tan(-\theta) = -\tan\theta$. Assume that

17. $\frac{\pi}{2} < \theta < \pi$.

18. $\pi < \theta < \frac{3\pi}{2}$.

19. Show that $\sin(\beta + 6\pi) = \sin\beta$. Assume that $\pi/2 < \beta < \pi$.

20. Show that $\tan(\beta + \pi) = \tan\beta$. Assume that $3\pi/2 < \beta < 2\pi$.

Find each function value. Do not use a calculator or the trig table. You may use tables 7.1 and 7.2.

21. $\sin\left(\frac{-7\pi}{4}\right)$

22. $\sin\left(\frac{13\pi}{3}\right)$

23. $\cos(9\pi)$

24. $\cos\left(\frac{-17\pi}{6}\right)$

25. $\tan\left(\frac{-11\pi}{4}\right)$

26. $\tan(7\pi)$

27. $\sec\left(\frac{9\pi}{4}\right)$

28. $\sec\left(\frac{-5\pi}{2}\right)$

29. $\cot\left(\frac{19\pi}{6}\right)$

30. $\cot\left(\frac{-7\pi}{3}\right)$

31. $\csc\left(\frac{-3\pi}{2}\right)$

32. $\csc\left(\frac{11\pi}{3}\right)$

Find each function value to three decimal places in two ways. First, use a calculator, and then use the trig table.

33. $\sin 6.21$

34. $\sin(-5.32)$

35. $\tan(-11.63)$

36. $\tan 2.94$

37. $\cos 12.94$

38. $\cos(-4.75)$

39. $\cot(-7.86)$

40. $\cot 3.66$

41. $\sec 9.83$

42. $\sec(-13.15)$

43. $\csc(-8.72)$

44. $\csc 10.56$

(B)

Use figure 7.26 to argue that

45. $\tan\theta = \cot\left(\frac{\pi}{2} - \theta\right)$, $0 < \theta < \frac{\pi}{2}$.

46. $\sec\theta = \csc\left(\frac{\pi}{2} - \theta\right)$, $0 < \theta < \frac{\pi}{2}$.

47. Construct a figure similar to figure 7.26 to argue that $\sin\theta = \cos\left(\frac{\pi}{2} - \theta\right)$, $\frac{\pi}{2} < \theta < \pi$.

48. Given $\sin\theta = \cos\left(\frac{\pi}{2} - \theta\right)$, show that $\cos\beta = \sin\left(\frac{\pi}{2} - \beta\right)$, where $\beta = \frac{\pi}{2} - \theta$.

49. Given $\tan\theta = \cot\left(\frac{\pi}{2} - \theta\right)$, show that $\cot\beta = \tan\left(\frac{\pi}{2} - \beta\right)$.

7.5 Graphs of the Trigonometric Functions

The Basic Sine Curve

The depiction of one period of a trigonometric function in a graph produces *one cycle* of the graph. Two cycles of a sine graph, over the interval $[-2\pi, 2\pi]$, are shown in figure 7.27. The graph is obtained by brute force, that is, by carefully plotting many points and then connecting them. The portion of the graph that is to the right of the origin shall be called **the basic sine curve.** To realize a graph of $y = \sin x$ over R, one need only extend the basic sine curve left and right over intervals that are 2π units in length. It will be easy to generate variations of the sine graph as translations, reflections, expansions, or contractions of the basic sine curve. We shall also graph $y = \csc x$ from the basic sine curve because $\csc x = 1/\sin x$.

x	0	$\frac{\pi}{6}$	$\frac{\pi}{4}$	$\frac{\pi}{3}$	$\frac{\pi}{2}$	$\frac{2\pi}{3}$	$\frac{3\pi}{4}$	$\frac{5\pi}{6}$	π	$\frac{7\pi}{6}$	$\frac{5\pi}{4}$	$\frac{4\pi}{3}$	$\frac{3\pi}{2}$	$\frac{5\pi}{3}$	$\frac{7\pi}{4}$	$\frac{4\pi}{6}$	2π
sin x	0	$\frac{1}{2}$	$\frac{\sqrt{2}}{2}$	$\frac{\sqrt{3}}{2}$	1	$\frac{\sqrt{3}}{2}$	$\frac{\sqrt{2}}{2}$	$\frac{1}{2}$	0	$-\frac{1}{2}$	$-\frac{\sqrt{2}}{2}$	$-\frac{\sqrt{3}}{2}$	-1	$-\frac{\sqrt{3}}{2}$	$-\frac{\sqrt{2}}{2}$	$-\frac{1}{2}$	0

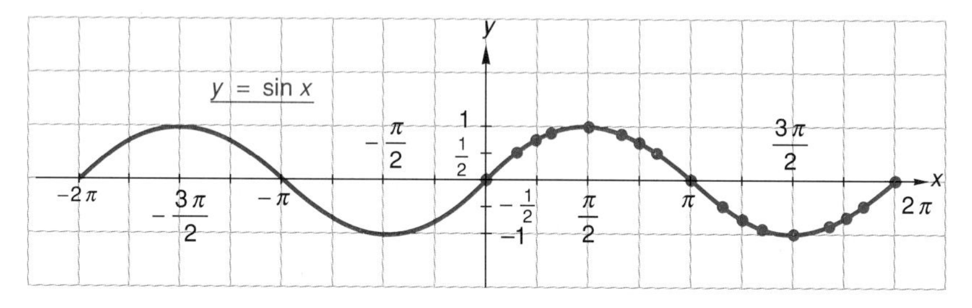

Figure 7.27

A segment of arbitrary length, with left endpoint at the origin, is chosen on the positive x-axis. The right endpoint is labeled 2π, and the quarter points are labeled $\pi/2$, π, and $3\pi/2$. So as not to distort the graph, vertical and horizontal scales are made equal. Thus, the length of the segment on the positive y-axis between 0 and 1 is about $\frac{2}{3}$ the length of the segment between 0 and $\pi/2$ on the positive x-axis. A set of ordered pairs, $(x,\sin x)$, are generated in a table, and the points determined by the ordered pairs are plotted and then connected. The graph in figure 7.27 clearly shows that the range of the sine function is $[-1,1]$. It should also be noted that the graph is symmetric with respect to the origin. This fact might have been anticipated because of the discussion on symmetry in section 2.4.

$$f(x) = \sin x = -\sin(-x) = -f(-x)$$

The Basic Cosine Curve

Two cycles of a graph of $y = \cos x$ are shown in figure 7.28. The portion of the graph over $[-2\pi,0]$ is a copy of the portion of the graph over $[0,2\pi]$ because the cosine function has a period of 2π units. We shall call the cycle over $[0,2\pi]$ **the basic cosine curve.** Figure 7.28 shows that the range of $\cos x$ is $[-1,1]$ and that a graph of $y = \cos x$ is symmetric with respect to the y-axis. The symmetry results from the fact that

$$f(x) = \cos x = \cos(-x) = f(-x).$$

x	0	$\frac{\pi}{6}$	$\frac{\pi}{4}$	$\frac{\pi}{3}$	$\frac{\pi}{2}$	$\frac{2\pi}{3}$	$\frac{3\pi}{4}$	$\frac{5\pi}{6}$	π	$\frac{7\pi}{6}$	$\frac{5\pi}{4}$	$\frac{4\pi}{3}$	$\frac{3\pi}{2}$	$\frac{5\pi}{3}$	$\frac{7\pi}{4}$	$\frac{11\pi}{6}$	2π
cos x	1	$\frac{\sqrt{3}}{2}$	$\frac{\sqrt{2}}{2}$	$\frac{1}{2}$	0	$-\frac{1}{2}$	$-\frac{\sqrt{2}}{2}$	$-\frac{\sqrt{3}}{2}$	-1	$-\frac{\sqrt{3}}{2}$	$-\frac{\sqrt{2}}{2}$	$-\frac{1}{2}$	0	$\frac{1}{2}$	$\frac{\sqrt{2}}{2}$	$\frac{\sqrt{3}}{2}$	1

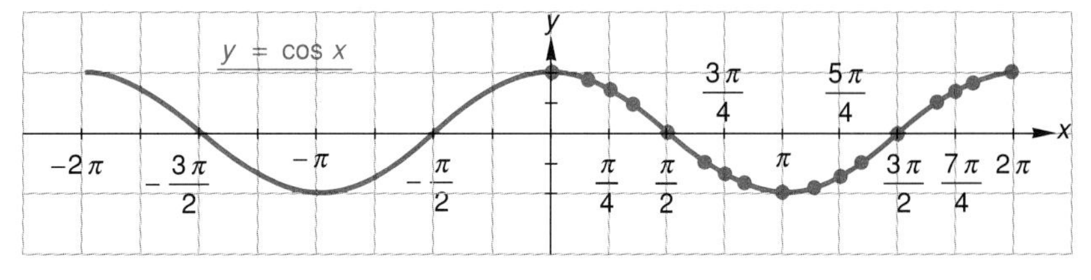

Figure 7.28

The Basic Tangent Curve

Because the tangent function has a period of π units, we are able to display two cycles of $y = \tan x$ in the interval $[-\pi,\pi]$ (figure 7.29). Also,

$$f(x) = \tan x = -\tan(-x) = -f(-x),$$

so that the graph of the tangent function is symmetric with respect to the origin. We call the cycle of a tangent graph in $[0,\pi]$ **the basic tangent curve.** A vertical asymptote occurs at $x = \pi/2$ of the basic tangent curve because the tangent function is not defined there. As x assumes values closer and closer to $\pi/2$, $|\tan x|$ tends to ∞ (see table below). The periodicity of the tangent function dictates that a graph of $y = \tan x$ will have vertical asymptotes at $x = \pi/2 + k \cdot \pi$, $k \varepsilon J$.

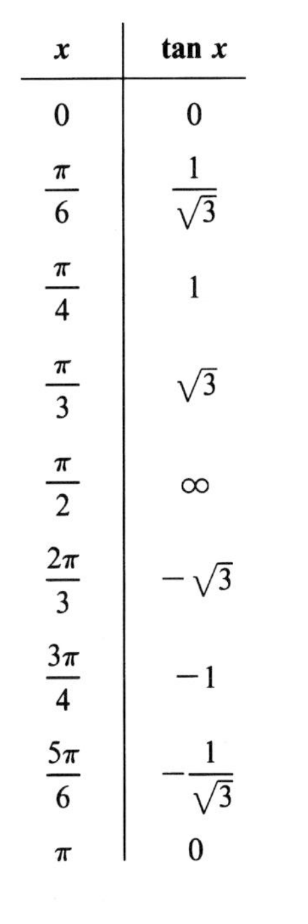

x	tan x
0	0
$\frac{\pi}{6}$	$\frac{1}{\sqrt{3}}$
$\frac{\pi}{4}$	1
$\frac{\pi}{3}$	$\sqrt{3}$
$\frac{\pi}{2}$	∞
$\frac{2\pi}{3}$	$-\sqrt{3}$
$\frac{3\pi}{4}$	-1
$\frac{5\pi}{6}$	$-\frac{1}{\sqrt{3}}$
π	0

x	$\frac{5\pi}{12}$	$\frac{4\pi}{9}$	$\frac{17\pi}{36}$	$\frac{22\pi}{45}$	$\frac{7\pi}{12}$	$\frac{5\pi}{9}$	$\frac{19\pi}{36}$	$\frac{23\pi}{45}$
tan x	3.732	5.671	11.43	28.64	−3.732	−5.671	−11.43	−28.64

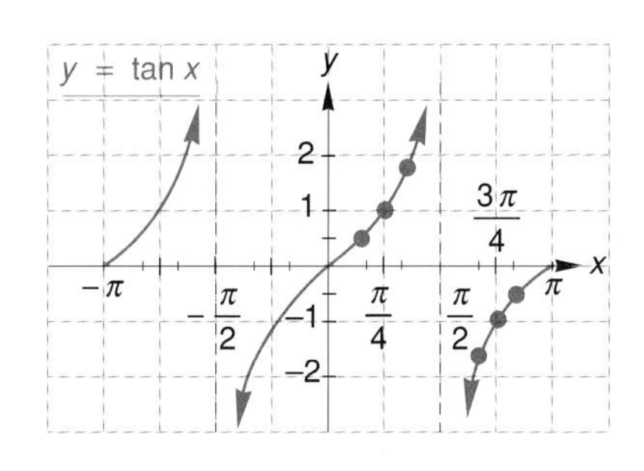

Figure 7.29

You shall see that sketching the graphs of trigonometric functions involves little more than sketching some variation of one of the above basic graphs.

Graphs of the Reciprocal Functions

A graph of $y = \csc x$ can be sketched directly from a graph of $y = \sin x$. Since $\csc x = 1/\sin x$, and $-1 \leq \sin x \leq 1$, then $|\csc x| > 1$. Also, $|\csc x|$ assumes greater and greater values as $\sin x$ approaches closer and closer to zero. Therefore, a cosecant graph has vertical asymptotes at those value of x for which $\sin x = 0$ (see figure 7.30).

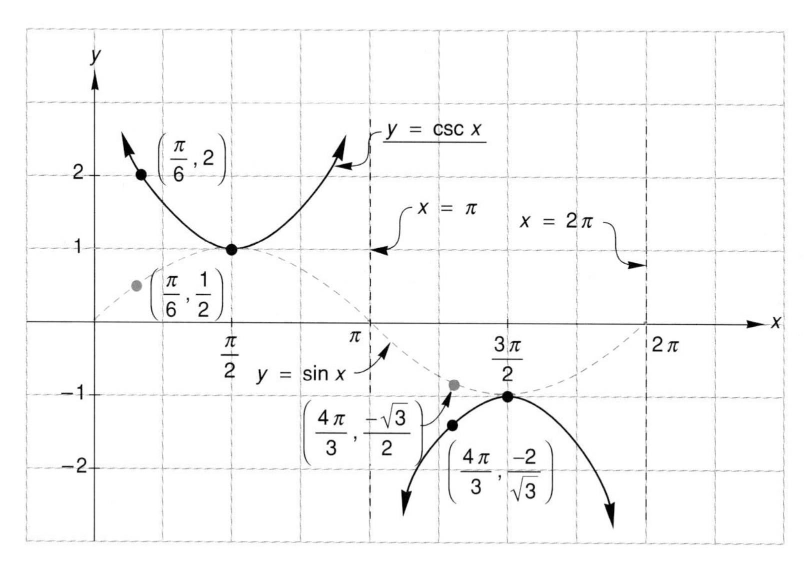

Figure 7.30

Graphs of $y = \sec x$ and $y = \cot x$ are sketched in figure 7.31 from the cosine and tangent graphs, respectively. Note that

$$|\sec x| \geq 1 \quad \text{and} \quad -\infty < \cot x < \infty.$$

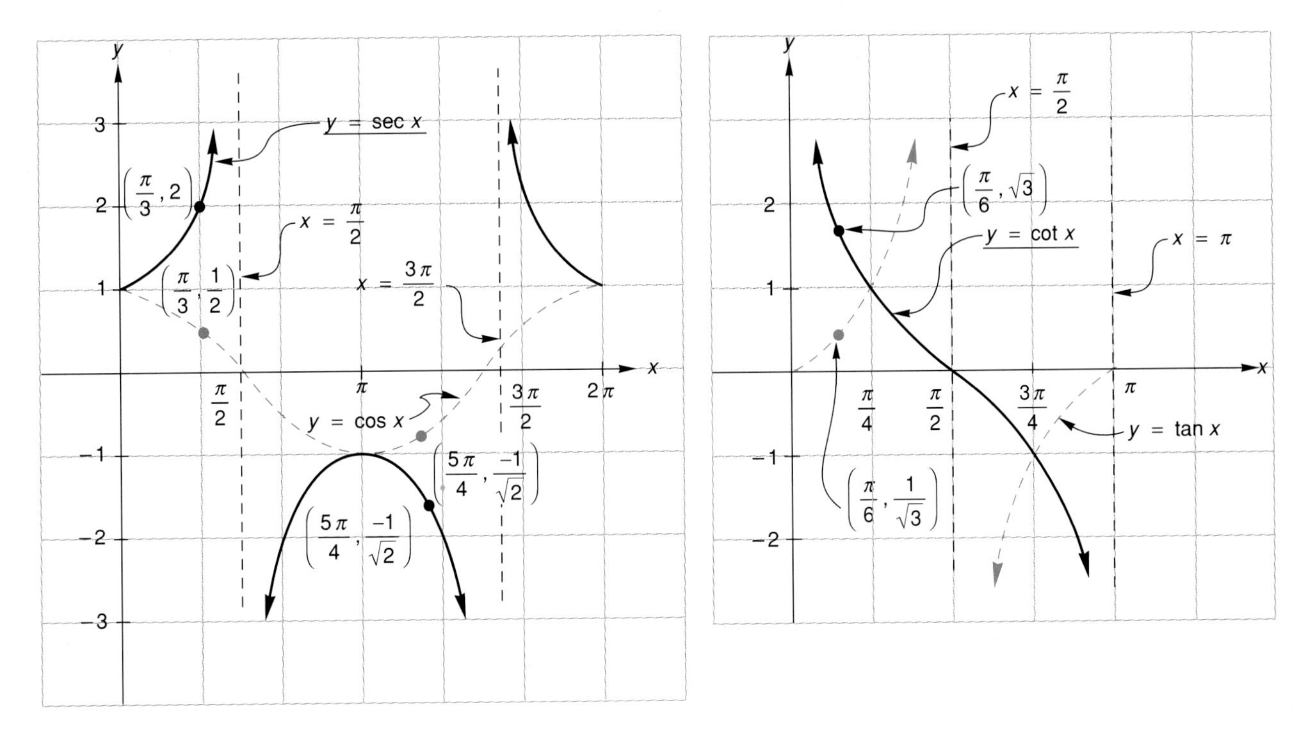

Figure 7.31

Graphs of $y = A\,f[B(x + C)]$

One is more likely to encounter an equation of the type

$$y = A \sin B(x + C) \tag{1}$$

in applications than $y = \sin x$. Equation $y = \sin 2\pi kt$ (page 276), which describes the motion of a pump piston, has the form of (1) with $C = 0$. A graph of (1) can be easily constructed by noting the effect that each of the constants, A, B, and C has on the basic sine curve.

It is shown in section 2.4 that the graph of $y = A\,f(x)$ is a stretching away from, or a contraction towards, the x-axis of the graph of $y = f(x)$. A stretching occurs if $|A| > 1$. There is a contraction if $0 < |A| < 1$.

Example 1 Graph one circle each of $y = \sin x$, $y = 3 \sin x$, $y = -3 \sin x$, and $y = \frac{1}{2} \sin x$, on the same coordinates axes. Start each graph at the origin.

Solution If $f(x) = 3 \sin x$, then $-f(x) = -3 \sin x$, and a graph of $y = -3 \sin x$ is a reflection, with respect to the x-axis, of a graph of $y = 3 \sin x$ (see section 2.4). The required graphs are shown in figure 7.32.

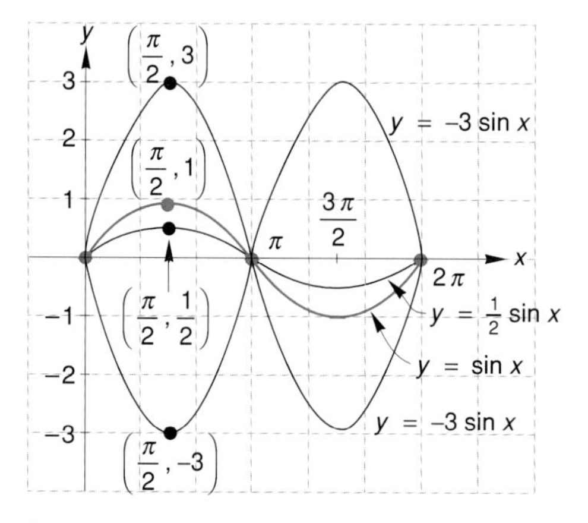

Figure 7.32

In section 2.4, a graph of $y = f(Bx)$ is shown to be a stretching or a contraction, with respect to the y-axis, of a graph of $y = f(x)$. There is a stretching if $0 < |B| < 1$, and a contraction if $|B| > 1$. To sketch one cycle of $y = \sin Bx$, $B > 0$, starting at the origin, Bx must assume all values from 0 to 2π. That is,

$$0 \leq Bx \leq 2\pi$$

or

$$0 \leq x \leq \frac{2\pi}{B}.$$

Thus, one cycle of the curve can be graphed in the interval $[0,2\pi/B]$.

Example 2 Graph one cycle each of $y = \sin x$, $y = \sin 2x$, $y = \sin(-2x)$, and $y = \sin(x/2)$, on the same coordinate axes. Start each graph at the origin.

Solution Since $\sin(-2x) = -\sin 2x$, the graphs of $y = \sin 2x$ and $y = \sin(-2x)$ are reflections of each other with respect to the x-axis. The periods of the graphs of $y = \sin x$, $y = \sin 2x$, and $y = \sin(x/2)$ are 2π, $2\pi/2 = \pi$, and $2\pi/(\frac{1}{2}) = 4\pi$. The required graphs are shown in figure 7.33.

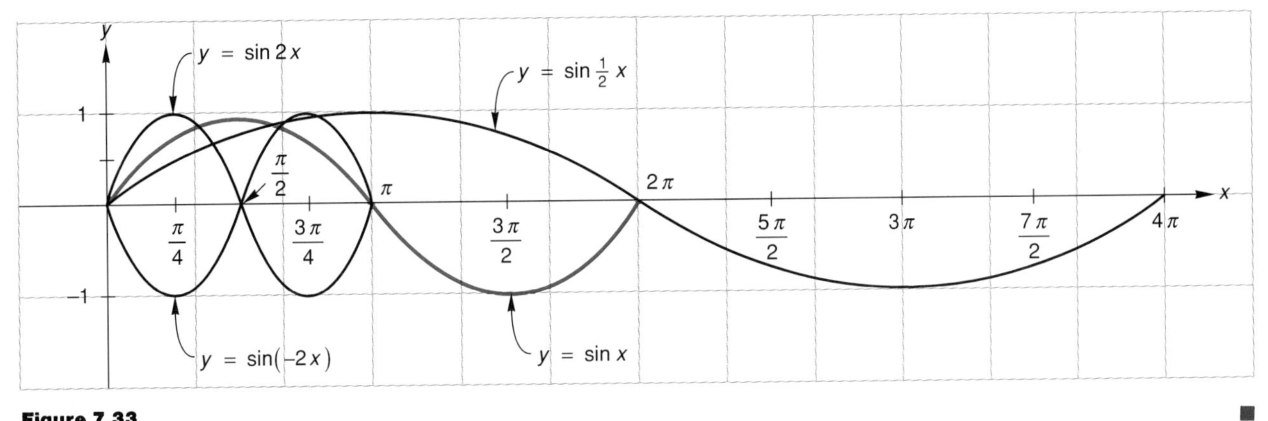

Figure 7.33

■

A graph of $y = f(x + C)$ can be obtained by shifting a graph of $y = f(x)$ C units right or left, depending upon whether C is negative or positive (see section 2.4). Thus, a graph of $y = \sin x$ is shifted $\pi/6$ units to the left to produce a graph of $y = \sin(x + \pi/6)$.

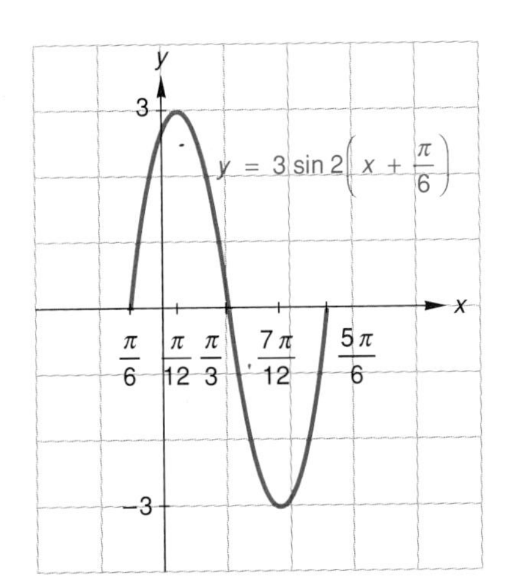

Figure 7.34

Example 3 Graph one cycle of $y = 3 \sin 2(x + \pi/6)$.

Solution By comparing the given equation to (1) we see that $A = 3$, $B = 2$, and $C = \pi/6$. Consequently, $-3 \leq y \leq 3$, and one cycle of the graph can be sketched over a period of $2\pi/2 = \pi$ units. Since the graph is shifted $\pi/6$ units to the left with respect to the basic sine curve, it is possible to sketch one cycle of the graph in the interval $[-\pi/6, 5\pi/6]$. The quarter points of the interval are located at $-\pi/6 + \pi/4 = \boldsymbol{\pi/12}$, $\pi/12 + \pi/4 = \boldsymbol{\pi/3}$, and $\pi/3 + \pi/4 = \boldsymbol{7\pi/12}$. It is clear from the basic sine curve that every sine graph has its maximum and minimum values at the first and third quarter points of its period. Its x-intercepts occur at the endpoints and midpoint of the period (figure 7.34). ■

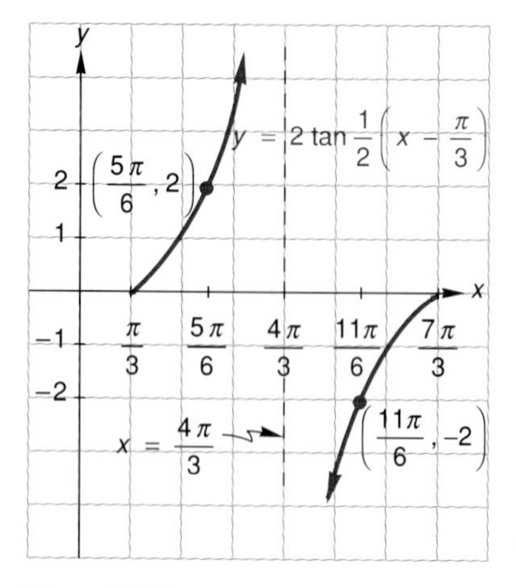

Figure 7.35

Example 4 Graph one cycle of $y = 2 \tan \frac{1}{2}(x - \pi/3)$.

Solution $A = 2$, $B = \frac{1}{2}$, and $C = -\pi/3$. The graph has a period of $\pi/(\frac{1}{2}) = 2\pi$ units, and it is shifted $\pi/3$ units to the right with respect to the basic tangent curve. Hence, one cycle of the graph can be sketched in the interval $[\pi/3, 7\pi/3]$, and, as illustrated by the basic tangent curve, the graph has a vertical asymptote at the midpoint of its cycle (figure 7.35). ■

Example 5 Graph one cycle of $y = \sec(3x + 3\pi/4)$.

Solution The equation has the form $y = A \sec B(x + C)$ when it is written $y = \sec 3(x + \pi/4)$. Thus $A = 1$, $B = 3$, and $C = \pi/4$. We graph $y = \cos 3(x + \pi/4)$ and then use the reciprocal relationship between cosine and secant to sketch the secant curve. The basic cosine curve is shifted $\pi/4$ units to the left and squeezed into a period of $2\pi/3$ units, to produce a graph of $y = \cos 3(x + \pi/4)$ (figure 7.36). The quarter points of the interval $[-\pi/4, 5\pi/12]$, to which the graph is constrained, are $-\pi/4 + \pi/6 = \mathbf{-\pi/12}$, $-\pi/12 + \pi/6 = \mathbf{\pi/12}$, and $\pi/12 + \pi/6 = \mathbf{\pi/4}$.

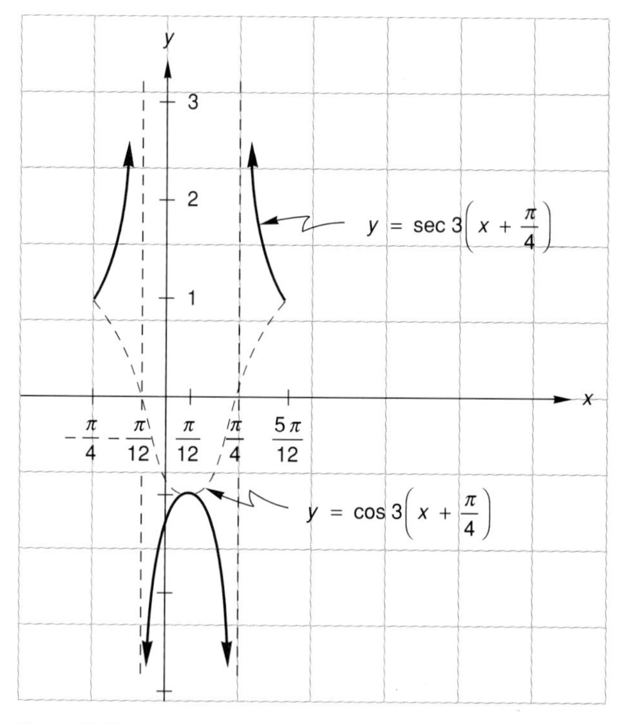

Figure 7.36

■

Sums of Trigonometric Functions

The work of the French mathematician Fourier (1768–1830) enables us to approximate any periodic function, as closely as we desire, with a sum of sine and cosine functions.

Consider the function $y = \sin 3x + \cos 2x$. It can be shown that the sum of two periodic functions is periodic only if the quotient of their periods is a rational number. The quotient of the periods of $\sin 3x$ and $\cos 2x$ is $(2\pi/3)/\pi = \frac{2}{3}$. Hence, the sum function is periodic. Its period is 2π, which is the least common multiple of $2\pi/3$ and π.

One cycle of the sum function can be graphed by plotting a sufficient number of points in $[0,2\pi]$ and then connecting those points with a continuous curve. Another approach that yields a rough sketch (figure 7.37) of $y = \sin 3x + \cos 2x$, takes advantage of our ability to make quick sketches of $y_1 = \sin 3x$ and $y_2 = \cos 2x$. Each graph is sketched over $[0,2\pi]$ on the same coordinate system. Corresponding ordinates of the two graphs are then added to produce points in the graph of the sum function. We usually add ordinates at those x-values where

1. one of the graphs crosses the x-axis ($x = x_2$ and $x = 2\pi$ are two such points).
2. one of the graphs has a high or a low point (at $x = x_3$ and $x = x_4$).
3. the two graphs intersect (at $x = x_1$ and $x = x_5$).

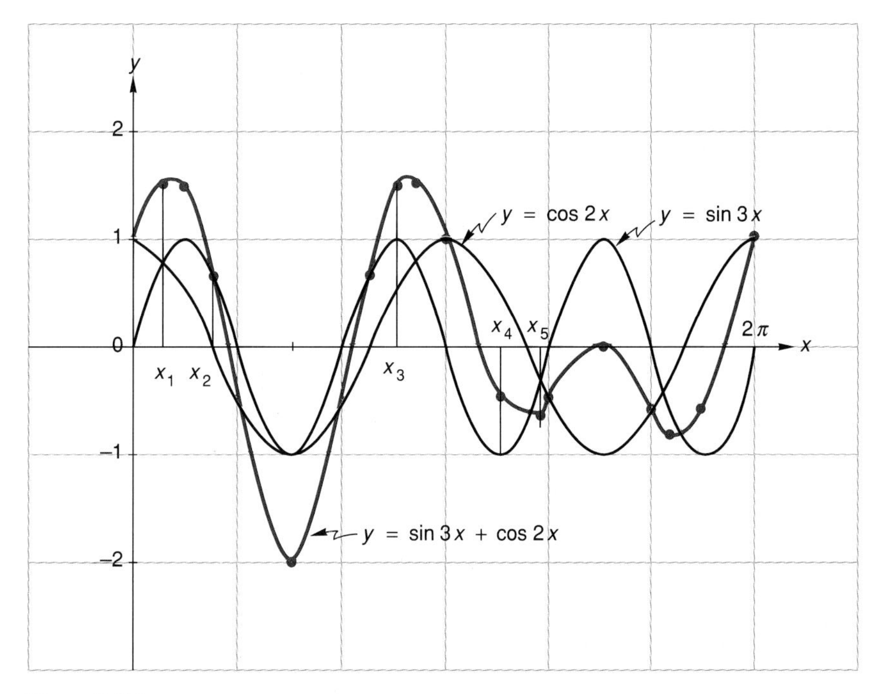

Figure 7.37

Exercise Set 7.5

(A)

Sketch one cycle of a graph of each of the three functions on the same coordinate axes.

1. $y = \sin x$, $y = -2 \sin x$, $y = 4 \sin x$
2. $y = 3 \sin x$, $y = -3 \sin x$, $y = \frac{1}{2} \sin x$
3. $y = \cos x$, $y = 2 \cos x$, $y = -2 \cos x$
4. $y = -\cos x$, $y = -\frac{1}{2} \cos x$, $y = \frac{3}{2} \cos x$
5. $y = \tan x$, $y = -\tan x$, $y = 3 \tan x$
6. $y = -2 \tan x$, $y = \frac{1}{2} \tan x$, $y = 4 \tan x$

Sketch one cycle of a graph of each of the two functions on the same coordinate axes.

7. $y = \sin 2x$, $y = \sin \dfrac{x}{2}$
8. $y = \sin \dfrac{2x}{3}$, $y = \sin 3x$
9. $y = \cos \dfrac{x}{3}$, $y = \cos 3x$
10. $y = \cos 4x$, $y = \cos \dfrac{4x}{3}$
11. $y = \tan \dfrac{x}{2}$, $y = \tan 2x$
12. $y = \tan 3x$, $y = \tan \dfrac{3x}{4}$
13. $y = \sin\left(x + \dfrac{\pi}{3}\right)$, $y = \sin\left(x - \dfrac{\pi}{3}\right)$
14. $y = \sin\left(x - \dfrac{\pi}{2}\right)$, $y = \sin\left(x + \dfrac{\pi}{4}\right)$
15. $y = \cos\left(x + \dfrac{\pi}{2}\right)$, $y = \cos\left(x - \dfrac{\pi}{6}\right)$
16. $y = \cos\left(x - \dfrac{\pi}{2}\right)$, $y = \cos\left(x + \dfrac{\pi}{3}\right)$
17. $y = \tan\left(x + \dfrac{\pi}{4}\right)$, $y = \tan\left(x - \dfrac{\pi}{4}\right)$
18. $y = \tan\left(x - \dfrac{\pi}{3}\right)$, $y = \tan\left(x + \dfrac{\pi}{6}\right)$

Sketch one cycle of a graph of each function.

19. $y = 2 \sec 3x$
20. $y = \frac{1}{2} \sec 2x$
21. $y = -\csc \dfrac{x}{2}$
22. $y = 3 \csc \dfrac{4x}{3}$
23. $y = -2 \cot \dfrac{x}{3}$
24. $y = \frac{1}{3} \cot 2x$
25. $y = -3 \sin 2\left(x + \dfrac{\pi}{4}\right)$
26. $y = 2 \sin \dfrac{1}{2}\left(x - \dfrac{\pi}{3}\right)$
27. $y = \cos 3\left(x - \dfrac{\pi}{2}\right)$
28. $y = -2 \cos 2\left(x + \dfrac{\pi}{4}\right)$
29. $y = 4 \tan \dfrac{1}{3}\left(x - \dfrac{\pi}{6}\right)$
30. $y = -\tan 2\left(x + \dfrac{\pi}{4}\right)$

Given an amplitude A, period P, and shift C, write an equation for the indicated function.

31. $A = 2$, $P = \dfrac{\pi}{2}$, $C = \dfrac{\pi}{3}$ right; sine function
32. $A = 3$, $P = 4\pi$, $C = \dfrac{\pi}{2}$ left; sine function
33. $A = \dfrac{1}{2}$, $P = \pi$, $C = \dfrac{\pi}{4}$ left; cosine function
34. $A = 4$, $P = 3\pi$, $C = \dfrac{\pi}{6}$ right; cosine function

(B)

Sketch one cycle of a graph of each function.

35. $y = \sin(-3x)$
36. $y = 2 \tan(-2x)$
37. $y = 2 \cos\left(-2x + \dfrac{\pi}{3}\right)$
38. $y = \sin\left(-3x - \dfrac{\pi}{2}\right)$
39. $y = -3 \tan\left[-\dfrac{2}{3}\left(x - \dfrac{\pi}{6}\right)\right]$
40. $y = \cos\left[-2\left(x + \dfrac{\pi}{4}\right)\right]$

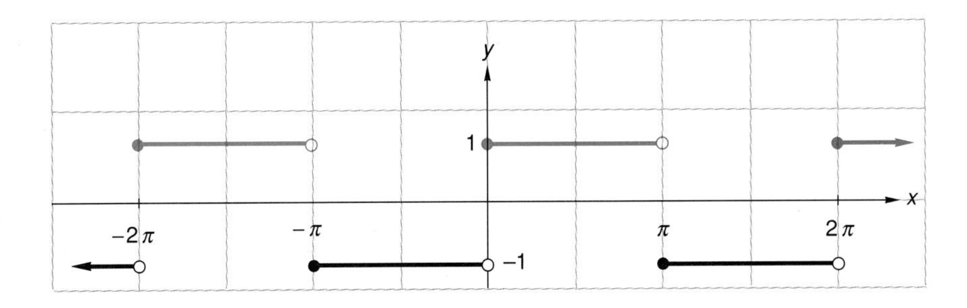

Sketch one cycle of a graph of each function, starting at the origin.

41. $y = 3 \cos \frac{2}{3}\left(x - \frac{\pi}{6}\right)$

42. $y = 2 \tan 2\left(x + \frac{\pi}{3}\right)$

Sketch one cycle of a graph of each function by the method of adding ordinates as discussed on page 295. Use a calculator to check your graphs.

43. $y = \sin 2x + \cos x$

44. $y = 2 \sin \frac{2x}{3} + 3 \cos 2x$

45. $y = 2 \sin 3x + \sin 2x$

46. $y = \cos x + \cos 3x$

47. $y = 2 \sin x - \cos \frac{x}{2}$

48. $y = 3 \cos \frac{2x}{3} - 2 \sin 2x$

49. Let $f(x) = \sin 3x + \cos 2x$. Show that $f(x + 2\pi) = f(x)$. Can you conclude that the period of f is 2π?

50. Graph $y = \sin^2 x$, $y = \sin^4 x$, and $y = \sin^6 x$ on the same coordinate axes, and over the interval $[0,\pi]$.

Sketch a graph of each function over the interval $[-2\pi,2\pi]$.

51. $y = \sin x - 1$

52. $y = x + \sin x$

53. $y = x - \sin x$

54. $y = x \sin x$

55. $y = \frac{\sin x}{x}$

56. A graph of the square wave function

$$f(x) = \begin{cases} -1, & -\pi \le x < 0 \\ 1, & 0 \le x < \pi \end{cases}$$

is shown in the figure. The function has period 2π.

The Fourier series corresponding to $f(x)$ is

$$\frac{4}{\pi}\left(\sin x + \frac{1}{3}\sin 3x + \frac{1}{5}\sin 5x + \ . \ . \ .\right).$$

Use a calculator to help you graph

a. $g(x) = \frac{4}{\pi}\left(\sin x + \frac{1}{3}\sin 3x + \frac{1}{5}\sin 5x\right)$, $0 \le x < \pi$.

b. $h(x) = \frac{4}{\pi}\left(\sin x + \frac{1}{3}\sin 3x + \frac{1}{5}\sin 5x + \frac{1}{7}\sin 7x\right)$, $0 \le x < \pi$.

c. $p(x) = \frac{4}{\pi}\left(\sin x + \frac{1}{3}\sin 3x + \frac{1}{5}\sin 5x + \frac{1}{7}\sin 7x + \frac{1}{9}\sin 9x + \frac{1}{11}\sin 11x\right)$, $0 \le x < \pi$.

Observe that the graphs of g, h, and p approach closer and closer to the graph of f in $[0,\pi)$.

7.6 The Inverse Trigonometric Functions

The horizontal line test, which is used to determine if a function is one-to-one, is applied to sin x in figure 7.38. The test shows that a particular y-value has many x-values associated with it. Consequently, the sine function is not one-to-one, and it cannot have an inverse function. However, if we restrict the domain of $y = \sin x$ to $[-\pi/2, \pi/2]$ (figure 7.38), then the function is one-to-one, and it does have an inverse function. We are stymied when we try to find the x that is associated with a particular y by solving $y = \sin x$ for x in terms of y. Algebraic methods alone are insufficient to enable us to produce such a solution. Yet, as we see in figure 7.38, for each y there is associated exactly one x. We name that x-value

$$\mathbf{x = \sin^{-1} y,}$$

to be consistent with the inverse function notation, f^{-1}, introduced in section 2.5. There is a tendency for the student to confuse $\sin^{-1} y$ with $1/\sin y$, so we offer the admonition

$$\mathbf{\sin^{-1} y \neq \frac{1}{\sin y}}.$$

(a)

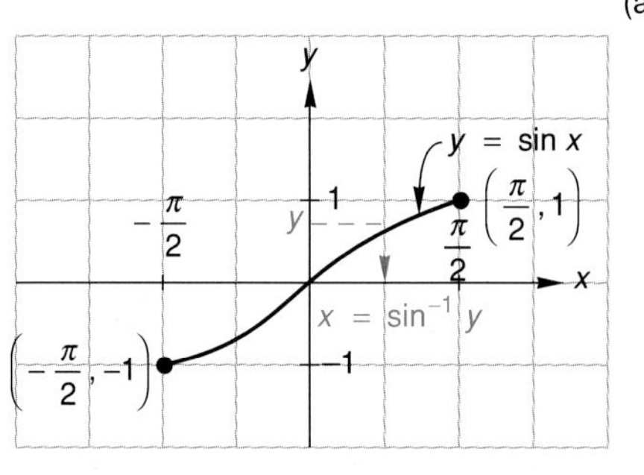

(b)

Figure 7.38

We have established that if

$$y = \sin x, \quad -\frac{\pi}{2} \le x \le \frac{\pi}{2}, \quad \text{then} \quad x = \sin^{-1} y.$$

But, to relate the inverse sine function to all functions written in the standard form $y = f(x)$, we give the following definition.

Definition

$$y = \sin^{-1} x \quad \text{if and only if} \quad x = \sin y, \quad -\frac{\pi}{2} \le y \le \frac{\pi}{2} \tag{1}$$

From the discussion on inverse functions in section 2.5, we know that a function and its inverse function interchange domain and range. We also know that graphs of a function and its inverse function are reflections of each other with respect to the line $y = x$. Thus, in figure 7.39, *the domain of* $\sin^{-1} x$, **[−1, 1]**, is the range of $\sin x$ and *the range of* $\sin^{-1} x$, $[-\pi/2, \pi/2]$, is the restricted domain of $\sin x$.

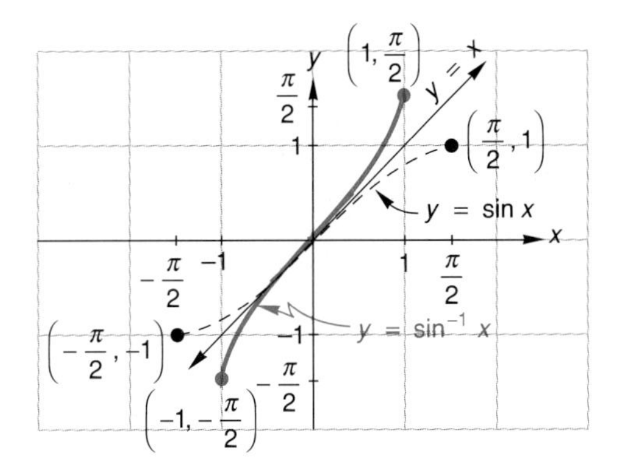

Figure 7.39

By restricting the domains of the cosine and tangent functions, it is possible to construct an inverse cosine function and an inverse tangent function.

Definitions

$$y = \cos^{-1} x \quad \text{if and only if} \quad x = \cos y, \quad 0 \le y \le \pi \tag{2}$$

$$y = \tan^{-1} x \quad \text{if and only if} \quad x = \tan y, \quad -\frac{\pi}{2} < y < \frac{\pi}{2} \tag{3}$$

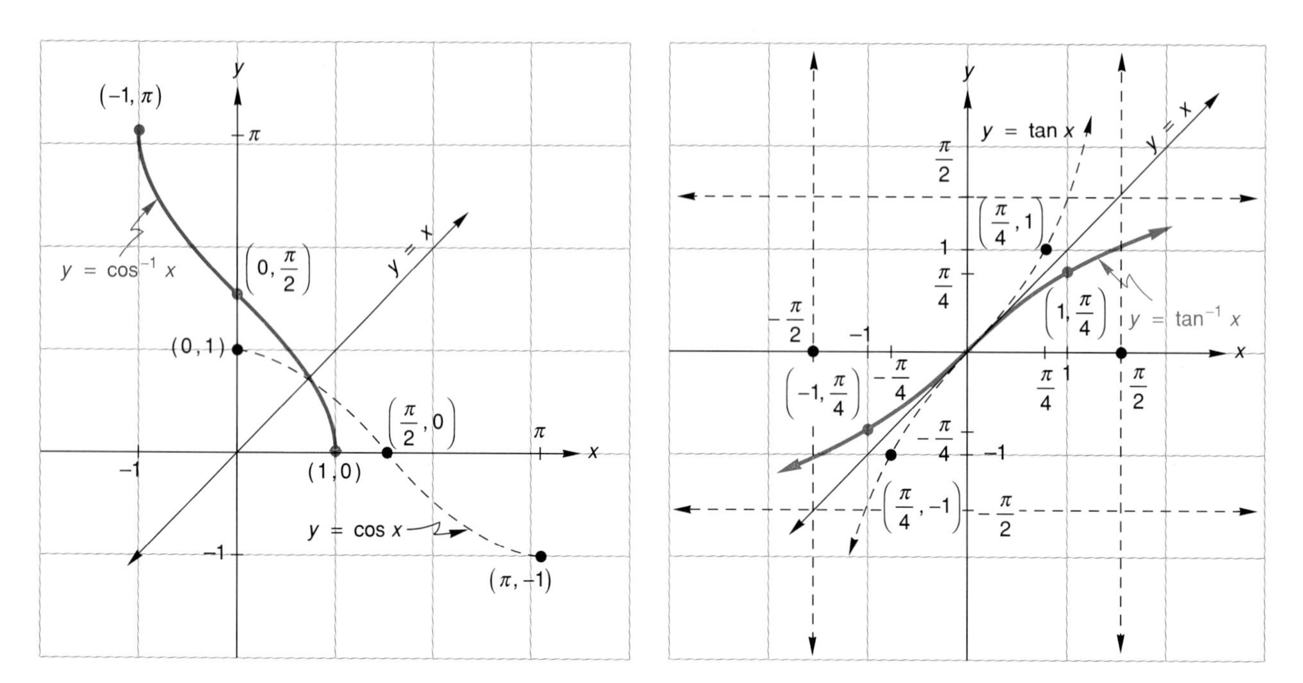

Figure 7.40

A graph of $y = \cos^{-1} x$ is constructed from the graph of $y = \cos x$, $0 \le x \le \pi$, in figure 7.40 in much the same way as the graph of $y = \sin^{-1} x$ is constructed in figure 7.39. The coordinates of the ordered pairs $(0, 1)$, $(\pi/2, 0)$, and $(\pi, -1)$ in $\cos x$ are interchanged to become ordered pairs in $\cos^{-1} x$. *The domain of* $\cos^{-1} x$, $\mathbf{[-1, 1]}$, is the range of $\cos x$ and *the range of* $\cos^{-1} x$, $\mathbf{[0, \pi]}$, is the restricted domain of $\cos x$.

In the graph of $y = \tan^{-1} x$ (figure 7.40) x tends to infinity as y approaches $\pi/2$ because, in the graph of $y = \tan x$, y tends to infinity as x approaches $\pi/2$. *The domain of* $\tan^{-1} x$, $\mathbf{(-\infty, \infty)}$, is the range of $\tan x$, and *the range of* $\tan^{-1} x$, $\mathbf{(-\pi/2, \pi/2)}$, is the restricted domain of $\tan x$.

Evaluating $\sin^{-1} x$, $\cos^{-1} x$, and $\tan^{-1} x$

The definition given in (1) can often be used to evaluate the inverse sine directly. We know that

$$y = \sin^{-1} 1 \quad \text{if and only if} \quad 1 = \sin y.$$

Consequently, $y = \pi/2$. We do not say that $\sin^{-1} 1 = 5\pi/2$, despite the fact that $\sin 5\pi/2 = 1$. The range of $\sin^{-1} x$ does not contain $5\pi/2$.

$$\sin^{-1}\left(\frac{1}{2}\right) = \frac{\pi}{6} \quad \text{because} \quad \sin\frac{\pi}{6} = \frac{1}{2}$$

$$\sin^{-1}\left(-\frac{\sqrt{2}}{2}\right) = -\frac{\pi}{4} \quad \text{because} \quad \sin\left(-\frac{\pi}{4}\right) = -\frac{\sqrt{2}}{2}$$

The number whose sine is 0.27, symbolized as *sin*$^{-1}$ *0.27,* is approximated in the trigonometric table by 0.2734. We look for 0.27 in the sine column and associate with it the corresponding number, 0.2734, in the radian column. A calculator used to evaluate $\sin^{-1} 0.27$ gives the following result.

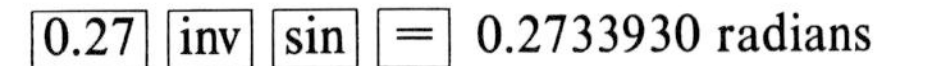
0.27 inv sin = 0.2733930 radians

Example 1 Evaluate

a. $\tan^{-1} 1$

b. $\cos^{-1}(-\frac{1}{2})$

c. $\sin^{-1} 0$

d. $\tan^{-1}(-\sqrt{3})$

e. $\cos^{-1}\left(\dfrac{\sqrt{3}}{2}\right)$

f. $\cos^{-1}(-0.74)$

g. $\tan^{-1}(-0.42)$

Solution Remember that the ranges of $\sin^{-1} x$, $\cos^{-1} x$, and $\tan^{-1} x$ are $[-\pi/2,\pi/2]$, $[0,\pi]$, and $(-\pi/2,\pi/2)$, respectively. (See figures 7.39 and 7.40.)

a. $\tan^{-1} 1 = \dfrac{\pi}{4}$ because $\tan \dfrac{\pi}{4} = 1$

b. $\cos^{-1}\left(-\dfrac{1}{2}\right) = \dfrac{2\pi}{3}$ because $\cos \dfrac{2\pi}{3} = -\dfrac{1}{2}$

c. $\sin^{-1} 0 = 0$ because $\sin 0 = 0$

d. $\tan^{-1}(-\sqrt{3}) = -\dfrac{\pi}{3}$ because $\tan\left(-\dfrac{\pi}{3}\right) = -\sqrt{3}$

e. $\cos^{-1}\left(\dfrac{\sqrt{3}}{2}\right) = \dfrac{\pi}{6}$ because $\cos \dfrac{\pi}{6} = \dfrac{\sqrt{3}}{2}$

f. $\cos^{-1}(-0.74) = 2.4038667$, [0.74] [+/−] [inv] [cos]

g. $\tan^{-1}(-0.42) = -0.39762799$, [0.42] [+/−] [inv] [tan] ■

Some Relationships

Recall that

$$f^{-1}(f(a)) = a \quad \text{and} \quad f(f^{-1}(b)) = b.$$

Therefore,

$$\left.\begin{array}{lll} \sin^{-1}(\sin a) = a & \text{and} & \sin(\sin^{-1} b) = b \\ \cos^{-1}(\cos a) = a & \text{and} & \cos(\cos^{-1} b) = b \\ \tan^{-1}(\tan a) = a & \text{and} & \tan(\tan^{-1} b) = b. \end{array}\right\} \qquad (4)$$

In equations (4), a is limited by the range restrictions of equations (1), (2), and (3), and b is limited by the domain restrictions of the inverse trigonometric functions.

Example 2 Find the exact value of

a. $\cos^{-1}(\cos \pi/3)$
b. $\tan(\tan^{-1} \frac{1}{2})$.

Solution

a. Either use (4) to obtain $\pi/3$ or read $\cos^{-1}(\cos \pi/3)$ as, "the number whose cosine is the cosine of $\pi/3$," and obtain $\pi/3$.
b. From (4), $\tan(\tan^{-1} \frac{1}{2}) = \frac{1}{2}$. We can also read b. as, "the tangent of the number whose tangent is $\frac{1}{2}$," and obtain $\frac{1}{2}$. ■

Example 3 Find the exact value of

a. $\sin^{-1}(\cos \pi/5)$,
b. $\cos(\tan^{-1} \frac{1}{2})$.

Then use a calculator to find **a** and **b,** correct to four decimal places.

Solution

a. Recall that $\cos x = \sin\left(\frac{\pi}{2} - x\right)$. Thus,

$$\sin^{-1}\left(\cos \frac{\pi}{5}\right) = \sin^{-1}\left(\sin\left(\frac{\pi}{2} - \frac{\pi}{5}\right)\right) = \sin^{-1}\left(\sin \frac{3\pi}{10}\right) = \frac{3\pi}{10}.$$

By calculator:

[π] [÷] [5] [=] [cos] [inv] [sin] [=] 0.9425.

b. Let $\theta = \tan^{-1} \frac{1}{2}$, then $\tan \theta = \frac{1}{2}$. In figure 7.41, $r = \sqrt{1^2 + 2^2} = \sqrt{5}$. Therefore,

$$\cos(\tan^{-1} \tfrac{1}{2}) = \cos \theta = \frac{x}{r} = \frac{2}{\sqrt{5}}.$$

By calculator:

[0.5] [inv] [tan] [cos] [=] 0.8944. ■

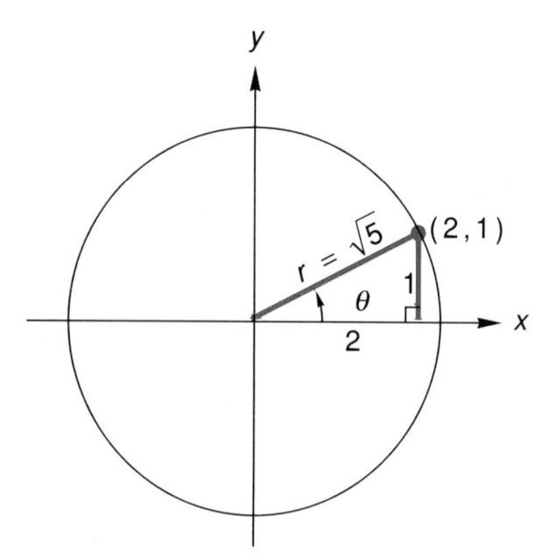

Figure 7.41

$\text{Csc}^{-1}\,x$, $\text{Sec}^{-1}\,x$, $\text{Cot}^{-1}\,x$

If the domain of the secant function is restricted to $[0,\pi/2) \cup (\pi/2,\pi]$, then it has an inverse function, $\mathbf{y = sec^{-1}\,x}$. In figure 7.42, a graph of $y = \sec^{-1} x$ is drawn as a reflection, with respect to the line $y = x$, of the graph of $y = \sec x$. The domain of the inverse secant function is $(-\infty,-1] \cup [1,\infty)$, and its range is $[0,\pi/2) \cup (\pi/2,\pi]$. Where the secant has a vertical asymptote at $x = \pi/2$, the inverse secant has a horizontal asymptote at $y = \pi/2$.

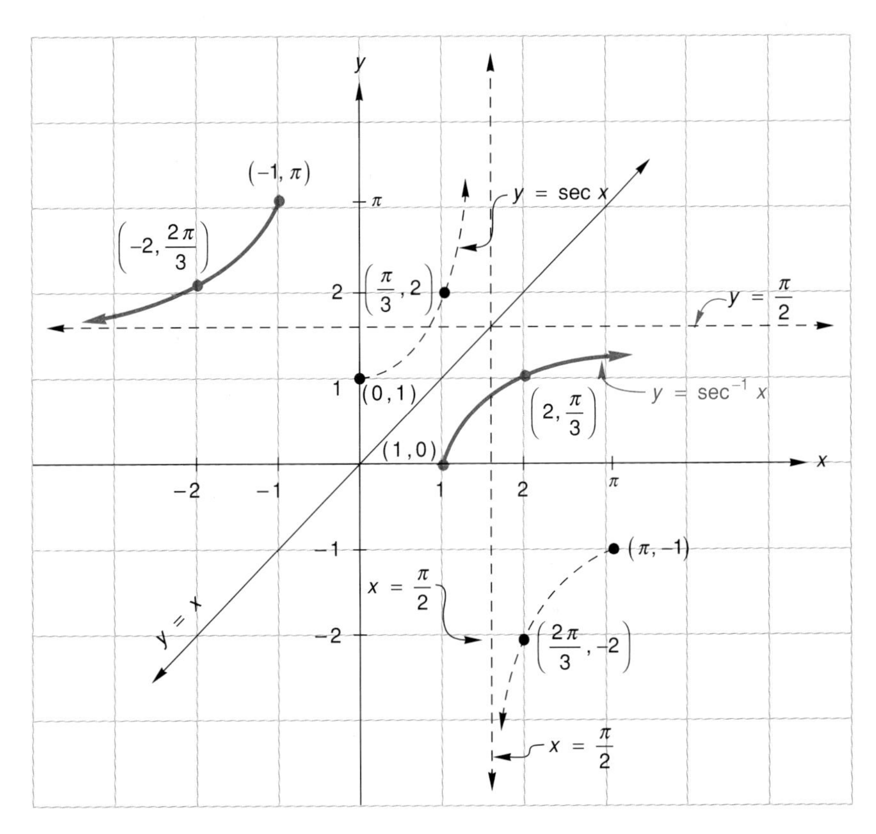

Figure 7.42

Let $y = \sec^{-1} x$, then

$$\sec y = x,$$

$$\frac{1}{\cos y} = x,$$

$$\cos y = \frac{1}{x},$$

and

$$y = \cos^{-1}\frac{1}{x}.$$

Thus,

$$\mathbf{sec^{-1}\,x = cos^{-1}\frac{1}{x}}.$$

By repeating much of the above discussion, we can establish the existence of the inverse cosecant and the inverse cotangent functions, and the relations

$$\csc^{-1} x = \sin^{-1} \frac{1}{x}$$

$$\cot^{-1} x = \tan^{-1} \frac{1}{x}.$$

Rather than construct a graph of $y = \csc^{-1} x$ as a reflection with respect to $y = x$, of $y = \csc x$, we use the relationship $\csc^{-1} x = \sin^{-1} \frac{1}{x}$ to construct the graph. The domain of the inverse sine function is $[-1,1]$. Therefore, $\sin^{-1} \frac{1}{x}$ is defined if

$$-1 \leq \frac{1}{x} \leq 1 \quad \text{or} \quad |x| \geq 1.$$

A graph of $y = \csc^{-1} x$ is sketched in figure 7.43 from the plotted points $(x, \sin^{-1} x)$, where $|x| \geq 1$. We note that as $|x|$ tends to infinity, $\frac{1}{x}$ tends to 0. As a result, $\csc^{-1} x$, which equals $\sin^{-1} \frac{1}{x}$, tends to 0, and the x-axis serves as a horizontal asymptote for a graph of $y = \csc^{-1} x$.

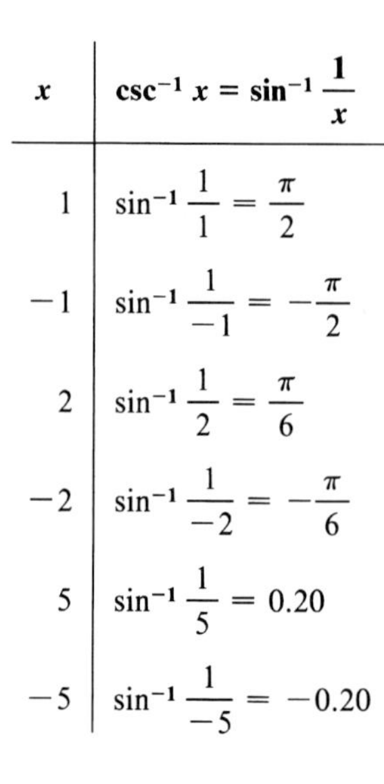

x	$\csc^{-1} x = \sin^{-1} \frac{1}{x}$
1	$\sin^{-1} \frac{1}{1} = \frac{\pi}{2}$
-1	$\sin^{-1} \frac{1}{-1} = -\frac{\pi}{2}$
2	$\sin^{-1} \frac{1}{2} = \frac{\pi}{6}$
-2	$\sin^{-1} \frac{1}{-2} = -\frac{\pi}{6}$
5	$\sin^{-1} \frac{1}{5} = 0.20$
-5	$\sin^{-1} \frac{1}{-5} = -0.20$

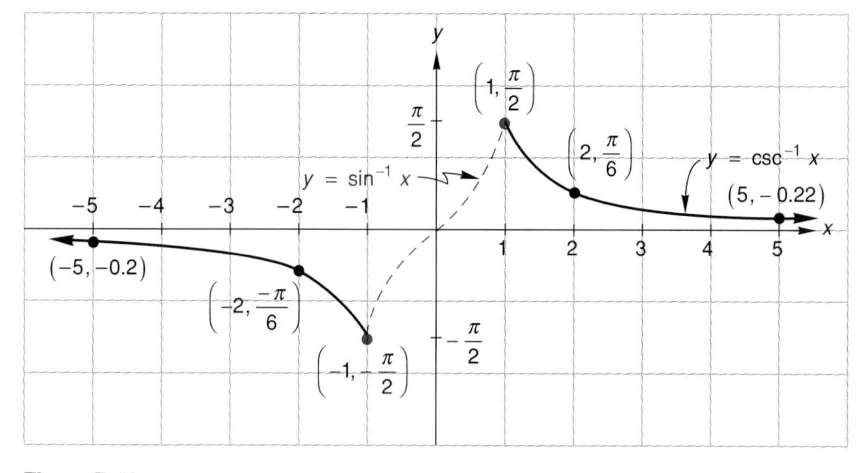

Figure 7.43

Example 4 Use the relationship $\sec^{-1} x = \cos^{-1} \dfrac{1}{x}$ to sketch a graph of $y = \sec^{-1} x$.

Solution Because $\sec^{-1} x = \cos^{-1} \dfrac{1}{x}$, the same y-value in the graphs of $y = \sec^{-1} x$ and $y = \cos^{-1} x$ are paired with x-values that are *reciprocals* of each other. Thus,

$$\frac{\pi}{3} = \sec^{-1} 2 = \cos^{-1} \frac{1}{2}, \text{ and } \frac{3\pi}{4} = \sec^{-1} -\sqrt{2} = \cos^{-1} -\frac{1}{\sqrt{2}}.$$

The ordered pairs $\left(\dfrac{1}{2}, \dfrac{\pi}{3}\right)$ and $\left(-\dfrac{1}{\sqrt{2}}, \dfrac{3\pi}{4}\right)$ in the inverse cosine function correspond to the ordered pairs $\left(2, \dfrac{\pi}{3}\right)$ and $\left(-\sqrt{2}, \dfrac{3\pi}{4}\right)$ in the inverse secant function (figure 7.44). A graph of $y = \sec^{-1} x$ is constructed through the points (x,θ) which are obtained from the corresponding points $\left(\dfrac{1}{x}, \theta\right)$ in the graph of $y = \cos^{-1} x$. As $|x|$ tends to ∞, $\dfrac{1}{x}$ tends to 0. Since $\cos^{-1} 0 = \dfrac{\pi}{2}$, the graph of $y = \sec^{-1} x$, in figure 7.44, has $y = \dfrac{\pi}{2}$ as a horizontal asymptote.

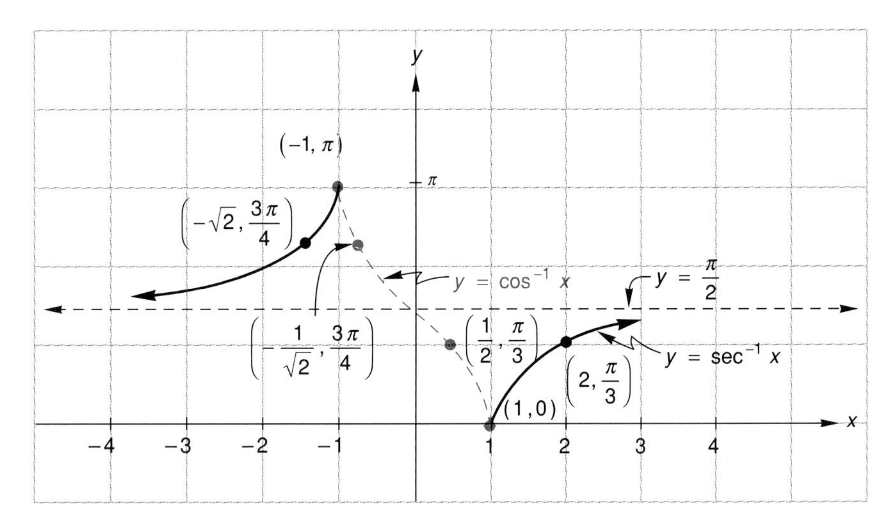

Figure 7.44 ■

Example 5 Find the exact value of

a. $\csc^{-1} \sqrt{2}$
b. $\tan(\cot^{-1} \frac{3}{4})$
c. $\cos(\sec^{-1} \frac{5}{2})$
d. $\sin(\sec^{-1} \frac{7}{3})$.

Then use a calculator to find **a, b, c,** and **d** correct to four decimal places.

Solution

a. $\csc^{-1} \sqrt{2} = \sin^{-1} \dfrac{1}{\sqrt{2}} = \dfrac{\pi}{4}$ because $\sin \dfrac{\pi}{4} = \dfrac{1}{\sqrt{2}}$.
By calculator: Calculators do not have inverse cosecant functions.
You must express $\csc^{-1} x$ as $\sin^{-1} \dfrac{1}{x}$.

$$\boxed{2}\ \boxed{\sqrt{\ }}\ \boxed{1/x}\ \boxed{\text{inv}}\ \boxed{\sin}\ \boxed{=}\ 0.7854$$

b. $\tan(\cot^{-1} \frac{3}{4}) = \tan(\tan^{-1} \frac{4}{3}) = \frac{4}{3}$.
By calculator:

$$\boxed{4}\ \boxed{\div}\ \boxed{3}\ \boxed{=}\ \boxed{\text{inv}}\ \boxed{\tan}\ \boxed{\tan}\ \boxed{=}\ 1.3333.$$

c. $\cos(\sec^{-1} \frac{5}{2}) = \cos(\cos^{-1} \frac{2}{5}) = \frac{2}{5}$.
By calculator:

$$\boxed{2}\ \boxed{\div}\ \boxed{5}\ \boxed{=}\ \boxed{\text{inv}}\ \boxed{\cos}\ \boxed{\cos}\ \boxed{=}\ 0.4000.$$

d. Let $\theta = \sec^{-1} \frac{7}{3}$, then $\sec \theta = \frac{7}{3}$. But $\sec \theta = r/x$. Thus, if $r = 7$ and $x = 3$, then $3^2 + y^2 = 49$ and $y = \sqrt{40} = 2\sqrt{10}$. Now,

$$\sin\left(\sec^{-1} \frac{7}{3}\right) = \sin \theta = \frac{y}{r} = \frac{2\sqrt{10}}{7}.$$

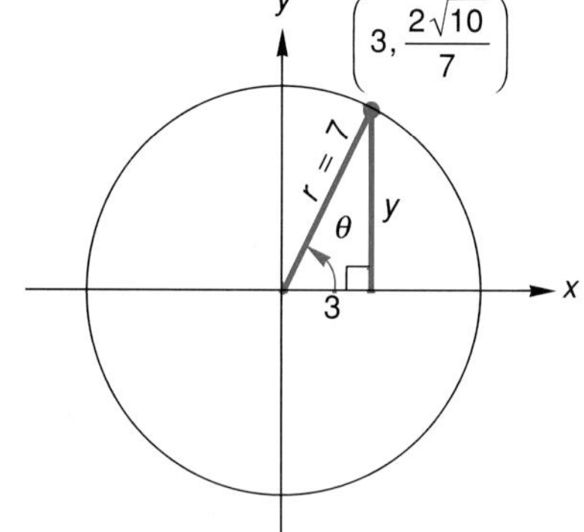

By calculator:

3 ÷ 7 = inv cos sin = 0.9035. ■

Function	Domain	Range
$\sin^{-1} x$	$[-1,1]$	$\left[-\frac{\pi}{2}, \frac{\pi}{2}\right]$
$\cos^{-1} x$	$[-1,1]$	$[0,\pi]$
$\tan^{-1} x$	R	$\left(-\frac{\pi}{2}, \frac{\pi}{2}\right)$
$\csc^{-1} x$	$(-\infty,-1] \cup [1,\infty)$	$\left[-\frac{\pi}{2},0\right) \cup \left(0, \frac{\pi}{2}\right]$
$\sec^{-1} x$	$(-\infty,-1] \cup [1,\infty)$	$\left[0,\frac{\pi}{2}\right) \cup \left(\frac{\pi}{2},\pi\right]$
$\cot^{-1} x$	R	$\left[-\frac{\pi}{2},0\right) \cup \left(0,\frac{\pi}{2}\right]$

Domains and Ranges of the Inverse Trigonometric Functions

Exercise Set 7.6

(A)

Evaluate exactly.

1. $\cos^{-1}(-1)$ **2.** $\cos^{-1}(\frac{1}{2})$

3. $\sin^{-1}\left(\frac{\sqrt{3}}{2}\right)$ **4.** $\sin^{-1}(-\frac{1}{2})$

5. $\tan^{-1}(\sqrt{3})$ **6.** $\tan^{-1}(-1)$

7. $\sec^{-1}(-2)$ **8.** $\csc^{-1}(1)$

9. $\cot^{-1}\left(-\frac{1}{\sqrt{3}}\right)$ **10.** $\sin^{-1}(0)$

11. $\cos^{-1}(0)$ **12.** $\sec^{-1}\left(\frac{2}{\sqrt{3}}\right)$

Use a calculator to find each function value correct to three decimal places.

13. $\tan^{-1}(1.34)$ **14.** $\tan^{-1}(-6.71)$

15. $\sin^{-1}(-0.94)$ **16.** $\sin^{-1}(0.3)$

17. $\cos^{-1}(-0.77)$ **18.** $\cos^{-1}(0.06)$

19. $\sec^{-1}(12.38)$ **20.** $\sec^{-1}(-4.26)$

21. $\cot^{-1}(5.12)$ **22.** $\csc^{-1}(23.46)$

Evaluate exactly.

23. $\sin^{-1}\left(\sin\frac{\pi}{6}\right)$ **24.** $\sin^{-1}\left(\cos\frac{\pi}{8}\right)$

25. $\cos(\sec^{-1} 2)$ **26.** $\sin(\sin^{-1}\frac{1}{2})$

27. $\tan(\cot^{-1} -\frac{2}{3})$ **28.** $\tan^{-1}\left(\cos\frac{\pi}{2}\right)$

29. $\cos(\tan^{-1}\frac{1}{5})$ **30.** $\cos^{-1}\left(\sin\frac{\pi}{10}\right)$

31. $\sin(\cos^{-1}\frac{5}{13})$ **32.** $\tan(\cos^{-1} -\frac{3}{5})$

33. $\sin(\sec^{-1} -6)$ **34.** $\sin(\csc^{-1} 3)$

35. $\cos(\sin^{-1}\frac{1}{8})$ **36.** $\sin^{-1}(\cos\frac{1}{8})$

Sketch a graph of the inverse function for

37. $y = \sin x, \frac{\pi}{2} \leq x \leq \frac{3\pi}{2}$.

38. $y = \sin x, -\frac{3\pi}{2} \leq x \leq -\frac{\pi}{2}$.

39. $y = \cos x, -\pi \leq x \leq 0$.

40. $y = \cos x, \pi \leq x \leq 2\pi$.

41. $y = \tan x, \frac{\pi}{2} < x < \frac{3\pi}{2}$.

42. $y = \tan x, -\frac{3\pi}{2} < x < -\frac{\pi}{2}$.

Graph each pair of functions on the same coordinate axes.

43. $y = \sin^{-1}(x - 2), y = \sin^{-1} 2x$

44. $y = -2\cos^{-1} x, y = 3 + \cos^{-1} x$

45. $y = \tan^{-1}(x - 1), y = -\pi + \tan^{-1} x$

46. $y = \sec^{-1} 2x, y = \sec^{-1} x + \frac{\pi}{4}$

Express in terms of x. (Hint: In exercise 47 let θ be the angle whose sine is x and find $\tan \theta$ in terms of x. Draw a sketch of θ.)

47. $\tan(\sin^{-1} x)$

48. $\sin(\sec^{-1} 2x)$

49. $\sin(\tan^{-1} x)$

50. $\cos(\cot^{-1} 3x)$

51. Show that $\sin^{-1} x = \cos^{-1} \sqrt{1 - x^2}$ $= \tan^{-1}(x/\sqrt{1 - x^2}), 0 < x < 1$.

52. Show that $\sin^{-1}(-x) = -\sin^{-1} x$, and that $\tan^{-1}(-x) = -\tan^{-1} x$.

53. What property may be attributed to graphs of $y = \sin^{-1} x$ and $y = \tan^{-1} x$ because of the statements in exercise 52?

Chapter 7 Review

Major Points for Review

Section 7.1 Ratios of the sides of 30°–60°–90° triangles and 45°–45°–90° triangles

Definition of a periodic function

Section 7.2 The relationship between radian and degree measures

Angles in standard position

Positive and negative angles

Section 7.3 Definitions of the six trigonometric functions

The sign attached to each trigonometric function of θ when the terminal side of θ is in the first, second, third, or fourth quadrant

Section 7.4 The reciprocal relationships among the trigonometric functions

Tan $\theta = \sin \theta/\cos \theta$, $\sin^2 \theta + \cos^2 \theta = 1$

The association between $f(-\theta)$ and $f(\theta)$ when f is a trigonometric function

The period of each trigonometric function

Co-functions

Section 7.5 The basic graphs of the trigonometric functions

Graphs of the form $y = A f(B(x + C))$ where f is a trigonometric function

Graphs of sums of trigonometric functions

Section 7.6 Definitions of the inverse trigonometric functions

Evaluating the inverse trigonometric functions

Graphing the inverse trigonometric functions

Review Exercises

Convert each degree measure to radian measure.

1. $48°$

2. $560°$

Convert each radian measure to a degree measure.

3. $\frac{2\pi}{3}$

4. 9.5

The first number is the radius of a circle. The second number is the measure of a central angle in the circle. Calculate the arc length that is intercepted by the central angle

5. 8 cm, 1.32.

6. 2 ft. 100°.

Evaluate exactly.

7. $\sin \frac{\pi}{2}$

8. $\cos \frac{11\pi}{6}$

9. $\tan\left(-\frac{\pi}{3}\right)$

10. $\sin\left(-\frac{\pi}{6}\right)$

11. $\cos \frac{3\pi}{4}$

12. $\tan \frac{8\pi}{3}$

13. $\sec\left(-\frac{7\pi}{6}\right)$

14. $\csc\left(\frac{5\pi}{3}\right)$

15. $\tan^{-1} 1$

16. $\cos^{-1} \frac{\sqrt{3}}{2}$

17. $\sin^{-1} \frac{1}{2}$

18. $\cot^{-1} \frac{1}{\sqrt{3}}$

19. $\sec^{-1}(-\sqrt{2})$

20. $\csc^{-1}(-1)$

21. $\sin^{-1}\left(\sin \frac{\pi}{3}\right)$

22. $\cos(\sec^{-1} 3)$

23. $\tan^{-1}\left(\sin \frac{\pi}{2}\right)$

24. $\tan(\cos^{-1} -\frac{1}{2})$

25. $\sin(\csc^{-1} -5)$

26. $\cot(\sin^{-1} \frac{12}{13})$

Assume that θ is in standard position in a unit circle. What is the quadrant position of the terminal side of θ if:

27. $\sin \theta < 0$ and $\tan \theta > 0$

28. $\cos \theta > 0$ and $\sin \theta < 0$

29. $\tan \theta > 0$ and $\cos \theta < 0$

30. $\sec \theta < 0$ and $\cot \theta > 0$

The value of a trigonometric function and the domain of θ are given. Evaluate the other five trigonometric functions exactly.

31. $\sin \theta = \frac{3}{5}, 0 < \theta < \frac{\pi}{2}$

32. $\cos \theta = \frac{5}{13}, 0 < \theta < \frac{\pi}{2}$

33. $\tan \theta = \frac{15}{8}, \pi < \theta < \frac{3\pi}{2}$

34. $\csc \theta = \frac{5}{4}, \frac{\pi}{2} < \theta < \pi$

35. $\sec \theta = \frac{17}{15}, \frac{3\pi}{2} < \theta < 2\pi$

36. $\tan \theta = -3, \frac{\pi}{2} < \theta < \pi$

37. $\cos \theta = -\frac{2}{3}, -\pi < \theta < -\frac{\pi}{2}$

38. $\sin \theta = -\frac{1}{5}, 3\pi < \theta < \frac{7\pi}{2}$

Use a calculator to find each function value correct to three decimal places.

39. $\sin(-7.18)$
40. $\sin(10.43)$
41. $\cos(1.40)$
42. $\cos(-4.23)$
43. $\tan(-9.12)$
44. $\tan(6.10)$
45. $\sec(11.27)$
46. $\cot(-5.49)$
47. $\sin^{-1}(0.72)$
48. $\tan^{-1}(-3.64)$
49. $\cos^{-1}(-0.28)$
50. $\sec^{-1}(8.47)$

Graph one cycle of each function.

51. $y = 2 \sin x$
52. $y = \sin 3x$
53. $y = \cos\left(x - \frac{\pi}{3}\right)$
54. $y = -\cos\left(x + \frac{\pi}{4}\right)$
55. $y = \tan 2\left(x - \frac{\pi}{6}\right)$
56. $y = \cot 3\left(x + \frac{\pi}{4}\right)$
57. $y = 3 \csc\left(2x + \frac{\pi}{3}\right)$
58. $y = \sec\left[-2\left(x + \frac{\pi}{2}\right)\right]$
59. $y = 3 \sin\left[-\frac{1}{2}\left(x - \frac{\pi}{3}\right)\right]$
60. $y = \cos x + \sin 2x$
61. $y = 2 \sin x - \cos 2x$
62. $y = \cos x + \cos 2x$

Graph on the same coordinate axes.

63. $y = \cos^{-1} x$ and $y = \cos^{-1} 2x$
64. $y = \tan^{-1} x$ and $y = \cot^{-1} x$
65. $y = \sin^{-1}(x + 1)$ and $y = \sin^{-1}(x - 1)$
66. $y = \tan^{-1} x + 2$ and $y = \tan^{-1} x - 2$

Express in terms of x.

67. $\sin(\tan^{-1} 3x)$
68. $\cot\left(\cos^{-1} \frac{x}{2}\right)$

69. Extend f, where $f(x) = x + 1, 0 \leq x < 1$, so that f is periodic over the interval $[0,4)$.

70. A gear makes 400 revolutions per minute (rpm). Through how many radians does a tooth of the gear turn in three minutes?

71. Given that $\frac{\pi}{2} < \theta < \pi$, why does $\cos \theta = -\cos(\pi - \theta)$?

72. Show that $\cos \theta = \sin\left(\frac{\pi}{2} - \theta\right)$ when $\frac{\pi}{2} < \theta < \pi$

Graph.

73. $y = 2x \cos x, \ -2\pi \leq x \leq 2\pi$
74. $y = x^2 \sin x, \ -2\pi \leq x \leq 2\pi$

Chapter 7 Test 1

1. Convert
   a. 60° to radian measure.
   b. $5\pi/6$ radians to degree measure.

2. A circle with a 5 cm radius contains a central angle of 0.63 radians.
   a. What arc length is intercepted by the central angle?
   b. The 0.63 radian central angle and the arc it intercepts determine a sector of the given circle. The ratio of the areas of the sector and the circle is equal to $0.63/2\pi$. In a circle with radius r find the formula for the area of a sector with central angle θ.

3. Let θ be in standard position in a unit circle. Find the quadrant position of its terminal side if:
   a. $\tan \theta > 0$ and $\sin \theta < 0$.
   b. $\sec \theta < 0$ and $\cot \theta < 0$.

4. Evaluate exactly.
   a. $\sin \frac{7\pi}{6}$
   b. $\tan\left(-\frac{\pi}{3}\right)$

5. Evaluate exactly.
   a. $\cos^{-1}\left(\frac{1}{2}\right)$
   b. $\sin^{-1}\left(-\frac{\sqrt{3}}{2}\right)$

6. Evaluate exactly.
   a. $\tan(\cot^{-1} 3)$
   b. $\sin^{-1}\left(\cos \frac{3\pi}{4}\right)$

7. Evaluate the other five trigonometric functions if $\cos \theta = \frac{8}{17}$ and $\frac{3\pi}{2} < \theta < 2\pi$.

8. Graph one cycle of $y = 3 \sin 2\left(x - \frac{\pi}{4}\right)$.
9. Graph $y = \cos^{-1}(x - 1)$.
10. Express $\sin(\cos^{-1} 2x)$ in terms of x.

Chapter 7 Test 2

1. Convert
   a. 248° to radian measure.
   b. −2.6 radians to degree measure.
2. An arc ten inches in length is intercepted by a central angle of measure θ in a circle with a four-inch radius.
   a. What is the measure of θ?
   b. What is the area of the circular sector determined by θ and the ten-inch arc that it intercepts?
3. Evaluate exactly.
   a. $4 \sin^{-1}\left(-\frac{1}{\sqrt{2}}\right)$ b. $\sec^{-1}(-2)$
4. Evaluate exactly.
   a. $\sin(\csc^{-1} 5)$ b. $\tan^{-1}(\tan 6.4)$
5. Use a calculator to evaluate each function value correct to three decimal places.
   a. $\sec 7.38$ b. $\cot^{-1}(-2.68)$
6. Express $\tan(\sin^{-1}(x + 2))$ in terms of x.
7. Evaluate the other five trigonometric functions if $\sec \theta = -\frac{13}{12}$ and $\frac{\pi}{2} < \theta < \pi$.
8. Graph one cycle of $y = -\cos\left(3x + \frac{\pi}{2}\right)$.
9. Graph $y = \sin^{-1} \frac{x}{2}$.
10. Graph $y = 2x + \sin x$, $-2\pi \leq x \leq 2\pi$.

Chapter 8

Trigonometric Identities and Equations

Analytical and numerical techniques for solving equations containing trigonometric expressions are developed in the latter part of the chapter. The analytical techniques rely heavily on the recognition of a number of frequently encountered trigonometric identities. Those identities are constructed in the first two sections of the chapter.

8.1 Identities

The equation, $\tan x = \sin x/\cos x$, is called an identity because *the replacement set for x equals the solution set of the equation* (see section 1.1). More simply put, the equation is an identity because equality holds for all values of the variable for which both sides of the equation are defined. Left and right members of the equation are equal as long as x is not replaced by a real number for which $\cos x = 0$. Thus, if $x \neq (2k - 1)\frac{\pi}{2}$, $k \varepsilon J$, then $\tan x$ and $\sin x/\cos x$ are *interchangeable expressions.*

Replacing an expression with another because the expressions form an identity may enable one to solve an equation that appears intractable. Also, with the use of identities, one may convert a formidable expression to a simpler and more easily interpretable form. In short, the rewards are worth the effort to learn a number of the more often encountered identities. We begin with a listing of some identities generated in section 7.4.

$$\tan x = \frac{\sin x}{\cos x} \tag{1}$$

$$\csc x = \frac{1}{\sin x} \tag{2}$$

$$\sec x = \frac{1}{\cos x} \tag{3}$$

$$\cot x = \frac{1}{\tan x} \tag{4}$$

$$\sin(-x) = -\sin x \tag{5}$$

$$\cos(-x) = \cos x \tag{6}$$

$$\tan(-x) = -\tan x \tag{7}$$

$$\sin^2 x + \cos^2 x = 1 \tag{8}$$

$$\cos\left(\frac{\pi}{2} - x\right) = \sin x \tag{9}$$

$$\sin\left(\frac{\pi}{2} - x\right) = \cos x \tag{10}$$

If each member of equation (8) is divided in turn by $\cos^2 x$ and then $\sin^2 x$, the following identities are produced.

$$\frac{\sin^2 x + \cos^2 x}{\cos^2 x} = \frac{1}{\cos^2 x}$$

$$\frac{\sin^2 x + \cos^2 x}{\sin^2 x} = \frac{1}{\sin^2 x}$$

Pythagorean Identities

$$\tan^2 x + 1 = \sec^2 x \tag{11}$$

$$1 + \cot^2 x = \csc^2 x \tag{12}$$

Verifying That an Equation Is an Identity

Suppose we want to show that

$$\frac{\sin x}{\csc x} = 1 - \cos^2 x$$

is an identity. We do not assume at the outset that the left and right members of the equation are equal. Instead

1. we use known identities to convert the left member of the equation to $1 - \cos^2 x$, or
2. we use known identities to convert the right member of the equation to $\sin x/\csc x$, or
3. we convert both members of the equation to the same expression.

Whichever technique is used, if done correctly, should produce an equation equivalent to the original equation in which the left and right members are identical. Certainly, in such an equation, the replacement set and the solution set are equal.

1.
$$\begin{aligned}\frac{\sin x}{\csc x} &= \frac{\sin x}{\frac{1}{\sin x}} && \text{[from (2)]}\\ &= \sin^2 x\\ &= 1 - \cos^2 x && \text{[from (8)]}\end{aligned}$$

2.
$$\begin{aligned}1 - \cos^2 x &= \sin^2 x && \text{[from (8)]}\\ &= \sin x \cdot \sin x\\ &= \sin x \cdot \frac{1}{\csc x} && \text{[from (2)]}\\ &= \frac{\sin x}{\csc x}\end{aligned}$$

3.
$$\begin{aligned}\frac{\sin x}{\csc x} &= \frac{\sin x}{\frac{1}{\sin x}} && \text{[from (2)]}\\ &= \sin^2 x\\ 1 - \cos^2 x &= \sin^2 x && \text{[from (8)]}\end{aligned}$$

Since the left and right members of the equation both equal $\sin^2 x$, the equation is an identity.

Example 1 Verify that $\cos^4 x + 2 \sin^2 x = 1 + \sin^4 x$ is an identity.

Solution

$$\begin{aligned}1 + \sin^4 x &= 1 + (\sin^2 x)^2\\ &= 1 + (1 - \cos^2 x)^2 && \text{[from (8)]}\\ &= 1 + 1 - 2\cos^2 x + \cos^4 x\\ &= 2(1 - \cos^2 x) + \cos^4 x\\ &= 2 \sin^2 x + \cos^4 x && \text{[from (8)]}\end{aligned}$$

■

Example 2 Verify that $\tan^2 x(1 + \cot^2 x) = \dfrac{1}{\cos^2 x}$ is an identity.

Solution

$$\begin{aligned}\tan^2 x(1 + \cot^2 x) &= \tan^2 x \cdot \csc^2 x && \text{[from (12)]}\\ &= \frac{\sin^2 x}{\cos^2 x} \cdot \frac{1}{\sin^2 x}\\ &= \frac{1}{\cos^2 x}\end{aligned}$$

■

Sum and Difference Formulas

The circle in figure 8.1 is a unit circle. Therefore, the x and y coordinates of point T are, by definition, $\cos \alpha$ and $\sin \alpha$, respectively. Also by definition, the coordinates of points S and P are $(\cos(\alpha + \beta), \sin(\alpha + \beta))$ and $(\cos(-\beta), \sin(-\beta))$. Chords $\overline{SQ}$ and $\overline{TP}$ are equal because equal central angles in the same circle intercept equal chords. Applying the distance formula to the equal chords gives us

$$\sqrt{(\cos(\alpha + \beta) - 1)^2 + (\sin(\alpha + \beta) - 0)^2}$$
$$= \sqrt{(\cos \alpha - \cos(-\beta))^2 + (\sin \alpha - \sin(-\beta))^2}$$

$$(\cos(\alpha + \beta) - 1)^2 + \sin^2(\alpha + \beta)$$
$$= (\cos \alpha - \cos \beta)^2 + (\sin \alpha + \sin \beta)^2$$

$$\cos^2(\alpha + \beta) - 2\cos(\alpha + \beta) + 1 + \sin^2(\alpha + \beta)$$
$$= \cos^2 \alpha - 2\cos \alpha \cos \beta + \cos^2 \beta + \sin^2 \alpha + 2\sin \alpha \sin \beta + \sin^2 \beta$$

$$2 - 2\cos(\alpha + \beta)$$
$$= 2 - 2\cos \alpha \cos \beta + 2\sin \alpha \sin \beta$$

$$\cos(\alpha + \beta)$$
$$= \cos \alpha \cos \beta - \sin \alpha \sin \beta.$$

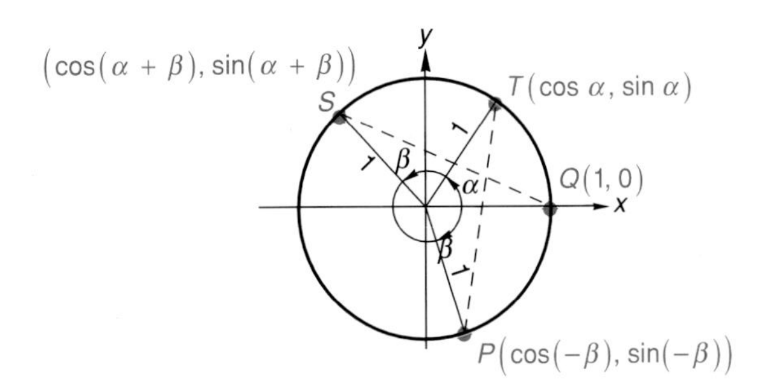

Figure 8.1

The last equation is called the **cosine sum formula.** We shall refer to it in the form

Cosine Sum Formula

$$\cos(x + y) = \cos x \cos y - \sin x \sin y \tag{13}$$

to remind us that in this chapter, angle measures are in radians.

Example 3

a. Use the cosine sum formula to show that $\cos(x + 2k\pi) = \cos x$, $k \in J$.

b. Use (13) to evaluate $\cos \frac{5\pi}{12}$ exactly.

Solution

a.
$$\begin{aligned}\cos(x + 2k\pi) &= \cos x \cos 2k\pi - \sin x \sin 2k\pi \\ &= \cos x \cdot 1 - \sin x \cdot 0 \\ &= \cos x\end{aligned}$$

b.
$$\begin{aligned}\cos \frac{5\pi}{12} &= \cos\left(\frac{\pi}{4} + \frac{\pi}{6}\right) \\ &= \cos\frac{\pi}{4} \cos \frac{\pi}{6} - \sin \frac{\pi}{4} \sin \frac{\pi}{6} \\ &= \frac{\sqrt{2}}{2} \cdot \frac{\sqrt{3}}{2} - \frac{\sqrt{2}}{2} \cdot \frac{1}{2} \\ &= \frac{\sqrt{6} - \sqrt{2}}{4}\end{aligned}$$

■

The following identities follow directly from (13). They are called the **cosine difference formula** and the **sine sum and difference formulas.**

Sum and Difference Formulas

$$\cos(x - y) = \cos x \cos y + \sin x \sin y \tag{14}$$
$$\sin(x + y) = \sin x \cos y + \cos x \sin y \tag{15}$$
$$\sin(x - y) = \sin x \cos y - \cos x \sin y \tag{16}$$

$$\begin{aligned}\cos(x - y) &= \cos(x + (-y)) \\ &= \cos x \cos(-y) - \sin x \sin(-y) && \text{[from (13)]} \\ &= \cos x \cos y + \sin x \sin y && \text{[from (5) and (6)]}\end{aligned}$$

$$\begin{aligned}\sin(x + y) &= \cos\left(\frac{\pi}{2} - (x + y)\right) && \text{[from (9)]} \\ &= \cos\left(\left(\frac{\pi}{2} - x\right) - y\right) \\ &= \cos\left(\frac{\pi}{2} - x\right)\cos y + \sin\left(\frac{\pi}{2} - x\right)\sin y && \text{[from (14)]} \\ &= \sin x \cos y + \cos x \sin y && \text{[from (9) and (10)]}\end{aligned}$$

The derivation of (16) follows the same pattern as the derivation of (14). Try it.

Example 4 If $\sin x = \frac{1}{2}$, $\cos y = -\frac{3}{4}$, $0 \le x \le \pi/2$, and $\pi/2 \le y \le \pi$, find

a. $\sin(x + y)$

b. $\cos(x + y)$.

Also, determine if $\pi/2 \le x + y \le \pi$ or if $\pi \le x + y \le 3\pi/2$. Use a calculator to find $x + y$ correct to three decimal places.

Solution From (8), $\cos^2 x = 1 - \sin^2 x$. Since $0 \le x \le \pi/2$, $\cos x > 0$. Thus, $\cos x = \sqrt{1 - \frac{1}{4}} = \dfrac{\sqrt{3}}{2}$. Because $\pi/2 \le y \le \pi$, $\sin y > 0$. In addition, $\sin^2 y = 1 - \cos^2 y$. Therefore, $\sin y = \sqrt{1 - \frac{9}{16}} = \dfrac{\sqrt{7}}{4}$.

a.
$$\begin{aligned}\sin(x + y) &= \sin x \cos y + \cos x \sin y \\ &= \left(\frac{1}{2}\right)\left(-\frac{3}{4}\right) + \left(\frac{\sqrt{3}}{2}\right)\left(\frac{\sqrt{7}}{4}\right) \\ &= \frac{-3 + \sqrt{21}}{8}.\end{aligned}$$

b.
$$\begin{aligned}\cos(x + y) &= \cos x \cos y - \sin x \sin y \\ &= \left(\frac{\sqrt{3}}{2}\right)\left(-\frac{3}{4}\right) - \left(\frac{1}{2}\right)\left(\frac{\sqrt{7}}{4}\right) \\ &= -\frac{3\sqrt{3} + \sqrt{7}}{8}.\end{aligned}$$

Because $\sin(x + y) > 0$ and $\cos(x + y) < 0$, $\pi/2 \leq x + y \leq \pi$. The calculator determined radian measure of $x + y$ is

$$x + y = \cos^{-1}\left(-\frac{3\sqrt{3} + \sqrt{7}}{8}\right) \approx 2.942.$$

Check: $\sin 2.942 \approx \dfrac{-3 + \sqrt{21}}{8}$. ■

Example 5 Verify that the angle sum formula for the tangent,

Tangent Sum Formula

$$\tan(x + y) = \frac{\tan x + \tan y}{1 - \tan x \tan y}, \qquad (17)$$

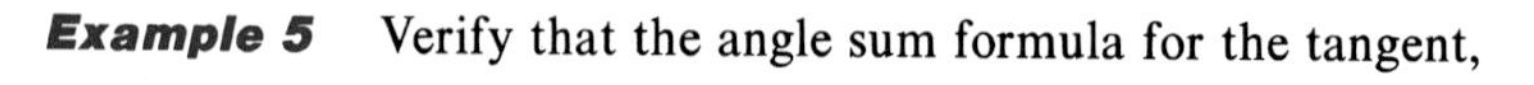

is an identity.

Solution

$$\begin{aligned}\tan(x + y) &= \frac{\sin(x + y)}{\cos(x + y)}\\ &= \frac{\sin x \cos y + \cos x \sin y}{\cos x \cos y - \sin x \sin y}\\ &= \frac{\dfrac{\sin x \cos y + \cos x \sin y}{\cos x \cos y}}{\dfrac{\cos x \cos y - \sin x \sin y}{\cos x \cos y}}\\ &= \frac{\tan x + \tan y}{1 - \tan x \tan y}\end{aligned}$$

■

Example 6 Establish the identity,

$$\frac{\cos(x + y) + \cos(x - y)}{2} = \cos x \cos y.$$

Solution Both (13) and (14) contain the product $\cos x \cos y$. Hence,

$$\begin{aligned}\cos(x + y) &= \cos x \cos y - \sin x \sin y\\ \cos(x - y) &= \cos x \cos y + \sin x \sin y\\ \hline \cos(x + y) + \cos(x - y) &= 2 \cos x \cos y\\ \frac{\cos(x + y) + \cos(x - y)}{2} &= \cos x \cos y.\end{aligned}$$

■

Exercise Set 8.1

(A)

Use the identities

$$\sin^2 x + \cos^2 x = 1,$$
$$\cos(x + y) = \cos x \cos y - \sin x \sin y,$$

and

$$\sin(x + y) = \sin x \cos y + \cos x \sin y$$

to *derive* the identities in exercises 1 through 8. Do not refer to the text.

1. $\tan^2 x + 1 = \sec^2 x$
2. $\cos(x - y) = \cos x \cos y + \sin x \sin y$
3. $\sin(x - y) = \sin x \cos y - \cos x \sin y$
4. $1 + \cot^2 x = \csc^2 x$
5. $\tan(x + y) = \dfrac{\tan x + \tan y}{1 - \tan x \tan y}$
6. $\tan(x - y) = \dfrac{\tan x - \tan y}{1 + \tan x \tan y}$
7. $\cos x \cos y = \dfrac{\cos(x + y) + \cos(x - y)}{2}$
8. $\cos x \sin y = \dfrac{\sin(x + y) - \sin(x - y)}{2}$

Verify that each equation is an identity.

9. $\sin x \sec x = \tan x$
10. $\sin x \cot x \sec x = 1$
11. $\dfrac{\cos x}{1 - \sin x} = \dfrac{1 + \sin x}{\cos x}$
12. $\cot^2 x = \dfrac{1 - \sin^2 x}{\sin^2 x}$
13. $\dfrac{\sec^2 x - 1}{\sec^2 x} = \sin^2 x$
14. $3 \tan^2 x - 2 \sec^2 x + 5 = \sec^2 x + 2$
15. $(\csc^2 x - 1)(1 - \cos^2 x) = \dfrac{1}{\sec^2 x}$
16. $\dfrac{1}{\csc x - \sin x} = \tan x \sec x$
17. $\dfrac{1 - \cos x}{1 + \cos x} = (\cot x - \csc x)^2$
18. $\cos(x + \pi) = -\cos x$
19. $\sin(x - \pi) = -\sin x$
20. $\sin(x + y) + \sin(x - y) = 2 \sin x \cos y$
21. $\cos(x + y) - \cos(x - y) = -2 \sin x \sin y$
22. $\sin 8x \cos 3x - \cos 8x \sin 3x = \sin 5x$
23. $\cos 2x \cos 5x + \sin 2x \sin 5x = \cos 3x$
24. $\sin x \cos 2x + \cos x \sin 2x = \sin 3x$
25. $\dfrac{\sin(x + y)}{\sin(x - y)} = \dfrac{\tan x + \tan y}{\tan x - \tan y}$
26. $\dfrac{\cos(x + y)}{\cos(x - y)} = \dfrac{1 - \tan x \tan y}{1 + \tan x \tan y}$
27. $\cos\left(\dfrac{x}{2}\right)\cos\left(\dfrac{x}{3}\right) - \sin\left(\dfrac{x}{2}\right)\sin\left(\dfrac{x}{3}\right) = \cos\left(\dfrac{5x}{6}\right)$
28. $\tan(x + \pi) = \tan x$
29. $\csc(x - y) = \dfrac{\csc x \csc y}{\cot y - \cot x}$
30. $\sin(x - y) \sin(x + y) = \cos^2 y - \cos^2 x$

Find $\sin(x + y)$ and $\cos(x + y)$ under the given conditions. Use a calculator to evaluate $x + y$ correct to three decimal places.

31. $\sin x = -\dfrac{2}{3}$, $\pi < x < \dfrac{3\pi}{2}$ and $\cos y = \dfrac{1}{5}$, $\dfrac{3\pi}{2} < y < 2\pi$
32. $\sin x = \dfrac{7}{10}$, $\dfrac{\pi}{2} < x < \pi$ and $\cos y = -\dfrac{7}{10}$, $\dfrac{\pi}{2} < y < \pi$
33. $\sin x = \dfrac{3}{4}$, $0 < x < \dfrac{\pi}{2}$ and $\cos y = -\dfrac{5}{8}$, $\dfrac{\pi}{2} < y < \pi$
34. $\sin x = -\dfrac{1}{4}$, $\dfrac{3\pi}{2} < x < 2\pi$ and $\cos y = \dfrac{3}{10}$, $\dfrac{3\pi}{2} < y < 2\pi$

Verify that each equation is an identity.

35. $\sin(x + y + z) = \sin x \cos y \cos z - \sin x \sin y \sin z + \cos x \sin y \cos z + \cos x \cos y \sin z$

36. $\sin(x + y - z) = \sin x \cos y \cos z + \cos x \sin y \cos z - \cos x \cos y \sin z + \sin x \sin y \sin z$

37. $\dfrac{\sin x + \cos x - 1}{\sin x - \cos x + 1} = \dfrac{1 - \sin x}{\cos x}$

38. $\dfrac{\cos x \cot x}{\cot x - \cos x} = \dfrac{\cot x + \cos x}{\cos x \cot x}$

(B)

Determine which of the following equations are identities and which are not identities. An equation is not an identity if there is at least one number in its replacement set for which the left and right members of the equation are *not* equal.

39. $6 \cos 4x \cos x + 6 \sin 4x \sin x = 3 \cos 3x$

40. $\dfrac{2}{\sec x} = \cos^2 x - \sin^2 x - 1$

41. $\dfrac{\tan(x + y) - \tan z}{1 + \tan(x + y) \tan z} = \dfrac{\tan x + \tan(y - z)}{1 - \tan x \tan (y - z)}$

42. $3 \cos^2 x + \cos x - 2 = 0$

43. $\dfrac{\cos(x + y)}{\sin(x + y)} = \dfrac{\cot x + \cot y}{\cot x \cot y - 1}$

44. $\sqrt{\sin^2 x + \cos^2 x} = \sin x + \cos x$

45. In exercise 20, let $w = x + y$ and $z = x - y$. Then show that

$$\sin w + \sin z = 2 \sin\left(\frac{w + z}{2}\right) \cos\left(\frac{w - z}{2}\right).$$

46. In exercise 21, let $w = x + y$ and $z = x - y$. Then show that

$$\cos w - \cos z = -2 \sin\left(\frac{w + z}{2}\right) \sin\left(\frac{w - z}{2}\right).$$

47. Use exercises 45 and 46 to show that

$$\frac{\sin 5x + \sin 9x}{\cos 5x - \cos 9x} = \cot 2x.$$

48. Verify the identity.

$$\frac{\sin 6x + \sin 2x}{\sin 6x - \sin 2x} = \frac{\tan 4x}{\tan 2x}$$

49. Verify the identity.

$$\frac{\sin 3x + \sin x}{\cos 3x + \cos x} = \tan 2x$$

50. Show that $\sin^2 x - \sin^2 y = \sin(x + y) \sin(x - y)$.

51. Show that $\cos^2 x - \sin^2 y = \cos(x + y) \cos(x - y)$.

52. Show that $\cos 348°/\cos 102° = -\tan 78°$.

8.2 More Identities

It must seem to the reader that the list of identities is endless. But relatively few identities occur frequently enough to warrant memorizing them. They are:

The Pythagorean Formulas

$$\sin^2 x + \cos^2 x = 1$$
$$\tan^2 x + 1 = \sec^2 x$$
$$1 + \cot^2 x = \csc^2 x,$$

and

The Sum and Difference Formulas

$$\cos(x + y) = \cos x \cos y - \sin x \sin y$$
$$\cos(x - y) = \cos x \cos y + \sin x \sin y$$
$$\sin(x + y) = \sin x \cos y + \cos x \cos y$$
$$\sin(x - y) = \sin x \cos y - \cos x \sin y,$$

which are derived in section 8.1, and the *Double Angle* and *Half Angle Formulas,* which are derived in this section. It is shown in section 8.1 and in this section that ten of the thirteen identities suggested for memorization can be derived from the three identities,

$$\sin^2 x + \cos^2 x = 1$$
$$\cos(x + y) = \cos x \cos y - \sin x \sin y$$
$$\sin(x + y) = \sin x \cos y + \cos x \sin y.$$

Memorize three, and the other ten can be derived.

Double Angle Formulas

The cosine sum and sine sum formulas lead directly to the double angle formulas.

$$\begin{aligned} \cos 2x &= \cos(x + x) \\ &= \cos x \cos x - \sin x \sin x \end{aligned}$$

Cosine Double Angle Formula

$$\cos 2x = \cos^2 x - \sin^2 x \qquad (1)$$

$$\begin{aligned} \sin 2x &= \sin(x + x) \\ &= \sin x \cos x + \cos x \sin x \end{aligned}$$

Sine Double Angle Formula

$$\sin 2x = 2 \sin x \cos x \qquad (2)$$

Useful variations of (1) are obtained if $\cos^2 x$ is replaced by $1 - \sin^2 x$ or $\sin^2 x$ is replaced by $1 - \cos^2 x$. Thus,

$$\begin{aligned}\cos 2x &= \cos^2 x - \sin^2 x \\ &= (1 - \sin^2 x) - \sin^2 x\end{aligned}$$

Second Cosine Double Angle Formula

$$\cos 2x = 1 - 2\sin^2 x, \tag{3}$$

$$\begin{aligned}\cos 2x &= \cos^2 x - \sin^2 x \\ &= \cos^2 x - (1 - \cos^2 x)\end{aligned}$$

Third Cosine Double Angle Formula

$$\cos 2x = 2\cos^2 x - 1. \tag{4}$$

Example 1 Express $\cos 4y$ in terms of $\cos y$.

Solution

$$\begin{aligned}\cos 4y &= \cos 2(2y) \\ &= 2\cos^2 2y - 1 && \text{[from (4)]} \\ &= 2(2\cos^2 y - 1)^2 - 1 && \text{[from (4)]} \\ &= 8\cos^4 y - 8\cos^2 y + 1\end{aligned}$$

■

Example 2 Evaluate $\cos 2x$ if $\sin x = \frac{1}{3}$.

Solution From (3), $\cos 2x = 1 - 2(\frac{1}{3})^2 = \frac{7}{9}$. ■

Example 3 Evaluate $\sin 2x$ if $\sin x = \frac{1}{3}$ and $\pi/2 \leq x \leq \pi$.

Solution Knowing that $\cos^2 x = 1 - \sin^2 x$ and that $\pi/2 \leq x \leq \pi$, then

$$\cos x = -\sqrt{1 - \sin^2 x} = -\sqrt{\frac{8}{9}} = -\frac{2\sqrt{2}}{3}.$$

From (2),

$$\sin 2x = 2\left(\frac{1}{3}\right)\left(-\frac{2\sqrt{2}}{3}\right) = -\frac{4\sqrt{2}}{9}.$$

■

Example 4 Verify the identity, $\cos 3x = 4\cos^3 x - 3\cos x$.

Solution

$$\begin{aligned}\cos 3x &= \cos(2x + x)\\ &= \cos 2x \cos x - \sin 2x \sin x\\ &= (2\cos^2 x - 1)(\cos x) - 2\sin x \cos x \sin x\\ &= 2\cos^3 x - \cos x - 2(1 - \cos^2 x)\cos x\\ &= 4\cos^3 x - 3\cos x\end{aligned}$$

■

In calculus, one often needs to convert $\sin^n x$ or $\cos^n x$, where n is an even natural number, into a sum of terms of the form $\cos kx$, where k is a natural number. Equations (3) and (4) allow us to do this.

Example 5 Express $\sin^4 x$ as a sum of functions of the form $\cos kx$, $k \varepsilon N$.

Solution

$$\begin{aligned}\sin^4 x &= (\sin^2 x)^2\\ &= \left(\frac{1 - \cos 2x}{2}\right)^2 && \text{[from (3)]}\\ &= \frac{1}{4} - \frac{\cos 2x}{2} + \frac{\cos^2 2x}{4}\\ &= \frac{1}{4} - \frac{\cos 2x}{2} + \frac{\left(\frac{1 + \cos 4x}{2}\right)}{4} && \text{[from (4)]}\\ &= \frac{1}{4} - \frac{\cos 2x}{2} + \frac{1}{8} + \frac{\cos 4x}{8}\\ &= \frac{3}{8} - \frac{\cos 2x}{2} + \frac{\cos 4x}{8}\end{aligned}$$

■

Half Angle Formulas

Equation (3) can be written in the equivalent form

$$\sin^2 x = \frac{1 - \cos 2x}{2},$$

from which it follows that

$$\sin x = \pm\sqrt{\frac{1 - \cos 2x}{2}}.$$

The sign before the radical is determined by the magnitude of x. For example, if $0 < x < \pi$, the sign is $+$. If $\pi < x < 2\pi$, then the sign is $-$. Now, if we let $w = 2x$, the last equation becomes **the half angle formula for the sine.**

The half angle formula for the cosine, is developed in the same way except that we start with equation (4).

Half Angle Formulas

$$\sin\left(\frac{w}{2}\right) = \pm\sqrt{\frac{1 - \cos w}{2}} \tag{5}$$

$$\cos\left(\frac{w}{2}\right) = \pm\sqrt{\frac{1 + \cos w}{2}}, \tag{6}$$

Example 6 If $\tan 2x = 2$ and $0 < x < \pi/4$, find $\sin x$, $\cos x$, and $\tan x$.

Solution

$$\begin{aligned}
\sec^2 2x &= 1 + \tan^2 2x \\
\sec^2 2x &= 5 \\
\sec 2x &= \sqrt{5} \\
\cos 2x &= \frac{1}{\sqrt{5}} \\
\sin x &= \sqrt{\frac{1 - 1/\sqrt{5}}{2}} && \text{[from (5)]} \\
&= \sqrt{\frac{\sqrt{5} - 1}{2\sqrt{5}}} \\
\cos x &= \sqrt{\frac{1 + 1/\sqrt{5}}{2}} && \text{[from (6)]} \\
&= \sqrt{\frac{\sqrt{5} + 1}{2\sqrt{5}}} \\
\tan x &= \frac{\sin x}{\cos x} \\
&= \sqrt{\frac{\sqrt{5} - 1}{\sqrt{5} + 1}}
\end{aligned}$$

■

Example 7 Determine the exact value of $\sin(5\pi/24)$.

Solution In example 3b of section 8.1 it is shown that $\cos 5\pi/12$ is equal to $(\sqrt{6} - \sqrt{2})/4$. Since $5\pi/24$ is one half of $5\pi/12$, we can use the sine half angle formula to find

$$\begin{aligned}\sin\left(\frac{5\pi}{24}\right) &= \sqrt{\frac{1 - \cos(5\pi/12)}{2}}\\ &= \sqrt{\frac{1 - (\sqrt{6} - \sqrt{2})/4}{2}}\\ &= \sqrt{\frac{4 - \sqrt{6} + \sqrt{2}}{8}}.\end{aligned}$$

You might wish to check this result against a calculator-generated value for $\sin(5\pi/24)$. ■

Evaluating Expressions That Contain Inverse Trigonometric Functions

Example 8 Evaluate the expression $\sin^{-1}\frac{4}{5} + \sin^{-1}\frac{12}{13}$ exactly.

Solution Let $x = \sin^{-1}\frac{4}{5}$. The domain and range of the inverse sine function are $[-1,1]$ and $[-\pi/2,\pi/2]$, respectively. Therefore x is an angle between 0 and $\pi/2$ whose sine is $\frac{4}{5}$. Similarly, if $y = \sin^{-1}\frac{12}{13}$, then y is the angle in $[0,\pi/2]$ whose sine is $\frac{12}{13}$ (see figure 8.2). The expression we seek to evaluate is the sum, $x + y$. It can be obtained as follows:

$$\begin{aligned}\sin(x + y) &= \sin x \cos y + \cos x \sin y\\ &= \sin(\sin^{-1}\tfrac{4}{5})\cos(\sin^{-1}\tfrac{12}{13}) + \cos(\sin^{-1}\tfrac{4}{5})\sin(\sin^{-1}\tfrac{12}{13})\\ &= \tfrac{4}{5}\cdot\tfrac{5}{13} + \tfrac{3}{5}\cdot\tfrac{12}{13} \qquad \text{(figure 8.2)}\\ x + y &= \sin^{-1}\tfrac{56}{65}.\end{aligned}$$

We note that $x + y$ can also be found as the inverse cosine of a number if we begin our computation with $\cos(x + y)$. Try it.

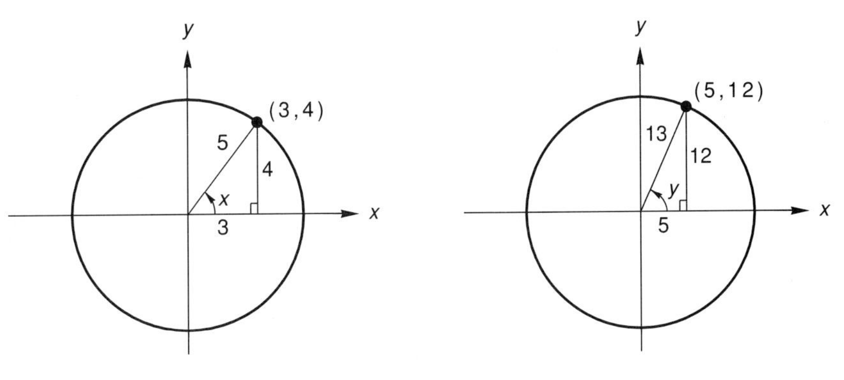

Figure 8.2 ■

Example 9 Evaluate $\cos(\sin^{-1} -\frac{1}{5} + \cos^{-1} \frac{1}{4})$ exactly. Then use a calculator to check your result.

Solution Let $x = \sin^{-1} -\frac{1}{5}$ and $y = \cos^{-1} \frac{1}{4}$. Figure 8.3 shows us that the inverse sine of a number in the interval $[-1,0]$ is an angle between 0 and $-\pi/2$, and that the inverse cosine of a number in $[0,1]$ is an angle between 0 and $\pi/2$. Consequently, $\cos x = 2\sqrt{6}/5$, and $\sin y = \sqrt{15}/4$. Also, $\sin x = \sin(\sin^{-1} -\frac{1}{5}) = -\frac{1}{5}$, and $\cos y = \cos(\cos^{-1} \frac{1}{4}) = \frac{1}{4}$.

$$\begin{aligned}\cos\left(\sin^{-1} -\frac{1}{5} + \cos^{-1} \frac{1}{4}\right) &= \cos(x + y)\\ &= \cos x \cos y - \sin x \sin y\\ &= \left(\frac{2\sqrt{6}}{5}\right)\left(\frac{1}{4}\right) - \left(-\frac{1}{5}\right)\left(\frac{\sqrt{15}}{4}\right)\\ &= \frac{2\sqrt{6} + \sqrt{15}}{20}\end{aligned}$$

Check by calculator: $\sin^{-1} -\frac{1}{5} \approx -0.2014$, $\cos^{-1} \frac{1}{4} \approx 1.3181$, $\cos(-0.2014 + 1.3181) \approx$ **0.4386,** $(2\sqrt{6} + \sqrt{15})/20 \approx$ **0.4386.**

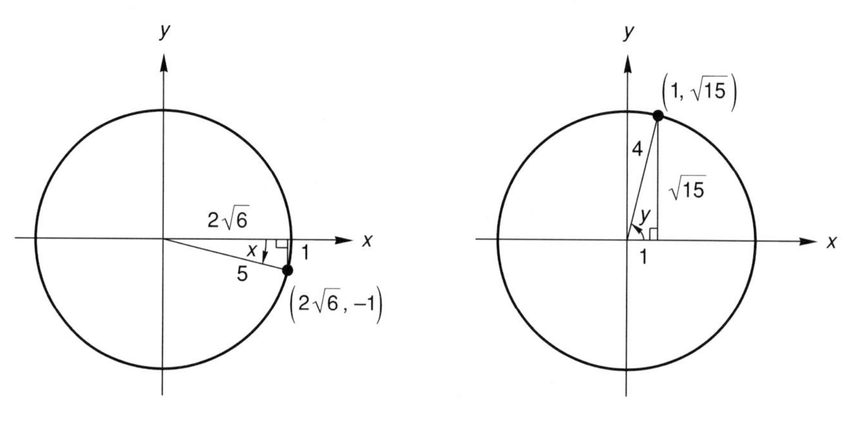

Figure 8.3

■

Example 10 Evaluate $\sin(2 \sin^{-1} \frac{5}{13})$ exactly. Check your result with a calculator.

Solution Let $x = \sin^{-1} \frac{5}{13}$ then, $\sin x = \sin(\sin^{-1} \frac{5}{13}) = \frac{5}{13}$ and $\cos x = \frac{12}{13}$ (see figure 8.4).

$$\begin{aligned}\sin(2 \sin^{-1} \tfrac{5}{13}) &= \sin 2x\\ &= 2 \sin x \cos x \qquad \text{[from (2)]}\\ &= 2 \cdot \tfrac{5}{13} \cdot \tfrac{12}{13}\\ &= \tfrac{120}{169}\end{aligned}$$

Calculator check: $2 \sin^{-1} \frac{5}{13} \approx 0.7896$, $\sin(2 \sin^{-1} \frac{5}{13}) \approx$ **0.7101,** $\frac{120}{169} \approx$ **0.7101.**

■

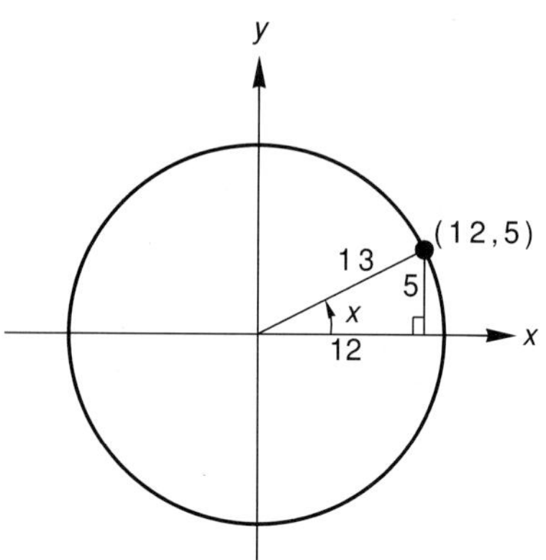

Figure 8.4

Exercise Set 8.2

(A)

Use the identities

$$\sin^2 x + \cos^2 x = 1,$$
$$\cos(x + y) = \cos x \cos y - \sin x \sin y,$$
and
$$\sin(x + y) = \sin x \cos y + \cos x \sin y$$

to derive the identities in exercises 1 through 8. Do not refer to the text.

1. $\sin 2x = 2 \sin x \cos x$

2. $\cos 2x = \cos^2 x - \sin^2 x$

3. $\cos 2x = 2 \cos^2 x - 1$

4. $\cos 2x = 1 - 2 \sin^2 x$

5. $\sin \frac{x}{2} = \pm \sqrt{\frac{1 - \cos x}{2}}$

6. $\cos \frac{x}{2} = \pm \sqrt{\frac{1 + \cos x}{2}}$

7. $\tan 2x = \frac{2 \tan x}{1 - \tan^2 x}$

8. $\tan^2 x = \frac{1 - \cos 2x}{1 + \cos 2x}$

Verify that each equation is an identity.

9. $\sin 3x = 3 \sin x - 4 \sin^3 x$

10. $\sin 4x = 4 \sin x \cos x(1 - 2 \sin^2 x)$

11. $\sin^4 x = \frac{1}{8} \cos 4x - \frac{1}{2} \cos 2x + \frac{3}{8}$

12. $\frac{2 \tan x}{1 + \tan^2 x} = \sin 2x$

13. $\tan x + \cot x = 2 \csc 2x$

14. $\frac{1 - \tan x}{1 + \tan x} = \frac{1 - \sin 2x}{\cos 2x}$

15. $\cot^2 \frac{x}{2} = \frac{\sec x + 1}{\sec x - 1}$

16. $\cos^4 x - \sin^4 x = \cos 2x$

17. $\sec 2x = \frac{\sec^2 x}{2 - \sec^2 x}$

18. $\frac{2}{1 + \cos 2x} = \sec^2 x$

19. $\cos 4x = 8 \cos^4 x - 8 \cos^2 x + 1$

20. $\frac{1 - \tan^2 x}{1 + \tan^2 x} = \cos 2x$

21. $2 \sin^2 3x = \sin 6x \tan 3x$

22. $\sin^2 2x \cos^2 2x = \frac{1 - \cos 8x}{8}$

23. $2 \sin^2 \frac{x}{2} \tan x = \tan x - \sin x$

24. $\sec^2 \left(\frac{x}{2}\right) = 2 \tan \frac{x}{2} \csc x$

25. $\tan \frac{x}{2} = \frac{\sin x}{1 + \cos x}$ $\left(\text{Hint: Start with } \frac{\sin \frac{x}{2}}{\cos \frac{x}{2}} \text{ and multiply top and bottom by } 2 \cos \frac{x}{2}.\right)$

26. $\tan \frac{x}{2} = \frac{1 - \cos x}{\sin x}$ (Hint: See exercise 25.)

Write an identity for the first function in terms of the second function.

27. $\sec 2x$; $\sec x$

28. $\cot 2x$; $\cot x$

29. $\tan 3x$; $\tan x$

30. $\sin 5x$; $\sin x$

31. $\cos 5x$; $\cos x$

32. $\tan 4x$; $\tan x$

(B)

Find each function value exactly. (See example 7.)

33. $\cos \frac{5\pi}{24}$

34. $\sin \frac{7\pi}{48}$

35. $\cos \frac{35\pi}{96}$

36. $\cos \frac{7\pi}{24}$

Evaluate the indicated functions under the given conditions.

37. $\sin 2x$; $\cos x = \frac{1}{6}$, $0 < x < \frac{\pi}{2}$

38. $\cos 2x$; $\cos x = \frac{3}{4}$, $\frac{3\pi}{2} < x < 2\pi$

39. $\cos x$; $\cos 2x = -0.4$, $\frac{\pi}{2} < x < \frac{3\pi}{4}$

40. $\sin x$; $\cos 2x = 0.28$, $0 < x < \frac{\pi}{2}$

41. $\sin x$, $\cos x$, and $\tan x$; $\tan 2x = -2, \frac{\pi}{2} < x < \pi$

42. $\sin x$, $\cos x$, and $\tan x$; $\csc 2x = -5, \frac{3\pi}{4} < x < \pi$

Evaluate each expression exactly.

43. $\cos(\sin^{-1} \frac{1}{2} - \sin^{-1} \frac{-3}{10})$

44. $\sin(\cos^{-1} \frac{1}{5} + \tan^{-1} 4)$

45. $\sin(\sin^{-1} -\frac{5}{8} - \cos^{-1} - \frac{2}{3})$

46. $\tan(\cos^{-1} \frac{3}{4} + \tan^{-1} - 3)$

47. $\sin(2 \sin^{-1} \frac{1}{5})$

48. $\sin(2 \cos^{-1} - \frac{1}{2})$

49. $\cos(2 \sin^{-1} - \frac{9}{10})$

50. $\cos(2 \cos^{-1} \frac{3}{8})$

Determine which of the following equations are identities and which are not identities. An equation is not an identity if there is at least one element in its replacement set for which the left and right members of the equation are not equal.

51. $\tan x = \dfrac{1 - \cos 2x}{\sin x}$

52. $\sec 2x = \dfrac{\csc^2 x}{\csc^2 x - 1}$

53. $\dfrac{1 - \sin 2x}{\cos 2x} = \dfrac{1 - \tan x}{1 + \tan x}$

54. $\dfrac{(1 + \tan x)^2 - 2}{\sec^2 x} = \sin 2x - \cos 2x$

Express in terms of x.

55. $\cos(2 \sin^{-1} x)$

56. $\cos(2 \cos^{-1} x)$

57. $\sin(\frac{1}{2} \cos^{-1} x)$

58. $\sin(\frac{1}{2} \sin^{-1} x)$

59. Show that $\sin^{-1} x + \cos^{-1} x = \pi/2$ if $-1 \le x \le 1$.

60. Express $\sin^6 x$ as a sum of functions of the form $\cos kx$ where $k \in N$.

61. Let $z = \tan \frac{x}{2}$. Show that $\cos x = \dfrac{1 - z^2}{1 + z^2}$ and that $\sin x = \dfrac{2z}{1 + z^2}$. (Hint: For $\cos x$, start with $\tan^2 \frac{x}{2} + 1 = \sec^2 \frac{x}{2}$. For $\sin x$, start with $\sin x = 2 \sin \frac{x}{2} \cos \frac{x}{2} = 2 \tan \frac{x}{2} \cos^2 \frac{x}{2}$.)

8.3 Trigonometric Equations

Equations whose members include trigonometric functions can often be solved with algebraic techniques. To illustrate this fact we obtain a solution of

$$\sin x + 3 \sin x \cos x = 0, \quad 0 \le x < 2\pi$$

after factoring $\sin x$ from the two terms in the left member of the equation.

$$\sin x(1 + 3 \cos x) = 0$$
$$\sin x = 0 \quad \text{or} \quad 1 + 3 \cos x = 0$$
$$\cos x = -\tfrac{1}{3}$$

The values of x in $[0,2\pi)$, for which $\sin x$ is zero, are 0 and π. If we don't remember that $\sin 0 = 0$, we can use a calculator to find x from the equation $x = \sin^{-1} 0$.

The calculator gives only the zero root. But it might sufficiently jog the memory to enable one to remember that if $x = k \cdot \pi$, $k \in J$, then $\sin x = 0$.

The values of x in $[0,2\pi)$, for which $\cos x = -\frac{1}{3}$, must be the measures of second and third quadrant angles. A calculator gives us the second quadrant root from the relationship $x = \cos^{-1}(-\frac{1}{3})$.

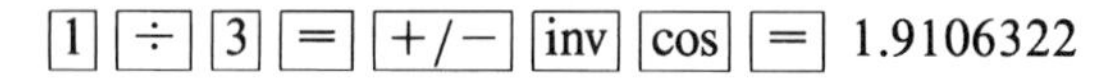

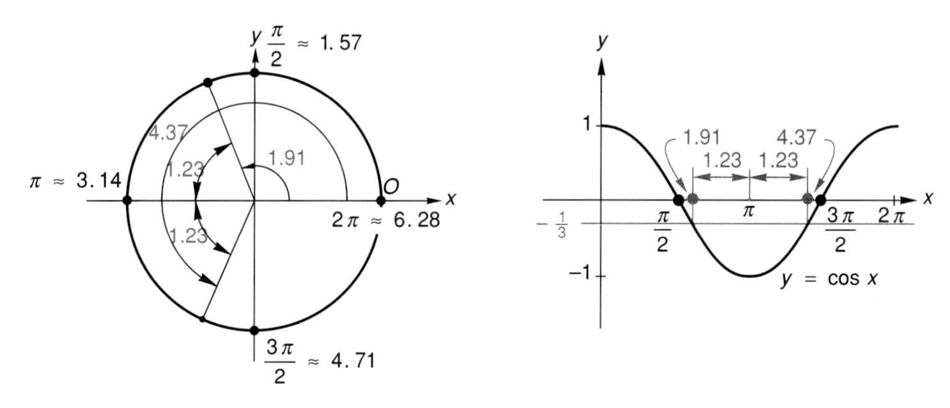

Figure 8.5

The third quadrant root is found as follows. We know that

$$\cos(1.91) = -\cos(\pi - 1.91) = -\cos(1.23) \qquad \text{(figure 8.5)}.$$

As a result, the terminal side of the third quadrant angle must make an angle of 1.23 radians with the negative x-axis. Thus,

$$\cos(1.91) = \cos(\pi + 1.23) = \cos(4.37),$$

and the solution set of

$$\sin x + 3 \sin x \cos x = 0, \quad 0 \le x < 2\pi,$$

is $\{0, 1.91, \pi, 4.37\}$.

Example 1 Solve $\sin 2x + \cos x = 0$, $0 \le x < 2\pi$.

Solution Substitute $2 \sin x \cos x$ for $\sin 2x$ and factor.

$$2 \sin x \cos x + \cos x = 0$$
$$\cos x(2 \sin x + 1) = 0$$
$$\cos x = 0 \quad \text{or} \quad \sin x = -\frac{1}{2}$$
$$x = \frac{\pi}{2}, \frac{3\pi}{2} \quad \text{or} \quad x = \frac{7\pi}{6}, \frac{11\pi}{6}$$

The solution set is $\{\pi/2, 7\pi/6, 3\pi/2, 11\pi/6\}$. ■

An equation that contains more than one trigonometric function can often be solved by rewriting it so that it contains only one trig function. Factoring helps us isolate cosine and sine in example 1. We end up solving two equations, each containing a single trig function.

There is nothing about the equation

$$2 \tan x + \sec x = 2, \quad 0 \leq x < 2\pi$$

that would suggest factoring as a key to its solution. Rather we might be attracted to the possibility of using the identity

$$1 + \tan^2 x = \sec^2 x$$

to change the equation into one that contains only the tangent function. All we have to do is some squaring.

$$\begin{aligned} (2 \tan x - 2)^2 &= (-\sec x)^2 \\ 4 \tan^2 x - 8 \tan x + 4 &= \sec^2 x \\ 4 \tan^2 x - 8 \tan x + 4 &= 1 + \tan^2 x \\ 3 \tan^2 x - 8 \tan x + 3 &= 0 \end{aligned}$$

The last equation has the quadratic form $3y^2 - 8y + 3 = 0$, where $y = \tan x$. Using the quadratic formula gives us

$$\begin{aligned} \tan x &= 8 \pm \frac{\sqrt{(-8)^2 - 4(3)(3)}}{2(3)} \\ &= 2.215 \text{ or } 0.451 \end{aligned}$$

$$\begin{aligned} x &= \tan^{-1} 2.215 \quad &\text{or} \quad x &= \tan^{-1} 0.451 \\ x &= 1.15 \quad &\text{or} \quad x &= 0.42. \end{aligned}$$

A calculator gives only one value each for $\tan^{-1} 2.215$ and $\tan^{-1} 0.451$. But, as figure 8.6 shows, there are two x-values in $[0,2\pi)$ for which $\tan x = 2.215$, and two x-values for which $\tan x = 0.451$. The apparent solution set is

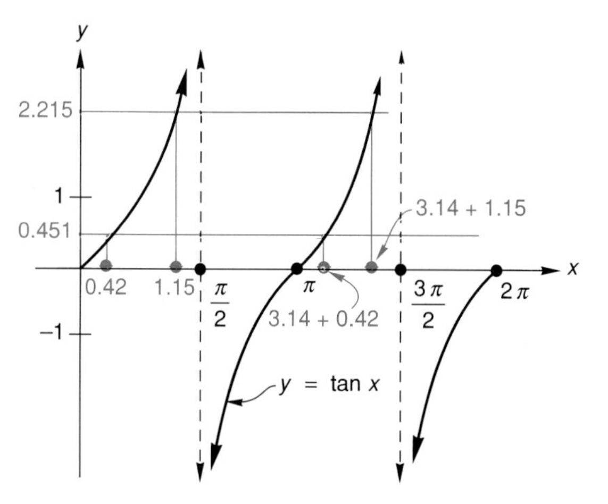

Figure 8.6

$\{0.42, 1.15, 3.56, 4.29\}$. However, squaring an equation can produce extraneous roots. It is therefore necessary to try each possible root in the original equation. Only 4.29 and 0.42 satisfy the original equation. Thus, the solution set of

$$2 \tan x + \sec x = 2, \quad 0 \leq x < 2\pi,$$

is $\{0.42, 4.29\}$.

Example 2 Solve $2 \sin^2 x + \cos^2 x = 1, 0 \leq x < 2\pi$.

Solution Substituting $1 - \cos^2 x$ for $\sin^2 x$ produces an equation containing only cosine functions.

$$\begin{aligned} 2(1 - \cos^2 x) + \cos^2 x + \cos x &= 1 \\ -\cos^2 x + \cos x + 1 &= 0 \\ \cos x &= \frac{-1 \pm \sqrt{1 - 4(-1)(1)}}{-2} \\ \cos x &= -0.618, 1.618 \end{aligned}$$

We must reject 1.618 as a possible value for $\cos x$, as the range of the cosine function is $[-1,1]$.

$$\begin{aligned} x &= \cos^{-1} -0.618, \quad 0 \leq x < 2\pi, \\ x &= 2.24 \end{aligned}$$

From figure 8.7 we see that the solution set is $\{2.24, 4.04\}$.

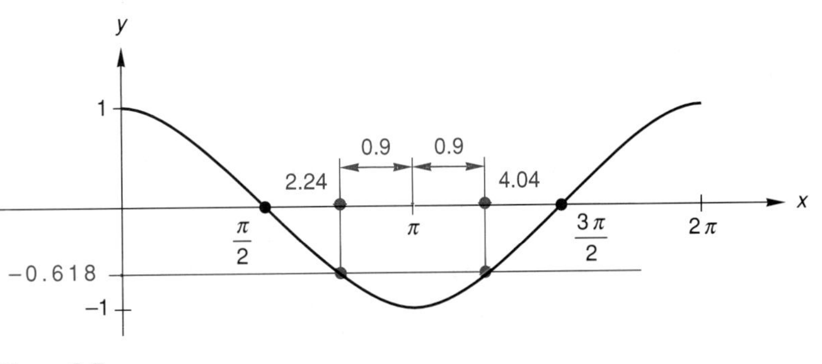

Figure 8.7

■

Example 3 Solve $\cos 4x + 3\cos 2x = -2, 0 \le x < 2\pi$.

Solution $\cos 4x = \cos 2(2x) = 2\cos^2 2x - 1$. Substituting $2\cos^2 2x - 1$ for $\cos 4x$ in the given equation yields

$$\begin{aligned} 2\cos^2 2x - 1 + 3\cos 2x &= -2 \\ 2\cos^2 2x + 3\cos 2x + 1 &= 0 \\ (2\cos 2x + 1)(\cos 2x + 1) &= 0 \\ \cos 2x = -\tfrac{1}{2} \quad \text{or} \quad \cos 2x &= -1. \end{aligned}$$

Recall that for $k \in J$, $\cos(2\pi/3 + k \cdot 2\pi) = -\frac{1}{2}$, $\cos(4\pi/3 + k \cdot 2\pi) = -\frac{1}{2}$, and $\cos(\pi + k \cdot 2\pi) = -1$. Therefore, the solutions of the equation are those x in $[0,2\pi)$ that satisfy each of the equations

$$2x = \frac{2\pi}{3} + k \cdot 2\pi,$$

$$2x = \frac{4\pi}{3} + k \cdot 2\pi,$$

and $$2x = \pi + k \cdot 2\pi.$$

The solution set is $\{\pi/3, \pi/2, 2\pi/3, 4\pi/3, 3\pi/2, 5\pi/3\}$. ■

Equations with Inverse Trigonometric Functions

Let us solve

$$2\sin^{-1} x = \sin^{-1} 2x.$$

Because the domain of the inverse sine function is $[-1,1]$, $2x$ must be a number in the interval $[-1,1]$. Thus, the replacement set for x is $[-\frac{1}{2}, \frac{1}{2}]$. Let $\alpha = \sin^{-1} x$ and $\beta = \sin^{-1} 2x$ (see figure 8.8). Taking the sine of both sides of the given equation yields

$$\begin{aligned} \sin(2\alpha) &= \sin \beta \\ 2\sin\alpha\cos\alpha &= \sin\beta \\ 2\sin(\sin^{-1} x)\cos(\sin^{-1} x) &= \sin(\sin^{-1} 2x) \\ 2x\sqrt{1 - x^2} &= 2x \\ 2x(\sqrt{1 - x^2} - 1) &= 0 \\ 2x = 0 \quad \text{or} \quad \sqrt{1 - x^2} - 1 &= 0 \\ x = 0 \quad \text{or} \quad 1 - x^2 &= 1 \\ x &= 0. \end{aligned}$$

Zero is in the replacement set for x. Hence, the solution set of the equation is $\{0\}$.

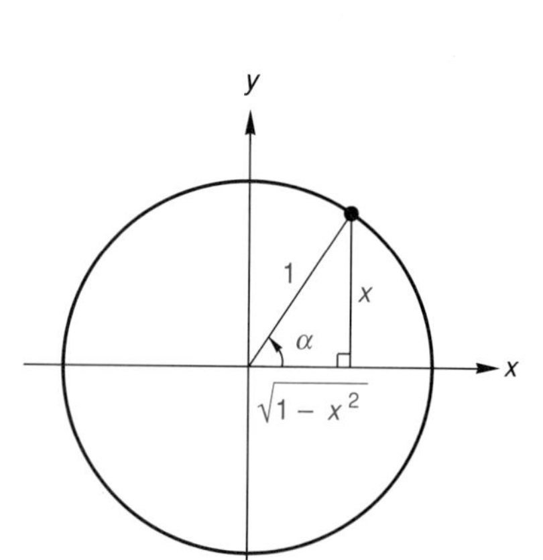

Figure 8.8

Example 4 Solve $2 \sin^{-1} x + \cos^{-1} x = \pi$.

Solution The replacement set for x is $[-1,1]$. Let $\alpha = \sin^{-1} x$ and $\beta = \cos^{-1} x$. Taking the sine of both sides of the equation gives us

$$\sin(2\alpha + \beta) = \sin \pi.$$

Expanding the left side of this equation with the sine sum formula produces

$$\sin 2\alpha \cos \beta + \cos 2\alpha \sin \beta = 0.$$

If we replace $\sin 2\alpha$ with $2 \sin \alpha \cos \alpha$, and $\cos 2\alpha$ with $1 - 2 \sin^2 \alpha$, the last equation can be rewritten as an equation without double angles.

$$2 \sin \alpha \cos \alpha \cos \beta + (1 - 2 \sin^2 \alpha)\sin \beta = 0$$

We solve for x by substituting $\sin^{-1} x$ for α and $\cos^{-1} x$ for β.

$$\begin{aligned} 2 \sin(\sin^{-1}x)\cos(\sin^{-1} x)\cos(\cos^{-1}x) + [1 - 2(\sin^2(\sin^{-1}x))]\sin(\cos^{-1}x) &= 0 \\ (2x)(\sqrt{1 - x^2})(x) + (1 - 2x^2)(\sqrt{1 - x^2}) &= 0 \\ 2x^2\sqrt{1 - x^2} - 2x^2\sqrt{1 - x^2} + \sqrt{1 - x^2} &= 0 \\ \sqrt{1 - x^2} &= 0 \\ x &= \pm 1 \end{aligned}$$

You can refer to figure 8.8 to see that $\cos(\sin^{-1} x) = \sqrt{1 - x^2}$. A similar figure can be drawn to show that $\sin(\cos^{-1}x) = \sqrt{1 - x^2}$. It is a good idea to check the apparent solutions of equations containing inverse trig functions because of the limited range of these functions. Hence, we check 1 and -1 in the original equation.

$$2 \sin^{-1} 1 + \cos^{-1} 1 = 2\left(\frac{\pi}{2}\right) + 0 = \pi$$

$$2 \sin^{-1} - 1 + \cos^{-1} - 1 = 2\left(-\frac{\pi}{2}\right) + \pi = 0 \neq \pi$$

Thus, -1 is an extraneous root, and the solution set of the equation is $\{1\}$. ■

An Application

Ms. J can get to her island home at C from A by cycling along the beach road at 15 mph to a point P. There she gets into a small boat, which takes her to C at the rate of 5 mph.

1. Express the time (T) that it takes Ms. J to travel from A to C as a function of θ.
2. What is the domain of the function?
3. Where is P located if it takes Ms. J one hour to go from A to C?
4. Graph the function in 1. From the graph, estimate the least amount of time that it should take Ms. J to travel from A to C. Estimate also the value of θ associated with the least time of travel.

5. With the aid of calculus, the value of θ that yields the least time of travel is gotten as the solution of the equation

$$0 = 0.1 \csc^2 \theta - 0.3 \csc \theta \cot \theta.$$

Compare the values of θ obtained in parts 4 and 5.

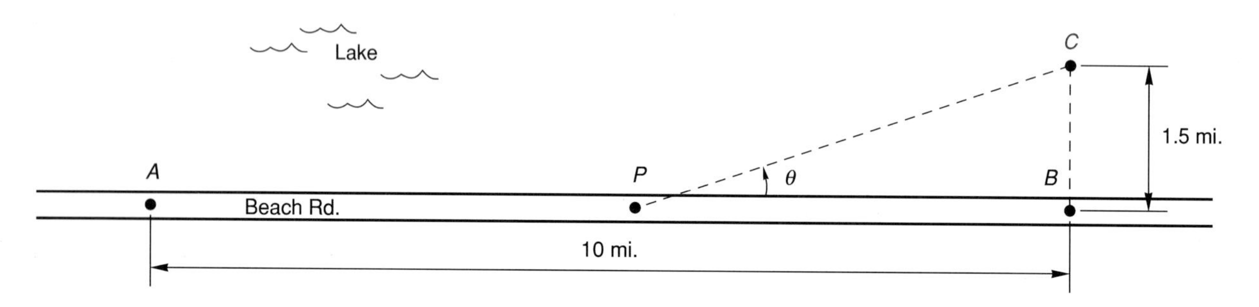

1. The distance from P to B is $1.5 \cot \theta$ miles. Therefore, the distance from A to P is $10 - 1.5 \cot \theta$ miles. The distance from P to C is $1.5 \csc \theta$ miles. Since time is distance divided by rate, we have

$$T = \frac{10 - 1.5 \cot \theta}{15} + \frac{1.5 \csc \theta}{5}.$$

2. The smallest angle that θ can be is angle CAB, when P is at A. The measure of that angle is $\tan^{-1} 0.15$. θ can be 90° when P is at B. The domain of the function is $\tan^{-1} 0.15 \leq \theta \leq \pi/2$.

3.
$$\begin{aligned}
1 &= \frac{10 - 1.5 \cot \theta}{15} + \frac{1.5 \csc \theta}{5} \\
15 &= 10 - 1.5 \cot \theta + 4.5 \csc \theta \\
5 &= 4.5 \csc \theta - 1.5 \cot \theta \\
0 &= 0.9 \csc \theta - 0.3 \cot \theta - 1 \\
0 &= \frac{9}{\sin \theta} - \frac{3 \cos \theta}{\sin \theta} - 10 \\
0 &= \frac{9 - 3 \cos \theta - 10 \sin \theta}{\sin \theta} \\
0 &= 9 - 3\sqrt{1 - \sin^2 \theta} - 10 \sin \theta \\
3\sqrt{1 - \sin^2 \theta} &= 9 - 10 \sin \theta \\
9 - 9 \sin^2 \theta &= 81 - 180 \sin \theta + 100 \sin^2 \theta \\
0 &= 109 \sin^2 \theta - 180 \sin \theta + 72 \\
\sin \theta &= \frac{180 \pm \sqrt{(-180)^2 - 4(109)(72)}}{218} \\
\sin \theta &= 0.9713 \quad \text{or} \quad 0.6800 \\
\theta &= \sin^{-1} 0.9713 \quad \text{or} \quad \theta = \sin^{-1} 0.68
\end{aligned}$$

4.

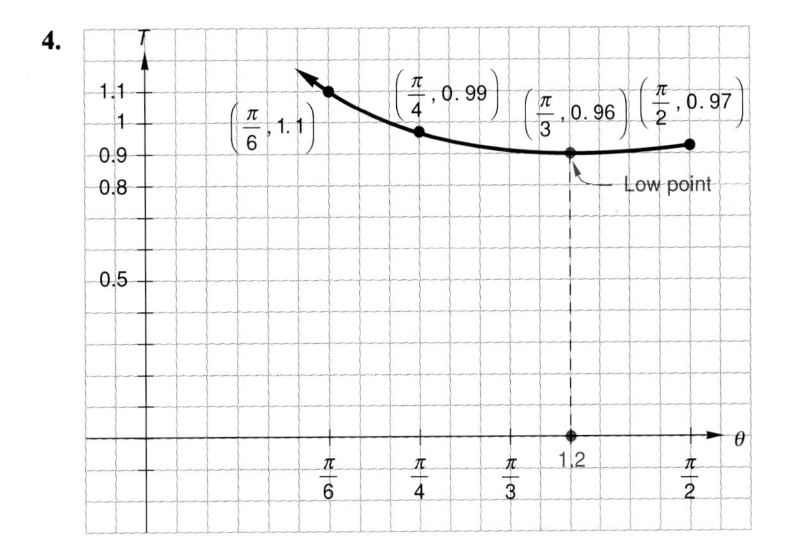

5.

$$0 = \csc^2 \theta - 3 \csc \theta \cot \theta$$
$$0 = \csc \theta(\csc \theta - 3 \cot \theta)$$

$\csc \theta = 0$ or $\csc \theta - 3 \cot \theta = 0$

no solution (Why?) or $\dfrac{1}{\sin \theta} - \dfrac{3 \cos \theta}{\sin \theta} = 0$

$$\frac{1 - 3 \cos \theta}{\sin \theta} = 0$$
$$1 - 3 \cos \theta = 0$$
$$\cos \theta = \frac{1}{3}$$
$$\theta = \cos^{-1} \frac{1}{3}$$
$$\theta = 1.23 \text{ radians}$$

Exercise Set 8.3

(A)

Solve for x, $0 \leq x < 2\pi$.

1. $2 \sin x = \sqrt{3}$

2. $\cos^2 x = 1$

3. $\sin 3x = \frac{1}{2}$

4. $\cos 4x = 1$

5. $\sqrt{3} \cos x = \sin x$

6. $\sin 2x + 1 = 0$

7. $(2 \sin x + 1)(\cos x + 3) = 0$

8. $(2 \cos x - \sqrt{3})(\cos x - 2) = 0$

9. $2 \cos^2 x + \cos x \sin x = 0$

10. $\sin 2x - 3 \sin x = 0$

11. $2 \cos^2 x - \cos x - 1 = 0$

12. $2 \sin^2 x + \sin x = 1$

13. $3 \sin^2 x - \cos x = 2$

14. $2 \cos^2 x + 7 \sin x - 5 = 0$

15. $\tan^2 x + \sec x - 5 = 0$

16. $\csc^2 x - 3 \cot x = 4$
17. $\cos 2x - 5 \cos^2 x + 2 = 0$
18. $6 \sin^2 x - 2 \cos 2x + 1 = 0$
19. $\sin x + \cos x = 1$
20. $3 \tan x - 2 \sec x = 2$
21. $6 \cos x + 3 = \sec x$
22. $2 \sin x - \cot x \cos x = 1$
23. $\sin 4x - 3 \cos 2x = 0$
24. $\cos 4x - \cos 2x = 2$
25. $15 \sin^2 2x - 25 \sin 2x + 10 = 0$
26. $8 \cos^2 3x + 2 \cos 3x - 3 = 0$
27. $\sec x - 2 \cos x - 1 = 0$
28. $2 \sin x + \csc x - 3 = 0$

(B)

Solve for x.

29. $2 \cos^{-1} x = \cos^{-1} 2x$
30. $\tan^{-1} 2x - 2 \tan^{-1} x = 0$
31. $\cos^{-1} x + \cos^{-1} 2x = \frac{\pi}{2}$
32. $\sin^{-1} x + \cos^{-1} x = \frac{\pi}{2}$
33. $\cos^{-1}(2x - 3) = \sin^{-1}(x - 1)$
34. $\sec^{-1} x + \csc^{-1} x = \frac{\pi}{2}$

Solve for x, $0 \leq x < 2\pi$.

35. $\cos 3x + 4 \sin^2 x - 2 = 0$
36. $2 \cos 6x + \sin 3x = -1$
37. $\cos^2(2x - 3) - 5 \sin(2x - 3) + 2 = 0$
38. $\sec 3x + 3 \tan 3x = 1$
39. $\sin 5x + \sin 3x = \frac{1}{2} \sin 4x$. (Hint: Let $u + v = 5x$ and $u - v = 3x$. Then use the identity in exercise 20 of Exercise Set 8.1 to replace $\sin 5x + \sin 3x$ with $2 \sin 4x \cos x$.)
40. $\cos 7x - \cos x + 2 \sin 3x = 0$. (Hint: Let $u + v = 7x$ and $u - v = x$. Then use the identity in exercise 21 of Exercise Set 8.1 to replace $\cos 7x - \cos x$ with $2 \sin 4x \sin 3x$.)
41. Show that $\sin^{-1} x + \tan^{-1} x = 2\pi/3$ has a root in the interval (0.9,1). Convert the equation to a fourth degree polynomial equation. Start with $\sin^{-1} x = 2\pi/3 - \tan^{-1} x$, and then take the sine of both sides of the equation. Use the techniques of chapter 4 to find the root.
42. Solve $2 \sin^{-1} x - \cos^{-1} x = \pi/3$. (Hint: See exercise 41.)
43. Solve $\cos x = 2 \cos \frac{x}{2}$, $0 \leq x < 2\pi$.
44. Solve $\sin x + \cos x = \frac{\sqrt{6}}{2}$, $0 \leq x \leq \pi$.
45. If $\tan x + \cot x = A$, show that $|A| \geq 2$. (Hint: Convert the left side to sines and cosines.)
46. Express the area of the rectangle inscribed in the semicircle of radius 9″ as a function of θ. What value of θ produces the rectangle of maximum area? What are the dimensions of the maximum rectangle?

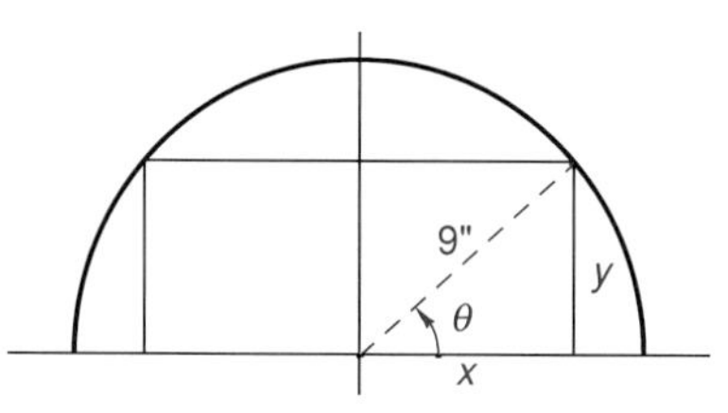

47. A rain gutter is to be made from a one-foot wide strip of sheet metal by bending, as shown in the figure, though $\theta°$. The base and sides of the gutter are to be of equal length. Express the cross sectional area of the gutter as a function of θ. Graph the function. From the graph, estimate the value of θ that produces the maximum cross sectional area. With calculus, the maximum area can be found by solving the equation $\cos^2 \theta - \sin^2 \theta + \cos \theta = 0$. Compare your solution of this equation with your graph estimate of θ.

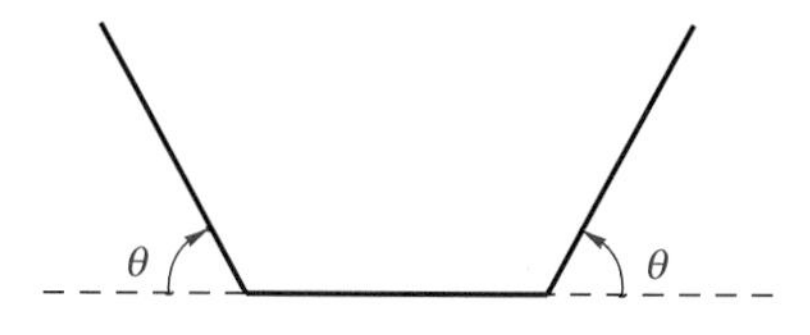

48. A metallic rod is part of a device that must be moved from a six-foot-wide hall into a ten-foot-wide hall. The rod must remain horizontal as it is being moved. Express the total length (L) of the rod as a function of θ. Is it possible to move the device and its attached rod into the ten-foot hallway if the rod is twenty feet long? Answer the question by graphing the length function and estimating, from the graph, the maximum length rod that can be moved into the ten-foot hallway. The maximum rod length can be found analytically, with the aid of calculus, by solving the equation

$$6 \sin^3 \theta - 10 \cos^3 \theta = 0, 0 < \theta < \frac{\pi}{2}.$$

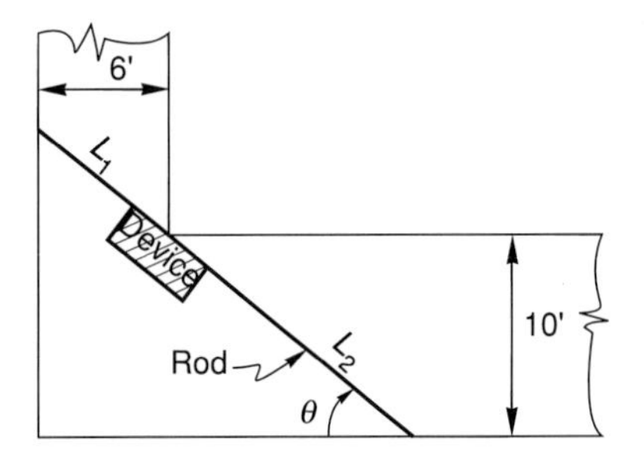

Substitute the solution of the equation into the length function to find the maximum length and compare this number with the maximum length obtained from the graph.

8.4 Numerical Techniques

If an equation resists solution by algebraic methods, we can often approximate its solution with numerical methods.

Let us find a root of $x - 2 \sin x = 0$ to three decimal places using the iteration technique introduced in section 4.5. The equation is written in the form $x = 2 \sin x$. A solution of this equation is an x-coordinate of one of the points of intersection between the graphs of $y = x$ and $y = 2 \sin x$ (figure 8.9). We choose the x-coordinate that appears to be somewhat greater than $\pi/2$ and begin the iteration with $x_0 = 2.0000$.

$$\begin{aligned}
x_1 &= 2 \sin 2.0000 = 1.8186 \\
x_2 &= 2 \sin 1.8186 = 1.9389 \\
x_3 &= 2 \sin 1.9389 = 1.8660 \\
x_4 &= 2 \sin 1.8660 = 1.9134 \\
x_5 &= 2 \sin 1.9134 = 1.8837 \\
x_6 &= 2 \sin 1.8837 = 1.9029 \\
x_7 &= 2 \sin 1.9029 = 1.8907 \\
x_8 &= 2 \sin 1.8907 = 1.8985 \\
x_9 &= 2 \sin 1.8985 = 1.8935 \\
x_{10} &= 2 \sin 1.8935 = 1.8967 \\
x_{11} &= 2 \sin 1.8967 = 1.8947 \\
x_{12} &= 2 \sin 1.8947 = 1.8960 \\
x_{13} &= 2 \sin 1.8960 = 1.8952 \\
x_{14} &= 2 \sin 1.8952 = 1.8957 \\
x_{15} &= 2 \sin 1.8957 = 1.8954
\end{aligned}$$

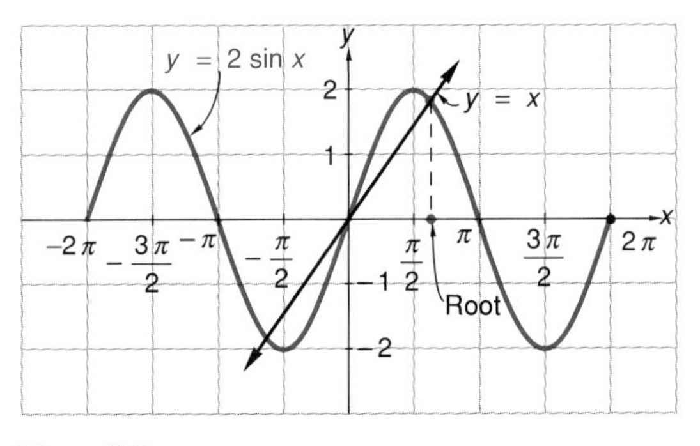

Figure 8.9

Since $|x_{15} - x_{14}| = |1.8954 - 1.8957| = 0.0003 < 0.0005$, then 1.895 is a root correct to three decimal places.

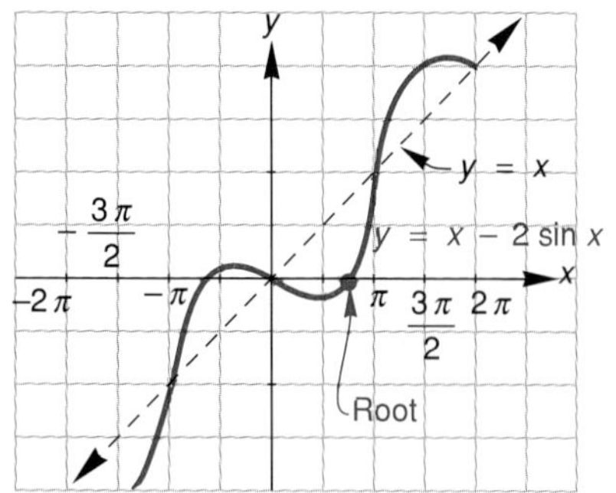

Figure 8.10

Linear interpolation is another numerical technique that can be used to find a root of $x - 2 \sin x = 0$ to any desired precision. The method attempts to locate the x-intercept of a graph of $y = x - 2 \sin x$ ever more precisely through a sequence of simple calculations called interpolations.

A graph of $y = x - 2 \sin x$ (figure 8.10) can be constructed either by point plotting or by subtraction of corresponding ordinates of the graphs in figure 8.9. The graph has an x-intercept between $\pi/2$ and π. We attempt to narrow the location of the intercept between successive tenths. Since y is negative at $x = 1.8$ and positive at $x = 1.9$, an x-intercept must lie between 1.8 and 1.9.

x	1.7	1.8	1.9
y	−0.283	−0.148	0.007

Figure 8.11

We approximate the x-intercept of the curve with the x-intercept of the line between points P and Q (figure 8.11). Similar triangles PBQ and PAR help us find the x-coordinate of R. The calculation below yields a *linear interpolation* of the distance between point A and the root of the equation.

$$\overline{AR}/\overline{BQ} = \overline{PA}/\overline{PB}$$

$$\overline{AR} = \frac{\overline{BQ} \cdot \overline{PA}}{\overline{PB}}$$

$$\overline{AR} = \frac{(0 - (-0.148))(1.9 - 1.8)}{0.007 - (-0.148)}$$

$$\overline{AR} = \left(\frac{0.148}{0.155}\right)(0.1)$$

$$\overline{AR} = \left(\frac{148}{155}\right)(0.1) = 0.095$$

The last result is obtained directly from the table above as

$$\frac{148}{148 + 7} = 0.95,$$

where 0.95 is the ratio $\overline{AR}/\overline{BQ}$. Hence, the x-coordinate of R is

$$1.8 + (0.95)(0.1) = 1.895.$$

The interpolation process begins anew between x-values 1.895 and 1.896. Those values substituted into $y = x - 2 \sin x$ yield y-values −0.0008 and 0.0008, respectively.

x	1.895	1.896
y	−0.0008	0.0008

The second interpolation is gotten from the ratio $\frac{8}{8+8} = 0.5$. It is

$$1.895 + (0.5)(0.001) = 1.8955.$$

Example 1 Estimate a root of $\sin^{-1} x + \tan^{-1} x = 2\pi/3$ correct to six decimal places, using

a. iteration and

b. interpolation.

Solution We first try to capture a root between successive tenths by evaluating $y = \sin^{-1} x + \tan^{-1} x - 2\pi/3$.

x	0.8	0.9	1.0
y	−0.492	−0.241	0.262

a. A root exists between 0.9 and 1.0. Write the given equation so that x is isolated on one side.

$$\sin^{-1} x + \tan^{-1} x = \frac{2\pi}{3}$$

$$\sin^{-1} x = \frac{2\pi}{3} - \tan^{-1} x$$

$$\sin(\sin^{-1} x) = \sin\left(\frac{2\pi}{3} - \tan^{-1} x\right)$$

$$x = \sin\left(\frac{2\pi}{3} - \tan^{-1} x\right)$$

Let $x_0 = 0.95$.

$$x_1 = \sin\left(\frac{2\pi}{3} - \tan^{-1} 0.95\right) = 0.9722427$$

$$x_2 = \sin\left(\frac{2\pi}{3} - \tan^{-1} 0.9722427\right) = 0.9694724$$

$$x_3 = \sin\left(\frac{2\pi}{3} - \tan^{-1} 0.9694724\right) = 0.9698211$$

$$x_4 = \sin\left(\frac{2\pi}{3} - \tan^{-1} 0.9698211\right) = 0.9697773$$

$$x_5 = \sin\left(\frac{2\pi}{3} - \tan^{-1} 0.9697773\right) = 0.9697828$$

$$x_6 = \sin\left(\frac{2\pi}{3} - \tan^{-1} 0.9697828\right) = 0.9697821$$
$$x_7 = \sin\left(\frac{2\pi}{3} - \tan^{-1} 0.9697821\right) = 0.9697821$$

A root correct to six decimal places is 0.969782.

b. Given the y-values obtained above for $x = 0.9$ and $x = 1.0$, a first interpolation, x_1, is calculated as follows:

$$\frac{241}{241 + 262} \approx 0.5$$
$$x_1 = 0.9 + (0.5)(0.1) = 0.95.$$

New y-values are found from $y = \sin^{-1} x + \tan^{-1} x - 2\pi/3$ for x-values 0.95, 0.96, and 0.97.

x	0.95	0.96	0.97
y	−0.0814	−0.0393	0.0010

A second interpolation, x_2, can now be calculated.

$$\frac{393}{393 + 10} \approx 0.975$$
$$x_2 = 0.96 + (0.975)(0.01) = 0.9697$$

The process continues:

x	0.9697	0.9698
y	−0.000379	0.000082

$$\frac{379}{379 + 82} \approx 0.822$$
$$x_3 = 0.9697 + (0.822)(0.0001) = 0.969782.$$

x	0.969782	0.969783
y	−0.00000084	0.00000377

The last calculations for y indicate that the root is closer to 0.969782 than 0.969783. Linear interpolation gives us 0.969782 as a root that is correct to six decimal places. ■

Exercise Set 8.4

Estimate a root, other than 0, to three decimal places. Use (a) iteration, and (b) interpolation.

1. $x^{-1} - \sin x = 0$
2. $\sin x = x^3$
3. $2 \cos x - x^2 = 0$
4. $x \tan x = 0.5$
5. $\cos 2x = \dfrac{x^{3/2}}{6}$
6. $2x^2 + 2 \sin^2 x = 5$
7. $2 + 7\sqrt{2}x - 14 \sin x = 0$
8. $x^2 - 3x - 4 \sin^2 x = 0$
9. $\tan^{-1} x + \cos^{-1} x = 1.75$
10. $2 \sin^{-1} x - \cos^{-1} x = \dfrac{\pi}{3}$
11. Solve $30 = 6 \sec \theta + 10 \csc \theta$ by interpolation. Find a root correct to two decimal places. This equation is associated with exercise 48 in section 8.3. That the equation has a solution implies that a thirty-foot rod could not pass from the six-foot hallway into the ten-foot hallway.

Chapter 8 Review

Major Points for Review

Section 8.1 When an equation is an identity

How to verify that an equation is an identity

Section 8.2 The Pythagorean Identities

The cosine sum and difference formulas

The sine sum and difference formulas

The cosine double angle formulas

The sine double angle formula

The half angle formulas

Section 8.3 Algebraic techniques for solving equations with trigonometric functions

Section 8.4 Using iteration to solve trigonometric equations

Using interpolation to solve trigonometric equations

Review Exercises

Verify that each equation is an identity.

1. $\sec x \cot x = \csc x$
2. $\sin x + \cot x \cos x = \csc x$
3. $\dfrac{\tan x}{1 + \tan^2 x} = \sin x \cos x$
4. $\sin^4 x - \cos^4 x = \sin^2 x - \cos^2 x$
5. $\sin 2x \cos 5x + \cos 2x \sin 5x = \sin 7x$
6. $\dfrac{\sin(x + y)}{\cos x \cos y} = \tan x + \tan y$
7. $2 \cos x \cos y = \cos(x + y) + \cos(x - y)$
8. $\sin\left(x + \dfrac{3\pi}{2}\right) = -\cos x$
9. $\dfrac{\sin 2x}{1 + \cos 2x} = \tan x$
10. $\cot 2x + \csc 2x = \cot x$
11. $(\tan x - \cot x)\tan 2x = -2$
12. $\dfrac{\sin x - \sin y}{\cos x + \cos y} = \tan\left(\dfrac{x - y}{2}\right)$
13. $\dfrac{\sin 3x + \sin x}{\cos 3x + \cos x} = \tan 2x$
14. $\dfrac{1 - \cos 2x}{1 + \cos 2x} = \tan^2 x$

15. Find $\sin(x + y)$ if $\sin x = \frac{3}{4}$, $\pi/2 < x < \pi$ and $\cos y = \frac{2}{3}$, $0 < y < \pi/2$. Evaluate $x + y$ correct to three decimal places.
16. Find $\cos(x + y)$ if $\cos x = -\frac{4}{5}$, $\pi < x < 3\pi/2$ and $\cos y = \frac{1}{5}$, $0 < y < \pi/2$. Evaluate $x + y$ correct to three decimal places.
17. Evaluate $\sin 2x$ if $\cos x = \frac{1}{4}$, $0 < x < \pi/2$.
18. Evaluate $\cos 2x$ if $\cos x = -\frac{3}{5}$, $\pi/2 < x < \pi$.
19. Evaluate $\sin x$ if $\cos 2x = 0.35$, $0 < x < \pi/2$.
20. Evaluate $\sin x$, $\cos x$, and $\tan x$ if $\tan 2x = 1.8$, $\pi < x < 3\pi/2$.

Evaluate each expression exactly.

21. $\cos(\sin^{-1} \frac{4}{5} - \cos^{-1}(-\frac{5}{13}))$
22. $\sin(\cos^{-1} \frac{15}{17} + \cos^{-1}(-\frac{8}{17}))$
23. $\tan(2 \cos^{-1} \frac{1}{6})$
24. $\sin(2 \tan^{-1} \frac{12}{5})$

Solve for x, $0 \le x < 2\pi$.

25. $\tan 2x = \cot x$
26. $6 \sin^2 x - \cos x = 4$
27. $-3 + 2 \cos^2 x = 3 \sin x$
28. $\sin 3x = \cos 2x$
29. $\cos 4x + 3 \sin 2x + 4 = 0$
30. $\tan 2x - 2 \cos x = 0$
31. $\sin x - 3 \cos x = 1$
32. $\sec x + \tan x = 3$

Solve for x.

33. $\sin^{-1} x = \sin^{-1} 2x - \dfrac{\pi}{6}$
34. $\cos^{-1}(x + 4) = \sin^{-1}(2x - 1)$

Either verify that the equation is an identity or prove that it is not an identity.

35. $\sec(x + y) = \sec x + \sec y$
36. $\cos x + \cos 2x = \cos 3x$
37. $\csc 2x = \dfrac{1 + \tan^2 x}{2 \tan x}$
38. $\dfrac{1 + \sin x}{1 - \sin x} = (\sec x + \tan x)^2$
39. $\dfrac{\sin 3x}{\sin x} - \dfrac{\sin 4x}{\sin 2x} = 1$
40. $\cos 2x \cos 3x = \cos 6x$

Solve for x, $0 \le x < 2\pi$.

41. $\sin 6x = \sin 3x$
42. $\cos^2 x + \cos x = 1$
43. $4 \sin^2 \dfrac{x}{2} + \tan^2 \dfrac{x}{2} = 6$
44. $\sin x^2 \cos x - \cos x^2 \sin x = 0$

Find a root for each equation by (a) iteration and (b) interpolation. Obtain results correct to three decimal places.

45. $\cos x = \dfrac{x}{5}$ **46.** $\sin x - x^2 + 1 = 0$

47. Power is to be brought by cable from a power station at A to an offshore platform at P. The cable is laid underground from A to S and then under water from S to P.

a. Express the cost (C) of laying the cable from A to S to P as a function of θ. (See the figure.) Let k be the cost, in dollars per mile, of laying cable underground. Assume that it costs twice as much to lay cable under water as it does to lay it underground.

b. Where is S if the cost for laying the cable from A to S to P is $10k$.

c. Graph the cost function in a. From the graph, estimate the minimum cost for laying the cable from A to P. Estimate the position of S for minimum laying cost.

d. The value of θ that produces the minimum laying cost is gotten, with the aid of calculus, as the solution of the equation

$$\csc^2 \theta - 2 \csc \theta \cot \theta = 0.$$

Compare the solution of the equation with the value of θ obtained in c.

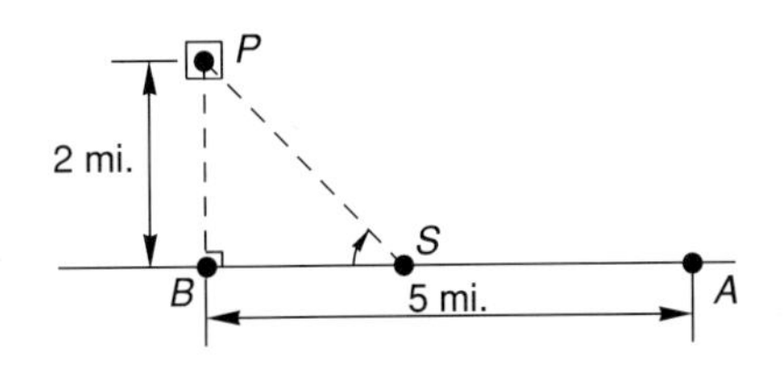

Chapter 8 Test 1

In Exercises 1–4, verify that each equation is an identity.

1. $\cos\left(2x - \dfrac{\pi}{2}\right) = \sin 2x$

2. $\dfrac{\sin 2x \cos x}{\cot^2 x} = 2 \sin^3 x$

3. $\dfrac{2 \tan x}{1 + \tan^2 x} = \tan 2x$

4. $\cot \dfrac{x}{2} = \dfrac{\sin x}{1 + \cos x}$

5. Find $\sin 2x$ if $\sin x = \frac{2}{3}$ and $\pi/2 < x < \pi$.

6. Evaluate $\cos(\cos^{-1}(\frac{4}{5}) + \sin^{-1}(\frac{5}{13}))$.

Solve each equation.

7. $\sin 2x - 3 \cos x = 0, 0 \leq x < 2\pi$

8. $2 \tan^2 x - \sec^2 x + 2 = 2 \tan x, 0 < x < \dfrac{\pi}{2}$

9. $\cos 3x = 0.6, 0 \leq x < 2\pi$

10. Either verify that the equation

$$\sin x + \sin 2x = \sin 3x$$

is an identity or prove that it is not an identity.

Chapter 8 Test 2

Verify that each equation is an identity.

1. $\dfrac{(\cos^2 x - \sin^2 x)^2}{\cos^4 x - \sin^4 x} = 2 \cos^2 x - 1$

2. $\cot(x + y) = \dfrac{\cot x \cot y - 1}{\cot x + \cot y}$

3. $\dfrac{\sin x + \sin 3x}{\cos x - \cos 3x} = \cot x$

4. $\dfrac{(\tan x + 1)^2}{2} = \dfrac{1 + \csc 2x}{\cot x}$

5. Find $\sin(x + y)$ if $\sin x = -\frac{3}{5}$, $3\pi/2 < x < 2\pi$, and $\cos y = \frac{12}{13}$, $3\pi/2 < y < 2\pi$. Evaluate $x + y$ correct to three decimal places.

6. Evaluate $\tan(2 \sin^{-1}(-\frac{15}{17}))$ exactly.

Solve each equation.

7. $2 \cos^4 x = 2 - 3 \sin^2 x, 0 \leq x < 2\pi$

8. $\sec x - \tan^2 x + 3 = 0, 0 < x < \pi$

9. $2 \sin^{-1} x = \cos^{-1} 0$

10. Either verify that the equation

$$\cot x = \frac{1 + \cos x}{\sin 2x}$$

is an identity or prove that it is not an identity.

Chapter 9

Applications of the Trigonometric Functions

The applications in this chapter range from the relationship between the speed of a bicycle and the rate at which the bike pedals are rotated, to land measuring problems in surveying, to the way radio signals are constructed so that they can be received by our radios. Some aspect of trigonometry is at the center of each application.

9.1 Angular and Linear Velocity

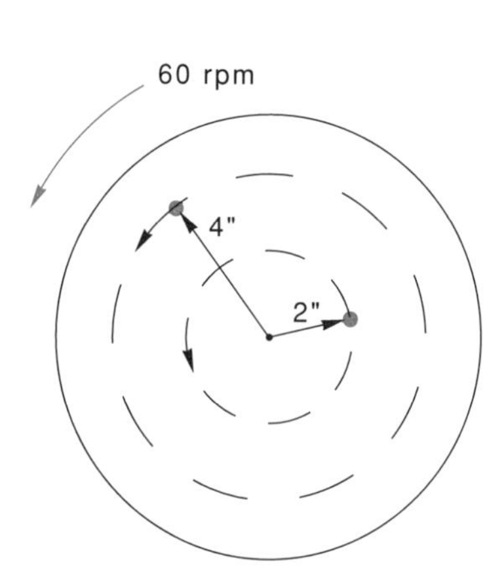

Figure 9.1

Two flies are sitting on a one-foot diameter potter's wheel that is rotating at the rate of 60 revolutions per minute (rpm). One fly is two inches from the center of the wheel. The other fly is four inches from the center (figure 9.1). Both flies rotate once around the center of the wheel every second. They each experience an **angular velocity** of *1 revolution per second* or *2π radians per second.* But they experience different **linear velocities.** The fly that is two inches from the center of the wheel travels over the circumference of a two-inch radius circle in one second. It experiences a linear velocity of 4π in./sec. The other fly has a linear velocity of 8π in./sec. The flies may not recognize it, but there is a relationship between their linear and angular velocities.

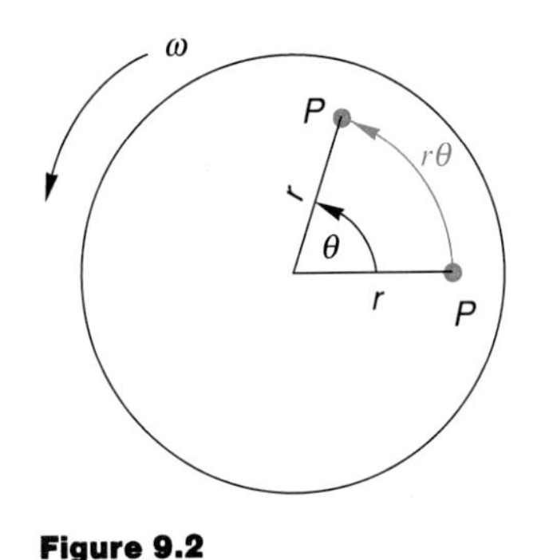

Figure 9.2

Let P be a point r units from the center of a wheel that is rotating uniformly. If P rotates through θ radians in t seconds (figure 9.2), its angular velocity, ω (omega), is

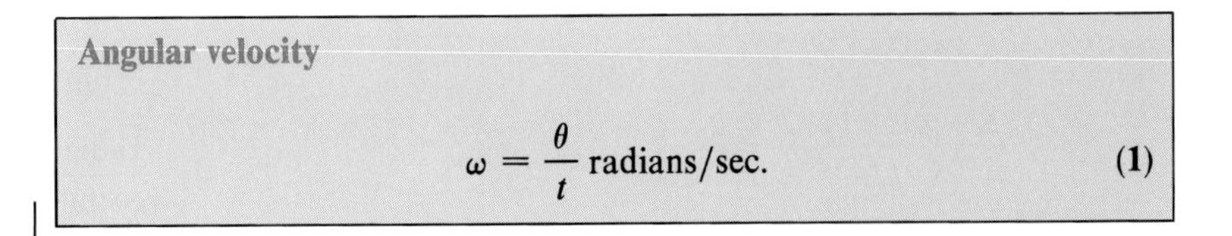

Angular velocity

$$\omega = \frac{\theta}{t} \text{ radians/sec.} \tag{1}$$

Since P travels through an arc length of $r \cdot \theta$ units in t seconds, it has a linear velocity

$$v = r\frac{\theta}{t}. \tag{2}$$

Combining (1) and (2) yields

The Relationship between v and ω

$$v = \left(\frac{\theta}{t}\right)r = \omega r. \tag{3}$$

Example 1 What linear velocity, in miles per hour (mph), does a person at the equator experience? Assume that the radius of the earth is 3,960 miles.

Solution The angular velocity of the earth is

$$\omega = 2\pi \frac{\text{radians}}{\text{day}} \cdot \frac{1 \text{ day}}{24 \text{ hours}} = \frac{\pi}{12} \frac{\text{radians}}{\text{hour}}.$$

The linear velocity of a person at the equator is

$$v = \frac{\pi}{12} \cdot 3{,}960 = 1{,}037 \text{ mph.}$$

■

Example 2 A belt around the rim of a 12″ diameter turntable drives it at $33\frac{1}{3}$ rpm. What is the linear velocity of the belt in ft./min.?

Solution

$$\begin{aligned}\omega &= 33\frac{1}{3}\text{ rpm}\\ &= 2\pi\frac{\text{radians}}{\text{revolution}}\cdot 33\frac{1}{3}\text{ rpm}\\ &= 200\frac{\pi}{3}\frac{\text{radians}}{\text{min.}}\\ v &= 200\frac{\pi}{3}\frac{\text{radians}}{\text{min.}}\cdot\frac{1}{2}\text{ ft.}\\ &= 100\frac{\pi}{3}\frac{\text{ft.}}{\text{min.}}\\ &= 104.72\frac{\text{ft.}}{\text{min.}}\end{aligned}$$

■

Example 3 A cyclist is traveling 20 mph. His bike has 26″ diameter wheels. What is the angular velocity of the wheels?

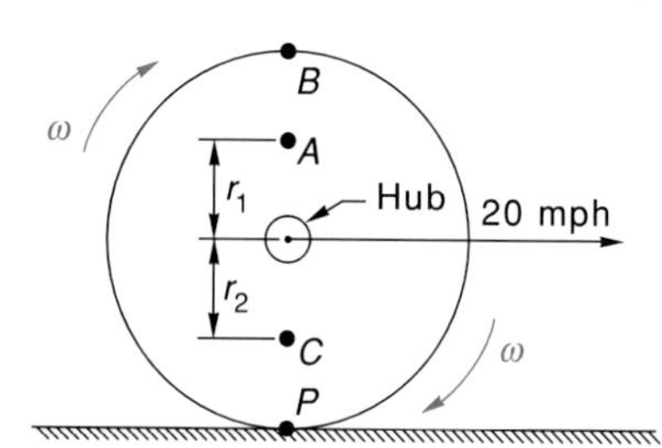

Figure 9.3

Solution Figure 9.3 shows one bike wheel. The wheelhub as well as all points on the wheel have a linear velocity of 20 mph or

$$20\frac{\text{mi.}}{\text{hr.}}\cdot 5{,}280\frac{\text{ft.}}{\text{mi.}}\cdot 12\frac{\text{in.}}{\text{ft.}}\cdot\frac{1\text{ hr.}}{60\text{ min.}}\cdot\frac{1\text{ min.}}{60\text{ sec.}}$$
$$= 352\frac{\text{in.}}{\text{sec.}}$$

However, all points on the wheel except the hub have an additional linear velocity due to the angular velocity (ω) of the wheel. A has the linear velocity $v_A = 352 + \omega r_1$ inches per second. B has the linear velocity $v_B = 352 + \omega\cdot 13$ inches per second, and C has the linear velocity $v_C = 352 - \omega r_2$ inches per second. The instant that P is on the ground, its linear velocity is 0. Thus,

$$v_P = 0 = 352 - \omega\cdot 13,$$

and we see that
$$\omega = \frac{352}{13} = 27.08\frac{\text{radians}}{\text{sec.}}$$

The angular velocity in rpm is

$$\omega = \frac{27.08 \dfrac{\text{radians}}{\text{sec.}}}{2\pi \dfrac{\text{radians}}{\text{revolution}}} \cdot 60 \frac{\text{sec.}}{\text{min.}} = 258.59 \text{ rpm.}$$

■

One may conclude from example 3 that the angular velocity of the wheels of a vehicle traveling with linear velocity v is given by $\omega = v/r$, where r is the radius of the wheels.

Example 4 How many rpm are made by the wheels of a truck that is traveling 60 mph? The truck wheels are 36 inches in diameter.

Solution

$$v = 60 \text{ mph} \cdot 5{,}280 \frac{\text{ft.}}{\text{mi.}} \cdot \frac{1 \text{ hr.}}{60 \text{ min.}} = 5{,}280 \frac{\text{ft.}}{\text{min.}}$$

$$\omega = \frac{5{,}280 \dfrac{\text{ft.}}{\text{min.}}}{1.5 \text{ ft.}}$$

$$= 3{,}520 \frac{\text{radians}}{\text{min.}}$$

$$= \frac{3{,}520 \dfrac{\text{radians}}{\text{min.}}}{2\pi \dfrac{\text{radians}}{\text{revolution}}}$$

$$= \frac{1{,}760}{\pi} \text{ rpm}$$

■

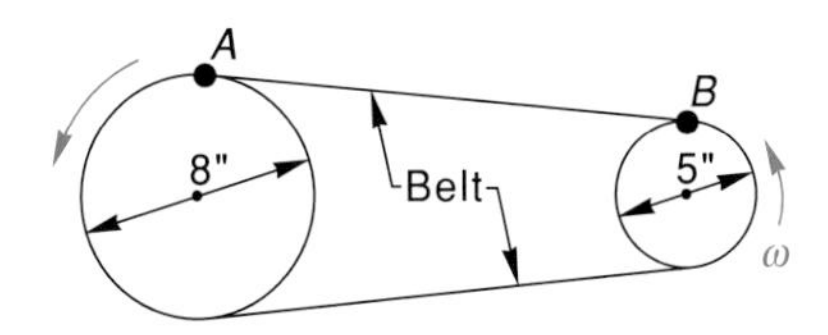

Figure 9.4

Example 5 An 8″ diameter pulley rotating at 300 rpm drives a 5″ diameter pulley (figure 9.4). What is the angular velocity, in rpm, of the smaller pulley?

Solution Every point on the belt moves with the same linear velocity. Therefore, the points of contact between the belt and the pulleys, points A and B, must have the same linear velocity.

large pulley:

$$v = \omega \cdot r$$

$$= 300 \cdot 4''$$

$$= 1{,}200 \frac{\text{in.}}{\text{min.}}$$

small pulley:

$$1{,}200\,\frac{\text{in.}}{\text{min.}} = \omega \cdot \frac{5}{2}\text{ inches}$$
$$\frac{2{,}400}{5} = \omega$$
$$480\text{ rpm} = \omega$$

Exercise Set 9.1

(A)

Convert each angular velocity to rpm.

1. 100π radians/sec.
2. $25{,}000\pi$ radians/hr.
3. 50,000 radians/hr.
4. 80 radians/sec.

5. The crankshaft of an engine rotates at 4,000 rpm. What is its angular velocity?
6. A carousel makes one complete revolution in 10 seconds. What is its angular velocity in radians/sec.?
7. If a wooden horse stands 9 feet from the center of the carousel in exercise 6, what is its linear velocity in ft./sec., and in mph?
8. A lawnmower engine is started by pulling a rope wound around a 4″ diameter drum. With what speed, in ft./sec., must the rope be pulled to give the drum an angular velocity of 600 rpm?
9. A belt-driven pulley turns through 300 rpm. The pulley has an 8″ diameter. What is the linear velocity of the belt (ft./sec.)?
10. One-inch stock is being turned in a lathe at 250 rpm. What is the linear speed of cutting (ft./sec.)?
11. A car is traveling 70 mph. Its wheels have a 22″ diameter. What is the angular velocity of the wheels in rpm?
12. A truck's speed is 100 km/hr. What is the angular velocity of its 88 cm diameter wheels in rpm?
13. The 8″ diameter front sprocket gear of a bicycle is being rotated at 80 rpm. The bicycle chain is attached to a 3″ diameter rear sprocket gear. What is the rpm of the rear gear?
14. If the rpm of the bicycle wheels in exercise 13 is the same as the rpm of the rear sprocket gear, what is the speed (mph) of the bicycle? The bike has 26″ diameter wheels.
15. A gear rotates at 10 rpm. It meshes with and drives another gear at 25 rpm. The distance between the gear centers is 35 cm. Determine the radii of the gears.

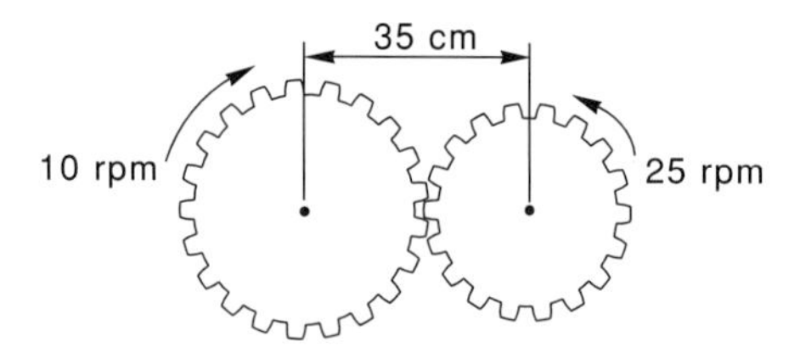

16. A 10″ diameter pulley, rotating at 400 rpm, drives a belt that drives a 6″ diameter pulley. What is the angular velocity, in rpm, of the smaller pulley?

(B)

17. A bullet is shot through a pair of rotating discs that are mounted 4 feet apart on an axle that is rotating at 2,000 rpm. The bullet is fired parallel to the axle. It produces holes in the discs that are displaced 24°. What is the speed (mph) of the bullet?

18. A cyclist is traveling 15 mph on a bicycle that has 27″ diameter wheels. What is the linear velocity (mph) of a reflector that is 8″ from the center of the wheel, when the reflector is directly above the wheel's center?

19. The 3″ crank in the figure has an angular velocity of 2,500 rpm. Find the average piston speed in ft./min. (Hint: How far does the piston travel in one revolution of the crank?)

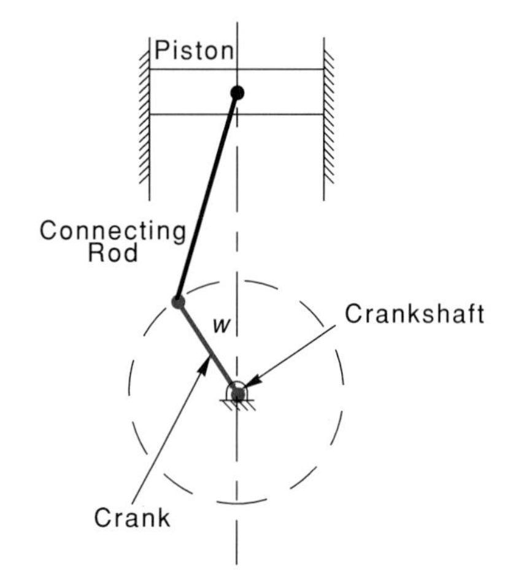

9.2 Right Triangle Trigonometry

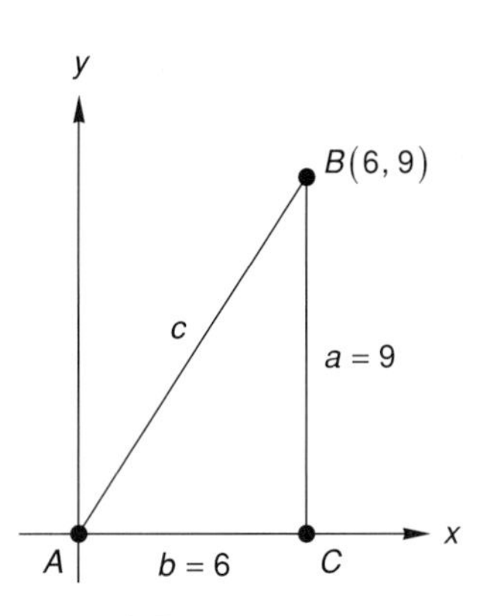

Figure 9.5

There is an enormous number of applied problems that are solved only after a side or an angle of a triangle is determined. Methods for finding lengths of the sides and measures of the angles of a triangle when only some of those six quantities are known are discussed in this section and the next. Degree measure is used, as it is the measure commonly used in engineering instrumentation.

We start our triangle computations with the right triangle in figure 9.5. Given side lengths a and b and angle C, we want to find c and angles A and B. Imagine a coordinate system through A so that side $\overline{AC}$ is contained in the x-axis. Because the coordinates of B are (6,9),

$$\tan A = \tfrac{9}{6}$$

$$A = \tan^{-1}\tfrac{9}{6} = 56.31°.$$

Also,

$$B = 180 - (90 + 56.31) = 33.69°,$$

and

$$c = \sqrt{9^2 + 6^2} = \sqrt{117}\,.$$

Imagining a coordinate system through A enabled us to use the definition of the tangent of an angle, $\tan A = y/x$, to find A. The definition can be paraphrased to give us the tangent of angle A without reference to a coordinate system.

$$\tan A = \frac{\text{measure of side opposite } A}{\text{measure of side adjacent to } A} = \frac{\text{opp}}{\text{adj}}$$

A similar paraphrasing for sine and cosine gives us the following function definitions associated with an *acute angle A* in a *right triangle.*

Definitions

$$\sin A = \frac{\text{opp}}{\text{hyp}}$$
$$\cos A = \frac{\text{adj}}{\text{hyp}}$$
$$\tan A = \frac{\text{opp}}{\text{adj}}$$

Can you construct the right triangle definitions for the reciprocal trigonometric functions?

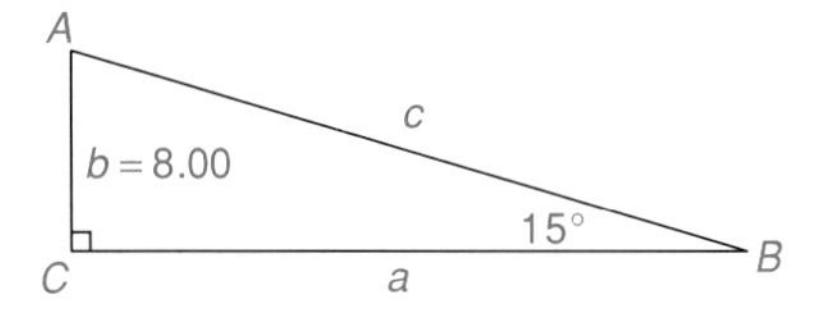

Example 1 Solve triangle ABC.

Solution To solve a triangle means to find its three sides and its three angles. Thus, we must find A, a, and c.

$$A = 180 - (90 + 15) = 75°$$
$$\sin 15° = \frac{\text{opp}}{\text{hyp}} = \frac{8.00}{c}$$
$$c = \frac{8.00}{\sin 15°} = 30.91$$
$$\tan 15° = \frac{\text{opp}}{\text{adj}} = \frac{8.00}{a}$$
$$a = \frac{8.00}{\tan 15°} = 29.86$$

■

The computations for a and c may be checked by using the measures to compute A, which is known to be 75°.

$$\sin A = \frac{\text{opp}}{\text{hyp}} = \frac{a}{c}$$
$$A = \sin^{-1}\frac{29.86}{30.91} = 75.02°$$

The difference in the actual and computed measures of A is due to the rounding off of the computed measures of a and c.

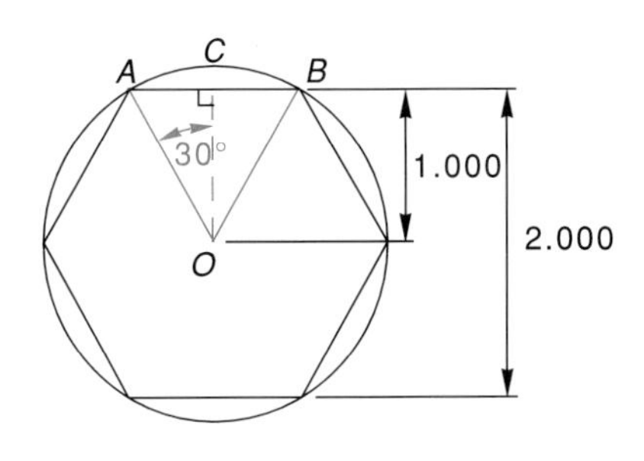

Example 2 A machinist must shape a hexagonal bar from circular bar stock. The distance between opposite sides of the hexagon must be 2.000 inches. What is the smallest diameter bar that can be used by the machinist?

Solution Central angle AOB intercepts one of the six equal sides of the hexagon. Its measure must therefore be 60°. Because triangle AOB is isosceles, altitude $\overline{OC}$ is an angle bisector of AOB. In triangle AOC,

$$\cos 30° = \frac{\text{adj}}{\text{hyp}} = \frac{1.000}{r}$$

$$r = \frac{1.000}{\cos 30°} = 1.1547 \text{ inches.}$$

The smallest diameter bar that can be used by the machinist is

$$2r = 2.309 \text{ inches.}$$

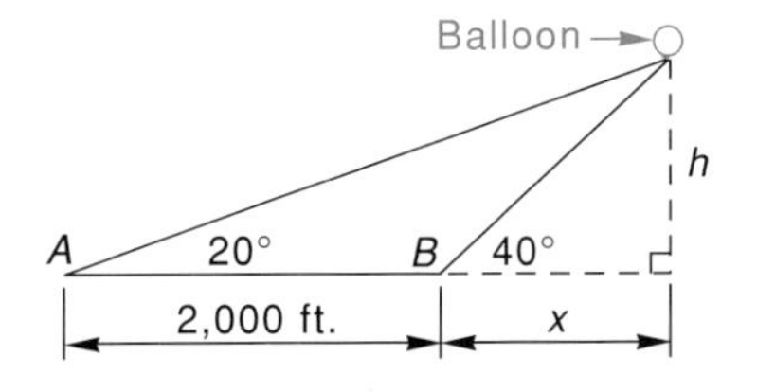

Example 3 A hot air balloon is observed at the same instant from points A and B that are on level ground, and are 2,000 feet apart. The angle of elevation (an angle measured up from the horizontal) from A to the balloon is 20°. The angle of elevation from B to the balloon is 40°. How high is the balloon above the ground at the moment of observation?

Solution

Referring to the figure, we can write

$$\tan 20° = \frac{\text{opp}}{\text{adj}} = \frac{h}{2{,}000 + x},$$

$$\tan 40° = \frac{h}{x}.$$

Solving the second equation for x in terms of h and substituting the result into the first equation yields,

$$\tan 20° = \frac{h}{2{,}000 + \dfrac{h}{\tan 40°}}$$

$$2{,}000 \tan 20° + \frac{h \tan 20°}{\tan 40°} = h$$

$$2{,}000 \tan 20° = h\left(1 - \frac{\tan 20°}{\tan 40°}\right)$$

$$1{,}286 \text{ ft.} = h.$$

Example 4 Two cities, A and B, are at 42° N latitude. The longitudes of A and B are 110° W and 160° W. What is the distance along the 42° N latitude circle between A and B? Assume that the radius of the earth is 3,960 miles.

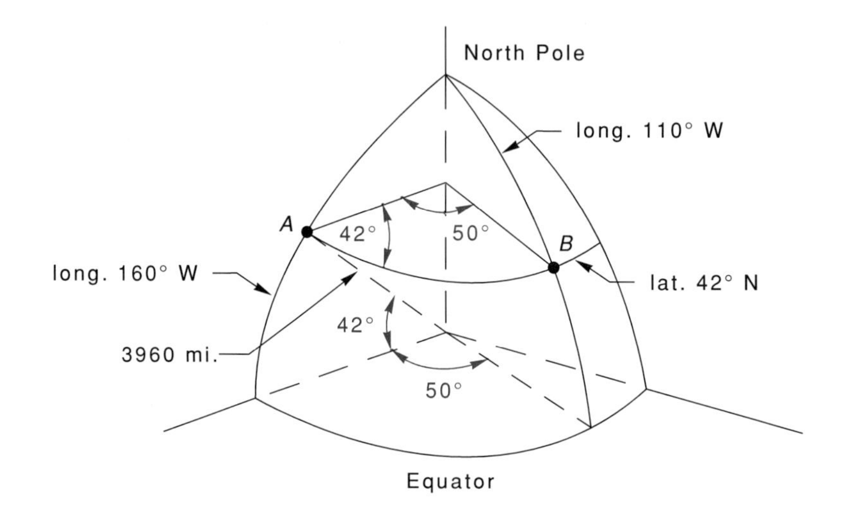

Solution

We must find the length of arc $\widehat{AB}$. Its length is given by equation (4) of section 7.2 as $r\,\theta$, where θ is in radian measure.

$$\frac{\theta}{\pi} = \frac{50°}{180°}$$

$$\theta = \frac{5\pi}{18}$$

The radius r is found as

$$r = 3{,}960 \cos 42° = 2{,}943 \text{ miles.}$$

Now,

$$\widehat{AB} = 2{,}943 \cdot \frac{5\pi}{18}$$

$$= 2{,}568 \text{ miles.}$$

■

Exercise Set 9.2

(A)

Exercises 1–6 pertain to a right triangle such as the one depicted here. Solve each right triangle from the given data.

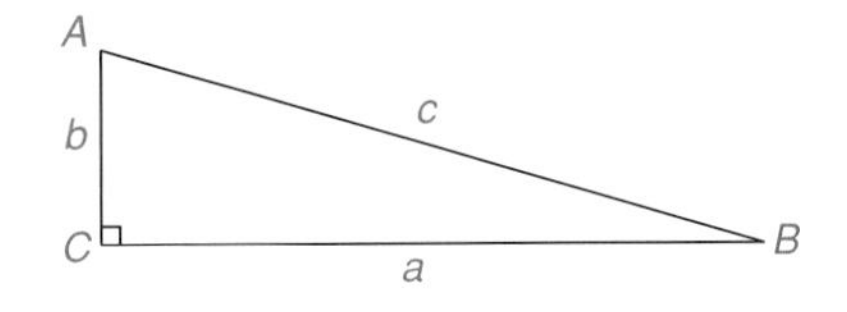

1. $A = 22°$, $b = 12.5$
2. $B = 65°$, $c = 17$
3. $a = 13.6$, $b = 15.1$
4. $c = 29.8$, $b = 15.3$
5. $B = 46°$, $c = 120$
6. $a = 1{,}830$, $c = 1{,}950$
7. Write definitions, such as those given at the beginning of this section for sine, cosine, and tangent, for the reciprocal trigonometric functions.
8. At a point 30 feet from the base of a redwood tree, the angle of elevation to the top of the tree is 82°. How high is the tree?
9. The angle of depression (an angle measured down from the horizontal) from a 1,200-foot vertical cliff to a sailboat is 28°. How far is the sailboat from the base of the cliff?
10. An airplane is flying at an approximate elevation of 2,000 feet. Ten seconds after it is directly overhead, the angle of elevation to the plane is 45°. How fast (in mph) is the plane flying?
11. From their positions in a level field, two observers spot a balloon that is 200 feet above the field. If the angles of elevation to the balloon from the observers are 26° and 43°, and if the balloon and observers are in the same vertical plane, how far apart are the observers? (There are two possible answers.)
12. While walking on a level street a tourist stops and estimates that the angle of elevation to the top of a distant skyscraper is 42°. The tourist walks 500 feet toward the skyscraper and then estimates the angle of elevation to the top of the skyscraper to be 55°. Approximately how high above the level street is the top of the skyscraper?
13. Find the distance between two cities that are located on latitude 36° N and longitudes 15° E and 35° W. Assume that the radius of the earth is 3,960 miles.
14. Find the distance between two cities that are located on latitude 48° S and longitudes 52° W and 86° W. The radius of the earth is 3,960 miles.
15. Find the perimeter of a regular polygon of 50 sides that is inscribed in a 10″ radius circle.
16. Find the linear velocity (mph) of a point on earth that is on latitude 70° S. The earth's radius is 3,960 miles.
17. What time in the morning does a 10-foot tree cast an 8-foot shadow on level ground if on the day in question, the sun rises at 5:30 A.M. and is directly overhead at 12:30 P.M.?
18. On a given day the sun rose at 5:30 A.M. and set at 8:30 P.M. Sometime during the day an 8-foot vertical pole cast a 15-foot horizontal shadow in an easterly direction. What time of day was it?
19. A surveying traverse is a sequence of connected straight line segments whose lengths and bearings are determined. A bearing is an acute angle measured from either the north or the south. Find the distance between A and D and the bearing of AD, using the data for the open traverse in the figure. The bearing N 68°-20′ E is read "North 68 degrees 20 minutes East." A minute is $\frac{1}{60}$ of a degree.

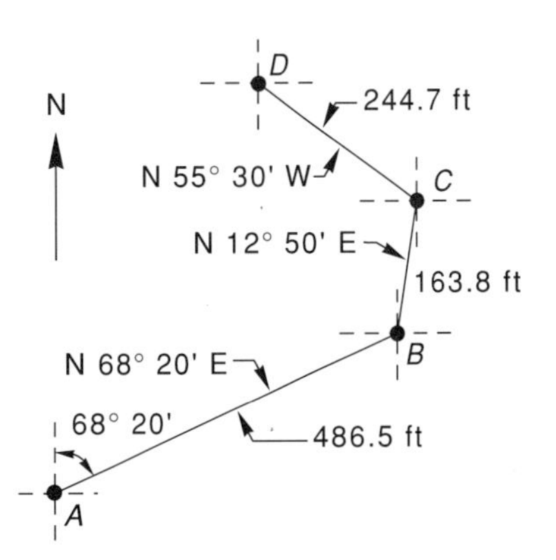

20. A closed traverse has the same beginning and end point. Use the traverse data in the table to determine the distance between the beginning and end positions of point A. This distance is called the error of closure of the traverse. In practice, the error of closure is distributed amongst the lengths and bearings of the traverse so that the beginning and end point positions are the same.

Line	Distance (ft.)	Bearing
$\overline{AB}$	200.0	N 70°-30′ E
$\overline{BC}$	144.1	S 38°-45′ E
$\overline{CD}$	171.3	S 29°-52′ W
$\overline{DA}$	273.7	N 44°-42′ W

21. Find the lengths of $\overline{AB}$ and $\overline{AD}$.

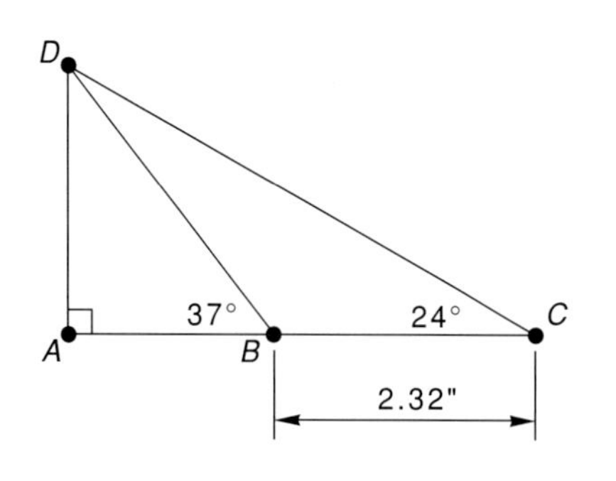

(B)

22. Find the length of the continuous belt. (See figure.)

23. Find the length of the crossed belt. (See figure.)

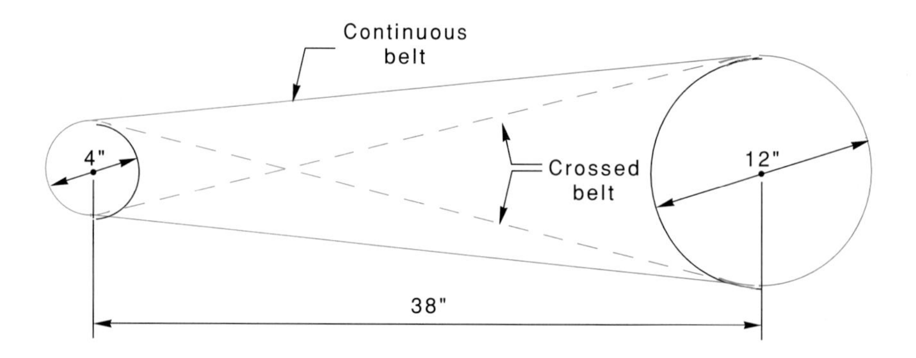

24. Show that the perimeter of an n-sided regular polygon that is inscribed in a circle of radius r, is $2rn \sin \pi/n$. Given that $\sin \pi/n \approx \pi/n$ when n is large and π/n is therefore small, what conclusion can you draw about the perimeters of inscribed regular polygons?

25. A large radio transmitting tower stands on the roof of an office building. The angles of elevation to the bottom and top of the tower from a point on level ground that is x meters from the building are $\alpha°$ and $\theta°$ respectively. Express the height, h, of the tower in terms of x, α, and θ.

26. The surface area of a sphere of radius r is $4\pi r^2$. The surface area of a spherical cap of height h is $2\pi rh$. Show that the percent of the earth's surface that can be seen from s miles above the surface of the earth is given by $50s/(3{,}960 + s)$, where 3,960 miles is assumed to be the radius of the earth.

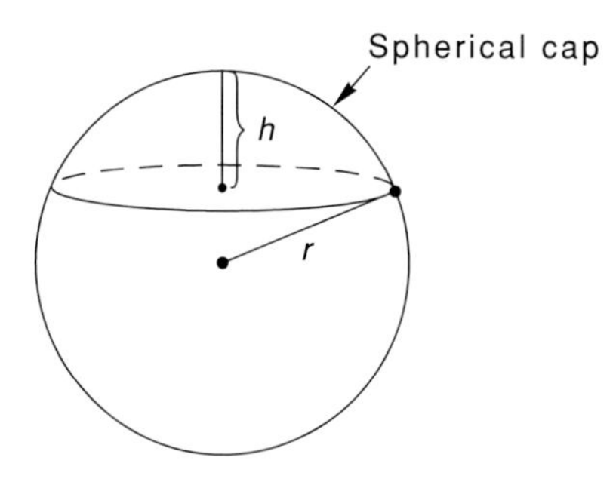

27. The surface speed of the grindstone in the figure is 300 ft./sec. What is the average speed of point A on the treadle? The connecting rod is vertical at the end positions of A.

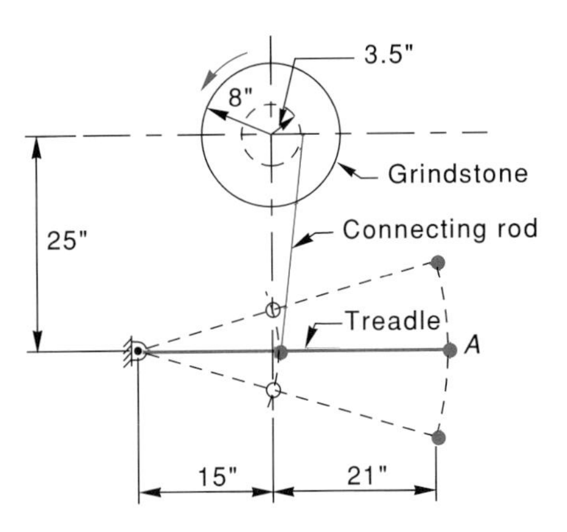

28. A block wall that is h meters high stands on a north-south property line. The bearing of the sun is S $\theta°$ E, and its angle of elevation is $\alpha°$. The wall stands on level ground. Find the width of its shadow in terms of h, α, and θ.

29. A three-foot high painting is hung so that its base is one foot above an observer's eye. If the observer is x feet from the wall on which the painting is hung, what is the angle θ, at the observer's eye, between the top and bottom of the painting? Express θ in terms of x and determine what happens to θ as x tends to ∞, and as x tends to 0. The observer best views the painting when θ assumes its maximum value. Evaluate θ for different values of x to see if you can approximate how far the observer has to be from the wall to view the painting best.

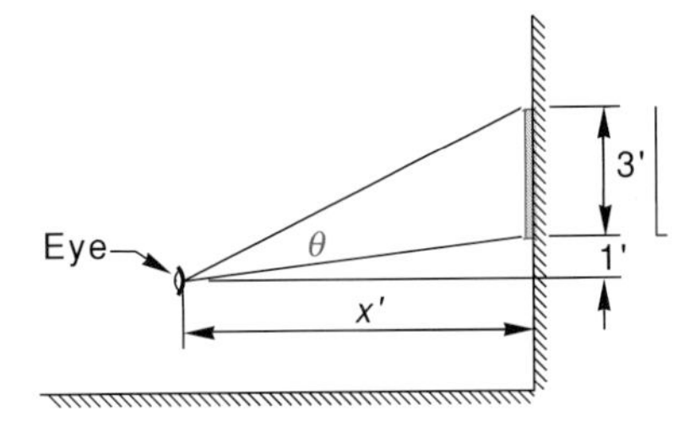

30. In the figure, $\alpha + \beta = \theta$. Why?

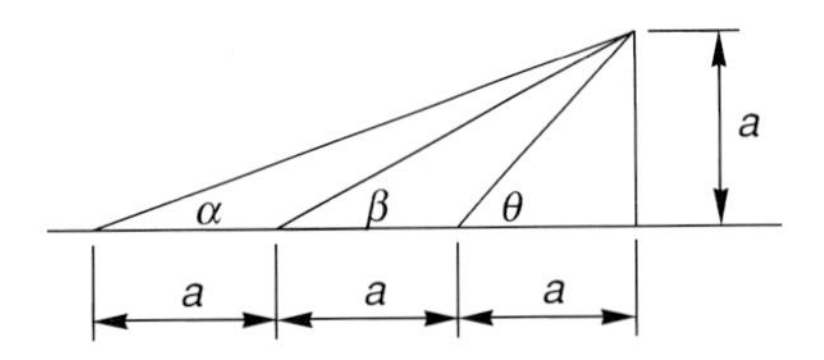

31. Compute θ from the dimensions given in the figure.

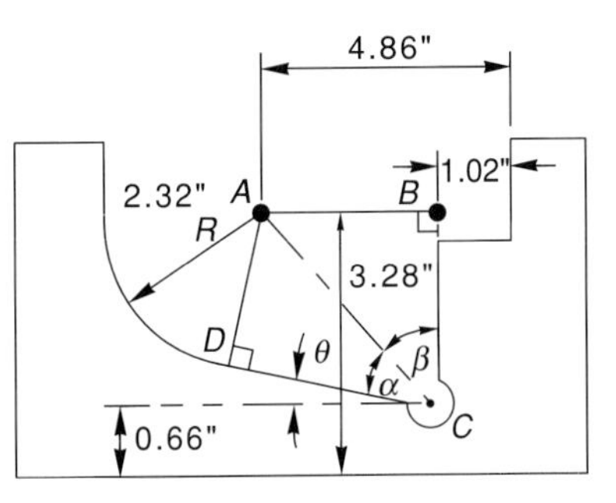

For the triangle given below,

32. show that $b = x[\tan(\alpha + \beta) - \tan \alpha]$.

33. if $\alpha = \beta$, and $b > a$, show that $x = a\sqrt{(b + a)/(b - a)}$.

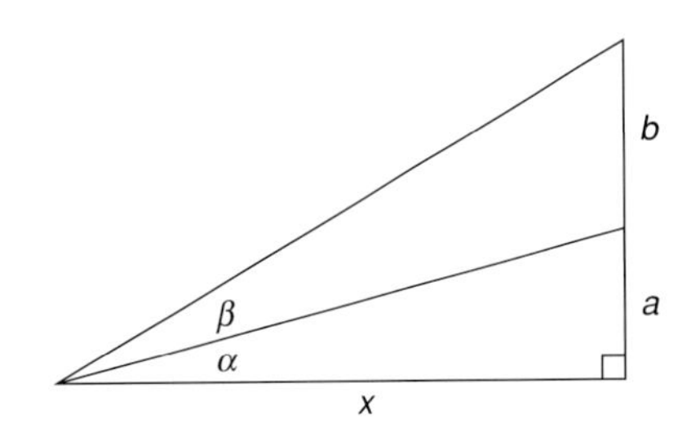

9.3 The Sine and Cosine Laws

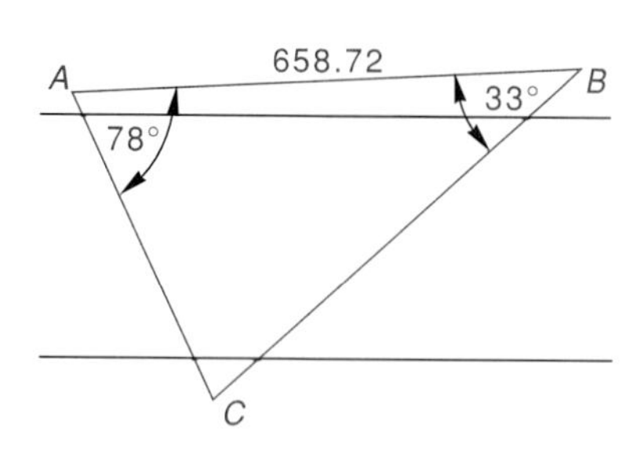

Figure 9.6

A highway bridge is to span a ravine along a line through points A and C on opposite sides of the ravine (figure 9.6). The distance between A and C is required for the design of the bridge. A surveying team chooses a point B on the same side of the ravine as A from which there are clear lines of sight to A and C. Angles CAB and ABC are measured with a transit and length $\overline{AB}$ is taped. With these field measurements and the "Law of Sines," $\overline{AC}$ can be computed.

Law of Sines

Consider a triangle that is not a right triangle (figure 9.7). Let h be the length of altitude CD. Triangles ACD and BCD are right triangles. Therefore,

$$\sin A = \frac{h}{b} \quad \text{and} \quad \sin B = \frac{h}{a}.$$

Figure 9.7

From these equations we have that

$$b \sin A = a \sin B = h,$$

and

$$\frac{\sin A}{a} = \frac{\sin B}{b}. \tag{1}$$

Repeating the above argument for right triangles ABE and CBE, it is possible to show that

$$\frac{\sin A}{a} = \frac{\sin C}{c}. \tag{2}$$

If (1) and (2) are combined, one form of the **Law of Sines** is obtained. A second form is obtained if each of the fractions in (1) and (2) are inverted.

Law of Sines

$$\frac{\sin A}{a} = \frac{\sin B}{b} = \frac{\sin C}{c} \tag{3}$$

$$\frac{a}{\sin A} = \frac{b}{\sin B} = \frac{c}{\sin C} \tag{4}$$

We are now able to do the surveyor's calculation for finding the distance $\overline{AC}$ across the ravine (figure 9.6). Substituting into (4),

$$\frac{\overline{AC}}{\sin 33°} = \frac{658.72}{\sin 69°}$$
$$\overline{AC} = 384.29 \text{ ft.}$$

Example 1 Solve triangle ABC.

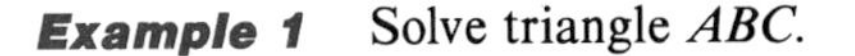

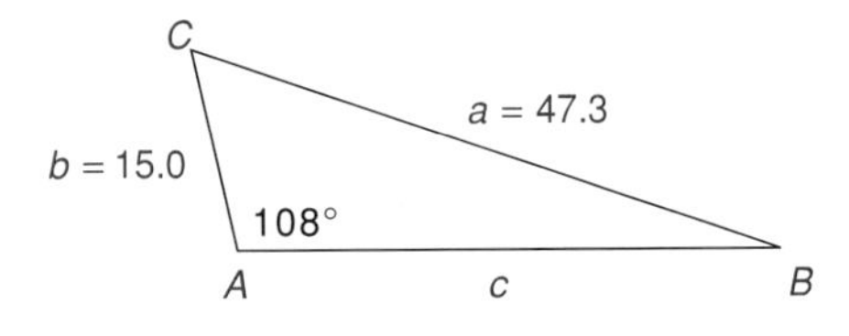

Solution Substituting into (3),

$$\frac{\sin 108°}{47.3} = \frac{\sin B}{15.0}$$
$$0.301604 = \sin B$$
$$\sin^{-1} 0.301604 = B$$
$$17.6° = B.$$
$$C = 180 - (108 + 17.6) = 54.4°$$
$$\frac{c}{\sin 54.4°} = \frac{47.3}{\sin 108°}$$
$$c = 40.4.$$

■

Had we not been given a sketch of the triangle in example 1, the computation for the measure of angle B would have been ambiguous. After all, an angle in a triangle whose sine is 0.301604 is either 17.6° or 152.4°. The range of the inverse sine function is $[-90°, 90°]$ so that our calculator will always give us a positive or negative acute angle when the inverse and sin keys are depressed. However, in solving triangles we must be aware of the possibility that the triangle may have an *obtuse* angle that is the *supplement* of the acute angle displayed by our calculator. The sketch shows angle B to be acute, but a sketch is not necessary to see that B must be acute. Since the sum of the angles in a triangle is 180°, and $A = 108°$, B cannot be 152.4°.

Example 2 An engineer wishes to sketch a triangular piece of land from the field data, $B = 32.0°$, $b = 1{,}216.8$ ft., and $c = 2{,}005.1$ ft. Can he draw the sketch?

Solution From the Law of Sines,

$$\frac{\sin 32.0°}{1{,}216.8} = \frac{\sin C}{2{,}005.1}$$
$$0.87323 = \sin C$$
$$C = 60.8° \text{ or } C = 119.2°.$$

The engineer can sketch one of the two triangles, ABC_1 or ABC_2, shown in figure 9.8. In the absence of additional information, he does not know which of the triangles is the correct representation of the piece of land.

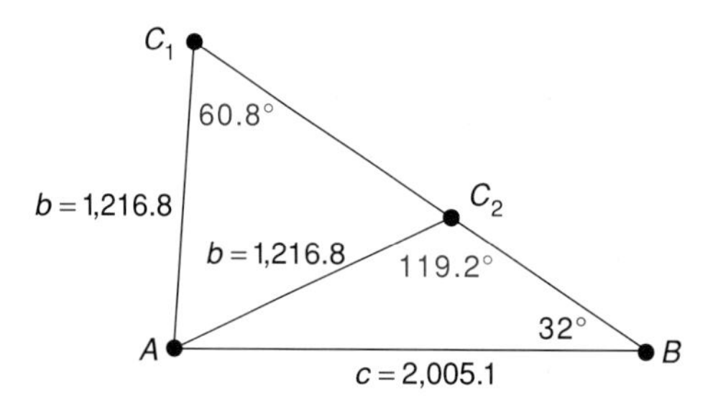

Figure 9.8

■

The congruence postulates of plane geometry tell us that exactly one triangle can be constructed if we know

a. its three sides (S.S.S.), or

b. two sides and an included angle (S.A.S.), or

c. two angles and a side (A.S.A.).

It would seem then that there must be an analytic tool that would enable us to solve a triangle, given **a,** or **b,** or **c.**

By associating the Law of Sines with **a, b,** or **c,**

a. S.S.S. $\quad \dfrac{\sin A}{a^{\checkmark}} = \dfrac{\sin B}{b^{\checkmark}} = \dfrac{\sin C}{c^{\checkmark}}$

b. S.A.S. $\quad \dfrac{\sin A}{a^{\checkmark}} = \dfrac{\sin B}{b^{\checkmark}} = \dfrac{\sin C^{\checkmark}}{c}$

c. A.S.A. $\quad \dfrac{\sin A^{\checkmark}}{a^{\checkmark}} = \dfrac{\sin B^{\checkmark}}{b} = \dfrac{\sin C^{\checkmark}}{c},$

we see that it enables us to solve a triangle only in the A.S.A. case. Given two angles of a triangle, we can compute the third angle, knowing that the sum of the angles must be 180°. If we also know a side, then a proportion can be formed from the Law of Sines in which three of the four quantities of the proportion are known. In **c,** we can solve for b from

$$\frac{\sin A^{\checkmark}}{a^{\checkmark}} = \frac{\sin B^{\checkmark}}{b},$$

and then solve for c from

$$\frac{\sin A^{\checkmark}}{a^{\checkmark}} = \frac{\sin C^{\checkmark}}{c}.$$

Neither the S.S.S. case nor the S.A.S. case can be solved with only the Law of Sines. But the Sine Law alone can sometimes provide a unique solution of a triangle in which *2 sides and an angle not included between the sides are known* (S.S.A.). Such is the case in example 1 in which A, a, and b are known.

$$\frac{\sin A^{\checkmark}}{a^{\checkmark}} = \frac{B}{b^{\checkmark}} = \frac{\sin C}{c}$$

It is also possible in such cases, as example 2 illustrates, that the Sine Law produces two solutions. A unique solution is obtainable only with additional information. A third possibility for the S.S.A. case is that the Sine Law does not generate a solution (see exercises 17 and 18).

Law of Cosines

The S.S.S. and S.A.S. cases can be solved with an analytic tool called the Law of Cosines.

A railroad right-of-way is to be constructed from A to B through a mountain (figure 9.9).

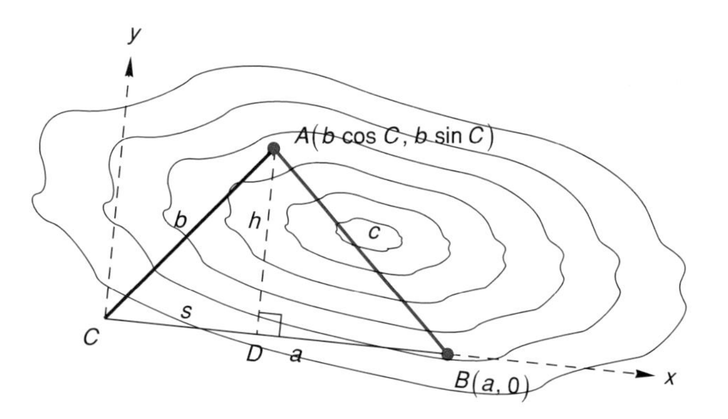

Figure 9.9

A point C is selected that has a clear line of sight to A and B. Angle C and a and b are determined. With these measurements c can be calculated. We imagine a coordinate system in the plane of A, B, and C that contains $\overline{CB}$ in the x-axis. The coordinates of B are $(a,0)$. The coordinates of A may be determined as the lengths of the sides of right triangle CDA.

$$\frac{s}{b} = \cos C \quad \text{and} \quad \frac{h}{b} = \sin C$$
$$s = b \cos C \quad \text{and} \quad h = b \sin C$$

The distance formula between two points in the plane is now used to find c.

$$\begin{aligned} c &= \sqrt{(b \cos C - a)^2 + (b \sin C)^2} \\ c^2 &= b^2 \cos^2 C - 2ab \cos C + a^2 + b^2 \sin^2 C \\ c^2 &= a^2 + b^2(\sin^2 C + \cos^2 C) - 2ab \cos C \\ c^2 &= a^2 + b^2 - 2ab \cos C \end{aligned}$$

The last equation is one form of the Law of Cosines. Alternate forms can be derived in the same way.

Law of Cosines

$$c^2 = a^2 + b^2 - 2ab \cos C \tag{5}$$
$$a^2 = b^2 + c^2 - 2bc \cos A \tag{6}$$
$$b^2 = a^2 + c^2 - 2ac \cos B \tag{7}$$

Example 3 Solve triangle ABC.

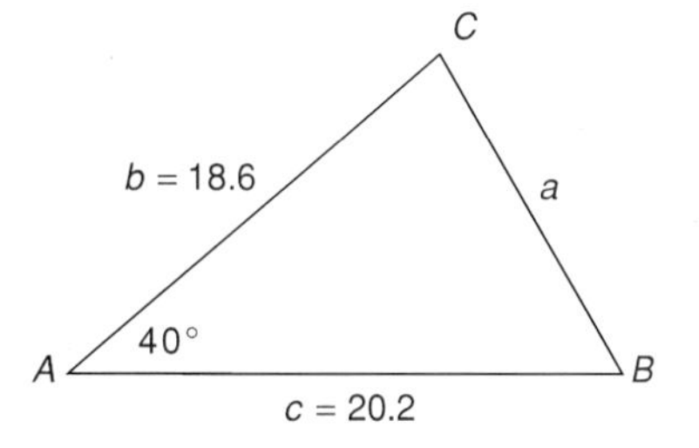

Solution Using (6) we have,

$$\begin{aligned} a^2 &= (18.6)^2 + (20.2)^2 - 2(18.6)(20.2)\cos 40° \\ a &= 13.4. \end{aligned}$$

Having a, b, and c, we can use (5) to find C.

$$(20.2)^2 = (13.4)^2 + (18.6)^2 - 2(13.4)(18.6)\cos C$$
$$\cos C = \frac{(13.4)^2 + (18.6)^2 - (20.2)^2}{2(13.4)(18.6)}$$
$$C = 76.4°$$

Finally, $B = 180 - (40 + 76.4) = 63.6°$. ■

The calculation for angle C in example 3 illustrates how the Cosine Law can be used to find the angles of a triangle when the sides of the triangle are known. Thus, the Law of Cosines enables us to solve the S.A.S. and S.S.S. cases.

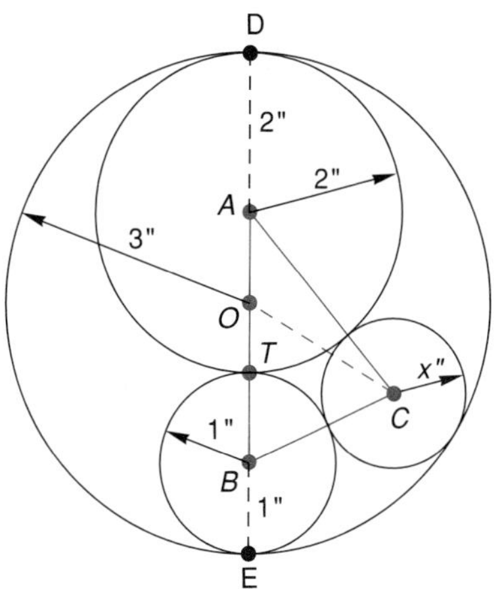

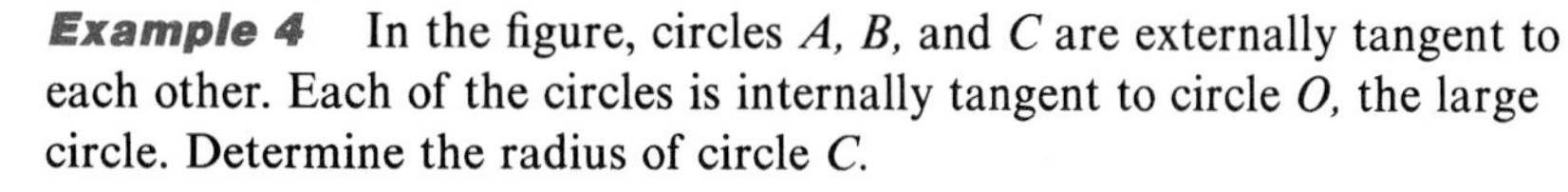

Example 4 In the figure, circles A, B, and C are externally tangent to each other. Each of the circles is internally tangent to circle O, the large circle. Determine the radius of circle C.

Solution Recall that the radius of a circle is perpendicular to a tangent of the circle at the point of tangency. Hence, $\overline{AT}$ and $\overline{BT}$ are perpendicular to a common tangent at T. As a result, points A, T, and B must lie on the same straight line. Therefore, $\overline{AB} = \overline{AT} + \overline{TB} = 3$. Similarly, $\overline{AC} = 2 + x$ and $\overline{BC} = 1 + x$. Applying the Law of Cosines to triangle ABC gives us

$$\cos A = \frac{\overline{AB}^2 + \overline{AC}^2 - \overline{BC}^2}{2(\overline{AB})(\overline{AC})}$$

$$\cos A = \frac{9 + (2 + x)^2 - (1 + x)^2}{2(3)(2 + x)}$$

$$= \frac{2x + 12}{6(2 + x)}.$$

Extending segment $\overline{AB}$ to D and E produces a diameter of the circle. (Why?) Therefore, the center O of the large circle must be on segment $\overline{AT}$. Now, applying the Law of Cosines to triangle AOC gives us

$$\cos A = \frac{\overline{AO}^2 + \overline{AC}^2 - \overline{OC}^2}{2(\overline{AO})(\overline{AC})}$$

$$= \frac{1 + (2 + x)^2 - (3 - x)^2}{2(1)(2 + x)}$$

$$= \frac{10x - 4}{2(2 + x)}.$$

The two expressions for $\cos A$ are equated.

$$\frac{10x - 4}{2(2 + x)} = \frac{2x + 12}{6(2 + x)}$$

$$30x - 12 = 2x + 12$$

$$x = \frac{6}{7}$$

■

Exercise Set 9.3

(A)

Use the Law of Sines or the Law of Cosines, or both laws, to solve each triangle.

1. $A = 42°, C = 76°, b = 15.2$
2. $A = 35°, B = 68°, c = 11.8$
3. $B = 76°, a = 12.1, c = 19.5$
4. $C = 108°, b = 26.3, a = 14.8$
5. $a = 8.7, B = 25°, C = 115°$
6. $b = 54.3, A = 115°, C = 28°$
7. $a = 15.8, b = 26.5, C = 19.6°$

8. $a = 103.6, b = 159.7, c = 241.9$
9. $A = 0.76, B = 1.32, c = 63.7$
10. $B = 2.14, C = 0.60, a = 74.6$
11. $C = 0.96, a = 12.8, b = 12.3$
12. $c = 111.8, A = 1.15, b = 93.2$
13. $A = 1.32, b = 63.2, a = 190.1$
14. $C = 1.03, a = 26.5, c = 71.8$
15. $B = 35°, c = 361.8, b = 240.3$
16. $A = 62°, c = 32.8, b = 30.5$
17. $C = 112°, c = 18.3, b = 25.7$
18. $B = 86°, a = 742.5, b = 628.3$
19. $a = 3{,}642, b = 5{,}810, c = 3{,}145$
20. $a = 31.8, b = 26.7, C = 18°$

21. Leaving at the same instant from the point where two straight highways intersect, two cars travel along the highways for two hours. The speed of one car is 10km/hr. faster than the other car. The highways make an angle of 40° with each other. At the end of two hours the cars are 120 km apart. What are the speeds of the cars?

22. One ship leaves port on a bearing of N 43° E, with a speed of 15 mph. Another ship leaves port two hours after the first with a speed of 18 mph and a course of S 82° E. How far apart are the ships five hours after the first ship leaves port?

23. A fifteen-foot pole stands on the side of a hill that inclines $\theta°$ with the horizontal. The pole leans 10° from the vertical towards the uphill side of the incline and casts a twelve-foot shadow uphill. The angle of elevation of the sun is 50°. Find θ.

24. A thirty-foot vertical flagpole is guyed by a fifty-foot guy wire that is attached to the pole ten feet from its top. If the pole stands on the side of a hill that inclines 15° with the horizontal, how far downhill does the guy wire extend?

25. Two mountain peaks P and Q can be sighted from points A and B in the valley below. Determine the distance between P and Q from the data in the figure.

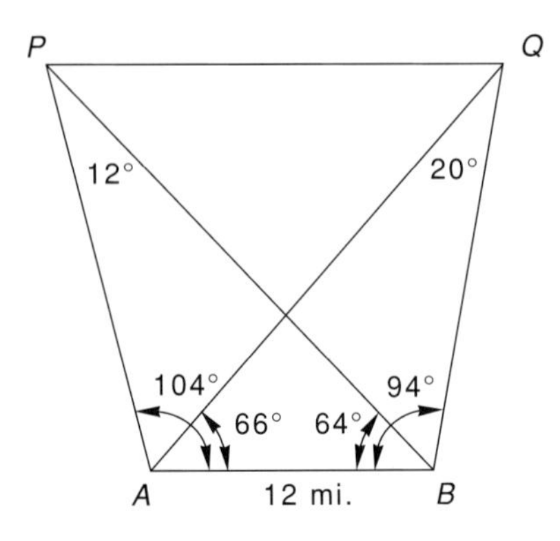

26. Two observers at points A and B, a mile apart, spot a plane in level flight. The angle at A from B to the plane is 56°, and the angle at B from A to the plane is 61°. Observations are made again ten seconds later. The angle at A from B to the plane is 48°, and the angle at B from A to the plane is 94°. What is the speed of the airplane if it and the points A and B are in the same vertical plane?

27. Thee circles with radii 4″, 3″, and x″ are externally tangent to each other. All three circles are internally tangent to a 7″ diameter circle (see example 4). Find x.

28. A straight portion of a road is to be constructed between A and C. There is no clear line of sight between A and C, but both points can be seen from D and B. Using the data in the figure, find the length and bearing of AC.

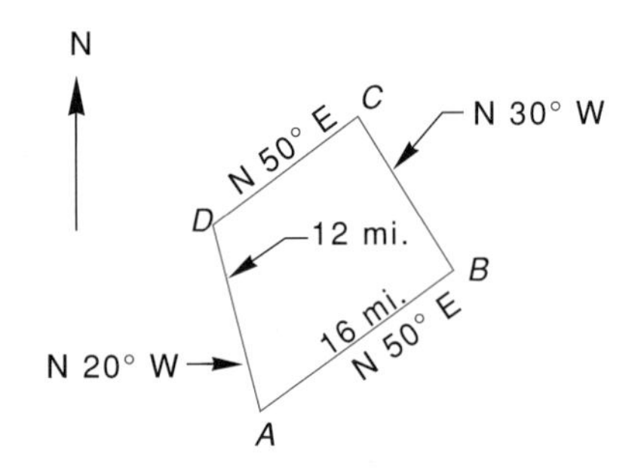

29. A satellite orbits the earth every 2.5 hours in a circular orbit that has a 6,000-mile radius. The satellite is to pass directly over a tracking station on earth at 1:00 P.M. The tracking station's antenna has a 50° angle of elevation. At what time will the satellite pass through the beam of the antenna? Assume the earth's radius is 4,000 miles.

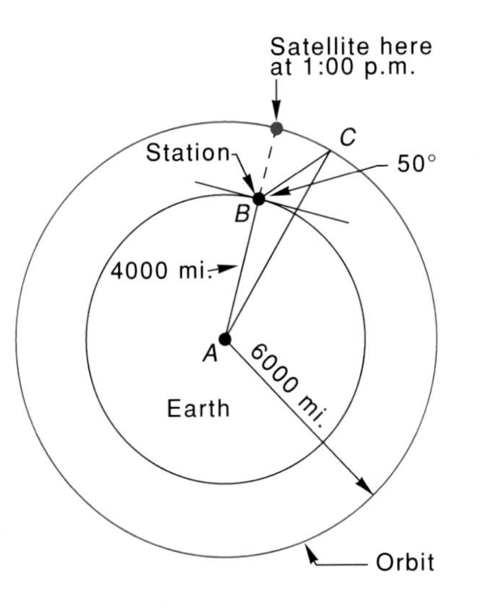

30. What is the distance between the satellite and the tracking station, referred to in exercise 29, at 12:50 P.M.?

(B)

31. Find the angle measures of triangle ABC in the figure. The triangle is contained in a cube that has 10-unit sides.

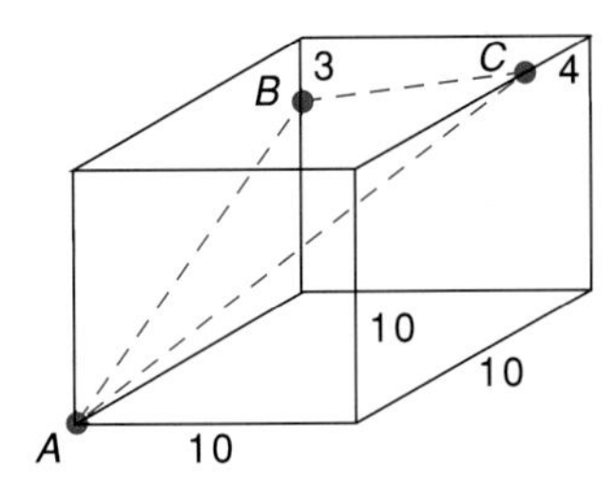

32. Find the angle measures of triangle ABC that is contained in the rectangular box shown in the figure.

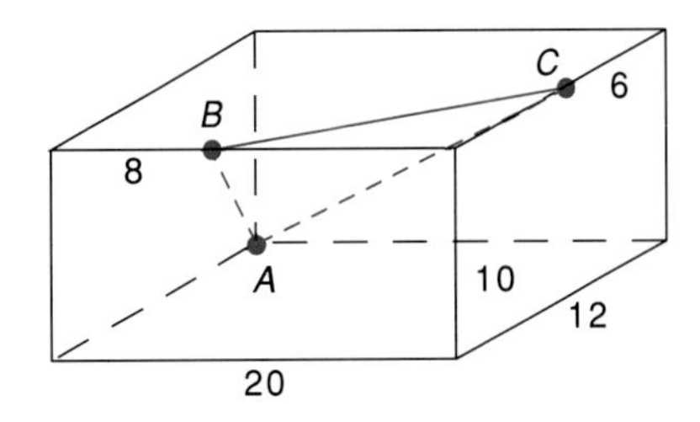

Exercises 33–36 are designed to lead the reader to Heron's formula for the area of a triangle.

$$\text{Area} = \sqrt{S(S-a)(S-b)(S-c)},$$

where a, b, and c are the lengths of the sides of a triangle and $S = (a + b + c)/2$.

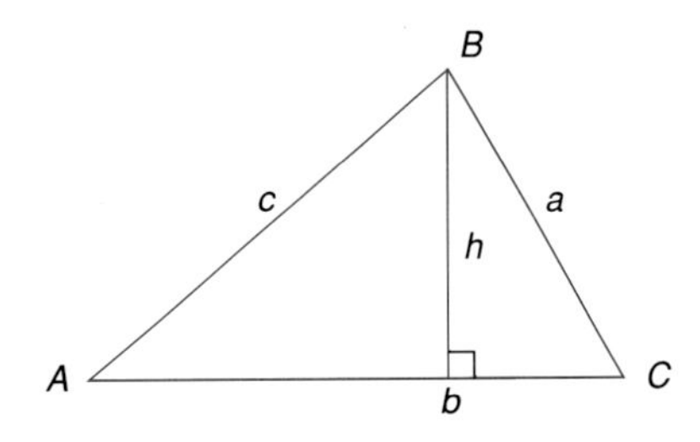

33. Show that the area of triangle ABC in the figure is given by the formula:

$$\text{Area} = \frac{(bc \sin A)}{2}. \tag{8}$$

34. Let a be the area of the triangle ABC. Show that

$$a^2 = (\tfrac{1}{2}bc)(\tfrac{1}{2}bc)(1 - \cos A)(1 + \cos A). \tag{9}$$

35. Use the law of cosines to substitute for $\cos A$ in (9) and obtain

$$\frac{1}{2}bc(1 + \cos A) = \frac{(b + c - a)(b + c + a)}{4} \tag{10}$$

and

$$\frac{1}{2}bc\,(1 - \cos A) = \frac{(a - b + c)(a + b - c)}{4}. \tag{11}$$

36. Substitute (10) and (11) in (9) and obtain

$$\text{Area} = \sqrt{S(S-a)(S-b)(S-c)}.$$

In a triangle with angles A, B, and C show that

37. $\cos A + \cos B + \cos C =$
$1 + 4 \sin \frac{A}{2} \sin \frac{B}{2} \sin \frac{C}{2}$. (Hint: Use the Law of Cosines to express $\cos A$, $\cos B$, and $\cos C$ in terms of the sides of the triangle.)

38. $\sin A + \sin B + \sin C = 4 \cos \frac{A}{2} \cos \frac{B}{2} \cos \frac{C}{2}$.
(Hint: Use equation 8 in exercise 33 and Heron's Formula.)

39. $\cot A \cot B + \cot A \cot C + \cot B \cot C = 1$. (Hint: $\cot A$ equals $(\cos A)/(\sin A)$. Substitute for $\sin A$ from (8) in exercise 33 and use the Law of Cosines to substitute for $\cos A$.)

9.4 Vectors

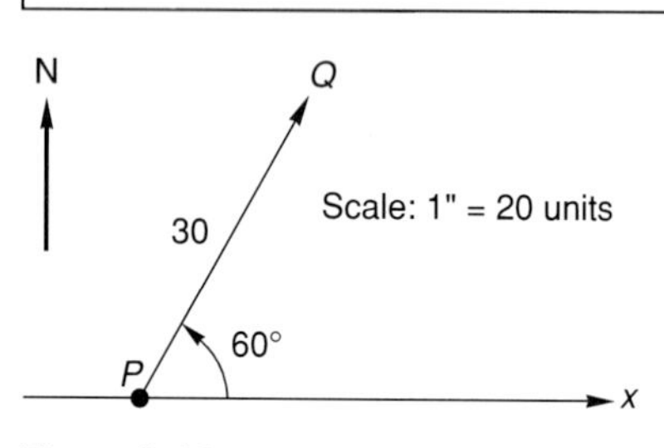

Figure 9.10

Some physical entities such as area, pressure, and temperature are completely described by their magnitudes. They are called **scalars.** Other physical entities such as velocity, force, and acceleration are completely described when their *magnitudes and directions* are known. They are called **vectors.**

A 30 mile per hour wind that has a bearing of N 30° E can be represented by the arrow $\overline{PQ}$ (figure 9.10). $\overline{PQ}$ can also represent a force of 30 pounds that makes a 60° angle with the positive horizontal axis. In fact $\overline{PQ}$ can represent *any vector of magnitude 30 that makes a counterclockwise 60° angle with the positive horizontal axis.*

If $\overline{PQ}$ is placed in a coordinate system (figure 9.11) with P at the origin, then the coordinates of Q are

$$x = 30 \cos 60° = 15$$
$$y = 30 \sin 60° = 15\sqrt{3}.$$

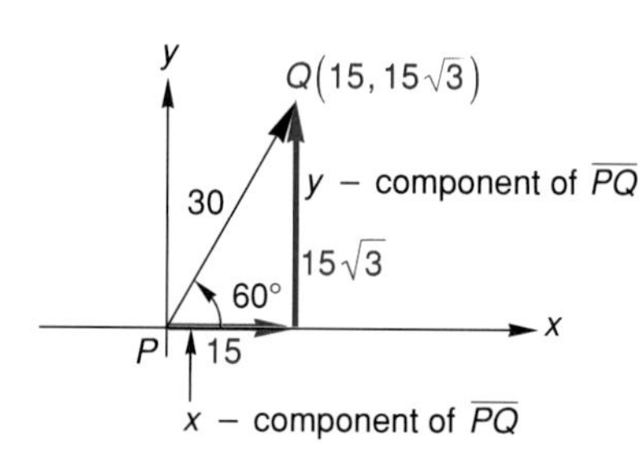

Figure 9.11

A vector represented by PQ is said to have an ***x*-component** of 15 units and a ***y*-component** of $15\sqrt{3}$ units, and it is often denoted simply as the ordered pair $(15,15\sqrt{3})$. The x and y components of a vector are themselves vectors that are parallel to the x and y axes, respectively.

Example 1

a. The coordinates of the points L and M are $(3,1)$ and $(-2,7)$, respectively. Find the components of the vector represented by the arrow $\overline{LM}$ and denote the vector as an ordered pair.

b. Draw an arrow representing vector $(5,-3)$.

Solution

a. The vector represented by $\overline{LM}$ is shown in figure 9.12. Its x-component is $-2 - 3 = -5$ and its y-component is $7 - 1 = 6$. It can therefore be denoted as $(-5,6)$.

b. Vector $(5,-3)$ can be represented by either of the arrows $\overline{A}$ or $\overline{B}$ in figure 9.12, or *any arrow that has the same magnitude and direction as* $\overline{A}$. ■

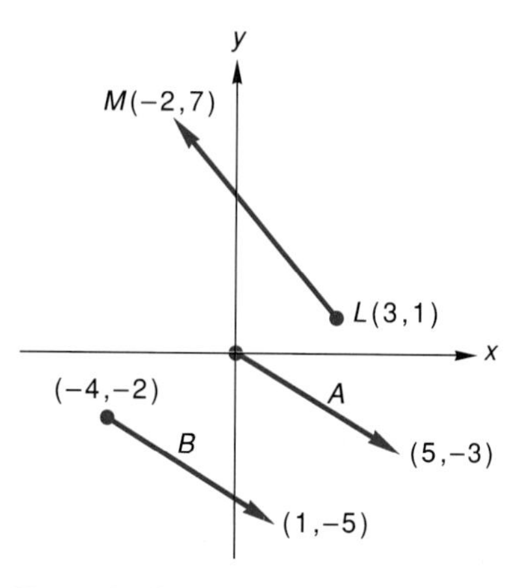

Figure 9.12

Magnitude of Vectors

Henceforth a vector shall be named with a single boldface letter. Thus the vector represented by either arrow $\overline{A}$ or arrow $\overline{B}$ in figure 9.12 can be named **A**.

We have seen that the magnitude of a vector **A** is simply the length of the arrow representing **A**. That length, or magnitude, will be denoted as $||\mathbf{A}||$.

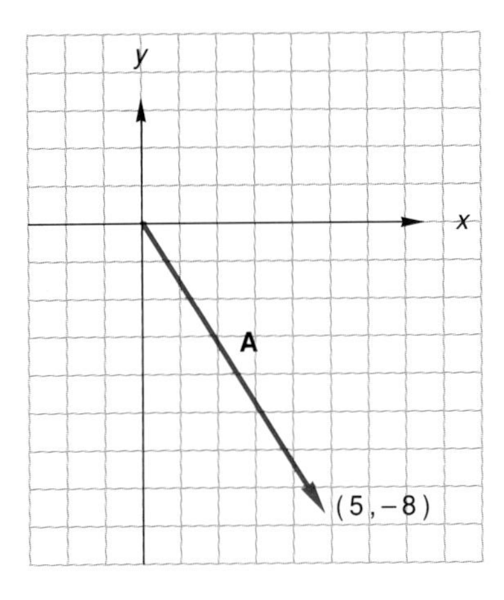

Example 2 Graph $\mathbf{A} = (5,-8)$, and find $||\mathbf{A}||$.

Solution Applying the Pythagorean Theorem to **A** in the figure gives us

$$||\mathbf{A}|| = \sqrt{5^2 + (-8)^2} = \sqrt{89}.$$ ■

Generalizing from example 2,

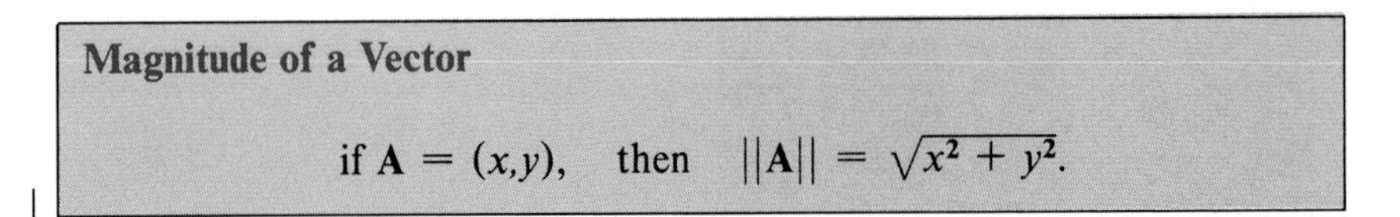

Magnitude of a Vector

$$\text{if } \mathbf{A} = (x,y), \quad \text{then} \quad ||\mathbf{A}|| = \sqrt{x^2 + y^2}.$$

Vector Addition

A small plane flies N 60° E from an airport with an airspeed of 120 mph. The plane is subjected to a 20-mph wind from the north. Where is the plane, relative to the airport, two hours after the flight begins?

The flight can be viewed as the two distinct displacements shown in figure 9.13. One can imagine the plane being displaced 240 miles from the airport on a N 60° E bearing because of a two-hour flight in *still* air. The second displacement is 40 miles south because of the wind. The actual displacement of the plane from the airport is shown as the vector **A** in figure 9.13. **A** is called the vector sum of **B** and **C**, where **C** is added to **B** by placing the tail of the arrow representing **C** to the head of the arrow representing **B.** The sum of **B** and **C** is then the vector **A**, which is represented by the arrow from the tail of **B** to the head of **C**.

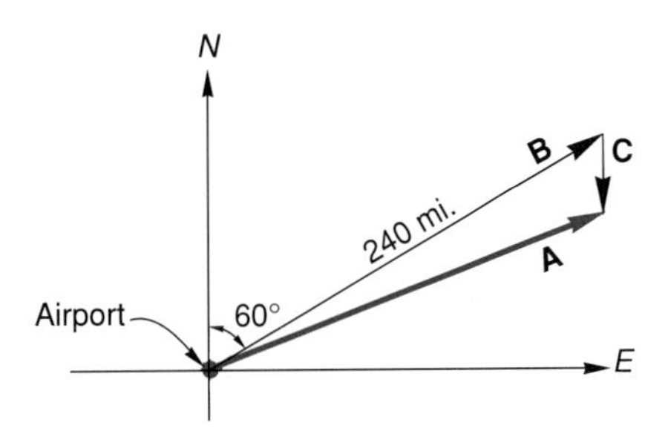

Figure 9.13

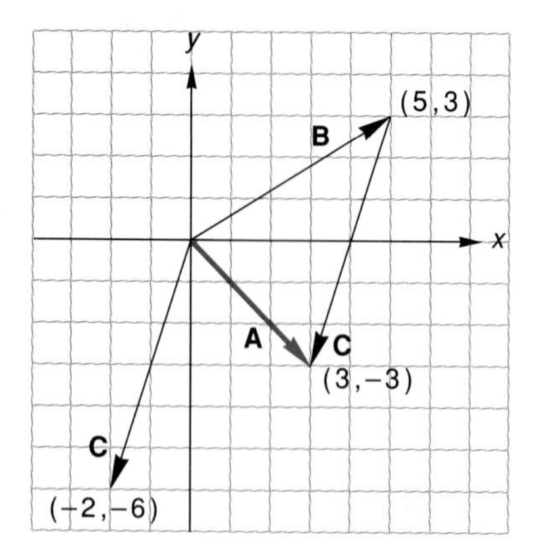

Example 3 Let $\mathbf{B} = (5,3)$, $\mathbf{C} = (-2,-6)$, and $\mathbf{A} = \mathbf{B} + \mathbf{C}$. Graph **B**, **C**, and **A** and denote **A** as an ordered pair.

Solution We represent **B** and **C** in the figure below as arrows whose tails are at the origin. To find **A** we construct another arrow representing **C** so that its tail is at the head of **B**. **A** is represented by the arrow from the tail of **B** to the head of **C**. We note that the head of **A** is at the point $(3,-3)$ since one arrives at that point from the head of **B** by traveling the length of **C**. That is, from the point $(5,3)$ one moves down six units and to the left two units.

$$\mathbf{A} = (5 + (-2), 3 + (-6)) = (3,-3)$$

The ordered pair that denotes **A** can be obtained *by adding the corresponding components of* **B** *and* **C.** ■

Example 3 suggests that we can view vector addition either geometrically or algebraically. The geometric point of view requires that we add two vectors by placing the tail of one vector to the head of the other vector. The algebraic point of view requires that we add the corresponding components of the two vectors.

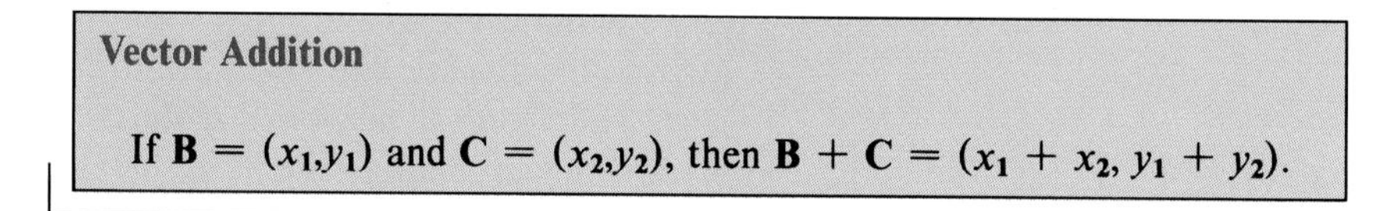
Vector Addition

If $\mathbf{B} = (x_1,y_1)$ and $\mathbf{C} = (x_2,y_2)$, then $\mathbf{B} + \mathbf{C} = (x_1 + x_2, y_1 + y_2)$.

Example 4 Denote **A** in figure 9.13 as an ordered pair.

Solution We denote **B** and **C** as ordered pairs and then add their corresponding components.

$$\mathbf{B} = (240 \cos 30°, 240 \sin 30°) = (120\sqrt{3},120)$$
$$\mathbf{C} = (0,-40)$$
$$\mathbf{A} = \mathbf{B} + \mathbf{C} = (120\sqrt{3} + 0, 120 + (-40)) = (120\sqrt{3}, 80)$$ ■

Vector Applications

The remaining examples in this section illustrate the use of vectors in constructing mathematical models of applied problems.

Example 5 A car travels N 10° W for 150 miles and then N 70° E for 100 miles. How far is the car from the starting point? What is the bearing of the car at the end of the trip, relative to its starting point?

Solution Figure 9.14 is a vector model of the problem.

We seek the magnitude and direction of **C**, where **C** = **A** + **B**. Vectors **A** and **B** represent the 150-mile and 100-mile displacements of the car. By the Law of Cosines,

$$\begin{aligned}\|\mathbf{C}\|^2 &= \|\mathbf{A}\|^2 + \|\mathbf{B}\|^2 - 2\,\|\mathbf{A}\|\,\|\mathbf{B}\| \cos 100^\circ \\ &= (150)^2 + (100)^2 - 2(150)(100)\cos 100^\circ \\ &= 37{,}709.45 \\ \|\mathbf{C}\| &= 194.2 \text{ miles.}\end{aligned}$$

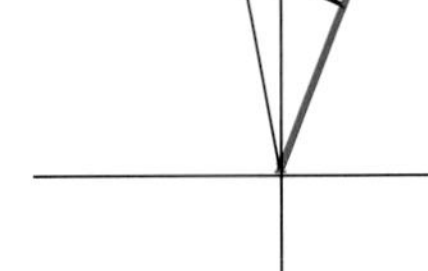

Figure 9.14

We now use the Law of Sines to find β, the angle between **A** and **C**.

$$\frac{\sin\beta}{100} = \frac{\sin 100^\circ}{194.2}$$

$$\beta = 30.5^\circ$$

The direction of **C** is N 20.5° E.

Alternate Solution Express **A** and **B** in terms of their components.

$$\begin{aligned}\mathbf{A} &= (150\cos 100^\circ,\ 150\sin 100^\circ) = (-26.05,\ 147.72) \\ \mathbf{B} &= (100\cos 20^\circ,\ 100\sin 20^\circ) = (93.97,\ 34.20) \\ \mathbf{C} &= \mathbf{A} + \mathbf{B} = (67.92,\ 181.92) \\ \|\mathbf{C}\| &= \sqrt{(67.92)^2 + (181.92)^2} = 194.2 \\ \tan\theta &= \frac{181.92}{67.92} \\ \theta &= 69.5^\circ\end{aligned}$$

The direction of **C** is N 20.5° E. ■

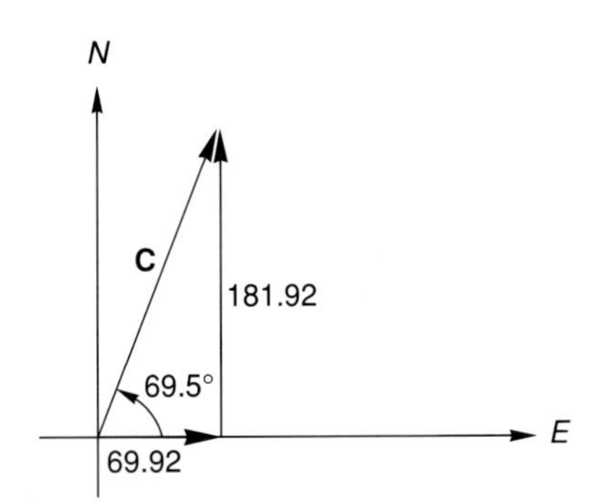

Example 6 An airplane has an airspeed of 550 mph and a heading of 160°. The plane encounters a 75-mph wind that has a heading of 60°. Find the speed and the heading of the plane relative to an observer on the ground.

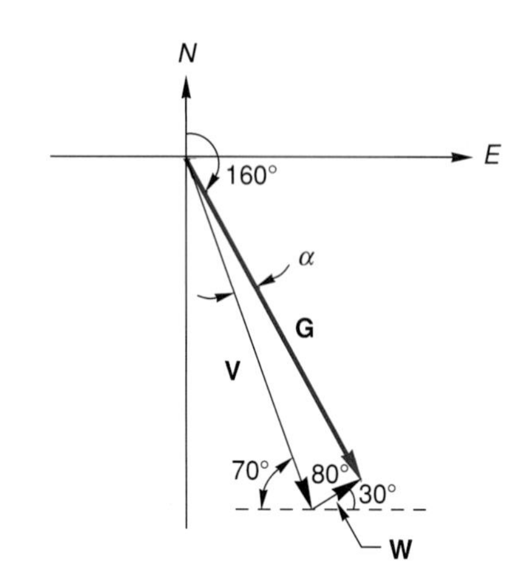

Solution A heading is an angle measured clockwise from the north. In the figure, **V** is the velocity of the plane in still air, **W** is the wind velocity, and **G** is the velocity of the plane relative to the ground.

$$\begin{aligned} ||\mathbf{G}||^2 &= ||\mathbf{V}||^2 + ||\mathbf{W}||^2 - 2\,||\mathbf{V}||\,||\mathbf{W}|| \cos 80° \\ &= (550)^2 + (75)^2 - 2(550)(75) \cos 80° \\ ||\mathbf{G}|| &= 542 \text{ mph} \end{aligned}$$

$$\frac{\sin \alpha}{75} = \frac{\sin 80°}{542}$$
$$\alpha = 7.8°$$

The heading of the plane, relative to an observer on the ground, is $160° - 7.8° = 152.2°$. ■

Example 7 A man wants to travel due west, from A to B, across a $\frac{1}{2}$- mile-wide river. His motorboat has a speed of 15 mph in still water. The river has a 3-mph north current. What heading should the boat take to go directly from A to B? How long does the trip take?

Solution The boat must head in a southwesterly direction to compensate for the river current. Let **M** represent the still water velocity of the boat, **R** the river current, and **V** the actual velocity of the boat as observed from point A on shore. **V** is perpendicular to **R.**

$$\begin{aligned} ||\mathbf{V}|| &= \sqrt{(15)^2 - 3^2} \qquad \sin\theta = \tfrac{3}{15} \\ &= 14.7 \text{ mph} \qquad\qquad \theta = 11.5° \end{aligned}$$

The motorboat heading is $270° - 11.5° = 258.5°$. The trip time is 0.5 mi/14.7 mph = 0.034 hr.

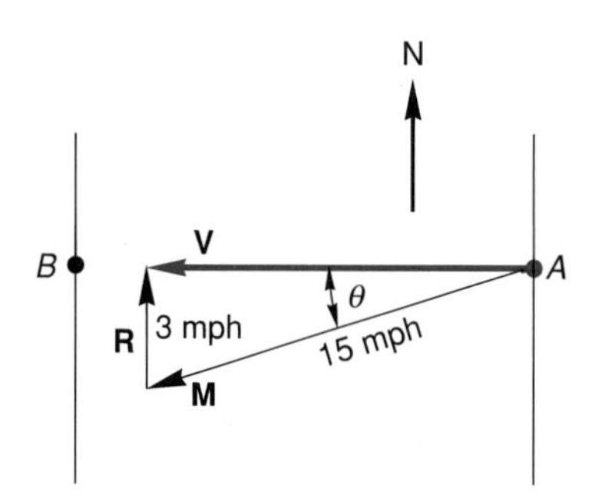

■

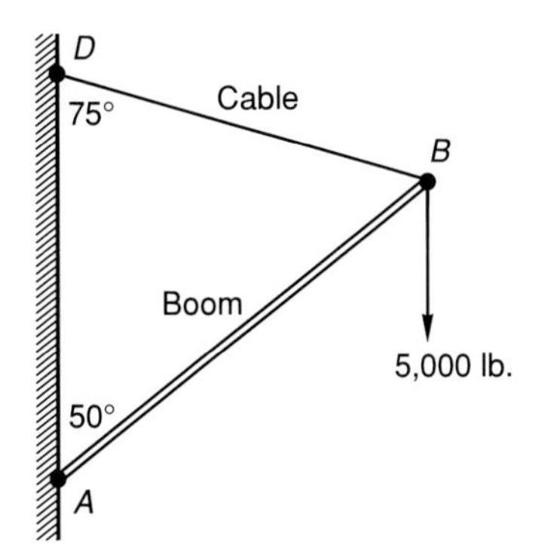

Example 8 In the figure, a 5,000-pound load is supported by a boom that is hinged at A, and a cable that is connected to a support at D. Assume that the weight of the boom is negligible compared to the load, and that the boom carries only a compressive load. Find the tension in the cable and the compression in the boom.

Solution The forces acting at B are represented in the figures below. $\mathbf{C}$ is the compression in the boom. $\mathbf{T}$ is the tension in the cable. The vertical vector is the 5,000-pound load. Point B does not move under its loading because the forces acting there are in equilibrium. Therefore, the sum of the vectors at B must be $\mathbf{0}$. This fact is reflected by the triangle that is formed when the vectors are added in the tail to head fashion. We see that $\mathbf{C} + \mathbf{T}$ is a vector that is equal in magnitude and opposite in direction to the 5,000-pound load.

$$\mathbf{C} + \mathbf{T} = \mathbf{-5{,}000} \quad \text{or} \quad \mathbf{C} + \mathbf{T} + \mathbf{5{,}000} = \mathbf{0}$$

Also,

$$\frac{\|\mathbf{C}\|}{\sin 75°} = \frac{5{,}000}{\sin 55°}, \qquad \frac{\|\mathbf{T}\|}{\sin 50°} = \frac{5{,}000}{\sin 55°}$$

$$\|\mathbf{C}\| = 5{,}896 \text{ pounds}, \qquad \|\mathbf{T}\| = 4{,}676 \text{ pounds}.$$

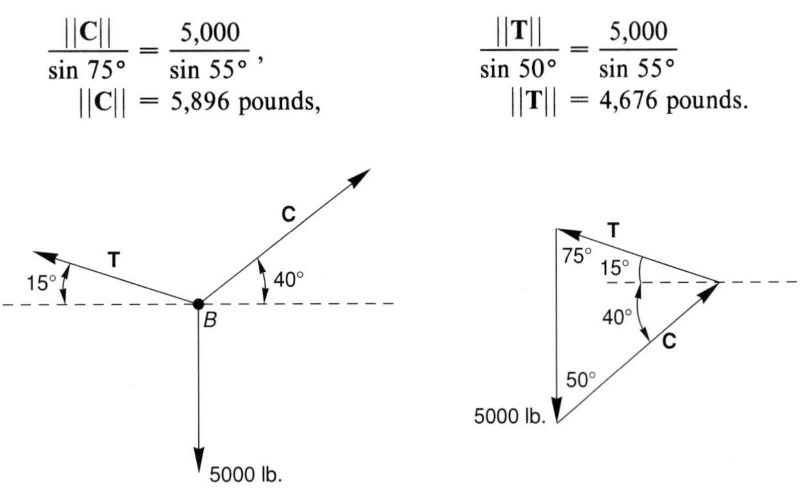

Alternate Solution Since point B is in equilibrium, the algebraic sum of the horizontal components at B must be 0. The algebraic sum of the vertical components must also be 0.

$$\|\mathbf{C}\| \cos 40° - \|\mathbf{T}\| \cos 15° = 0$$
$$\|\mathbf{C}\| \sin 40° + \|\mathbf{T}\| \sin 15° - 5{,}000 = 0$$

Solving the system of equations yields $\|\mathbf{C}\| = 5{,}896$ and $\|\mathbf{T}\| = 4{,}676$. ■

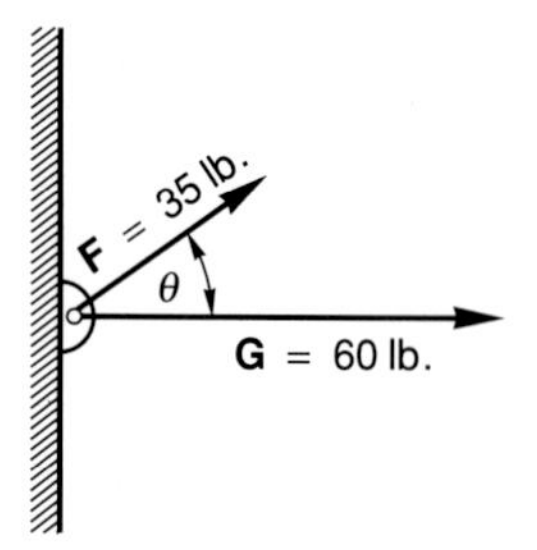

Example 9 A 35-pound force **F**, and a 60-pound force **G**, are applied at a pin support. What angle θ should **F** make with **G** so that $||\mathbf{F} + \mathbf{G}|| = 80$ pounds? What angle α, does $\mathbf{F} + \mathbf{G}$ make with **G** when $||\mathbf{F} + \mathbf{G}|| = 80$ pounds?

Solution In the figure below, $\mathbf{F} + \mathbf{G}$ is called the **resultant** of **F** and **G.** It is a single force that has the same effect on the pin as **F** and **G** acting together. To find θ, we apply the Law of Cosines to the force triangle shown below.

$$\cos(180 - \theta) = \frac{||\mathbf{F}||^2 + ||\mathbf{G}||^2 - ||\mathbf{F} + \mathbf{G}||^2}{2||\mathbf{F}||\,||\mathbf{G}||}$$

$$180 - \theta = \frac{\cos^{-1}(35^2 + 60^2 - 80^2)}{2 \cdot 35 \cdot 60}$$

$$180 - \theta = 112°$$

$$\theta = 68°$$

$$\frac{\sin \alpha}{35} = \frac{\sin 112°}{80}$$

$$\alpha = 23.9°$$

■

Exercise Set 9.4

(A)

In exercises 1–6, (a) graph a vector $\overline{PQ}$ given $||\overline{PQ}||$ and θ, the counterclockwise angle that $\overline{PQ}$ makes with the positive x-axis, and (b) determine the components of $\overline{PQ}$.

1. $||\overline{PQ}|| = 8, \theta = 45°$
2. $||\overline{PQ}|| = 10, \theta = 120°$
3. $||\overline{PQ}|| = 6, \theta = 150°$
4. $||\overline{PQ}|| = 50, \theta = 220°$
5. $||\overline{PQ}|| = 100, \theta = 300°$
6. $||\overline{PQ}|| = 25, \theta = 260°$

Given points P and Q, determine, (a) $\overline{PQ}$ as an ordered pair, and (b) $||\overline{PQ}||$.

7. $P(0,0), Q(5,8)$
8. $P(-6,2), Q(4,-11)$
9. $P(5,5), Q(2,-3)$
10. $P(-6,-1), Q(-8,7)$
11. $P(10,4), Q(-3,12)$
12. $P(7,3), Q(2,9)$

A + **B** is the vector from the tail of **A** to the head of **B** after the tail of **B** is placed on the head of **A**. Obtain **A** + **B** this way and then verify your results by adding the corresponding components of **A** and **B.**

13. $\mathbf{A} = (5,-2), \mathbf{B} = (3,4)$
14. $\mathbf{A} = (8,3), \mathbf{B} = (-2,5)$
15. $\mathbf{A} = (-6,-2), \mathbf{B} = (4,-3)$
16. $\mathbf{A} = (-1,1), \mathbf{B} = (7,6)$
17. $\mathbf{A} = (3,-2), \mathbf{B} = (-8,3)$
18. $\mathbf{A} = (4,5), \mathbf{B} = (3,2)$

19. A car travels 180 miles on a N 20° E bearing. It then travels 215 miles on a S 30° E bearing. How far is the car from the starting point? What is the bearing of the car from the starting point?

20. A driver is S 42° W and 600 mi. from the starting point of a two-day car trip. He traveled N 74° W and 350 miles the first day. How far, and in which direction, did he travel the second day?

21. An airplane has an airspeed of 525 mph and a heading of 55°. An 80-mph wind from the northwest with a heading of 160° affects the plane's course and speed. What is the plane's actual heading and its ground speed?

22. An airplane has an airspeed of 180 mph and a 250° heading. The plane's groundspeed is 200 mph because of a wind that has a 220° heading. Find the wind speed and the plane's actual course.

23. A swimmer starts across a river from its west bank. The river flows south at 2 mph. If the swimmer can swim 4 mph in still water, in what direction should he head to swim directly across the river? What is his actual speed as he crosses the river?

24. An ocean liner is traveling due east with a speed of 20 mph. A deckhand walks N 45° W across the deck at the rate of 3 mph. What is the deckhand's speed and direction relative to the earth?

25. A pilot wants to fly his plane due north. The plane has an airspeed of 200 mph. A 50-mph wind with a 265° heading alters the plane's course. What should the pilot's heading be so that his course is due north? What is his flying time for a 180-mi. trip?

26. A plane heads east with an airspeed of 160 mph. A 30-mph wind from the southeast forces the plane into an 85° course. Find the plane's groundspeed and the heading of the wind.

27. A boat heads S 25° E. Its speed in the water is 16 mph. After 2 hours the boat has traveled 37 miles and is S 31° E of its starting point. Find the current speed and direction.

28. A $\frac{1}{2}$-mile-wide river has a due east, 3-mph current. A boat leaves the south bank of the river and crosses diagonally to a point on the north bank that is $\frac{1}{4}$ mile downstream from the starting point. The trip takes 8 minutes. What is the heading of the boat, and what is its waterspeed?

29. A 10,000-lb. load is supported by a cable and a boom of negligible weight that carries only compression (see example 8). The boom makes an angle of 70° with the vertical, and the cable makes an angle of 55° with the vertical. Find the compression in the boom and the tension in the cable.

30. A 200-pound force, which makes an angle of 26° with the horizontal, is to be replaced by two forces, a horizontal force **F** and a 110-pound force. Find $||\mathbf{F}||$.

31. An object weighing 800 pounds is held in equilibrium by two cables that make angles of 30° and 40° with the horizontal. What is the tension in each cable?

32. An object weighing 1,200 pounds is held in equilibrium by two cables making angles of 47° and 38° with the vertical. What is the tension in each cable?

33. A 500-pound box is prevented from sliding down a frictionless, inclined plane by a cable parallel to the plane. Find the tension in the cable.

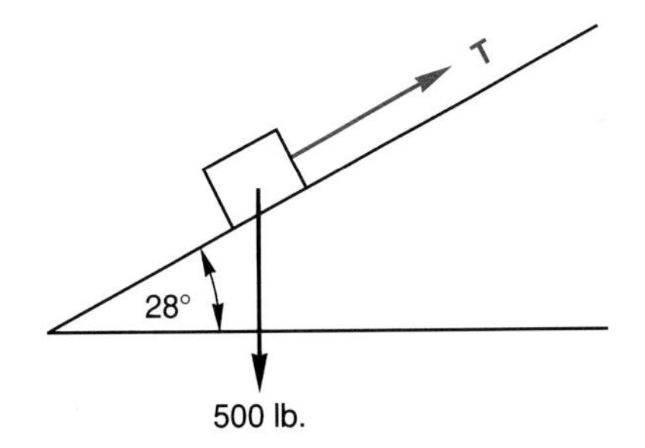

34. If in the figure for exercise 33 the incline is 35° and **T** = 420 pounds, find the weight of the box.

(B)

35. An astronaut can practice moon walking on the device shown in the figure. A sling holds the astronaut perpendicular to the inclined plane. The sling is attached to a cable, which is attached to a trolley that runs along an overhead track. The device allows the astronaut to move freely in a plane perpendicular to the inclined plane. If the device is to simulate lunar gravity, which is one-sixth earth's gravity, what must θ be?

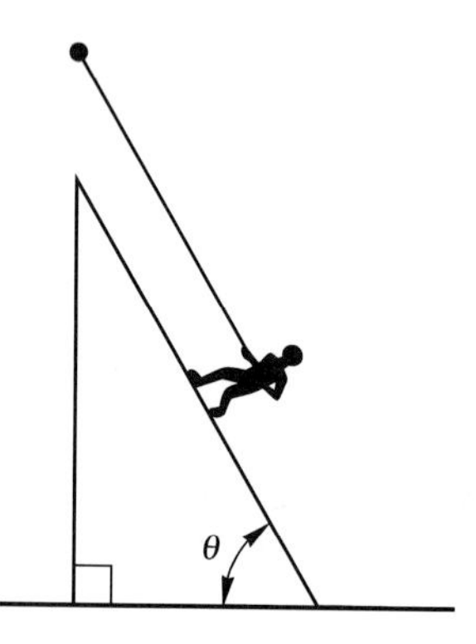

36. Resolve the 250-pound force along the lines AC and CB. These are the forces exerted on the pins at A and B.

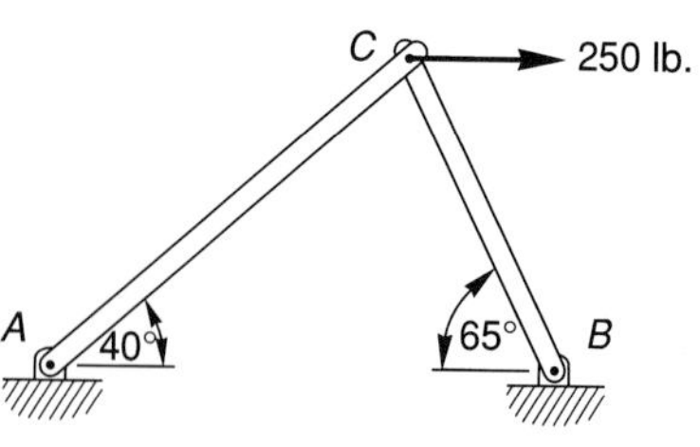

37. Find the tension in the guy wire and the compression in the compression strut.

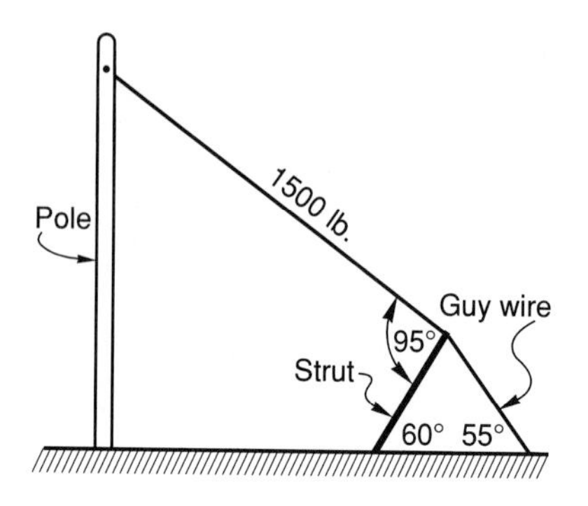

38. The tensile capacity of each cable in the figure is 10,000 pounds. Find the smallest angle θ that the cables can make with the horizontal before they are loaded beyond capacity.

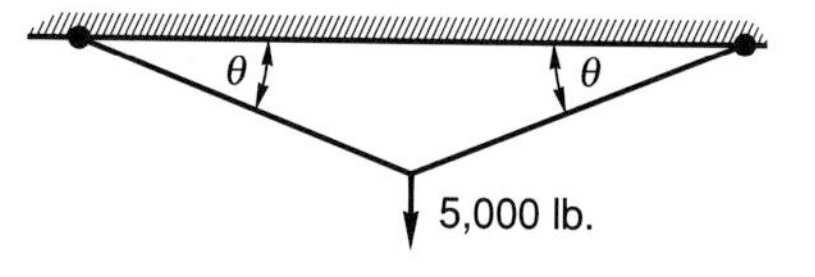

9.5 Trigonometric Models of Periodic Phenomena

Sound is propagated through air as a periodic pressure wave. A sound source, be it the human voice, a musical instrument, or a car engine, forces air molecules to move back and forth. Each vibrated molecule moves in essentially the same way as the piston of the water pump in section 7.1. The sound, or pressure wave, strikes the eardrum over some period of time, causing it to vibrate. That sensation is recorded by the brain as sound. The back and forth vibratory motions of the pump piston, the air molecules, and the eardrum all have the same mathematical description.

$$y = A \sin 2\pi kt \tag{1}$$

Intensity, or loudness of a sound, is indicated by the amplitude (A) of the pressure wave given by (1). Because $2\pi/2\pi k = 1/k$, a graph of (1) exhibits one cycle every $1/k$ seconds. Therefore, the pressure wave represented by (1) has a **frequency** of k cycles per second, or k **hertz** (***HZ***).

Beats

The air around us is scrambled with pressure waves of varying amplitudes and frequencies. As a result, the sound we hear at any instant is produced by a pressure wave of the form,

$$y = A_1 \sin 2\pi k_1 t + A_2 \sin 2\pi k_2 t + A_3 \sin 2\pi k_3 t + \cdots. \tag{2}$$

Two sounds that have pressure waves with the same amplitude and frequency can cancel each other and produce silence (figure 9.15).

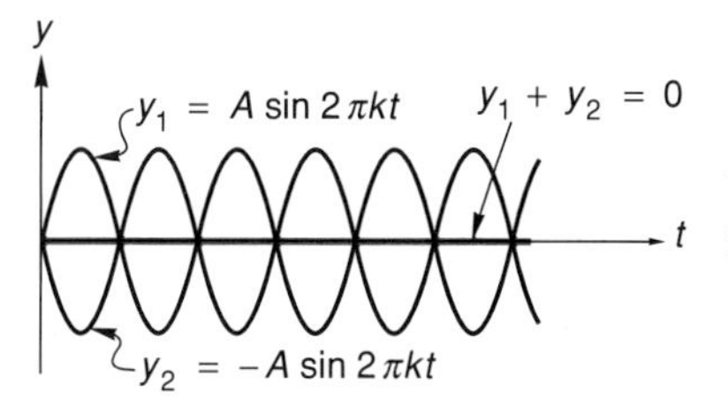

Figure 9.15

$$y = A \sin 2\pi kt - A \sin 2\pi kt = 0$$

A more interesting phenomenon occurs when a point in space is disturbed by two pressure waves that have the same amplitude but slightly different frequencies.

Assume that pressure waves of amplitude A and frequencies 28 HZ and 32 HZ are heard over a time interval. The equation,

$$y = A \sin 64\pi t + A \sin 56\pi t, \tag{3}$$

describes the sound generated by the pressure waves. We shall analyze that sound after constructing another form of (3).

Since

$$\sin(x + y) = \sin x \cos y + \cos x \sin y$$

and

$$\sin(x - y) = \sin x \cos y - \cos x \sin y,$$

then

$$\sin(x + y) + \sin(x - y) = 2 \sin x \cos y. \tag{4}$$

If

$$x + y = 64\pi t$$

and

$$x - y = 56\pi t,$$

then

$$2x = 120\pi t, \quad 2y = 8\pi t$$

and

$$x = 60\pi t, \quad y = 4\pi t.$$

Now, using (4) and the obtained values for x and y, (3) can be written as

$$\begin{aligned} y &= A(\sin 64\pi t + \sin 56\pi t) \\ &= A(2 \sin 60\,\pi t \cos 4\pi t) \\ &= 2A \cos 4\pi t \sin 60\,\pi t. \end{aligned} \tag{5}$$

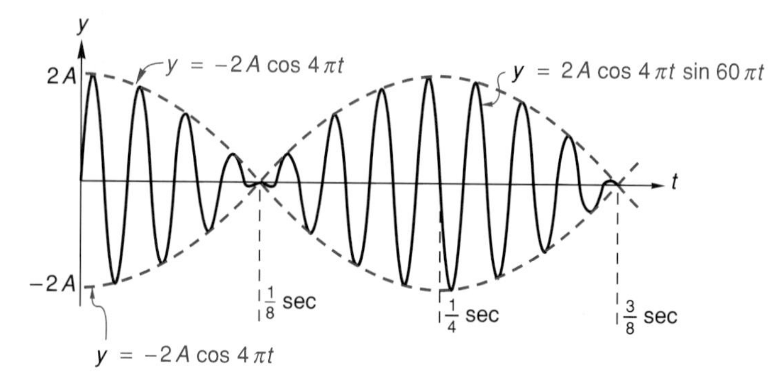

Figure 9.16

A graph of (5) is simply a graph of $y = \sin 60\pi t$ *modulated* by the factor $2A \cos 4\pi t$. The rapidly oscillating sine curve is constrained to lie between the curves

$$y = 2A \cos 4\pi t \quad \text{and} \quad y = -2A \cos 4\pi t \text{ (figure 9.16).}$$

The sound described by (5) is alternately loud and soft, four times per second. One hears four beats a second.

A concert violinist may tune his violin's A string by altering its tension until he hears no beats between the vibrating A string and a standard A note struck on the piano.

Modulation Theory

Information can be transmitted by electromagnetic waves in the form of sound (radio), or visual images (television). Because the audible range of the human ear, 20–20,000 *HZ,* is too low for practical radio transmission, audible frequencies are transmitted "piggyback" on very high frequency radio waves called carrier waves. Low frequency sound information is imposed on a carrier wave, thereby modulating the shape of the wave. A radio receiver demodulates the altered carrier to recover the audible information. Figure 9.17 illustrates amplitude modulation (AM) and frequency modulation (FM).

Let $y = A \sin 2\pi Kt$ represent a carrier wave and let $y = C \sin 2\pi kt$ represent an audio signal to be transmitted by AM. The modulated wave has the form

$$y = A[1 + C \sin 2\pi kt] \sin 2\pi Kt, \quad 0 < C < 1 \qquad \textbf{(6)}$$

The constant A is related to the intensity of the radio station's signal. The constant C bears on the quality of the signal transmitted.

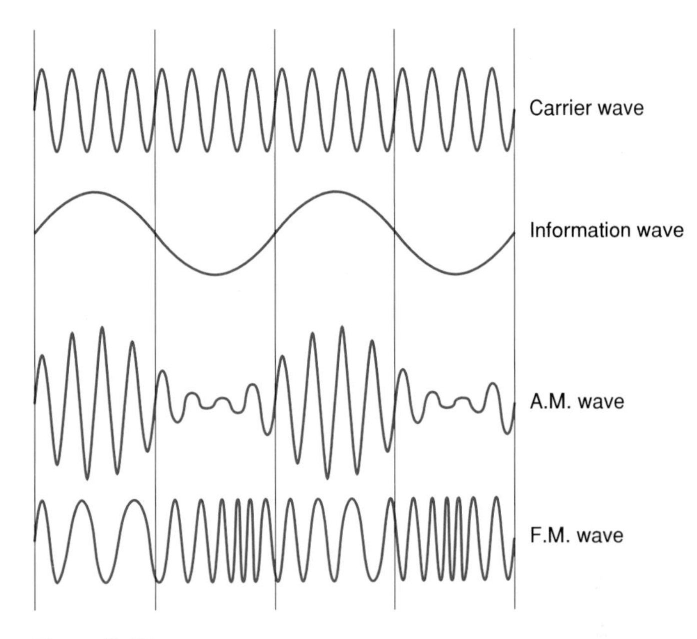

Figure 9.17

Equation (6) can be shown to represent three distinct signals, the carrier wave oscillating at a particular radio station's frequency and two signals, each of which carries the audio information being transmitted.

Rewriting (6) as

$$A[\sin 2\pi Kt + C \sin 2\pi kt \sin 2\pi Kt],$$

and noting (from exercise 21, section 8.1) that

$$\sin 2\pi kt \sin 2\pi Kt = \tfrac{1}{2} \cos 2\pi(K - k)t - \tfrac{1}{2} \cos 2\pi(K + k)t,$$

we are able to represent the AM wave as

$$y = A[\sin 2\pi Kt + \frac{C}{2} \cos 2\pi(K - k)t - \frac{C}{2} \cos 2\pi(K + k)t]. \qquad (7)$$

An AM receiver has resonance circuitry that enables one to tune the receiver to the frequency of the carrier wave $\sin 2\pi Kt$. Other circuitry in the receiver converts the information carrying signals, $C/2 \cos 2\pi(K - k)t$ and $C/2 \cos 2\pi(K + k)t$, into mechanical motion of the speaker.

Example 1 Write an equation for the signal generated by radio station 105 AM when it broadcasts a sound with frequency 400 HZ. Determine the frequencies of the two audio carrying signals.

Solution The form of (6) gives us an equation for the signal generated by station 105 AM.

$$y = A[1 + C\sin(2\pi \cdot 400)t]\sin(2\pi \cdot 105{,}000)t$$

The audio carrying signals are obtained from (7). They are

$$\frac{C}{2}\cos 2\pi(105{,}000 - 400)t$$

and

$$\frac{C}{2}\cos 2\pi(105{,}000 + 400)t.$$

Thus, the frequencies of the audio carrying signals are 104,600 HZ and 105,400 HZ. ■

Exercise Set 9.5

(A)

Follow the sequence of steps that leads from equation (3) to equation (5) in this section, and convert each sine sum to a sine-cosine product.

1. $\sin 6t + \sin 10t$
2. $\sin 30t + \sin 20t$
3. $10\sin 24\pi t + 10\sin 28\pi t$
4. $15\sin 20\pi t + 15\sin 14\pi t$
5. $\sin 120\pi t + \sin 100\pi t$
6. $\sin 264\pi t + \sin 132\pi t$

7. Use the results of exercise 3 to sketch a portion of the graph of $y = 10\sin 24\pi t + 10\sin 28\pi t$ (see figure 9.16). How many beats per second are denoted by the graph?
8. Use the results of exercise 5 to sketch a portion of the graph of $y = \sin 120\pi t + \sin 100\pi t$. How many beats per second are denoted by the graph?
9. Write an equation for the signal generated by radio station 100 AM when it broadcasts a sound with frequency 500 HZ. Determine the frequencies of the two audio carrying signals. (See example 1.)
10. Write an equation for the signal generated by radio station 98 AM when it broadcasts a sound with frequency 260 HZ. Determine the frequencies of the two audio carrying signals.

(B)

11. Assume that the periodic motion of a piston in a car engine can be described by equation (1) of this section, $y = A\sin 2\pi kt$. Write an equation that describes the motion of a piston with a 10-cm stroke in an engine that is operating at 3,000 rpm. (Hint: Figure 7.1 might help you determine A.)

12. In city M daylight lasts 12 hours on March 21 and September 21. The city experiences 14.5 hours of daylight on June 21 and 9.5 hours of daylight on December 21. Assume that the variation from 12 hours in the number of hours of daylight in city M is a periodic function of the form

$$y = A \sin 2\pi kt.$$

Write an equation for the number of hours of daylight in city M for any day of the year. (Hint: Express t in days. March 21 is the 80th day of the year.)

13. A pendulum vibrates 25 times a second. It swings through an angle of 10° from the vertical. Use equation (1) of this section to express the angular displacement of the pendulum as a function of time.

14. The energy released at impact when a car goes over a bump in the road is absorbed in the springs of the car. With no constraints on them, the springs would alternately contract and expand, causing the car to bounce all over the road. Shock absorbers constrain the springs by "dampening" their oscillations. A mathematical description of such a dampening process is given by

$$y = e^{-t} \cos 10\pi t.$$

Graph the function for $0 \leq t \leq 1$.

Chapter 9 Review

Major Points for Review

Section 9.1 The relationship between angular and linear velocity

Applications

Section 9.2 Definitions of the trigonometric functions in terms of the sides and the hypotenuse of a right triangle

Applications of right triangle models

Section 9.3 Law of sines

Law of cosines

Applications of non-right triangle models

Section 9.4 Representation of a vector as an arrow

The x and y components of a vector

A vector as an ordered pair

Magnitude of a vector

Geometric addition of two vectors

Algebraic addition of two vectors

Applications of vector models

Review Exercises

Convert each angular velocity to rpm.

1. 200π radians/sec.
2. 30,000 radians/hr.

Solve each right triangle from the given data. Assume that the triangle has angles A, B, and C where C is a right angle. Assume also that the lengths of the sides opposite angles A, B, and C are a, b, and c respectively.

3. $A = 30°, b = 25$
4. $B = 58°, c = 18$
5. $a = 1{,}485, c = 1{,}720$
6. $b = 15.3, c = 24.8$

Use the Law of Sines or the Law of Cosines or both laws to solve each triangle.

7. $A = 45°, C = 60°, b = 12.0$
8. $A = 100°, B = 40°, c = 32.5$
9. $c = 106.4, A = 83°, b = 98.3$
10. $C = 130°, b = 42.8, a = 18.9$
11. $a = 94.2, b = 28.6, c = 74.7$
12. $B = 68°, b = 241, c = 218$

The magnitude and direction, θ, of vector **A** are given. Determine the components of **A**.

13. $||\mathbf{A}|| = 140, \theta = 162°$
14. $||\mathbf{A}|| = 27, \theta = 235°$

In exercises 15–16, the ordered pair representation of vectors **A** and **B** are given. Graph **A**, **B**, and **A** + **B**. Then determine **A** + **B** as an ordered pair.

15. $\mathbf{A} = (8,5), \mathbf{B} = (-3,3)$
16. $\mathbf{A} = (-2,-7), \mathbf{B} = (6,-1)$
17. A belt-driven pulley turns through 750 rpm. The pulley has a 10-inch diameter. What is the linear velocity, in ft./sec., of the belt?
18. A car with 30-inch diameter wheels is moving along the highway at 65 mph. What is the angular velocity, in rpm, of the wheels?
19. A gear rotating at 60 rpm meshes with and drives another gear at 85 rpm. The distance between the centers of the gears is 26 cm. Find the radii of the gears.
20. A 9-inch diameter pulley rotating at 500 rpm drives a belt that is attached to a 5-inch diameter pulley. What is the angular velocity, in rpm, of the smaller pulley?
21. Assuming that the radius of the earth is 4,000 miles, what is the linear velocity of a point on earth (a) at the equator and (b) at 40° north latitude?
22. At a point on level ground that is 200 feet from the base of a building, the angle of elevation to the top of the building is 64°. How high is the building?
23. A surveyor takes sightings on a point M that is on a distant mountaintop, from two points, A and B, that are on level ground. Points M, A, and B are in the same vertical plane. The angle of elevation from A to M is 48°–20′. The angle of elevation from B to M is 36°–30′. The distance between A and B is 612 feet. How high above A and B is point M?
24. An aerial photograph of a mountain crater shows that the side of the crater cast a shadow $\frac{1}{2}$ inch wide when the elevation of the sun was 42°. If the scale of the photograph is $1'' = \frac{1}{2}$ mile, how deep is the crater?

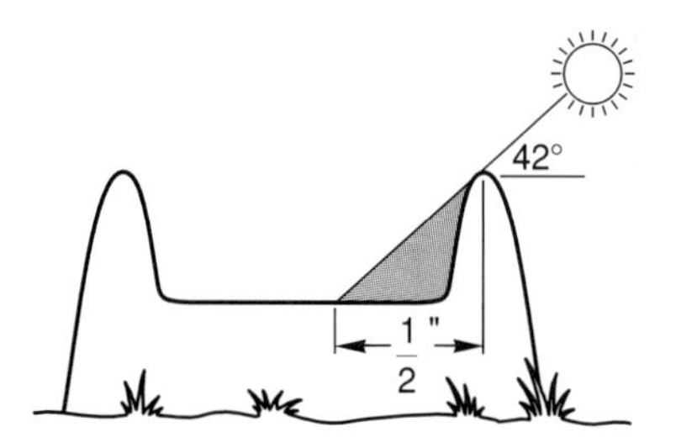

25. A helicopter is flying at 600 feet elevation. Five seconds after it is directly overhead, the angle of elevation to the helicopter is 40°. How fast is the helicopter flying (mph)?
26. Two points on earth, at latitude 58° N, are at longitudes 28° E and 14° W. Find the distance between the points. Assume that the radius of the earth is 4,000 mi.
27. Two cars leave at the same instant from a point where two highways intersect. The highways make an angle of 70° with each other. The cars speed along the two highways at 60 and 55 mph. How far apart are the cars two hours after they leave the highway intersection?

vibrate. That sensation is recorded by the brain as sound. The back and forth vibratory motions of the pump piston, the air molecules, and the eardrum all have the same mathematical description.

$$y = A \sin 2\pi kt \tag{1}$$

Intensity, or loudness of a sound, is indicated by the amplitude (A) of the pressure wave given by (1). Because $2\pi/2\pi k = 1/k$, a graph of (1) exhibits one cycle every $1/k$ seconds. Therefore, the pressure wave represented by (1) has a **frequency** of k cycles per second, or k **hertz** (***HZ***).

Beats

The air around us is scrambled with pressure waves of varying amplitudes and frequencies. As a result, the sound we hear at any instant is produced by a pressure wave of the form,

$$y = A_1 \sin 2\pi k_1 t + A_2 \sin 2\pi k_2 t + A_3 \sin 2\pi k_3 t + \cdots. \tag{2}$$

Two sounds that have pressure waves with the same amplitude and frequency can cancel each other and produce silence (figure 9.15).

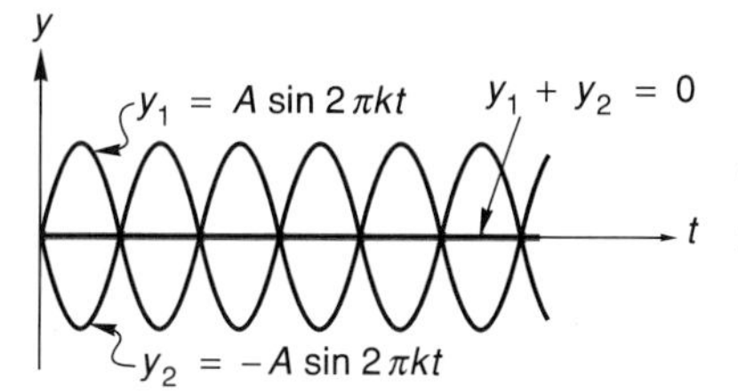

Figure 9.15

$$y = A \sin 2\pi kt - A \sin 2\pi kt = 0$$

A more interesting phenomenon occurs when a point in space is disturbed by two pressure waves that have the same amplitude but slightly different frequencies.

Assume that pressure waves of amplitude A and frequencies 28 HZ and 32 HZ are heard over a time interval. The equation,

$$y = A \sin 64\pi t + A \sin 56\pi t, \tag{3}$$

describes the sound generated by the pressure waves. We shall analyze that sound after constructing another form of (3).

Since

$$\sin(x + y) = \sin x \cos y + \cos x \sin y$$

and

$$\sin(x - y) = \sin x \cos y - \cos x \sin y,$$

then

$$\sin(x + y) + \sin(x - y) = 2 \sin x \cos y. \tag{4}$$

If

$$x + y = 64\pi t$$

and

$$x - y = 56\pi t,$$

then

$$2x = 120\pi t, \quad 2y = 8\pi t$$

and

$$x = 60\pi t, \quad y = 4\pi t.$$

Now, using (4) and the obtained values for x and y, (3) can be written as

$$\begin{aligned} y &= A(\sin 64\pi t + \sin 56\pi t) \\ &= A(2 \sin 60\, \pi t \cos 4\pi t) \\ &= 2A \cos 4\pi t \sin 60\, \pi t. \end{aligned} \tag{5}$$

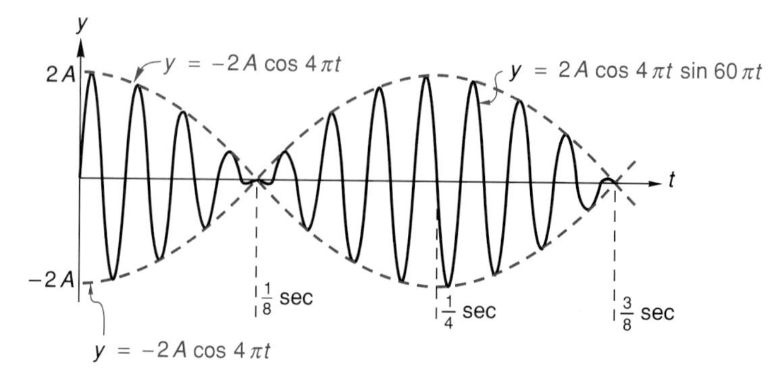

Figure 9.16

A graph of (5) is simply a graph of $y = \sin 60\pi t$ *modulated* by the factor $2A \cos 4\pi t$. The rapidly oscillating sine curve is constrained to lie between the curves

$$y = 2A \cos 4\pi t \quad \text{and} \quad y = -2A \cos 4\pi t \text{ (figure 9.16).}$$

The sound described by (5) is alternately loud and soft, four times per second. One hears four beats a second.

A concert violinist may tune his violin's A string by altering its tension until he hears no beats between the vibrating A string and a standard A note struck on the piano.

Modulation Theory

Information can be transmitted by electromagnetic waves in the form of sound (radio), or visual images (television). Because the audible range of the human ear, 20–20,000 *HZ*, is too low for practical radio transmission, audible frequencies are transmitted "piggyback" on very high frequency radio waves called carrier waves. Low frequency sound information is imposed on a carrier wave, thereby modulating the shape of the wave. A radio receiver demodulates the altered carrier to recover the audible information. Figure 9.17 illustrates amplitude modulation (AM) and frequency modulation (FM).

Let $y = A \sin 2\pi Kt$ represent a carrier wave and let $y = C \sin 2\pi kt$ represent an audio signal to be transmitted by AM. The modulated wave has the form

$$y = A[1 + C \sin 2\pi kt] \sin 2\pi Kt, \quad 0 < C < 1 \tag{6}$$

The constant A is related to the intensity of the radio station's signal. The constant C bears on the quality of the signal transmitted.

28. A 10-foot pole stands on the side of a hill that inclines 20° with the horizontal. The pole inclines 5° from the vertical toward the uphill side of the incline. The angle of elevation of the sun is 60°. What is the length of the downhill shadow cast by the pole?

29. Points P and Q are on opposite sides of a lake. To find the distance from P to Q, points R and S are located as shown in the figure and angles SRQ and RSP are measured with a transit. The lengths $\overline{PS}$, $\overline{RS}$, and $\overline{RQ}$ are determined. With the measurements shown in the figure, find the distance between P and Q.

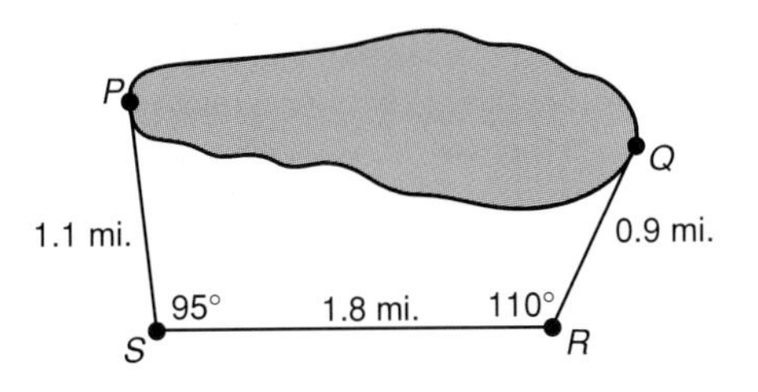

30. The sides of a triangular piece of land measure 246.8 ft., 391.2 ft., and 355.6 ft. Find the angles at the corners of the property.

31. Use the measurements in the figure to estimate the diameter of the moon.

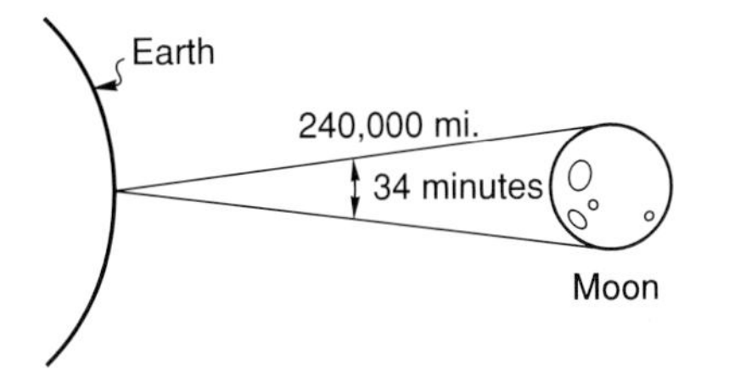

32. A plane flies 340 miles on a N 15° W bearing and then 280 miles on a S 80° W bearing. How far is the plane from its starting point, and what is its bearing from the starting point?

33. A plane has an airspeed of 450 mph and a bearing of S 72° E. The plane's course and speed are affected by a 55 mph wind that has a bearing of N 48° E. What is the plane's actual bearing and its ground speed?

34. A boat heads N 68° W. Its engine speed is 20 mph. After three hours the boat has traveled 70 miles and is N 71° W of its starting point. Find the speed and direction of the current.

35. An object weighing 1,000 pounds is held in equilibrium by two cables making angles of 35° and 42° with the horizontal. What is the tension in each cable?

36. An 850-pound box is prevented from sliding down a frictionless, inclined plane by a cable parallel to the plane. What is the tension in the cable?

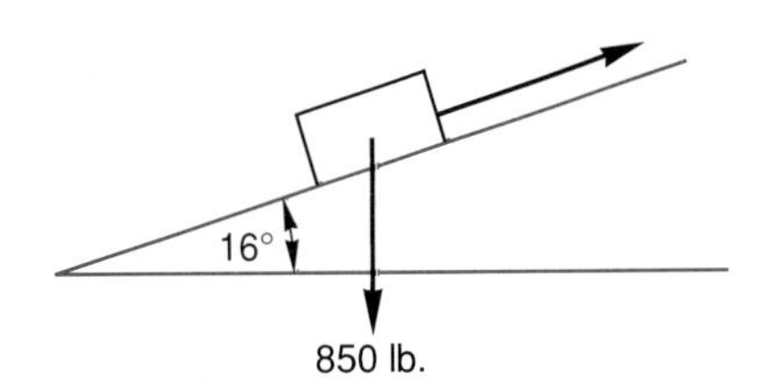

37. Convert the sum, $\sin 8\pi t + \sin 12\pi t$, into a sine-cosine product. Sketch a portion of the graph of $y = \sin 8\pi t + \sin 12\pi t$. How many beats per second are indicated by your sketch?

38. Use the equation $y = A \sin 2\pi kt$ to describe the motion of a car engine piston that has an 8-cm stroke. The engine is operating at 4,000 rpm.

39. Find the length of the continuous belt.

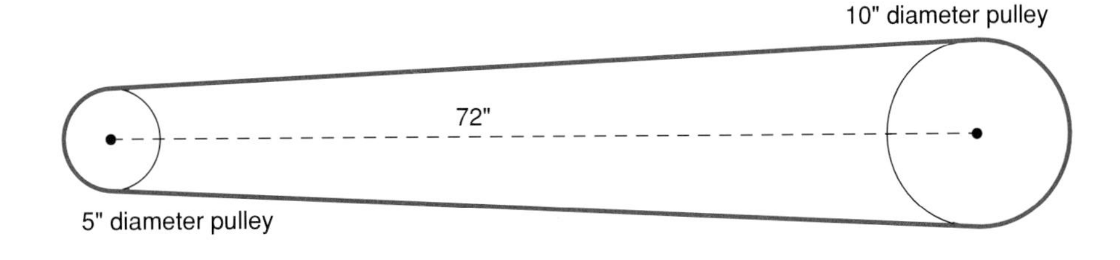

40. In the figure, D is the top of a tower. Points A, B, and C are on level ground, and C is directly below D. Find the height of D above C from the information in the diagram.

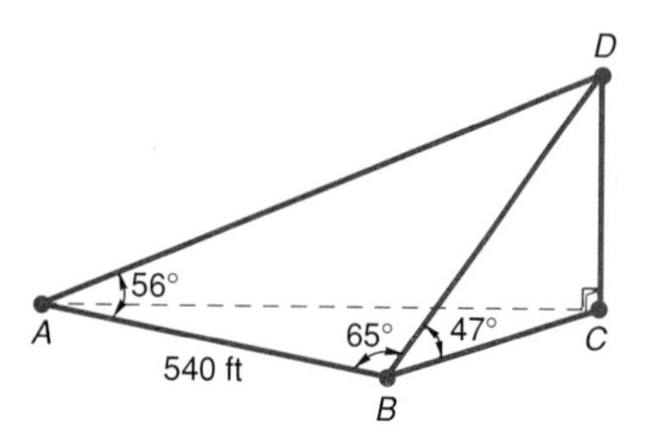

41. Find the angle measures of triangle ABC.

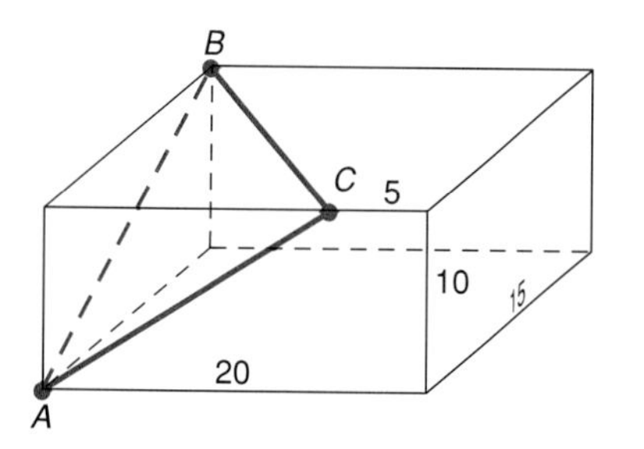

42. Find the compressive load in the boom and the tensile load in the cable.

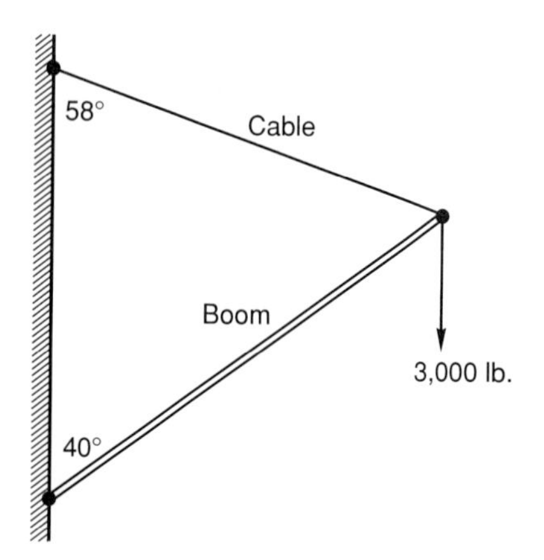

Chapter 9 Test 1

1. One angle of a right triangle is 36°. The side opposite the right angle measures 245 cm. Solve the right triangle.

The angles of a triangle are A, B, and C, and the sides opposite those angles have measures a, b, and c, respectively. Solve the triangle with the given information.

2. $B = 108°$, $C = 23°$, $b = 19.5$
3. $c = 12.8$, $B = 72°$, $a = 15.1$

4. An 8-inch diameter pulley rotating at 400 rpm drives other pulleys with a belt. What is the linear speed of the belt?
5. An observer on a cliff that is 800 feet above a body of water sights a boat on the water below. The angle of depression of the line of sight is 26°. How far is the boat from the observer?
6. Two cars start at the same point. One car travels 65 mph on a road that bears N 8° W from the starting point. The second car travels 55 mph due east. How far apart are the cars after four hours?
7. A tunnel is to be dug through the side of a mountain between points P and Q. A point C is established from which there are clear lines of sight to P and Q. The distances from C to P and from C to Q are measured to be 826 ft. and 641 ft. Angle PCQ is measured to be 62°. What is the length of the tunnel?
8. Determine the angle that the line $x + 2y = 6$ makes with the x-axis.
9. A power boat has a still water speed of 10 mph. In what direction should the boat be headed to move due east across a $\frac{1}{4}$ mile river that has a 2-mph south current? How long does it take the boat to cross the river?

10. A 50-pound force $\mathbf{F}_1$, and a horizontal 70-pound force $\mathbf{F}_2$, are applied at a pin support. What angle, θ, should $\mathbf{F}_1$ make with $\mathbf{F}_2$ so that $||\mathbf{F}_1 + \mathbf{F}_2|| = 100$ pounds?

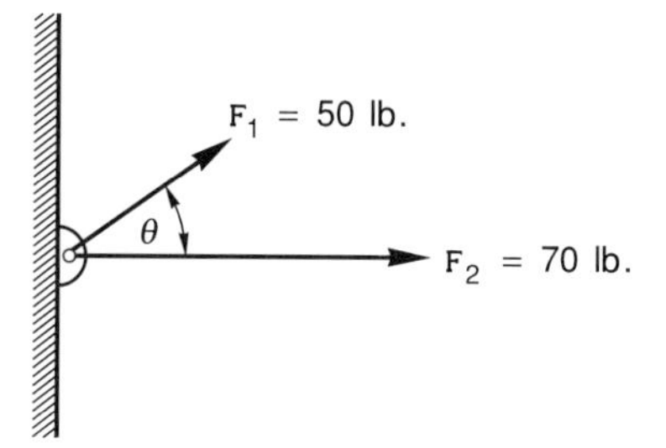

Chapter 9 Test 2

1. The sides of a right triangle measure 5 and 12 units. Solve the triangle.

The angles of a triangle are A, B, and C, and the lengths of the sides opposite those angles are a, b, and c, respectively. Solve the triangle with the given information.

2. $A = 54°$, $a = 106.8$, $b = 115.4$, B is obtuse.
3. $a = 26.3$, $b = 85.1$, $c = 70.5$

4. What is the linear velocity of a point on earth that is on the 30° N circle of latitude? Assume that the radius of the earth is 4,000 miles.
5. The angles of elevation from points A and B, in a level meadow, to the top of a distant hill are 24° and 31°. The top of the hill and points A and B are in the same vertical plane. The distance between A and B is 460 feet. How high is the top of the hill above the meadow?
6. On a certain day the sun rose at 5:00 A.M. and set at 8:00 P.M. Some time during the day a 12-foot pole cast an 8-foot shadow in the westerly direction. What time of day was it?
7. From observation points A and B that are three miles apart, a plane is spotted in level flight. The angle at A from B to the plane is 48°, and the angle at B from A to the plane is 66°. A second pair of observations are made 20 seconds later. At that time the angle at A from B to the plane is 40°, and the angle at B from A to the plane is 96°. What is the speed of the plane?
8. Find the angle measures of triangle ABC.

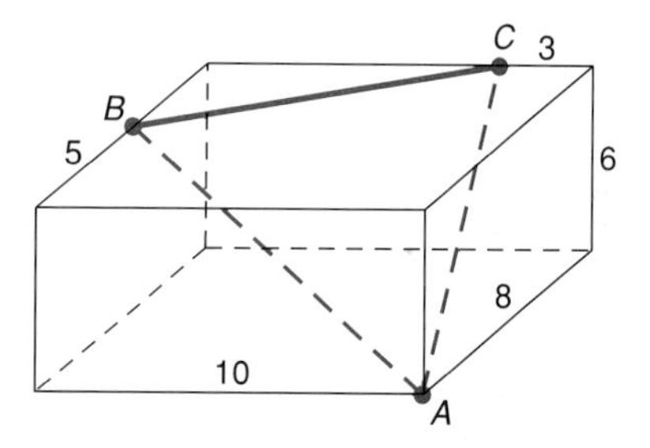

9. A plane has a bearing of N 84° E and an air speed of 500 mph. In flight the plane encounters a 50-mph wind that has a bearing of S 70° E. Find the true bearing of the plane and its ground speed.
10. An object weighing 2,000 pounds is held in equilibrium by two cables making angles of 32° and 26° with the horizontal. Find the tension in each cable.

Chapter 10

Polar Coordinate Graphs and Parametric Equations

Certain relationships can be expressed more simply in a coordinate system that is different from the rectangular coordinate system. The rectangular coordinate equation of a circle of radius a, centered at the origin, is $x^2 + y^2 = a^2$. The same circle has the equation $r = a$ in the **polar coordinate system.** That system is introduced in this chapter.

We learn how to sketch polar coordinate graphs. Polar coordinates and the notion of a parameter, which was introduced in chapter 3, are employed to help us construct a number of relationships that are either very difficult or impossible to construct in the rectangular coordinate system. Finally, we use polar coordinates to extend the "Theory of Equations" of chapter 4 to find complex roots of polynomial equations.

10.1 Polar Coordinates

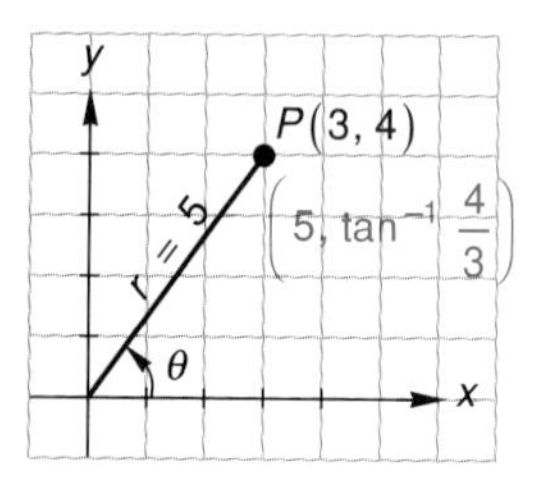

Figure 10.1

A point P can be located in a plane relative to the origin of a rectangular coordinate system imposed in the plane (figure 10.1). Thus, $P(3,4)$ is three units to the right of and four units above the origin. P may also be located relative to the origin by assigning to it a pair of numbers (r,θ). The distance from P to the origin is given by r, and θ is an angle between the positive x-axis and a ray through P. In figure 10.1,

$$r = \sqrt{3^2 + 4^2} = \mathbf{5} \quad \text{and} \quad \theta = \mathbf{tan^{-1}\tfrac{4}{3}}.$$

The numbers 5 and $\tan^{-1}\frac{4}{3}$ are called the **polar coordinates** of P.

Every point in the plane can be located relative to the origin by imposing a polar coordinate system in the plane. In such a system (figure 10.2) the positive x-axis is called the **polar axis** and the origin is called the **pole.** Also, each point in the plane has *infinitely* many ordered number pairs associated with it. $P(5,\pi/6)$ is located five units from the pole on a ray through the pole that makes an angle of $\pi/6$ radians with the polar axis. P may also be located five units from the pole in the direction *opposite* to the ray that makes an angle of $7\pi/6$ radians with the polar axis. So located, P has the polar coordinates $(-5,7\pi/6)$. Other polar coordinate descriptions of P are $(5,13\pi/6)$, $(-5,-5\pi/6)$, $(-5,19\pi/6)$, $(5,-11\pi/6)$, and $(5,25\pi/6)$.

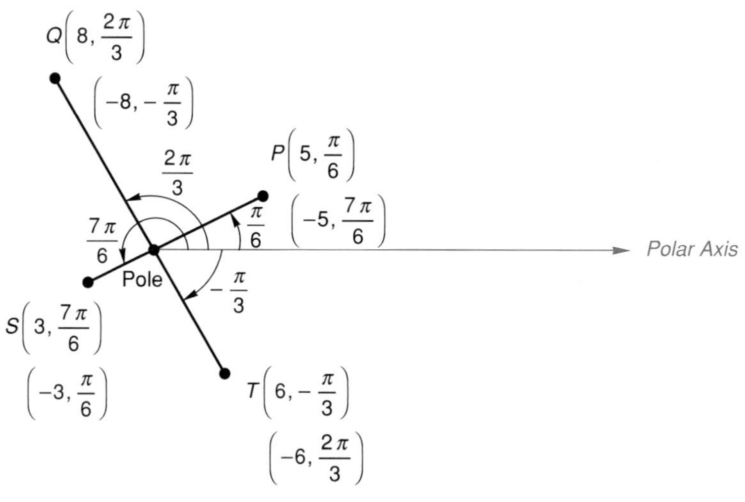

Figure 10.2

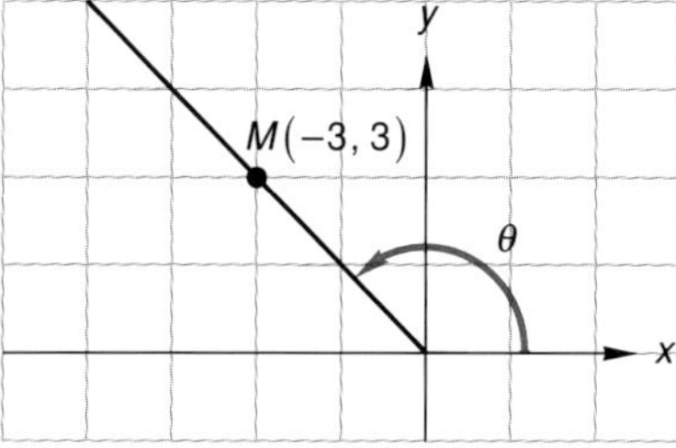

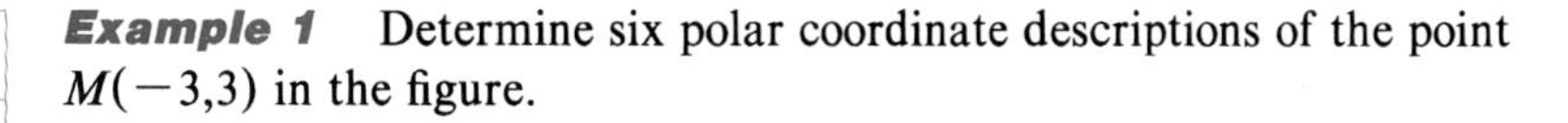

Example 1 Determine six polar coordinate descriptions of the point $M(-3,3)$ in the figure.

Solution From the figure we see that $\theta = \dfrac{3\pi}{4}$ because

$\tan\theta = \dfrac{3}{-3} = -1$, and the ray through M is in the second quadrant.

Also,
$$r = \sqrt{(-3)^2 + 3^2} = 3\sqrt{2}.$$

Thus, one polar coordinate description of M is $(3\sqrt{2}, 3\pi/4)$. Five other polar coordinate descriptions of the same point are $(3\sqrt{2},-5\pi/4)$, $(3\sqrt{2},11\pi/4)$, $(-3\sqrt{2},-\pi/4)$, $(-3\sqrt{2},7\pi/4)$, and $(-3\sqrt{2},-9\pi/4)$. ■

Polar and Rectangular Equations

We have shown that if a point in the plane has the rectangular description (x,y), then its polar coordinate description is (r,θ), where

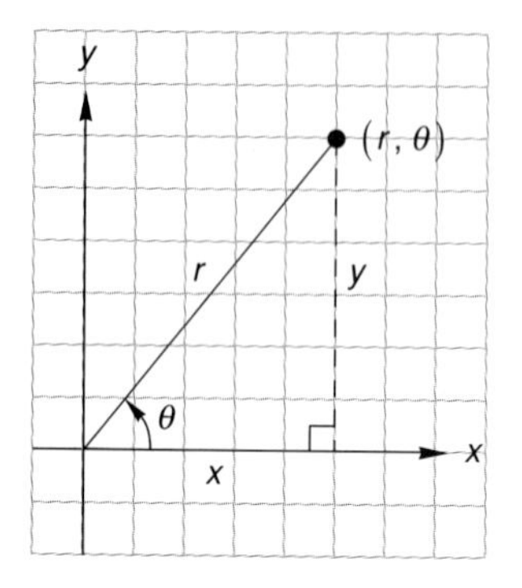

Figure 10.3

Transformation Equations

$$r = \pm\sqrt{x^2 + y^2}$$
$$\tan\theta = \frac{y}{x}. \tag{1}$$

Similarly (figure 10.3) if (r,θ) is a polar coordinate description of a point in the plane, then the point's rectangular coordinates are

Transformation Equations

$$x = r\cos\theta$$
$$y = r\sin\theta. \tag{2}$$

Example 2 Determine the rectangular coordinates of the points $P(2,\pi/3)$, $Q(-5,5\pi/6)$, and $T(-8,3\pi/4)$.

Solution Using (2), the rectangular coordinates of P, Q, and T are determined as follows:

$$x = 2\cos\frac{\pi}{3} = 1, \quad y = 2\sin\frac{\pi}{3} = \sqrt{3} \qquad P(1,\sqrt{3})$$

$$x = -5\cos\frac{5\pi}{6} = \frac{5\sqrt{3}}{2}, \quad y = -5\sin\frac{5\pi}{6} = -\frac{5}{2} \qquad Q\left(\frac{5\sqrt{3}}{2}, -\frac{5}{2}\right)$$

$$x = -8\cos\frac{3\pi}{4} = 4\sqrt{2}, \quad y = -8\sin\frac{3\pi}{4} = -4\sqrt{2} \qquad T(4\sqrt{2},-4\sqrt{2})$$

■

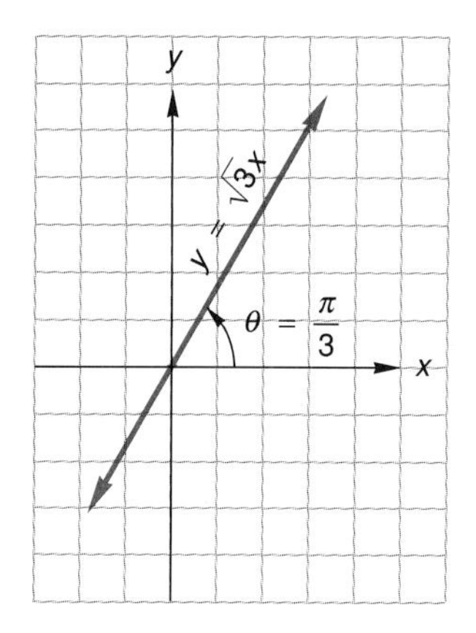

Figure 10.4

With (1) and (2), polar coordinate equations can be converted to rectangular coordinate equations and vice versa. The polar coordinate equation of the line $y = \sqrt{3}x$ (figure 10.4) is obtained with the aid of (2).

$$
\begin{aligned}
r \sin \theta &= \sqrt{3}\, r \cos \theta \\
\tan \theta &= \sqrt{3} \\
\theta &= \tan^{-1}\sqrt{3} \\
\theta &= \frac{\pi}{3} + k\pi,\ k \in J
\end{aligned}
$$

In each of the equations,

$$\theta = \frac{\pi}{3}, \quad \theta = \frac{\pi}{3} + \pi, \quad \theta = \frac{\pi}{3} - \pi, \quad \theta = \frac{\pi}{3} + 2\pi, \text{ etc.,}$$

r assumes all real values. Each set,

$$\left\{\left(r, \frac{\pi}{3}\right) \middle| r \in R\right\}, \quad \left\{\left(r, \frac{4\pi}{3}\right) \middle| r \in R\right\}, \quad \left\{\left(r, -\frac{2\pi}{3}\right) \middle| r \in R\right\}, \text{ etc.,}$$

determines the line $y = \sqrt{3}x$. Therefore,

$$\theta = \frac{\pi}{3} \quad \text{and} \quad y = \sqrt{3}x$$

are equations of the same straight line. The above argument can be made for any oblique line $y = mx$. Thus,

$$\boldsymbol{\theta = \tan^{-1} m \quad \text{and} \quad y = mx}$$

are equations of the same straight line.

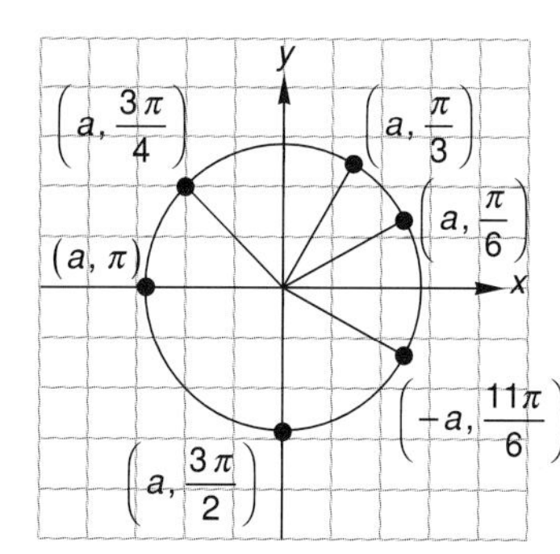

Example 3 Find a polar coordinate equation that corresponds to $x^2 + y^2 = a^2$.

Solution From (1), $x^2 + y^2 = r^2$. Hence, the given equation has the polar form

$$r^2 = a^2,$$

or

$$r = \pm a.$$

There is no θ in the polar equation $r = a$, implying that no matter what value θ assumes, r is equal to a. Some point plotting for $r = a$ and $r = -a$ (figure 10.5) should convince you that each equation determines the circle with radius a that is centered at the origin. Thus, $r = a$ or $r = -a$ corresponds to $x^2 + y^2 = a^2$. ■

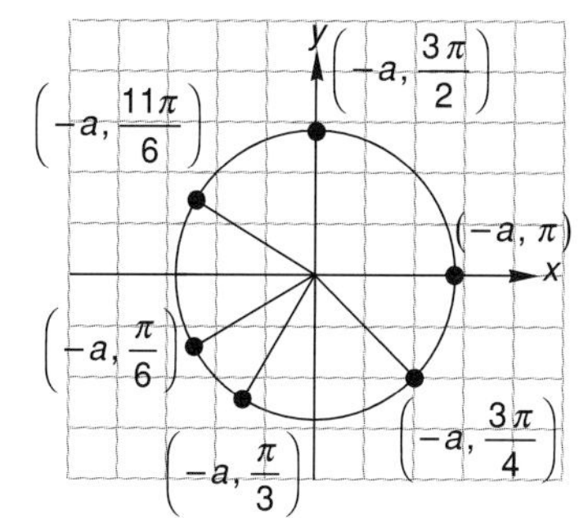

Figure 10.5

Example 4 Find a polar coordinate equation that corresponds to $x^2 + y^2 - 4x = 0$.

Solution Letting $r^2 = x^2 + y^2$ and $r \cos \theta = x$, the given equation becomes

$$r^2 - 4r \cos \theta = 0$$

or

$$r(r - 4 \cos \theta) = 0.$$

The solution set of $r = 0$ is $\{(0,\theta) | \theta \varepsilon R\}$. No matter what value θ assumes, $(0,\theta)$ is a description of the pole. But the pole is a point in the graph of $r - 4 \cos \theta = 0$ because $(0,\pi/2)$ solves the equation. Hence, $r - 4 \cos \theta = 0$ and $x^2 + y^2 - 4x = 0$ determine the same set of points in the plane. ■

Example 5 Find a rectangular coordinate equation that corresponds to $r = 2 \sin \theta$.

Solution Multiplying both sides of the given equation by r yields

$$r^2 = 2r \sin \theta$$
$$r^2 - 2r \sin \theta = 0$$
$$r(r - 2 \sin \theta) = 0.$$

The solution set of $r^2 = 2r \sin \theta$ is the union of the solution sets of

$$r = 0 \quad \text{and} \quad r - 2 \sin \theta = 0.$$

But the solution set of $r = 0$ is the pole that is included in the solution set of $r - 2 \sin \theta = 0$. (Why?) Hence,

$$r = 2 \sin \theta \quad \text{and} \quad r^2 = 2r \sin \theta$$

determine the same set of points in the plane. In $r^2 = 2r \sin \theta$, substitute $x^2 + y^2$ for r^2, and y for $r \sin \theta$. The resulting equation, $x^2 + y^2 = 2y$, corresponds to $r = 2 \sin \theta$. ■

Example 6 Find a rectangular coordinate equation that corresponds to $r = \dfrac{2}{1 - \sin \theta}$.

Solution Write the given equation in the form

$$r - r \sin \theta = 2$$

and replace r with $\pm\sqrt{x^2 + y^2}$, and $r \sin \theta$ with y. In the resulting equation,

$$\pm\sqrt{x^2 + y^2} - y = 2.$$

Transpose y and square to obtain

$$\begin{aligned}(\pm\sqrt{x^2+y^2})^2 &= (2+y)^2\\ x^2+y^2 &= 4+4y+y^2\\ \frac{x^2}{4}-1 &= y.\end{aligned}$$

■

You may be unwilling to accept $y = \dfrac{x^2}{4} - 1$ as the rectangular coordinate equation that corresponds to $r = \dfrac{2}{1-\sin\theta}$. After all, it was obtained by squaring both sides of an equation, a process that can produce extraneous roots. To convince you that $y = \dfrac{x^2}{4} - 1$ and $r = \dfrac{2}{1-\sin\theta}$ determine the same set of points in the plane, we work back from the rectangular coordinate equation to the polar equation.

$$\begin{aligned}y &= \frac{x^2}{4}-1\\ 4y+4 &= x^2\\ 4r\sin\theta+4 &= r^2\cos^2\theta\\ 4r\sin\theta+4 &= r^2(1-\sin^2\theta)\\ r^2\sin^2\theta+4r\sin\theta+4 &= r^2\\ (r\sin\theta+2)^2 &= r^2\end{aligned}$$

$$r\sin\theta+2=r \quad\text{or}\quad r\sin\theta+2=-r$$

$$2=r-r\sin\theta \quad\text{or}\quad 2=-r-r\sin\theta$$

$$\frac{2}{1-\sin\theta}=r \quad\text{or}\quad \frac{2}{1+\sin\theta}=-r$$

Equation $\dfrac{2}{1+\sin\theta} = -r$ may also be written in the form $\dfrac{2}{1-\sin(\theta+\pi)} = -r$ because $\sin(\theta+\pi) = -\sin\theta$. If (r,θ) is a solution of $r = \dfrac{2}{1-\sin\theta}$, then $(-r,\theta+\pi)$ is a solution of $-r = \dfrac{2}{1-\sin(\theta+\pi)}$ or $-r = \dfrac{2}{1+\sin\theta}$. But (r,θ) and $(-r,\theta+\pi)$ are polar coordinates of the same point. (Why?) Therefore,

$$r = \frac{2}{1-\sin\theta} \quad\text{and}\quad -r = \frac{2}{1+\sin\theta}$$

determine the set of points that constitute the graph of $y = x^2/4 - 1$.

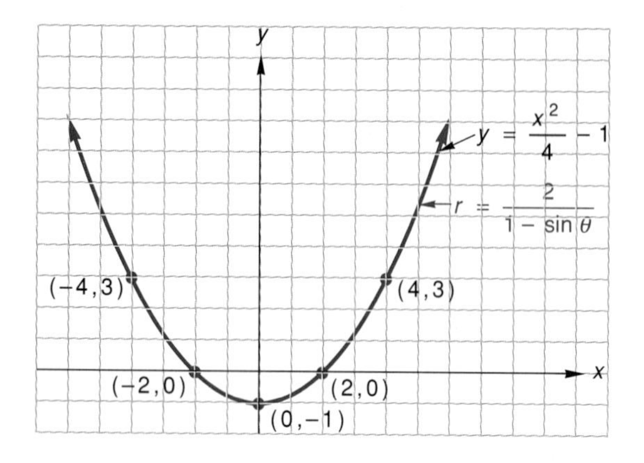

Exercise Set 10.1

(A)

Given the rectangular coordinates of a point, write four pairs of polar coordinates of the point, two with positive r and two with negative r.

1. $(3,3)$ **2.** $(-2,2)$ **3.** $(-\sqrt{3},-1)$

4. $(\sqrt{3},-1)$ **5.** $(-5,0)$ **6.** $(0,-4)$

Write the pair of rectangular coordinates that correspond to the given pair of polar coordinates.

7. $\left(1, \frac{\pi}{6}\right)$ **8.** $\left(-3, \frac{5\pi}{6}\right)$

9. $\left(-2, -\frac{4\pi}{3}\right)$ **10.** $\left(10, -\frac{\pi}{4}\right)$

11. $\left(-8, \frac{3\pi}{4}\right)$ **12.** $\left(-6, -\frac{5\pi}{4}\right)$

Write each equation in polar coordinates.

13. $y = x$ **14.** $y = 2x$

15. $x^2 + y^2 = 9$ **16.** $x^2 + y^2 = 4$

17. $y = -3x$ **18.** $-y = \frac{1}{2}x$

19. $x^2 + y^2 = 12$ **20.** $x^2 + y^2 = 45$

Write each equation in rectangular coordinates.

21. $\theta = \frac{\pi}{6}$ **22.** $\theta = \frac{2\pi}{3}$

23. $r = 1$ **24.** $r = 3$

25. $\theta = -\frac{\pi}{4}$ **26.** $\theta = -\frac{3\pi}{2}$

27. $r = -1$ **28.** $r = -3$

Write each equation in polar coordinates.

29. $x^2 + y^2 - 2x = 0$

30. $x^2 + y^2 + 4y = 0$

31. $x + 2y = 5$

32. $y = -3x + 1$

33. $x = 2$

34. $y = -5$

35. $x^2 + y^2 - 6x + 4y = 0$

36. $x^2 + y^2 - 8x - 10y = 0$

37. $x^2 = 2y$

38. $x^2 - y^2 = 1$

Write each equation in rectangular coordinates.

39. $r = 5 \sec\theta$ **40.** $r = \csc\theta$

41. $r = 4 \sin\theta$ **42.** $r = -\sin\theta$

43. $r = 4 \cos\theta$ **44.** $r = -\cos\theta$

45. $r \cos\left(\theta + \frac{\pi}{6}\right) = 1$ **46.** $r \sin\left(\theta - \frac{\pi}{4}\right) = 3$

47. $r = \frac{6}{2 - \cos\theta}$ **48.** $r = \frac{1}{1 - \cos\theta}$

49. $r = \frac{6}{2 - 3\sin\theta}$ **50.** $r = \frac{4}{2 - 2\sin\theta}$

51. $r^2 \sin 2\theta = 1$ **52.** $r^2 \cos 2\theta = 9$

(B)

53. The distance between the points (x_1,y_1) and (x_2,y_2) is given by the formula

$$d = \sqrt{(x_1 - x_2)^2 + (y_1 - y_2)^2}.$$

Write a corresponding polar coordinate distance formula. Use the formula to find the distance between the points $\left(1, \frac{\pi}{6}\right)$ and $\left(3, \frac{3\pi}{4}\right)$.

Write each equation in polar coordinates. (Hint: See example 6.)

54. $y + 2 = \frac{x^2}{8}$

55. $y^2 = -12(x - 3)$

10.2 Polar Coordinate Graphs

A relation $r = f(\theta)$ can be graphed by plotting a sufficient number of ordered pairs (r,θ) and connecting the points obtained.

Example 1 Sketch a graph of $r = 1 + \cos\theta$.

Solution Let θ assume values that are multiples of $\pi/6$ and $\pi/4$, $0 \leq \theta \leq 2\pi$. Compute the resulting r values and plot the ordered pairs (r,θ).

r	θ
2	0
$1 + \frac{\sqrt{3}}{2}$	$\frac{\pi}{6}$
$1 + \frac{\sqrt{2}}{2}$	$\frac{\pi}{4}$
$\frac{3}{2}$	$\frac{\pi}{2}$
$\frac{1}{2}$	$\frac{2\pi}{3}$
$1 - \frac{\sqrt{2}}{2}$	$\frac{3\pi}{4}$
$1 - \frac{\sqrt{3}}{2}$	$\frac{5\pi}{6}$
0	π

r	θ
$1 - \frac{\sqrt{3}}{2}$	$\frac{7\pi}{6}$
$1 - \frac{\sqrt{2}}{2}$	$\frac{5\pi}{4}$
$\frac{1}{2}$	$\frac{4\pi}{3}$
1	$\frac{3\pi}{2}$
$\frac{3}{2}$	$\frac{5\pi}{3}$
$1 + \frac{\sqrt{2}}{2}$	$\frac{7\pi}{4}$
$1 + \frac{\sqrt{3}}{2}$	$\frac{11\pi}{6}$
2	2π

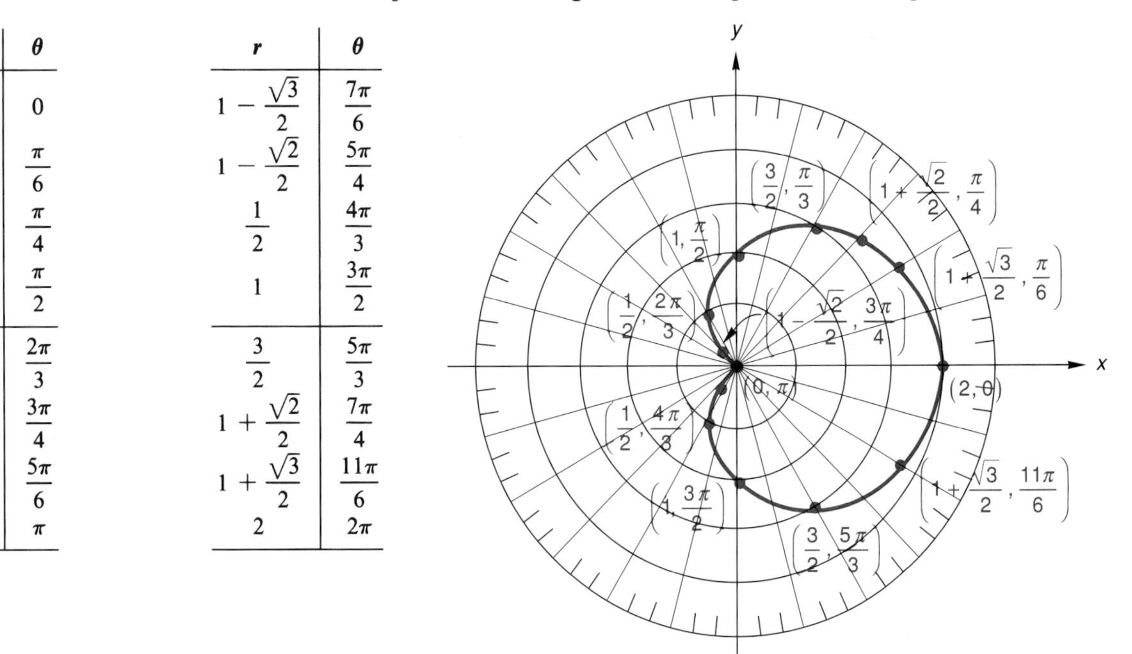

Figure 10.6

The graph in figure 10.6 is called a cardioid. Had we established with symmetry tests that the graph is symmetric to the x-axis, we could have avoided plotting all (r,θ) for which $\pi < \theta \leq 2\pi$. ■

Symmetry Tests

Because a point in the plane has more than one polar coordinate description, there is more than one polar coordinate symmetry test for each of the coordinate axes and the pole. It is possible (see example 3) for a relation $r = f(\theta)$ to satisfy one y-axis symmetry test and not another. *As long as the relation satisfies any one y-axis symmetry test, its graph is symmetric to the y-axis.*

There are many different tests for symmetry with respect to the pole, or symmetry with respect to either of the coordinate axes. We shall utilize only a few of those tests for graphing. It is too time consuming, when constructing a graph, to search through a catalogue of symmetry tests for one that works.

(I) In figure 10.7, points $P(r,\theta)$ and $Q(r,-\theta)$ are reflections of each other with respect to the x-axis. Therefore, *a graph of $r = f(\theta)$ is symmetric to the x-axis if the same or an equivalent equation is obtained when θ is replaced by $-\theta$ in $r = f(\theta)$.*

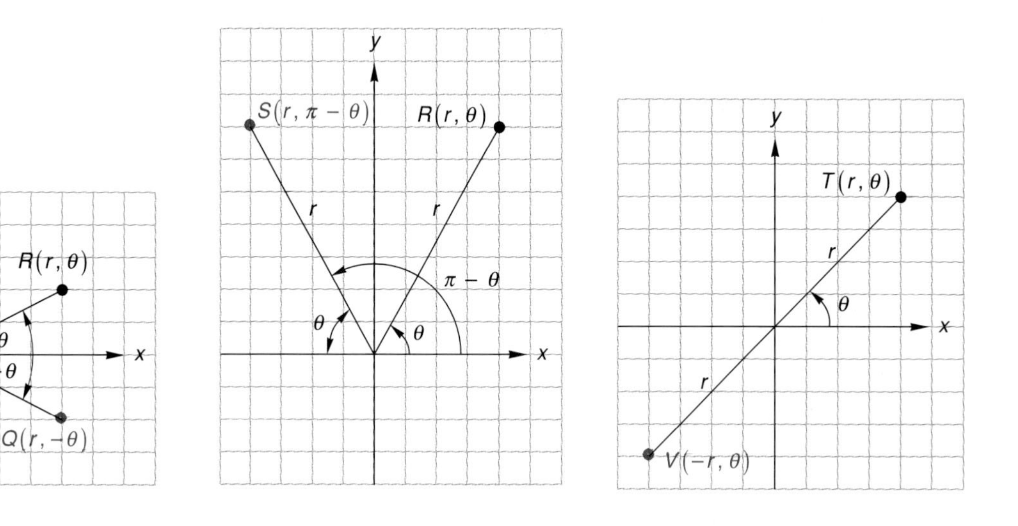

Figure 10.7

(II) Points $R(r,\theta)$ and $S(r, \pi - \theta)$ are reflections of each other with respect to the y-axis. Therefore, *a graph of $r = f(\theta)$ is symmetric to the y-axis if the same or an equivalent equation is obtained when θ is replaced by $\pi - \theta$ in $r = f(\theta)$.*

(III) Points $T(r,\theta)$ and $V(-r,\theta)$ are reflections of each other with respect to the pole. Therefore, *a graph of $r = f(\theta)$ is symmetric to the pole if the same or an equivalent equation is obtained when r is replaced by $-r$.*

Example 2 Show that the graph of $r = 1 + \cos\theta$ is symmetric to the x-axis.

Solution Replacing θ by $-\theta$ yields

$$r = 1 + \cos(-\theta) = 1 + \cos\theta,$$

the same equation as the given equation. Hence, the graph of $r = 1 + \cos\theta$ is symmetric to the x-axis. ■

The y-axis symmetry test fails when applied to $r = \sin 2\theta$. If θ is replaced by $\pi - \theta$, the resulting equation,

$$\begin{aligned} r &= \sin 2(\pi - \theta) \\ &= \sin 2\pi \cos 2\,\theta - \cos 2\pi \sin 2\,\theta \\ &= -\sin 2\,\theta, \end{aligned}$$

is not equivalent to $r = \sin 2\,\theta$. Nevertheless, a graph of $r = \sin 2\,\theta$ is symmetric to the y-axis because the given equation satisfies the y-axis symmetry test suggested by figure 10.8.

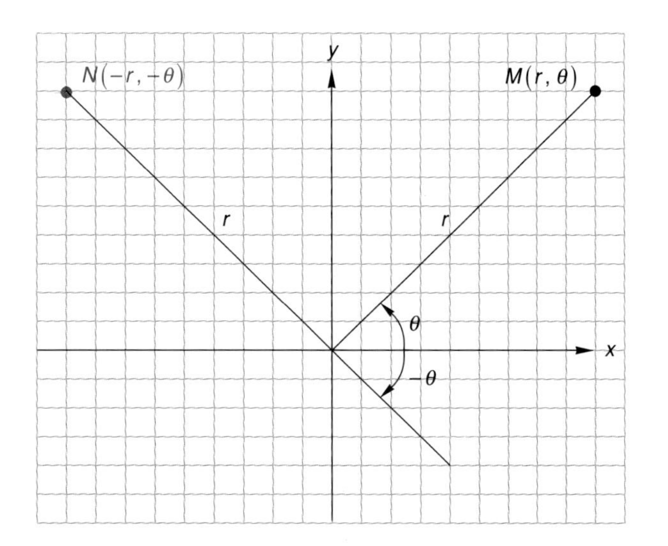

Figure 10.8

Points $M(r,\theta)$ and $N(-r,-\theta)$ are reflections of each other with respect to the y-axis. Therefore, the graph of $r = f(\theta)$ is symmetric to the y-axis if the same or an equivalent equation is obtained when r is replaced by $-r$ and θ is replaced by $-\theta$ in $r = f(\theta)$.

$$\begin{aligned} -r &= \sin 2(-\theta) \\ -r &= \sin(-2\,\theta) \\ -r &= -\sin 2\,\theta \end{aligned}$$

But $-r = -\sin 2\,\theta$ is equivalent to $r = \sin 2\,\theta$. Therefore, the graph of $r = \sin 2\,\theta$ is symmetric to the y-axis.

The test developed for symmetry with respect to the pole fails when applied to $r = \sin 2\,\theta$ (try it). However, the test suggested by figure 10.9 does not fail. Points $K(r,\theta)$ and $J(r,\pi + \theta)$ are reflections of each other with respect to the pole. Hence, the graph of $r = f(\theta)$ is symmetric to the pole if the same or an equivalent equation is obtained when θ is replaced by $\pi + \theta$ in $r = f(\theta)$.

$$\begin{aligned} r &= \sin 2(\pi + \theta) \\ &= \sin 2\pi \cos 2\,\theta + \cos 2\pi \sin 2\,\theta \\ &= \sin 2\,\theta \end{aligned}$$

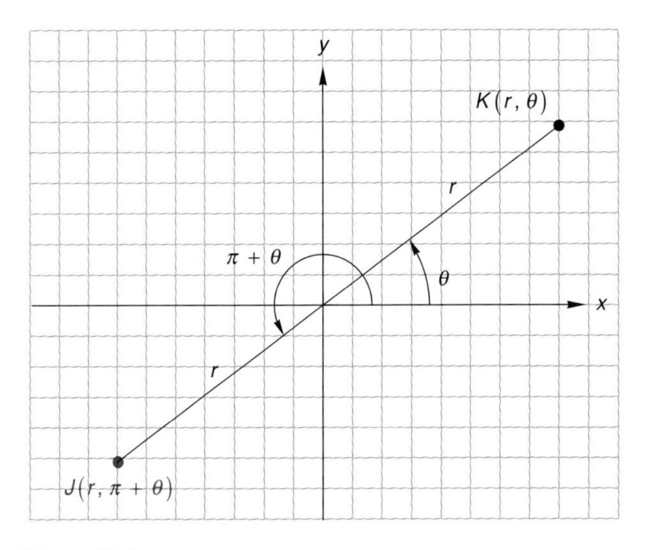

Figure 10.9

Example 3 Graph $r = \sin 2\theta$.

Solution As shown on page 392, the graph of the given relation is symmetric to the y-axis and the pole. It is therefore also symmetric to the x-axis (see section 2.4). We need only plot points in the first quadrant and let symmetry do the rest (figure 10.10).

r	θ
0	0
$\frac{\sqrt{3}}{2}$	$\frac{\pi}{6}$
1	$\frac{\pi}{4}$
$\frac{\sqrt{3}}{2}$	$\frac{\pi}{3}$
0	$\frac{\pi}{2}$

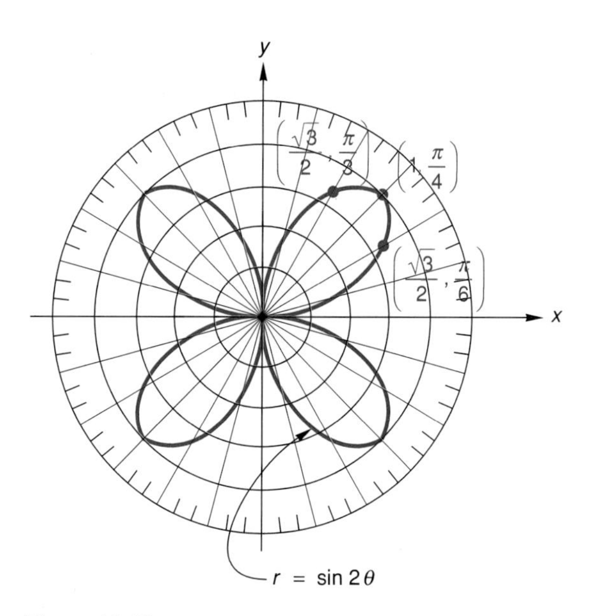

Figure 10.10

The graph is called a four-leaf rose. ■

Example 4 Graph $r^2 = 4 \cos 2\theta$.

Solution Test the relation for symmetry.

> Pole: The graph is symmetric to the pole because the equation obtained when r is replaced by $-r$ is the same as the given equation.
>
> x-axis: Replace θ by $-\theta$.

$$r^2 = 4 \cos 2(-\theta) = r \cos(-2\theta) = 4 \cos 2\theta$$

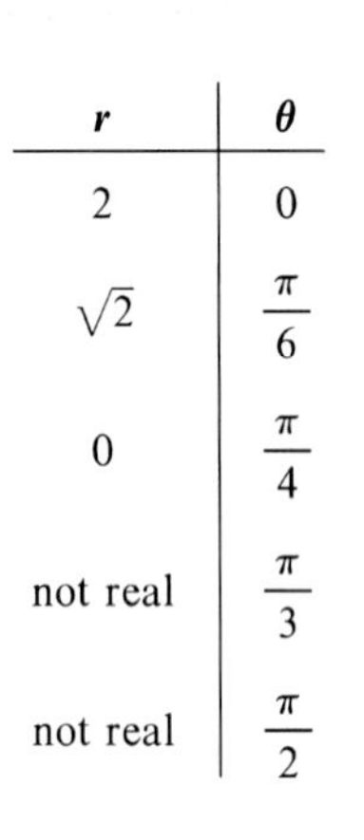

r	θ
2	0
$\sqrt{2}$	$\frac{\pi}{6}$
0	$\frac{\pi}{4}$
not real	$\frac{\pi}{3}$
not real	$\frac{\pi}{2}$

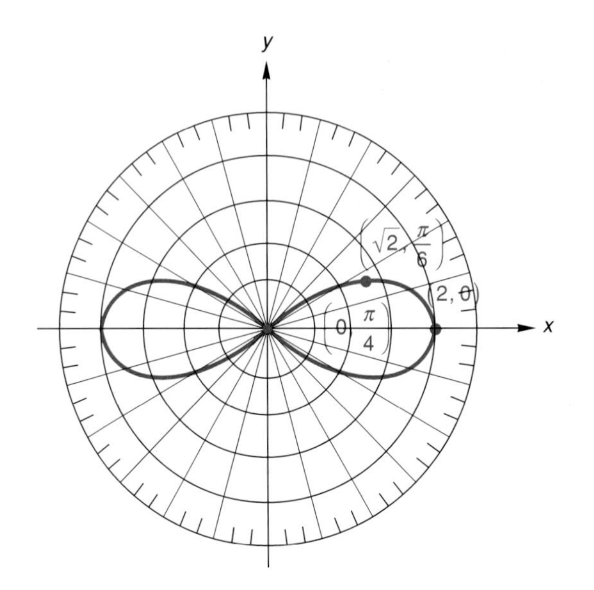

Figure 10.11

Since the resulting equation is the same as the original, the graph is symmetric to the x-axis.

> y-axis: Symmetry to the pole and to the x-axis imply symmetry to the y-axis.

As in example 3, we need only plot points in the first quadrant and then take advantage of symmetry (figure 10.11).

The graph is called a lemniscate. ■

Exercise Set 10.2

(A)

Sketch a graph of each equation. Apply the symmetry tests of this section to aid in the graphing.

1. $r = 2$

2. $r = -3$

3. $\theta = \frac{\pi}{6}$

4. $\theta = \frac{2\pi}{3}$

5. $r = 5 \sec \theta$

6. $r = -2 \csc \theta$

7. $r = 3 \cos \theta$

8. $r = 2 \sin \theta$

9. $r = 3 - 3 \sin \theta$

10. $r = 2 + 2 \cos \theta$

11. $r = 2 - \cos \theta$

12. $r = 4 + 2 \sin \theta$

13. $r = 1 - 3 \sin \theta$

14. $r = 2 + 4 \cos \theta$

15. $r = \cos 3\theta$

16. $r = \sin 3\theta$

17. $r = 2 \cos 2\theta$

18. $r = 2 \sin 4\theta$

19. $r^2 = 9 \sin 2\theta$

20. $r^2 = -4 \cos 2\theta$

21. $r^2 = \cos \theta$

22. $r^2 = \sin \theta$

23. $r = \theta$

24. $r\theta = 10$

(B)

25. Without graphing, argue that the graphs of $r = 2 \cos 4\theta$ and $-r = 2 \cos 4(\theta + \pi)$ must be the same. (Hint: See example 6, section 10.1.)

26. Without graphing, argue that the graphs of $r = \dfrac{1}{2 - \cos\theta}$ and $-r = \dfrac{1}{2 + \cos\theta}$ must be the same. (Hint: See example 6, section 10.1.)

27. Devise a symmetry test for each axis and the pole that is different from any of the symmetry tests developed in the text.

Graph.

28. $r = e^{\theta}$ (logarithmic spiral)

29. $r = \sin\theta \cos^2\theta$ (bifolium)

30. $r = 3 \sec\theta - 2$ (conchoid)

31. $r = \sin\theta \tan\theta$ (cissoid)

10.3 Parametric Equations

Let s be the distance from the origin to any point on the line $y = 3x$ (figure 10.12). Then

$$\begin{aligned} s^2 &= x^2 + y^2 \\ s^2 &= x^2 + (3x)^2 \\ s^2 &= 10x^2 \\ \pm\frac{s}{\sqrt{10}} &= x. \end{aligned}$$

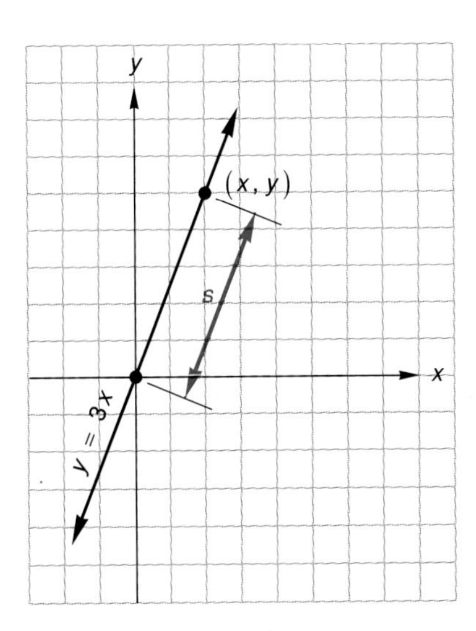

Figure 10.12

Since $y = 3x$ it follows that $y = \pm 3s/\sqrt{10}$. For $s \in R$, the equations

$$x = \frac{s}{\sqrt{10}} \quad \text{and} \quad y = \frac{3s}{\sqrt{10}} \tag{1}$$

are called **parametric equations** of the line $y = 3x$, and s is called a **parameter.** Negative s can be interpreted as the distance from the origin to a third quadrant point that is on the line.

Why would we choose equations (1) to describe the line in figure 10.12 instead of the single equation $y = 3x$? One answer is that equations (1) give more information. If we are studying the linear motion of a particle, then (1) gives us the position of the particle as well as its distance from the starting point. The equation $y = 3x$ gives us only position.

Example 1 The motion of a particle is given by the parametric equations

$$x = 10t \quad \text{and} \quad y = 10t - t^2,$$

in which the parameter t is time. Plot the path of the particle for $0 \le t \le 10$. Also, establish a direct relationship between y and x by eliminating the parameter from the given equations.

Solution We assume values for t, compute corresponding x and y values, and plot the resulting (x,y) pairs.

t	x	y
0	0	0
2	20	16
5	50	25
8	80	16
10	100	0

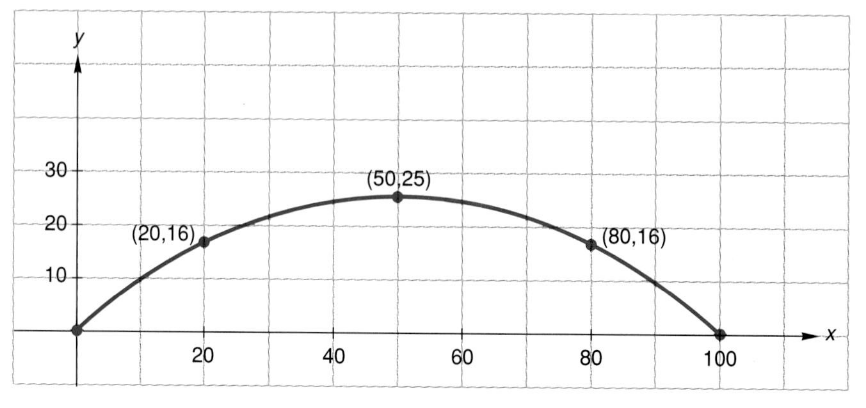

Figure 10.13

The parameter t is eliminated from the given equations by solving $x = 10t$ for t, and then substituting in $y = 10t - t^2$.

$$y = 10\left(\frac{x}{10}\right) - \left(\frac{x}{10}\right)^2$$

$$y = x - \frac{x^2}{100}$$

A plot of $y = x - x^2/100$, $0 \leq x \leq 100$, yields the graph in figure 10.13. ■

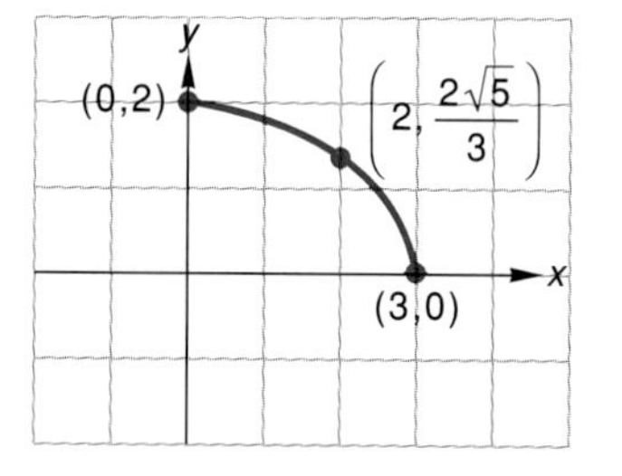

Example 2 Eliminate the parameter from the equations

$$x = 3 \cos t \quad y = 2 \sin t,$$

and sketch a graph of the resulting equation.

Solution

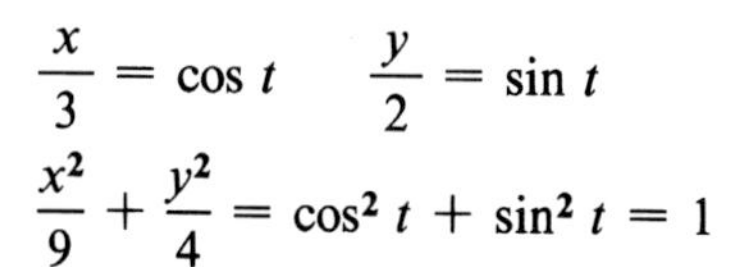

$$\frac{x}{3} = \cos t \quad \frac{y}{2} = \sin t$$

$$\frac{x^2}{9} + \frac{y^2}{4} = \cos^2 t + \sin^2 t = 1$$

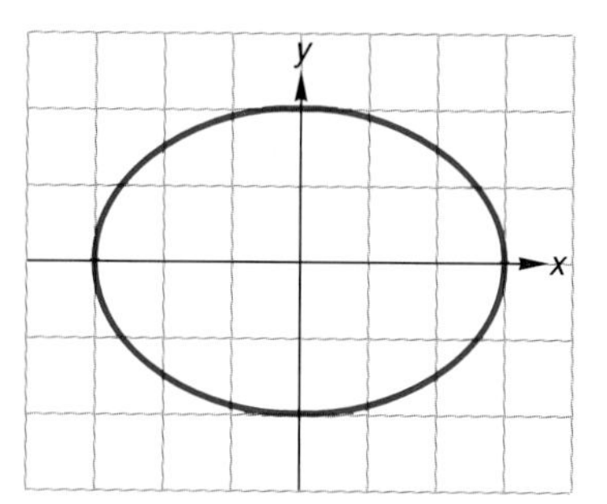

A graph of $x^2/9 + y^2/4 = 1$ is symmetric to the x-axis, y-axis, and the origin (see section 2.4). We sketch a first quadrant portion of the graph from a set of plotted points and then use symmetry to generate the rest of the graph. ■

Example 3 Write a pair of parametric equations, $x = f(\theta)$ and $y = g(\theta)$, for the relation $r = \sin 2\,\theta$.

Solution In the transformation equations

$$x = r \cos \theta \quad y = r \sin \theta$$

that are introduced in section 10.1, $\sin 2\,\theta$ is substituted for r. The resulting equations are the desired parametric equations.

$$x = \sin 2\,\theta \cos \theta \quad y = \sin 2\,\theta \sin \theta$$ ■

It is not always the case that a parameter such as t in example 1 is introduced into a relationship, $y = f(x)$, for the purpose of obtaining additional information. Sometimes a parameter is necessary to establish a relationship.

Let us try to write the equation of the curve generated by point P on the rolling wheel in figure 10.14. Assume the wheel has radius a. After the wheel has turned through θ radians, the distance between points O and Q is equal to the length of arc QP. Hence, the coordinates of P can be determined as follows.

$$x = \overline{OQ} - \overline{CP} \cos\left(\theta - \frac{\pi}{2}\right) \quad y = \overline{QC} + \overline{CP} \sin\left(\theta - \frac{\pi}{2}\right)$$

$$x = a\,\theta - a \sin \theta \qquad y = a - a \cos \theta \tag{2}$$

Equations (2) are the parametric equations of a curve called a **cycloid.** You might get some sense of the difficulty of obtaining an equation of the cycloid in the form $y = f(x)$ by trying to eliminate the parameter θ from equations (2).

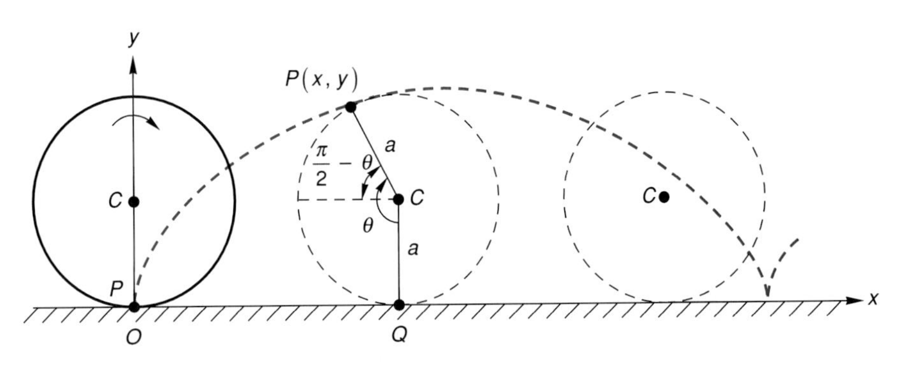

Figure 10.14

Exercise Set 10.3

(A)

(a) Sketch a graph directly from the given pair of parametric equations, and (b) eliminate the parameter from the given pair of parametric equations. Verify that the graph of the resulting rectangular coordinate equations is the graph sketched in part a.

1. $x = t - 2,\ y = 2t + 1$

2. $x = 3t,\ y = -5t$

3. $x = t - 1,\ y = 2t + 3$

4. $x = t^2 + 3,\ y = t - 1$

5. $x = e^t,\ y = e^{-2t}$

6. $x = t^2,\ y = t^3$

7. $x = \ln\frac{t}{2},\ y = \ln(3t)$

8. $x = \frac{1}{t},\ y = t^2$

9. $x = \cos\theta,\ y = \sin\theta$

10. $x = \sec\theta,\ y = \tan\theta$

11. $x = 3\cos\theta,\ y = 4\sin\theta$

12. $x = 5\sec\theta,\ y = 2\tan\theta$

13. $x = -3 + 2\cos\theta,\ y = 1 + 2\sin\theta$

14. $x = -2 + \cos\theta,\ y = -1 + 4\sin\theta$

Write a pair of parametric equations for each line using s, the distance of each point on the line from the y-intercept, as a parameter.

15. $y = 3x + 5$

16. $y = 6x - 2$

17. $y = -2x - 8$

18. $y = -5x + 4$

Write a pair of parametric equations, $x = f(\theta)$, $y = g(\theta)$, for each polar coordinate relation.

19. $r = 2\cos\theta$

20. $r = -3\sin\theta$

21. $r = 1 - \cos\theta$

22. $r = 2 + \cos\theta$

23. $\frac{1}{r} = \sin 2\theta$

24. $\frac{1}{r} = \cos 2\theta$

(B)

Eliminate the parameter and obtain a rectangular coordinate relation. Sketch a graph using either the parametric equations or the obtained x-y equation.

25. $x = t^2 + 3t - 1$, $y = t^2 - t + 2$

26. $x = 2t^2 + t + 1$, $y = t^2 + 2t + 2$

27. $x = 1 + \sec\theta$, $y = -2 + 3\tan\theta$

28. $x = \sin^{-1} t$, $y = \cos^{-1} t$

29. Prove that the parametric equations $x = 2(t + 1)$ and $y = t(t + 2)$ and the polar coordinate equation $r(1 - \sin\theta) = 2$ determine the same curve.

30. Write a pair of parametric equations for the curve generated by a point P on a wheel as the wheel rolls along a straight line on a horizontal surface. Assume that the radius of the wheel is a units and that P is b units from the center of the wheel, $b < a$.

31. In the figure, a string is unwound from a circular spool while held taut in the plane of the circle. The curve traced out by the end of the string, P, is called an involute of the circle. Use θ, the angle to the point of tangency T of the taut string as a parameter, and write a pair of parametric equations that determine the involute. The radius of the circle is a.

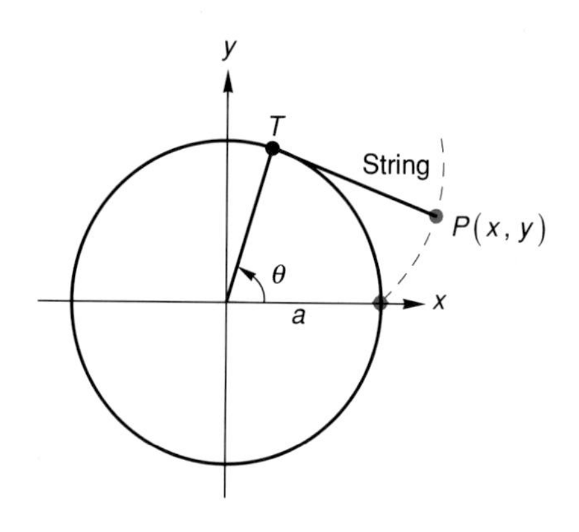

32. In the figure, a circle of radius b rolls inside a circle of radius a, $b < a$. A point P on the circumference of the small circle generates a curve called a hypocycloid. Let θ be the angle that a line through the centers of the circles makes with the positive x-axis. With θ as parameter, develop a pair of parametric equations that determine the hypocycloid.

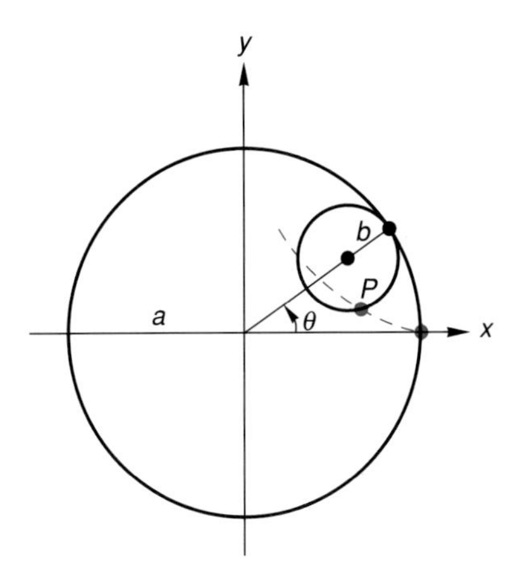

10.4 Complex Roots and deMoivre's Theorem

"Theory of Equations" in chapter 4 tells us that the equation $x^5 - 32 = 0$ has five roots in C. It also tells us that no negative number can be a root. Synthetic division of $x^5 - 32$ by 2, which is a zero of the polynomial

$$\begin{array}{r|rrrrrr} & 1 & 0 & 0 & 0 & 0 & -32 \\ \hline 2 & 1 & 2 & 4 & 8 & 16 & 0 \end{array}$$

enables us to write

$$x^5 - 32 = (x - 2)(x^4 + 2x^3 + 4x^2 + 8x + 16).$$

Descarte's Rule of Signs guarantees that equation

$$x^4 + 2x^3 + 4x^2 + 8x + 16 = 0$$

has no positive roots. Thus, $x^5 - 32 = 0$ has the real root 2 and four complex roots. We will find the complex roots with the aid of polar coordinates.

Polar Form of a Complex Number

A complex number $x + iy$, can be graphed in what is called the *complex plane* (figure 10.15). If $r \cos \theta$ and $r \sin \theta$ are substituted for x and y, respectively, $x + iy$ takes on the **polar form**

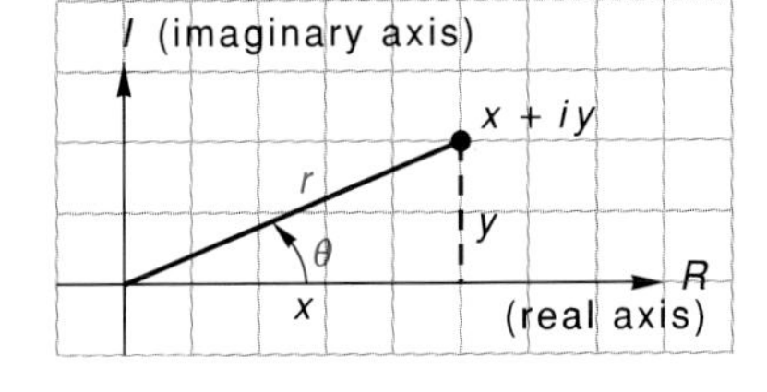

Figure 10.15

Polar Form of a Complex Number

$$r(\cos \theta + i \sin \theta). \tag{1}$$

In (1), r is a *positive* number called the **modulus.** The number θ, called the **argument,** is determined to a multiple of 2π.

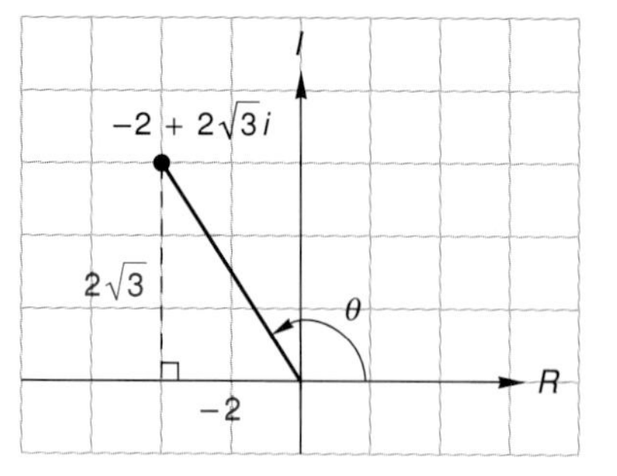

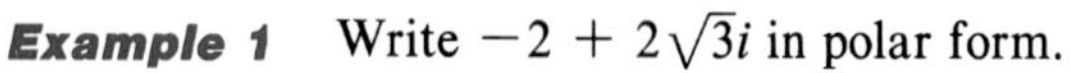

Example 1 Write $-2 + 2\sqrt{3}i$ in polar form.

Solution

$$r = \sqrt{(-2)^2 + (2\sqrt{3})^2} = 4$$
$$\tan\theta = \frac{2\sqrt{3}}{-2} = -\sqrt{3}$$

Since θ is a second quadrant angle, it must be $2\pi/3$.

$$-2 + 2\sqrt{3}i = 4\left(\cos\frac{2\pi}{3} + i\sin\frac{2\pi}{3}\right),$$

or $$-2 + 2\sqrt{3}i = 4\left(\cos\left(\frac{2\pi}{3} + 2\pi\right) + i\sin\left(\frac{2\pi}{3} + 2\pi\right)\right),$$

or $$-2 + 2\sqrt{3}i = 4\left(\cos\left(\frac{2\pi}{3} + k\cdot 2\pi\right) + i\sin\left(\frac{2\pi}{3} + k\cdot 2\pi\right)\right),\ k \in J$$ ■

Product of Complex Numbers

If $z_1 = r_1(\cos\theta_1 + i\sin\theta_1)$ and $z_2 = r_2(\cos\theta_2 + i\sin\theta_2)$, then

$$\begin{aligned} z_1 \cdot z_2 &= r_1 \cdot r_2(\cos\theta_1 + i\sin\theta_1)(\cos\theta_2 + i\sin\theta_2) \\ &= r_1 \cdot r_2[(\cos\theta_1\cos\theta_2 - \sin\theta_1\sin\theta_2) + i(\sin\theta_1\cos\theta_2 + \cos\theta_1\sin\theta_2)] \\ &= r_1 \cdot r_2[\cos(\theta_1 + \theta_2) + i\sin(\theta_1 + \theta_2)]. \end{aligned}$$

Product of Complex Numbers in Polar Form

The product of two complex numbers is the number whose modulus is the product of the moduli of the given numbers, and whose argument is the sum of the arguments of the given numbers.

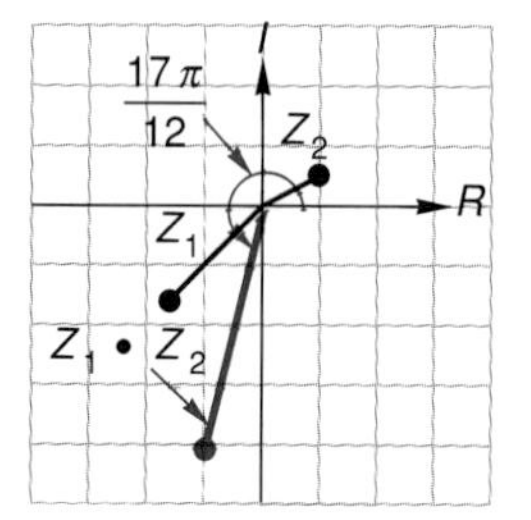

Example 2 Compute the product of $-3 - 3i$ and $\sqrt{3} + i$ in polar form.

Solution

$$z_1 = -3 - 3i = 3\sqrt{2}\left(\cos\frac{5\pi}{4} + i\sin\frac{5\pi}{4}\right)$$
$$z_2 = \sqrt{3} + i = 2\left(\cos\frac{\pi}{6} + i\sin\frac{\pi}{6}\right)$$
$$\begin{aligned} z_1 \cdot z_2 &= 3\sqrt{2}(2)\left[\cos\left(\frac{5\pi}{4} + \frac{\pi}{6}\right) + i\sin\left(\frac{5\pi}{4} + \frac{\pi}{6}\right)\right] \\ &= 6\sqrt{2}\left(\cos\frac{17\pi}{12} + i\sin\frac{17\pi}{12}\right) \end{aligned}$$ ■

If $z = r(\cos\theta + i\sin\theta)$, then $z^2 = r^2(\cos 2\theta + i\sin 2\theta)$. It is not difficult to show that

$$z^3 = r^3(\cos 3\theta + i\sin 3\theta)$$
$$z^4 = r^4(\cos 4\theta + i\sin 4\theta), \text{ and so forth.}$$

The sequence of formulas leads to:

deMoivre's Theorem[1]

If $z = r(\cos\theta + i\sin\theta)$, then $z^n = r^n(\cos n\theta + i\sin n\theta)$, $n \varepsilon N$. **(2)**

We use the theorem to find all the complex roots of $x^5 - 32 = 0$. Let $R(\cos\alpha + i\sin\alpha)$ be one of the complex roots of the equation. Then,

$$R^5(\cos 5\alpha + i\sin 5\alpha) - 32 = 0. \tag{3}$$

Because 32 has the polar form

$$32[\cos(0 + k\cdot 2\pi) + i\sin(0 + k\cdot 2\pi)],\ k \varepsilon J,$$

(3) can be written as

$$R^5(\cos 5\alpha + i\sin 5\alpha) = 32[\cos(0 + k\cdot 2\pi) + i\sin(0 + k\cdot 2\pi)],\ k \varepsilon J.$$

It follows from the last equation that

$$R^5 = 32 \quad \text{and} \quad \alpha = \frac{0 + k\cdot 2\pi}{5}.$$

Therefore, the five complex roots of the given equation are

$$\begin{aligned}
&2(\cos 0 + i\sin 0), && k = 0\\
&2\left(\cos\frac{2\pi}{5} + i\sin\frac{2\pi}{5}\right), && k = 1\\
&2\left(\cos\frac{4\pi}{5} + i\sin\frac{4\pi}{5}\right), && k = 2\\
&2\left(\cos\frac{6\pi}{5} + i\sin\frac{6\pi}{5}\right), && k = 3\\
&2\left(\cos\frac{8\pi}{5} + i\sin\frac{8\pi}{5}\right), && k = 4.
\end{aligned}$$

If k assumes any integral value other than 0, 1, 2, 3, or 4, one of the above roots is obtained.

The complex roots of $x^5 - 32 = 0$ are shown in figure 10.16.

[1]Abraham deMoivre (1667–1754)

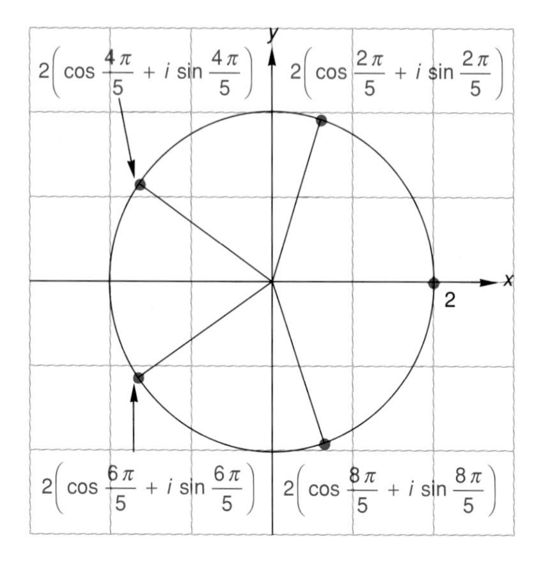

Figure 10.16

Example 3 Solve $x^4 + 16 = 0$.

Solution The equation has no real roots. Each of the four complex roots has the polar form $R(\cos \alpha + i \sin \alpha)$. Thus

$$\begin{aligned} R^4(\cos 4\,\alpha + i \sin 4\,\alpha) &= -16 \\ &= 16(\cos(\pi + k \cdot 2\pi) + i \sin(\pi + k \cdot 2\,\pi)) \end{aligned}$$

where $k = 0, 1, 2$ or 3. It follows that

$$R^4 = 16 \quad \text{and} \quad \alpha = \frac{\pi + k \cdot 2\pi}{4}, \quad k = 0, 1, 2, 3.$$

The four roots are

$$\begin{aligned} 2\left(\cos\frac{\pi}{4} + i\sin\frac{\pi}{4}\right) &= \sqrt{2} + \sqrt{2}i, & k &= 0 \\ 2\left(\cos\frac{3\pi}{4} + i\sin\frac{3\pi}{4}\right) &= -\sqrt{2} + \sqrt{2}i, & k &= 1 \\ 2\left(\cos\frac{5\pi}{4} + i\sin\frac{5\pi}{4}\right) &= -\sqrt{2} - \sqrt{2}i, & k &= 2 \\ 2\left(\cos\frac{7\pi}{4} + i\sin\frac{7\pi}{4}\right) &= \sqrt{2} - \sqrt{2}i, & k &= 3. \end{aligned}$$

■

Example 4 Determine the cube roots of i.

Solution The polar form of i is

$$1\left[\cos\left(\frac{\pi}{2} + k \cdot 2\pi\right) + i \sin\left(\frac{\pi}{2} + k \cdot 2\pi\right)\right].$$

Hence,

$$R^3(\cos 3\alpha + i \sin 3\alpha) = 1\left[\cos\left(\frac{\pi}{2} + k \cdot 2\pi\right) + i \sin\left(\frac{\pi}{2} + k \cdot 2\pi\right)\right]$$

$$R^3 = 1 \quad \text{and} \quad \alpha = \frac{\frac{\pi}{2} + k \cdot 2\pi}{3}, \quad k = 0, 1, 2.$$

The cube roots are

$$\cos\frac{\pi}{6} + i \sin\frac{\pi}{6} = \frac{\sqrt{3}}{2} + \frac{1}{2}i$$

$$\cos\frac{5\pi}{6} + i \sin\frac{5\pi}{6} = -\frac{\sqrt{3}}{2} + \frac{1}{2}i$$

$$\cos\frac{3\pi}{2} + i \sin\frac{3\pi}{2} = -i.$$

■

Exercise Set 10.4

(A)

Graph each complex number in the complex plane. Then write the number in polar form.

1. $2 + 2i$

2. $-3 - 3i$

3. $1 - \sqrt{3}i$

4. $-4\sqrt{3} + 4i$

5. 6

6. -8

7. $5i$

8. $-2i$

Use deMoivre's Theorem to express each number in the form $a + bi$.

9. $(2 + 2i)^6$

10. $(1 - \sqrt{3}i)^{10}$

11. $(\sqrt{3} - i)^5$

12. $(-3 + 3i)^8$

13. $(-4\sqrt{3} + 4i)^4$

14. $(-2i)^9$

Solve each equation.

15. $x^5 - 1 = 0$

16. $x^5 + 1 = 0$

17. $x^4 + 8 = 0$

18. $x^4 - 16 = 0$

19. $x^6 + 64 = 0$

20. $x^6 - 27 = 0$

Find the indicated roots.

21. Fourth roots of i

22. Cube roots of $8i$.

23. Cube roots of $1 - \sqrt{3}i$

24. Cube roots of $2 + 2i$

25. Fifth roots of $16\sqrt{3} - 16i$

26. Cube roots of 64.

(B)

27. Let $z_1 = r_1(\cos\theta_1 + i \sin\theta_1)$ and $z_2 = r_2(\cos\theta_2 + i \sin\theta_2)$. Show that

$$\frac{z_1}{z_2} = \frac{r_1}{r_2}[\cos(\theta_1 - \theta_2) + i \sin(\theta_1 - \theta_2)].$$

28. Prove that if $z = r(\cos\theta + i \sin\theta)$, then

$$z^3 = r^3(\cos 3\theta + i \sin 3\theta).$$

(This result was assumed to be true in the derivation of deMoivre's Theorem.)

Chapter 10 Review

Major Points for Review

Section 10.1 Determining polar coordinates of a point in the plane

The equations relating polar and rectangular coordinates

Converting a rectangular coordinate equation to a polar coordinate equation and vice versa

Section 10.2 Symmetry tests for graphs in polar coordinates

Graphing in polar coordinates

Section 10.3 Eliminating the parameter from a pair of parametric equations

Graphing a relation defined by a pair of parametric equations

Section 10.4 Polar form of a complex number

Product of complex numbers in polar form

deMoivre's Theorem

Finding the complex roots of a number

Review Exercises

Write two polar coordinate pairs, one with positive r and one with negative r, for each point given in rectangular coordinates.

1. $(4,-4)$ **2.** $(-\sqrt{3},1)$

Write the pair of rectangular coordinates that correspond to the given pair of polar coordinates.

3. $(3,\pi/3)$ **4.** $(-5,-\pi/4)$

Write each number in polar form.

5. $\sqrt{3} - i$ **6.** $-4 + 4i$

Use deMoivre's Theorem to express each number in the form $a + bi$.

7. $(3 + 3i)^5$ **8.** $(2 - 2\sqrt{3}i)^8$

Write each equation in polar coordinates.

9. $y = -5x$ **10.** $x^2 + y^2 = 16$

11. $x + y = 2$ **12.** $y = 4$

13. $x^2 + y^2 + 6x = 0$

14. $x^2 + y^2 - 4x + 10y = 0$

15. $x^2 = 9y$ **16.** $x^2 + 2y^2 = 4$

Write each equation in rectangular coordinates.

17. $\theta = \dfrac{7\pi}{6}$ **18.** $r = -4$

19. $r = -2 \csc \theta$ **20.** $r = 3 \cos \theta$

21. $r = \dfrac{4}{1 - \cos \theta}$ **22.** $r = \dfrac{1}{1 - 2 \cos \theta}$

23. $2r^2 \cos 2\theta = 5$ **24.** $r = 5 \sec \theta \tan \theta$

Sketch a graph of each equation.

25. $r = 4$

26. $\theta = \frac{\pi}{4}$

27. $r = 6 \sin \theta$

28. $r = 2 \cos \theta$

29. $r = 2 - 2 \cos \theta$

30. $r = 5 - 3 \sin \theta$

31. $r = 1 + 2 \cos \theta$

32. $r = \cos 2\theta$

33. $r = \sin 5\theta$

34. $r^2 = 4 \cos \theta$

Sketch a graph of the relation defined by the given pair of parametric equations.

35. $x = -2t, y = 6t$

36. $x = e^{2t}, y = e^{6t}$

37. $x = 3 \sin \theta, y = 3 \cos \theta$

38. $x = 2 \cos \theta, y = 3 \sin \theta$

39. $x = -2 \sec \theta, y = 3 \tan \theta$

40. $x = 2t - 1, y = t^2 + 1$

Solve each equation. $x \in C$.

41. $x^3 + 8 = 0$

42. $x^5 - 32 = 0$

Write a pair of parametric equations for each straight line using *s*, the distance of any point on the line from the origin, as a parameter.

43. $y = x$

44. $y = -2x$

45. Find the fourth roots of -16.

46. Find the cube roots of $-27i$.

Write a pair of parametric equations, $x = f(\theta)$ and $y = g(\theta)$, for each equation.

47. $r = \sin 2\theta$

48. $r - 1 = \cos \theta$

49. Eliminate the parameter t from the equations, $x = t^2 - 5t$ and $y = 2t^2 + t$.

50. Write the equation, $4y^2 = 1 - 4x$, in polar coordinates.

Chapter 10 Test 1

1. Determine

a. the pair of rectangular coordinates that correspond to the polar coordinates $(1,\pi/6)$.

b. a polar coordinate pair that corresponds to the rectangular coordinates $(\sqrt{3},3)$.

2. Write the equation, $y = 2x^2$, in polar coordinates.

3. Write the equation, $r = 5 \sin \theta$, in rectangular coordinates.

Graph.

4. $r = 3 \sec \theta$

5. $r = 1 + 2 \sin \theta$

6. Sketch a graph of the relation defined by the equations, $x = 2t - 2, y = 4t + 5$.

7. Write a pair of parametric equations, $x = f(\theta)$ and $y = g(\theta)$, for the equation $r = \cos 3\theta$.

8. Find the cube roots of -125.

Chapter 10 Test 2

1. Determine

a. the pair of rectangular coordinates that correspond to the polar coordinates $(-5,3\pi/4)$.

b. a pair of polar coordinates that correspond to the rectangular coordinates $(4,-5)$.

2. Write the equation $x^2 + y^2 + 2x - 5y = 0$ in polar coordinates.

3. Write the equation $r = \frac{2}{1 - 3 \cos \theta}$ in rectangular coordinates.

Graph.

4. $r = 3 + 4 \sin \theta$

5. $r^2 = 4 \cos 2\theta$

6. Sketch a graph of the relation defined by the equations, $x = \sin \theta, y = 2 \cos \theta$.

7. Write a pair of parametric equations, $x = f(\theta)$ and $y = g(\theta)$, for the equation $r = 1 - 2 \sin \theta$.

8. Find the fourth roots of $-1 + i$.

Chapter 11

The Conic Sections

A cone is generated by lines that make the same angle with a fixed line called the axis of the cone (see figure 11.1) and that pass through the same point on the axis. The lines are called *generators* of the cone. The point through which all generators pass is called the *vertex* of the cone.

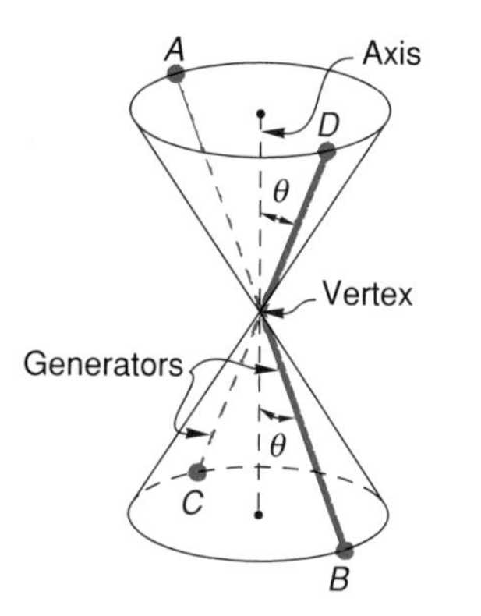

Figure 11.1

If a cone such as that shown in figure 11.1 is intersected by a plane that is

1. perpendicular to the axis of the cone,
2. not parallel to any generator of the cone or perpendicular to the axis of the cone,
3. parallel to exactly one generator of the cone,
4. parallel to exactly two generators of the cone,

then the resulting curve of intersection is

1. a **circle** or a point.
2. an **ellipse** or a point.
3. a **parabola** or a line.
4. a **hyperbola** or two straight lines that are generators of the cone.

These curves of intersection are called conics (figure 11.2).

Many properties of the conics were discovered by the Greek geometer Apollonius about 200 B.C. Pappus, another ancient Greek geometer, was able to show that a conic is a plane curve that can be generated by a point that

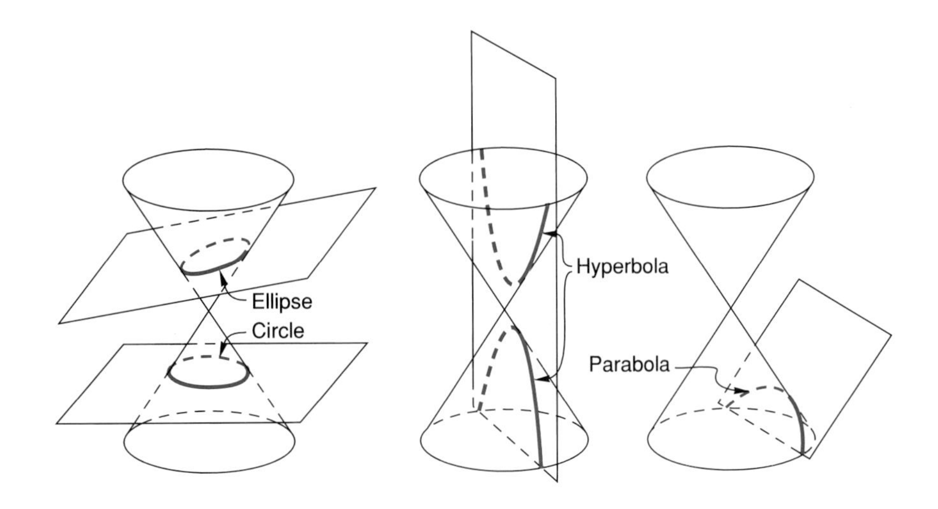

Figure 11.2

moves according to some prescribed constraints. Thus, a circle can be generated by a point that moves in a plane so that its distance from a fixed point in the plane remains constant. A parabola is generated by a point that moves in a plane and is always equidistant from a fixed point and a fixed line in the plane. There are similar definitions for the ellipse and hyperbola.

The French mathematician Descartes (1596–1650) used his "Analytic Geometry" to convert Pappus' geometric descriptions of the conics to algebraic descriptions. Descartes was able to show that each conic section is a graph of a second degree polynomial of the form

General Second Degree Equation

$$Ax^2 + Bxy + Cy^2 + Dx + Ey + F = 0. \qquad (1)$$

We shall take the path of Pappus and Descartes in studying the conics, and in the process become acquainted with some of their many applications in science and engineering.

11.1 Circles

A circle is the set of all points in the plane that are equidistant from a fixed point in the plane. The fixed point is the center.

Let (h,k) be the coordinates of the center of a circle with radius r. Let $P(x,y)$ be a point on the circle (figure 11.3). Then

$$\sqrt{(x-h)^2 + (y-k)^2} = r$$

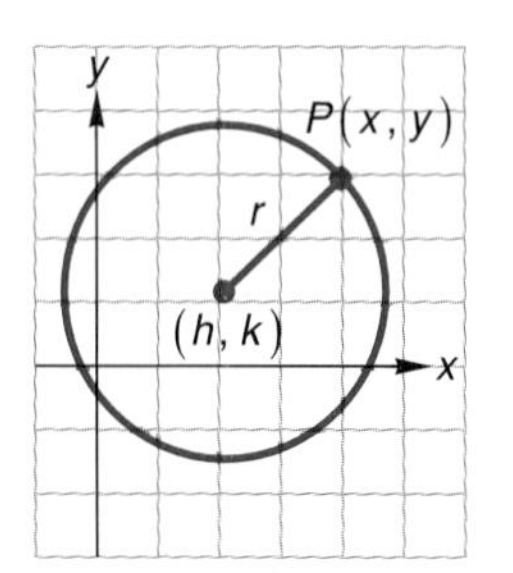

Figure 11.3

Standard Form Equation of a Circle

$$(x-h)^2 + (y-k)^2 = r^2. \tag{2}$$

Equation (2) is called the *standard form* of the equation of a circle with center (h,k) and radius r.

Expanding (2) gives

$$x^2 + y^2 - 2hx - 2ky + h^2 + k^2 - r^2 = 0$$

or

General Form Equation of a Circle

$$x^2 + y^2 + Dx + Ey + F = 0 \tag{3}$$

where $D = -2h$, $E = -2k$ and $F = h^2 + k^2 - r^2$. Equation (3) is called the *general form* of the equation of a circle. Note that (3) is a special instance of (1) in which $B = 0$ and $A = C$.

Example 1 Write the standard form and the general form of the equation of a circle with center $(-2,5)$ and radius 4.

Solution The standard form is

$$(x + 2)^2 + (y - 5)^2 = 16.$$

Expanding the standard equation

$$x^2 + 4x + 4 + y^2 - 10y + 25 = 16$$

gives us the general equation

$$x^2 + y^2 + 4x - 10y + 13 = 0.$$

■

Example 2 Find the center and radius of the circle whose equation is $x^2 + y^2 - 6x + 14y + 8 = 0$.

Solution We can use the method of completing the square to change the given general form equation to standard form.

$$(x^2 - 6x + 9) + (y^2 + 14y + 49) = -8 + 9 + 49$$
$$(x - 3)^2 + (y + 7)^2 = 50$$

The center of the circle is $(3,-7)$ and its radius is $\sqrt{50}$. ■

The coefficients of x^2 and y^2 in the general equation (3) of a circle need not be 1. Both sides of that equation can be multiplied by any non-zero constant to produce an equivalent equation. Conversely, an equation such as

$$4x^2 + 4y^2 + 12x + 4y + 20 = 0$$

can be put in the form of (3) by dividing through by 4. The result is

$$x^2 + y^2 + 3x + y + 5 = 0.$$

Now we might ask the following question. Since each circle in the plane has an equation that is a specific instance of (3), does every equation of the form

$$x^2 + y^2 + DX + Ey + F = 0$$

determine a circle?

Example 3 Show that equations $x^2 + y^2 - 6x + 14y + 58 = 0$ and $4x^2 + 4y^2 + 12x + 4y + 20 = 0$ do not determine circles.

Solution We change both equations to standard form.

$$(x^2 - 6x + 9) + (y^2 + 14y + 49) = 9 + 49 - 58$$
$$(x - 3)^2 + (y + 7)^2 = 0$$

$$(x^2 + 3x + \tfrac{9}{4}) + (y^2 + y + \tfrac{1}{4}) = \tfrac{10}{4} - 5$$
$$(x + \tfrac{3}{2})^2 + (y + \tfrac{1}{2})^2 = -\tfrac{5}{2}$$

The only ordered pair that satisfies the first equation is $(3,-7)$. We say the circle degenerates to a single point, its center. No ordered pair of real numbers satisfies the second equation. The left side of the equation is nonnegative for any real values of x and y, but the right side is negative. Since the equation has no real solutions, there is no graph. ■

Circles Determined by Three Conditions

The undetermined constants D, E, and F in the general equation of a circle can be evaluated if we know three conditions that the circle must satisfy. The circle can be determined if three points on the circle are known (example 4) or if we know a point on the circle and the circle's center and radius (example 5). Other sets of three conditions are given in example 6 and in the exercises.

Example 4 Write an equation of the circle containing points (2,7), (3,−4), and (−1,5).

Solution The coordinates of the three points are substituted into the general form equation (3) of a circle to produce three equations in D, E, and F.

$$\begin{aligned} 4 + 49 + 2D + 7E + F &= 0 \\ 9 + 16 + 3D - 4E + F &= 0 \\ 1 + 25 - D + 5E + F &= 0 \end{aligned}$$

$$\begin{aligned} 2D + 7E + F &= -53 \\ 3D - 4E + F &= -25 \\ -D + 5E + F &= -26 \end{aligned}$$

The system of linear equations is solved by the method of linear combinations, as discussed in chapter 3. The solution is

$$D = -\tfrac{241}{35} \quad E = -\tfrac{111}{35} \quad F = -\tfrac{596}{35}.$$

Substituting these values into (3) gives us

$$\begin{aligned} x^2 + y^2 - \tfrac{241}{35}x - \tfrac{111}{35}y - \tfrac{596}{35} &= 0, \\ 35x^2 + 35y^2 - 241x - 111y - 596 &= 0. \end{aligned}$$

The last equation is the general form equation of the circle that contains (2,7), (3,−4), and (−1,5). ■

Example 5 Find the center and radius of the circle that contains (2,5) and is tangent to both the y-axis and $y = x$.

Solution We make use of the formula for the distance from a point (x_0,y_0) to the line $Ax + By + C = 0$ as given in equation (9) of section 3.1. That formula is

$$s = \frac{|Ax_0 + By_0 + C|}{\sqrt{A^2 + B^2}}. \tag{4}$$

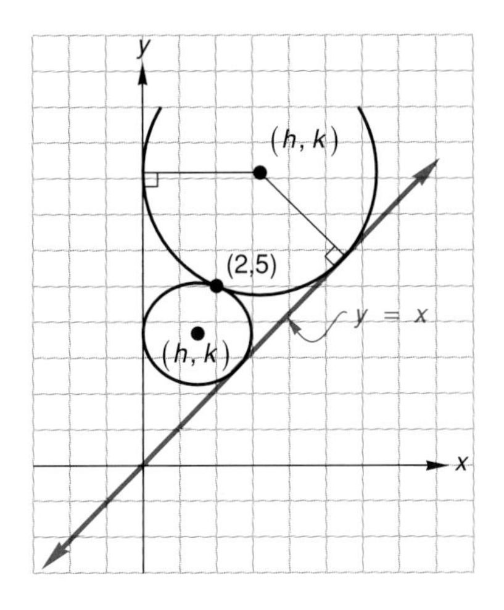

Figure 11.4

If we write the equation of the line $y = x$ (figure 11.4) in the equivalent form, $-x + y = 0$, then (4) gives us the distance from (h,k) to the line as $|-h + k|/\sqrt{2}$. Since the center of the circle is equidistant from the tangents to the circle, and since the distance of (h,k) from the y-axis is h, we can write the equation

$$h = \frac{|-h + k|}{\sqrt{2}}.$$

Squaring both sides of the equation yields

$$2h^2 = h^2 - 2hk + k^2$$

or

$$h^2 + 2kh - k^2 = 0.$$

Using the quadratic formula to solve for h in terms of k gives $h = (\pm\sqrt{2} - 1)k$. But h and k must both be positive. Therefore, $h = (\sqrt{2} - 1)k$. If we substitute into the standard form equation (2) $x = 2$, $y = 5$, and $r = (\sqrt{2} - 1)k$, we obtain

$$(2 - (\sqrt{2} - 1)k)^2 + (5 - k)^2 = ((\sqrt{2} - 1)k)^2,$$

or

$$k^2 - (6 + 4\sqrt{2})k + 29 = 0.$$

It follows that $k = (3 + 2\sqrt{2}) \pm \sqrt{12(\sqrt{2} - 1)} \approx 8.06$ or 3.60 and that $h = (\sqrt{2} - 1)k \approx 3.34$ or 1.49. Thus two circles satisfy the given conditions. Their centers and radii are

$$(3.34,8.06),\ 3.34 \quad \text{and} \quad (1.49,3.60),\ 1.49.$$

■

Example 6 Find an equation for the circle that contains the origin and the points of intersection of the circles

$$x^2 + y^2 + 2y - 36 = 0 \quad \text{and} \quad x^2 + y^2 + 2x + 4y - 48 = 0.$$

Solution We find the points of intersection of the given circles by solving their equations simultaneously.

$$\begin{array}{r} x^2 + y^2 \qquad\quad + 2y = 36 \\ x^2 + y^2 + 2x + 4y = 48 \\ \hline -2x - 2y = -12 \end{array}$$

If the squared terms are eliminated by subtraction, the resulting equation, $-2x - 2y = -12$, determines a straight line that contains the points common to the circles (see figure 11.5). To find those points obtain the solution of the system

$$\begin{aligned} -2x - 2y &= -12 \\ x^2 + y^2 + 2y &= 36. \end{aligned}$$

In the first equation $x = 6 - y$. Substituting for x in the second equation produces

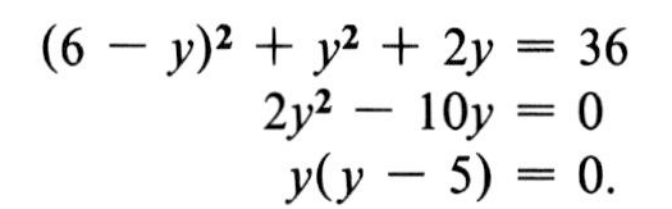

$$\begin{aligned} (6 - y)^2 + y^2 + 2y &= 36 \\ 2y^2 - 10y &= 0 \\ y(y - 5) &= 0. \end{aligned}$$

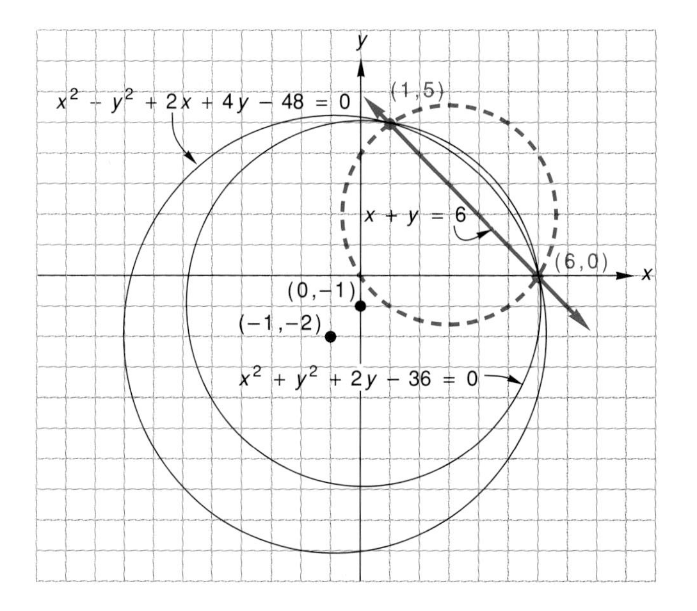

Figure 11.5

The points of intersection are (6,0) and (1,5). Now repeat the steps taken in example 4 and find the general equation of the circle containing (0,0), (6,0), and (1,5). That equation is $x^2 + y^2 - 6x - 4y = 0$. (See exercise 39 for a different solution of this problem.) ■

Parametric Equations

The basic trigonometric identity $\cos^2\theta + \sin^2\theta = 1$ suggests a natural parametrization of the circle $x^2 + y^2 = a^2$. Let $x = a\cos\theta$ and $y = a\sin\theta$; then

$$x^2 + y^2 = a^2\cos^2\theta + a^2\sin^2\theta = a^2.$$

You can verify for yourself that the parametric equations $x = a\cos\theta$ and $y = a\sin\theta$ trace out a circle of radius a, counterclockwise, as θ assumes all the values in the interval $[0,2\pi]$.

Exercise Set 11.1

(A)

Write standard form and general form equations of the circle with center C and radius r.

1. $C(1,-5)$, $r = 4$ **2.** $C(0,0)$, $r = 5$

3. $C(-7,-3)$, $r = 2$ **4.** $C(6,2)$, $r = 7$

Find the center and radius of the given circle.

5. $x^2 + y^2 + 6x - 8y = 0$

6. $x^2 + y^2 - 12x + 4y + 4 = 0$

7. $5x^2 + 5y^2 + 20x + 40y - 25 = 0$

8. $3x^2 + 3y^2 - 18x - 15y + 12 = 0$

9. $x^2 + y^2 - 4x - 10y + 29 = 0$

10. $2x^2 + 2y^2 + 8x + 14y + 35 = 0$

Write the general form equation of the circle that satisfies the given conditions.

11. Contains the points (0,0), (2,3), (5,−1).

12. Contains the points (−6,−5), (−8,3), (2,4).

13. Has center (3,1) and contains (2,5).

14. Has center (5,−4) and contains the origin.

15. Center is (−7,2), tangent to the x-axis.

16. Center is (3,8), tangent to y-axis.

17. Center is (4,−3), tangent to $5x + 2y = 10$.

18. Center is (0,6), tangent to $2x - 3y = 6$.

19. Center is on the x-axis, contains (2,5) and (6,3).

20. Center is on the y-axis, contains (4,2) and (−3,7).

21. Concentric with $x^2 + y^2 + 4x - 6y - 3 = 0$ and tangent to $y - x = 1$.

22. Concentric with $x^2 + y^2 - 8x + 2y + 1 = 0$ and tangent to $x + 2y = 10$.

Write a standard form equation of the circle whose parametric equations are given.

23. $x = 2\cos\theta$, $y = 2\sin\theta$

24. $x = 4\sin\theta$, $y = 4\cos\theta$

25. $x = -1 + 3\cos\theta$, $y = 2 + 3\sin\theta$

26. $x = 5 + 4\sin\theta$, $y = -3 + 4\cos\theta$

(B)

27. Find the center and radius of the circle that contains (4,3) and is tangent to $y = x$ and the x-axis. (See example 5.)

28. Find the center and radius of the circle that contains (−6,4) and is tangent to $x + y = 2$ and the x-axis. (See example 5.)

29. The parametrization $x = 2\sin\theta$, $y = 2\cos\theta$, yields the same standard form equation of a circle as does the parametrization in exercise 23. What is the significance of the different parametrizations?

30. Use the fact that the perpendicular bisectors of two non-parallel chords in a circle intersect at the center of the circle as the basis for doing example 4 another way.

31. The equation $(x - 1)^2 + (y + 2)^2 = r^2$ represents a family of concentric circles centered at $(1,-2)$. Graph three members of this family in the same coordinate system.

32. The equation $(x - h)^2 + y^2 = 16$ represents a family of circles, each of which has radius 4 and center on the x-axis. Graph three members of this family in the same coordinate system.

Write an equation for the family of circles satisfying the given conditions. (Hint: See exercises 31 and 32.)

33. Center is $(6,2)$.

34. Center on y-axis, radius 3.

35. Center on the line $x + y = 5$, radius 2.

36. Center on the line $y = x$, tangent to coordinate axis.

37. Center on line $y = 2x$, tangent to line $y = -\frac{1}{2}x$.

38. Argue that the equation $(x^2 + y^2 + 4x - 8y - 7) + t(x^2 + y^2 + 6x + 10y - 2) = 0$ represents a family of circles containing the points of intersection of the circles $x^2 + y^2 + 4x - 8y - 7 = 0$ and $x^2 + y^2 + 6x + 10y - 2 = 0$. (Hint: See example 6.)

39. In example 6 you are shown how to find the circle containing the origin and the points of intersection of the circles $x^2 + y^2 + 2y - 36 = 0$ and $x^2 + y^2 + 2x + 4y - 48 = 0$. In exercise 38 it is shown that an equation of the form $(x^2 + y^2 + 2y - 36) + t(x^2 + y^2 + 2x + 4y - 48) = 0$ represents a family of circles, each of which contains the points of intersection of the circles given in example 6. Find that member of the family that contains the origin.

11.2 Parabolas

A parabola is a set of all points in a plane that are equidistant from a fixed point in the plane and a fixed line in the plane. The fixed point is called the ***focus,*** *and the fixed line is called the* ***directrix.***

An algebraic description of the parabola can be constructed directly from the given geometric description in the following way.

Let point $F(0,c)$ be the focus and line $y = -c$ be the directrix of a parabola (figure 11.6). Then the origin, which is equidistant from the focus and the directrix, is a point on the parabola. It is in fact the **vertex** of the parabola. The vertex is the *only* point of the parabola that is on a line called the *axis of the parabola.* The axis must contain the focus and be *perpendicular to the directrix.*

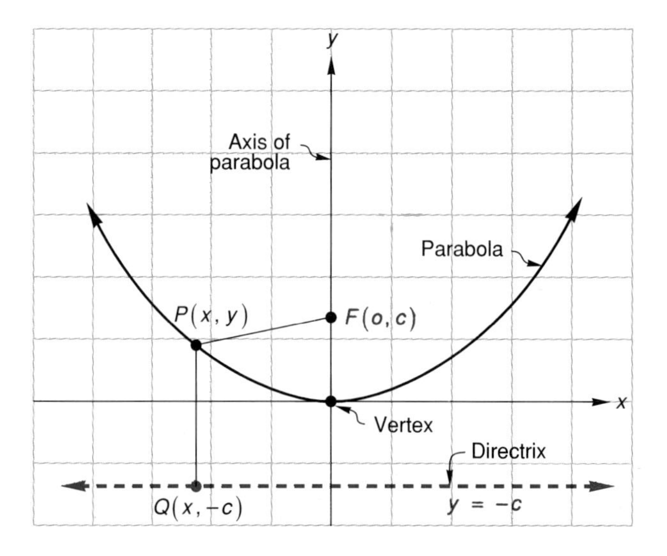

Figure 11.6

The distances from any point $P(x,y)$ on the parabola to the directrix and focus must be equal. That is, $\overline{PQ} = \overline{PF}$. Thus, we may write

$$\sqrt{x^2 + (y - c)^2} = |y - (-c)|$$
$$x^2 + y^2 - 2cy + c^2 = y^2 + 2cy + c^2$$
$$x^2 = 4cy$$

Vertical Parabola—Vertex at Origin

$$x^2 = 4cy. \qquad (1)$$

Equation (1) is often called the *standard equation of a parabola with vertical axis and vertex at the origin.* The equation tells us that the axis of the parabola shown in figure 11.6 is an axis of symmetry.

A "horizontal" parabola with vertex at the origin has

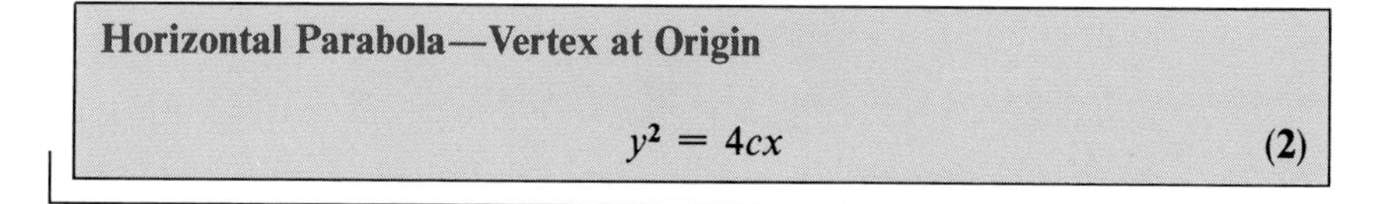

Horizontal Parabola—Vertex at Origin

$$y^2 = 4cx \qquad (2)$$

as its standard equation. In both (1) and (2), $|c|$ gives the distance between the vertex and focus of the parabola.

Example 1 Determine the coordinates of the focus, and write an equation of the directrix of the parabola $y^2 = -12x$.

Solution The given equation has the form of a horizontal parabola with vertex at the origin (2). Furthermore, the parabola opens to the left, as $-12x$ must be nonnegative. (Why?) From (2) we know that $4c = -12$. Hence, the coordinates of the focus are $(-3,0)$. The directrix must be perpendicular to the x-axis, which is the axis of the parabola, and it must be three units from the vertex. It is the line $x = 3$. ■

Example 2 Write an equation of the "vertical" parabola, with vertex at the origin, that contains (2,5).

Solution Substituting 2 for x and 5 for y in (1) yields

$$(2)^2 = 4c\,(5)$$
$$\tfrac{1}{5} = c.$$

Hence, the equation sought is $x^2 = \frac{4}{5}y$. ■

Translated Parabolas

A parabola with vertex not at the origin and axis parallel to one of the coordinate axes is said to be *translated* with respect to the origin.

In figure 11.7 a coordinate axis pair, x'-y', is passed through the vertex (h,k), of the translated parabola. A point on the parabola now has two coordinate descriptions, (x',y') or (x,y). Which description is used depends upon the coordinate axes to which the parabola is referred. In any case it is possible to shift from one coordinate description to the other via the set of transformation equations (3), that come directly out of figure 11.7.

Transformation Equations between the x-y and x'-y' Axes

$$\begin{cases} x = x' + h \\ y = y' + k \end{cases} \qquad \begin{cases} x' = x - h \\ y' = y - k \end{cases} \tag{3}$$

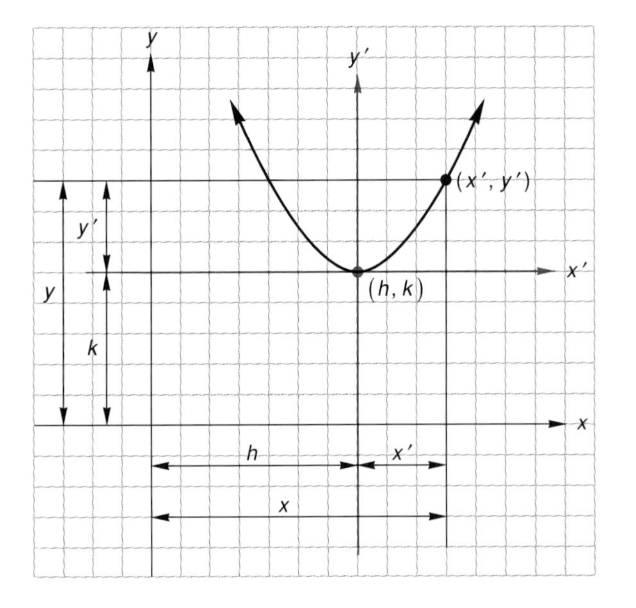

Figure 11.7

If the parabola in figure 11.7 is referred to the x'-y' axes, its equation has the form of equation (1),

$$x'^2 = 4cy'.$$

This last equation can be converted to

Vertical Parabola—Vertex at (h,k)

$$(x - h)^2 = 4c(y - k) \qquad (4)$$

with equations (3). Thus, in (4) we have the *standard form of the equation of a vertical parabola with vertex at (h,k).*

A parallel development would give us

Horizontal Parabola—Vertex at (h,k)

$$(y - k)^2 = 4c(x - h) \qquad (5)$$

as the standard form of the horizontal parabola with vertex at (h,k).

Example 3 Write an equation of the horizontal parabola that contains $(5,-2)$ and has $(-3,2)$ as its vertex. Locate the focus of the parabola.

Solution We substitute for h and k in (5) to obtain

$$(y - 2)^2 = 4c(x + 3).$$

Now substitute the coordinates of the given point in this last equation to find c.

$$(-2 - 2)^2 = 4c(5 + 3)$$
$$\tfrac{1}{2} = c$$

An equation of the parabola is $(y - 2)^2 = 2(x + 3)$. Its focus is $(-\frac{5}{2},2)$ (figure 11.8).

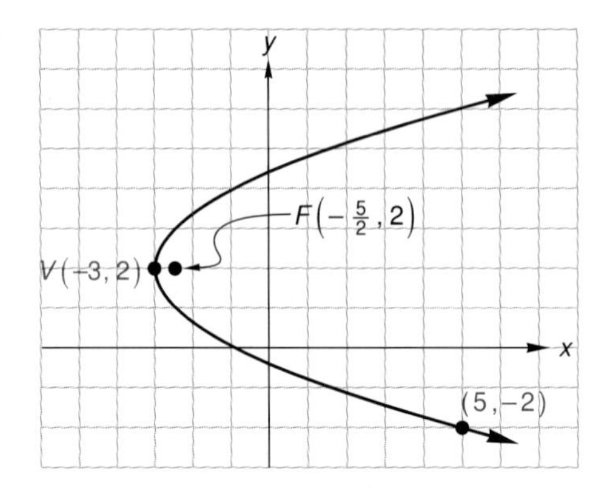

Figure 11.8 ■

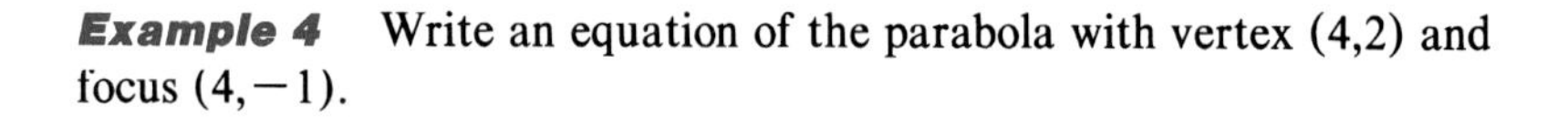
Example 4 Write an equation of the parabola with vertex (4,2) and focus (4, −1).

Solution The axis of our parabola contains (4,2) and (4, −1). Hence, the axis is $x = 4$, and the parabola must have an equation of the form $(x - h)^2 = 4c(y - k)$. We know that $h = 4$, $k = 2$, and $|c| = 2 - (-1) = 3$. As the focus is below the vertex, the parabola opens downward (figure 11.9), and c must be negative. (Why?)

The parabola's equation is

$$(x - 4)^2 = -12(y - 2).$$

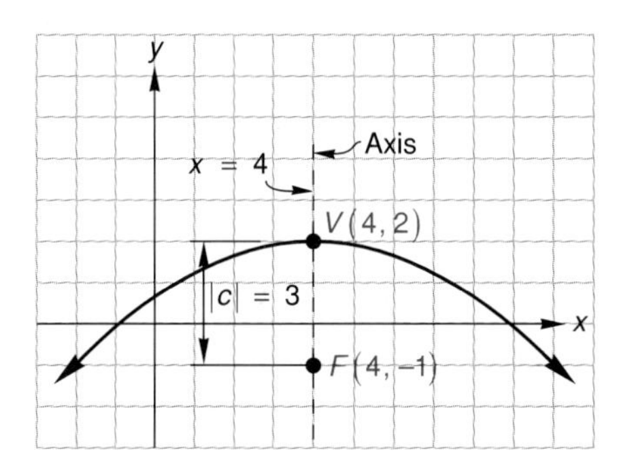

Figure 11.9 ■

Equation (4) can be rewritten as

$$x^2 - 2hx - 4cy + h^2 + 4ck = 0,$$

or as

$$Ax^2 + Dx + Ey + F = 0, \tag{6}$$

where $A = 1$, $D = -2h$, $E = -4c$ and $F = h^2 + 4ck$. In the same manner (5) can be written as

$$Cy^2 + Dx + Ey + F = 0. \tag{7}$$

From equations (6) and (7) we conclude that the equation of a translated parabola is of the form

$$Ax^2 + Bxy + Cy^2 + Dx + Ey + F = 0,$$

in which *B and either A or C are zero.*

Example 5 Determine the vertex and focus of the parabola whose equation is $y^2 + 8x - 6y - 31 = 0$.

Solution

$$\begin{aligned} y^2 - 6y &= -8x + 31 \\ y^2 - 6y + 9 &= -8x + 40 \\ (y - 3)^2 &= -8(x - 5) \end{aligned}$$

Comparing this last equation with (5), we see the parabola is horizontal with vertex (5,3). Since $4c = -8$ and $c = -2$, the parabola opens to the left, and its focus is (3,3). ■

A parabola may have an equation that has more than one second degree term if its axis is oblique.

Example 6 Write an equation of the parabola whose focus is the origin and whose directrix is $x + 2y - 4 = 0$.

Solution The distances from any point $P(x,y)$ on the parabola to the focus and directrix are equal. Using the distance formula from a point to a line (equation (4), section 11.1) we write

$$d_1 = \sqrt{x^2 + y^2} = \frac{|x + 2y - 4|}{\sqrt{5}} = d_2$$

$$x^2 + y^2 = \frac{x^2 + 4xy + 4y^2 - 8x - 16y + 16}{5}$$

$$4x^2 - 4xy + y^2 + 8x + 16y - 16 = 0.$$

The last equation defines our parabola. Note the coincidence of a parabola whose axis is not parallel to one of the coordinate axes and the xy term in the equation of the parabola (figure 11.10).

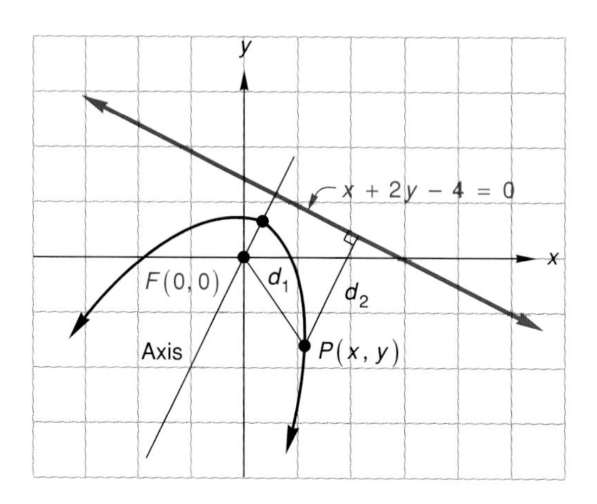

Figure 11.10

■

Summary

The standard equation of a parabola with vertex at (h,k) and axis parallel to one of the coordinate axes is

Vertical Parabola:

$$(x - h)^2 = 4c(y - k) \begin{cases} \text{opens up if } c > 0 \\ \text{opens down if } c < 0 \end{cases}$$

Horizontal Parabola:

$$(y - k)^2 = 4c(x - h) \begin{cases} \text{opens right if } c > 0 \\ \text{opens left if } c < 0 \end{cases}$$

$|c|$ is the distance from the focus to the vertex.

The general equation of a parabola whose axis is parallel to one of the coordinate axes has the form

$$Ax^2 + Cy^2 + Dx + Ey + F = 0,$$

where either A or C is zero.

Exercise Set 11.2

(A)

Graph each pair of parabolas using the same set of coordinate axes. Locate the focus and write an equation of the directrix of each parabola.

1. $x^2 = 4y$, $x^2 = -4y$
2. $x^2 = 20y$, $x^2 = -2y$
3. $y^2 = 24x$, $y^2 = 8x$
4. $y^2 = 12x$, $y^2 = -12x$

Write an equation of the parabola with vertex at the origin that satisfies the given condition(s).

5. Focus (5,0)
6. Focus (0,−6)
7. Directrix $y = -3$
8. Directrix $x = 2$
9. Contains (2,8), axis is y-axis
10. Contains (−5,−3), axis is x-axis.

Sketch a graph of each parabola. Locate the focus, vertex, and axis of the parabola.

11. $(x - 3)^2 = 6(y + 1)$
12. $x^2 = 8(y - 3)$
13. $(y + 4)^2 = 4(x - 1)$
14. $(y - 3)^2 = 9x$
15. $(x + 6)^2 = 5(y - 2)$
16. $y^2 = 7(x + 4)$

Write an equation of the parabola that satisfies the given conditions.

17. vertical axis, vertex (4,−1), contains (1,3)
18. horizontal axis, vertex (−2,3), contains (0,5)
19. axis $y = -3$, focus (0,−3), directrix $x = -4$
20. axis $x = 2$, focus (2,−1), directrix $y = -5$
21. vertex (3,1), focus (3,−2)
22. vertex (2,5), focus (−2,5)
23. axis $x = 1$, contains (3,1) and (−2,3)
24. axis $y = 4$, contains (6,0) and (1,5)

Find the vertex and focus of the parabola determined by each equation. Sketch a graph.

25. $x^2 + 6x + 2y - 8 = 0$
26. $4x^2 + 12x + 3y + 2 = 0$
27. $2y^2 - 2x + 8y + 10 = 0$
28. $y^2 + x + 4y - 3 = 0$
29. $3x^2 - 9x + 2y - 8 = 0$
30. $5y^2 + 4x - 20y + 12 = 0$

Write an equation of the parabola with the given focus and directrix.

31. focus (0,0), directrix $x + y = 1$
32. focus (0,0), directrix $2x - 3y = 6$
33. focus (3,2), directrix $x - 3y - 9 = 0$
34. focus (−4,1), directrix $2x - y + 10 = 0$

(B)

35. The support cables of a suspension bridge hang from the bridge towers in a parabolic arc. If the ends of one such cable are 200 feet above the lowest point of the cable, and if the horizontal distance between the ends of the cable is 1,000 feet, how high above the low point of the cable are the horizontal quarter points of the cable? (See figure 11.11.)

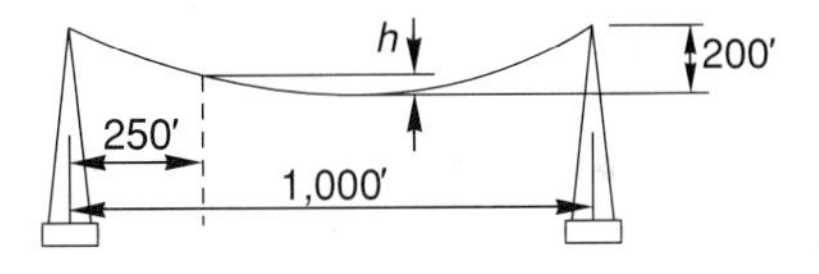

Figure 11.11

36. A ball is thrown from a point that is 7 feet above a level field. Its flight path is assumed to be parabolic. Besides the point (0,7), points (20,30) and (50,50) are observed to be in the ball's path. How far down the field does the ball land? (Hint: Assume the parabolic path has an equation of the form $y = ax^2 + bx + c$. Write three linear equations in a, b, and c.)

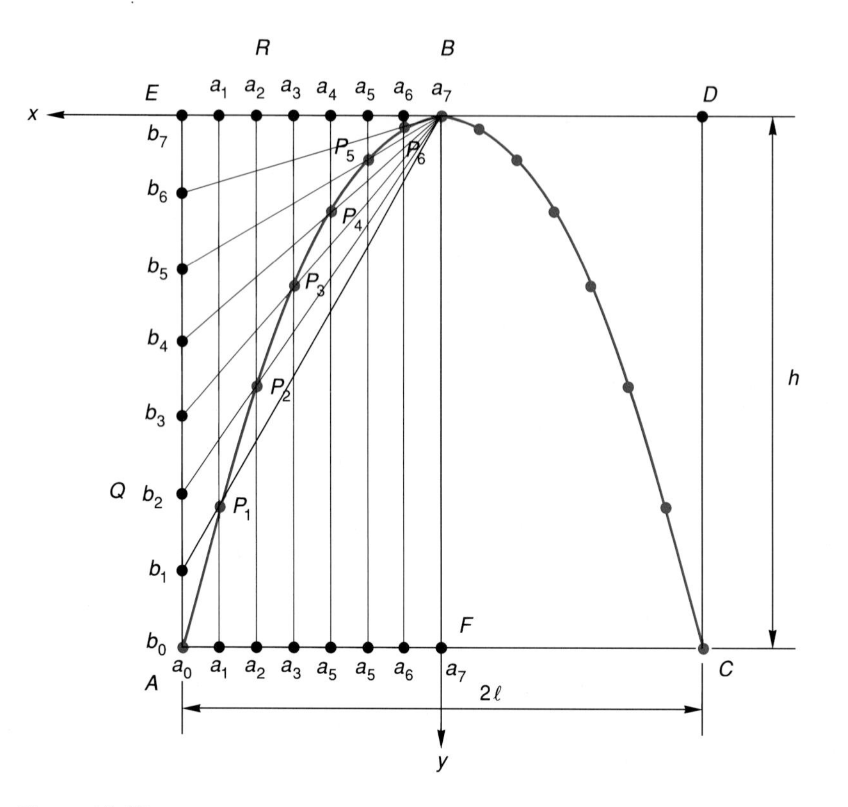

Figure 11.12

37. A parabolic arch of height h and base width 2ℓ (figure 11.12) can be drawn in the following way. Let a_0, a_1, a_2, . . . , a_n be the end points of n equal subdivisions of $\overline{AF}$ and $\overline{EB}$. Let b_0, b_1, . . . , b_n be the end points of n equal subdivisions of $\overline{AE}$. The intersections of $\overline{Bb_1}$ and $\overline{a_1a_1}$, $\overline{Bb_2}$ and $\overline{a_2a_2}$, $\overline{Bb_3}$ and $\overline{a_3a_3}$ are points on the parabola. Explain why. (Hint: Consider a coordinate system through B as shown in figure 11.12. Use the similar triangles BRP_2 and BEQ to show that the coordinates of P_2 satisfy the equation $x^2 = (\ell^2/h)y$.

38. In the beginning of the section we showed that if a point (x_0,y_0) is equidistant from the point $(0,c)$ and the line $y = -c$, then (x_0,y_0) satisfies equation (1), $x^2 = 4cy$. Show the converse; that if (x_0,y_0) satisfies (1), it must be equidistant from point $(0,c)$ and line $y = -c$.

11.3 Ellipses

An ellipse is a set of all points in a plane for which the sum of the distances from two fixed points in the plane is a constant. The two fixed points are called ***foci.***

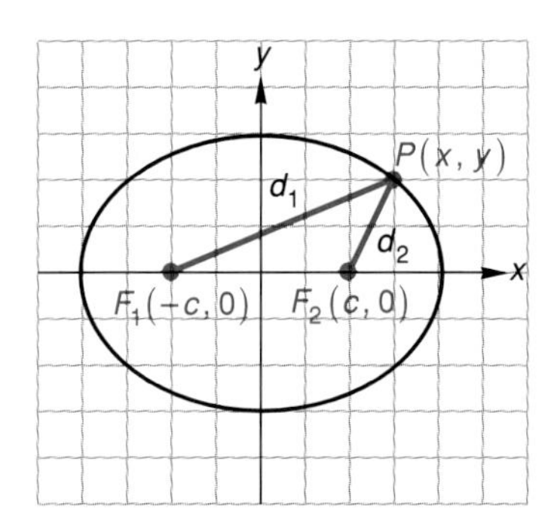

Figure 11.13

Let $F_1(-c,0)$ and $F_2(c,0)$ be the foci of an ellipse and let $P(x,y)$ be a point on the ellipse (figure 11.13). Then, according to the given definition, $d_1 + d_2 = 2a$, where $2a$ is a constant.

It follows that

$$\begin{aligned}
\sqrt{(x+c)^2+y^2} + \sqrt{(x-c)^2+y^2} &= 2a \\
\sqrt{(x+c)^2+y^2} &= 2a - \sqrt{(x-c)^2+y^2} \\
x^2 + 2cx + c^2 + y^2 &= 4a^2 - 4a\sqrt{(x-c)^2+y^2} \\
&\quad + x^2 - 2cx + c^2 + y^2 \\
a\sqrt{(x-c)^2+y^2} &= a^2 - cx \\
a^2x^2 - 2a^2cx + a^2c^2 + a^2y^2 &= a^4 - 2a^2cx + c^2x^2 \\
(a^2-c^2)x^2 + a^2y^2 &= a^2(a^2-c^2).
\end{aligned}$$

The sum of the lengths of two sides of a triangle must be greater than the length of the third side. Thus, in figure 11.13

$$\begin{aligned}
d_1 + d_2 &> \overline{F_1F_2} \\
2a &> 2c \\
a &> c.
\end{aligned}$$

Because $a > c$, $a^2 - c^2 > 0$, and we can replace $a^2 - c^2$ by a positive number, b^2. Since $a^2 - c^2 = b^2$, *a must be greater than b.* Substituting b^2 for $a^2 - c^2$ in the last equation above, we get

$$b^2x^2 + a^2y^2 = a^2b^2.$$

We divide both sides of this equation by a^2b^2 and obtain one of the standard form equations of the ellipse.

Horizontal Ellipse—Centered at the Origin

$$\frac{x^2}{a^2} + \frac{y^2}{b^2} = 1 \qquad \textbf{(1)}$$

A graph of (1) (figure 11.14) is symmetric to the x and y axes and the origin. (Why?) Letting $y = 0$ gives us $x^2/a^2 = 1$ or $x^2 = a^2$. Consequently, the x-intercepts are a and $-a$. The segment on the x-axis between the x-intercepts is called the **major axis** of the ellipse. The **minor axis** of the ellipse is the segment on the y-axis between the y-intercepts, which are obtained by setting $x = 0$ in (1). Note that the lengths of the major and minor axes are $2a$ and $2b$, respectively.

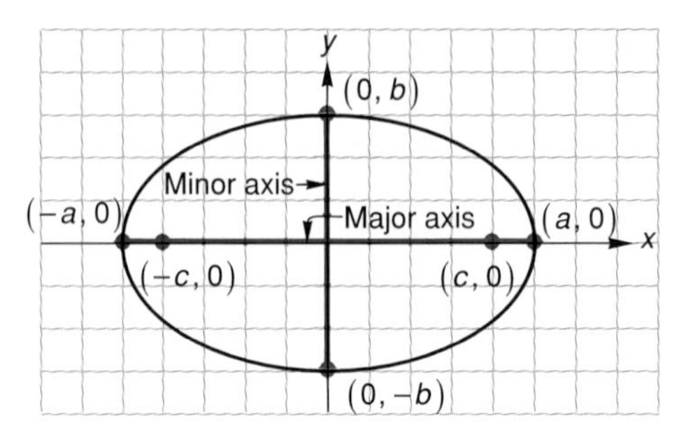

Figure 11.14

The derivation of (1) proceeds from the assumption that the foci of the ellipse are on the x-axis and equidistant from the origin. If it is assumed that the foci are on the y-axis and equidistant from the origin, then the major axis would lie on the y-axis, and the derivation that produced (1) would yield

Vertical Ellipse—Centered at the Origin

$$\frac{x^2}{b^2} + \frac{y^2}{a^2} = 1. \tag{2}$$

Equations (1) and (2) are called **standard equations** of ellipses centered at the origin. The center of an ellipse is the point of intersection of its major and minor axes. The ends of its major axis are called the **vertices** of an ellipse.

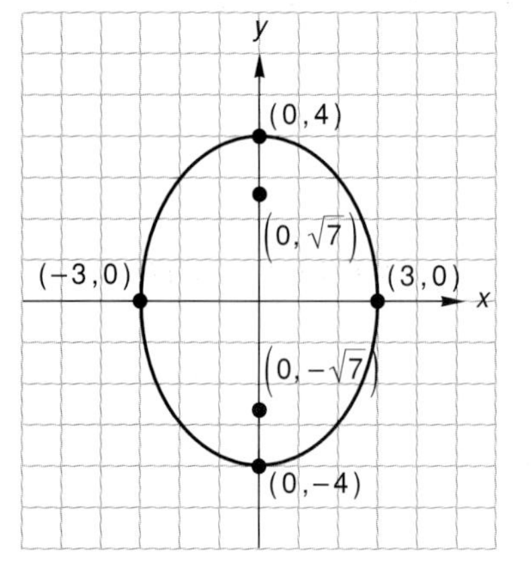

Figure 11.15

Example 1 Show that $16x^2 + 9y^2 = 144$ is the equation of an ellipse. Determine the coordinates of the vertices and foci, and sketch a graph of the ellipse.

Solution Divide both sides of the given equation by 144 to obtain

$$\frac{x^2}{9} + \frac{y^2}{16} = 1.$$

The equation has the form of (2) so that it represents an ellipse that is centered at the origin with its major axis on the y-axis. Since $a^2 = 16$, the coordinates of the vertices are $(0,4)$ and $(0,-4)$ (figure 11.15). The coordinates of the foci can be found from the relationship $a^2 - c^2 = b^2$ where $a^2 = 16$ and $b^2 = 9$. Thus, $c = \pm\sqrt{7}$ and the coordinates of the foci are $(0,\sqrt{7})$ and $(0,-\sqrt{7})$. ■

Example 2 Write an equation of the ellipse centered at the origin with axes on the coordinate axes, whose graph contains the points (5,1) and (−2,4). Determine the foci of the ellipse and sketch its graph.

Solution The equation sought may have either form (1) or (2). Let us assume that it has the form

$$\frac{x^2}{b^2} + \frac{y^2}{a^2} = 1.$$

Substituting the coordinates of the given points into this equation yields

$$\frac{25}{b^2} + \frac{1}{a^2} = 1$$
$$\frac{4}{b^2} + \frac{16}{a^2} = 1.$$

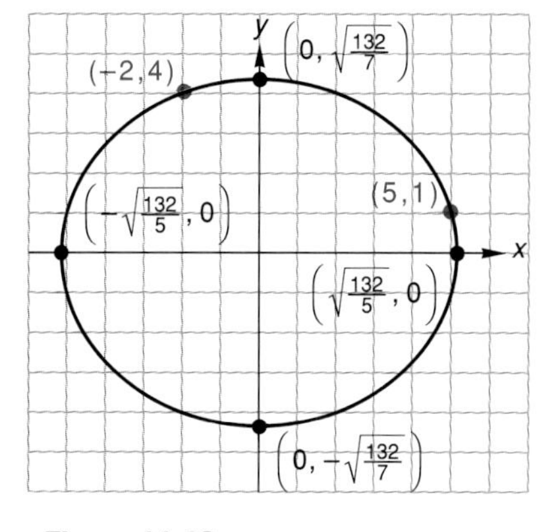

Figure 11.16

Multiplying the first equation by −16 and the second equation by 1 and adding gives

$$\frac{-396}{b^2} = -15.$$

We find that $b^2 = \frac{132}{5}$ and $a^2 = \frac{132}{7}$. But a must be greater than b. It is not. Thus, our assumption that the ellipse has its major axis on the y-axis is incorrect. The ellipse (figure 11.16) equation must have the form

$$\frac{x^2}{a^2} + \frac{y^2}{b^2} = 1.$$

Repeating the above computations for a and b produces the system

$$\frac{25}{a^2} + \frac{1}{b^2} = 1$$
$$\frac{4}{a^2} + \frac{16}{b^2} = 1,$$

in which $a^2 = \frac{132}{5}$ and $b^2 = \frac{132}{7}$. As a result an equation for the ellipse is $5x^2 + 7y^2 = 132$. ■

Translated Ellipses

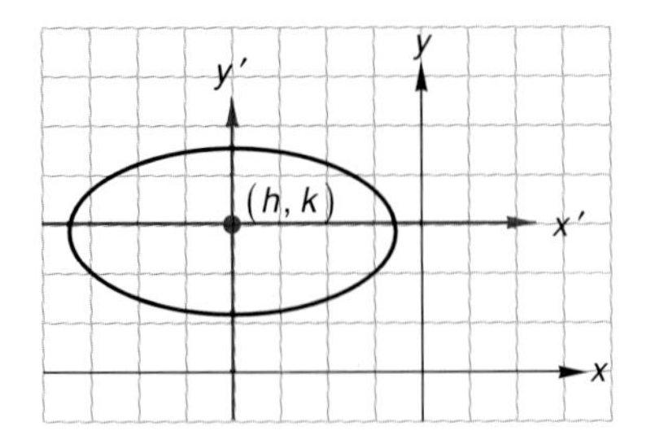

Figure 11.17

If an ellipse is centered at some point other than the origin and its axes are parallel to the coordinate axes, we say the ellipse is *translated* relative to the origin. Such a translated ellipse is depicted in figure 11.17. Since it is centered with respect to the x'-y' axes, its equation relative to those axes is $(x')^2/a^2 + (y')^2/b^2 = 1$.

Applying the transformation equations

$$x' = x - h \quad \text{and} \quad y' = y - k,$$

to $x'^2/a^2 + y'^2/b^2 = 1$ yields the standard equation of a translated ellipse with horizontal major axis.

Horizontal Ellipse—Center (*h*,*k*)

$$\frac{(x-h)^2}{a^2} + \frac{(y-k)^2}{b^2} = 1 \tag{3}$$

The standard equation of a translated ellipse with vertical major axis is

Vertical Ellipse—Center (*h*,*k*)

$$\frac{(x-h)^2}{b^2} + \frac{(y-k)^2}{a^2} = 1. \tag{4}$$

Example 3 The vertices of an ellipse are located at $(-2,3)$ and $(-2,-5)$. Its foci are at $(-2,2)$ and $(-2,-4)$. Write an equation of the ellipse.

Solution The center of the ellipse, $(-2,-1)$, is the midpoint of the line segment connecting the vertices. Its major axis, contained in the line $x = -2$, is vertical. Hence the ellipse has an equation of the form $(x + 2)/b^2 + (y + 1)/a^2 = 1$. Since $a = 4$, $c = 3$, and $b^2 = a^2 - c^2 = 7$, we can write

$$\frac{(x+2)^2}{7} + \frac{(y+1)^2}{16} = 1$$

as an equation of the ellipse. ■

Equation (3) is equivalent to

$$b^2x^2 + a^2y^2 - 2b^2hx - 2a^2ky + b^2h^2 + a^2k^2 - a^2b^2 = 0,$$

which is an equation of the form

$$Ax^2 + Cy^2 + Dx + Ey + F = 0 \tag{5}$$

where $A = b^2$, $C = a^2$, $D = -2b^2h$, $E = -2a^2k$, and $F = b^2h^2 + a^2k^2 - a^2b^2$. Equation (4) can also be written as (5) with $A = a^2$ and $C = b^2$. Thus, we are tempted to call (5) the general equation of a translated ellipse. However, translated circles and parabolas also have (5) as their general equation.

It was shown in section 11.1 that if $A = C$, then (5) determines a circle. In section 11.2 we saw that if A or C is zero, then (5) determines a parabola. In this section we see that $A = b^2$ and $C = a^2$ or vice versa. Therefore (5) will determine an ellipse *if A and C are of the same sign and unequal.*

Example 4 Identify the conic determined by $-3x^2 - 2y^2 + 12x - 16y + 5 = 0$ and sketch its graph.

Solution Comparing the given equation to (5), we see that A and C are of the same sign and unequal. The equation determines an ellipse. Now we convert the given equation to either (3) or (4) by completing the square.

$$-3(x^2 - 4x + 4) - 2(y^2 + 8y + 16) = -5 - 12 - 32$$

$$-3(x - 2)^2 - 2(y + 4)^2 = -49$$

$$\frac{(x-2)^2}{\frac{49}{3}} + \frac{(y+4)^2}{\frac{49}{2}} = 1$$

$$\frac{(x')^2}{\frac{49}{3}} + \frac{(y')^2}{\frac{49}{2}} = 1$$

We see that $(2,-4)$ is the center of the ellipse (figure 11.18), that $\frac{49}{2}$, being the greater of $\frac{49}{2}$ and $\frac{49}{3}$, is a^2 and that b^2 then is $\frac{49}{3}$. Because a^2 is under the y-term, the major axis of the ellipse is vertical. ■

Parametric Equations

As with the circle, the trigonometric identity $\cos^2\theta + \sin^2\theta = 1$ provides us with a parametrization of the ellipse. Let $x = a\cos\theta$ and $y = b\sin\theta$, then

$$\frac{x^2}{a^2} + \frac{y^2}{b^2} = \cos^2\theta + \sin^2\theta = 1.$$

Thus, as θ assumes all real values between 0 and 2π, the parametric equations

$$x = a\cos\theta \text{ and } y = b\sin\theta$$

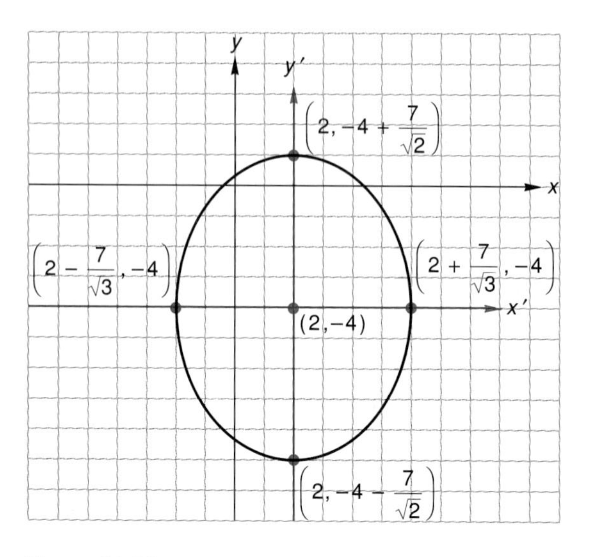

Figure 11.18

generate the coordinates of all points on the ellipse

$$\frac{x^2}{a^2} + \frac{y^2}{b^2} = 1.$$

Summary

The standard equation of an ellipse with center at (h,k) and axes parallel to the coordinate axes is

Horizontal Ellipse:

$$\frac{(x - h)^2}{a^2} + \frac{(y - k)^2}{b^2} = 1, \quad a > b$$

Vertical Ellipse:

$$\frac{(x - h)^2}{b^2} + \frac{(y - k)^2}{a^2} = 1, \quad a > b$$

The distance from the center to each focus is $|c|$, where c is obtained from $c^2 = a^2 - b^2$.

The general equation of an ellipse whose axes are parallel to the coordinate axes has the form

$$Ax^2 + Cy^2 + Dx + Ey + F = 0$$

where $A \neq C$ and A and C have the same sign.

Exercise Set 11.3

(A)

Graph each ellipse. Label foci and vertices.

1. $\dfrac{x^2}{25} + \dfrac{y^2}{9} = 1$ **2.** $\dfrac{x^2}{16} + \dfrac{y^2}{9} = 1$

3. $\dfrac{x^2}{4} + \dfrac{y^2}{9} = 1$ **4.** $\dfrac{x^2}{16} + \dfrac{y^2}{36} = 1$

Write an equation of the centered ellipse, with axes on the coordinate axes, that satisfies the given conditions.

5. focus (4,0), vertex (6,0)

6. focus (0,2), vertex (0,4)

7. major axis is vertical, $c = 3$, $a = 5$

8. major axis is horizontal, $c = 2$, $b = 2$

9. its graph contains (6,3) and (4,5)

10. its graph contains (3,8) and (−1,9)

Graph each ellipse. Locate center, foci, and vertices.

11. $\dfrac{(x-2)^2}{4} + \dfrac{(y+1)^2}{16} = 1$

12. $\dfrac{(x+8)^2}{9} + \dfrac{(y-3)^2}{25} = 1$

13. $\dfrac{(x-9)^2}{49} + \dfrac{(y-5)^2}{4} = 1$

14. $\dfrac{(x+4)^2}{16} + \dfrac{(y+3)^2}{9} = 1$

Write an equation of the translated ellipse that satisfies the following conditions.

15. foci are (3,2) and (3,8); vertices are (3,1) and (3,9)

16. vertices are (2,5) and (10,5); one focus is at (7,5)

17. center is (−4,6); graph contains (0,2) and (−9,7)

18. center is (5,−2); graph contains (4,1) and (1,−4)

19. foci are (−9,−2) and (1,−2); length of major axis is 14

20. foci are (−2,−6) and (−2,2); length of major axis is 10

Graph each ellipse. Locate center, foci, and vertices.

21. $x^2 + 2y^2 + 6x + 4y - 3 = 0$

22. $x^2 + 3y^2 - 4x - 12y = 0$

23. $2x^2 + y^2 + 8x - 8y - 48 = 0$

24. $9x^2 + 4y^2 - 18x + 24y - 4 = 0$

25. $-9x^2 - 16y^2 + 36x + 32y + 92 = 0$

26. $-4x^2 - 9y^2 - 8x + 36y - 4 = 0$

(B)

27. Derive equation (2) of this section.

28. In the beginning of this section we showed that if the sum of the distances from a point (x_0,y_0) to points $(c,0)$ and $(-c,0)$ equals $2a$, then (x_0,y_0) satisfies equation (1),

$$\frac{x^2}{a^2} + \frac{y^2}{b^2} = 1.$$

Show the converse, that if (x_0,y_0) satisfies equation (1), then the sum of its distances from points $(c,0)$ and $(-c,0)$ is equal to $2a$. (Hint: Start with an expression for the sum of the distances from (x_0,y_0) to the points $(c,0)$ and $(-c,0)$. Then, from equation (1), substitute for y in that expression.)

29. The center of the earth is at one focus of the elliptical orbit of a research satellite. The minimum distance (the perigee of the orbit) of the satellite from the earth's center is 80 miles. The maximum distance (the apogee of the orbit) of the satellite from the Earth's center is 320 miles. Write an equation of the orbit. Assume a set of coordinate axes through the center of the earth that are in the plane of the satellite's orbit.

30. An ellipse is to be drawn on a 12″ by 20″ rectangular board as shown in figure 11.19. A string of certain length is attached to two nails that are hammered into the board at points on the major axis of the ellipse. A pencil, holding the string taut, is moved over the board and sketches the ellipse. How long is the string? Where are the nails located? Why does the process work?

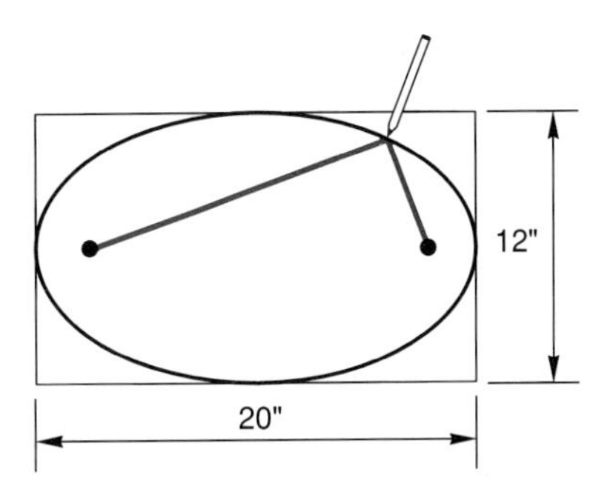

Figure 11.19

Write a standard form equation of the ellipse determined by each pair of parametric equations.

31. $x = 2 \cos \theta$, $y = 3 \sin \theta$

32. $x = 6 \cos \theta$, $y = 4 \sin \theta$

33. $x = 9 \sin \theta$, $y = 4 \cos \theta$

34. $x = 6 \sin \theta$, $y = 4 \cos \theta$

35. The different parametrizations in exercises 32 and 34 produce the same ellipse. Can you explain this?

36. The ellipse $x^2/a^2 + y^2/b^2 = 1$ can be constructed in the following way. Draw concentric circles with radii a and b (see figure 11.20). Draw a radius OA of the large circle. Label as B the point of intersection of OA and the small circle. Draw a perpendicular from A to the x-axis. Draw a line through B parallel to the x-axis, and label as P the point of intersection of this line and the perpendicular through A. P is a point on the ellipse $x^2/a^2 + y^2/b^2 = 1$. Why?

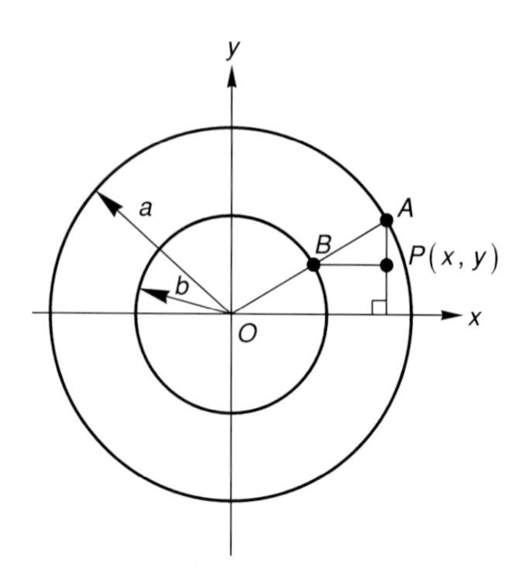

Figure 11.20

11.4 Hyperbolas

A hyperbola is a set of all points in a plane for which the difference of the distances from two fixed points in the plane is a positive constant. The two fixed points are called the ***foci.***

Let $P(x,y)$ be a point on a hyperbola whose foci are $F_1(-c,0)$ and $F_2(c,0)$ (figure 11.21). Then $|d_1 - d_2| = 2a$. This absolute value equation is equivalent to the equations

$$d_1 - d_2 = 2a \quad \text{or} \quad d_1 - d_2 = -2a.$$

Substituting the distance formula between points in the plane into the first of these equations gives

$$\begin{aligned}
\sqrt{(x+c)^2 + y^2} - \sqrt{(x-c)^2 + y^2} &= 2a \\
\sqrt{(x+c)^2 + y^2} &= 2a + \sqrt{(x-c)^2 + y^2} \\
x^2 + 2cx + c^2 + y^2 &= 4a^2 + 4a\sqrt{(x-c)^2 + y^2} \\
&\quad + x^2 - 2cx + c^2 + y^2 \\
4cx &= 4a^2 + 4a\sqrt{(x-c)^2 + y^2} \\
cx - a^2 &= a\sqrt{(x-c)^2 + y^2} \\
c^2x^2 - 2a^2cx + a^4 &= a^2x^2 - 2a^2cx + a^2c^2 + a^2y^2 \\
x^2(c^2 - a^2) - a^2y^2 &= a^2(c^2 - a^2) \\
\frac{x^2}{a^2} - \frac{y^2}{c^2 - a^2} &= 1.
\end{aligned}$$

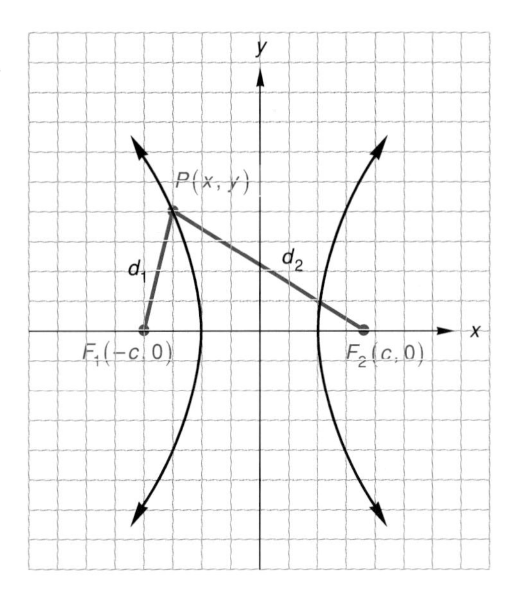

Figure 11.21

We now show that $c^2 - a^2$ is a positive number so that it can be replaced with the positive number b^2. The following inequalities (see figure 11.21) state that the sum of the lengths of two sides of a triangle is greater than the length of the third side.

$$2c + d_1 > d_2, \qquad 2c + d_2 > d_1$$
$$2c > d_2 - d_1, \qquad 2c > d_1 - d_2$$

Combining the last two inequalities gives us

$$2c > |d_1 - d_2| = 2a.$$

As a result, $c > a$ and $c^2 - a^2 > 0$. If we set $b^2 = c^2 - a^2$ in

$$\frac{x^2}{a^2} - \frac{y^2}{c^2 - a^2} = 1,$$

we produce the standard form equation of a hyperbola whose foci are on the x-axis and equidistant from the origin.

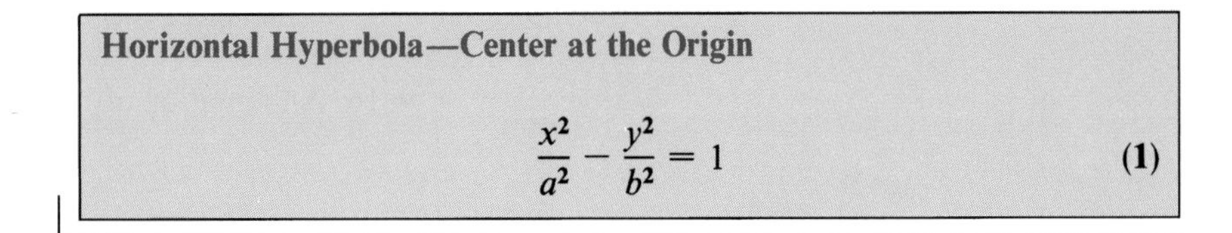

Horizontal Hyperbola—Center at the Origin

$$\frac{x^2}{a^2} - \frac{y^2}{b^2} = 1 \qquad \textbf{(1)}$$

Had we started the derivation of (1) with $d_1 - d_2 = -2a$, we would have gotten the same result. (Try it.)

A graph of (1) is symmetric to the coordinate axes and the origin. (Why?) The x-intercepts of the graph, obtained by setting $y = 0$ in (1), are $\pm a$. If x is set equal to zero in (1), the resulting equation, $-y^2/b^2 = 1$, has no real number solutions. Hence, a graph of (1) has no y-intercepts. Points $V_1(-a,0)$ and $V_2(a,0)$ are the **vertices** of the hyperbola. The line segment determined by the vertices is called the **transverse axis** of the hyperbola. The midpoint of the transverse axis is called the **center** of the hyperbola (figure 11.22).

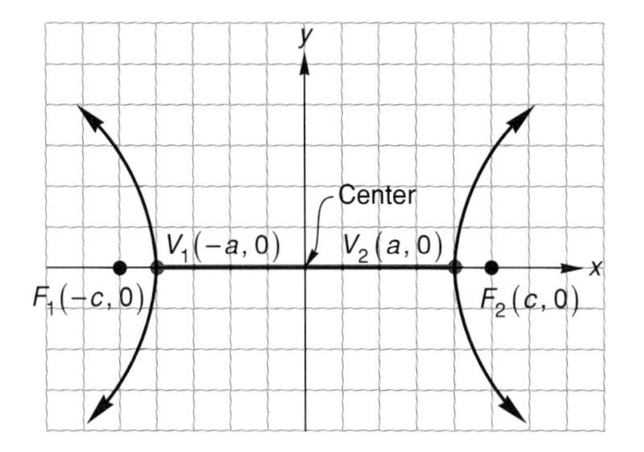

Figure 11.22

If the transverse axis is horizontal we say the hyperbola is horizontal. Thus, (1) is the equation of a horizontal hyperbola centered at the origin. A vertical hyperbola centered at the origin has the equation

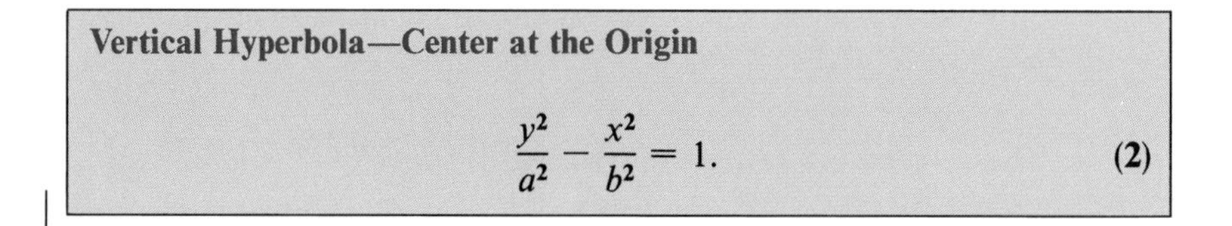

Vertical Hyperbola—Center at the Origin

$$\frac{y^2}{a^2} - \frac{x^2}{b^2} = 1. \tag{2}$$

Example 1 Write an equation of the hyperbola with vertices (0,3), (0,−3), and foci (0,5), (0,−5). Sketch a graph of the hyperbola.

Solution The transverse axis is vertical, and its midpoint is (0,0). Thus we have a vertical hyperbola centered at the origin with $a = 3$, $c = 5$, and $b^2 = 25 - 9 = 16$. An equation of the hyperbola is

$$\frac{y^2}{9} - \frac{x^2}{16} = 1.$$

We solve the last equation for y and obtain

$$y = \pm\tfrac{3}{4}\sqrt{x^2 + 16}.$$

This result is used to find some first quadrant points in the graph of the hyperbola. Symmetry gives us a sufficient number of points in the other quadrants to sketch the graph (figure 11.23).

x	y
0	3
1	$\frac{3\sqrt{17}}{4}$
3	$\frac{15}{4}$
6	$\frac{3\sqrt{13}}{2}$

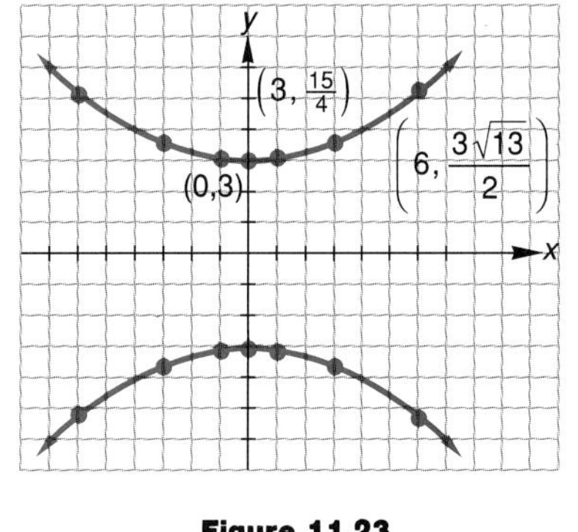

Figure 11.23

■

Asymptotes

If (1) and (2) are solved for y in terms of x, the equations

$$y = \pm\frac{b}{a}\sqrt{x^2 - a^2} \quad \text{and} \quad y = \pm\frac{a}{b}\sqrt{x^2 + b^2}$$

are obtained. Because a and b are constants, the expressions $\sqrt{x^2 - a^2}$ and $\sqrt{x^2 + b^2}$ each approach closer and closer to $\sqrt{x^2}$ as $|x|$ becomes larger and larger. Therefore, for large $|x|$, the differences

$$\pm\frac{b}{a}\sqrt{x^2 - a^2} - \left(\pm\frac{b}{a}x\right) \quad \text{and} \quad \pm\frac{a}{b}\sqrt{x^2 + b^2} - \left(\pm\frac{a}{b}x\right)$$

become very small. The graph determined by (1) or by (2) approaches closer and closer to the pairs of lines

Asymptotes of a Hyperbola

$$y = \pm\frac{b}{a}x, \quad \text{or} \quad y = \pm\frac{a}{b}x, \tag{3}$$

as $|x|$ becomes larger and larger.

The first of equations (3) gives the oblique asymptotes of a horizontal hyperbola centered at the origin. The second of equations (3) gives the oblique asymptotes of a vertical hyperbola centered at the origin.

Example 2 Show that $y = \pm\frac{2}{3}x$ are the asymptotes of the hyperbola $x^2/36 - y^2/16 = 1$. Sketch a graph of the hyperbola.

Solution An equivalent equation of the hyperbola is $y = \pm\frac{2}{3}\sqrt{x^2 - 36}$. Let us compare y-values on the hyperbola with the corresponding y-values on the lines $y = \pm\frac{2}{3}x$.

x	$y = \pm\frac{2}{3}\sqrt{x^2 - 36}$	$y = \pm\frac{2}{3}x$
± 10	± 5.3333	± 6.6667
± 100	± 66.5466	± 66.6667
$\pm 1{,}000$	± 666.6547	± 666.6667
$\pm 10{,}000$	$\pm 6{,}666.6655$	$\pm 6{,}666.6667$

It is clear from the table that as $|x|$ becomes larger and larger, the difference between the corresponding y-values on the hyperbola and the lines becomes smaller and smaller. Therefore, lines $y = \pm\frac{2}{3}x$ are the asymptotes of the hyperbola.

Having determined the asymptotes, it is a simple matter to fit the hyperbola between its vertices and asymptotes (figure 11.24).

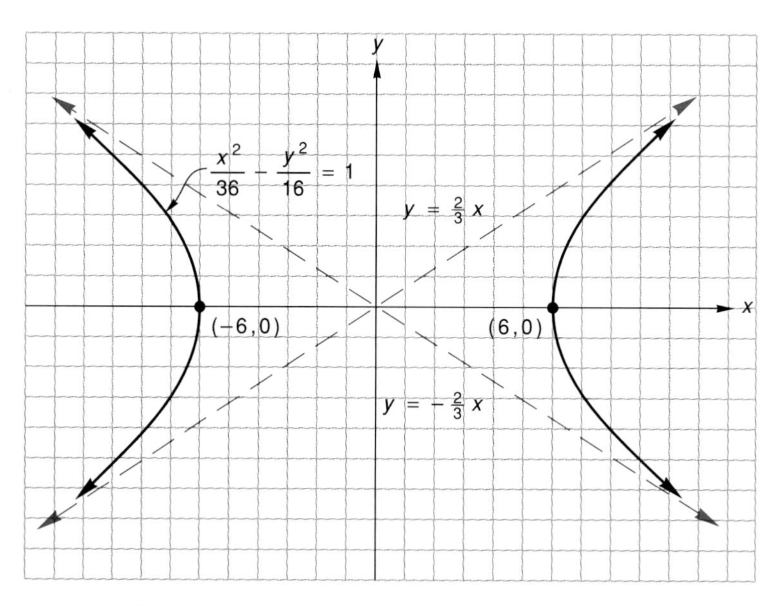

Figure 11.24

■

Translated Hyperbolas

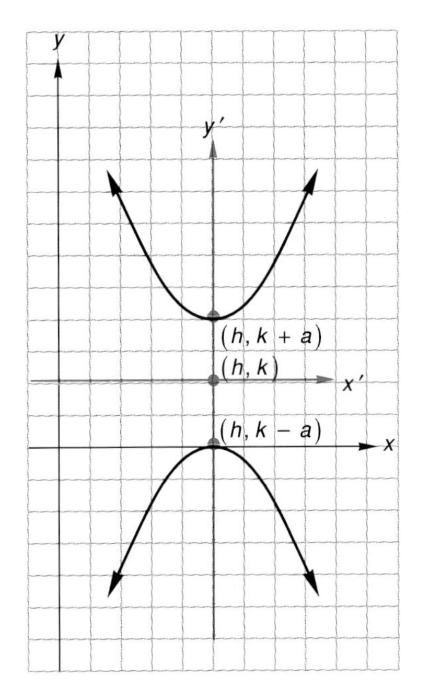

Figure 11.25

If the transverse axis of a hyperbola is parallel to one of the coordinate axes and its center is not at the origin, we say the hyperbola is *translated* with respect to the origin.

The translated hyperbola in figure 11.25 is centered at (h,k). Its equation relative to the x'-y' axes is $y'^2/a^2 - x'^2/b^2 = 1$. Substituting for x' and y', where

$$x' = x - h \quad \text{and} \quad y' = y - k,$$

one obtains an equation of the translated vertical hyperbola relative to the x-y axes

Vertical Hyperbola—Center at (h,k)

$$\frac{(y-k)^2}{a^2} - \frac{(x-h)^2}{b^2} = 1. \tag{4}$$

A translated horizontal hyperbola has the equation

Horizontal Hyperbola—Center at (h,k)

$$\frac{(x-h)^2}{a^2} - \frac{(y-k)^2}{b^2} = 1. \tag{5}$$

Example 3 Determine an equation of the hyperbola centered at $(-3,-2)$ that contains $(0,0)$ and $(-7,1)$.

Solution The equation sought may have the form of (5) or (6). We guess that it has the form of (5).

$$\frac{(x+3)^2}{a^2} - \frac{(y+2)^2}{b^2} = 1$$

When $(0,0)$ and $(-7,1)$ are substituted into this equation, we generate the system

$$\frac{9}{a^2} - \frac{4}{b^2} = 1$$
$$\frac{16}{a^2} - \frac{9}{b^2} = 1.$$

If the first equation is multiplied by 9 and the second by -4, and if the resulting equations are added, we get

$$\frac{17}{a^2} = 5$$
$$a^2 = \frac{17}{5}.$$

Multiplying the first equation by 16 and the second by -9, and adding the resulting equations yields

$$\frac{17}{b^2} = 7$$
$$b^2 = \frac{17}{7}.$$

The equation sought is

$$\frac{(x+3)^2}{\frac{17}{5}} - \frac{(y+2)^2}{\frac{17}{7}} = 1,$$

or
$$5(x+3)^2 - 7(y+2)^2 = 17.$$ ■

The solution in example 3 is a horizontal hyperbola centered at $(-3,-2)$. It is reasonable to ask if there is also a vertical hyperbola centered at $(-3,-2)$ that contains $(0,0)$ and $(-7,1)$. We need only substitute $(0,0)$ and $(-7,1)$ into $(y+2)^2/a^2 - (x+3)^2/b^2 = 1$ to find out. (Try it.)

Finally, we want to show that the standard forms of the translated hyperbolas, equations (4) and (5), have the same general form

$$Ax^2 + Cy^2 + Dx + Ey + F = 0, \tag{6}$$

as all the previously discussed conics. We expand (4) and (5) to get

$$-a^2x^2 + b^2y^2 + 2a^2hx - 2b^2ky - a^2h^2 + b^2k^2 - a^2b^2 = 0$$

and
$$b^2x^2 - a^2y^2 - 2b^2hx + 2a^2ky - a^2k^2 + b^2h^2 - a^2b^2 = 0,$$

respectively. Both equations have the form of (6), in which A and C are of opposite sign.

We see now that all translated conics are represented by (6) and that the equation determines

1. a circle if $A = C$,
2. a parabola if A or C is zero,
3. an ellipse if $A \neq C$ and A and C are of the same sign,
4. a hyperbola if A and C do not have the same sign.

Example 4 Identify and sketch the conic determined by

$$5x^2 - y^2 - 30x + 40 = 0.$$

Solution Since A and C are of opposite sign, the equation determines a hyperbola. Now put the equation in standard form to determine the center, the vertices, and the foci of the hyperbola.

$$5(x^2 - 6x + 9) - y^2 = -40 + 45$$
$$5(x - 3)^2 - (y - 0)^2 = 5$$
$$\frac{(x - 3)^2}{1} - \frac{(y - 0)^2}{5} = 1$$

Because of the form of the standard equation, the transverse axis of the hyperbola is horizontal. The center of the hyperbola is (3,0). Since $a^2 = 1$, the vertices are (2,0) and (4,0). We note that $b^2 = 5$ and recall that $c^2 - a^2 = b^2$. Therefore, $c^2 = a^2 + b^2 = 6$, and the foci are $(3 - \sqrt{6},0)$ and $(3 + \sqrt{6},0)$. If we let $x' = x + 3$ and $y' = y$, the standard form equation becomes an equation of a hyperbola centered with respect to the x'-y' axes.

$$\frac{x'^2}{1} - \frac{y'^2}{5} = 1$$

Equations of the asymptotes of a horizontal hyperbola that is centered at the origin are given in (3) as $y = (\pm b/a)x$. Therefore, the asymptotes of this hyperbola, relative to the x'-y' axes, are $y' = \pm\sqrt{5}x'$. The equations of the asymptotes relative to the x-y axes are $y = \pm\sqrt{5}(x + 3)$. (See figure 11.26.)

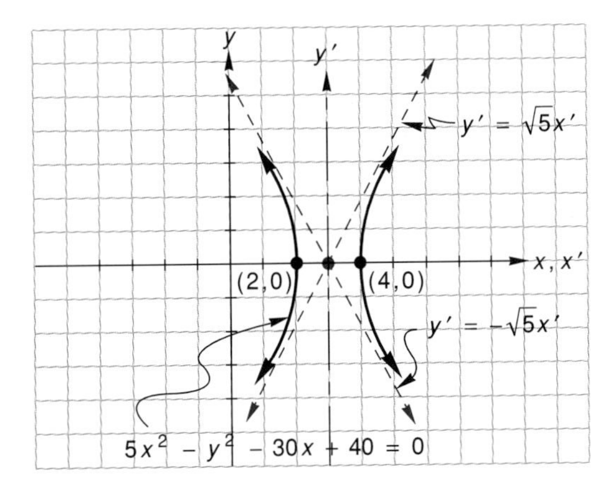

Figure 11.26

■

Parametric Equations

A parametrization of $x^2/a^2 - y^2/b^2 = 1$ is suggested by the trigonometric identity $\sec^2 \theta - \tan^2 \theta = 1$. Let $x = a \sec \theta$ and $y = b \tan \theta$. Then

$$\frac{x^2}{a^2} - \frac{y^2}{b^2} = \sec^2 \theta - \tan^2 \theta = 1.$$

As θ assumes all values in the interval $[0,2\pi]$, except $\pi/2$ and $3\pi/2$, the parametric equations generate all the points on the hyperbola (see exercise 37).

Summary

The standard equation of a hyperbola with vertex at (h,k) and its transverse axes parallel to one of the coordinate axes is

Horizontal Hyperbola:

$$\frac{(x-h)^2}{a^2} - \frac{(y-k)^2}{b^2} = 1$$

Vertical Hyperbola:

$$\frac{(y-k)^2}{a^2} - \frac{(x-h)^2}{b^2} = 1.$$

Equations of the asymptotes, relative to the x'-y' axes through (h,k) are

$$y' = \pm\left(\frac{b}{a}\right)x' \quad \text{horizontal hyperbola}$$

$$y' = \pm\left(\frac{a}{b}\right)x' \quad \text{vertical hyperbola.}$$

The distance from the center to each focus is $|c|$, where $c^2 = a^2 + b^2$.

The general equation of a hyperbola whose transverse axis is parallel to one of the coordinate axes has the form

$$Ax^2 + Cy^2 + Dx + Ey + F = 0,$$

in which A and C are of opposite sign.

Exercise Set 11.4

(A)

Graph each hyperbola. Label vertices, foci, and asymptotes.

1. $\frac{x^2}{9} - y^2 = 1$

2. $\frac{x^2}{4} - \frac{y^2}{16} = 1$

3. $\frac{y^2}{25} - \frac{x^2}{9} = 1$

4. $\frac{y^2}{9} - \frac{x^2}{36} = 1$

Write an equation of the hyperbola that satisfies the given conditions.

5. vertices $(4,0)$, $(-4,0)$ and foci $(5,0)$, $(-5,0)$

6. foci on y axis, $a = 6$, $c = 9$, center is $(0,0)$

7. vertices are $(0,3)$ and $(0,-3)$, and $b = 4$

8. vertex $(-1,0)$, corresponding focus $(-3,0)$, center $(0,0)$

9. vertical hyperbola, asymptotes $y = \pm 2x$, $c = \pm 2\sqrt{5}$

10. horizontal hyperbola, asymptotes $y = \pm x$, $c = \pm 2\sqrt{2}$

Graph each hyperbola. Label vertices, foci, and asymptotes.

11. $\frac{(x-5)^2}{36} - \frac{(y-1)^2}{4} = 1$

12. $\frac{(x+3)^2}{9} - \frac{(y-4)^2}{16} = 1$

13. $\frac{(y+3)^2}{9} - \frac{(x+2)^2}{25} = 1$

14. $\frac{(y+1)^2}{49} - \frac{(x-2)^2}{36} = 1$

Write an equation of the translated hyperbola that satisfies the following conditions.

15. center $(4,1)$, contains $(6,2)$ and $(8,-3)$

16. center $(-2,-5)$, contains $(0,0)$ and $(-5,1)$

17. vertices $(1,2)$ and $(6,2)$, contains $(0,3)$

18. foci $(-3,-4)$ and $(-3,6)$, contains $(-1,-5)$

19. Asymptotes $x - 2y + 6 = 0$ and $x + 2y - 10 = 0$, contains $(6,5)$

20. Asymptotes $3x + 4y + 17 = 0$ and $3x - 4y + 1 = 0$, contains $(-4,5)$

Graph each hyperbola. Show vertices, foci, and asymptotes. Write equations of the asymptotes relative to the x-y axes.

21. $9x^2 - 4y^2 + 18x + 16y + 29 = 0$

22. $4x^2 - 3y^2 - 32x + 6y + 73 = 0$

23. $4y^2 - 3x^2 + 8y - 12x - 16 = 0$

24. $25x^2 - 9y^2 - 100x - 54y + 10 = 0$

25. $4y^2 - x^2 + 32y - 8x + 49 = 0$

26. $x^2 - y^2 + 6x + 4y + 5 = 0$

Identify the conic determined by each equation.

27. $4x^2 + y^2 + 24x - 10y + 45 = 0$

28. $9x^2 - y^2 - 36x + 12y - 9 = 0$

29. $3x^2 + 3y^2 + 6x + y - 1 = 0$

30. $x^2 + 4x - y + 1 = 0$

31. $5y^2 + 6x - 30y - 8 = 0$

32. $-x^2 - y^2 + 6x + 4y + 4 = 0$

33. $2y^2 - x^2 + 4x - 6y + 5 = 0$

34. $-3y^2 - 2x^2 - 6x + 9y - 10 = 0$

(B)

35. Derive equation (2) of this section.

36. At the beginning of this section we showed that if the absolute value of the difference of the distances from (x_0,y_0) to the points $(c,0)$ and $(-c,0)$ is equal to $2a$, then (x_0,y_0) satisfies equation (1),

$$\frac{x^2}{a^2} - \frac{y^2}{b^2} = 1.$$

Show the converse; that if (x_0,y_0) satisfies equation (1), then the absolute value of the difference of its distances from the points $(c,0)$ and $(-c,0)$ is $2a$. (Hint: See exercise 28, exercise set 11.3.)

Write a standard form equation of the hyperbola determined by each pair of parametric equations.

37. $x = 2 \sec \theta$, $y = 3 \tan \theta$

38. $x = 4 \sec \theta$, $y = \tan \theta$

39. $x = 2 \tan \theta$, $y = 3 \sec \theta$

40. $x = 4 \tan \theta$, $y = \sec \theta$

41. A hyperbola may be constructed as follows: choose two points F_1 and F_2 as centers of circles with radii r_1 and r_2 such that $r_1 - r_2 = \pm 2a$, $a > 0$. The points of intersection of the circles are points on the hyperbola. Why does the method work?

42. Use the method described in exercise 41 to sketch a hyperbola in which $a = 5$ units.

43. Three observers are located in a mountainous region at A, B, and C as shown in figure 11.27. Each observer hears the sound of an avalanche at the times indicated on the figure. Assume sound travels at the rate of 1,100 ft./sec. and that the distances in figure 11.27 are in feet. Where does the avalanche occur? Construct a graphical solution. (Hint: The avalanche occurs at the intersection of two hyperbolas. One hyperbola has foci at A and B. The other hyperbola has foci at B and C.)

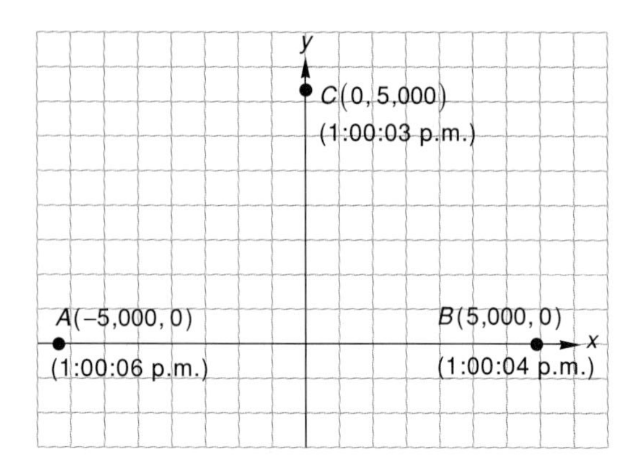

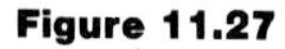
Figure 11.27

11.5 Rotation of Axes

The ellipse in figure 11.28 has the equation $x'^2/a^2 + y'^2/b^2 = 1$ relative to the x'-y' coordinate axes, which are rotated θ degrees with respect to the x-y axes. To find an equation of the ellipse relative to the x-y axes, we must establish a pair of transformation equations between x' and y' and x and y. Toward that end let r be the distance from the origin to a point P on the ellipse. The

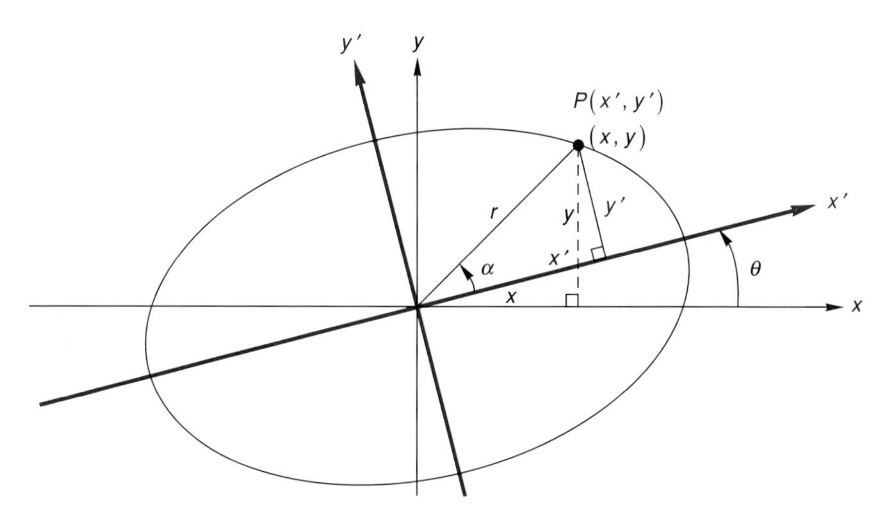

Figure 11.28

coordinates of P are either (x,y) or (x',y'), depending upon to which coordinate system P is referred. Let α be the angle between the x' axis and the line segment between the origin and P. It is clear from figure 11.28 that

$$x' = r\cos\alpha \quad \text{and} \quad y' = r\sin\alpha, \tag{1}$$

and that

$$x = r\cos(\alpha + \theta) \quad \text{and} \quad y = r\sin(\alpha + \theta). \tag{2}$$

Expanding equations (2) and substituting from (1) yields

$$\begin{aligned} x &= r(\cos\alpha\cos\theta - \sin\alpha\sin\theta) & y &= r(\sin\alpha\cos\theta + \cos\alpha\sin\theta) \\ x &= x'\cos\theta - y'\sin\theta & y &= y'\cos\theta + x'\sin\theta. \end{aligned}$$

Transformation Equations: x-y Axes to x'-y' Axes

$$x = x'\cos\theta - y'\sin\theta \tag{3}$$
$$y = y'\cos\theta + x'\sin\theta \tag{4}$$

Equations (3) and (4) can be solved for x' and y' in terms of x and y. To solve for x', multiply (3) by $\cos\theta$ and (4) by $\sin\theta$ and add the resulting equations. We multiply (3) by $-\sin\theta$ and (4) by $\cos\theta$ to find y'.

$$\begin{array}{rll} x'\cos\theta - y'\sin\theta = x & (\cos\theta & (-\sin\theta \\ \underline{x'\sin\theta + y'\cos\theta = y} & \underline{(\sin\theta} & \underline{(\cos\theta} \\ x'\cos^2\theta + x'\sin^2\theta = x\cos\theta + y\sin\theta & & \\ x' = x\cos\theta + y\sin\theta & & \\ y'\sin^2\theta + y'\cos^2\theta = -x\sin\theta + y\cos\theta & & \\ y' = -x\sin\theta + y\cos\theta & & \end{array}$$

Transformation Equations: x'-y' Axes to x-y Axes

$$x' = x\cos\theta + y\sin\theta \tag{5}$$
$$y' = -x\sin\theta + y\cos\theta \tag{6}$$

With equations (5) and (6), the equation of the ellipse whose major and minor axes are in the x' and y' axes respectively,

$$\frac{(x')^2}{a^2} + \frac{(y')^2}{b^2} = 1,$$

becomes an equation of an ellipse centered at the origin and rotated through an angle of $\theta°$ with respect to the x-y axes (figure 11.28)

$$\frac{(x \cos \theta + y \sin \theta)^2}{a^2} + \frac{-x \sin \theta + y \cos \theta}{b^2} = 1.$$

Expanding this equation gives us

$$b^2(x^2 \cos^2 \theta + 2xy \sin \theta \cos \theta + y^2 \sin^2 \theta) \\ + a^2(x^2 \sin^2 \theta - 2xy \sin \theta \cos \theta + y^2 \cos^2 \theta) = a^2b^2$$

$$(a^2 \sin^2 \theta + b^2 \cos^2 \theta)x^2 \\ + ((b^2 - a^2)\sin 2\,\theta)xy + (a^2 \cos^2 \theta + b^2 \sin^2 \theta)y^2 - a^2b^2 = 0.$$

The last equation has the general form

$$Ax^2 + Bxy + Cy^2 + F = 0.$$

Example 1 A parabola is both rotated and translated with respect to the x-y axes as shown in figure 11.29. If $y'' = (x'')^2$ defines the parabola relative to the x''-y'' axes, write a general form equation of the parabola relative to the x-y axes. Also find the x-y coordinates of the parabola's focus.

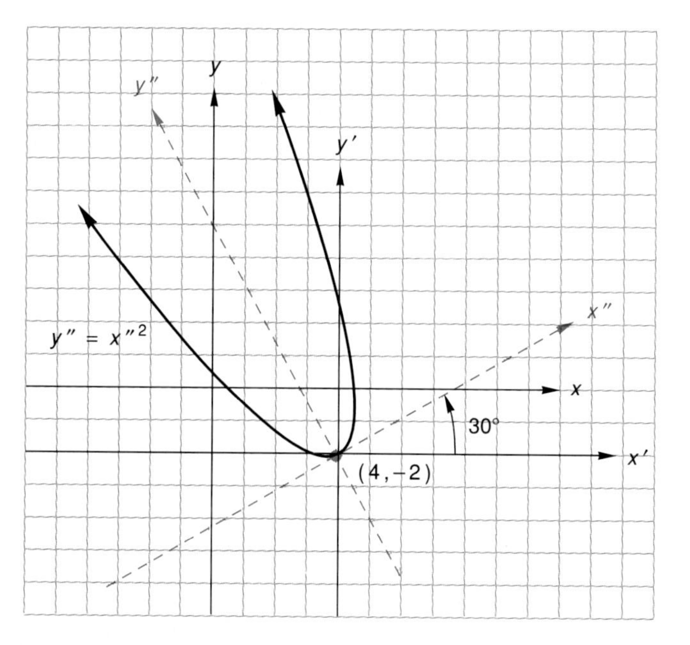

Figure 11.29

Solution We use equations (5) and (6) to transform $y'' = (x'')^2$ into an x'-y' equation.

$$-x' \sin 30° + y' \cos 30° = (x' \cos 30° + y' \sin 30°)^2$$
$$\frac{-x'}{2} + \frac{\sqrt{3}}{2}y' = \frac{3}{4}(x')^2 + \frac{\sqrt{3}}{2}x'y' + \frac{(y')^2}{4}$$
$$3(x')^2 + 2\sqrt{3}x'y' + (y')^2 + 2x' - 2\sqrt{3}y' = 0$$

Now the transformation equations $x' = x - 4$ and $y' = y + 2$ (equations (3), section 11.2) transform the x'-y' equation of the parabola into

$$3(x - 4)^2 + 2\sqrt{3}(x - 4)(y + 2) + (y + 2)^2 + 2(x - 4) - 2\sqrt{3}(y + 2) = 0$$
$$3x^2 + 2\sqrt{3}xy + y^2 + (4\sqrt{3} - 22)x + (4 - 10\sqrt{3})y + (44 - 20\sqrt{3}) = 0.$$

By comparing $y'' = (x'')^2$ to the standard form equation of a parabola with vertex at the origin, $4cy'' = (x'')^2$, we compute c to be $\frac{1}{4}$. Therefore the x''-y'' coordinates of the focus are $(0, \frac{1}{4})$. The x'-y' coordinates of the focus are obtained from equations (3) and (4).

$$x' = 0 \cdot \cos 30° - \frac{1}{4} \sin 30°, \quad y' = 0 \cdot \sin 30° + \frac{1}{4} \cos 30°$$
$$x' = -\frac{1}{8}, \quad y' = \frac{\sqrt{3}}{8}$$

Finally, the x-y coordinates are found using the transformation equations $x = x' + h$ and $y = y' + k$.

$$x = -\frac{1}{8} + 4 = \frac{31}{8}, \quad y = \frac{\sqrt{3}}{8} - 2 = \frac{-16 + \sqrt{3}}{8}$$ ■

Were it not repetitious and tedious, we would generalize the results of example 1, and the discussion preceding it, to show that the equation of a rotated and translated conic has the general form

$$Ax^2 + Bxy + Cy^2 + Dx + Ey + F = 0. \tag{7}$$

It is more interesting to show that (7) determines a conic or some degenerate form of a conic such as a pair of straight lines, or a point, etc. (See exercises 13 and 14.)

If equations (3) and (4) are substituted into (7), the result is the equation

$$A'(x')^2 + B'x'y' + C'(y')^2 + D'x' + E'y' + F' = 0,$$

in which

$$\begin{aligned}
A' &= A\cos^2\theta + B\sin\theta\cos\theta + C\sin^2\theta \\
B' &= B(\cos^2\theta - \sin^2\theta) - 2(A - C)\sin\theta\cos\theta \\
C' &= A\sin^2\theta - B\sin\theta\cos\theta + C\cos^2\theta \\
D' &= D\cos\theta + E\sin\theta \\
E' &= -D\sin\theta + E\cos\theta \\
F' &= F.
\end{aligned}$$

A proper choice of θ can make $B' = 0$. In the case that $A = C$

$$B' = B(\cos^2\theta - \sin^2\theta) = B\cos 2\theta$$

and $B' = 0$ if $\theta = 45°$. If $A \neq C$; then $B' = 0$ if

$$B\cos 2\theta - (A - C)\sin 2\theta = 0$$

or

$$\tan 2\theta = \frac{B}{A - C}. \tag{8}$$

We conclude that if in (7), $B \neq 0$, it is possible to find an x'-y' coordinate axes pair that is rotated θ degrees with respect to the x-y axes so that (7) becomes

$$A'(x')^2 + C'(y')^2 + D'x' + E'y' + F' = 0 \tag{9}$$

relative to the x'-y' axes. Equation (9) has been shown to be the equation of a conic or some degenerate form of a conic.

Example 2 Graph the conic determined by $x^2 - 4xy + y^2 + 10x - 8y + 7 = 0$.

Solution Since $A = C$, a rotation through 45° will transform the given equation into one having the form of (9). Substituting 45° for θ in (3) and (4) yields

$$x = x'\cos 45° - y'\sin 45° = \frac{x' - y'}{\sqrt{2}}$$

and

$$y = x'\sin 45° + y'\cos 45° = \frac{x' + y'}{\sqrt{2}}.$$

Now we substitute for x and y in the given equation to get

$$\left(\frac{x' - y'}{\sqrt{2}}\right)^2 - 4\left(\frac{x' - y'}{\sqrt{2}}\right)\left(\frac{x' + y'}{\sqrt{2}}\right) + \left(\frac{x' + y'}{\sqrt{2}}\right)^2 + 10\left(\frac{x' - y'}{\sqrt{2}}\right) - 8\left(\frac{x' + y'}{\sqrt{2}}\right) + 7 = 0$$

or

$$-x'^2 + 3y'^2 + \sqrt{2}x' - 9\sqrt{2}y' + 7 = 0.$$

Since the coefficients of $(x')^2$ and $(y')^2$ are opposite in sign, we have a hyperbola. The vertices, foci, and asymptotes of the hyperbola can be easily determined after completing the square of the x'-y' equation.

$$-\left[(x')^2 - \sqrt{2}x + \frac{1}{2}\right] + 3\left[(y')^2 - 3\sqrt{2}y' + \frac{9}{2}\right] = -7 + \frac{26}{2}$$

$$-\left(x' - \frac{\sqrt{2}}{2}\right)^2 + 3\left(y' - \frac{3\sqrt{2}}{2}\right)^2 = 6$$

Let $x'' = x' - \dfrac{\sqrt{2}}{2}$ and $y'' = y' - \dfrac{3\sqrt{2}}{2}$. Then the equation of the hyperbola relative to an x''-y'' coordinate system is

$$\frac{(y'')^2}{2} - \frac{(x'')^2}{6} = 1.$$

Also $a^2 = 2$, $b^2 = 6$, $c^2 = 8$, and the equations of the asymptotes are $y'' = \pm(1/\sqrt{3})x''$. To sketch a graph of the hyperbola we locate the origin of the x''-y'' system at the point $(\sqrt{2}/2, 3\sqrt{2}/2)$ of the x'-y' coordinate system. The asymptotes are drawn through the origin, and the vertices are located $\sqrt{2}$ units from the origin on the y'' axis (see figure 11.30).

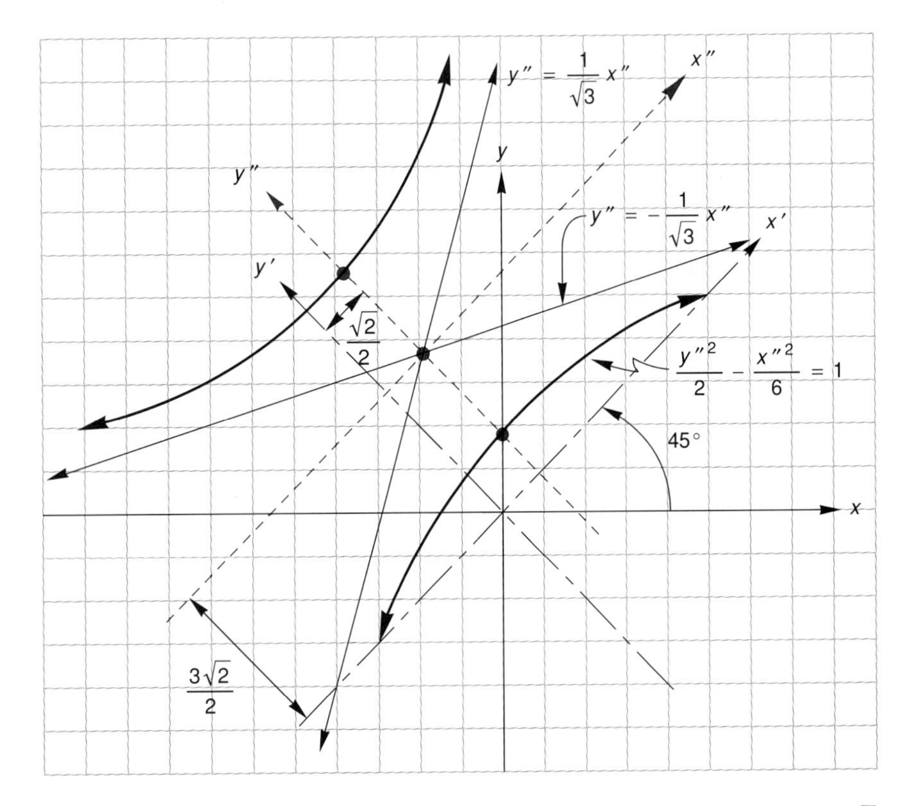

Figure 11.30

Example 3 Determine the x-y coordinates of the center and the vertices of the hyperbola in example 2. Also determine the x-y equations of the hyperbola's asymptotes.

Solution Points located by pairs of x''-y'' coordinates are easily renamed x-y coordinates with the aid of the transformation equations for rotation,

$$x = \frac{x' - y'}{\sqrt{2}} \quad \text{and} \quad y = \frac{x' + y'}{\sqrt{2}},$$

and the transformation equations for translation

$$x' = x'' + \frac{\sqrt{2}}{2} \quad \text{and} \quad y' = y'' + \frac{3\sqrt{2}}{2}.$$

(x'',y'')	(x',y')	(x,y)	
$(0,0)$	$\left(\frac{\sqrt{2}}{2},\frac{3\sqrt{2}}{2}\right)$	$(-1,2)$	Center
$(0,\sqrt{2})$	$\left(\frac{\sqrt{2}}{2},\frac{5\sqrt{2}}{2}\right)$	$(-2,3)$	Vertex
$(0,-\sqrt{2})$	$\left(\frac{\sqrt{2}}{2},\frac{\sqrt{2}}{2}\right)$	$(0,1)$	Vertex

The equations of the asymptotes, $y'' = \pm\frac{\sqrt{3}}{3}x''$, become

$$y' - \frac{3\sqrt{2}}{2} = \pm\frac{\sqrt{3}}{3}\left(x' - \frac{\sqrt{2}}{2}\right)$$

in the x'-y' coordinate system. But

$$x' = \frac{\sqrt{2}}{2}(x + y) \quad \text{and} \quad y' = \frac{\sqrt{2}}{2}(y - x). \text{ (Why?)}$$

Hence, the x-y equations of the asymptotes are

$$-(3 + \sqrt{3})x + (3 - \sqrt{3})y = 9 - \sqrt{3} \quad \text{and}$$
$$(\sqrt{3} - 3)x + (3 + \sqrt{3})y = 9 + \sqrt{3}.$$

■

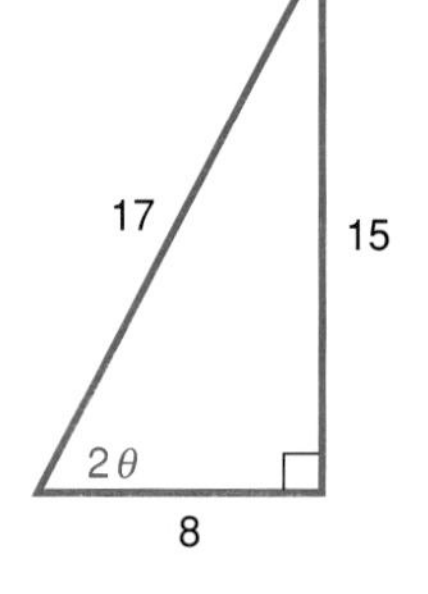

Figure 11.31

Example 4 Remove the xy term from $6x^2 + 15xy - 2y^2 - 10 = 0$ with a rotation of axes transformation.

Solution

$$\tan 2\theta = \frac{B}{A - C} = \frac{15}{8}$$

From the right triangle in figure 11.31 we see that $\cos 2\theta = \frac{8}{17}$. Recall that $\cos 2\theta = 2\cos^2\theta - 1 = 1 - 2\sin^2\theta$. Therefore

$$2\cos^2\theta - 1 = \frac{8}{17} \quad \text{and} \quad 1 - 2\sin^2\theta = \frac{8}{17}$$

$$\cos\theta = \frac{5}{\sqrt{34}} \quad \text{and} \quad \sin\theta = \frac{3}{\sqrt{34}}.$$

Substituting the sine and cosine values into equations (3) and (4) gives

$$x = \frac{5x' - 3y'}{\sqrt{34}} \quad \text{and} \quad y = \frac{3x' + 5y'}{\sqrt{34}}.$$

Now if we substitute for x and y in the given equation, we obtain the equation

$$357(x')^2 - 221(y')^2 - 340 = 0.$$

■

Identifying Conics

We now know that the general second degree equation in x and y,

$$Ax^2 + Bxy + Cy^2 + Dx + Ey + F = 0, \tag{7}$$

determines a conic or some degenerate form of a conic. Moreover, we have seen how transformation equations (3) and (4) convert (7) into

$$A'(x')^2 + B'x'y' + C'(y')^2 + D'x' + E'y' + F' = 0. \tag{10}$$

The transformation from (7) to (10) leaves the quantity $B^2 - 4AC$ invariant (see exercise 16). That is

$$(B')^2 - 4A'C' = B^2 - 4AC. \tag{11}$$

If in transformation equations (3) and (4), θ is chosen to make $B' = 0$, then (10) becomes

$$A'(x')^2 + C'(y')^2 + D'x' + E'y' + F' = 0, \tag{9}$$

an equation of a *translated conic*. Also, (11) becomes

$$-4A'C' = B^2 - 4AC.$$

Therefore the conic determined by (7) is

1. a parabola if $B^2 - 4AC = 0$, since (9) is a parabola if $A' = 0$ or $C' = 0$.
2. an ellipse if $B^2 - 4AC < 0$, since (9) is an ellipse if $A' \neq C'$ and A' and C' are of the same sign.
3. a hyperbola if $B^2 - 4AC > 0$, since (9) is a hyperbola if A' and C' are of opposite signs.

Example 5 Identify the conic $x^2 - 4xy + y^2 + 10x - 8y + 7 = 0$.

Solution $B^2 - 4AC = (-4)^2 - 4(1)(1) = 12 > 0$. Hence the conic is a hyperbola. Note that the conics in examples 2 and 5 are the same. ■

Exercise Set 11.5

(A)

Remove the xy term in each equation with a rotation of axes transformation. Sketch a graph of each equation showing vertices, foci, and asymptotes. Locate the vertices and foci, and name the asymptotes in the x-y coordinate system.

1. $x^2 - xy + y^2 - 6 = 0$
2. $5x^2 - 8xy + 5y^2 - 9 = 0$
3. $3x^2 + 2\sqrt{3}xy + y^2 + 2x + 2\sqrt{3}y = 0$
4. $3x^2 - 2\sqrt{3}xy + y^2 + 4x + 4\sqrt{3}y = 0$
5. $xy - 1 = 0$
6. $11x^2 - 10\sqrt{3}xy + y^2 + 4 = 0$

(B)

Sketch a graph of each equation that is centered at the origin in an x''-y'' coordinate system that is rotated and translated with respect to the x-y coordinate system. Write the given equation in x''-y'' coordinates. (See example 2.)

7. $x^2 - xy + y^2 + 2x - y - 5 = 0$
8. $-4x^2 + 6xy + 4y^2 - 8\sqrt{10}x + 6\sqrt{10}y - 45 = 0$
9. $16x^2 - 24xy + 9y^2 - 60x - 80y + 100 = 0$
10. $x^2 - 2xy + y^2 + 2x + 4y - 4 = 0$
11. $x^2 + 24xy - 6y^2 + 4x + 48y + 34 = 0$
12. $7x^2 + 12xy + 2y^2 + 140x + 76y + 548 = 0$
13. $3x^2 + 2\sqrt{3}xy + y^2 + 8y - 4 = 0$
14. $43x^2 + 48xy + 57y^2 + 124x - 18y + 133 = 0$

Locate the vertices and foci, in the x-y coordinate system, of the graph in the indicated exercise. If the graph is a hyperbola, name its asymptotes in the x-y coordinate system.

15. Exercise 7
16. Exercise 9
17. Exercise 11
18. Exercise 13

19. In the process of deriving equation (8) of this section the claim is made that

$$A' = A\cos^2\theta + B\sin\theta\cos\theta + C\sin^2\theta,$$
$$B' = B(\cos^2\theta - \sin^2\theta) - 2(A - C)\sin\theta\cos\theta,$$
and
$$C' = A\sin^2\theta - B\sin\theta\cos\theta + C\cos^2\theta.$$

Verify this claim.

20. Show that $B^2 - 4AC$ is invariant under rotation. You must show that

$$(B')^2 - 4A'C' = B^2 - 4AC$$

for any θ.

11.6 Eccentricity and Polar Coordinate Forms

An equation for each conic form has been derived from a distinct geometric definition of the conic. It is also possible and useful to define *all* of the conics as we did the parabola. Each conic can be viewed as a set of points whose distances from a focus and a directrix determine a ratio called the **eccentricity** e of the conic. In the case of a parabola, $e = 1$ because each point on the parabola is equidistant from its focus and directrix. Ellipses and hyperbolas also have focus-directrix pairs and eccentricities, as we shall demonstrate below.

Consider the ellipse in figure 11.32. Its equation, $x^2/a^2 + y^2/b^2 = 1$, is derived in section 11.3. The derivation begins with the distance relationship

$$\sqrt{(x - c)^2 + y^2} + \sqrt{(x + c)^2 + y^2} = 2a. \tag{1}$$

If the first radical in (1) is transposed, and both sides of the resulting equation are squared, we get

$$x^2 + 2cx + c^2 + y^2 = 4a^2 - 4a\sqrt{(x - c)^2 + y^2} + x^2 - 2cx + c^2 + y^2$$

$$\sqrt{(x - c)^2 + y^2} = a - \frac{c}{a}x$$

$$\sqrt{(x - c)^2 + y^2} = \frac{c}{a}\left(\frac{a^2}{c} - x\right).$$

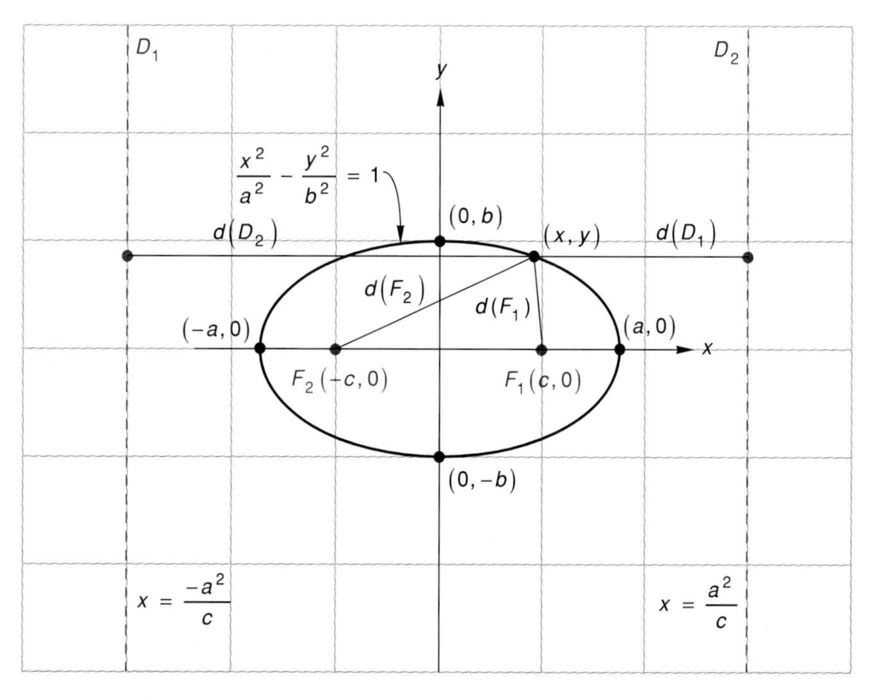

Figure 11.32

You can read the last result as, "the distance of a point (x,y) on the ellipse to the focus $(c,0)$ is a constant multiple, c/a, of the distance of the point from the line $x = a^2/c$." Let us label the focus $(c,0)$ as F_1 and the line $x = a^2/c$ as D_1. Also, let $e = c/a$ and $d(F_1)$ and $d(D_1)$ be the respective distances of (x,y) from F_1 and D_1. Then every point on the ellipse satisfies the relationship

$$d(F_1) = ed(D_1), \quad 0 < e < 1. \tag{2}$$

Equation (2) is a focus-directrix definition of an ellipse in which e is the eccentricity of the ellipse. You shall see that the closer e is to zero, the closer is the shape of the ellipse to that of a circle. As e draws closer to 1, the shape of the ellipse becomes more cigarlike.

Example 1 Find equations of the ellipses centered at the origin that have focus (2,0) and

a. $e = \frac{1}{4}$,

b. $e = \frac{7}{8}$.

Sketch each ellipse.

Solution

a. $e = \dfrac{c}{a} = \dfrac{1}{4}$ and $c = 2$. Hence, $\dfrac{2}{a} = \dfrac{1}{4}$ and $a = 8$. Also

$$b^2 = a^2 - c^2 = 64 - 4 = 60.$$

An equation of the ellipse in part **a** is

$$\frac{x^2}{64} + \frac{y^2}{60} = 1.$$

b. $\dfrac{2}{a} = \dfrac{7}{8}$, $a = \dfrac{16}{7}$

$$b^2 = \left(\frac{16}{7}\right)^2 - 4 = \frac{60}{49}$$

An equation of the ellipse in part **b** is

$$\frac{x^2}{\frac{256}{49}} + \frac{y^2}{\frac{60}{49}} = 1.$$

Graphs of the ellipses are shown in figure 11.33. ■

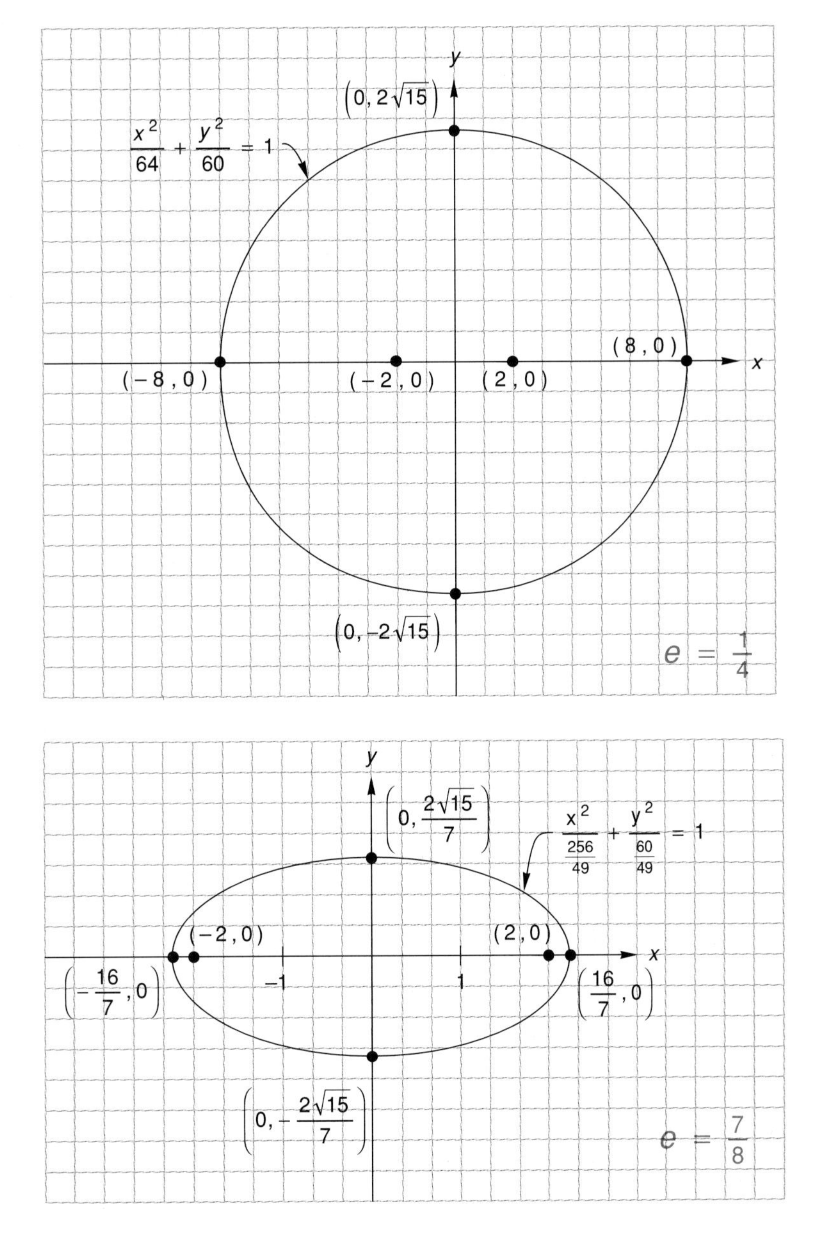

$\frac{x^2}{64} + \frac{y^2}{60} = 1$
$(0, 2\sqrt{15})$
$(-8, 0)$
$(-2, 0)$
$(2, 0)$
$(8, 0)$
$(0, -2\sqrt{15})$
$e = \frac{1}{4}$
$\left(0, \frac{2\sqrt{15}}{7}\right)$
$\frac{x^2}{\frac{256}{49}} + \frac{y^2}{\frac{60}{49}} = 1$
$(-2, 0)$
$(2, 0)$
$\left(-\frac{16}{7}, 0\right)$
$\left(\frac{16}{7}, 0\right)$
-1
1
$\left(0, -\frac{2\sqrt{15}}{7}\right)$
$e = \frac{7}{8}$

Figure 11.33

If we begin the discussion that led to equation (2) by transposing the second radical in equation (1), we produce

$$\sqrt{(x + c)^2 + y^2} = \frac{c}{a}\left(\frac{a^2}{c} + x\right)$$

or

$$d(F_2) = ed(D_2). \tag{3}$$

In other words, *an ellipse can be generated from either its right or left focus-directrix pair.*

Example 2 Write an equation of the centered ellipse whose major axis is 6 units long and whose left directrix is $x = -6$.

Solution

$a = \frac{6}{2} = 3$. Also $-a^2/c = -6$. Hence $c = \frac{3}{2}$.
$b^2 = a^2 - c^2 = \frac{27}{4}$ and $e = c/a = \frac{1}{2}$.
An equation of the ellipse is $x^2/9 + y^2/(27/4) = 1$. ■

You will have an opportunity in the exercise set to show that the hyperbola $x^2/a^2 - y^2/b^2 = 1$ also has directrices $x = a^2/c$ and $x = -a^2/c$ (figure 11.34). You will also be asked to show that for any point (x,y) on the hyperbola, $d(F_1) = ed(D_1)$ with $e > 1$ (see figure 11.35).

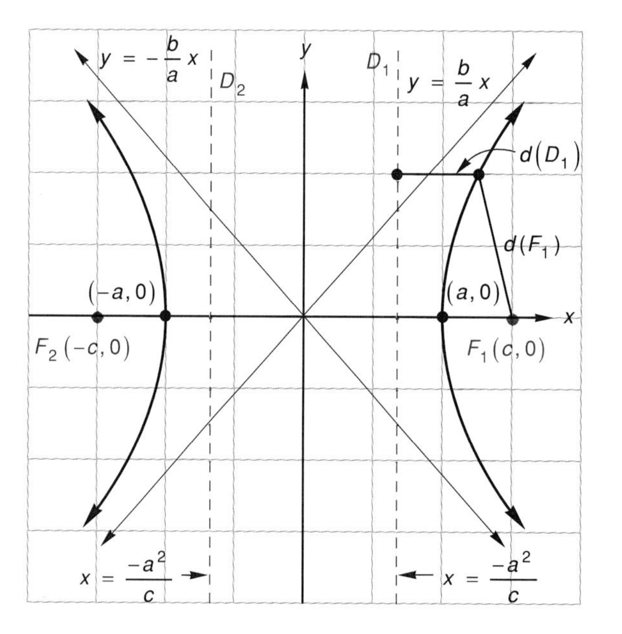

Figure 11.34

Example 3 Find equations of the hyperbolas centered at the origin that have focus (3,0) and

a. $e = \frac{5}{4}$,

b. $e = 4$.

Sketch each hyperbola.

Solution

a. $\dfrac{3}{a} = \dfrac{5}{4}$, $a = \dfrac{12}{5}$.

$$b^2 = c^2 - a^2 = 9 - \frac{144}{25} = \frac{81}{25}$$

An equation of the hyperbola in part **a** is

$$\frac{x^2}{\frac{144}{25}} - \frac{y^2}{\frac{81}{25}} = 1.$$

b. $\dfrac{3}{a} = 4$, $a = \dfrac{3}{4}$

$$b^2 = 9 - \frac{9}{16} = \frac{135}{16}$$

An equation of the hyperbola in part **b** is

$$\frac{x^2}{\frac{9}{16}} - \frac{y^2}{\frac{135}{16}} = 1.$$

The hyperbolas are sketched in figure 11.35.

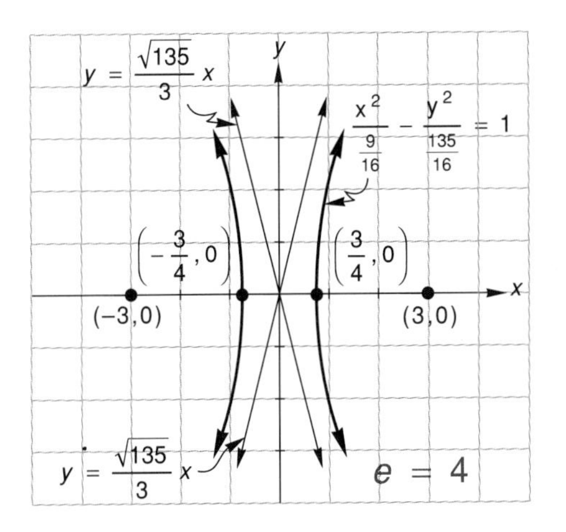

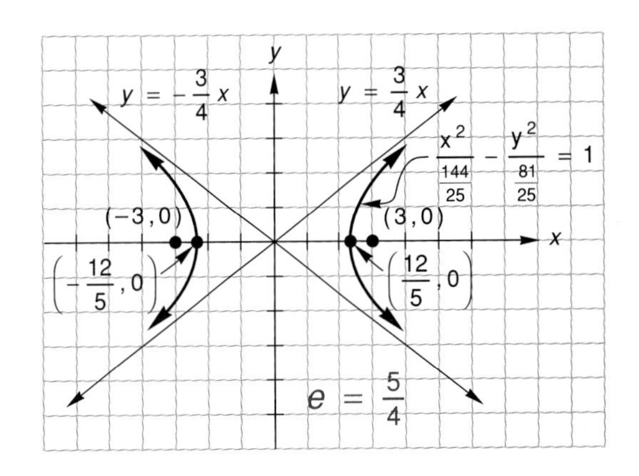

Figure 11.35

■

Example 4 Write an equation for the ellipse with center $(1,-2)$, focus $(5,-2)$, and directrix $x = \frac{29}{4}$.

Solution The ellipse has an equation of the form $(x')^2/a^2 + (y')^2/b^2 = 1$ where $x' = x - 1$ and $y' = y + 2$ (see figure 11.36). The equation of the directrix relative to the x'-y' axes is $x' + 1 = \frac{29}{4}$ or $x' = \frac{25}{4}$. Since the general equation of the right directrix of a centered ellipse is $x = a^2/c$, we have $a^2/4 = \frac{25}{4}$ or $a^2 = 25$. Now $b^2 = a^2 - c^2 = 25 - 16 = 9$. Thus an x'-y' equation of the ellipse is $(x')^2/25 + (y')^2/9 = 1$. An equation of the ellipse relative to the x-y axes is

$$\frac{(x-1)^2}{25} + \frac{(y+2)^2}{9} = 1.$$

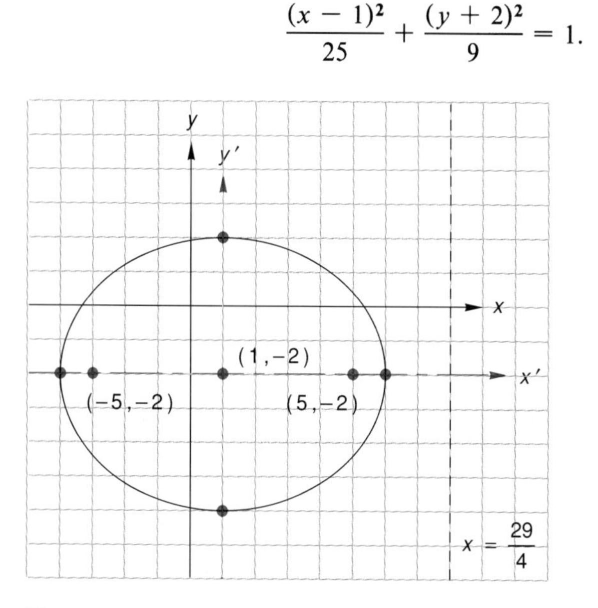

Figure 11.36 ■

Polar Coordinate Forms of the Conics

Each planet has an elliptic orbit with the sun at one focus of that orbit. Comets whose orbits are wholly contained in the solar system travel an elliptic path about the sun, which is located at a focus of the ellipse. Many comets enter and leave the solar system in hyperbolic or parabolic orbits about the sun, which is located at a focus of each of those orbits. Most artificial satellites are hurled into elliptic orbit around the earth, which sits at a focus of the orbit.

The study of planetary, comet, or satellite orbits is facilitated by describing those orbits with polar coordinates rather than rectangular coordinates.

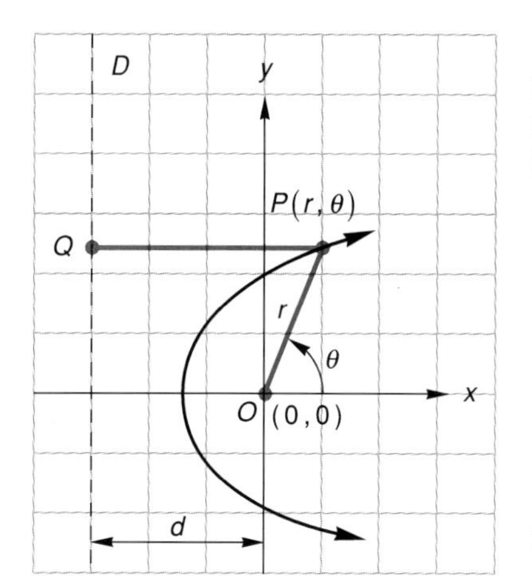

Figure 11.37

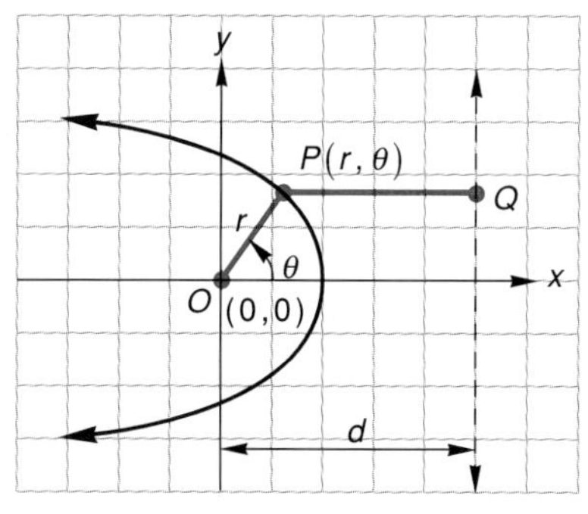

Figure 11.38

Imagine that a focus of a conic with eccentricity e is at the pole, as in figure 11.37. Let D, a directrix of the conic, be d units from the pole. The focus-directrix description of a conic tells us that the ratio of the distances between P and O and P and Q is e. That is,

$$\begin{aligned} \frac{r}{d + r\cos\theta} &= e \\ r &= e(d + r\cos\theta) \\ r - re\cos\theta &= ed \\ r &= \frac{ed}{1 - e\cos\theta}. \end{aligned}$$

If the directrix is to the right of the focus of the conic, as in figure 11.38, then the polar equation of the conic is obtained in the following way

$$\begin{aligned} \frac{r}{d - r\cos\theta} &= e \\ r &= e(d - r\cos\theta) \\ r(1 + e\cos\theta) &= ed \\ r &= \frac{ed}{1 + e\cos\theta}. \end{aligned}$$

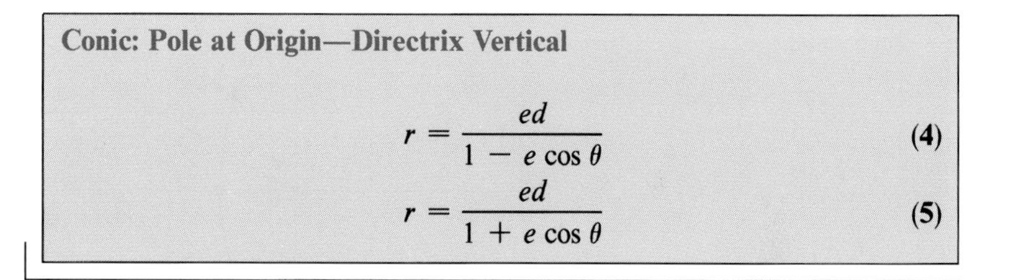

Conic: Pole at Origin—Directrix Vertical

$$r = \frac{ed}{1 - e\cos\theta} \tag{4}$$

$$r = \frac{ed}{1 + e\cos\theta} \tag{5}$$

The polar equations of a conic whose focus is at the pole, but whose directrix is horizontal are (see exercise 27)

Conic: Pole at Origin—Directrix Horizontal

$$r = \frac{ed}{1 - e\sin\theta} \tag{6}$$

$$r = \frac{ed}{1 + e\sin\theta}. \tag{7}$$

Example 5 Name and sketch the conic $r = 15/2 + 3 \cos \theta$. Also write a rectangular coordinate equation for the conic.

Solution The given equation is equivalent to

$$r = \frac{\frac{15}{2}}{1 + \frac{3}{2}\cos\theta}.$$

Comparing this equation to (5) we see that $ed = \frac{15}{2}$ and $e = \frac{3}{2}$. The conic is a hyperbola since $e > 1$. *Its graph is symmetric to the x-axis,* (Why?), so we need plot only a few points to obtain a good representation of the hyperbola.

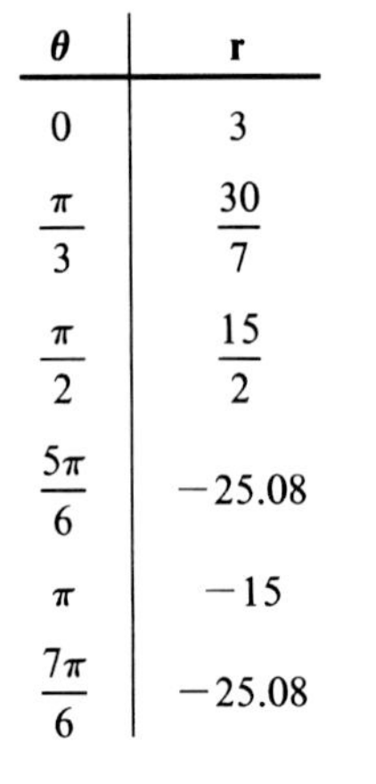

θ	r
0	3
$\frac{\pi}{3}$	$\frac{30}{7}$
$\frac{\pi}{2}$	$\frac{15}{2}$
$\frac{5\pi}{6}$	−25.08
π	−15
$\frac{7\pi}{6}$	−25.08

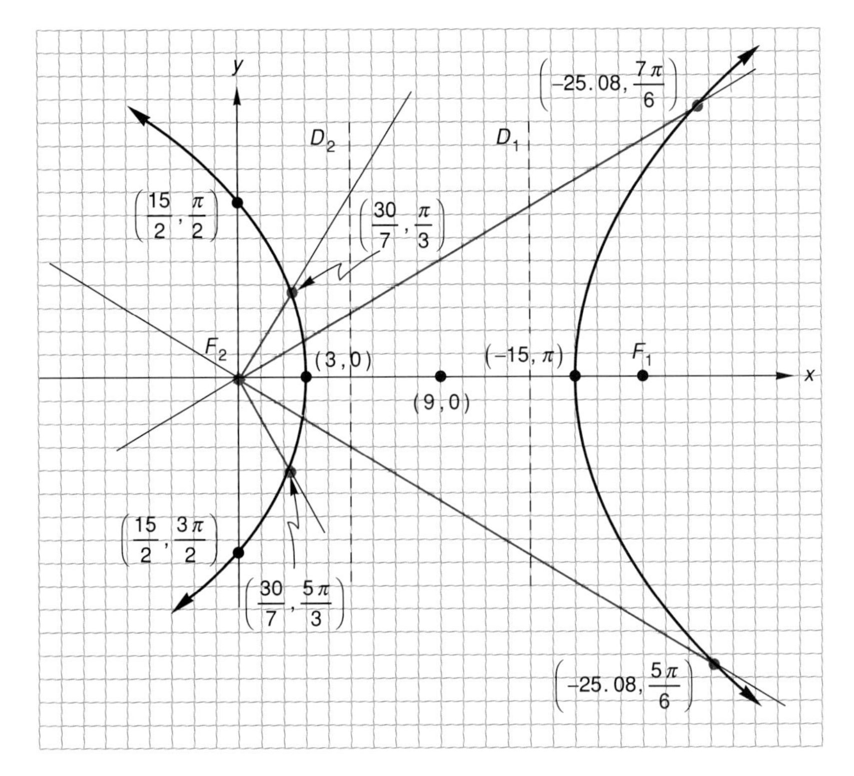

Figure 11.39

We shall discuss two methods for finding a rectangular coordinate equation of the conic.

Method I: Use the transformation equations between rectangular and polar coordinates.

$$\begin{aligned} r &= \frac{15}{2 + 3\cos\theta} \\ 2r + 3r\cos\theta &= 15 \\ 2\sqrt{x^2 + y^2} + 3x &= 15 \\ (2\sqrt{x^2 + y^2})^2 &= (15 - 3x)^2 \\ -5x^2 + 90x + 4y^2 &= 225 \\ -5(x^2 - 18x + 81) + 4y^2 &= 225 - 405 \\ \frac{(x-9)^2}{36} - \frac{y^2}{45} &= 1 \end{aligned}$$

Method II: Since $ed = \frac{15}{2}$ and $e = \frac{3}{2}$, then $d = 5$. The equation has form (5); therefore the directrix of the conic is 5 units to the right of the pole (see figure 11.39). By setting θ equal to 0 and to π, we find that the vertices of the hyperbola are (3,0) and (15,0). Hence, the center of the hyperbola is (9,0), and directrix D_2 is 4 units to the left of the hyperbola's center. As a result we can write $-a^2/c = -4$. But $a^2/c = a/e$. Therefore,

$$\frac{a}{\frac{3}{2}} = 4 \text{ or } a = 6.$$

Finally, $b^2 = c^2 - a^2 = 9^2 - 6^2 = 45$. The equation sought is

$$\frac{(x-9)^2}{36} - \frac{y^2}{45} = 1.$$

■

Exercise Set 11.6

(A)

Write an equation of the conic that satisfies the given conditions.

1. focus (4,0), $e = \frac{1}{2}$, conic is centered at the origin
2. focus $(0,-3)$, $e = \frac{4}{3}$, conic is centered at the origin
3. $e = 2$, directrices $x = \pm 1$, conic is centered at the origin
4. $e = \frac{1}{4}$, directrices $y = \pm 6$, conic is centered at the origin
5. left directrix is $x = -8$, $a = 5$, ellipse is centered at the origin
6. upper directrix is $y = 2$, $a = 3$, hyperbola is centered at the origin
7. directrices $y = \pm 4$, asymptotes $y = \pm\frac{3}{2}x$
8. focus $(-5,0)$, directrix $x = -7$, center (0,0), ellipse
9. focus (2,1), directrix $y = -1$, $e = 1$
10. focus $(-5,-2)$, directrix $x = -8$, $e = \frac{3}{5}$
11. focus (3,0), directrix $x = 1$, $e = \frac{3}{2}$
12. focus $(-3,4)$, directrix $x = 2$, $e = 1$

Find a polar coordinate equation of the conic, with focus at the pole, that satisfies the following conditions.

13. $e = 1$, directrix $x = 2$

14. $e = 1$, directrix $y = -1$

15. $e = \frac{1}{2}$, directrix $x = 5$

16. $e = \frac{7}{8}$, directrix $y = 6$

17. $e = \frac{4}{3}$, directrix $y = 1$

18. $e = 3$, directrix $x = -5$

Write a rectangular coordinate equation for each conic. Sketch a graph of the conic and locate the graph's foci or focus.

19. $r = \dfrac{6}{1 - \cos\theta}$

20. $r = \dfrac{2}{1 - \sin\theta}$

21. $r = \dfrac{1}{3 - \sin\theta}$

22. $r = \dfrac{10}{5 + 3\cos\theta}$

23. $r = \dfrac{15}{2 - 3\cos\theta}$

24. $r = \dfrac{3}{1 - 2\sin\theta}$

(B)

25. Derive equation (3) of this section.

26. Show that the hyperbola $x^2/a^2 - y^2/b^2 = 1$ satisfies the equation $d(F_1) = ed(D_1)$ with $e > 1$. (Hint: Review the derivation of equation (2) of this section.)

27. Derive equation (6) of this section.

Use the directrix-focus definition of a conic, $d(F) = ed(D)$, to write a Cartesian equation of the conic with the given eccentricity and focus-directrix pair.

28. $e = \frac{1}{4}$, $(1,0)$, $x + y = 5$

29. $e = \frac{2}{3}$, $(-3,1)$, $x - 2y + 8 = 0$

30. $e = 2$, $(0,0)$, $x + y - 1 = 0$

31. $e = 3$, $(-2,-4)$, $x + 2y = 0$

32. A satellite is placed in elliptic orbit about the earth. The minimum (perigee) and maximum (apogee) distances of the satellite from the surface of the earth are 200 miles and 600 miles, respectively. Write a polar equation that describes the orbit of the satellite. Place the center of the earth at the pole, and assume that the radius of the earth is 4,000 miles.

33. The length of the major axis of the earth's orbit about the sun is approximately 1.9×10^8 miles. The eccentricity of the orbit is approximately 0.02. Write a polar equation of the orbit with the pole at the center of the sun. Make a scale sketch of the orbit.

34. The elliptic orbit of Halley's Comet about the sun has a major axis that is approximately 34×10^8 miles and a minor axis that is approximately 8.5×10^8 miles. Write a polar equation of the orbit with the pole at the center of the sun. Make a scale sketch of the orbit.

11.7 Tangents and the Reflecting Property of the Conics

The angle at which a light ray strikes a reflecting surface equals the angle of reflection of the ray from the surface. Thus, in figure 11.40 the **angle of incidence** α, equals β, **the angle of reflection.** This physical law enables us to show that a light ray emanating from the focus of a parabola will be reflected from the parabola along a line parallel to the axis of the parabola.

We shall need two results to demonstrate the reflecting property of the parabola—one, a formula for the angle between two lines and two, a formula for the slope of the tangent to a parabola at any point on the parabola.

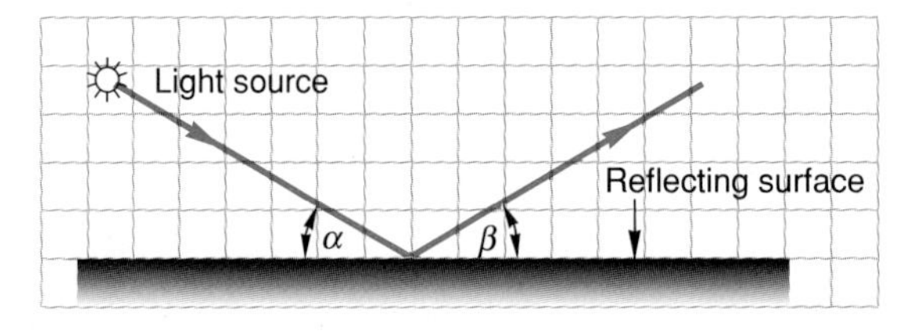

Figure 11.40

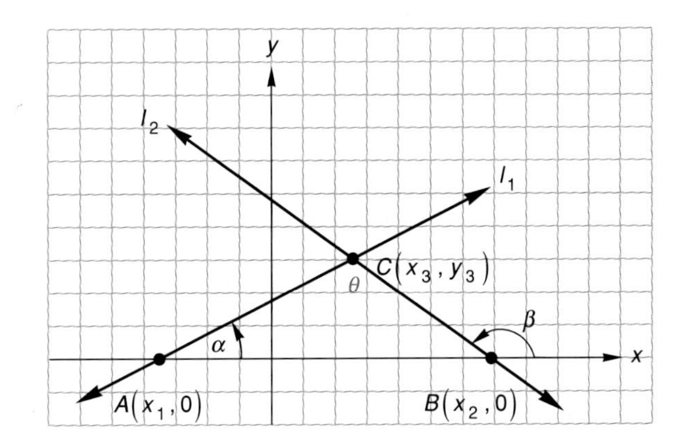

Figure 11.41

The Angle between Two Lines

The angle from line l_1 to line l_2 in figure 11.41 is θ. The *positive angles* that the lines make with the x-axis, which we shall call *angles of inclination,* are α and β. Because β is an exterior angle of triangle ABC, it is equal to the sum of the opposite two interior angles. That is,

$$\beta = \alpha + \theta.$$

It follows that

$$\begin{aligned} \theta &= \beta - \alpha \\ \tan \theta &= \tan (\beta - \alpha) \\ \tan \theta &= \frac{\tan \beta - \tan \alpha}{1 + \tan \beta \tan \alpha}. \end{aligned} \tag{1}$$

Equation (1) has another, more useful, form for our purposes. We see in figure 11.41 that

$$\begin{aligned} \tan \alpha &= \frac{y_3 - 0}{x_3 - x_1} \\ &= m_1 \end{aligned} \quad \text{and} \quad \begin{aligned} \tan \beta &= \frac{y_3 - 0}{x_3 - x_2} \\ &= m_2 \end{aligned}$$

where m_1 and m_2 are the respective slopes of lines l_1 and l_2. Substituting m_1 for $\tan \alpha$ and m_2 for $\tan \beta$ in equation (1) transforms that equation into

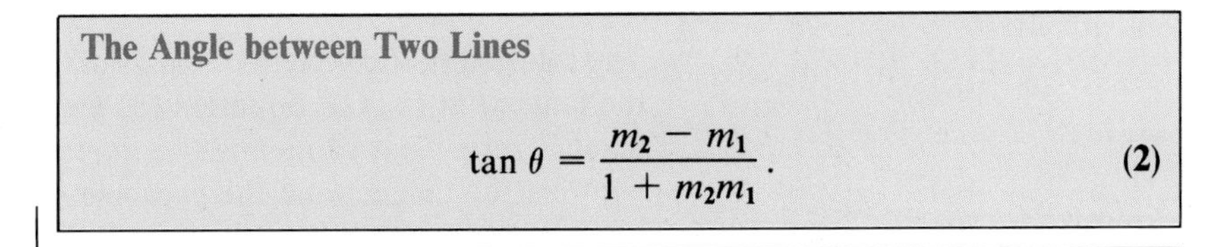

The Angle between Two Lines

$$\tan \theta = \frac{m_2 - m_1}{1 + m_2 m_1}. \tag{2}$$

We note that the development of (2) makes use of the fact that *the slope of a line equals the tangent of the line's angle of inclination.*

Equations (1) and (2) hold only when the angle of inclination (β) of l_2 is greater than the angle of inclination (α) of l_1. In other words, when finding the angle between two lines be sure to *label the line with the greater angle of inclination as* $\boldsymbol{l_2}$.

Equations (1) and (2) fail if the lines are perpendicular because then $m_1m_2 = -1$. The equations also fail if one of the lines is vertical, as the slope of a vertical line is undefined.

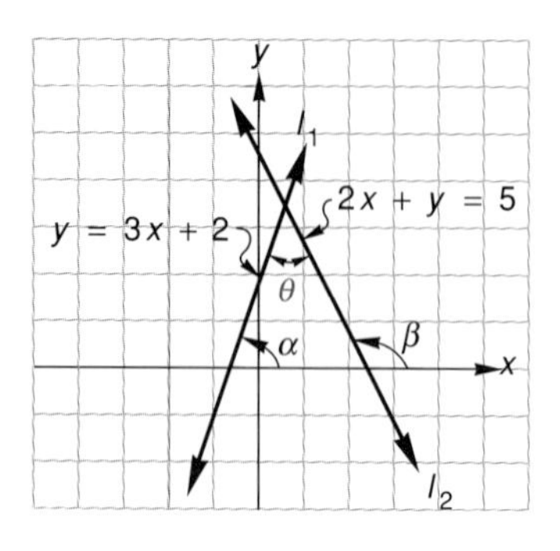

Figure 11.42

Example 1 Find the angle between the lines $2x + y = 5$ and $y = 3x + 2$. Graph the lines in the same coordinate system.

Solution The slopes of the lines are -2 and 3 (figure 11.42). Since the line with the negative slope has the greater angle of inclination (why?), it is labeled l_2. Substituting the slopes into equation (2) yields

$$\begin{aligned} \tan\theta &= \frac{-2-3}{1+(-2)(3)} \\ \tan\theta &= 1 \\ \theta &= 45^\circ. \end{aligned}$$

■

Tangents

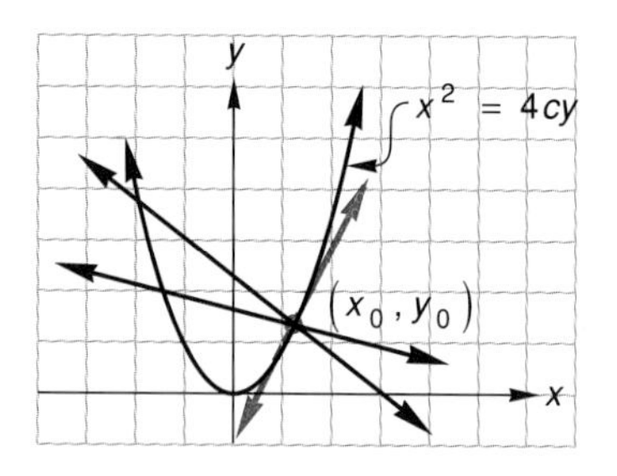

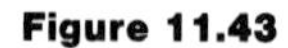
Figure 11.43

Let (x_0,y_0) be a point on the parabola $x^2 = 4cy$ in figure 11.43. A tangent of the parabola through (x_0,y_0) is one line in a pencil of lines through the point, all of which have an equation of the form $y - y_0 = m(x - x_0)$. Each line in the pencil of lines, except the tangent, intersects the parabola in two points. The tangent line has only (x_0,y_0) in common with the parabola.

To find an equation of the tangent we seek a solution of the system of equations

$$\begin{aligned} y - y_0 &= m(x - x_0) \\ x^2 &= 4cy. \end{aligned}$$

We replace y in the second equation with $mx + y_0 - mx_0$, the expression that, in the first equation, is found to be equal to y.

$$\begin{aligned} x^2 &= 4c(mx + y_0 - mx_0) \\ x^2 - 4cmx + 4cmx_0 - 4cy_0 &= 0 \\ x &= \frac{4cm \pm \sqrt{16c^2m^2 - 16cmx_0 + 16cy_0}}{2} \end{aligned} \tag{3}$$

The solution of the system of equations is $\{(x_0,y_0)\}$ if $y - y_0 = m(x - x_0)$ is the tangent at (x_0,y_0). Equation (3) yields only one value for x if the discriminant is zero. That value must be x_0, the x-coordinate of the point of contact between the tangent and the parabola. Hence, from (3) we have

$$x_0 = \frac{4cm}{2}$$

$$m = \frac{x_0}{2c}. \tag{4}$$

An equation of the tangent line at (x_0,y_0) is

$$y - y_0 = \frac{x_0}{2c}(x - x_0). \tag{5}$$

Had the above analysis been carried through for a horizontal parabola $y^2 = 4cx$, we would have gotten

$$m = \frac{2c}{y_0} \tag{6}$$

and

$$y - y_0 = \frac{2c}{y_0}(x - x_0). \tag{7}$$

Example 2 Determine the equation of the tangent line to the parabola $y^2 - 4x - 6y + 1 = 0$ at $(2,-1)$.

Solution It is easy to verify that $(2,-1)$ is a point on the parabola. Thus, we can use the results above if we can write the parabola in the one of the forms

$$(y')^2 = 4cx' \quad \text{or} \quad (x')^2 = 4cy'.$$

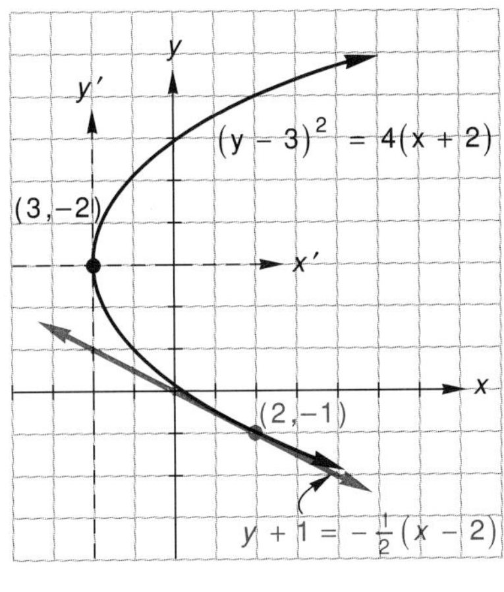

Figure 11.44

Completing the square gives us

$$(y^2 - 6y + 9) = 4x - 1 + 9$$
$$(y - 3)^2 = 4(x + 2).$$

We have a horizontal parabola with vertex at $(-2,3)$. If we pass x'-y' coordinate axes through the vertex (figure 11.44), and if we let

$$y' = y - 3 \quad \text{and} \quad x' = x + 2,$$

then the equation of the parabola, relative to the x'-y' axes, is

$$(y')^2 = 4x'.$$

The point $(2,-1)$ becomes $(4,-4)$ in the x'-y' system. Now, using (7) with $c = 1$, $x_0 = 4$, and $y_0 = -4$, the equation of the tangent line is

$$y' + 4 = -\tfrac{1}{2}(x' - 4).$$

In the x-y system, the equation of the tangent line is

$$y + 1 = -\tfrac{1}{2}(x - 2).$$ ■

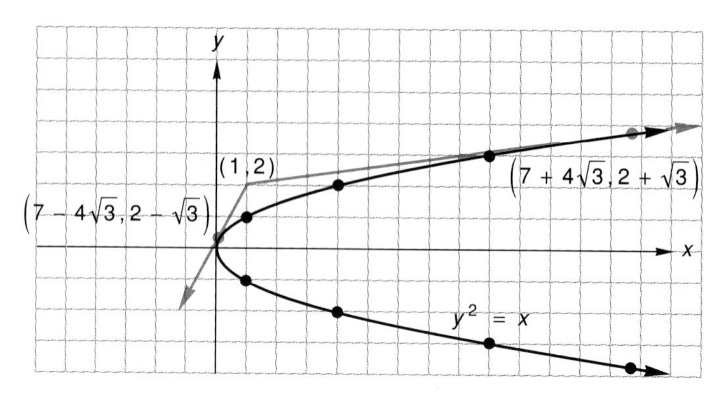

Figure 11.45

Example 3 Determine equations of the lines from (1,2) that are tangent to the parabola $y^2 = x$ in figure 11.45. What are the coordinates of the points of tangency?

Solution We solve the system

$$\begin{aligned} y - 2 &= m(x - 1) \\ y^2 &= x \end{aligned}$$

to find the points of intersection between the tangent lines and the parabola. Replace x in the second equation by $(y - 2 + m)/m$, the expression for x obtained from the first equation.

$$\begin{aligned} y^2 &= \frac{y - 2 + m}{m} \\ my^2 - y + 2 - m &= 0 \\ y &= \frac{1 \pm \sqrt{1 - 4m(2 - m)}}{2m} \end{aligned}$$

A tangent line through (1,2) intersects the parabola in exactly one point. Therefore, the discriminant must be zero. (Why?)

$$\begin{aligned} 4m^2 - 8m + 1 &= 0 \\ m &= \frac{2 \pm \sqrt{3}}{2} \\ y &= \frac{1 \pm \sqrt{4m^2 - 8m + 1}}{2m} \\ &= \frac{1 \pm \sqrt{0}}{\dfrac{2(2 \pm \sqrt{3})}{2}} \\ &= \frac{1}{2 \pm \sqrt{3}} \\ &= 2 \pm \sqrt{3} \text{ (Why?)} \end{aligned}$$

The corresponding x-coordinates can be computed from $y^2 = x$. They are $7 + 4\sqrt{3}$ and $7 - 4\sqrt{3}$. The points of tangency are $(7 + 4\sqrt{3}, 2 + \sqrt{3})$ and $(7 - 4\sqrt{3}, 2 - \sqrt{3})$. The equations of the tangent lines are

$$y - 2 = \frac{2 - \sqrt{3}}{2}(x - 1) \quad \text{and} \quad y - 2 = \frac{2 + \sqrt{3}}{2}(x - 1).$$

Reflecting Property of the Parabola

Assume that a light ray from the focus of parabola $y^2 = 4cx$ strikes the parabola at $P(x_0, y_0)$ (figure 11.46). The angle of incidence of the ray, which is the angle between the ray and the tangent at P, is α. The slope of the tangent line to the parabola at P is $2c/y_0$ (equation (7)). We want to show that β, the angle between a horizontal line through P and the tangent line through P, is equal to α. We will then have shown that β is the angle of reflection of the light ray. It will follow that any light ray emanating from the focus of the parabola is reflected along a line that is parallel to the axis of the parabola. Using (2), we can write

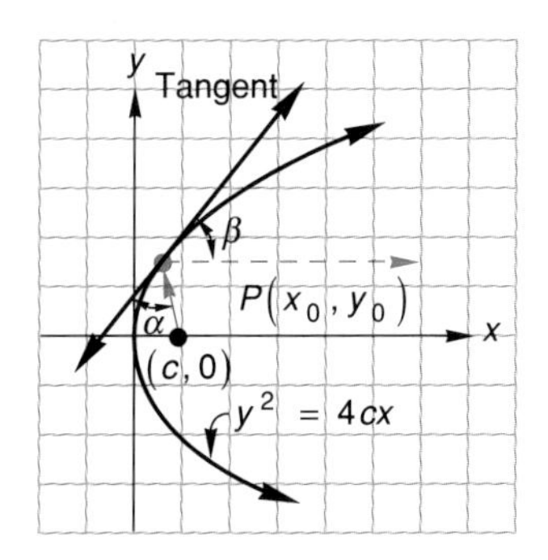

Figure 11.46

$$\begin{aligned} \tan\alpha &= \frac{\dfrac{y_0}{x_0 - c} - \dfrac{2c}{y_0}}{1 + \left(\dfrac{y_0}{x_0 - c}\right)\left(\dfrac{2c}{y_0}\right)} \\ &= \frac{y_0^2 - 2cx_0 + 2c^2}{x_0y_0 - cy_0 + 2cy_0} \\ &= \frac{4cx_0 - 2cx_0 + 2c^2}{y_0(x_0 + c)} \\ &= \frac{2c}{y_0}. \end{aligned}$$

Since β is the angle of inclination of the tangent line, the slope of the tangent line is $\tan\beta$. Hence,

$$\tan\beta = \frac{2c}{y_0},$$

and $\beta = \alpha$.

The reflecting property of the parabola is used in the construction of sealed beam headlights. The light source in the headlight unit is placed at the focus of the unit's paraboloid reflecting surface (figure 11.47), so that light from the source is reflected as a concentrated beam on the road. A paraboloid is the surface generated when a parabola is revolved around its axis.

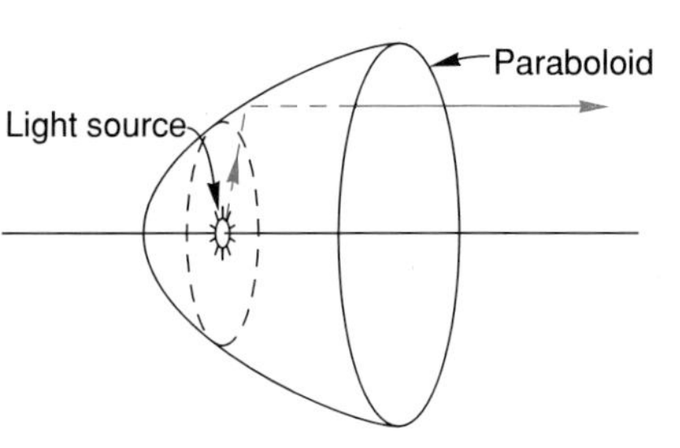

Figure 11.47

If a parabolic dish (a paraboloid), is pointed at an energy source such as the sun or a more distant star, it collects the incoming energy at its focus. This is the principle underlying the construction of radio telescopes, solar furnaces, radar, and other energy collectors.

Reflecting Property of the Ellipse and the Hyperbola

If an energy source is placed at one focus of an ellipse, the energy it radiates is reflected by the ellipse to the other focus (see figure 11.48). You will be asked to prove this fact in the exercises.

If an ellipse is rotated through 180° about its major axis, it generates a three-dimensional surface called an ellipsoid (figure 11.48). There are many buildings throughout the world that contain so-called whispering galleries because some large room has a ceiling in the shape of an ellipsoid. Persons whispering at one focus of the ellipsoid can be heard clearly at the other focus.

There is a medical technique for breaking up kidney stones into fragments small enough to be passed. The method obviates the need for painful and debilitating surgery. A subject is placed in an ellipsoidal bath so that the region containing the kidney stones is at one focus of the ellipsoid. An ultrasound source is at the other focus.

If a light ray is directed at one focus of a hyperbola, it is reflected by the hyperbola to its other focus (see figure 11.49). You will be asked to prove this fact in the exercises. The reflecting property of the hyperbola is applied in the design of certain telescopes that utilize a collection of hyperbolic and parabolic mirrors to focus the light from distant celestial objects.

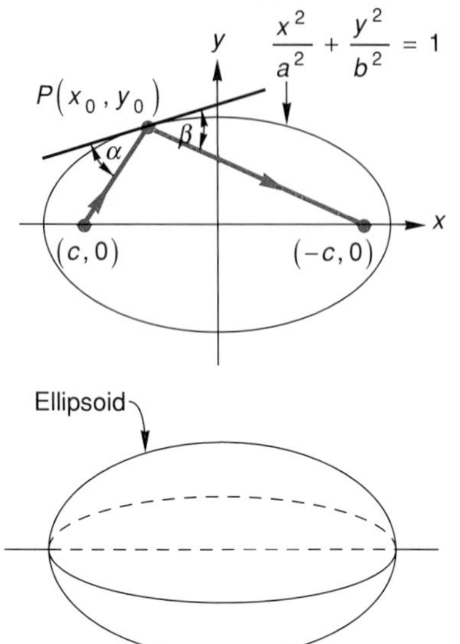

Figure 11.48

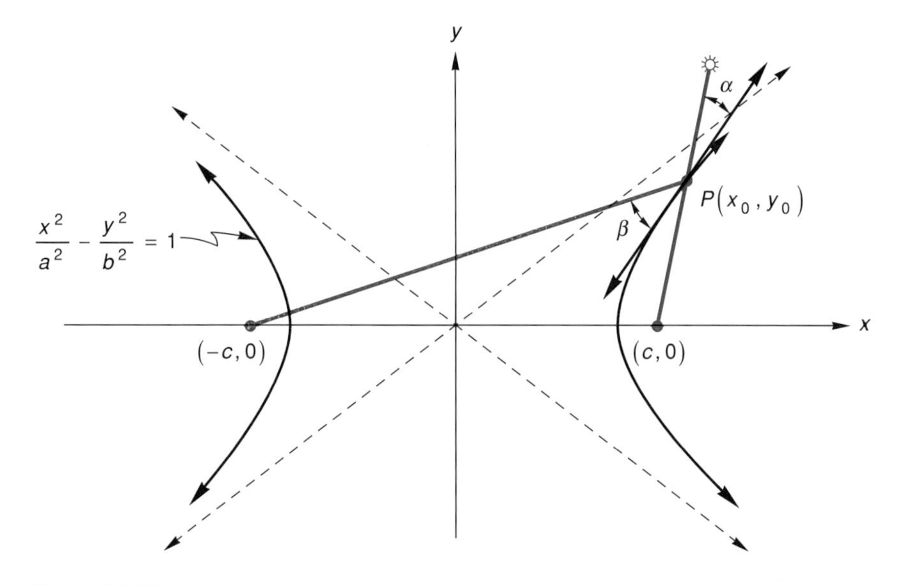

Figure 11.49

Exercise Set 11.7

(A)

Find the angle between the given lines.

1. $x + 3y = 5$, $3x + 5y = 5$

2. $y = x - 2$, $y = 2x + 1$

3. $4y = 3x$, $2y = 5x$

4. $y = -\frac{3}{2}x - 1$, $4x + y = 2$

5. $6x - 2y = 5$, $y = 1$

6. $y = -3$, $2x + y = 4$

Find the angles of the triangle whose vertices are given.

7. $(0,0)$, $(2,5)$, $(-1,3)$

8. $(-4,6)$, $(-1,-2)$, $(5,1)$

9. $(0,1)$, $(1,8)$, $(4,4)$

10. $(2,0)$, $(10,0)$, $(6,3)$

Find the equations of the lines that make the indicated angle with the given line through the given point.

11. $y = x$, $\theta = 30°$, $(2,2)$

12. $2x + y = 3$, $\theta = 45°$, $(0,3)$

13. $-x + 4y = 4$, $\theta = 60°$, $(-4,0)$

14. $3x + 4y = 7$, $\theta = 30°$, $(1,1)$

Write an equation for the tangent line through the given point that is tangent to the given parabola.

15. $y = x^2$, $(2,4)$

16. $y^2 = x$, $(9,3)$

17. $(x - 2)^2 = 4(y - 1)$, $(-2,5)$

18. $(y + 1)^2 = -2(x - 2)$, $(0,1)$

19. $x^2 + 4x + 5y - 2 = 0$, $(-1,1)$

20. $y^2 - 2x + 2y + 1 = 0$, $(2,1)$

Write an equation for the tangent line to the given circle through the given point. (Hint: The tangent line is perpendicular to the radius at the point of tangency.)

21. $x^2 + y^2 = 9$, $(1,2\sqrt{2})$

22. $x^2 + y^2 = 16$, $(-3,\sqrt{7})$

23. $(x + 2)^2 + (y + 2)^2 = 13$, $(0,1)$

24. $x^2 + y^2 - 4x + 10y = 0$, $(0,0)$

Write equations for the tangent lines to the given parabola through the given point.

25. $y = x^2$, $(-3,-2)$

26. $y^2 = x$, $(1,-5)$

27. $x^2 + 4x + 5y - 2 = 0$, $(0,2)$

28. $(x - 2)^2 = 4(y - 1)$, $(-1,2)$

(B)

Write equations for the tangent lines to the given circle through the given point. (Hint: The distance from the center of the circle to the point of tangency is equal to the length of the radius. Use the formula for the distance from a point to a line, equation (4) section 11.1.)

29. $(x + 3)^2 + (y - 1)^2 = 25$, $(5,-3)$

30. $x^2 + y^2 + 8x - 6y + 9 = 0$, $(-8,4)$

Find the equations of the angle bisectors of the given pair of lines.

31. $x - 3y = 5$, $3x - y = 5$

32. $y = x - 2$, $y = 2x + 1$

33. Show that the slope of the tangent to the ellipse $x^2/a^2 + y^2/b^2 = 1$ at a point (x_0,y_0) on the ellipse is $m = (-b^2x_0)/(a^2y_0)$.

34. Use the result in exercise 33 to prove the reflecting property of the ellipse.

35. Show that the slope of the tangent to the hyperbola $x^2/a^2 - y^2/b^2 = 1$ at a point (x_0,y_0) on the hyperbola is $m = (b^2x_0)/(a^2y_0)$.

36. Use the result in exercise 35 to prove the reflecting property of the hyperbola.

Chapter 11 Review

Major Points for Review

Section 11.1 The geometric description of a circle

The standard and general forms of the equation of a circle

Determining the center and radius of a circle from its standard and general forms

Determining the general equation of a circle from three conditions that the circle must satisfy

Parametric equations of a circle

Graphing circles

Section 11.2 The geometric description of a parabola

The standard form equations for vertical and horizontal parabolas with vertices at the origin and with vertices at (h,k)

Graphing parabolas

The transformation equations for the translation of the coordinate axes

Section 11.3 The geometric description of an ellipse

The standard form equations for vertical and horizontal ellipses centered at the origin, and centered at (h,k)

Parametric equations for the ellipse

Graphing ellipses

Section 11.4 The geometric description of a hyperbola

The standard form equations for vertical and horizontal hyperbolas centered at the origin and centered at (h,k)

Determining the asymptotes of a hyperbola

Parametric equations of a hyperbola

Graphing hyperbolas

Determine which conic is defined by an equation $Ax^2 + Cy^2 + Dx + Ey + F = 0$

Section 11.5 The transformation equations between coordinate systems that are rotated with respect to each other

The formula that determines the angle of rotation that is necessary to eliminate the xy term from the equation $Ax^2 + Bxy + Cy^2 + Dx + Ey + F = 0$

The use of the discriminant, $B^2 - 4AC$, in identifying the conic defined by the general second degree equation in two variables

Section 11.6 The focus-directrix definition of a conic

The eccentricity associated with each conic

Polar coordinate equation forms of the conics

Section 11.7 Formulas for the angle between two lines

Finding the equation of the tangent line through a point on a parabola

Finding the equations of the tangent lines to a parabola from a point not on the parabola

The reflecting properties of the parabola, ellipse, and hyperbola

Review Exercises

Write a standard form equation of the indicated conic that satisfies the given conditions.

1. circle: center (1,2), radius 4
2. circle: center (0,0), radius 5
3. parabola: vertex (0,0), directrix $x = 1$
4. parabola: vertex $(2,-5)$, focus $(2,-1)$
5. ellipse: center (0,0), focus (2,0), vertex (4,0)
6. ellipse: vertices $(-3,2)$ and $(-3,8)$, focus $(-3,7)$
7. hyperbola: vertices $(0,-4)$ and (0,4), foci $(0,-6)$ and (0,6)
8. horizontal hyperbola: asymptotes $y = \pm 3x/2$, $a = 4$

Graph each conic.

9. $\dfrac{(x-3)^2}{4} - \dfrac{(y+2)^2}{9} = 1$
10. $(x-1)^2 + (y-3)^2 = 16$
11. $(y+1)^2 = 8(x-5)$
12. $(x+2)^2 = 6(y+3)$
13. $\dfrac{(x+4)^2}{25} + \dfrac{y^2}{16} = 1$
14. $\dfrac{(y-2)^2}{9} + (x-1)^2 = 1$

Write an equation of the indicated conic that satisfies the given conditions.

15. circle: contains the points (1,2), (0,5), $(-2,7)$
16. circle: center $(-3,8)$, tangent to the y-axis
17. parabola: horizontal axis, vertex (1,3), contains (5,0)
18. parabola: axis $y = 4$, focus (2,4), directrix $x = 5$
19. ellipse: center $(3,-1)$, contains (6,2) and $(-1,-3)$
20. ellipse: foci are $(-1,-1)$ and $(-1,9)$, length of major axis is 12
21. hyperbola: vertices (8,2) and $(-2,2)$, contains $(-4,6)$
22. hyperbola: asymptotes $2x - y = 7$ and $2x + y = 5$, contains (5,2)

Identify and graph the conic determined by each equation.

23. $x^2 + 2y^2 + 6x - 8y = 0$
24. $2x^2 + 2y^2 - 8x - 12y + 1 = 0$
25. $3y^2 + 2x - 6y + 7 = 0$
26. $5x^2 + y^2 - 4y - 6 = 0$
27. $3y^2 - 4x^2 - 12y + 8x - 16 = 0$
28. $x^2 + 8x - y + 12 = 0$

Eliminate the parameter and write a standard form equation of the conic determined by each pair of parametric equations.

29. $x = 2 \sec \theta, y = 3 \tan \theta$
30. $x = 1 + 4 \sin \theta, y = -2 + 4 \cos \theta$
31. $x = 2 \cos \theta, y = 5 + \sin \theta$
32. $x = -3 + 4 \tan \theta, y = 2 \sec \theta$

Determine the angle through which the coordinate axes must be rotated to remove the xy term from the equation.

33. $x^2 - 3xy + y^2 - 5 = 0$
34. $2xy + 5 = 0$
35. $3x^2 + 2\sqrt{3}xy + y^2 + 2x + 2\sqrt{3}y = 0$
36. $6x^2 + 24xy - y^2 - 12x + 26y + 11 = 0$

Find the angle between the given lines.

37. $2x + y = 4, x - y = 2$
38. $x + y = 1, 3x + 4y = 6$

Write a rectangular coordinate equation of the conic that satisfies the given conditions.

39. focus $(0,1)$, $e = 2$, conic is centered at the origin
40. $e = 1$, directrix $x = 3$, focus $(0,0)$
41. focus $(2,3)$, directrix $y = 5$, $e = \frac{4}{5}$
42. directrices $x = \pm 3$, asymptotes $y = \pm 2x$

Find a polar coordinate equation of the conic, with focus at the pole, that satisfies the given conditions.

43. $e = \frac{1}{4}$, directrix $x = 4$
44. $e = 1$, directrix $y = 2$

Graph the conic from its polar equation. Then convert the polar equation to a rectangular coordinate equation.

45. $r = \dfrac{2}{4 - 3 \cos \theta}$
46. $r = \dfrac{3}{2 + 5 \sin \theta}$

Write equations for the tangent line(s) to the given conic through the indicated point.

47. $x^2 = 8y$, $(4,2)$
48. $y^2 - 4x + 4y - 8 = 0$, $(1,3)$

49. Find the center and radius of the circle that contains $(1,4)$ and is tangent both to the y-axis and to the line $y = 2x$.
50. Write an equation of the parabola with focus $(-1,2)$ and directrix $x + y = 6$.

Sketch a graph of each equation so that it is centered at the origin of an x''-y'' coordinate system that is rotated and translated relative to the x-y coordinate system. Write the given equation in x''-y'' coordinates. Name vertices, foci, and asymptotes relative to the x-y coordinate system.

51. $3x^2 + 2\sqrt{3}xy + y^2 + 2x + 2\sqrt{3}y = 0$
52. $6x^2 + 24xy - y^2 - 12x + 26y + 11 = 0$

53. Find the equations of the angle bisectors of the angle formed by the lines $2x + y = 4$ and $x - y = 2$.
54. Use the focus-directrix definition of a conic to write a Cartesian equation of the conic with $e = \frac{3}{4}$ and the focus-directrix pair $(-2,1)$, $y - x = 6$.

Chapter 11 Test 1

Write an equation of the indicated conic that satisfies the given conditions.

1. circle: radius 2, center $(4,5)$
2. parabola: focus $(-3,4)$, vertex $(1,4)$
3. ellipse: center $(-2,-2)$, focus $(2,-2)$, vertex $(4,-2)$
4. hyperbola: vertices $(3,2)$ and $(3,8)$, foci $(3,0)$ and $(3,10)$

Graph each conic. Label foci and vertices.

5. $(y - 3)^2 = 2x$
6. $(x + 2)^2 + (y - 5)^2 = 16$
7. $x^2 + 4y^2 + 2x - 8y - 4 = 0$
8. $2x^2 - y^2 + 12x + 4y - 11 = 0$

9. Write an equation of the parabola with focus (0,0) and directrix $x + 2y = 6$.

10. Find the angle through which the coordinate axes must be rotated to remove the xy term from the equation $x^2 + 2xy + 3y^2 - 6x + 4y - 10 = 0$.

Chapter 11 Test 2

Write an equation of the indicated conic that satisfies the given conditions.

1. circle: center (7,4), tangent to the x-axis
2. parabola: vertical axis, vertex $(2,-5)$, contains (4,0)
3. ellipse: center $(-5,4)$, contains $(-4,-2)$ and $(-7,6)$

Graph each conic. Label foci and vertices.

4. $x^2 + 2x - y = 0$
5. $y^2 - x^2 + 4x + 2y - 2 = 0$
6. $2x^2 + y^2 - 4x - 2y - 1 = 0$

7. Write the Cartesian coordinate equation of the conic defined by the parametric equations $x = 2 + 3 \cos \theta$, $y = -1 + 2 \sin \theta$
8. Write the equations of the tangent lines to the parabola $y^2 = x - 2$ from the point (3,4).
9. Find the angle between the lines $2x - y = 4$ and $x + y = 5$.
10. Write the Cartesian equation of the conic with directrix $x = 4$, focus (0,2), and eccentricity $e = \frac{2}{3}$.

Chapter 12
Vectors

Vectors are introduced in section 9.4 as a tool for solving displacement problems such as determining the bearing of a plane whose course is altered by wind. Static loading problems, such as determining the magnitude and direction of the total force acting on a pin connection, are also solved with vectors. In this chapter we use the ordered pair representation of a vector introduced in section 9.4 to construct rules for adding and multiplying vectors. We develop a vector algebra and with it are able to construct, in a very simple way, formulas for the angle between two lines and the distance from a point to a line. Our vector algebra also enables us to generate easily equations for lines and planes in three dimensions.

12.1 Vector Algebra

Equations (1) and (2), where $\mathbf{A} = (x_1,y_1)$ and $\mathbf{B} = (x_2,y_2)$, are developed in section 9.4.

$$\mathbf{A} + \mathbf{B} = (x_1 + x_2, y_1 + y_2) \qquad (1)$$

$$||\mathbf{A}|| = \sqrt{x_1^2 + y_1^2} \qquad (2)$$

The equations state that

Summary

1. the sum of two vectors is a third vector whose components are the sums of the corresponding components of the original two vectors, and
2. the *magnitude or length* of a vector $\mathbf{A}$ ($||\mathbf{A}||$) is the square root of the sum of the squares of the components.

The zero vector, **0**, is defined by the ordered pair (0,0). Hence, $||\mathbf{0}|| = \sqrt{0^2 + 0^2} = 0$. Also, the sum of any nonzero vector **A** and **0** is **A**. If $\mathbf{A} = (x_1,y_1)$, then

$$\mathbf{A} + \mathbf{0} = (x_1,y_1) + (0,0) = (x_1 + 0, \quad y_1 + 0) = (x_1,y_1) = \mathbf{A}.$$

Scalar Multiplication

The positive angles that $\mathbf{A} = (x,y)$ and $\mathbf{B} = (3x,3y)$ make with the x-axis (figure 12.1) are equal since

$$\tan^{-1}\frac{y}{x} = \tan^{-1}\frac{3y}{3x}.$$

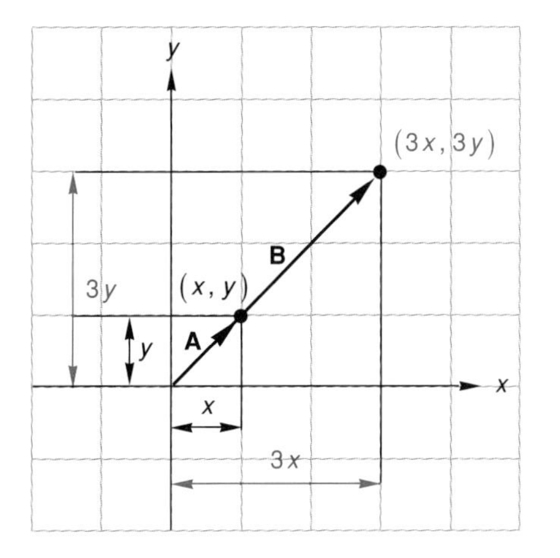

Figure 12.1

Therefore, **A** and **B** have the same direction but different magnitudes.

$$||\mathbf{A}|| = \sqrt{x^2 + y^2}, \quad ||\mathbf{B}|| = \sqrt{(3x)^2 + (3y)^2} = 3\sqrt{x^2 + y^2}$$

We say **B** is a **scalar multiple** of **A** and write $\mathbf{B} = 3\mathbf{A}$. In general

Multiplying a Vector by a Real Number

if a vector $\mathbf{A} = (x,y)$ is multiplied by a real number k, called a scalar, then $\mathbf{A} = (kx,ky)$.

Example 1 Graph $\mathbf{A} = (6,4)$, $\mathbf{B} = \mathbf{A}/2$, and $\mathbf{C} = -2\mathbf{A}$. Evaluate $||\mathbf{A}||$, $||\mathbf{B}||$, and $||\mathbf{C}||$.

Solution In figure 12.2,

$$||\mathbf{A}|| = \sqrt{6^2 + 4^2} = 2\sqrt{13},$$
$$\mathbf{B} = \tfrac{1}{2}(6,4) = (3,2), \quad ||\mathbf{B}|| = \sqrt{3^2 + 2^2} = \sqrt{13},$$
$$\mathbf{C} = -2(6,4) = (-12,-8), \text{ and } ||\mathbf{C}|| = \sqrt{144 + 64} = 4\sqrt{13}.$$

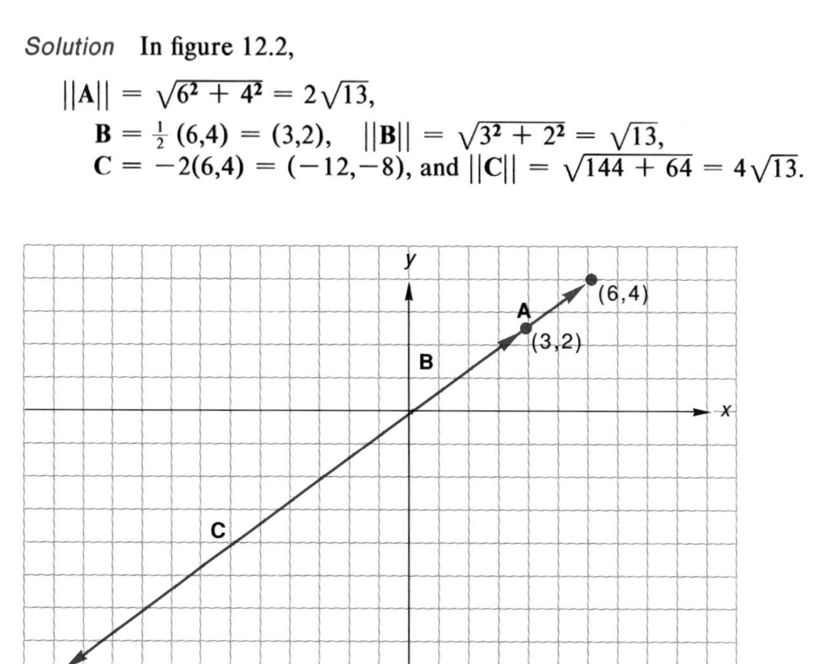

Figure 12.2 ■

Example 2 Find a unit vector $\mathbf{U}$ having the same direction as $\mathbf{A} = (3,4)$.

$$1 = ||\mathbf{U}|| = ||k\mathbf{A}|| = \sqrt{(3k)^2 + (4k)^2} = 5k, \quad \text{and} \quad k = \frac{1}{5}.$$

As a result $\mathbf{U} = \left(\frac{3}{5}, \frac{4}{5}\right)$. ■

Example 2 suggests the following generalization.

Unit Vector

A unit vector having the same direction as (x,y) is

$$(x/\sqrt{x^2 + y^2}, y/\sqrt{x^2 + y^2}).$$

Properties of Vector Addition and Scalar Multiplication

Each property in the following list can be derived directly from the definitions of vector addition and scalar multiplication.

1. $\mathbf{A} + \mathbf{B} = \mathbf{B} + \mathbf{A}$ (commutativity)
2. $\mathbf{A} + (\mathbf{B} + \mathbf{C}) = (\mathbf{A} + \mathbf{B}) + \mathbf{C}$ (associativity)
3. $\mathbf{A} + \mathbf{O} = \mathbf{A}$, $\mathbf{O} = (0,0)$ (additive identity)
4. If $\mathbf{A} = (x,y)$, then $-\mathbf{A} = (-x,-y)$ and $\mathbf{A} + (-\mathbf{A}) = \mathbf{O}$. (additive inverse)

Letting k and t be scalars,

5. $(kt)\mathbf{A} = k(t\mathbf{A})$ (associativity)
6. $k(\mathbf{A} + \mathbf{B}) = k\mathbf{A} + k\mathbf{B}$, $(k + t)\mathbf{A} = k\mathbf{A} + t\mathbf{A}$ (distributive properties)
7. $1(\mathbf{A}) = \mathbf{A}$ (multiplicative identity)

Proof of Property 2: Let $\mathbf{A} = (a_1,a_2)$, $\mathbf{B} = (b_1,b_2)$, and $\mathbf{C} = (c_1,c_2)$.

$$\begin{aligned}
\mathbf{A} + (\mathbf{B} + \mathbf{C}) &= (a_1,a_2) + ((b_1,b_2) + (c_1,c_2)) \\
&= (a_1,a_2) + (b_1 + c_1, b_2 + c_2) \\
&= (a_1 + (b_1 + c_1), a_2 + (b_2 + c_2)) \\
&= ((a_1 + b_1) + c_1, (a_2 + b_2) + c_2) \\
&= (a_1 + b_1, a_2 + b_2) + (c_1,c_2) \\
&= ((a_1,a_2) + (b_1,b_2)) + (c_1,c_2) \\
&= (\mathbf{A} + \mathbf{B}) + \mathbf{C}
\end{aligned}$$

Proof of Property 5: Let $\mathbf{A} = (x,y)$.

$$\begin{aligned}
(kt)\mathbf{A} &= kt(x,y) \\
&= (ktx,kty) \\
&= k(tx,ty) \\
&= k(t\mathbf{A})
\end{aligned}$$

Basis Vectors

Let $\mathbf{A} = (7,2)$ and $\mathbf{B} = (-3,4)$. Is it possible to express $\mathbf{C} = (12,-2)$ in terms of **A** and **B**, using only scalar multiplication and vector addition? In other words, can we stretch or contract **A** and **B** and add the resulting vectors to obtain **C**? Figure 12.3 suggests that the answer is yes and that

$$\mathbf{C} = k\mathbf{A} + t\mathbf{B}.$$

Vector **C** is called a **linear combination** of **A** and **B**. To find k and t, we write

$$\begin{aligned}
(12,-2) &= k(7,2) + t(-3,4) \\
12 &= 7k - 3t \\
-2 &= 2k + 4t.
\end{aligned}$$

The solution of the system of equations is $k = \frac{21}{17}$ and $t = -\frac{19}{17}$. Thus, both **A** and **B** are stretched, and the direction of **B** is reversed.

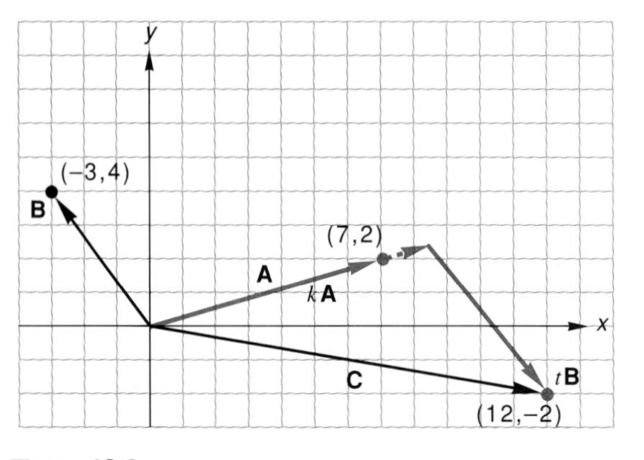

Figure 12.3

We leave it as an exercise to prove that a vector **C** can always be written as a linear combination of any two non-parallel, nonzero vectors **A** and **B.** It follows that *if* **B** *is not a scalar multiple of* **A,** *then* **A** *and* **B** *can generate any other vector in the plane.* As such, **A** and **B** are called **basis vectors** in the plane. One pair of basis vectors that has great utility consists of the unit vectors $\mathbf{i} = (1,0)$ and $\mathbf{j} = (0,1)$.

Example 3 Write $\mathbf{C} = (12,-2)$ as a linear combination of **i** and **j.**

Solution

$$\begin{aligned}(12,-2) &= k(1,0) + t(0,1)\\ 12 &= k\\ -2 &= t\\ (12,-2) &= 12(1,0) - 2(0,1)\end{aligned}$$ ■

Example 3 suggests that

Representing a Vector in the i–j Basis

Each vector (x,y) can be written as a linear combination of basis vectors **i** and **j,** as $x\mathbf{i} + y\mathbf{j}$.

Example 4 Use vector methods to show that a line segment joining the midpoints of two sides of a triangle is parallel to the third side and has a length that is one-half the length of the third side.

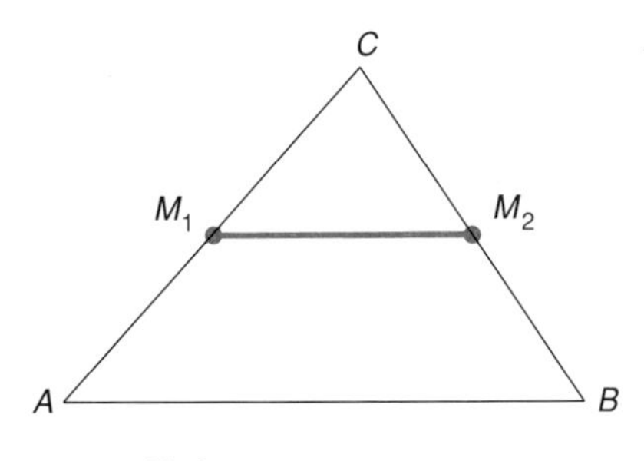

Figure 12.4

Solution In the solution, vectors are denoted as directed line segments (figure 12.4). Because M_1 and M_2 are the respective midpoints of sides AC and BG, we have

$$\overline{M_1M_2} = \overline{M_1C} + \overline{CM_2} = \frac{\overline{AC}}{2} + \frac{\overline{CB}}{2} = \frac{\overline{AC} + \overline{CB}}{2} = \frac{\overline{AB}}{2}.$$

■

Exercise Set 12.1

(A)

Find the x and y components of **A** if θ is the positive angle that **A** makes with the x-axis.

1. $\|\mathbf{A}\| = 6, \theta = 60°$
2. $\|\mathbf{A}\| = 200, \theta = 45°$
3. $\|\mathbf{A}\| = 520, \theta = 75°$
4. $\|\mathbf{A}\| = 75, \theta = 135°$
5. $\|\mathbf{A}\| = 15, \theta = 300°$
6. $\|\mathbf{A}\| = 180, \theta = 210°$

Let **A** be a vector determined by points P and Q. (a) Write **A** as an ordered pair. (See examples 1 and 2 in section 9.4.) (b) Find $\|\mathbf{A}\|$.

7. $P(0,0), Q(2,5)$
8. $P(-2,8), Q(6,4)$
9. $P(1,4), Q(7,12)$
10. $P(3,-4), Q(-2,8)$
11. $P(-5,-3), Q(-13,12)$
12. $P(6,-1), Q(2,8)$

Find x and y given that **A** = **B**. Recall that two vectors are equal if and only if their corresponding components are equal.

13. $\mathbf{A} = (4,8), \mathbf{B} = (2x + y, x - 3y)$
14. $\mathbf{A} = (-7,6), \mathbf{B} = (x + 4y, -2x + y)$
15. $\mathbf{A} = (-5x - 2y, x + y), \mathbf{B} = 12\mathbf{i} - 3\mathbf{j}$
16. $\mathbf{A} = (8x + 3y, 7x - 5y), \mathbf{B} = -10\mathbf{i} + 6\mathbf{j}$

Graph **A**, **B**, and **C** in the same coordinate system.

17. $\mathbf{A} = (3,5), \mathbf{B} = 2\mathbf{A}, \mathbf{C} = -3\mathbf{A}$
18. $\mathbf{A} = (6,4), \mathbf{B} = -\dfrac{\mathbf{A}}{2}, \mathbf{C} = \dfrac{2\mathbf{A}}{3}$
19. $\mathbf{A} = 4\mathbf{i} - 2\mathbf{j}, \mathbf{B} = -2\mathbf{A}, \mathbf{C} = \dfrac{5\mathbf{A}}{4}$
20. $\mathbf{A} = -2\mathbf{i} - 7\mathbf{j}, \mathbf{B} = 3\mathbf{A}, \mathbf{C} = -\dfrac{5\mathbf{A}}{2}$

Find a unit vector having the same direction as **A**.

21. $\mathbf{A} = \mathbf{i} + 6\mathbf{j}$
22. $\mathbf{A} = -3\mathbf{i} + 2\mathbf{j}$
23. $\mathbf{A} = (9,5)$
24. $\mathbf{A} = (-15,-8)$

Given $\mathbf{A} = (2,5)$, $\mathbf{B} = (3,6)$, and $\mathbf{C} = (-8,1)$, find

25. $\mathbf{A} + \mathbf{B}$
26. $\mathbf{B} - \mathbf{A}$
27. $\mathbf{A} + \mathbf{B} + \mathbf{C}$
28. $\mathbf{A} - \mathbf{B} - \mathbf{C}$
29. $2\mathbf{A} - 5\mathbf{C}$
30. $12(\mathbf{A} + \mathbf{C})$
31. $\|\mathbf{A} + \mathbf{B} + \mathbf{C}\|$
32. $-2\mathbf{A} + 3\mathbf{B} - 5\mathbf{C}$

(B)

Given **A**, **B**, and **C**, write **C** as a linear combination of **A** and **B**.

33. $\mathbf{A} = 6\mathbf{i} - 2\mathbf{j}, \mathbf{B} = (5,8), \mathbf{C} = (-2,4)$
34. $\mathbf{A} = (-8,-8), \mathbf{B} = -4\mathbf{i} + 4\mathbf{j}, \mathbf{C} = (5,5)$
35. $\mathbf{A} = (10,4), \mathbf{B} = (2,6), \mathbf{C} = 15\mathbf{i} + 12\mathbf{j}$
36. $\mathbf{A} = (9,2), \mathbf{B} = (-3,12), \mathbf{C} = (-20,-22)$

37. Prove Property 1, the commutative property of vector addition.
38. Prove Property 6, the distributive properties of scalar multiplication.
39. Prove that any vector $\mathbf{C} = (x,y)$ can be written as a linear combination of two non-parallel vectors, $\mathbf{A} = (x_1,y_1)$ and $\mathbf{B} = (x_2,y_2)$.

40. Use vector addition and figure 12.5 to derive the translation equations of section 11.2;

$$x = x' + h, \quad y = y' + k.$$

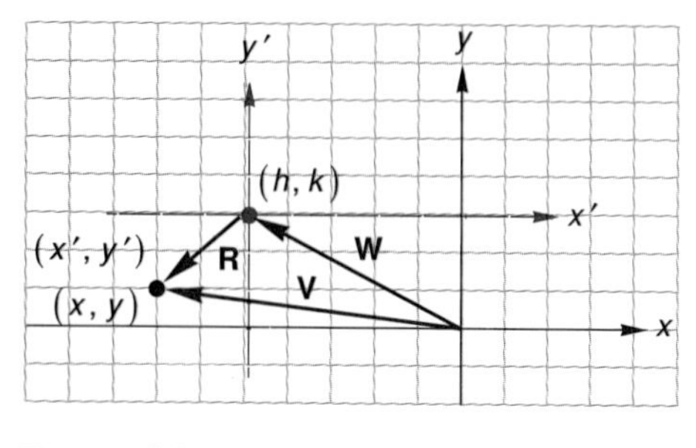

Figure 12.5

41. Show that the vector equation $||\mathbf{P} - \mathbf{C}|| = ||\mathbf{R}||$ produces the coordinate equation of the circle in figure 12.6.

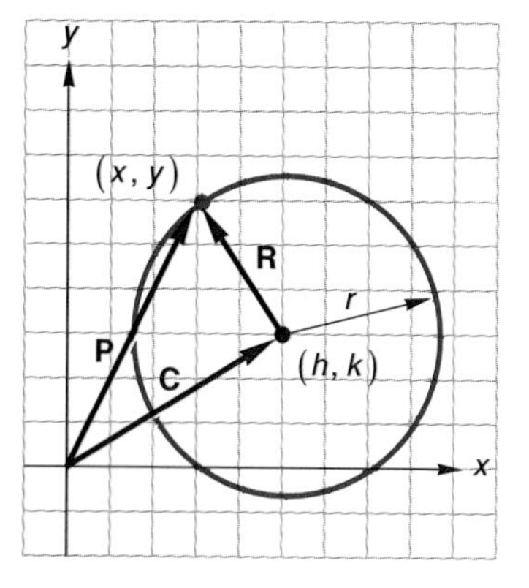

Figure 12.6

42. Write a vector equation for the parabola in figure 12.7. Use vectors **F**, **P**, and **D**. Convert the vector equation into a coordinate equation.

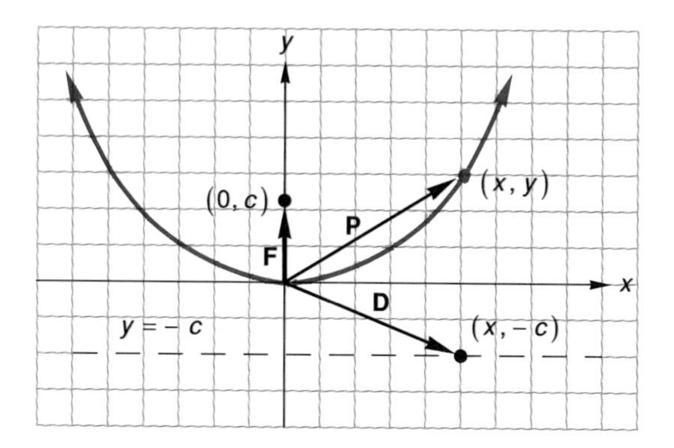

Figure 12.7

43. Show that the vector equation $||\mathbf{P} - \mathbf{F}_1|| + ||\mathbf{P} - \mathbf{F}_2|| = 2a$, for the ellipse in figure 12.8, generates the coordinate equation $x^2/a^2 + y^2/b^2 = 1$.

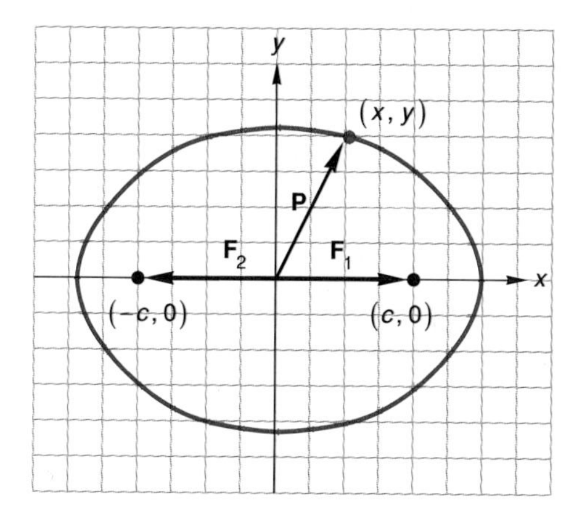

Figure 12.8

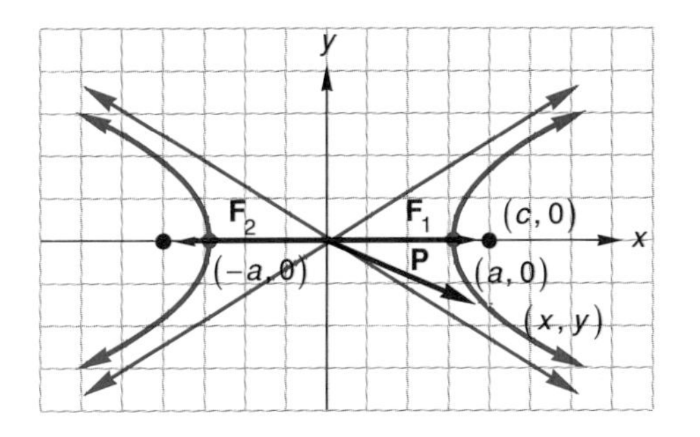

Figure 12.9

44. Show that the vector equation $\|\mathbf{P} - \mathbf{F}_1\| - \|\mathbf{P} - \mathbf{F}_2\| = 2a$, for the hyperbola in figure 12.9, generates the coordinate equation $x^2/a^2 - y^2/b^2 = 1$.
45. Use vector methods to prove that the line joining the midpoints of the sides of a quadrilateral form a parallelogram. (Hint: See example 4 of this section.)

12.2 The Dot Product

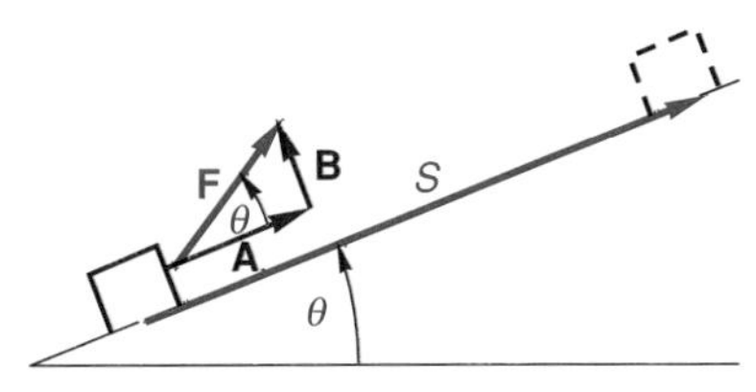

Figure 12.10

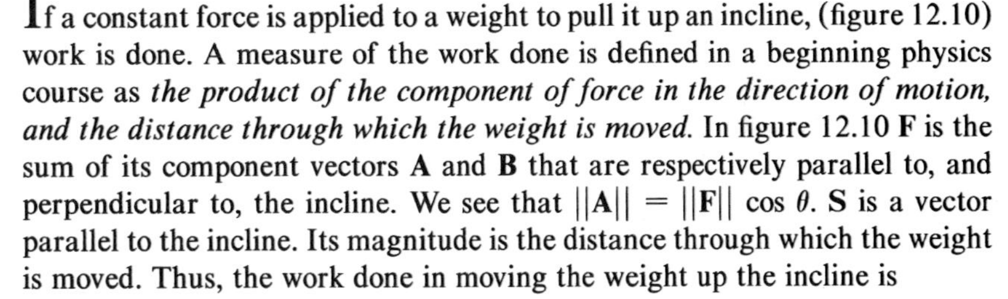

If a constant force is applied to a weight to pull it up an incline, (figure 12.10) work is done. A measure of the work done is defined in a beginning physics course as *the product of the component of force in the direction of motion, and the distance through which the weight is moved.* In figure 12.10 **F** is the sum of its component vectors **A** and **B** that are respectively parallel to, and perpendicular to, the incline. We see that $\|\mathbf{A}\| = \|\mathbf{F}\| \cos \theta$. **S** is a vector parallel to the incline. Its magnitude is the distance through which the weight is moved. Thus, the work done in moving the weight up the incline is

$$\begin{gathered} \text{force} \times \text{distance} \\ \|\mathbf{F}\| \cos \theta \, \|\mathbf{S}\|. \end{gathered} \tag{1}$$

Expression (1) is called the *inner product* or *dot product* between vectors **F** and **S.** We write

Definition

$$\mathbf{F} \cdot \mathbf{S} = \|\mathbf{F}\| \, \|\mathbf{S}\| \cos \theta, \quad 0 \leq \theta \leq \pi. \tag{2}$$

It is worth noting that vector addition and scalar multiplication yield vectors. *The dot product yields a scalar* (a real number).

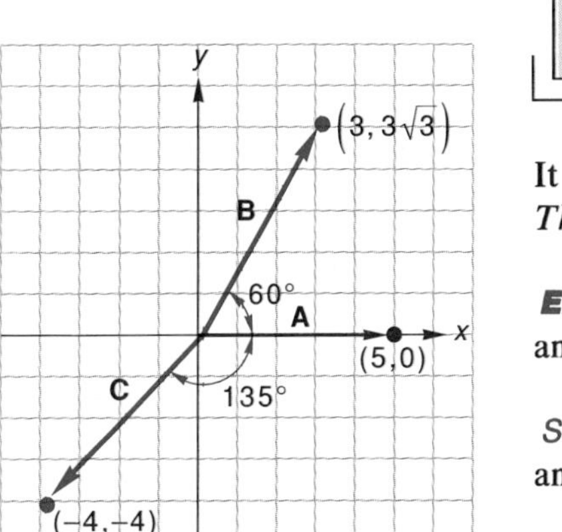

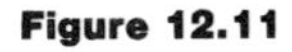

Figure 12.11

Example 1 Find **A** · **B** and **A** · **C**, where $\mathbf{A} = (5,0)$, $\mathbf{B} = (3,3\sqrt{3})$, and $\mathbf{C} = (-4,-4)$.

Solution In figure 12.11, the angles that **B** and **C** make with **A** are 60° and 135°, respectively. Hence,

$$\mathbf{A} \cdot \mathbf{B} = \sqrt{5^2 + 0^2}\,\sqrt{3^2 + (3\sqrt{3})^2} \cos 60° = 15$$
$$\mathbf{A} \cdot \mathbf{C} = \sqrt{5^2 + 0^2}\,\sqrt{(-4)^2 + (-4)^2} \cos 135° = -20.$$

■

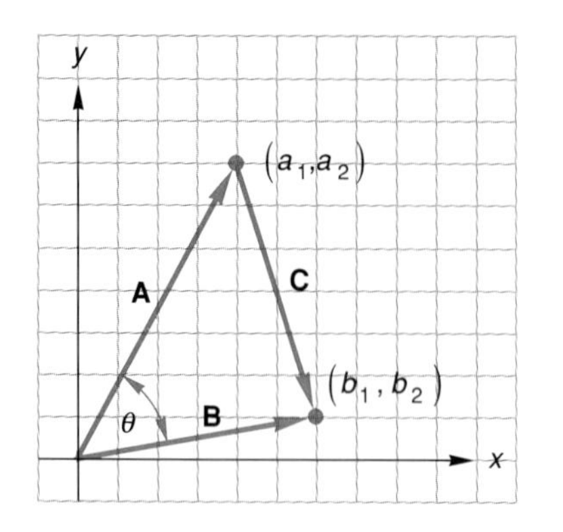

Figure 12.12

It is shown in example 1 that the dot product between two vectors is a positive or negative number, as the angle between the vectors is acute or obtuse. Example 1 also shows us that the dot product is easily obtained from (2) if θ is known. But θ is usually not known and must be computed. Angle θ, between vectors $\mathbf{A} = (a_1,a_2)$ and $\mathbf{B} = (b_1,b_2)$ can be found by applying the law of cosines to the triangle formed by **A**, **B**, and **C** (figure 12.12). Since $\mathbf{A} + \mathbf{C} = \mathbf{B}$, then $\mathbf{C} = \mathbf{B} - \mathbf{A} = (b_1 - a_1, b_2 - a_2)$. Now

$$||\mathbf{C}||^2 = ||\mathbf{A}||^2 + ||\mathbf{B}||^2 - 2||\mathbf{A}|| \, ||\mathbf{B}|| \cos\theta, \tag{3}$$

and equation (3) can be solved for $\cos\theta$.

A useful result is obtained if (3) is solved for the expression $||\mathbf{A}|| \, ||\mathbf{B}|| \cos\theta$, which is defined by (2) as the dot product between **A** and **B**.

$$\begin{aligned}
\mathbf{A} \cdot \mathbf{B} &= ||\mathbf{A}|| \, ||\mathbf{B}|| \cos\theta \\
&= \frac{||\mathbf{A}||^2 + ||\mathbf{B}||^2 - ||\mathbf{C}||^2}{2} \\
&= \frac{a_1^2 + a_2^2 + b_1^2 + b_2^2 - [(b_1 - a_1)^2 + (b_2 - a_2)^2]}{2} \\
&= a_1b_1 + a_2b_2
\end{aligned}$$

We now have a second definition for the dot product.

Definition

If $\mathbf{A} = (a_1,a_2)$ and $\mathbf{B} = (b_1,b_2)$, then

$$\mathbf{A} \cdot \mathbf{B} = a_1b_1 + a_2b_2. \tag{4}$$

Applying (4) to **A**, **B**, and **C** in example 1 gives us

$$\mathbf{A} \cdot \mathbf{B} = (5)(3) + (0)(3\sqrt{3}) = 15$$
$$\mathbf{A} \cdot \mathbf{C} = (5)(-4) + (0)(-4) = -20.$$

The Angle between Two Vectors

Taken together, (2) and (4) enable one to easily find the angle between two vectors. From (2)

$$\mathbf{A} \cdot \mathbf{B} = ||\mathbf{A}|| \, ||\mathbf{B}|| \cos\theta$$

and

The Angle between Two Vectors

$$\cos\theta = \frac{\mathbf{A} \cdot \mathbf{B}}{||\mathbf{A}|| \, ||\mathbf{B}||}. \tag{5}$$

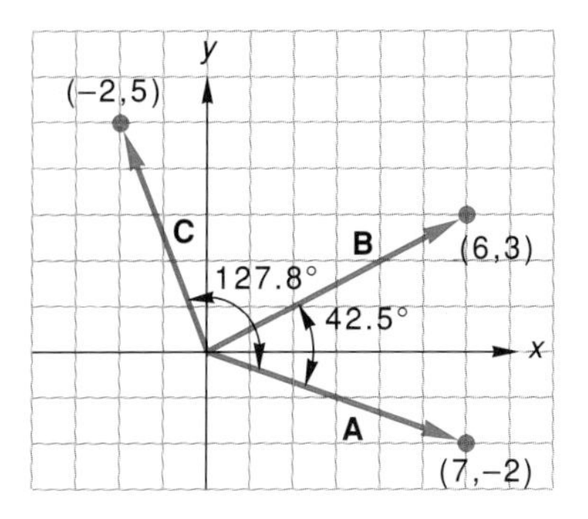

Figure 12.13

The coordinate form of (5) is

$$\cos \theta = \frac{a_1 b_1 + a_2 b_2}{\sqrt{a_1^2 + a_2^2}\,\sqrt{b_1^2 + b_2^2}}. \tag{6}$$

Example 2 Find the angles between **A** and **B,** and **A** and **C,** where **A** = (7,−2), **B** = (6,3), and **C** = (−2,5) (figure 12.13).

Solution Using (6), the angle between **A** and **B** is

$$\cos^{-1} \frac{(7)(6) + (-2)(3)}{\sqrt{53}\,\sqrt{45}} \approx 42.5°.$$

The angle between **A** and **C** is

$$\cos^{-1} \frac{(7)(-2) + (-2)(5)}{\sqrt{53}\,\sqrt{29}} \approx 127.8°.$$

■

Example 3 Find the acute angle between the lines L_1: $y = 3x + 3$ and L_2: $y = -x/2 + 2$.

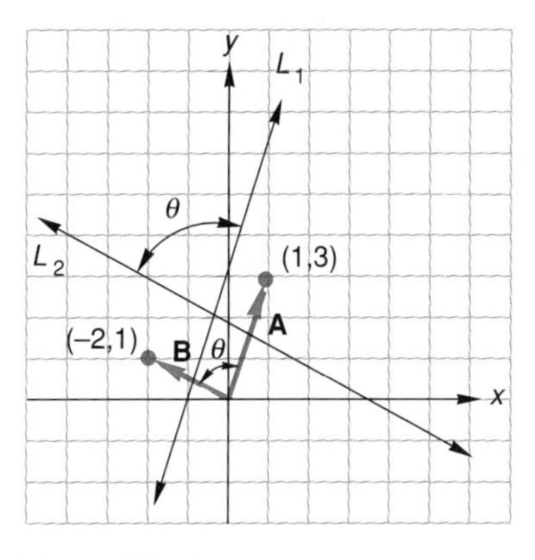

Figure 12.14

Solution We use the slope of L_1, 3/1, to find a vector parallel to L_1 (see figure 12.14). As slope is the change in y divided by the change in x, we let the components of **A** be 1, which is the change in x, and 3, which is the change in y. Thus, **A** = (1,3). Because the slope of L_2 is $-\frac{1}{2}$, **B** can be (−2,1) or (2,−1). We choose (−2,1). Now from (5)

$$\begin{aligned} \theta &= \cos^{-1} \frac{\mathbf{A} \cdot \mathbf{B}}{\|\mathbf{A}\|\,\|\mathbf{B}\|} \\ &= \cos^{-1} \frac{1}{\sqrt{50}} \\ &\approx 81.9°. \end{aligned}$$

Had we let **B** = (2,−1), we would have found θ to be 108.1°. (Try it.) ■

If **A** and **B** are any two vectors and **A** · **B** = 0, then it follows from (2) that either **A** = 0, or **B** = 0, or $\cos \theta = 0$.

Perpendicular Vectors

If neither **A** nor **B** is the zero vector and **A** · **B** = 0, then the vectors are *perpendicular.*

Example 4 Find a non-zero vector **B** perpendicular to **A** = (6,5).

Solution Let **B** = (x,y). Since **A** · **B** = 0, it follows that

$$6x + 5y = 0$$
$$y = -\frac{6x}{5},$$

and that **B** = $(x, -6x/5)$. If $x = 5$, then **B** = (5,−6). If $x = 1$, then **B** = $(-1, \frac{6}{5})$, etc. ■

The magnitude of a vector **A** can be found with the dot product because

$$\mathbf{A} \cdot \mathbf{A} = ||\mathbf{A}||\ ||\mathbf{A}|| \cos 0° = ||\mathbf{A}||^2.$$

Therefore,

Vector Magnitude

$$||\mathbf{A}|| = (\mathbf{A} \cdot \mathbf{A})^{1/2}.$$

Example 5 Use the dot product to find the magnitude of **A** = (−3,8).

Solution $||\mathbf{A}|| = (\mathbf{A} \cdot \mathbf{A})^{1/2} = \sqrt{(-3)(-3) + (8)(8)} = \sqrt{73}$ ■

The Distance from a Point to a Line

Assume that we are given a point $P(x_1,y_1)$ not on line L: $ax + by + c = 0$. Let us find the distance from P to L (see figure 12.15). Construct a vector **V** between (x_1,y_1) and any point (x_0,y_0) on L. $\mathbf{V} = (x_0-x_1, y_0-y_1)$. Also construct a vector **N** that is perpendicular to L. The slope of L is $-a/b$. Therefore, the slope of a line parallel to **N** is b/a, and $\mathbf{N} = (a,b)$. The vector **N** is called a *normal vector* to L. The distance sought is

$$d = ||\mathbf{V}||\ |\cos \theta|,$$

where θ is the angle between **V** and **N**. The absolute value of $\cos \theta$ is used because $\cos \theta$ is negative if θ is obtuse. Now, if we multiply and divide the right member of the distance equation by $||\mathbf{N}||$, we obtain

$$d = \frac{||\mathbf{V}||\ ||\mathbf{N}||\ |\cos \theta|}{||\mathbf{N}||}.$$

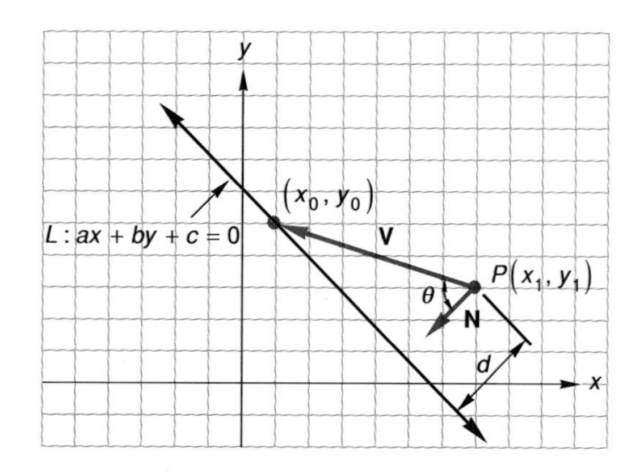

Figure 12.15

The numerator of the right member of the last equation is the dot product between **V** and **N**. Hence,

Distance from a Point to a Line—Vector Form

$$d = \frac{|\mathbf{V} \cdot \mathbf{N}|}{\|\mathbf{N}\|}. \tag{7}$$

Equation (7) is the *vector form* of the formula for the distance from a point to a line. We convert (7) to coordinate form as follows,

$$\begin{aligned} d &= \frac{|(x_0 - x_1)a + (y_0 - y_1)b|}{\sqrt{a^2 + b^2}} \\ &= \frac{|-ax_1 - by_1 + ax_0 + by_0|}{\sqrt{a^2 + b^2}}. \end{aligned}$$

Expression $ax_0 + by_0$ can be replaced by $-c$ because $ax_0 + by_0 + c = 0$. Finally, we have

Distance from a Point to a Line—Coordinate Form

$$d = \frac{|ax_1 + by_1 + c|}{\sqrt{a^2 + b^2}}, \tag{8}$$

the *coordinate form* of (7).

The geometric and trigonometric relationships suggested by figure 12.15 make the derivation of (7) quite intuitive. Contrast this derivation with that developed in exercise 57 of section 3.1, in which a great amount of algebraic manipulation is required to obtain (8).

Example 6 Find the distance from $(4,-7)$ to $y = 2x + 3$.

Solution To use (8) we write the given equation in the form $-2x + y - 3 = 0$.

$$\begin{aligned} d &= \frac{|-2x_1 + y_1 - 3|}{\sqrt{(-2)^2 + (1)^2}} \\ &= \frac{|-2(4) + (-7) - 3|}{\sqrt{5}} \\ &= \frac{18}{\sqrt{5}} \end{aligned}$$

■

Example 7 Find the distance from the origin to $-4x + 3y = 12$.

Solution

$$\begin{aligned} d &= \frac{|-4 \cdot 0 + 3 \cdot 0 - 12|}{\sqrt{(-4)^2 + (3)^2}} \\ &= \frac{12}{5} \end{aligned}$$

■

Properties of the Dot Product

We end this section with a list of properties of the dot product and an example in which some of the properties are used in the construction of a proof.

If **A**, **B**, and **C** are vectors and k is a scalar, then

1. $\mathbf{A} \cdot \mathbf{B} = \mathbf{B} \cdot \mathbf{A}$ (commutativity)
2. $\mathbf{A} \cdot (\mathbf{B} + \mathbf{C}) = \mathbf{A} \cdot \mathbf{B} + \mathbf{A} \cdot \mathbf{C}$ (distributive property)
3. $k(\mathbf{A} \cdot \mathbf{B}) = (k\mathbf{A}) \cdot \mathbf{B} = \mathbf{A} \cdot (k\mathbf{B})$ (associativity)
4. $\mathbf{0} \cdot \mathbf{A} = 0$
5. $\mathbf{A} \cdot \mathbf{A} = ||\mathbf{A}||^2$
6. $|\mathbf{A} \cdot \mathbf{B}| \le ||\mathbf{A}||\,||\mathbf{B}||$ (Cauchy-Schwartz Inequality)
7. $||\mathbf{A} + \mathbf{B}|| \le ||\mathbf{A}|| + ||\mathbf{B}||$ (Triangle Inequality)

Proof of Property 2: Let $\mathbf{A} = (a_1,a_2)$, $\mathbf{B} = (b_1,b_2)$, and $\mathbf{C} = (c_1,c_2)$.

$$\begin{aligned} \mathbf{A} \cdot (\mathbf{B} + \mathbf{C}) &= (a_1,a_2) \cdot (b_1 + c_1, b_2 + c_2) \\ &= a_1(b_1 + c_1) + a_2(b_2 + c_2) \\ &= (a_1b_1 + a_2b_2) + (a_1c_1 + a_2c_2) \\ &= \mathbf{A} \cdot \mathbf{B} + \mathbf{A} \cdot \mathbf{C} \end{aligned}$$

Proof of Property 6: The definition of the dot product permits us to write

$$\frac{|\mathbf{A} \cdot \mathbf{B}|}{||\mathbf{A}|| \ ||\mathbf{B}||} = |\cos \theta| \leq 1.$$

Therefore, $|\mathbf{A} \cdot \mathbf{B}| \leq ||\mathbf{A}|| \ ||\mathbf{B}||$.

You are asked to prove some of the other properties in the exercises.

Example 8 Use vector methods to prove that the altitudes of a triangle meet in a point.

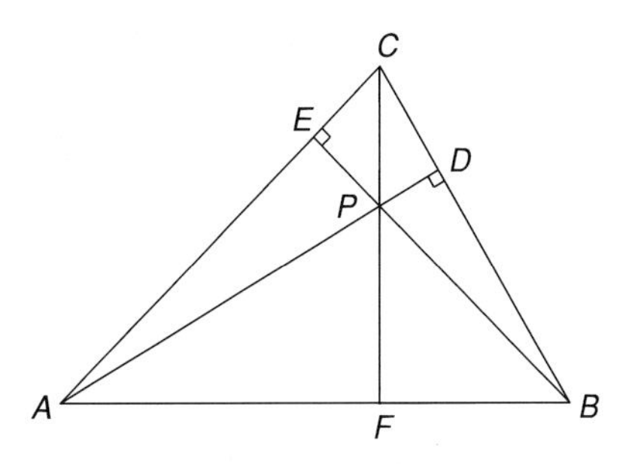

Figure 12.16

Solution Directed line segments are viewed as vectors in what follows. Altitudes $\overline{AD}$ and $\overline{BE}$ of triangle ABC meet at P (figure 12.16). Extend segment $\overline{CP}$ to F. We want to show that $\overline{CF}$ is perpendicular to $\overline{AB}$. We can do this by showing that the dot product of $\overline{CF}$ and $\overline{AB}$ is zero.

$$\begin{aligned}
\overline{CP} \cdot \overline{AB} &= \overline{CP} \cdot (\overline{AC} + \overline{CB}) \\
&= \overline{CP} \cdot \overline{AC} + \overline{CP} \cdot \overline{CB} \\
&= (\overline{CB} + \overline{BP}) \cdot \overline{AC} + (\overline{CA} + \overline{AP}) \cdot \overline{CB} \\
&= \overline{CB} \cdot \overline{AC} + \overline{BP} \cdot \overline{AC} + \overline{CA} \cdot \overline{CB} + \overline{AP} \cdot \overline{CB} \\
&= \overline{CB} \cdot (\overline{AC} + (-\overline{AC})) + \overline{BP} \cdot \overline{AC} + \overline{AP} \cdot \overline{CB} \\
&= 0 + \overline{BP} \cdot \overline{AC} + \overline{AP} \cdot \overline{CB}.
\end{aligned}$$

But $\overline{BP} \cdot \overline{AC} = 0$ and $\overline{AP} \cdot \overline{CB} = 0$ because AD and BE are altitudes. Hence, $\overline{CP} \cdot \overline{AB} = 0$. As $\overline{CP}$ is a scalar multiple of $\overline{CF}$, we can write

$$0 = \overline{CP} \cdot \overline{AB} = k\overline{CF} \cdot \overline{AB},$$

or

$$0 = k(\overline{CF} \cdot \overline{AB}).$$

Since $k \neq 0$, it follows that $\overline{CF}$ is perpendicular to $\overline{AB}$. ■

Exercise Set 12.2

(A)

Find $\mathbf{A} \cdot \mathbf{B}$

1. $||\mathbf{A}|| = 5$, $||\mathbf{B}|| = 8$, $\theta = 30°$

2. $||\mathbf{A}|| = 12$, $||\mathbf{B}|| = 6$, $\theta = 120°$

3. $\mathbf{A} = (8,-2)$, $\mathbf{B} = (7,3)$

4. $\mathbf{A} = (-10,-6)$, $\mathbf{B} = (-4,11)$

Find a unit vector parallel to the given line.

5. $x + 2y = 6$

6. $3x - y = 12$

7. $y = \frac{1}{2}x + 4$

8. $y = -5x - 2$

9. $4x + 9y - 1 = 0$

10. $5x - 12y + 15 = 0$

Find the acute angle between the given lines.

11. $y = 2x - 5$, $y = 5x + 2$

12. $y = \frac{1}{3}x$, $y = -2x$

13. $8x - 3y + 12 = 0$, $2x + 5y + 4 = 0$

14. $x + 6y - 5 = 0$, $10x + 4y + 9 = 0$

15. $5x + y = 5$, $3x - 7y = 6$

16. $3x - y = 9$, $2x - 3y = 12$

Find a unit vector perpendicular to $\mathbf{A}$.

17. $\mathbf{A} = (6,2)$

18. $\mathbf{A} = (-5,8)$

19. $\mathbf{A} = -3\mathbf{i} - 10\mathbf{j}$

20. $\mathbf{A} = 4\mathbf{i} - 15\mathbf{j}$

21. $\mathbf{A} = (10,10)$

22. $\mathbf{A} = (-\sqrt{3},1)$

Find the distance from point P to the given line.

23. $x + 2y = 6$, $P(9,5)$

24. $3x - y = 12$, $P(1,0)$

25. $8x - 3y + 12 = 0$, $P(4,-7)$

26. $10x + 4y + 9 = 0$, $P(-2,6)$

27. $y = \frac{1}{3}x$, $P(0,10)$

28. $y = -2x + 5$, $P(-3,-4)$

29. $3x - y = 9$, $P(0,0)$

30. $x + 6y - 5 = 0$, $P(0,0)$

Review equations (1) and (2) of this section before doing exercises 31–34. Assume that forces are measured in pounds and distances in feet.

31. A force $\mathbf{F} = 6\mathbf{i} + 3\mathbf{j}$ moves an object from $P(0,0)$ to $Q(10,0)$. Find the work done.

32. A force $\mathbf{F} = 2\mathbf{i} + 10\mathbf{j}$ moves an object from $P(0,0)$ to $Q(0,12)$. Find the work done.

33. Forces $\mathbf{F}_1 = (5,8)$ and $\mathbf{F}_2 = (6,4)$ acting together move an object from $P(3,1)$ to $Q(15,6)$. Find the work done.

34. Forces $\mathbf{F}_1 = (2,-8)$ and $\mathbf{F}_2 = (-3,-10)$ acting together move an object from $P(1,1)$ to $Q(-2,-18)$. Find the work done.

(B)

Find the distance between each pair of parallel lines.

35. $x + 2y = 6$, $x + 2y = -5$

36. $y = -3x + 8$, $y = -3x + 1$

37. Find the distance from $P(4,-7)$ to the line $8x - 3y + 12 = 0$ without using (8).

38. Write a formula for the distance from the origin to a line $ax + by + c = 0$. (See example 7.)

39. Write equations for the lines that are three units from the line $4x - 3y + 6 = 0$.

40. Write an equation for the line that is equidistant from lines $5x + 12y - 2 = 0$ and $5x + 12y + 14 = 0$.

41. Find the angle bisectors of the angles formed by the lines $4x - 3y + 6 = 0$ and $5x + 12y - 2 = 0$. (Hint: The points of an angle bisector are equidistant from the sides of the angle.)

42. Find the angle bisectors of the angles formed by the lines $5x + 12y + 14 = 0$ and $15x - 8y - 10 = 0$.

43. Find the area of the triangle determined by $P(-4,2)$, $Q(5,-1)$, and $R(3,8)$.

44. Show that the area of a triangle determined by $P(x_1,y_1)$, $Q(x_2,y_2)$, and $R(x_3,y_3)$ is

$$\tfrac{1}{2}|x_1(y_2 - y_3) + x_2(y_3 - y_1) + x_3(y_1 - y_2)|.$$

45. Show that $\mathbf{A} \cdot \mathbf{B} = \mathbf{A} \cdot \mathbf{C}$, $\mathbf{A} \neq 0$, does not imply $\mathbf{B} = \mathbf{C}$.

46. Show that $(\mathbf{A} + \mathbf{B}) \cdot (\mathbf{A} + \mathbf{B}) = ||\mathbf{A}||^2 + 2\mathbf{A} \cdot \mathbf{B} + ||\mathbf{B}||^2$.

47. Prove the triangle inequality

$$||\mathbf{A} + \mathbf{B}|| \leq ||\mathbf{A}|| + ||\mathbf{B}||.$$

(Hint: Use the result of exercise 46 and the Cauchy-Schwartz inequality on page 482).

48. Show that if $\mathbf{A} - \mathbf{B}$ is perpendicular to $\mathbf{A} + \mathbf{B}$, then $||\mathbf{A}|| = ||\mathbf{B}||$.

Use vector methods to prove the following statements. (See example 8.)

49. The median to the base of an isosceles triangle is perpendicular to the base.

50. The diagonals of a rhombus are perpendicular.

51. The perpendicular bisectors of the sides of a triangle meet in a point.

52. An angle inscribed in a semicircle is a right angle.

12.3 Parametric Equations of a Line

In figure 12.17, L: $ax + by + c = 0$ has slope $m = -a/b$. $\mathbf{V} = (b,-a)$ is a vector parallel to L. $\mathbf{R} = (x_0,y_0)$ is a vector from the origin to a known point (x_0,y_0) on L. $\mathbf{P} = (x,y)$ is a vector from the origin to a point (x,y) on L. The relationship

Vector Parametric Equation of a Line

$$\mathbf{P} = \mathbf{R} + t\mathbf{V}, \qquad (1)$$

in which t is a scalar, is called the **vector parametric equation** of L. Any point (x,y) on L can be generated by adding a scalar multiple of $\mathbf{V}$ to $\mathbf{R}$. Points on L to the right of (x_0,y_0) are generated by (1) if $t > 0$. If $t < 0$, then (1) produces points to the left of (x_0,y_0). $\mathbf{P}$ is called the **position vector,** and t is called a **parameter.**

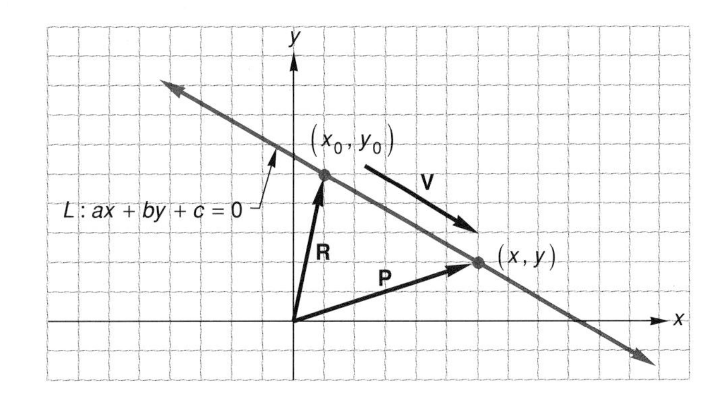

Figure 12.17

The coordinate form of (1),

$$(x,y) = (x_0,y_0) + t(b,-a),$$

yields a pair of **coordinate parametric equations** of L.

Coordinate Parametric Equations of a Line

$$x = x_0 + tb$$
$$y = y_0 - ta \qquad (2)$$

If t is eliminated from (2), the point-slope equation of L is obtained.

$$\frac{y - y_0}{-a} = t = \frac{x - x_0}{b}$$

$$y - y_0 = -\frac{a}{b}(x - x_0)$$

Example 1 Write a vector parametric equation for the line L: $y = -2x + 8$. Also determine a pair of coordinate parametric equations for L.

Solution $\mathbf{V} = (1,-2)$ is a vector parallel to L because the line has slope -2. As $(0,8)$ is a point on L, we let $\mathbf{R} = (0,8)$. Substituting for $\mathbf{V}$ and $\mathbf{R}$ in (1) produces the vector parametric equation

$$(x,y) = (0,8) + t(1,-2).$$

Equating corresponding components in the vector parametric equation gives us the following pair of coordinate parametric equations,

$$x = 0 + t \quad \text{and} \quad y = 8 - 2t. \qquad ■$$

Example 2 Write a pair of coordinate parametric equations for the line determined by the points $(1,-3)$ and $(5,2)$.

Solution A vector $\mathbf{V} = (5 - 1, 2 - (-3)) = (4,5)$, determined by the two points, is parallel to the line (figure 12.18). Thus,

$$\mathbf{P} = \mathbf{R} + t\mathbf{V},$$

or

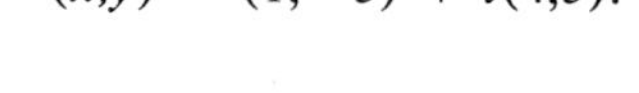

$$(x,y) = (1,-3) + t(4,5).$$

It follows that

$$x = 1 + 4t \quad \text{and} \quad y = -3 + 5t. \qquad ■$$

Figure 12.18

Example 3 Find points on the line $y = x + 1$ that are six units from (0,1).

Solution

Method 1 (without vectors): The distance between a point (x,y) and (0,1) is $\sqrt{(x - 0)^2 + (y - 1)^2}$. If (x,y) is a point on the line that is six units from (0,1), we can write

$$\begin{aligned} \sqrt{x^2 + ((x + 1) - 1)^2} &= 6 \\ 2x^2 &= 36 \\ x &= \pm 3\sqrt{2}. \end{aligned}$$

Points on the line that are six units from (0,1) are $(3\sqrt{2}, 1 + 3\sqrt{2})$ and $(-3\sqrt{2}, 1 - 3\sqrt{2})$.

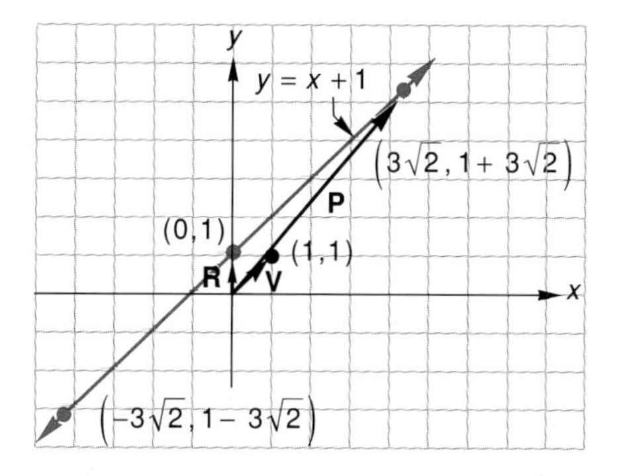

Figure 12.19

Method 2 (with vectors): In vector notation, the distance between (x,y) and (0,1) (figure 12.19) is

$$||\mathbf{P} - \mathbf{R}|| = ||t\mathbf{V}|| = |t|\ ||\mathbf{V}||.$$

If **V** is made a unit vector, then the distance is simply $|t|$. A unit vector in the direction of **V** is $(1/\sqrt{2}, 1/\sqrt{2})$. (Why?) Therefore, points on the line that are six units from (0,1) are

$$(x,y) = (0,1) + 6\left(\frac{1}{\sqrt{2}}, \frac{1}{\sqrt{2}}\right) = (3\sqrt{2}, 1 + 3\sqrt{2})$$

and $$(x,y) = (0,1) - 6\left(\frac{1}{\sqrt{2}}, \frac{1}{\sqrt{2}}\right) = (-3\sqrt{2}, 1 - 3\sqrt{2}).$$ ■

Example 4 Use vector methods to find the point of intersection between the lines L_1: $x - y = 11$ and L_2: $2x + 3y = 12$.

Solution Equations

$$(x,y) = (11,0) + t_1(1,1) \quad \text{and} \quad (x,y) = (0,4) + t_2(-3,2)$$

are vector parametric equations of lines L_1 and L_2. (Why?) Since the coordinates of the point of intersection must satisfy both equations, we can write

$$(11,0) + t_1(1,1) = (0,4) + t_2(-3,2).$$

Parameter t_1 can be found by forming the dot product of both sides of the last equation with (2,3). Because vector (2,3) is perpendicular to $(-3,2)$, t_2 will be eliminated. The dot product of perpendicular vectors is 0.

$$\begin{aligned}(2,3) \cdot [(11,0) + t_1(1,1)] &= (2,3) \cdot [(0,4) + t_2(-3,2)]\\ 22 + 5t_1 &= 12\\ t_1 &= -2\end{aligned}$$

The point of intersection is $(x,y) = (11,0) + (-2)(1,1) = (9,-2)$. ■

Subdivision of a Line Segment

A point (x,y) on the line segment determined by the points (x_1,y_1) and (x_2,y_2) can be located by vector **P** (figure 12.20), where

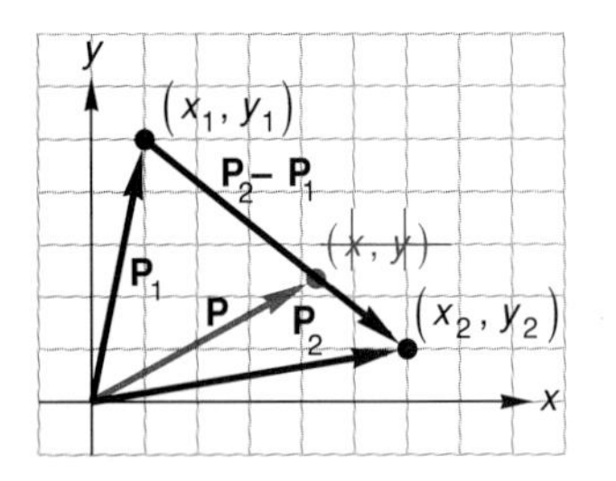

Figure 12.20

Subdivision Point—Vector Form

$$\mathbf{P} = \mathbf{P}_1 + t(\mathbf{P}_2 - \mathbf{P}_1), \quad 0 \le t \le 1. \tag{3}$$

Vector $\mathbf{P}_2 - \mathbf{P}_1$ is parallel to the line segment. Its magnitude is the length of the line segment. Thus, a point on the segment that is $\frac{1}{3}$ the distance from (x_1,y_1) to (x_2,y_2) is located by **P** as

$$\begin{aligned}\mathbf{P} &= (x_1,y_1) + \frac{1}{3}(x_2 - x_1, y_2 - y_1)\\ &= \left(x_1 + \frac{1}{3}(x_2 - x_1), y_1 + \frac{1}{3}(y_2 - y_1)\right)\\ &= \left(\frac{2x_1}{3} + \frac{x_2}{3}, \frac{2y_1}{3} + \frac{y_2}{3}\right).\end{aligned}$$

Hence, $x = 2x_1/3 + x_2/3$ and $y = 2y_1/3 + y_2/3$.

If $t = 0$, then (3) locates the point (x_1,y_1). If $t = 1$, (3) locates (x_2,y_2). A coordinate parametric form of (3) is found as follows.

$$\begin{aligned}(x,y) &= (x_1,y_1) + t(x_2 - x_1, y_2 - y_1)\\ &= (x_1 + t(x_2 - x_1), y_1 + t(y_2 - y_1))\\ &= [(1 - t)x_1 + tx_2, (1 - t)y_1 + ty_2]\end{aligned}$$

Thus,

Subdivision Point—Coordinate Form

$$x = (1 - t)x_1 + tx_2 \quad \text{and} \quad y = (1 - t)y_1 + ty_2, \quad 0 \le t \le 1. \qquad \textbf{(4)}$$

Example 5 A line segment is determined by points $(-5,2)$ and $(3,8)$. Locate

a. the midpoint of the segment,

b. the point that is $\frac{3}{4}$ the distance from $(-5,2)$ towards $(3,8)$.

Solution

$$\textbf{a. } x = \left(1 - \frac{1}{2}\right)(-5) + \frac{1}{2}(3) = -1$$

$$y = \left(1 - \frac{1}{2}\right)(2) + \frac{1}{2}(8) = 5$$

$$\textbf{b. } x = \left(1 - \frac{3}{4}\right)(-5) + \frac{3}{4}(3) = 1$$

$$y = \left(1 - \frac{3}{4}\right)(2) + \frac{3}{4}(8) = \frac{13}{2}$$

■

Example 6 Prove that the medians of a triangle intersect in a point that is $\frac{2}{3}$ the distance along a median from its vertex.

Solution In the proof we shall view vectors as directed line segments. We shall also make repeated use of (3) and an equivalent form of (3). Both vector equations are written here for easy reference.

$$\mathbf{P} = \mathbf{P}_1 + t(\mathbf{P}_2 - \mathbf{P}_1), \quad \mathbf{P} = (1 - t)\mathbf{P}_1 + t\mathbf{P}_2, \quad 0 \le t \le 1 \qquad (3)$$

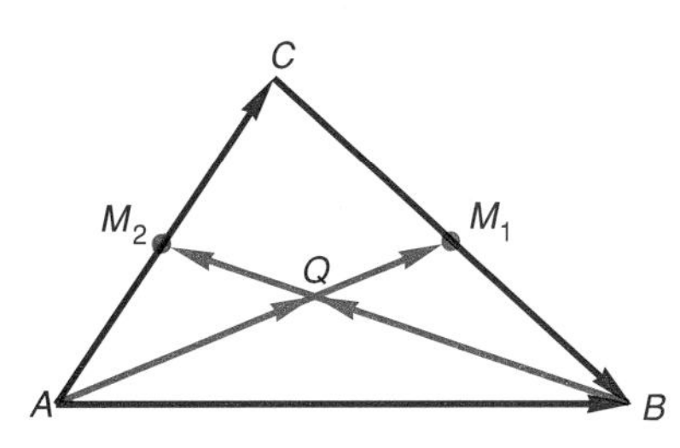

Figure 12.21

In figure 12.21, $\overline{AM_1}$ and $\overline{BM_2}$ are medians of triangle ABC. With $\overline{AC}$ as $\mathbf{P}_1$, $\overline{AB}$ as $\mathbf{P}_2$, $\overline{CB}$ as $\mathbf{P}_2 - \mathbf{P}_1$, and $t = \frac{1}{2}$, we write

$$\begin{aligned}\overline{AM_1} &= \frac{\overline{AC}}{2} + \frac{\overline{AB}}{2}\\ &= \overline{AM_2} + \frac{\overline{AB}}{2}.\end{aligned}$$

Since $\overline{AQ}$ is a scalar multiple of $\overline{AM_1}$, we have

$$\begin{aligned}\overline{AQ} &= k\,\overline{AM_1}\\ &= k\left(\overline{AM_2} + \frac{\overline{AB}}{2}\right)\\ &= k\,\overline{AM_2} + \frac{k}{2}\,\overline{AB}.\end{aligned}$$

The last result has the second form of (3) with $\overline{AQ}$ as $\mathbf{P}$, $\overline{AM_2}$ as $\mathbf{P}_1$, $\overline{AB}$ as $\mathbf{P}_2$, k as $1 - t$, and $k/2$ as t. In the second form of (3), the sum of the coefficients of $\mathbf{P}_1$ and $\mathbf{P}_2$ is 1. Hence,

$$\begin{aligned}k + \frac{k}{2} &= 1\\ k &= \frac{2}{3}.\end{aligned}$$

Therefore $\overline{AQ} = \frac{2}{3}\,\overline{AM_1}$.

We repeat the above argument to show that $\overline{BQ} = \frac{2}{3}\overline{BM_2}$.

$$\begin{aligned}\overline{BM_2} &= -\frac{\overline{AB}}{2} - \frac{\overline{CB}}{2}\\ &= \frac{\overline{BA}}{2} + \frac{\overline{BC}}{2}\\ &= \frac{\overline{BA}}{2} + \overline{BM_1}\\ \overline{BQ} &= k\overline{BM_2}\\ &= \frac{k}{2}\,\overline{BA} + k\overline{BM_1}\end{aligned}$$

Since $k/2 + k = 1$, then $k = \frac{2}{3}$, and $\overline{BQ} = \frac{2}{3}\,\overline{BM_2}$.

It is left to you in exercise 44 of this section to show that median $\overline{CM_3}$ passes through Q, and that $\overline{CQ} = \frac{2}{3}\,\overline{CM_3}$. ■

Exercise Set 12.3

(A)

Write (a) a vector parametric equation and (b) a pair of coordinate parametric equations for the following.

1. $y = 5x + 2$

2. $y = -\frac{3}{7}x + 8$

3. $2x + 6y = -15$

4. $11x - 4y = 11$

5. the line containing $(-6,1)$ and $(2,5)$

6. the line containing $(8,-3)$ and $(-4,-9)$

7. the line containing $(0,0)$ and $(7,10)$

8. the line containing $(-1,-3)$ and $(-12,6)$

Eliminate the parameter from the given pair of coordinate parametric equations and obtain a linear equation in point-slope form.

9. $x = 2 - 3t$
$y = 1 + t$

10. $x = -5 - 8t$
$y = -3 + 4t$

11. $x = 3t$
$y = 2 - 6t$

12. $x = 5t - 2$
$y = 6 - 2t$

Find points on line l that are the indicated number of units from point P on l.

13. l: $(x,y) = (2,-4) + t(3,1)$, 6 units, $P(5,-3)$

14. l: $(x,y) = (-5,5) + t(4,-6)$, 7 units, $P(3,-7)$

15. l: $10x + 3y = 12$, 2 units, $P(0,4)$

16. l: $\begin{cases} x = -3 + 2t \\ y = 5 - 5t \end{cases}$, 5 units, $P(-3,5)$

Use vector methods to find the point of intersection of lines l_1 and l_2.

17. l_1: $(x,y) = (2,-4) + t(3,1)$,
l_2: $(x,y) = (-5,5) + t(4,-6)$

18. l_1: $(x,y) = (3,8) + t(-2,2)$,
l_2: $(x,y) = (6,-1) + t(5,2)$

19. l_1: $(x,y) = (-9,3) + t(1,4)$,
l_2: $(x,y) = (0,0) + t(3,6)$

20. l_1: $(x,y) = (7,0) + t(2,-1)$,
l_2: $(x,y) = (0,-3) + t(3,2)$

Find the point on segment $\overline{PQ}$ that is the indicated distance from P to Q.

21. $P(8,1)$, $Q(2,10)$, $\frac{1}{3}$ **22.** $P(-5,2)$, $Q(7,5)$, $\frac{2}{3}$

23. $P(-3,-11)$, $Q(12,-2)$, $\frac{3}{4}$

24. $P(7,-4)$, $Q(2,-12)$, $\frac{5}{8}$

Given points P and Q, find a point R that divides segment $\overline{PQ}$ in the indicated ratio. Let $\overline{PR}$ be the shorter segment.

25. $P(8,1)$, $Q(2,10)$, 2:3 **26.** $P(-5,2)$, $Q(7,5)$, 1:3

27. $P(-3,-11)$, $Q(12,-2)$, 3:4

28. $P(7,-4)$, $Q(2,-12)$, 2:5

29. Write a vector parametric equation of the line containing the median from vertex A of triangle $A(1,3)$, $B(9,-2)$, $C(5,8)$.

30. Write a vector parametric equation of the line containing the altitude through A of triangle ABC of exercise 29.

31. Write a vector parametric equation of the line containing $(1,2)$ that is perpendicular to line $(x,y) = (4,-3) + t(3,5)$.

32. Write a vector parametric equation of the line that is perpendicular to the segment determined by $P(-10,2)$ and $Q(2,18)$ and contains the quarter point of $\overline{PQ}$ closest to Q.

(B)

33. The line $x - 2y - 3 = 0$ is tangent to a circle that has center $(2,-3)$. Find the point of tangency.

34. The line $3x - 4y + 22 = 0$ is tangent to a circle that has center $(-5,8)$. Find the point of tangency.

Segment subdivision formula (3) on page 488 (see figure 12.20) can be used to determine if points $P_1(x_1,y_1)$, $P(x,y)$, and $P_2(x_2,y_2)$ are collinear. Use (3) to determine if the following sets of points are collinear.

35. $P_1(0,6)$, $P(1,4)$, $P_2(3,0)$

36. $A(-2,20)$, $B(2,4)$, $C(1,8)$

37. $P(3,6)$, $Q(5,4)$, $R(8,-1)$

38. $S(12,-1)$, $T(8,3)$, $R(-4,6)$

39. Let $\overline{AB}$ be a line segment in the plane. Let P be any point in the plane and let M be the midpoint of $\overline{AB}$. Prove that $\overline{PA} + \overline{PB} = 2\overline{PM}$.

40. M is a midpoint of side $\overline{AB}$ of the parallelogram of figure 12.22. Show that P is a trisection point of diagonal $\overline{AC}$. (Hint: follow the method of proof used in example 6).

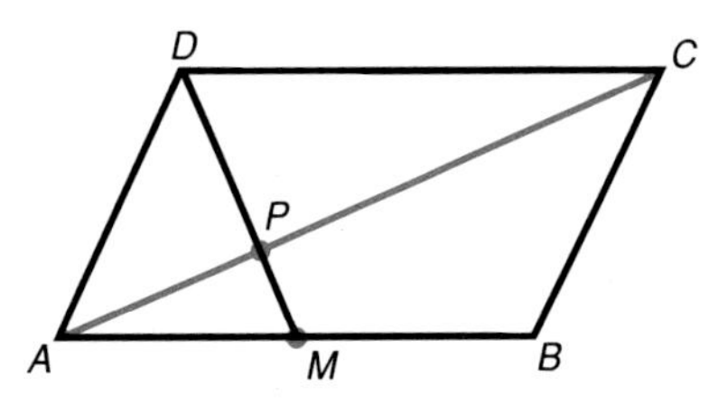

Figure 12.22

41. Prove that the diagonals of a parallelogram bisect each other. (Hint: Use the same procedure as in exercise 40.)

42. Prove that the median to $\overline{AB}$ (see figure 12.21) passes through Q. Call the median $\overline{CM_3}$ and show that $\overline{CQ} = \frac{2}{3}\,\overline{CM_3}$. (Hint: Repeat the argument in example 6 to show that $\overline{CM_3}$ and $\overline{AM_1}$ intersect in a point P so that $\overline{AP} = \frac{2}{3}\,\overline{AM_1}$ and $\overline{CP} = \frac{2}{3}\,\overline{CM_3}$. It is shown in example 6 that $\overline{AQ} = \frac{2}{3}\,\overline{AM_1}$. Thus P and Q are the same point.)

12.4 Lines and Planes in Three Dimensions

One of the virtues of vector notation is its compactness. Formulas expressed in vector notation usually have a simpler form than the coordinate form of the same formula. The distance from a point to a line is given in vector notation (section 12.2) as

$$d = \mathbf{V} \cdot \frac{\mathbf{N}}{||\mathbf{N}||}.$$

The coordinate form of the same formula is

$$d = \frac{|ax_1 + by_1 + c|}{\sqrt{a_2 + b_2}}.$$

In section 12.3, the vector parametric equation of a line,

$$\mathbf{P} = \mathbf{R} + t\mathbf{V},$$

yields a pair of coordinate parametric equations,

$$x = x_0 + tb$$
$$y = y_0 - ta.$$

The compactness of vector notation will again be displayed in this section where vector forms of the equations of a line and a plane in three dimensional space are generated.

A Three-Dimensional Coordinate System

There are three coordinate axes in 3-space, as shown in figure 12.23. The solid line portions of the axes point in the positive x, y, and z directions. Whereas the coordinate axes in the plane (2-space) divide the plane into four quadrants, the coordinate axes in 3-space divide the space into eight **octants.** Taken together, pairs of coordinate axes determine the **coordinate planes.** Thus, the x and z axes determine the x-z coordinate plane, and the y and z axes determine the y-z coordinate plane, etc. The first octant is bounded by the portions of the coordinate planes containing only the positive coordinate axes. The second octant is bounded by the portions of the coordinate planes that contain the positive y and z axes and the negative x-axis. The fifth octant is bounded by the portions of the coordinate planes containing the positive x and y axes and the negative z-axis. You can determine the boundaries of the other five octants.

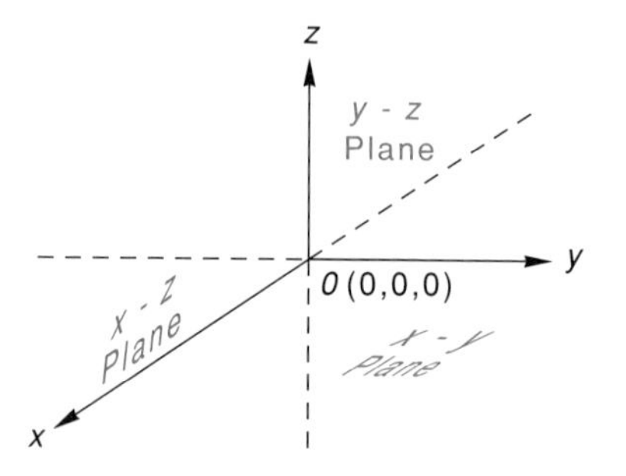

Figure 12.23

Each point in 3-space is uniquely described by an ordered triple of real numbers called its rectangular coordinates. The origin is the ordered triple (0,0,0). Any point in the x-y plane can be located with an ordered triple of the form $(x,y,0)$. Points in the y-z and x-z planes are described as $(0,y,z)$ and $(x, 0, z)$, respectively.

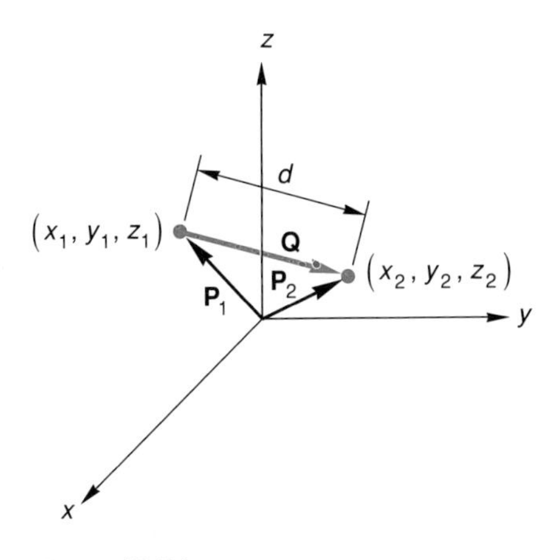

Figure 12.24

Vectors in 3-Space

As a two-dimensional vector is represented by its x and y components in the form of an ordered pair (x,y), so a three-dimensional vector is represented by its x, y, and z components in the form of an ordered triple (x,y,z). All the properties of vector addition and scalar multiplication given in section 12.1 apply to three-dimensional vectors, as do the dot product properties defined in section 12.2. We can illustrate the use of some of these properties in three dimensions by attempting to find the length, or magnitude, of vector **Q** in figure 12.24. It is clear from the figure that $\mathbf{P}_1 + \mathbf{Q} = \mathbf{P}_2$. Hence

$$\begin{aligned}
\mathbf{Q} &= \mathbf{P}_2 - \mathbf{P}_1 \\
&= (x_2,y_2,z_2) - (x_1,y_1,z_1) \\
&= (x_2 - x_1,\, y_2 - y_1,\, z_2 - z_1).
\end{aligned}$$

From section 12.2 we know that

$$\begin{aligned}
||\mathbf{Q}|| &= (\mathbf{Q} \cdot \mathbf{Q})^{1/2} \\
&= [(x_2 - x_1,\, y_2 - y_1,\, z_2 - z_1) \cdot (x_2 - x_1,\, y_2 - y_1,\, z_2 - z_1)]^{1/2}
\end{aligned}$$

Magnitude of Vector in 3-Space

$$||\mathbf{Q}|| = d = \sqrt{(x_2 - x_1)^2 + (y_2 - y_1)^2 + (z_2 - z_1)^2}. \tag{1}$$

Equation (1) not only gives us the length of **Q** in terms of its x, y, and z components, but it also gives us the distance between any two points (x_1,y_1,z_1) and (x_2,y_2,z_2) in 3-space.

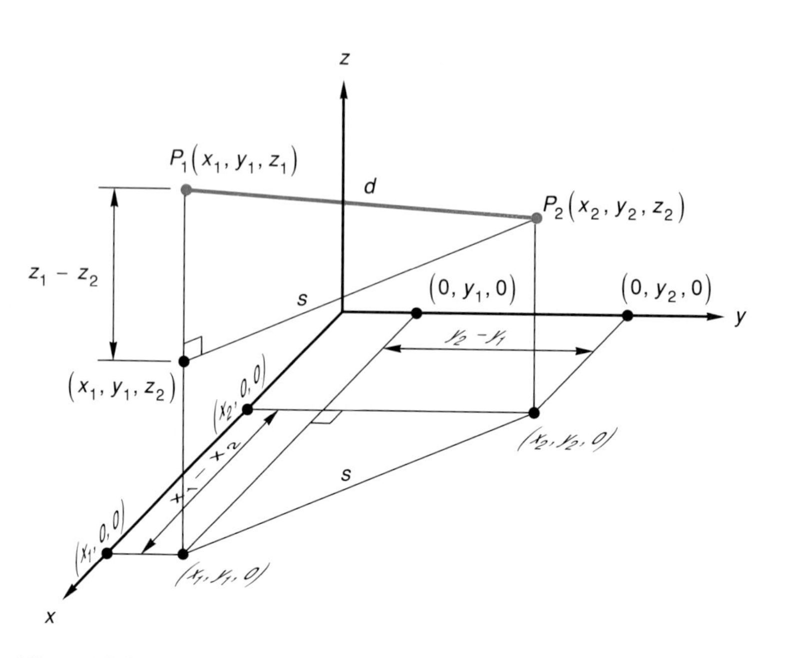

Figure 12.25

If we had attempted to develop (1) without the use of vectors, we could have proceeded as follows. In figure 12.25 we want to find d, the length of the segment between points P_1 and P_2. By the Pythagorean Theorem

$$d = \sqrt{(z_1 - z_2)^2 + s^2}.$$

But s is the length of the hypotenuse of a right triangle whose sides have lengths $x_1 - x_2$ and $y_2 - y_1$. Hence,

$$s^2 = (x_1 - x_2)^2 + (y_2 - y_1)^2.$$

Therefore,

$$\begin{aligned} d &= \sqrt{(z_1 - z_2)^2 + (x_1 - x_2)^2 + (y_2 - y_1)^2} \\ &= \sqrt{(x_2 - x_1)^2 + (y_2 - y_1)^2 + (z_2 - z_1)^2}. \end{aligned}$$

Lines in 3-Space

Let L be a line in 3-space and $\mathbf{V} = (a,b,c,)$, a vector parallel to L (figure 12.26). Let (x_0,y_0,z_0) be a known point on L and let (x,y,z) be another point on L. Then, as in section 12.3, a vector parametric equation of the line is

$$\mathbf{P} = \mathbf{R} + t\mathbf{V},$$

where t is a scalar. The equation states that you can reach any point on the line by adding to the vector $\mathbf{R}$ some scalar multiple of the vector $\mathbf{V}$.

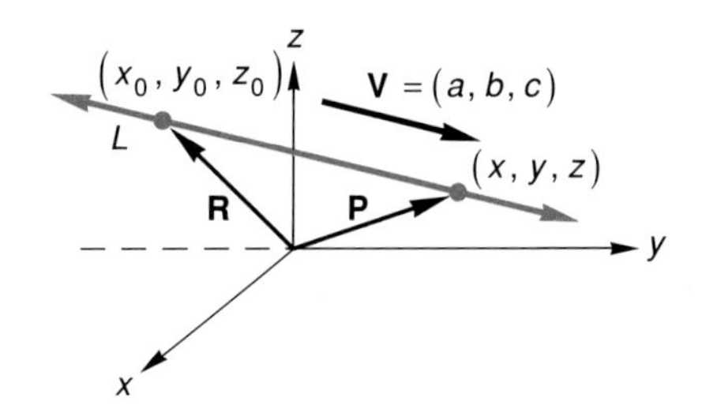

Figure 12.26

Rewriting the vector parametric equation of L as

$$(x,y,z) = (x_0,y_0,z_0) + t(a,b,c)$$

enables us to construct the coordinate parametric equations of the line.

Coordinate Parametric Equations of a Line

$$\begin{aligned} x &= x_0 + at \\ y &= y_0 + bt \\ z &= z_0 + ct \end{aligned} \qquad (2)$$

Eliminating t from equations (2) yields the **symmetric form** of the equation of a straight line in 3-space,

Symmetric Form of the Equation of a Line

$$\frac{x - x_0}{a} = \frac{y - y_0}{b} = \frac{z - z_0}{c}. \qquad (3)$$

Example 1 Write equations of the line containing the points $P_1(6,1,8)$ and $P_2(2,4,3)$, in parametric and symmetric forms.

Solution The two points on the line determine a vector **V** that is parallel to the line.

$$\begin{aligned} \mathbf{V} &= (2 - 6, 4 - 1, 3 - 8) \\ &= (-4,3,-5) \end{aligned}$$

Now, using the vector parametric form of the equation of a line with (6,1,8) playing the role of **R,** we have

$$(x,y,z) = (6,1,8) + t(-4,3,-5).$$

Hence the parametric equations of the line are

$$x = 6 - 4t$$
$$y = 1 + 3t$$
$$z = 8 - 5t.$$

Eliminating t from the parametric equations produces the symmetric equation

$$\frac{x-6}{-4} = \frac{y-1}{3} = \frac{z-8}{5}.$$

■

The numbers a, b, and c in equation (3) are called **direction numbers.** A line in 3-space has no slope. It has direction numbers instead. Thus, if a bug crawls along a line from point P_1 to point P_2 and experiences a change of **a** units in the x direction, it will also experience a change of **b** units in the y direction and a change of **c** units in the z direction. In example 1, our bug crawling from P_1 to P_2 experiences changes of -4 units, 3 units, and -5 units, respectively, in the x, y, and z directions.

Example 2 Find an equation of a line containing the point (3,7,1) that is perpendicular to the line $\frac{x+3}{2} = \frac{y-1}{-3} = \frac{z-6}{5}$.

Solution From example 1 we see that a vector **V** parallel to the given line has as its components the direction numbers of the line. That is, $\mathbf{V} = (2,-3,5)$. The line sought must be parallel to a vector $\mathbf{W} = (a,b,c)$ that is perpendicular to **V**. Consequently

$$\mathbf{V} \cdot \mathbf{W} = 0,$$

or

$$(2,-3,5) \cdot (a,b,c) = 0.$$

As a result

$$2a - 3b + 5c = 0,$$

or

$$a = \frac{3b - 5c}{2}.$$

Since we are solving for a in terms of b and c, there is an infinite number of solutions to our problem. One solution is obtained if we let $c = 1$ and $b = 3$. It follows that $a = 2$, and that

$$\frac{x-3}{2} = \frac{y-7}{3} = \frac{z-1}{1}$$

is the equation of a line that contains (3,7,1) and is perpendicular to the given line. Can you find another solution? ■

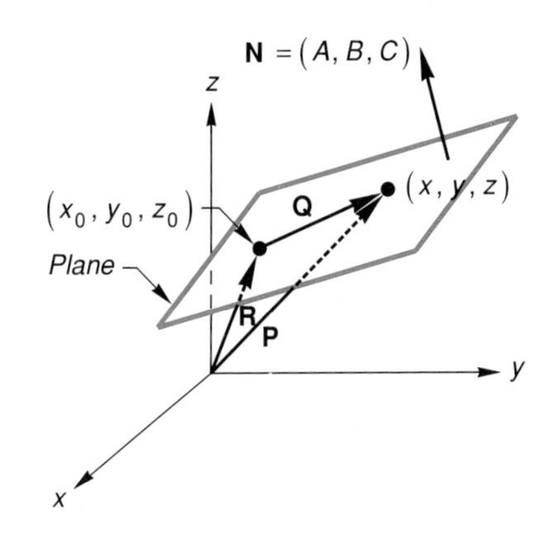

Figure 12.27

Planes in 3-Space

Let (x_0,y_0,z_0) be a known point in a plane and let (x,y,z) be any other point in the plane (figure 12.27). Name the vectors from the origin to (x_0,y_0,z_0) and (x,y,z) **R** and **P,** respectively. Also let $\mathbf{N} = (A,B,C)$ be a vector that is perpendicular to the plane. Hence **N** is perpendicular to any vector that is in the plane or is parallel to the plane. In particular, **N** is perpendicular to **Q** where $\mathbf{Q} = \mathbf{P} - \mathbf{R}$. Therefore

Vector Form of the Equation of a Plane

$$\mathbf{N} \cdot (\mathbf{P} - \mathbf{R}) = 0. \tag{4}$$

Equation (4) is a vector form of the equation of a plane. The corresponding coordinate form is obtained when (4) is written as

$$(A,B,C) \cdot (x - x_0, y - y_0, z - z_0) = 0.$$

Taking the dot product yields

$$A(x - x_0) + B(y - y_0) + C(z - z_0) = 0,$$

or

Coordinate Form of the Equation of a Plane

$$Ax + By + Cz + D = 0, \tag{5}$$

where $D = -(Ax_0 + By_0 + Cz_0)$.

Equation (5) is the coordinate form of the equation of a plane. Note that (5) is the three-dimensional analog of the equation of a line, $ax + by + c = 0$, in two dimensions. Note also that coefficients A, B, and C are the x, y, and z components of a vector that is perpendicular to the plane defined by (5). Such a vector is called a **normal vector** of the plane.

Example 3 Find an equation of the plane determined by the points $P_1(5,2,-2)$, $P_2(3,-6,8)$, and $P_3(-1,7,4)$.

Solution P_1 and P_2, and P_1 and P_3 determine vectors $\mathbf{V}$ and $\mathbf{W}$ in the plane whose equation we seek.

$$\begin{aligned}\mathbf{V} &= (3,-6,8) - (5,2,-2), \\ &= (-2,-8,10)\end{aligned} \qquad \begin{aligned}\mathbf{W} &= (-1,7,4) - (5,2,-2) \\ &= (-6,5,6)\end{aligned}$$

A normal vector $\mathbf{N}$ of the plane must be perpendicular to both $\mathbf{V}$ and $\mathbf{W}$. If $\mathbf{N} = (A,B,C)$, then $\mathbf{N} \cdot \mathbf{V} = 0$ and $\mathbf{N} \cdot \mathbf{W} = 0$. The dot products generate a pair of linear equations

$$\begin{aligned}-2A - 8B + 10C &= 0 \\ -6A + 5B + 6C &= 0\end{aligned}$$

which can be solved for A and B in terms of C to obtain

$$A = \frac{49}{29}C \quad \text{and} \quad B = \frac{24}{29}C.$$

If we let $C = 29$, then $A = 49$ and $B = 24$. Since A, B, and C are components of a normal vector, an equation of the plane through P_1, P_2, and P_3 has the form

$$49x + 24y + 29z + D = 0.$$

To find D we substitute the coordinates of any of the points P_1, P_2, or P_3 into the last equation.

$$\begin{aligned}49(5) + 24(2) + 29(-2) + D &= 0 \\ D &= -235\end{aligned}$$

An equation of the desired plane is $49x + 24y + 29z - 235 = 0$. ■

Example 4 Find an equation of the plane that contains the point $(4,3,-2)$ and is parallel to the plane whose equation is $x - 2y + 5z + 8 = 0$.

Solution If a plane through $(4,3,-2)$ is to be parallel to the given plane, it must have a normal vector that is parallel to a normal vector of the given plane. Such a vector has the form $k(1,-2,5)$, where k is any non-zero scalar and $(1,-2,5)$ is a normal vector of the given plane. If we let $k = 1$, then the equation sought has the form

$$x - 2y + 5z + D = 0.$$

Now substitute the coordinates of the given point into the equation to find D.

$$4 - 2(3) + 5(-2) + D = 0$$
$$D = 12$$

An equation of the desired plane is $x - 2y + 5z + 12 = 0$. ■

Exercise Set 12.4

(A)

Find the distance between the two points.

1. $(1,3,7), (2,-6,5)$ **2.** $(5,0,-9), (5,4,2)$
3. $(-8,-2,3), (-3,6,-3)$ **4.** $(2,-10,1), (-7,2,-8)$

Write equations in parametric and symmetric forms of the line that contains the two given points.

5. $(1,3,7), (2,-6,5)$ **6.** $(-8,-2,3), (-3,6,-3)$
7. $(2,-10,1), (-7,2,-8)$ **8.** $(5,0,-9), (-1,4,2)$
9. $(6,-3,11), (5,4,2)$ **10.** $(12,-5,-3), (-1,7,2)$

Parametric equations of a line are given. Write an equation of the line in symmetric form.

11. $x = 1 + 3t$, $y = 4 - t$, $z = 1 - 3t$
12. $x = -3 + 7t$, $y = 2 + 6t$, $z = 5 - 2t$
13. $x = -8 - 6t$, $y = 12 + 3t$, $z = -4 - 9t$
14. $x = 0 + 9t$, $y = 0 - 2t$, $z = 0 - t$

The symmetric form of the equation of a line is given. Write a set of parametric equations of the line.

15. $\frac{x-1}{2} = \frac{y+3}{4} = \frac{x+2}{9}$

16. $\frac{x}{3} = \frac{y}{3} = \frac{z}{-2}$

17. $x + 5 = \frac{y-5}{-5} = \frac{z-6}{2}$

18. $\frac{x-11}{-3} = y = -z$

Write, in symmetric form, an equation of the line that satisfies the given conditions.

19. contains the point $(1,4,1)$ and is perpendicular to the line $(x - 1)/2 = (y + 3)/4 = (z + 2)/9$

20. contains the point $(-3,2,5)$ and is perpendicular to the line $(x + 3)/2 = (y - 5)/-5 = (z - 6)/2$

21. contains the point $(2,2,2)$ and is perpendicular to each of the lines $x/3 = y/-3 = (z + 1)/7$ and $(x + 5)/-1 = (y - 2)/4 = z/6$

22. contains the point $(1,5,8)$ and is perpendicular to each of the lines $(x - 11)/-3 = y = -z$ and $(x - 8)/4 = (y - 5)/9 = (z + 1)/3$

Determine an equation of the plane containing the given points.

23. $(0,0,0), (1,2,3), (4,-5,6)$
24. $(6,-1,5), (2,0,-3), (-7,1,0)$
25. $(-2,-2,-2), (5,1,4), (8,-3,1)$
26. $(-6,-2,3), (1,-9,-4), (0,5,2)$
27. $(7,2,-9), (1,0,-12), (3,-2,7)$
28. $(1,8,-6), (7,0,3), (-5,10,4)$

Determine an equation of the plane satisfying the given conditions.

29. contains the point $(1,2,3)$ and is parallel to the plane $x + y - z + 5 = 0$

30. contains the point $(4,-5,6)$ and is perpendicular to the plane $2x - 7y - 4z - 3 = 0$

31. contains the point $(8,-3,1)$ and the line $x/3 = y/-3 = (z + 1)/7$

32. contains the point $(2,0,-3)$ and the line $(x + 5)/3 = (y - 2)/4 = z/6$

33. What portions of the coordinate axes determine the third octant?

34. What portions of the coordinate axes determine the eighth octant?

(B)

35. Write an equation of the x-y coordinate plane.

36. Write an equation of the x-z coordinate plane.

37. How would you interpret the fact that one of the direction numbers of a line is zero? If direction number c is zero in equation (3), how would you write a symmetric form equation of the line?

38. Graph the line containing the points $(8,2,3)$ and $(2,7,3)$. Write an equation of the line in symmetric form.

39. How do you interpret the fact that there is an infinite number of solutions to the problem posed in example 2?

40. The solution of the problem posed in example 3 suggests that any one of an infinite number of normal vectors could be used to find an equation of the plane containing the three points. For example, if we let $C = 58$, then $A = 98$ and $B = 48$. What is an equation of the plane if $\mathbf{N} = (98,48,58)$? How does the equation you obtained compare with the equation that was generated in example 3?

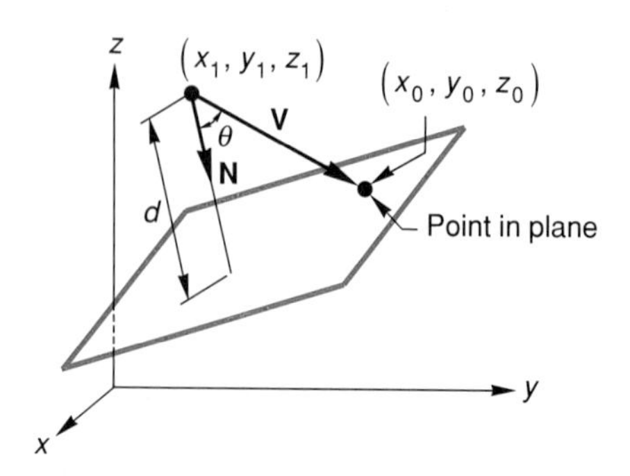

Figure 12.28

41. Use figure 12.28 and the derivation, in section 12.2, of the vector formula for the distance from a point to a line to show that the distance from point (x_1,y_1,z_1) to the plane $Ax + By + Cz + D = 0$ is

$$d = \frac{|\mathbf{V} \cdot \mathbf{N}|}{\|\mathbf{N}\|}.$$

Show how to convert this vector formula to the corresponding coordinate formula

$$d = \frac{|Ax_1 + By_1 + Cz_1 + D|}{\sqrt{A^2 + B^2 + C^2}}.$$

Use the results of exercise 41 to solve the following exercises.

42. Find the distance from the point $(6,3,-1)$ to the plane $5x - 2y + 4z + 2 = 0$.

43. Derive a formula for the distance from the origin to the plane $Ax + By + Cz + D = 0$.

44. Find the distance between the parallel planes $2x - y + 5z + 4 = 0$ and $2x - y + 5z - 11 = 0$.

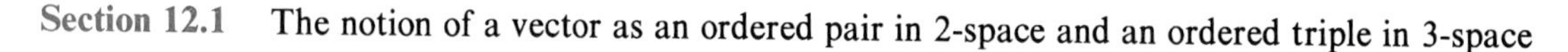

Chapter 12 Review

Major Points for Review

Section 12.1 The notion of a vector as an ordered pair in 2-space and an ordered triple in 3-space

Addition of vectors

The magnitude of a vector

Scalar multiplication

A unit vector

Basis vectors

Section 12.2 The dot product in geometric and algebraic forms

Using the dot product to find the angle between two vectors
Using the dot product to find the magnitude of a vector or to determine if two vectors are perpendicular

The distance formula from a point to a line in vector and coordinate forms

Properties of the dot product

Section 12.3 The vector parametric equation of a line in 2-space

The coordinate parametric equations of a line in 2-space

The vector and coordinate formulas for determining points of subdivision of line segments

Section 12.4 The vector parametric and coordinate parametric equations of a line in 3-space

The vector and coordinate equations of a plane in 3-space

Review Exercises

Let **V** be a vector determined by points P and Q. Write **V** as an ordered pair or an ordered triple and determine the magnitude of **V**.

1. $P(6,-2)$, $Q(-1,5)$ **2.** $P(-3,-8)$, $Q(2,4)$

3. $P(7,0,-3)$, $Q(5,-11,8)$ **4.** $P(-1,8,4)$, $Q(-6,-5,9)$

Graph **A**, **B**, and **C** in the same coordinate system.

5. $\mathbf{A} = 2\mathbf{i} - 5\mathbf{j}$, $B = 3\mathbf{A}$, $\mathbf{C} = -2\mathbf{A}$

6. $\mathbf{A} = (-4,3)$, $\mathbf{B} = \mathbf{A}/2$, $\mathbf{C} = 9\mathbf{A}/5$

Find a unit vector in the direction of **V**.

7. $\mathbf{V} = 5\mathbf{i} + 12\mathbf{j}$ **8.** $\mathbf{V} = (8,15)$

9. $\mathbf{V} = (2,-5,3)$ **10.** $\mathbf{V} = (-1,7,4)$

Given that $\mathbf{A} = (-3,8)$, $\mathbf{B} = (5,2)$, $\mathbf{C} = (4,-6)$, $\mathbf{V} = (-2,9,-3)$, and $\mathbf{W} = (5,-2,-1)$,

11. Find $3\mathbf{A} - 2\mathbf{B}$ **12.** Find $\mathbf{A} + 4\mathbf{B} - 3\mathbf{C}$

13. Find $||\mathbf{A} - \mathbf{B} + \mathbf{C}||$ **14.** Find $-4\mathbf{V} + 3\mathbf{W}$

15. Find $\mathbf{A} \cdot \mathbf{B}$ **16.** Find $\mathbf{V} \cdot \mathbf{W}$

17. Find $2\mathbf{A} \cdot 4\mathbf{C}$

18. Write **C** as a linear combination of **A** and **B.**

Find a unit vector that is

19. parallel to the line $2x - y = 5$.

20. parallel to the line $y = 3x - 1$.

21. perpendicular to $\mathbf{V} = (3,-2)$.

22. perpendicular to $\mathbf{V} = (-8,-5)$.

Find the acute angles between the given lines.

23. $2x - y = 5$, $y = 3x - 1$

24. $y = -2x + 5$, $3x + 4y = 12$

Find the distance from point P to the given line.

25. $P(0,0)$, $y = 3x + 5$

26. $P(4,7)$, $y = -2x - 8$

27. $P(-3,8)$, $5x - 2y = 10$

28. $P(5,0)$, $y = 2x$

Write a parametric vector equation and a pair of coordinate parametric equations for each given line or for the line determined by the given points.

29. $y = 7x - 4$

30. $x + 3y = 6$

31. (0,0), (2,4)

32. $(-5,3), (2,-6)$

Find points on the given line that are the indicated number of units from point P, which is on the line.

33. $(x,y) = (4,1) + t(3,4)$, 8 units, $P(1,-3)$

34. $5x - 2y = 10$, 5 units, $P(2,0)$

Use vector methods to find the point of intersection of each pair of lines.

35. $(x,y) = (-2,5) + t(1,2)$, $(x,y) = (3,-1) + t(-5,1)$

36. $3x + 4y = 12$, $y = -2x - 8$

Find $R(x,y)$, a point between P and Q on the segment $\overline{PQ}$, so that the indicated number is the ration $\overline{PR}/\overline{PQ}$.

37. $P(3,4)$, $Q(-1,-6)$, $\frac{2}{3}$

38. $P(-5,6)$, $Q(4,-6)$, $\frac{1}{8}$

Write equations, in parametric and symmetric forms, of the line that contains the given points.

39. $(2,0,6), (5,-3,4)$

40. $(-1,8,4), (-7,-10,9)$

The symmetric form of the equation of a line is given. Write a set of parametric equations of the line.

41. $\dfrac{x+2}{5} = \dfrac{y-1}{4} = \dfrac{z+5}{2}$

42. $x - 1 = y = \dfrac{z-3}{4}$

Determine an equation of the plane that contains the given points.

43. $(1,3,-2), (4,7,-1), (9,-3,5)$

44. $(-12,0,6), (-4,-8,1), (5,2,-9)$

45. Find the distance between the parallel lines $y = 2x - 5$ and $y = 2x + 3$.

46. Write equations for the lines that are two units from the line $3x + y + 9 = 0$.

47. Find the angle bisectors of the angles formed by the lines $y = x - 4$ and $y = -2x + 1$.

48. Find the point that divides the segment, determined by the points $P(6,2)$ and $Q(9,-4)$, into the ratio 1:5.

49. Write a parametric vector equation of the line containing $(5,-3)$ that is perpendicular to the line $(x,y) = (1,2) + t(-2,5)$.

50. Write a symmetric form equation of the line that contains $(0,3,-2)$ and is perpendicular to the line $(x + 2)/5 = (y - 1)/4 = (z + 5)/2$.

51. Write a symmetric form equation of the line that contains $(6,-1,-4)$ and is perpendicular to each of the lines $x/5 = y/2 = z/4$ and $x - 3 = (y + 1)/5 = (z - 2)/3$.

52. Determine an equation of the plane that contains (0,0,0) and is parallel to the plane $2x - y + 3z + 4 = 0$.

Chapter 12 Test 1

1. Graph $\mathbf{A} = (3,1)$, $\mathbf{B} = \dfrac{5\mathbf{A}}{2}$, and $\mathbf{C} = -3\mathbf{A}$ in the same coordinate system.

Given that $\mathbf{A} = (4,-3)$, $\mathbf{B} = (7,11)$, and $\mathbf{C} = (-2,-5)$, find

2. $6\mathbf{A} - 2\mathbf{B}$.

3. $\mathbf{A} \cdot (3\mathbf{C})$.

4. Find a unit vector that is perpendicular to $\mathbf{V} = (5,-12)$.

5. Find the acute angle between the lines $y = 2x + 1$ and $x + y = 4$.

6. Find the distance from the point (8,3) to the line $2x + 3y = 6$.

7. Write a pair of coordinate parametric equations for the line determined by the points $(-5,3)$ and $(2,7)$.

8. Find the point of intersection of the lines $(x,y) = (0,0) + t(1,3)$ and $(x,y) = (4,1) + t(-2,1)$.

9. Write a symmetric form equation of the line containing the points $(2,5,-1)$ and $(-4,7,3)$.

10. Determine an equation of the plane that contains the points $(1,2,3)$, $(1,5,-2)$, and $(4,1,6)$.

Chapter 12 Test 2

1. Find the acute angle between the lines $2x + y = 6$ and $3x - 4y = 12$.
2. Find the distance from the point $(-5,6)$ to the line $y = 2x - 1$.
3. Write a pair of coordinate parametric equations for the line $5x + 8y = 10$.
4. Find the point of intersection of the lines $(x,y) = (2,3) + t(5,1)$ and $(x,y) = (1,0) + t(1,-2)$.
5. Find the point that divides the segment, determined by the points $(1,4)$ and $(5,-4)$, into the ratio 1:3.
6. Write a symmetric form equation of the line that contains the points $(-2,5,0)$ and $(3,-4,-2)$.
7. Determine an equation of the plane that contains the points $(5,8,12)$, $(0,0,0)$, and $(-4,6,-1)$.
8. Write equations for the lines that are three units from the line $x + 4y - 8 = 0$.
9. Write a parametric vector equation of the line containing $(6,-2)$ that is perpendicular to the line $(x,y) = (0,3) + t(7,-4)$.
10. Write a symmetric form equation of the line that contains $(2,8,-3)$ and is perpendicular to the line $(x - 5)/2 = y/-2 = (z + 1)/5$.

Chapter 13

Induction, Series, and the Binomial Theorem

The formula $[n(n+1)]/2$ enables us to find the sum $1 + 2 + 3 + 4 + 5 + 6$, which is 21. With $n = 6$, the formula gives us $[6(6+1)]/2 = 42/2 = 21$. Does the formula work for $n = 10$, $n = 100$, $n = 1{,}000$? Mathematical induction, which is discussed in this chapter, gives us the answer. In this chapter we also show how to construct the formula $[n(n+1)]/2$ as well as other much used summation formulas. Finally, we develop some counting techniques that enable us to answer such questions as "How many different 5-card poker hands can be dealt from a 52-card deck" or "In how many ways can six numbers be chosen from the first 50 natural numbers".

13.1 Mathematical Induction

Suppose that your mind has been transported to a state of reverie by a boring lecture. You doodle

$$\begin{aligned} 1 &= 1 \\ 1 + 2 &= 3 \\ 1 + 2 + 3 &= 6 \\ 1 + 2 + 3 + 4 &= 10 \\ 1 + 2 + 3 + 4 + 5 &= 15. \end{aligned} \qquad \begin{matrix} \}\ 2 \\ \}\ 3 \\ \}\ 4 \\ \}\ 5 \end{matrix}$$

As you write down larger and larger sums, you begin to emerge from your reverie. Your scribblings have produced a clear pattern in which the differences between consecutive sums increase by 1. You excitedly ask yourself, "Is there a simple formula for these sums?" You look for another pattern that might suggest such a formula, and so you construct a table in which n is a natural number and $S(n)$ is the sum of the first n natural numbers. The ratios $S(n)/n$ each equal $(n + 1)/2$. There it is,

$$\frac{S(n)}{n} = \frac{n+1}{2}$$

and

$$S(n) = \frac{n(n+1)}{2}.$$

n	$S(n)$	$S(n)/n$
1	$1 = 1$	$1 = \frac{2}{2}$
2	$1 + 2 = 3$	$\frac{3}{2} = \frac{3}{2}$
3	$1 + 2 + 3 = 6$	$\frac{6}{3} = \frac{4}{2}$
4	$1 + 2 + 3 + 4 = 10$	$\frac{10}{4} = \frac{5}{2}$
5	$1 + 2 + 3 + 4 + 5 = 15$	$\frac{15}{5} = \frac{6}{2}$

You've done it. Your pulse quickens. You have created something. The formula works for $n = 1,2,3,4$, and 5. You test it for $n = 6$ and $n = 7$.

$$1 + 2 + 3 + 4 + 5 + 6 = 21 = \frac{6(6+1)}{2}$$

$$1 + 2 + 3 + 4 + 5 + 6 + 7 = 28 = \frac{7(7+1)}{2}$$

It works. Then you ask yourself the painful question, "Does it work for all *n?*" Must you test the formula for each natural number or is there a way to verify that it will work for all *n?*

The question, "Does it work for all *n,*" occurred to mathematicians whenever their work led them to a recursive pattern based on the natural numbers. A way of answering the question began appearing in the mathematics of the fifteenth century. Augustus DeMorgan (1807–1871) formalized the method of proof called **Mathematical Induction** in 1838. His method depends upon *The Axiom of Mathematical Induction.*

If S is a set of natural numbers that satisfies the conditions

1. $1 \in S$, and
2. whenever a natural number n belongs to S, the natural number $n + 1$ belongs to S,

then $S = \mathrm{N}$.

If a set S of natural numbers satisfies conditions **1** and **2,** then because 1 is in the set, $1 + 1 = 2$ must also be in the set. If 2 is in the set, then $2 + 1 = 3$ must be in the set, and so on. Clearly, all natural numbers must be in S.

Induction proofs are usually generated from the reformulated Axiom of Mathematical Induction given below.

Axiom of Mathematical Induction

Suppose we have a sequence of statements, $S_1, S_2, S_3, \ldots, S_n$ in which

1. S_1 is true, and
2. the truth of S_k implies the truth of S_{k+1},

then S_n must be true for all n.

Now we can answer the question, "Does it work for all n," about the pattern constructed by the bored student.

$$
\begin{array}{rl}
1 = 1 & S_1 \\
1 + 2 = 3 & S_2 \\
1 + 2 + 3 = 6 & S_3 \\
\vdots & \vdots \\
1 + 2 + 3 + \cdots + n = \dfrac{n(n+1)}{2} & S_n
\end{array}
$$

S_1 is a true statement. Assume S_k,

$$1 + 2 + 3 + \cdots + k = \frac{k(k+1)}{2},$$

is true. Now we must show that S_{k+1} is also a true statement: that is,

$$
\begin{aligned}
1 + 2 + 3 + \cdots + k + (k+1) &= \frac{(k+1)((k+1)+1)}{2} \\
&= \frac{(k+1)(k+2)}{2}.
\end{aligned}
$$

The kth statement is assumed to be true. If we add $k + 1$ to both sides of S_k, the left member becomes the sum of the first $k + 1$ natural numbers. It remains to show that the right side has the form $k(k + 1)/2$, where k is replaced by $k + 1$.

$$\underbrace{\overbrace{(1 + 2 + 3 + \cdots + k) + (k + 1)}^{S_{k+1}}}_{S_k} = \frac{k(k + 1)}{2} + (k + 1)$$

$$S_{k+1} = \frac{k(k + 1) + 2(k + 1)}{2}$$

$$S_{k+1} = \frac{(k + 1)(k + 2)}{2}$$

We have shown that if S_k is true, then S_{k+1} must be true. Thus, because S_1 is true, S_2 must be true. Now that S_2 is true, it follows that S_3 must be true, and so it goes for all S_n, n a natural number.

Example 1 Prove by induction that the statement

$$\frac{1}{2} + \frac{1}{2^2} + \frac{1}{2^3} + \cdots + \frac{1}{2^n} = 1 - \frac{1}{2^n}$$

is true for every natural number n.

Solution

$$S_1 \text{ is } \frac{1}{2} = 1 - \frac{1}{2}$$

$$S_2 \text{ is } \frac{1}{2} + \frac{1}{2^2} = 1 - \frac{1}{2^2}$$

$$\vdots$$

$$S_k \text{ is } \frac{1}{2} + \frac{1}{2^2} + \frac{1}{2^3} + \cdots + \frac{1}{2^k} = 1 - \frac{1}{2^k}.$$

$$S_{k+1} \text{ is } \frac{1}{2} + \frac{1}{2^2} + \frac{1}{2^3} + \cdots + \frac{1}{2^k} + \frac{1}{2^{k+1}} = 1 - \frac{1}{2^{k+1}}$$

We must show that

1. S_1 is true, and then
2. prove that S_{k+1} is true based on the assumption that S_k is true.

Clearly, S_1 is true. Now, assume S_k is true and add $\frac{1}{2^{k+1}}$ to both members of S_k.

$$\left(\frac{1}{2} + \frac{1}{2^2} + \cdots + \frac{1}{2^k}\right) + \frac{1}{2^{k+1}} = \left(1 - \frac{1}{2^k}\right) + \frac{1}{2^{k+1}}.$$

The left member is the sum of the first $k + 1$ terms. Hence we must show that the right member is equal to $1 - \frac{1}{2^{k+1}}$.

$$\left(1 - \frac{1}{2^k}\right) + \frac{1}{2^{k+1}} = \frac{2^{k+1} - 2 + 1}{2^{k+1}}$$

$$S_{k+1} = 1 - \frac{1}{2^{k+1}}$$

The proof is now complete, as we have shown that the assumption of the truth of S_k enables us to prove that S_{k+1},

$$\frac{1}{2} + \frac{1}{2^2} + \cdots + \frac{1}{2^{k+1}} = 1 - \frac{1}{2^{k+1}},$$

is true. ■

Proof by induction *requires* steps 1 and 2, as shown in example 1. Neither step can be left out of the proof. To illustrate this fact we consider two statements,

$$2^1 + 2^2 + 2^3 + \cdots + 2^n = 2 + 4\tan\left[\frac{(n-1)\pi}{4}\right]$$

and

$$1 + 2 + 3 + \cdots + n = 5 + \frac{n(n+1)}{2}.$$

The first statement is certainly not true for all natural numbers. Yet it is true for

$$S_1\colon 2^1 = 2 + 4\tan\left[\frac{(1-1)\pi}{4}\right]$$

$$2 = 2$$

and

$$S_2\colon 2^1 + 2^2 = 2 + 4\tan\left[\frac{(2-1)\pi}{4}\right]$$

$$6 = 6.$$

S_3 is false (try it). For the first statement we can verify step 1, but we cannot complete step 2.

The second statement is not true for $n = 1$, and therefore cannot be true for all n.

$$S_1\colon 1 \neq 5 + \frac{1(1 + 1)}{2}$$

Yet, if we assume the truth of S_k, we can prove the truth of S_{k+1}. Assume that

$$1 + 2 + \cdots + k = 5 + \frac{k(k + 1)}{2}$$

is true. Then

$$\begin{aligned}(1 + 2 + \cdots + k) + (k + 1) &= \left(5 + \frac{k(k + 1)}{2}\right) + (k + 1)\\ &= 5 + \frac{k(k + 1) + 2(k + 1)}{2}\\ &= 5 + \frac{(k + 1)(k + 2)}{2}.\end{aligned}$$

It is possible to verify step 2 for the second statement, but not step 1. The second statement is in fact false for all natural numbers.

We end with another pattern that can be generalized by mathematical induction. We know that $x - y$ is a factor of each of the expressions $x - y$, $x^2 - y^2$, and $x^3 - y^3$. It is not difficult to show with synthetic division that $x - y$ is also a factor of $x^4 - y^4$ and of $x^5 - y^5$. One might wonder if $x - y$ is a factor of $x^n - y^n$ for all natural numbers n. The following induction proof provides the answer.

Proof:

1. S_1 is true. That is, $x - y$ is a factor of $x - y$.
2. Assume the truth of S_k, that $x - y$ is a factor of $x^k - y^k$, and show that $x - y$ is a factor of $x^{k+1} - y^{k+1}$.

$$\begin{aligned}x^{k+1} - y^{k+1} &= (x^{k+1} - xy^k) + (xy^k - y^{k+1})\\ &= x(x^k - y^k) + y^k(x - y).\end{aligned}$$

Since $x - y$ is a factor of each of the expressions $x - y$ and $x^k - y^k$, then $x - y$ must also be a factor of $(x^{k+1} - y^{k+1})$.

By induction we conclude that $x - y$ is a factor of $x^n - y^n$ for all natural numbers n.

Exercise Set 13.1

(A)

Use Mathematical Induction to prove that each equation is an identity for all natural numbers n.

1. $1 + 3 + 5 + \cdots + (2n - 1) = n^2$
2. $2 + 4 + 6 + \cdots + 2n = n(n + 1)$
3. $1^2 + 2^2 + 3^2 + \cdots + n^2 = n(n + 1)(2n + 1)/6$
4. $1^3 + 2^3 + 3^3 + \cdots + n^3 = n^2(n + 1)^2/4$
5. $1 \cdot 3 + 2 \cdot 4 + 3 \cdot 5 + \cdots + n(n + 2)$ $= n(n + 1)(2n + 7)/6$
6. $\dfrac{1}{1 \cdot 2} + \dfrac{1}{2 \cdot 3} + \dfrac{1}{3 \cdot 4} + \cdots + \dfrac{1}{n(n + 1)} = \dfrac{n}{n + 1}$
7. $(ab)^n = a^n b^n$
8. $ln(b_1 b_2 \cdots b_n) = ln\, b_1 + ln\, b_2 + \cdots + ln\, b_n,\ b_i > 0$ for each i

9. Prove that $\sin(\theta + n\pi) = (-1)^n \sin \theta$.
10. Prove that $n^2 + n$ is even for all natural numbers n.
11. Prove that for all natural numbers n, $(1 + h)^n \geq 1 + nh$ if $h > -1$.
12. If m and n are natural numbers and b is a real number, prove that $(b^m)^n = b^{mn}$. (Hint: Use induction on n.)
13. Prove that $1 + 2n < 3^{n-1}$ if $n > 2$ and $n \varepsilon N$.
14. Prove that $2^n > 100n$ if $n > 10$ and $n \varepsilon N$.

(B)

15. Make a conjecture about the sum $2 + 7 + 12 + \cdots + (5n - 3)$, $n \varepsilon N$. Doodle as the bored student (at the beginning of the section) did, and then prove your conjecture.
16. Make a conjecture about the sum $1 + 4 + 7 + \cdots + (3n - 2)$, $n \varepsilon N$. Prove your conjecture.
17. Make a conjecture about the product $(1 + \frac{1}{1})(1 + \frac{1}{2}) \ldots (1 + 1/n)$, $n \varepsilon N$. Prove your conjecture.
18. Make a conjecture about the product $(1 - \frac{1}{4})(1 - \frac{1}{9}) \ldots (1 - 1/n^2)$, $n \varepsilon N$. Prove your conjecture.
19. Find a formula for the number of chords that can be constructed through n distinct points on a circle. Use mathematical induction to prove your assertion.
20. In the game "Towers of Hanoi" a stack of washers are placed on one of three spindles (see figure 13.1). The washers are arranged from bottom to top in order of decreasing radius. The object of the game is to transfer the stack of washers to one of the empty spindles, moving one washer at a time and never placing a larger washer above a smaller one. All three spindles may be used in transferring the stack from one spindle to another. Is it possible to make the transfer for a stack of any size? If so, how many moves are required? (Hint: Make the transfer with just two washers, then three washers, and then four washers. Count the number of moves in each case to see if you can discern a general pattern.)

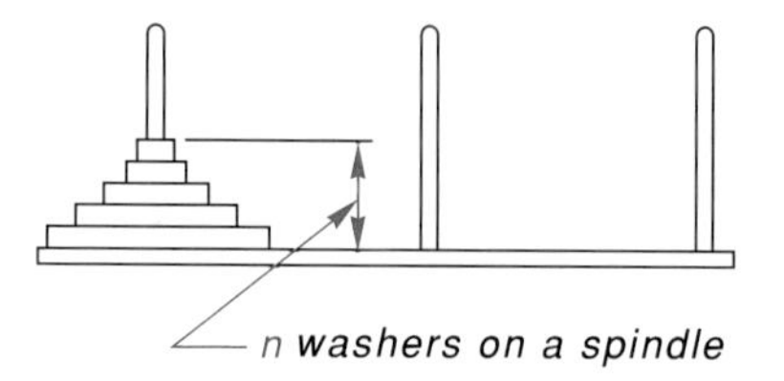

Figure 13.1

13.2 Sequences, Series, and Summation

Our bored student of section 13.1 showed us that the sum of the first n natural numbers is $n(n + 1)/2$. If a 1:1 correspondence is established between the natural numbers and the numbers $n(n + 1)/2$, a function is constructed whose domain is N and whose range is $\{n(n + 1)/2 | n \varepsilon N\}$. Any function whose domain is a subset of N is called a **sequence.**

Range	1	3	6	10	15	21	$\frac{n(n+1)}{2}$
	$\updownarrow$	$\updownarrow$	$\updownarrow$	$\updownarrow$	$\updownarrow$	$\updownarrow$	$\updownarrow$
Domain	1	2	3	4	5	6	n

It is common practice to call a sequence function simply a sequence and to denote it as $\{a_n\}$. The terms of the sequence are $a_1, a_2, a_3, \ldots, a_n, \ldots$, where a_n is called the **nth term.** The terms of the sequence $\{n(n+1)/2\}$ are $a_1 = 1, a_2 = 3, a_3 = 6, \ldots, a_n = n(n+1)/2$.

A sequence is said to be infinite if its domain is N, or an infinite subset of N. If a sequence has a domain that is a finite subset of N, then it is said to be a finite sequence.

Example 1 $\{3n + 2\}$, $n \varepsilon N$, is an infinite sequence whose terms are

$$a_1 = 3(1) + 2 = 5, a_2 = 3(2) + 2 = 8, a_3 = 3(3) + 2 = 11, \ldots,$$
$$a_n = 3n + 2.$$

Example 2 $\{(-1)^n 2^n\}$, $n \varepsilon \{1,2,3,4\}$, is a finite sequence whose terms are

$$a_1 = (-1)2 = -2, a_2 = (-1)^2 2^2 = 4, a_3 = (-1)^3 2^3 = -8,$$
$$a_4 = (-1)^4 2^4 = 16.$$

Series

Associated with each sequence is a *series that is an expression of the sum of the terms of the sequence.* If the sequence is infinite, so is the series. If the sequence is finite, then its series is finite.

Example 3 The infinite series associated with $\{3n + 2\}$, $n \varepsilon N$, is the expression

$$5 + 8 + 11 + 14 + \cdots + (3n + 2) + \cdots.$$

The finite series associated with $\{(-1)^n 2^n\}$, $n \varepsilon \{1,2,3,4\}$, is

$$(-2) + 4 + (-8) + 16.$$

The terms of the finite series of example 3 sum to 10. A finite series always has a sum. Such is not the case for an infinite series. If the terms of the infinite series in example 3 are added,

$$5 + 8 = 13, \quad 5 + 8 + 11 = 24, \quad 5 + 8 + 11 + 14 = 38, \text{ etc.,}$$

sums are gotten that grow large without bound and tend to infinity. Such unbounded growth is often symbolized as ∞. There are, however, infinite series whose terms sum to a real number. We shall encounter such series after discussing a shorthand notation for summing.

Sigma Notation

The greek letter Σ (sigma) is often used to indicate a sum. With sigma notation, the finite series in example 3 can be written

$$\sum_{i=1}^{4} (-1)^i 2^i,$$

where i is called the index of the sum. In this case i assumes, in sequence, all natural number values 1 through 4. The infinite series in example 3 can be written

$$\sum_{k=1}^{\infty} (3k + 2),$$

where the index k assumes, in sequence, all natural number values starting with 1.

Example 4 Express the series $\sum_{k=1}^{5} \left(1 + \frac{1}{k}\right)$ in expanded form.

Solution

$$(1 + \tfrac{1}{1}) + (1 + \tfrac{1}{2}) + (1 + \tfrac{1}{3}) + (1 + \tfrac{1}{4}) + (1 + \tfrac{1}{5})$$

or

$$2 + \tfrac{3}{2} + \tfrac{4}{3} + \tfrac{5}{4} + \tfrac{6}{5}$$

■

Properties of Σ Notation

There is always a price to be paid for short hand notation. One must learn the rules that govern the use of the notation. But the price is small, and the gain in an expanded ability for precise mathematical expression is great.

Summation Properties

If a_1, b_1, and c are real numbers and n is a natural number, then

1. $\sum_{i=1}^{n} (a_i + b_i) = \sum_{i=1}^{n} a_i + \sum_{i=1}^{n} b_i$
2. $\sum_{i=1}^{n} ca_i = c \sum_{i=1}^{n} a_i$
3. $\sum_{i=1}^{n} 1 = n$

Proof of Property 1:

Let S_k be the statement $\sum_{i=1}^{k} (a_i + b_i) = \sum_{i=1}^{k} a_i + \sum_{i=1}^{k} b_i$.

S_1 is true: $\sum_{i=1}^{1}(a_i + b_i) = a_1 + b_1 = \sum_{i=1}^{1} a_i + \sum_{i=1}^{1} b_i.$

Assume S_k is true and show that S_{k+1} is true.

$$\begin{aligned}
\sum_{i=1}^{k+1}(a_i + b_i) &= \sum_{i=1}^{k}(a_i + b_i) + (a_{k+1} + b_{k+1}) \\
&= \sum_{i=1}^{k} a_i + \sum_{i=1}^{k} b_i + a_{k+1} + b_{k+1} \\
&= \sum_{i=1}^{k} a_i + a_{k+1} + \sum_{i=1}^{k} b_i + b_{k+1} \\
&= \sum_{i=1}^{k+1} a_i + \sum_{i=1}^{k+1} b_i
\end{aligned}$$

Hence S_{k+1} is true. By induction S_n is true for all n.

Proof of Property 3:

$$\sum_{i=1}^{n} 1 = 1 + 1 + 1 + \cdots + 1 = n$$

The proof of Property 2 is left as an exercise.

Example 5 Find the sum of the series $\sum_{i=1}^{50}(2i - 3)$.

Solution

$$\begin{aligned}
\sum_{i=1}^{50}(2i - 3) &= \sum_{i=1}^{50}(2i + (-3)) \\
&= \sum_{i=1}^{50} 2i + \sum_{i=1}^{50}(-3) && \text{(Property 1)} \\
&= 2\sum_{i=1}^{50} i - 3\sum_{i=1}^{50} 1 && \text{(Property 2)} \\
&= 2\sum_{i=1}^{50} i - 3(50) && \text{(Property 3)} \\
&= \frac{2(50)(51)}{2} - 150 && \left(\sum_{i=1}^{n} i = \frac{n(n+1)}{2}\right) \\
&= 2{,}400
\end{aligned}$$

■

Exercise Set 13.2

(A)

For each infinite sequence, write the first four terms followed by the ellipsis symbol and the nth term.

1. $\{2^{n-1}\}$ **2.** $\{3^n\}$

3. $\left\{\frac{1}{3^n}\right\}$ **4.** $\{2n + 1\}$

5. $\{(-1)^n(2n)^{n-1}\}$ **6.** $\{(-1)^{n+1}(n)^n\}$

Write the indicated finite sequence.

7. $\left\{1 + \frac{1}{n}\right\}, n \in \{1,2,3,4,5\}$

8. $\left\{\frac{n}{2n-1}\right\}, n \in \{1,2,3,4,5,6\}$

9. $\left\{(-1)^n\frac{(n-2)}{n}\right\}, n \in \{1,2,3,4,5,6\}$

10. $\left\{\frac{n^2-3}{n+1}\right\}, n \in \{1,2,3,4,5\}$

11. $\left\{\frac{5}{n}\right\}, n \in \{1,2,3,4,5\}$

12. $\{3^{n-1}\}, n \in \{1,2,3,4,5\}$

Write the series associated with the indicated sequence. Use the ellipsis symbol as it is used in example 3 for those series that are infinite.

13. Exercise 7 **14.** Exercise 9

15. Exercise 1 **16.** Exercise 3

17. Exercise 11 **18.** Exercise 6

Write each series in expanded form.

19. $\sum_{k=1}^{8} (2k + 5)$ **20.** $\sum_{j=2}^{8} (-3j + 8)$

21. $\sum_{j=1}^{9} \left(\frac{j}{2} - 3\right)$ **22.** $\sum_{k=3}^{10} \left(\frac{2k}{3} - \frac{1}{9}\right)$

23. $\sum_{i=1}^{5} [(i + 2)(i - 1)]$ **24.** $\sum_{i=1}^{4} i(4i + 1)$

Use the properties of Σ notation to help you find the sum of each series.

25. Exercise 19 **26.** Exercise 20

27. Exercise 21 **28.** Exercise 22

29. Exercise 23 **30.** Exercise 24

Use summation notation to write each series.

31. $4 + 8 + 12 + 16$

32. $3 + 7 + 11 + 15 + 19$

33. $2 + 6 + 18 + 54$

34. $64 + 32 + 16 + 8 + 4 + 2$

A sequence can be generated with a **recursion formula.** The formula gives the first term of the sequence along with a rule for obtaining any term in the sequence, except the first, from the term that precedes it. To illustrate, let $a_1 = 6$ and let $a_{n+1} = a_n + 1/n$. The first four terms of the sequence generated by this recursion formula are 6, 7, $\frac{15}{2}$, and $\frac{47}{6}$. Now, you specify the first four terms of the sequence generated by each recursion formula.

35. $a_1 = 2, a_{n+1} = a_n + 5$

36. $a_1 = 3, a_{n+1} = 2a_n - 4$

37. $a_1 = \frac{1}{2}, a_{n+1} = \frac{a_n}{2}$

38. $a_1 = \frac{2}{3}, a_{n+1} = \frac{a_n}{3}$

(B)

39. In this exercise,

a. Write the first six terms of the sequence $\{3n\}$.

b. Write the first six terms of the sequence $\{3n - (n - 1)(n - 2)(n - 3)\}$.

c. Is a sequence uniquely determined by a finite number of its terms?

40. Write two sequences whose first five terms are the same, but which differ from the sixth term on. (See exercise 39.)

Find an expression for an nth term in each sequence.

41. $1, \frac{1}{2}, \frac{1}{3}, \frac{1}{4}, \ldots$

42. $2, 4, 8, 16, \ldots$

43. $\frac{x}{2}, \frac{2x}{3}, \frac{3x}{4}, \frac{4x}{5}, \ldots$

44. $\frac{x}{2}, -\frac{x^2}{4}, \frac{x^3}{8}, -\frac{x^4}{16}, \ldots$

Write a recursion formula that will generate the given sequence.

45. $\{n + 5\}$

46. $\{(-1)^{n+1}\}$

47. $\left\{\frac{1}{2^n}\right\}$

48. $\{2n + 3\}$

Write the nth term for each sequence.

49. $a_1 = 1, a_{n+1} = \left(\frac{n}{n+1}\right)a_n$

50. $a_1 = 2, a_{n+1} = 3a_n$

51. $a_1 = 3, a_{n+1} = a_n + 4$

52. $a_1 = 1, a_{n+1} = 2a_n + 3$

53. Prove Property 2 of Σ notation.

In exercises 54–55, show that each pair of expressions

54. $\sum_{i=1}^{10} (i + 2), \sum_{i=3}^{12} i$

55. $\sum_{i=4}^{9} \frac{i}{i-3}, \sum_{i=1}^{6} \frac{i+3}{i}$

56. Use mathematical induction to prove that the given assertion is true for all n, $n \in N$.

$$\sum_{i=1}^{n} (a_i + b_i)^2 = \sum_{i=1}^{n} a_i^2 + 2\sum_{i=1}^{n} a_i b_i + \sum_{i=1}^{n} b_i^2$$

57. The famous Fibonacci sequence is generated from the recursion formula $a_1 = 1, a_n = a_{n-1} + a_{n-2}$. Show that the sum of the first n terms of the series constructed from the Fibonacci sequence is $a_{n+2} - 1$. Assume $a_0 = 0$. (Hint: Use induction.)

58. Show that the sum of the first n terms of the series that is constructed from the Fibonacci numbers with odd subscripts is a_{2n}. (Hint: Show that $\Sigma\, a_{2i-1} = a_{2k}$ is true for $k = 1$. Then show that it must be true for $k + 1$ when it is assumed to be true for k.)

13.3 Arithmetic and Geometric Series

Consider the sequences

$$8, 15, 22, 29, \ldots; \quad 22, 19, 16, 13, \ldots;$$
$$x, 2x + 1, 3x + 2, 4x + 3, \ldots .$$

Each sequence has the property that *if a term is subtracted from the term following it, a common difference is obtained.* The common difference in the first sequence is 7 because

$$15 - 8 = 7, \quad 22 - 15 = 7, \quad \text{and} \quad 29 - 22 = 7.$$

In the second and third sequences the common differences are -3 and $x + 1$.

If in the sequence $\{a_n\}$, $a_i - a_{i-1} = d$ for all i, then the sequence is called an **arithmetic sequence,** and d is called its **common difference.** In such a sequence

$$a_2 = a_1 + d, a_3 = a_2 + d = (a_1 + d) + d = a_1 + 2d,$$
$$a_4 = a_3 + d = a_1 + 2d + d = a_1 + 3d,$$

and by induction (see exercise 53)

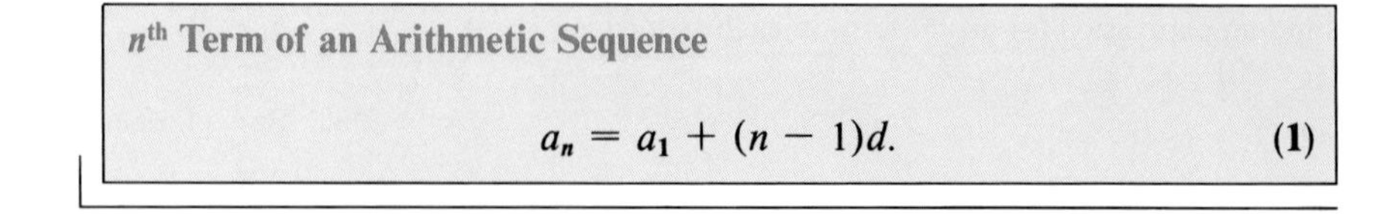
n^{th} Term of an Arithmetic Sequence

$$a_n = a_1 + (n - 1)d. \tag{1}$$

Example 1 Write the nth term for each arithmetic sequence.

a. 11,17,23, . . .

b. 16,12,8, . . .

c. $2y + 1, 3y + 3, 4y + 5, \ldots$.

Solution

a. $d = 6$, hence $a_n = 11 + (n - 1)6 = 5 + 6n$

b. $d = -4$, hence $a_n = 16 + (n - 1)(-4) = 20 - 4n$

c. $d = y + 2$, hence $a_n = 2y + 1 + (n - 1)(y + 2)$
$= (n + 1)y + 2n - 1.$ ■

Example 2 Find a_{12} in the arithmetic sequence $14, 6, -2, \ldots$.

Solution Since $d = -8$, $a_{12} = 14 + (12 - 1)(-8) = -74.$ ■

Arithmetic Series

The series associated with the arithmetic sequence $\{a_n\}$ is the expression

$$a_1 + (a_1 + d) + (a_1 + 2d) + \cdots + (a_1 + (n - 1)d) + \cdots .$$

The sum S of the first n terms of the series can be determined as follows.

$$\begin{aligned} S = \sum_{i=1}^{n} a_i &= \sum_{i=1}^{n} (a_1 + (i - 1)d) \\ &= \sum_{i=1}^{n} a_1 + \sum_{i=1}^{n} id + \sum_{i=1}^{n} (-d) \\ &= na_1 + d\sum_{i=1}^{n} i + (-d)\sum_{i=1}^{n} 1 \\ &= na_1 + \left(\frac{d}{2}\right)n(n + 1) - nd \\ &= \frac{n}{2}[2a_1 + d(n + 1) - 2d] \\ &= \frac{n}{2}[2a_1 + (n - 1)d] \\ &= \frac{n}{2}[a_1 + (a_1 + (n - 1)d)] \end{aligned}$$

Sum of an Arithmetic Series

$$S = \frac{n}{2}[a_1 + a_n] \tag{2}$$

Example 3 Find the sum of the first 12 terms of the arithmetic series associated with the sequence in example 2.

Solution $S = \frac{12}{2}[14 + (-74)] = -360$ ■

Geometric Sequences

Each of the sequences

$$3, 9, 27, 81, \ldots, \quad 4, -2, 1, -\tfrac{1}{2}, \ldots,$$
$$\text{and } x, x^2, x^3, x^4, \ldots,$$

enjoys the property that all ratios formed by dividing any term in the sequence by its predecessor term are the same. The **common ratio** of the first sequence is 3 because

$$\tfrac{9}{3} = 3, \quad \tfrac{27}{9} = 3, \quad \text{and } \tfrac{81}{27} = 3.$$

The common ratios for the second and third sequences are $-\frac{1}{2}$ and x, respectively.

A sequence $\{a_n\}$ is called a **geometric sequence** if $a_i/a_{i-1} = r$ for each i. If r, the common ratio of the sequence, is 1, then $a_i = a_{i-1}$ for all i, and each term of the sequence is 1. We shall avoid such sequences by insisting that r not be 1.

Because a geometric sequence has a common ratio, its terms can be written as follows

$$a_2 = a_1r, \quad a_3 = a_2r = (a_1r)r = a_1r^2, \quad a_4 = a_3r = (a_1r^2)r = a_1r^3, \text{ etc.}$$

Therefore, we have by induction

n^{th} Form of a Geometric Sequence

$$a_n = a_1r^{n-1}. \tag{3}$$

Example 4 Find the nth term of each sequence.

a. $6, 3, \frac{3}{2}, \ldots$

b. $\frac{y}{2}, 2, \frac{8}{y}, \ldots$

Solution

a. $r = \frac{3}{6} = \frac{1}{2}$. Hence, $a_n = 6\left(\frac{1}{2}\right)^{n-1} = \frac{3}{2^{n-2}}$.

b. $r = \frac{2}{y/2} = \frac{4}{y}$. Hence, $a_n = \left(\frac{y}{2}\right)\left(\frac{4}{y}\right)^{n-1} = \frac{2^{2n-3}}{y^{n-2}}$. ■

Example 5 Find a_9 in the geometric sequence $-9, 3, -1, \ldots$.

Solution $r = 3/(-9) = -\frac{1}{3}$. Therefore $a_9 = -9(\frac{1}{3})^8 = -9(\frac{1}{9})^4 = -\frac{1}{729}$. ■

Geometric Series

The series associated with a geometric sequence $\{a_n\}$ having common ratio r is

$$a_1 + a_1r + a_1r^2 + \cdots + a_1r^{n-1} + \cdots.$$

The sum S of the first n terms of the series can be found in the following way. Let

$$S = a_1 + a_1r + a_1r^2 + \cdots + a_1r^{n-1};$$

then

$$rS = a_1r + a_1r^2 + \cdots + a_1r^{n-1} + a_1r^n.$$

Subtracting the second equation from the first yields

$$S - rS = a_1 - a_1r^n$$

and

Sum of a Geometric Series

$$S = \frac{a_1 - a_1r^n}{1 - r}. \tag{4}$$

Example 6 Find the sum of the first nine terms of the geometric series associated with the sequence in example 5.

Solution Applying (4) with $a_1 = -9$, $r = (-\frac{1}{3})$, and $n = 9$,

$$S = \frac{-9 + 9\left(-\frac{1}{3}\right)^9}{1 - \left(-\frac{1}{3}\right)}$$

$$= \frac{-9\left(1 + \left(\frac{1}{19{,}683}\right)\right)}{\frac{4}{3}}$$

$$= -\frac{27(19{,}684)}{4(19{,}683)}$$

$$= -\frac{4{,}921}{729}.$$

■

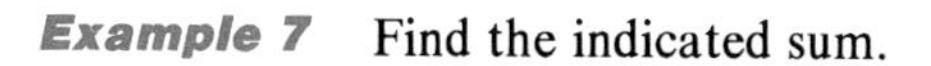

Example 7 Find the indicated sum.

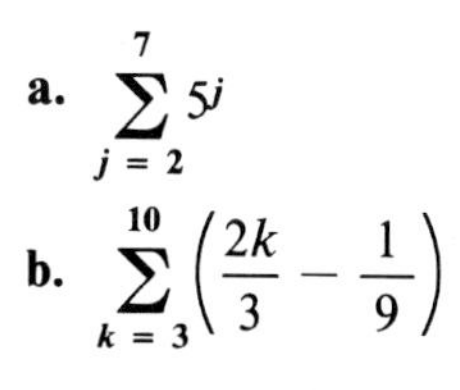

a. $\sum_{j=2}^{7} 5^j$

b. $\sum_{k=3}^{10} \left(\frac{2k}{3} - \frac{1}{9}\right)$

Solution

a. We seek the sum of a geometric series that has a common ratio of 5 and a first term of 25. Applying (4) gives us

$$S = \frac{25 - 25(5)^6}{1 - 5} = 97{,}650.$$

b. With the properties of Σ notation we can write the sum as

$$\left(\frac{2}{3}\right)\sum_{k=3}^{10} k - \left(\frac{1}{9}\right)\sum_{k=3}^{10} 1.$$

The first sum is an arithmetic series of eight terms that begins with 3 and has a common difference of 1. By (2)

$$\left(\frac{2}{3}\right)\sum_{k=3}^{10} k = \left(\frac{2}{3}\right)\left(\frac{8}{2}\right)(3 + 10) = \frac{104}{3}.$$

Also, $$\left(-\frac{1}{9}\right)\sum_{k=3}^{10} 1 = -\frac{8}{9}.$$

Hence, $$\sum_{k=3}^{10}\left(\frac{2k}{3} - \frac{1}{9}\right) = \frac{104}{3} - \frac{8}{9} = \frac{304}{9}.$$

The last result is also obtained by observing that

$$[(\tfrac{2}{3})(k + 1) - \tfrac{1}{9}] - [(\tfrac{2}{3})k - \tfrac{1}{9}] = \tfrac{2}{3}.$$

Thus, $\sum_{k=3}^{10} (2k/3 - \frac{1}{9})$ is an arithmetic series with eight terms and a common difference of $\frac{2}{3}$. The sum, using (2), is

$$\left(\frac{8}{2}\right)\left(\frac{17}{9} + \frac{59}{9}\right) = \frac{304}{9}.$$

■

Infinite Geometric Series

In grade school you determined the decimal fraction equivalent of $\frac{1}{3}$ by dividing 3 into 1,

$$\begin{array}{r} 0.333\ .\ .\ . \\ 3\overline{)\,1.000\ .\ .\ .} \\ \underline{9} \\ 10 \\ \underline{9} \\ 10 \end{array}$$

until it was apparent that the process continues to yield 3 after 3 after 3 ad nauseum. You willingly accepted that $\frac{1}{3}$ and 0.333 . . . were symbols for the same number, even though you never made it through to the end of the long division process. Simply put, is it possible to show that 0.3 followed by an endless string of threes equals $\frac{1}{3}$? If we look at 0.333 . . . as

$$\tfrac{3}{10} + \tfrac{3}{100} + \tfrac{3}{1{,}000} + \tfrac{3}{10{,}000} + \cdots, \tag{5}$$

then the question, "What number is represented by the symbol 0.333 . . ." becomes, "Does (5) have a sum?" Expression (5) is an **infinite geometric series** with common ratio $\frac{1}{10}$. It is a specific instance of an expression such as

$$\sum_{k=1}^{\infty} ar^{k-1} = a + ar + ar^2 + \cdots + ar^{n-1} + \cdots \tag{6}$$

that is called an **infinite geometric series with common ratio *r*.** Note that in expression (6) there is no last natural number value for the index k.

Having identified (5) as an infinite geometric series, we now find its sum with the aid of (4),

$$S = \frac{a - ar^n}{1 - r} = \frac{a(1 - r^n)}{1 - r} = \frac{a}{1 - r}(1 - r^n).$$

If $|r| < 1$, then r^n becomes very small and approaches 0 as n becomes very large (see the table below).

r^n

r \ n	1	2	5	10
$\frac{1}{10}$	0.1	0.01	0.00001	0.0000000001
$\frac{1}{4}$	0.25	0.0625	0.000997	0.000000954
$-\frac{2}{3}$	−0.67	0.44	−0.1317	0.01734

Hence, for large n,

$$\begin{aligned} S &= \frac{a}{1 - r}(1 - r^n) \\ &\approx \frac{a}{1 - r}(1 - 0) \\ &= \frac{a}{1 - r}. \end{aligned}$$

We conclude that if $|r| < 1$ and n is large, then the sum of the first n terms of (6) differs very little from $a/(1 - r)$. With the aid of calculus it is possible to show that if $|r| < 1$, then

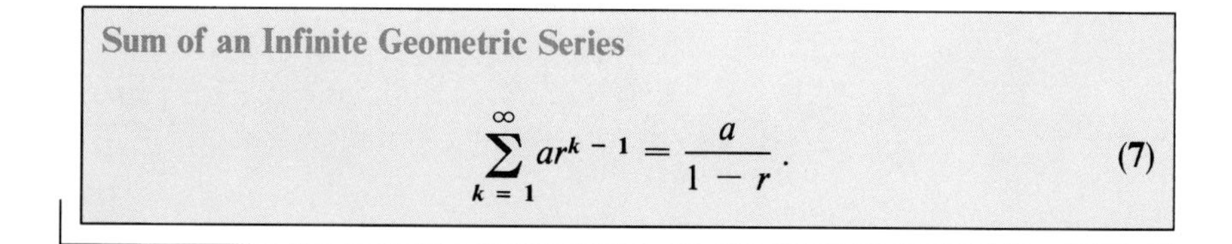

Sum of an Infinite Geometric Series

$$\sum_{k=1}^{\infty} ar^{k-1} = \frac{a}{1 - r}. \tag{7}$$

Example 8

a. Find the sum of the first 10 terms of (5).

b. Find the sum of the infinite geometric series denoted by (5).

Solution

a. $$S = \frac{a - ar^n}{1 - r} = \frac{a}{1 - r}(1 - r^n)$$

$$= \frac{\frac{3}{10}}{1 - \frac{1}{10}}\left(1 - \left(\frac{1}{10}\right)^{10}\right)$$

$$= \frac{1}{3}(0.9999999999)$$

$$\approx \frac{1}{3}$$

b. Applying (7) with $a = \frac{3}{10}$ and $r = \frac{1}{10}$

$$S = \frac{\frac{3}{10}}{1 - \frac{1}{10}}$$

$$= \frac{1}{3}.$$

■

Example 9 Find the rational number whose nonterminating decimal expansion is $0.12\overline{12}$. The bar over the last two digits indicates that the digit pair, 12, repeats ad infinitum.

Solution The nonterminating decimal, $0.12\overline{12}$, is seen as an infinite geometric series with $a = \frac{12}{100}$ and $r = \frac{1}{100}$ when it is written as

$$\frac{12}{100} + \frac{12}{10{,}000} + \frac{12}{1{,}000{,}000} + \cdots.$$

The sum of this series is

$$\frac{\frac{12}{100}}{1 - \frac{1}{100}} = \frac{\frac{12}{100}}{\frac{99}{100}}$$

$$= \frac{12}{99}$$

$$= \frac{4}{33}.$$

Hence, $0.12\overline{12} = \frac{4}{33}$.

■

Exercise Set 13.3

(A)

Determine if the sequence is arithmetic or geometric, and find the indicated term. Write an expression for the nth term.

1. $7, 14, 21, \ldots, a_{12}$
2. $27, 18, 9, \ldots, a_7$
3. $2, 4, 8, \ldots, a_9$
4. $3, 9, 27, \ldots, a_8$
5. $3, \frac{13}{5}, \frac{11}{5}, \ldots, a_{16}$
6. $-7, -3, 1, \ldots, a_7$
7. $\frac{1}{2}, \frac{3}{4}, \frac{9}{8}, \ldots, a_{10}$
8. $-\frac{1}{2}, \frac{1}{4}, -\frac{1}{8}, \ldots, a_{11}$
9. $9, -6, 4, \ldots, a_8$
10. $3, 0.03, 0.0003, \ldots, a_9$
11. $x, 2x + 1, 3x + 2, \ldots, a_{12}$
12. $y, 3, 6 - y, \ldots, a_8$

Use (2) or (4) and the properties of Σ notation to find the sum.

13. $\sum_{k=1}^{8} (2k + 5)$
14. $\sum_{j=2}^{8} (-3j + 8)$
15. $\sum_{j=2}^{6} \left(\frac{1}{3}\right)^j$
16. $\sum_{k=3}^{9} \left(\frac{1}{2}\right)^k$
17. $\sum_{k=2}^{7} (-1)^k \left(\frac{1}{2^k}\right)$
18. $\sum_{k=1}^{4} (-1)^{k+1} \left(\frac{1}{3^k}\right)$
19. $\sum_{j=1}^{8} \left(\frac{j}{2} - 3\right)$
20. $\sum_{k=3}^{10} \left(\frac{2k}{3} - \frac{1}{9}\right)$
21. $\sum_{k=3}^{8} (4 - 2^k)$
22. $\sum_{j=1}^{5} (3^j + j)$
23. $\sum_{i=1}^{10} 24\left(\frac{1}{2}\right)^{i-1}$
24. $\sum_{i=2}^{15} \left(\frac{4 - 3i}{10}\right)$
25. Find the sum of all even natural numbers between 11 and 93.
26. Find the sum of all integral multiples of 3 between 14 and 118.

Find the sum, if it exists, of each infinite geometric series.

27. $25 + 10 + 4 + \cdots$
28. $4 + 10 + 25 + \cdots$
29. $3 + 6 + 12 + \cdots$
30. $12 + 6 + 3 + \cdots$
31. $1 + \frac{2}{3} + \frac{4}{9} + \cdots$
32. $1 + \frac{2}{5} + \frac{4}{25} + \cdots$
33. $\sum_{j=1}^{10} \frac{1}{2^j}$
34. $\sum_{k=1}^{\infty} \frac{1}{3^{k-1}}$
35. $\sum_{j=1}^{\infty} (-1)^{j+1} 3^j$
36. $\sum_{k=1}^{\infty} (-1)^{k-1} 2^{k+1}$
37. $\sum_{i=1}^{\infty} \left(\frac{3}{4}\right)^i$
38. $\sum_{i=1}^{\infty} (-1)^i \left(\frac{1}{3}\right)^i$

Find the rational number associated with each nonterminating decimal.

39. $0.4\overline{4}$
40. $0.18\overline{18}$
41. $0.49\overline{9}$
42. $0.74\overline{4}$
43. $3.\overline{923076}$
44. $2.\overline{428571}$

(B)

45. Which term is 125 in the arithmetic sequence 2, 5, 8, . . . ?
46. How many even integers are there between 11 and 91?
47. How many integral multiples of 7 are there between 8 and 110?
48. Show that the sum of the first n positive odd integers is n^2.
49. Show that the sum of the first n positive even integers is $n(n + 1)$.
50. A ball is dropped from a height of 3 feet. Each time it strikes the floor, the ball bounces up $\frac{2}{3}$ of the distance it fell. Find the distance traveled by the ball up to the instant that it strikes the floor the fifth time. Find the distance traveled by the ball until it comes to rest.
51. From what height should the ball in exercise 50 be dropped so that the distance it travels until it comes to rest is 24 feet?
52. An auditorium has 20 rows of seats. The last row has 60 seats, and each row has 2 fewer seats than the row behind it. How many seats are in the auditorium?

53. Prove by induction that the nth term of an arithmetic sequence is given by the formula $a_n = a_1 + (n - 1)d$ where d is the common difference.

54. Prove by induction that the nth term of a geometric sequence is given by the formula $a_n = a_1r^{n-1}$ where r is the common ratio.

55. A \$3,600 loan is paid off in 18 months at \$200 a month plus interest, which is 1% a month on the unpaid balance. How much interest is paid over the life of the loan?

56. A new \$10,000 automobile depreciates at the rate of 20% a year. What is the market value of the car five years after it is purchased?

57. The population of a country increases at the rate of 1% a year. What will the country's population be 50 years from now if its current population is P?

58. A mixture that is 40% acid is to be diluted by replacing 25% of the mixture with pure water. What percentage of acid remains in the mixture if the dilution process is repeated 10 times? What is the minimum number of times that the dilution process must be repeated so that the mixture contains less than 20% acid?

59. A mysterious benefactor deposits P dollars at the end of each month into an account in your name. The account draws r% annual interest compounded monthly. How much money is in the account n years after it is opened? The first deposit, at the end of the first month, earns $P(1 + r/12)^{12n-1}$ dollars. The second deposit earns $P(1 + r/12)^{12n-2}$ dollars, . . . , and the last deposit, at the end of the last month, earns no interest. Show that the account has

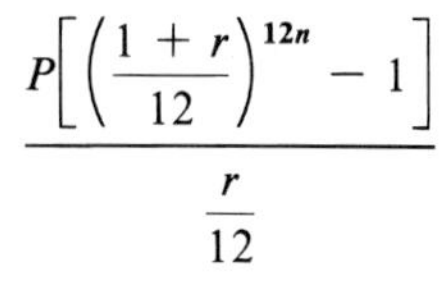

$$\frac{P\left[\left(\dfrac{1+r}{12}\right)^{12n} - 1\right]}{\dfrac{r}{12}}$$

dollars after n years.

60. Two hundred dollars is deposited into an annuity account at the end of each month. The account draws interest at the rate of 12% compounded monthly. How much money is in the account ten years after it is opened? (See exercise 59.)

61. A \$15,000 automobile is purchased on the following terms: 20% down, 12% annual interest rate compounded monthly, 48 equal monthly payments. What are the monthly payments? (Hint: Think of the monthly payments as building an annuity (see exercises 59 and 60) worth the amount of money that a principal of \$12,000 would accumulate were it drawing 12% annual interest compounded monthly over 48 months.)

13.4 Counting and the Binomial Theorem

This chapter began with an induction proof for the sum of an arithmetic series that was based on a pattern constructed out of the doodlings of a bored student. We consider another pattern in this section, one generated when the binomial $(a + b)$ is raised to positive integral powers starting with 1. The French mathematician Blaise Pascal (1623–1662) studied this pattern (see exercise 55), as did Isaac Newton.

$$\begin{aligned}
(a + b)^1 &= a + b \\
(a + b)^2 &= a^2 + 2ab + b^2 \\
(a + b)^3 &= a^3 + 3a^2b + 3ab^2 + b^3 \\
(a + b)^4 &= a^4 + 4a^3b + 6a^2b^2 + 4ab^3 + b^4 \\
(a + b)^5 &= a^5 + 5a^4b + 10a^3b^2 + 10a^2b^3 + 5ab^4 + b^5
\end{aligned} \tag{1}$$

A careful reading of equations (1) suggests that in the expansion of $(a + b)^n$, $n \varepsilon N$,

1. there are $n + 1$ terms.
2. the first and last terms are a^n and b^n.
3. the sum of the exponents on a and b in each term is n.
4. the exponent on a decreases by 1 and the exponent on b increases by 1 with each successive term.
5. there is a left–right symmetry in the appearance of a and b.

Taken together these observations suggest that

$$(a + b)^n = a^n + C_2a^{n-1}b + C_3a^{n-2}b^2 + \cdots + C_ra^{n-(r-1)}b^{r-1} + \cdots + C_nab^{n-1} + b^n, \qquad (2)$$

where C_r is the coefficient of the rth term of the expansion. A simple counting procedure enables us to find C_r. Let us find the coefficient of a^2b in the expansion of $(a + b)^3$.

Since $(a + b)^3 = (a + b)(a + b)(a + b)$,

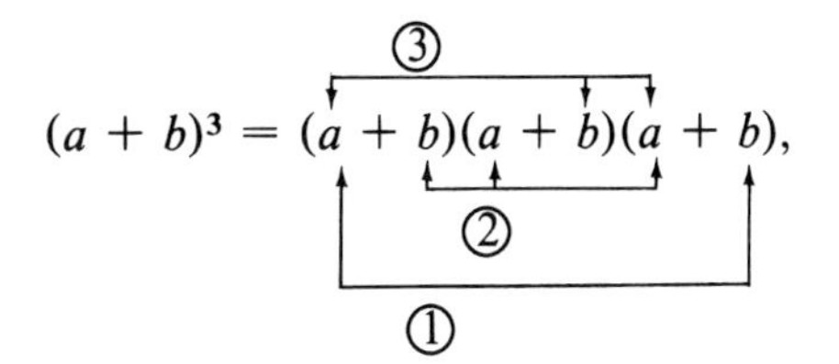

we see that the product a^2b is obtained when ① a's from the first two parentheses are combined with the b from the third parentheses, or when ② the b in the first parentheses is combined with the a's in the last two parentheses, or when ③ the b in the middle parentheses is combined with the a's in the outer parentheses. There are three ways a^2b can be formed, and so the coefficient of a^2b is 3. You might well ask what is so simple about this counting procedure. It is certainly no simple matter to pick and choose the a's and b's amongst the fifteen factors of $(a + b)$ in the expansion of $(a + b)^{15}$, to find the coefficient of a^7b^8. However, the procedure used to find the coefficient of a^2b is easily generalized so that a simple formula generates the coefficient of a^7b^8. Such a formula is derived below.

Permutations of n-Member Sets

A common counting problem is that of determining the number of ways that the elements of a set may be arranged without repeating an element in any arrangement. Such arrangements are called **permutations.**

If we want to find the number of permutations of the elements of the set $\{a,b,c,d\}$ we might consider the task of filling each of four blanks

____ ____ ____ ____

with one letter from the set, not permitting any repetitions. The first blank can be filled in four ways by any one of the four letters. The second blank can be filled in any one of three ways by any of the remaining three letters. Hence the first two blanks can be filled in 12, $(4 \cdot 3)$, ways. The third blank can be filled in two ways by either of the remaining two letters so that the first three blanks can be filled in 24 ways. Since there is only one letter left to fill the last blank, all four blanks can be filled in 24 ways.

$$\underline{\ 4\ } \cdot \underline{\ 3\ } \cdot \underline{\ 2\ } \cdot \underline{\ 1\ } = 24.$$

The indicated product $1 \cdot 2 \cdot 3 \cdot 4$ is often written in shorthand form as **4!**, and it is read as **four factorial.**

Example 1 Evaluate

a. $5!$

b. $6!$

c. $7!$.

Solution

a. $5! = 1 \cdot 2 \cdot 3 \cdot 4 \cdot 5 = 120$

b.
$$\begin{aligned} 6! &= 1 \cdot 2 \cdot 3 \cdot 4 \cdot 5 \cdot 6 \\ &= (1 \cdot 2 \cdot 3 \cdot 4 \cdot 5) \cdot 6 \\ &= 5! \cdot 6 \\ &= 120 \cdot 6 \\ &= 720 \end{aligned}$$

c. $7! = 6! \cdot 7 = 720 \cdot 7 = 5{,}040.$ ■

Example 2 How many different shelf arrangements of eight different books are possible?

Solution We must find the number of permutations of an eight-member set. Hence, we fill each of eight blanks (or positions) with one of the eight books.

$$\underline{\ 8\ } \cdot \underline{\ 7\ } \cdot \underline{\ 6\ } \cdot \underline{\ 5\ } \cdot \underline{\ 4\ } \cdot \underline{\ 3\ } \cdot \underline{\ 2\ } \cdot \underline{\ 1\ } = 8!$$

There are eight possibilities for the first position, seven for the second position, and so on. Thus, there are 8!, or 40,320 book arrangements. ■

The method for determining the number of permutations of an eight-member set generalizes to

Number of Permutations of n Things Taken n at a Time

$$P(n,n) = n!, \quad n \in N, \tag{3}$$

where $P(n,n)$ denotes the number of permutations of an n-member set. The symbol $P(n,n)$ is also often read as the *number of permutations of n things taken n at a time.*

It was shown in example 1 that $6! = 5! \cdot 6$ and that $7! = 6! \cdot 7$. If we attempt to generalize and say that

$$n! = (n - 1)! \cdot n$$

for all natural numbers n, we must consider the case

$$1! = (1 - 1)! \cdot 1$$

or

$$1! = 0! \cdot 1.$$

The last equation makes sense only if $0! = 1$. Hence **we define 0! to be 1.**

r-Member Subsets of n-Member Sets

Let us return to the book arrangement problem of example 2 and imagine that we have room on our bookshelf for only five of the eight books. How many different five-book arrangements are possible if we use all eight books? Our task now is to fill five positions with one of eight different book choices.

$$\underline{\ 8\ } \cdot \underline{\ 7\ } \cdot \underline{\ 6\ } \cdot \underline{\ 5\ } \cdot \underline{\ 4\ } = 6{,}720.$$

There are eight choices for the first position, seven for the second position, and so on. Note that

$$8 \cdot 7 \cdot 6 \cdot 5 \cdot 4 = \frac{(8 \cdot 7 \cdot 6 \cdot 5 \cdot 4)(3 \cdot 2 \cdot 1)}{3 \cdot 2 \cdot 1} = \frac{8!}{3!}.$$

It is not difficult to generalize the above result to show that the number of permutations of an r-member subset of an n-member set (or *the number of permutations of n things taken r at a time*) is

Number of Permutations of n Things Taken r at a Time

$$P(n,r) = \frac{n!}{(n - r)!}. \tag{4}$$

Example 3 How many four-digit numerals having no repeating digits can be formed from the elements in $\{1,2,3,4,5,6,7,8,9\}$?

Solution We must find the number of permutations of nine things taken four at a time. From (4)

$$P(9,4) = \frac{9!}{(9-4)!} = \frac{9 \cdot 8 \cdot 7 \cdot 6 \cdot (\cancel{5 \cdot 4 \cdot 3 \cdot 2 \cdot 1})}{\cancel{5 \cdot 4 \cdot 3 \cdot 2 \cdot 1}}$$
$$= 3{,}024.$$

■

Example 4 How many three-letter words that have no repeating letters can be formed from the letters $a,b,c,d,e,f,$ and g?

Solution We must find the number of permutations of seven things taken three at a time.

$$P(7,3) = \frac{7!}{(7-3)!}$$
$$= \frac{(7 \cdot 6 \cdot 5 \cdot \cancel{4}!)}{\cancel{4}!}$$
$$= 210$$

■

Combinations

Many counting problems require that we determine the number of r-element subsets of an n-element set. Such subsets are called **combinations.** Their number is denoted by the symbol $\binom{n}{r}$ which is often read, *"The number of combinations of n things taken r at a time."*

Example 5 Find the number of three-letter combinations of the four-letter set $\{a,b,c,d\}$.

Solution Because the number of letters in the given set is small, we are easily able to write all its three-letter combinations.

$$\{a,b,c\}, \quad \{a,b,d\}, \quad \{a,c,d\}, \quad \{b,c,d\}$$

Hence, $$\binom{4}{3} = 4.$$

■

If we associate with each combination in example 5 all of its three-letter permutations, then we generate all the permutations of four letters taken three at a time.

$\{a,b,c\}$	$\{a,b,d\}$	$\{a,c,d\}$	$\{b,c,d\}$
a,b,c	a,b,d	a,c,d	b,c,d
a,c,b	a,d,b	a,d,c	b,d,c
b,a,c	b,a,d	c,a,d	c,b,d
b,c,a	b,d,a	c,d,a	c,d,b
c,a,b	d,a,b	d,a,c	d,b,c
c,b,a	d,b,a	d,c,a	d,c,b

The above tabulation suggests equation (5),which determines the number of permutations of four things taken three at a time.

$$\binom{4}{3} \cdot P(3,3) = P(4,3) \tag{5}$$

But it is in general true that a combination of n things taken r at a time has $P(r,r)$ permutations associated with it. Therefore, the number of permutations of n things taken r at a time can be determined from equation (6), which is a generalization of (5).

$$\binom{n}{r} \cdot P(r,r) = P(n,r) \tag{6}$$

If (6) is solved for $\binom{n}{r}$, a useful formula for finding the number of combinations of n things taken r at a time is obtained.

$$\binom{n}{r} = \frac{P(n,r)}{P(r,r)}$$

$$\binom{n}{r} = \frac{\dfrac{n!}{(n-r)!}}{r!}$$

Number of Combinations of n Things Taken r at a Time

$$\binom{n}{r} = \frac{n!}{(n-r)!\,r!} \tag{7}$$

Example 6 How many different hands of five cards each can be dealt from an ordinary bridge deck of 52 cards?

Solution The number of five-card hands is the number of combinations of 52 things taken five at a time.

$$\binom{52}{5} = \frac{52!}{(52-5)!5!} = \frac{52 \cdot 51 \cdot 50 \cdot 49 \cdot 48 \cdot \cancel{47!}}{\cancel{47!} \cdot 5 \cdot 4 \cdot 3 \cdot 2 \cdot 1} = 2{,}598{,}960$$

■

Example 7 Find the coefficient of a^7b^8 in the expansion of $(a + b)^{15}$.

Solution The expansion of $(a + b)^{15}$ requires that we multiply fifteen binomial factors $(a + b)$. Thus products a^7b^8 can be formed by multiplying the a's from seven of the binomial factors and the b's from the remaining eight binomial factors. The number of such products is the number of combinations of fifteen things taken seven at a time or the number of combinations of fifteen things taken eight at a time. Because

$$\binom{15}{7} = \frac{15!}{(15-7)!7!} = \frac{15!}{(15-8)!8!} = \binom{15}{8},$$

the coefficient of a^7b^8 is

$$\frac{15 \cdot 14 \cdot 13 \cdot 12 \cdot 11 \cdot 10 \cdot 9 \cdot \cancel{8!}}{\cancel{8!} \cdot 7 \cdot 6 \cdot 5 \cdot 4 \cdot 3 \cdot 2 \cdot 1} = 6{,}435.$$

■

The Binomial Theorem

The result in example 7 is all we need to return to equation (2) at the beginning of the section and fill in the missing coefficients in the expansion of $(a + b)^n$. In example 7 we see that because there are eight factors of b in the term a^7b^8 of the expansion $(a + b)^{15}$, the coefficient of a^7b^8 is $\binom{15}{8}$. It follows from the same argument that is used in example 7 that the coefficient of the term $a^{n-(r-1)}b^{(r-1)}$ in the expansion of $(a + b)^n$ is $\binom{n}{r-1}$. Consequently, equation (2) becomes

The Binomial Theorem

$$(a + b)^n = \binom{n}{0}a^n + \binom{n}{1}a^{n-1}b + \binom{n}{2}a^{n-2}b^2 + \cdots + \binom{n}{r-1}a^{n-(r-1)}b^{(r-1)} + \cdots + \binom{n}{n}b^n. \tag{8}$$

Example 8 Write the expansion of $(2x - y)^7$.

Solution $(2x - y)^7 = (2x + (-y))^7$. Let $2x$ play the role of a, and let $-y$ play the role of b, in the expansion of $(a + b)^7$.

$$(a + b)^7 = \binom{7}{0}a^7 + \binom{7}{1}a^6b + \binom{7}{2}a^5b^2 + \binom{7}{3}a^4b^3 + \binom{7}{4}a^3b^4 + \binom{7}{5}a^2b^5 + \binom{7}{6}ab^6 + \binom{7}{7}b^7$$

$$\begin{aligned}(2x - y)^7 &= (2x)^7 + 7(2x)^6(-y) + 21(2x)^5(-y)^2 + 35(2x)^4(-y)^3 \\ &\quad + 35(2x)^3(-y)^4 + 21(2x)^2(-y)^5 + 7(2x)(-y)^6 + (-y)^7 \\ &= 128x^7 - 448x^6y + 672x^5y^2 - 560x^4y^3 + 280x^3y^4 - 84x^2y^5 \\ &\quad + 14xy^6 - y^7\end{aligned}$$

■

Exercise Set 13.4

(A)

Compute each of the following.

1. $P(6,6)$ **2.** $P(10,10)$ **3.** $P(7,5)$

4. $P(8,4)$ **5.** $\binom{5}{3}$ **6.** $\binom{5}{2}$

7. $P(9,3)$ **8.** $P(9,9)$ **9.** $\binom{10}{7}$

10. $\binom{10}{3}$ **11.** $\binom{12}{4}$ **12.** $\binom{12}{8}$

13. How many five-digit numerals can be written using the digits 1,2,3,4,5
a. without repetitions?
b. with repetitions?

14. How many six-digit numerals can be written using the digits 1,2,3,4,5, and 6
a. without repetitions?
b. with repetitions?

15. How many five-digit numerals can be written using the digits 1 through 9
a. without repetitions?
b. with repetitions?

16. How many six-digit numerals can be written using the digits 1 through 9
a. without repetitions?
b. with repetitions?

17. How many different three-letter "words" can be made by using three of the letters in the word TRIANGLE?

18. How many different four-letter "words" can be made using four of the letters in the word SQUARE?

19. In how many ways can a committee of five be chosen from a group of ten people?

20. In how many ways can a committee of three men and two women be chosen from a group of six men and six women?

21. In how many ways can a set of three books be selected from a set of eight different books?

22. Ten points lie on a circle. How many chords are determined by these points? How many triangles are determined by these points as vertices?

23. A man has four different kinds of coins and four bills of different denominations. How many different monetary combinations can he form if he selects three coins and two bills?

24. In how many ways can we select two mathematics texts and four English texts from a set of six mathematics texts and eight English texts?

25. In how many ways can a selection of six books be made from a set of thirteen different books? How many selections of six books can be made if two certain books must be chosen? How many selections of six books can be made if two certain books may not be chosen?

26. From a group of seven men and seven women, how many different committees of five can be chosen that contain

a. exactly three women?

b. at least one man?

Write and simplify the expansion of each binomial expression.

27. $(x + 3)^6$

28. $(y - 3)^5$

29. $(x + 2y)^7$

30. $(2y - 1)^5$

31. $(3x - 2y)^6$

32. $\left(x - \frac{y}{2}\right)^8$

33. $\left(2x - \frac{y}{3}\right)^4$

34. $\left(\frac{x}{3} + 2y\right)^5$

Write and simplify the first four terms of the expansion of each binomial expression.

35. $(x + 2y)^{48}$

36. $(x - \sqrt{3})^{12}$

37. $(x^2 - \sqrt{2})^{18}$

38. $\left(x - \frac{y}{3}\right)^{20}$

(B)

39. A combination of three marbles is to be chosen at random from an urn containing eight red, six white, and seven blue marbles. In how many ways can the chosen set contain at least one white marble?

40. Determine in how many ways we can choose four marbles at random from the urn in exercise 39 so that one marble is red and at least two marbles are blue.

41. In how many ways can a committee of four persons be selected from five married couples if

a. a couple may not serve together on a committee?

b. no two couples may serve on the same committee?

42. How many arrangements of the letters of the word FEINT can be made if the vowels are not to be separated?

43. How many five-digit numerals can be formed from the digits 1 through 9 if the odd digits must occupy the odd places and no digit is to be repeated in any numeral?

44. In how many ways can ten books be arranged on a shelf if two specified books must be side by side?

45. How many three-digit numbers are there that are less than 500 in which

a. no digit is repeated?

b. the digits may be repeated?

46. Show that $\binom{n}{r-r} = \binom{n}{r}$.

47. Show that $\binom{n+1}{r+1} = \binom{n}{r+1} + \binom{n}{r}$.

48. If $\binom{n}{2}/\binom{2n}{3} = \frac{3}{44}$, find n.

49. If $\binom{n}{7} = \binom{n}{9}$, find $\binom{n}{15}$ and $\binom{20}{n}$.

50. Show that $\binom{n}{0} + \binom{n}{1} + \cdots + \binom{n}{n} = 2^n$. (Hint: Think of the binomial expansion in which $a = b = 1$.)

51. Show that the number of ways of selecting at least one of n things is $2^n - 1$. (Hint: See exercise 50.)

52. How many different five-card hands containing four aces can be dealt from a 52-card deck?

53. How many different five-card hands, in which all the cards are from the same suit, can be dealt from a 52-card deck?

54. Prove $P(n,n) = n!$.

55. The triangle of numbers shown below is called Pascal's Triangle. The numbers in each row are the binomial coefficients for $n = 0$, $n = 1$, $n = 2$, etc. Generate the next two rows of the triangle. What is the rule for generating the numbers in any row?

$$
\begin{array}{ccccccccccc}
 & & & & & 1 & & & & & \\
 & & & & 1 & & 1 & & & & \\
 & & & 1 & & 2 & & 1 & & & \\
 & & 1 & & 3 & & 3 & & 1 & & \\
 & 1 & & 4 & & 6 & & 4 & & 1 & \\
1 & & 5 & & 10 & & 10 & & 5 & & 1
\end{array}
$$

Chapter 13 Review

Major Points for Review

Section 13.1 The Axiom of Mathematical Induction

How to construct proofs by induction

Section 13.2 What a sequence is

What a series is

The properties of sigma notation

Section 13.3 What an arithmetic sequence is

What a geometric sequence is

The formulas for the nth terms of arithmetic and geometric sequences

The summation formulas for finite arithmetic and geometric series

Under what conditions a geometric series has a sum

The formula for the sum of an infinite geometric series

Section 13.4 What is meant by the number of permutations of n things taken n at a time

The permutation formulas

What is meant by the number of combinations of n things taken r at a time

The formula for determining the number of combinations of n things taken r at a time

The Binomial Theorem

How to use the Binomial Theorem to expand a binomial that is raised to a natural number power

Review Exercises

Write the terms of the sequence.

1. $\{3n + 1\}, n \in \{1,2,3,4\}$

2. $\{(n^3 - 5)/n\}, n \in \{1,2,3,4\}$

Write the first four terms, followed by the ellipsis symbol and the nth term for each sequence.

3. $\left\{\frac{1}{2^n}\right\}$

4. $\{(-1)^{n+1} 3n\}$

Write the series associated with the sequence in the indicated exercises.

5. Exercise 4 **6.** Exercise 1

Write each series in expanded form.

7. $\sum_{i=1}^{8} (3i - 1)$ **8.** $\sum_{k=2}^{7} 5k^2$

9. $\sum_{j=1}^{6} (j + 3)(j - 1)$ **10.** $\sum_{k=3}^{11} \left(\frac{k}{4} - 3\right)$

Use sigma notation to write each series in compact form.

11. $6 + 12 + 24 + 48 + 96 + 192$

12. $14 + 19 + 24 + 29 + 34$

Write the nth term of each sequence.

13. $1, -\frac{1}{3}, \frac{1}{5}, -\frac{1}{7}, \frac{1}{9}, \ldots$

14. $10, 28, 82, 244, \ldots$

Specify the first four terms of the sequence given by each recursion formula.

15. $a_1 = -1, a_{n+1} = 2a_n - 3$

16. $a_1 = 4, a_{n+1} = (-1)^n \frac{a_n}{2^n}$

Determine if the sequence is arithmetic or geometric, and find the indicated term. Write an expression for the nth term.

17. $6, 18, 30, \ldots, a_9$ **18.** $34, 26, 18, \ldots, a_{12}$

19. $12, 16, \frac{64}{3}, \ldots, a_{10}$ **20.** $3, -12, 48, \ldots, a_7$

Find each sum.

21. $\sum_{i=1}^{9} (3i - 2)$ **22.** $\sum_{k=2}^{12} (-5k + 4)$

23. $\sum_{j=1}^{5} (-1)^j \left(\frac{3}{4}\right)^j$ **24.** $\sum_{k=3}^{10} (2^k - 3)$

25. $\sum_{i=2}^{8} (3^{i-1} - 2i)$ **26.** $\sum_{j=1}^{20} \left(\frac{5j}{2} + \frac{1}{8}\right)$

27. $\sum_{i=1}^{\infty} \left(\frac{2}{3}\right)^i$ **28.** $\sum_{k=2}^{\infty} \left(-\frac{5}{6}\right)^k$

Find the rational number associated with each nonterminating decimal.

29. $0.32\overline{32}$ **30.** $0.26\overline{6}$

Compute each number.

31. $P(7,7)$ **32.** $P(9,4)$

33. $\binom{9}{4}$ **34.** $\binom{11}{3}$

Write an expansion of each binomial expression.

35. $(x + 2)^7$ **36.** $(2y - 3)^6$

37. $(x - \frac{1}{2})^5$ **38.** $(3x + 2y)^8$

39. How many four-digit numerals can be written using the digits 5, 6, 7, and 8 so that no numeral has any repeating digits?

40. How many five-book shelf arrangements can be made from a set of eight books?

41. How many different license plates can be made if each license plate has three digits followed by three letters? No digit or letter may be repeated.

42. In how many ways can five books be selected from a set of eight books?

43. If you can choose three ingredients from a set of ten ingredients, how many different three-ingredient pizzas can you order?

44. How many seven-person committees can be selected from a group of ten people?

Use mathematical induction to prove each statement.

45. $\frac{1}{2} - \frac{1}{4} - \frac{1}{8} - \cdots - \frac{1}{2^n} = \frac{1}{2^n}$

46. $2 + 5 + 8 + \cdots + (3n - 1) = \frac{n(3n + 1)}{2}$

47. Write a recursion formula to generate the sequence $\{3n - 1\}$.

48. Write the nth term of the sequence generated by the recursion formula $a_1 = 5$, $a_{n+1} = 2a_n$.

49. Which term is 127 in the arithmetic sequence 3, 7, 11, . . . ?

50. Find a_{13} in the arithmetic sequence in which $a_4 = -11$ and $a_{19} = -21$.

51. Find the first term in the geometric sequence in which $a_4 = 48$ and the common ratio is 2.

52. Find the sum of all positive integers less than 180 that are multiples of 6.

53. A $9,600 loan is paid off in two years at the rate of $400 a month plus 2% interest on the unpaid balance. How much interest is paid over the life of the loan?

54. The current population of a country is P. What will the country's population be in 20 years if its population increases at the rate of 2% per year?

55. How many even numbers greater than 200 can be formed from the digits 0, 1, 2, 3, and 4 if no digit can be repeated?

56. How many different five-card hands that contain three kings can be dealt from a 52-card deck?

57. How many different committees of five can be chosen from a group of eight Democrats and five Republicans if each committee must contain exactly three Democrats?

58. Repeat exercise 57 with this change. Each committee must contain at least three Democrats.

Chapter 13 Test 1

1. Write the terms of the sequence $\{2n - 3\}$, $1 \leq n \leq 5$. Write the series associated with the sequence.

2. Write the nth term of the sequence, 2, 8, 32, ...

In exercises 3–4, find the sum.

3. $\sum_{n=1}^{15} (3n + 2)$ **4.** $\sum_{k=2}^{8} 3^{k-2}$

5. Find the rational number that has $0.4\overline{4}$ as its decimal expansion.

6. Compute each number.

a. $P(8,8)$

b. $\binom{8}{2}$

7. Write an expansion for the binomial expression $(x + 3)^7$.

8. How many three-digit numbers can be written using the digits 1, 2, 3, 4, and 5 if no number may have repeating digits?

9. How many different five-person committees can be formed from a group of nine people?

10. The cost of a new car is C dollars. The car depreciates 12% per year. What is the value of the car five years after it is purchased?

Chapter 13 Test 2

In exercises 1–2, find the sum.

1. $\sum_{k=3}^{26} \left(\frac{k}{5} - 3\right)$ **2.** $\sum_{k=2}^{10} (2^{k+1} - 3k)$

3. Find the rational number that has $0.56\overline{464}$ as its decimal expansion.

4. Write an expansion of the binomial expression

$$\left(3x - \frac{y}{2}\right)^8.$$

5. How many odd numbers less than 300 can be written using the digits 1, 2, 3, 4, and 5 if no number may have repeating digits?

6. How many five-person committees can be formed from a group of seven women and four men if no committee can have more than three men?

7. If in an arithmetic progression $a_3 = 17$ and $a_{13} = 43$, find a_{20}.

8. Write a recursion formula to generate the sequence $\{2n + 3\}$.

9. With each stroke, a vacuum pump removes $\frac{1}{5}$ of the volume of gas that remains in a container. How much of the original gas volume is left in the container after ten strokes?

10. A $5,000 loan is repaid in ten months with payments of $500 per month plus 3% interest on the unpaid balance. How much interest is paid on the loan?

Appendix A

Table 1 Exponential Functions

Table 2 Logarithms to Base 10

Table 3 Natural Logarithms (Base *e*)

Table 4 Values of Trigonometric Functions (Radian Measure)

Table 5 Values of Trigonometric Functions (Degree Measure)

Table 1 Exponential Functions

x	e^x	e^{-x}	x	e^x	e^{-x}
0.00	1.0000	1.0000	1.5	4.4817	0.2231
0.01	1.0101	0.9901	1.6	4.9530	0.2019
0.02	1.0202	0.9802	1.7	5.4739	0.1827
0.03	1.0305	0.9705	1.8	6.0496	0.1653
0.04	1.0408	0.9608	1.9	6.6859	0.1496
0.05	1.0513	0.9512	2.0	7.3891	0.1353
0.06	1.0618	0.9418	2.1	8.1662	0.1225
0.07	1.0725	0.9324	2.2	9.0250	0.1108
0.08	1.0833	0.9231	2.3	9.9742	0.1003
0.09	1.0942	0.9139	2.4	11.023	0.0907
0.10	1.1052	0.9048	2.5	12.182	0.0821
0.11	1.1163	0.8958	2.6	13.464	0.0743
0.12	1.1275	0.8869	2.7	14.880	0.0672
0.13	1.1388	0.8781	2.8	16.445	0.0608
0.14	1.1503	0.8694	2.9	18.174	0.0550
0.15	1.1618	0.8607	3.0	20.086	0.0498
0.16	1.1735	0.8521	3.1	22.198	0.0450
0.17	1.1853	0.8437	3.2	24.533	0.0408
0.18	1.1972	0.8353	3.3	27.113	0.0369
0.19	1.2092	0.8270	3.4	29.964	0.0334
0.20	1.2214	0.8187	3.5	33.115	0.0302
0.21	1.2337	0.8106	3.6	36.598	0.0273
0.22	1.2461	0.8025	3.7	40.447	0.0247
0.23	1.2586	0.7945	3.8	44.701	0.0224
0.24	1.2712	0.7866	3.9	49.402	0.0202
0.25	1.2840	0.7788	4.0	54.598	0.0183
0.30	1.3499	0.7408	4.1	60.340	0.0166
0.35	1.4191	0.7047	4.2	66.686	0.0150
0.40	1.4918	0.6703	4.3	73.700	0.0136
0.45	1.5683	0.6376	4.4	81.451	0.0123
0.50	1.6487	0.6065	4.5	90.017	0.0111
0.55	1.7333	0.5769	4.6	99.484	0.0101
0.60	1.8221	0.5488	4.7	109.95	0.0091
0.65	1.9155	0.5220	4.8	121.51	0.0082
0.70	2.0138	0.4966	4.9	134.29	0.0074
0.75	2.1170	0.4724	5.0	148.41	0.0067
0.80	2.2255	0.4493	5.5	244.69	0.0041
0.85	2.3396	0.4274	6.0	403.43	0.0025
0.90	2.4596	0.4066	6.5	665.14	0.0015
0.95	2.5857	0.3867	7.0	1,096.6	0.0009
1.0	2.7183	0.3679	7.5	1,808.0	0.0006
1.1	3.0042	0.3329	8.0	2,981.0	0.0003
1.2	3.3201	0.3012	8.5	4,914.8	0.0002
1.3	3.6693	0.2725	9.0	8,103.1	0.0001
1.4	4.0552	0.2466	10.0	22,026	0.00005

Table 2 Logarithms to Base 10

m	0	1	2	3	4	5	6	7	8	9
1.0	.0000	.0043	.0086	.0128	.0170	.0212	.0253	.0294	.0334	.0374
1.1	.0414	.0453	.0492	.0531	.0569	.0607	.0645	.0682	.0719	.0755
1.2	.0792	.0828	.0864	.0899	.0934	.0969	.1004	.1038	.1072	.1106
1.3	.1139	.1173	.1206	.1239	.1271	.1303	.1335	.1367	.1399	.1430
1.4	.1461	.1492	.1523	.1553	.1584	.1614	.1644	.1673	.1703	.1732
1.5	.1761	.1790	.1818	.1847	.1875	.1903	.1931	.1959	.1987	.2014
1.6	.2041	.2068	.2095	.2122	.2148	.2175	.2201	.2227	.2253	.2279
1.7	.2304	.2330	.2355	.2380	.2405	.2430	.2455	.2480	.2504	.2529
1.8	.2553	.2577	.2601	.2625	.2648	.2672	.2695	.2718	.2742	.2765
1.9	.2788	.2810	.2833	.2856	.2878	.2900	.2923	.2945	.2967	.2989
2.0	.3010	.3032	.3054	.3075	.3096	.3118	.3139	.3160	.3181	.3201
2.1	.3222	.3243	.3263	.3284	.3304	.3324	.3345	.3365	.3385	.3404
2.2	.3424	.3444	.3464	.3483	.3502	.3522	.3541	.3560	.3579	.3598
2.3	.3617	.3636	.3655	.3674	.3692	.3711	.3729	.3747	.3766	.3784
2.4	.3802	.3820	.3838	.3856	.3874	.3892	.3909	.3927	.3945	.3962
2.5	.3979	.3997	.4014	.4031	.4048	.4065	.4082	.4099	.4116	.4133
2.6	.4150	.4166	.4183	.4200	.4216	.4232	.4249	.4265	.4281	.4298
2.7	.4314	.4330	.4346	.4362	.4378	.4393	.4409	.4425	.4440	.4456
2.8	.4472	.4487	.4502	.4518	.4533	.4548	.4564	.4579	.4594	.4609
2.9	.4624	.4639	.4654	.4669	.4683	.4698	.4713	.4728	.4742	.4757
3.0	.4771	.4786	.4800	.4814	.4829	.4843	.4857	.4871	.4886	.4900
3.1	.4914	.4928	.4942	.4955	.4969	.4983	.4997	.5011	.5024	.5038
3.2	.5051	.5065	.5079	.5092	.5105	.5119	.5132	.5145	.5159	.5172
3.3	.5185	.5198	.5211	.5224	.5237	.5250	.5263	.5276	.5289	.5302
3.4	.5315	.5328	.5340	.5353	.5366	.5378	.5391	.5403	.5416	.5428
3.5	.5441	.5453	.5465	.5478	.5490	.5502	.5514	.5527	.5539	.5551
3.6	.5563	.5575	.5587	.5599	.5611	.5623	.5635	.5647	.5658	.5670
3.7	.5682	.5694	.5705	.5717	.5729	.5740	.5752	.5763	.5775	.5786
3.8	.5798	.5809	.5821	.5832	.5843	.5855	.5866	.5877	.5888	.5899
3.9	.5911	.5922	.5933	.5944	.5955	.5966	.5977	.5988	.5999	.6010
4.0	.6021	.6031	.6042	.6053	.6064	.6075	.6085	.6096	.6107	.6117
4.1	.6128	.6138	.6149	.6160	.6170	.6180	.6191	.6201	.6212	.6222
4.2	.6232	.6243	.6253	.6263	.6274	.6284	.6294	.6304	.6314	.6325
4.3	.6335	.6345	.6355	.6365	.6375	.6385	.6395	.6405	.6415	.6425
4.4	.6435	.6444	.6454	.6464	.6474	.6484	.6493	.6503	.6513	.6522
4.5	.6532	.6542	.6551	.6561	.6571	.6580	.6590	.6599	.6609	.6618
4.6	.6628	.6637	.6646	.6656	.6665	.6675	.6684	.6693	.6702	.6712
4.7	.6721	.6730	.6739	.6749	.6758	.6767	.6776	.6785	.6794	.6803
4.8	.6812	.6821	.6830	.6839	.6848	.6857	.6866	.6875	.6884	.6893
4.9	.6902	.6911	.6920	.6928	.6937	.6946	.6955	.6964	.6972	.6981
5.0	.6990	.6998	.7007	.7016	.7024	.7033	.7042	.7050	.7059	.7067
5.1	.7076	.7084	.7093	.7101	.7110	.7118	.7126	.7135	.7143	.7152
5.2	.7160	.7168	.7177	.7185	.7193	.7202	.7210	.7218	.7226	.7235
5.3	.7243	.7251	.7259	.7267	.7275	.7284	.7292	.7300	.7308	.7316
5.4	.7324	.7332	.7340	.7348	.7356	.7364	.7372	.7380	.7388	.7396

Table 2 *(continued)*

m	0	1	2	3	4	5	6	7	8	9
5.5	.7404	.7412	.7419	.7427	.7435	.7443	.7451	.7459	.7466	.7474
5.6	.7482	.7490	.7497	.7505	.7513	.7520	.7528	.7536	.7543	.7551
5.7	.7559	.7566	.7574	.7582	.7589	.7597	.7604	.7612	.7619	.7627
5.8	.7634	.7642	.7649	.7657	.7664	.7672	.7679	.7686	.7694	.7701
5.9	.7709	.7716	.7723	.7731	.7738	.7745	.7752	.7760	.7767	.7774
6.0	.7782	.7789	.7796	.7803	.7810	.7818	.7825	.7832	.7839	.7846
6.1	.7853	.7860	.7868	.7875	.7882	.7889	.7896	.7903	.7910	.7917
6.2	.7924	.7931	.7938	.7945	.7952	.7959	.7966	.7973	.7980	.7987
6.3	.7993	.8000	.8007	.8014	.8021	.8028	.8035	.8041	.8048	.8055
6.4	.8062	.8069	.8075	.8082	.8089	.8096	.8102	.8109	.8116	.8122
6.5	.8129	.8136	.8142	.8149	.8156	.8162	.8169	.8176	.8182	.8189
6.6	.8195	.8202	.8209	.8215	.8222	.8228	.8235	.8241	.8248	.8254
6.7	.8261	.8267	.8274	.8280	.8287	.8293	.8299	.8306	.8312	.8319
6.8	.8325	.8331	.8338	.8344	.8351	.8357	.8363	.8370	.8376	.8382
6.9	.8388	.8395	.8401	.8407	.8414	.8420	.8426	.8432	.8439	.8445
7.0	.8451	.8457	.8463	.8470	.8476	.8482	.8488	.8494	.8500	.8506
7.1	.8513	.8519	.8525	.8531	.8537	.8543	.8549	.8555	.8561	.8567
7.2	.8573	.8579	.8585	.8591	.8597	.8603	.8609	.8615	.8621	.8627
7.3	.8633	.8639	.8645	.8651	.8657	.8663	.8669	.8675	.8681	.8686
7.4	.8692	.8698	.8704	.8710	.8716	.8722	.8727	.8733	.8739	.8745
7.5	.8751	.8756	.8762	.8768	.8774	.8779	.8785	.8791	.8797	.8802
7.6	.8808	.8814	.8820	.8825	.8831	.8837	.8842	.8848	.8854	.8859
7.7	.8865	.8871	.8876	.8882	.8887	.8893	.8899	.8904	.8910	.8915
7.8	.8921	.8927	.8932	.8938	.8943	.8949	.8954	.8960	.8965	.8971
7.9	.8976	.8982	.8987	.8993	.8998	.9004	.9009	.9015	.9020	.9025
8.0	.9031	.9036	.9042	.9047	.9053	.9058	.9063	.9069	.9074	.9079
8.1	.9085	.9090	.9096	.9101	.9106	.9112	.9117	.9122	.9128	.9133
8.2	.9138	.9143	.9149	.9154	.9159	.9165	.9170	.9175	.9180	.9186
8.3	.9191	.9196	.9201	.9206	.9212	.9217	.9222	.9227	.9232	.9238
8.4	.9243	.9249	.9253	.9258	.9263	.9269	.9274	.9279	.9284	.9289
8.5	.9294	.9299	.9304	.9309	.9315	.9320	.9325	.9330	.9335	.9340
8.6	.9345	.9350	.9355	.9360	.9365	.9370	.9375	.9380	.9385	.9390
8.7	.9395	.9400	.9405	.9410	.9415	.9420	.9425	.9430	.9435	.9440
8.8	.9445	.9450	.9455	.9460	.9465	.9469	.9474	.9479	.9484	.9489
8.9	.9494	.9499	.9504	.9509	.9513	.9518	.9523	.9528	.9533	.9538
9.0	.9542	.9547	.9552	.9557	.9562	.9566	.9571	.9576	.9581	.9586
9.1	.9590	.9595	.9600	.9605	.9609	.9614	.9619	.9624	.9628	.9633
9.2	.9638	.9643	.9647	.9652	.9657	.9661	.9666	.9671	.9675	.9680
9.3	.9685	.9689	.9694	.9699	.9703	.9708	.9713	.9717	.9722	.9727
9.4	.9731	.9736	.9741	.9745	.9750	.9754	.9759	.9763	.9768	.9773
9.5	.9777	.9782	.9786	.9791	.9795	.9800	.9805	.9809	.9814	.9818
9.6	.9823	.9827	.9832	.9836	.9841	.9845	.9850	.9854	.9859	.9863
9.7	.9868	.9872	.9877	.9881	.9886	.9890	.9894	.9899	.9903	.9908
9.8	.9912	.9917	.9921	.9926	.9930	.9934	.9939	.9943	.9948	.9952
9.9	.9956	.9961	.9965	.9969	.9974	.9978	.9983	.9987	.9991	.9996

Table 3 Natural Logarithms (Base e)

	0	1	2	3	4	5	6	7	8	9
1.0	0.0000	0.0100	0.0198	0.0296	0.0392	0.0488	0.0583	0.0677	0.0770	0.0862
1.1	0.0953	0.1044	0.1133	0.1222	0.1310	0.1398	0.1484	0.1570	0.1655	0.1740
1.2	0.1823	0.1906	0.1989	0.2070	0.2151	0.2231	0.2311	0.2390	0.2469	0.2546
1.3	0.2624	0.2700	0.2776	0.2852	0.2927	0.3001	0.3075	0.3148	0.3221	0.3293
1.4	0.3365	0.3436	0.3507	0.3577	0.3646	0.3716	0.3784	0.3853	0.3920	0.3988
1.5	0.4055	0.4121	0.4187	0.4253	0.4318	0.4383	0.4447	0.4511	0.4574	0.4637
1.6	0.4700	0.4762	0.4824	0.4886	0.4947	0.5008	0.5068	0.5128	0.5188	0.5247
1.7	0.5306	0.5365	0.5423	0.5481	0.5539	0.5596	0.5653	0.5710	0.5766	0.5822
1.8	0.5878	0.5933	0.5988	0.6043	0.6098	0.6152	0.6206	0.6259	0.6313	0.6366
1.9	0.6419	0.6471	0.6523	0.6575	0.6627	0.6678	0.6729	0.6780	0.6831	0.6881
2.0	0.6931	0.6981	0.7031	0.7080	0.7129	0.7178	0.7227	0.7275	0.7324	0.7372
2.1	0.7419	0.7467	0.7514	0.7561	0.7608	0.7655	0.7701	0.7747	0.7793	0.7839
2.2	0.7885	0.7930	0.7975	0.8020	0.8065	0.8109	0.8154	0.8198	0.8242	0.8286
2.3	0.8329	0.8373	0.8416	0.8459	0.8502	0.8544	0.8587	0.8629	0.8671	0.8713
2.4	0.8755	0.8796	0.8838	0.8879	0.8920	0.8961	0.9002	0.9042	0.9083	0.9123
2.5	0.9163	0.9203	0.9243	0.9282	0.9322	0.9361	0.9400	0.9439	0.9478	0.9517
2.6	0.9555	0.9594	0.9632	0.9670	0.9708	0.9746	0.9783	0.9821	0.9858	0.9895
2.7	0.9933	0.9969	1.0006	1.0043	1.0080	1.0116	1.0152	1.0188	1.0225	1.0260
2.8	1.0296	1.0332	1.0367	1.0403	1.0438	1.0473	1.0508	1.0543	1.0578	1.0613
2.9	1.0647	1.0682	1.0716	1.0750	1.0784	1.0818	1.0852	1.0886	1.0919	1.0953
3.0	1.0986	1.1019	1.1053	1.1086	1.1119	1.1151	1.1184	1.1217	1.1249	1.1282
3.1	1.1314	1.1346	1.1378	1.1410	1.1442	1.1474	1.1506	1.1537	1.1569	1.1600
3.2	1.1632	1.1663	1.1694	1.1725	1.1756	1.1787	1.1817	1.1848	1.1878	1.1909
3.3	1.1939	1.1969	1.2000	1.2030	1.2060	1.2090	1.2119	1.2149	1.2179	1.2208
3.4	1.2238	1.2267	1.2296	1.2326	1.2355	1.2384	1.2413	1.2442	1.2470	1.2499
3.5	1.2528	1.2556	1.2585	1.2613	1.2641	1.2669	1.2698	1.2726	1.2754	1.2782
3.6	1.2809	1.2837	1.2865	1.2892	1.2920	1.2947	1.2975	1.3002	1.3029	1.3056
3.7	1.3083	1.3110	1.3137	1.3164	1.3191	1.3218	1.3244	1.3271	1.3297	1.3324
3.8	1.3350	1.3376	1.3403	1.3429	1.3455	1.3481	1.3507	1.3533	1.3558	1.3584
3.9	1.3610	1.3635	1.3661	1.3686	1.3712	1.3737	1.3762	1.3788	1.3813	1.3838
4.0	1.3863	1.3888	1.3913	1.3938	1.3962	1.3987	1.4012	1.4036	1.4061	1.4085
4.1	1.4110	1.4134	1.4159	1.4183	1.4207	1.4231	1.4255	1.4279	1.4303	1.4327
4.2	1.4351	1.4375	1.4398	1.4422	1.4446	1.4469	1.4493	1.4516	1.4540	1.4563
4.3	1.4586	1.4609	1.4633	1.4656	1.4679	1.4702	1.4725	1.4748	1.4771	1.4793
4.4	1.4816	1.4839	1.4861	1.4884	1.4907	1.4929	1.4951	1.4974	1.4996	1.5019
4.5	1.5041	1.5063	1.5085	1.5107	1.5129	1.5151	1.5173	1.5195	1.5217	1.5239
4.6	1.5261	1.5282	1.5304	1.5326	1.5347	1.5369	1.5390	1.5412	1.5433	1.5454
4.7	1.5476	1.5497	1.5518	1.5539	1.5560	1.5581	1.5602	1.5623	1.5644	1.5665
4.8	1.5686	1.5707	1.5728	1.5748	1.5769	1.5790	1.5810	1.5831	1.5851	1.5872
4.9	1.5892	1.5913	1.5933	1.5953	1.5974	1.5994	1.6014	1.6034	1.6054	1.6074
5.0	1.6094	1.6114	1.6134	1.6154	1.6174	1.6194	1.6214	1.6233	1.6253	1.6273
5.1	1.6292	1.6312	1.6332	1.6351	1.6371	1.6390	1.6409	1.6429	1.6448	1.6467
5.2	1.6487	1.6506	1.6525	1.6544	1.6563	1.6582	1.6601	1.6620	1.6639	1.6658
5.3	1.6677	1.6696	1.6715	1.6734	1.6752	1.6771	1.6790	1.6808	1.6827	1.6845
5.4	1.6864	1.6882	1.6901	1.6919	1.6938	1.6956	1.6974	1.6993	1.7011	1.7029

Table 3 *(continued)*

	0	1	2	3	4	5	6	7	8	9
5.5	1.7047	1.7066	1.7084	1.7102	1.7120	1.7138	1.7156	1.7174	1.7192	1.7210
5.6	1.7228	1.7246	1.7263	1.7281	1.7299	1.7317	1.7334	1.7352	1.7370	1.7387
5.7	1.7405	1.7422	1.7440	1.7457	1.7475	1.7492	1.7509	1.7527	1.7544	1.7561
5.8	1.7579	1.7596	1.7613	1.7630	1.7647	1.7664	1.7681	1.7699	1.7716	1.7733
5.9	1.7750	1.7766	1.7783	1.7800	1.7817	1.7834	1.7851	1.7868	1.7884	1.7901
6.0	1.7918	1.7934	1.7951	1.7967	1.7984	1.8001	1.8017	1.8034	1.8050	1.8066
6.1	1.8083	1.8099	1.8116	1.8132	1.8148	1.8165	1.8181	1.8197	1.8213	1.8229
6.2	1.8245	1.8262	1.8278	1.8294	1.8310	1.8326	1.8342	1.8358	1.8374	1.8390
6.3	1.8405	1.8421	1.8437	1.8453	1.8469	1.8485	1.8500	1.8516	1.8532	1.8547
6.4	1.8563	1.8579	1.8594	1.8610	1.8625	1.8641	1.8656	1.8672	1.8687	1.8703
6.5	1.8718	1.8733	1.8749	1.8764	1.8779	1.8795	1.8810	1.8825	1.8840	1.8856
6.6	1.8871	1.8886	1.8901	1.8916	1.8931	1.8946	1.8961	1.8976	1.8991	1.9006
6.7	1.9021	1.9036	1.9051	1.9066	1.9081	1.9095	1.9110	1.9125	1.9140	1.9155
6.8	1.9169	1.9184	1.9199	1.9213	1.9228	1.9242	1.9257	1.9272	1.9286	1.9301
6.9	1.9315	1.9330	1.9344	1.9359	1.9373	1.9387	1.9402	1.9416	1.9430	1.9445
7.0	1.9459	1.9473	1.9488	1.9502	1.9516	1.9530	1.9544	1.9559	1.9573	1.9587
7.1	1.9601	1.9615	1.9629	1.9643	1.9657	1.9671	1.9685	1.9699	1.9713	1.9727
7.2	1.9741	1.9755	1.9769	1.9782	1.9796	1.9810	1.9824	1.9838	1.9851	1.9865
7.3	1.9879	1.9892	1.9906	1.9920	1.9933	1.9947	1.9961	1.9974	1.9988	2.0001
7.4	2.0015	2.0028	2.0042	2.0055	2.0069	2.0082	2.0096	2.0109	2.0122	2.0136
7.5	2.0149	2.0162	2.0176	2.0189	2.0202	2.0215	2.0229	2.0242	2.0255	2.0268
7.6	2.0281	2.0295	2.0308	2.0321	2.0334	2.0347	2.0360	2.0373	2.0386	2.0399
7.7	2.0412	2.0425	2.0438	2.0451	2.0464	2.0477	2.0490	2.0503	2.0516	2.0528
7.8	2.0541	2.0554	2.0567	2.0580	2.0592	2.0605	2.0618	2.0631	2.0643	2.0656
7.9	2.0669	2.0681	2.0694	2.0707	2.0719	2.0732	2.0744	2.0757	2.0769	2.0782
8.0	2.0794	2.0807	2.0819	2.0832	2.0844	2.0857	2.0869	2.0882	2.0894	2.0906
8.1	2.0919	2.0931	2.0943	2.0956	2.0968	2.0980	2.0992	2.1005	2.1017	2.1029
8.2	2.1041	2.1054	2.1066	2.1078	2.1090	2.1102	2.1114	2.1126	2.1138	2.1150
8.3	2.1163	2.1175	2.1187	2.1199	2.1211	2.1223	2.1235	2.1247	2.1259	2.1270
8.4	2.1282	2.1294	2.1306	2.1318	2.1330	2.1342	2.1353	2.1365	2.1377	2.1389
8.5	2.1401	2.1412	2.1424	2.1436	2.1448	2.1459	2.1471	2.1483	2.1494	2.1506
8.6	2.1518	2.1529	2.1541	2.1552	2.1564	2.1576	2.1587	2.1599	2.1610	2.1622
8.7	2.1633	2.1645	2.1656	2.1668	2.1679	2.1691	2.1702	2.1713	2.1725	2.1736
8.8	2.1748	2.1759	2.1770	2.1782	2.1793	2.1804	2.1815	2.1827	2.1838	2.1849
8.9	2.1861	2.1872	2.1883	2.1894	2.1905	2.1917	2.1928	2.1939	2.1950	2.1961
9.0	2.1972	2.1983	2.1994	2.2006	2.2017	2.2028	2.2039	2.2050	2.2061	2.2072
9.1	2.2083	2.2094	2.2105	2.2116	2.2127	2.2138	2.2148	2.2159	2.2170	2.2181
9.2	2.2192	2.2203	2.2214	2.2225	2.2235	2.2246	2.2257	2.2268	2.2279	2.2289
9.3	2.2300	2.2311	2.2322	2.2332	2.2343	2.2354	2.2364	2.2375	2.2386	2.2396
9.4	2.2407	2.2418	2.2428	2.2439	2.2450	2.2460	2.2471	2.2481	2.2492	2.2502
9.5	2.2513	2.2523	2.2534	2.2544	2.2555	2.2565	2.2576	2.2586	2.2597	2.2607
9.6	2.2618	2.2628	2.2638	2.2649	2.2659	2.2670	2.2680	2.2690	2.2701	2.2711
9.7	2.2721	2.2732	2.2742	2.2752	2.2762	2.2773	2.2783	2.2793	2.2803	2.2814
9.8	2.2824	2.2834	2.2844	2.2854	2.2865	2.2875	2.2885	2.2895	2.2905	2.2915
9.9	2.2925	2.2935	2.2946	2.2956	2.2966	2.2976	2.2986	2.2996	2.3006	2.3016

$\ln 10 \approx 2.3026$

*The following example shows how to use the table to find $ln\ n$ if $n < 1$.

$$\begin{aligned} ln\ 0.06 &= ln(6 \times 10^{-2}) \\ &= ln\ 6 - 2\ ln\ 10 \\ &= 1.7918 - 2(2.3026) \\ &= -2.8134 \end{aligned}$$

Table 4 Values of Trigonometric Functions (Radian Measure)

Real Number x or θ radians	θ degrees	sin x or sin θ	csc x or csc θ	tan x or tan θ	cot x or cot θ	sec x or sec θ	cos x or cos θ
0.00	0°00′	0.0000	No value	0.0000	No value	1.000	1.000
.01	0°34′	.0100	100.0	.0100	100.0	1.000	1.000
.02	1°09′	.0200	50.00	.0200	49.99	1.000	0.9998
.03	1°43′	.0300	33.34	.0300	33.32	1.000	0.9996
.04	2°18′	.0400	25.01	.0400	24.99	1.001	0.9992
0.05	2°52′	0.0500	20.01	0.0500	19.98	1.001	0.9988
.06	3°26′	.0600	16.68	.0601	16.65	1.002	.9982
.07	4°01′	.0699	14.30	.0701	14.26	1.002	.9976
.08	4°35′	.0799	12.51	.0802	12.47	1.003	.9968
.09	5°09′	.0899	11.13	.0902	11.08	1.004	.9960
0.10	5°44′	0.0998	10.02	0.1003	9.967	1.005	0.9950
.11	6°18′	.1098	9.109	.1104	9.054	1.006	.9940
.12	6°53′	.1197	8.353	.1206	8.293	1.007	.9928
.13	7°27′	.1296	7.714	.1307	7.649	1.009	.9916
.14	8°01′	.1395	7.166	.1409	7.096	1.010	.9902
0.15	8°36′	0.1494	6.692	0.1511	6.617	1.011	0.9888
.16	9°10′	.1593	6.277	.1614	6.197	1.013	.9872
.17	9°44′	.1692	5.911	.1717	5.826	1.015	.9856
.18	10°19′	.1790	5.586	.1820	5.495	1.016	.9838
.19	10°53′	.1889	5.295	.1923	5.200	1.018	.9820
0.20	11°28′	0.1987	5.033	0.2027	4.933	1.020	0.9801
.21	12°02′	.2085	4.797	.2131	4.692	1.022	.9780
.22	12°36′	.2182	4.582	.2236	4.472	1.025	.9759
.23	13°11′	.2280	4.386	.2341	4.271	1.027	.9737
.24	13°45′	.2377	4.207	.2447	4.086	1.030	.9713
0.25	14°19′	0.2474	4.042	0.2553	3.916	1.032	0.9689
.26	14°54′	.2571	3.890	.2660	3.759	1.035	.9664
.27	15°28′	.2667	3.749	.2768	3.613	1.038	.9638
.28	16°03′	.2764	3.619	.2876	3.478	1.041	.9611
.29	16°37′	.2860	3.497	.2984	3.351	1.044	.9582
0.30	17°11′	0.2955	3.384	0.3093	3.233	1.047	0.9553
.31	17°46′	.3051	3.278	.3203	3.122	1.050	.9523
.32	18°20′	.3146	3.179	.3314	3.018	1.053	.9492
.33	18°54′	.3240	3.086	.3425	2.920	1.057	.9460
.34	19°29′	.3335	2.999	.3537	2.827	1.061	.9428
0.35	20°03′	0.3429	2.916	0.3650	2.740	1.065	0.9394
.36	20°38′	.3523	2.839	.3764	2.657	1.068	.9359
.37	21°12′	.3616	2.765	.3879	2.578	1.073	.9323
.38	21°46′	.3709	2.696	.3994	2.504	1.077	.9287
.39	22°21′	.3802	2.630	.4111	2.433	1.081	.9249

Table 4 *(continued)*

Real Number x or θ radians	θ degrees	sin x or sin θ	csc x or csc θ	tan x or tan θ	cot x or cot θ	sec x or sec θ	cos x or cos θ
0.40	22°55′	0.3894	2.568	0.4228	2.365	1.086	0.9211
.41	23°29′	.3986	2.509	.4346	2.301	1.090	.9171
.42	24°04′	.4078	2.452	.4466	2.239	1.095	.9131
.43	24°38′	.4169	2.399	.4586	2.180	1.100	.9090
.44	25°13′	.4259	2.348	.4708	2.124	1.105	.9048
0.45	25°47′	0.4350	2.299	0.4831	2.070	1.111	0.9004
.46	26°21′	.4439	2.253	.4954	2.018	1.116	.8961
.47	26°56′	.4529	2.208	.5080	1.969	1.122	.8916
.48	27°30′	.4618	2.166	.5206	1.921	1.127	.8870
.49	28°04′	.4706	2.125	.5334	1.875	1.133	.8823
0.50	28°39′	0.4794	2.086	0.5463	1.830	1.139	0.8776
.51	29°13′	.4882	2.048	.5594	1.788	1.146	.8727
.52	29°48′	.4969	2.013	.5726	1.747	1.152	.8678
.53	30°22′	.5055	1.978	.5859	1.707	1.159	.8628
.54	30°56′	.5141	1.945	.5994	1.668	1.166	.8577
0.55	31°31′	0.5227	1.913	0.6131	1.631	1.173	0.8525
.56	32°05′	.5312	1.883	.6269	1.595	1.180	.8473
.57	32°40′	.5396	1.853	.6410	1.560	1.188	.8419
.58	33°14′	.5480	1.825	.6552	1.526	1.196	.8365
.59	33°48′	.5564	1.797	.6696	1.494	1.203	.8309
0.60	34°23′	0.5646	1.771	0.6841	1.462	1.212	0.8253
.61	34°57′	.5729	1.746	.6989	1.431	1.220	.8196
.62	35°31′	.5810	1.721	.7139	1.401	1.229	.8139
.63	36°06′	.5891	1.697	.7291	1.372	1.238	.8080
.64	36°40′	.5972	1.674	.7445	1.343	1.247	.8021
0.65	37°15′	0.6052	1.652	0.7602	1.315	1.256	0.7961
.66	37°49′	.6131	1.631	.7761	1.288	1.266	.7900
.67	38°23′	.6210	1.610	.7923	1.262	1.276	.7838
.68	38°58′	.6288	1.590	.8087	1.237	1.286	.7776
.69	39°32′	.6365	1.571	.8253	1.212	1.297	.7712
0.70	40°06′	0.6442	1.552	0.8423	1.187	1.307	0.7648
.71	40°41′	.6518	1.534	.8595	1.163	1.319	.7584
.72	41°15′	.6594	1.517	.8771	1.140	1.330	.7518
.73	41°50′	.6669	1.500	.8949	1.117	1.342	.7452
.74	42°24′	.6743	1.483	.9131	1.095	1.354	.7385
0.75	42°58′	0.6816	1.467	0.9316	1.073	1.367	0.7317
.76	43°33′	.6889	1.452	.9505	1.052	1.380	.7248
.77	44°07′	.6961	1.436	.9697	1.031	1.393	.7179
.78	44°41′	.7033	1.422	.9893	1.011	1.407	.7109
.79	45°16′	.7104	1.408	1.009	.9908	1.421	.7038

Table 4 *(continued)*

Real Number x or θ radians	θ degrees	sin x or sin θ	csc x or csc θ	tan x or tan θ	cot x or cot θ	sec x or sec θ	cos x or cos θ
0.80	45°50′	0.7174	1.394	1.030	0.9712	1.435	0.6967
.81	46°25′	.7243	1.381	1.050	.9520	1.450	.6895
.82	46°59′	.7311	1.368	1.072	.9331	1.466	.6822
.83	47°33′	.7379	1.355	1.093	.9146	1.482	.6749
.84	48°08′	.7446	1.343	1.116	.8964	1.498	.6675
0.85	48°42′	0.7513	1.331	1.138	0.8785	1.515	0.6600
.86	49°16′	.7578	1.320	1.162	.8609	1.533	.6524
.87	49°51′	.7643	1.308	1.185	.8437	1.551	.6448
.88	50°25′	.7707	1.297	1.210	.8267	1.569	.6372
.89	51°00′	.7771	1.287	1.235	.8100	1.589	.6294
0.90	51°34′	0.7833	1.277	1.260	0.7936	1.609	0.6216
.91	52°08′	.7895	1.267	1.286	.7774	1.629	.6137
.92	52°43′	.7956	1.257	1.313	.7615	1.651	.6058
.93	53°17′	.8016	1.247	1.341	.7458	1.673	.5978
.94	53°51′	.8076	1.238	1.369	.7303	1.696	.5898
0.95	54°26′	0.8134	1.229	1.398	0.7151	1.719	0.5817
.96	55°00′	.8192	1.221	1.428	.7001	1.744	.5735
.97	55°35′	.8249	1.212	1.459	.6853	1.769	.5653
.98	56°09′	.8305	1.204	1.491	.6707	1.795	.5570
.99	56°43′	.8360	1.196	1.524	.6563	1.823	.5487
1.00	57°18′	0.8415	1.188	1.557	0.6421	1.851	0.5403
1.01	57°52′	.8468	1.181	1.592	.6281	1.880	.5319
1.02	58°27′	.8521	1.174	1.628	.6142	1.911	.5234
1.03	59°01′	.8573	1.166	1.665	.6005	1.942	.5148
1.04	59°35′	.8624	1.160	1.704	.5870	1.975	.5062
1.05	60°10′	0.8674	1.153	1.743	0.5736	2.010	0.4976
1.06	60°44′	.8724	1.146	1.784	.5604	2.046	.4889
1.07	61°18′	.8772	1.140	1.827	.5473	2.083	.4801
1.08	61°53′	.8820	1.134	1.871	.5344	2.122	.4713
1.09	62°27′	.8866	1.128	1.917	.5216	2.162	.4625
1.10	63°02′	0.8912	1.122	1.965	0.5090	2.205	0.4536
1.11	63°36′	.8957	1.116	2.014	.4964	2.249	.4447
1.12	64°10′	.9001	1.111	2.066	.4840	2.295	.4357
1.13	64°45′	.9044	1.106	2.120	.4718	2.344	.4267
1.14	65°19′	.9086	1.101	2.176	.4596	2.395	.4176
1.15	65°53′	0.9128	1.096	2.234	0.4475	2.448	0.4085
1.16	66°28′	.9168	1.091	2.296	.4356	2.504	.3993
1.17	67°02′	.9208	1.086	2.360	.4237	2.563	.3902
1.18	67°37′	.9246	1.082	2.427	.4120	2.625	.3809
1.19	68°11′	.9284	1.077	2.498	.4003	2.691	.3717

Table 4 *(continued)*

Real Number x or θ radians	θ degrees	sin x or sin θ	csc x or csc θ	tan x or tan θ	cot x or cot θ	sec x or sec θ	cos x or cos θ
1.20	68°45′	0.9320	1.073	2.572	0.3888	2.760	0.3624
1.21	69°20′	.9356	1.069	2.650	.3773	2.833	.3530
1.22	69°54′	.9391	1.065	2.733	.3659	2.910	.3436
1.23	70°28′	.9425	1.061	2.820	.3546	2.992	.3342
1.24	71°03′	.9458	1.057	2.912	.3434	3.079	.3248
1.25	71°37′	0.9490	1.054	3.010	0.3323	3.171	0.3153
1.26	72°12′	.9521	1.050	3.113	.3212	3.270	.3058
1.27	72°46′	.9551	1.047	3.224	.3102	3.375	.2963
1.28	73°20′	.9580	1.044	3.341	.2993	3.488	.2867
1.29	73°55′	.9608	1.041	3.467	.2884	3.609	.2771
1.30	74°29′	0.9636	1.038	3.602	0.2776	3.738	0.2675
1.31	75°03′	.9662	1.035	3.747	.2669	3.878	.2579
1.32	75°38′	.9687	1.032	3.903	.2562	4.029	.2482
1.33	76°12′	.9711	1.030	4.072	.2456	4.193	.2385
1.34	76°47′	.9735	1.027	4.256	.2350	4.372	.2288
1.35	77°21′	0.9757	1.025	4.455	0.2245	4.566	0.2190
1.36	77°55′	.9779	1.023	4.673	.2140	4.779	.2092
1.37	78°30′	.9799	1.021	4.913	.2035	5.014	.1994
1.38	79°04′	.9819	1.018	5.177	.1931	5.273	.1896
1.39	79°38′	.9837	1.017	5.471	.1828	5.561	.1798
1.40	80°13′	0.9854	1.015	5.798	0.1725	5.883	0.1700
1.41	80°47′	.9871	1.013	6.165	.1622	6.246	.1601
1.42	81°22′	.9887	1.011	6.581	.1519	6.657	.1502
1.43	81°56′	.9901	1.010	7.055	.1417	7.126	.1403
1.44	82°30′	.9915	1.009	7.602	.1315	7.667	.1304
1.45	83°05′	0.9927	1.007	8.238	0.1214	8.299	0.1205
1.46	83°39′	.9939	1.006	8.989	.1113	9.044	.1106
1.47	84°13′	.9949	1.005	9.887	.1011	9.938	.1006
1.48	84°48′	.9959	1.004	10.98	.0910	11.03	.0907
1.49	85°22′	.9967	1.003	12.35	.0810	12.39	.0807
1.50	85°57′	0.9975	1.003	14.10	0.0709	14.14	0.0707
1.51	86°31′	.9982	1.002	16.43	.0609	16.46	.0608
1.52	87°05′	.9987	1.001	19.67	.0508	19.69	.0508
1.53	87°40′	.9992	1.001	24.50	.0408	24.52	.0408
1.54	88°14′	.9995	1.000	32.46	.0308	32.48	.0308
1.55	88°49′	0.9998	1.000	48.08	0.0208	48.09	0.0208
1.56	89°23′	.9999	1.000	92.62	.0108	92.63	.0108
1.57	89°57′	1.000	1.000	1256	.0008	1256	.0008

Table 5 Values of Trigonometric Functions (Degree Measure)

radians	degrees θ	sin θ	cos θ	tan θ	cot θ	sec θ	csc θ		
.0000	**0°00′**	.0000	1.0000	.0000	—	1.000	—	**90°00′**	1.5708
.0029	10	.0029	1.0000	.0029	343.8	1.000	343.8	50	1.5679
.0058	20	.0058	1.0000	.0058	171.9	1.000	171.9	40	1.5650
.0087	30	.0087	1.0000	.0087	114.6	1.000	114.6	30	1.5621
.0116	40	.0116	.9999	.0116	85.94	1.000	85.95	20	1.5592
.0145	50	.0145	.9999	.0145	68.75	1.000	68.76	10	1.5563
.0175	**1°00′**	.0175	.9998	.0175	57.29	1.000	57.30	**89°00′**	1.5533
.0204	10	.0204	.9998	.0204	49.10	1.000	49.11	50	1.5504
.0233	20	.0233	.9997	.0233	42.96	1.000	42.98	40	1.5475
.0262	30	.0262	.9997	.0262	38.19	1.000	38.20	30	1.5446
.0291	40	.0291	.9996	.0291	34.37	1.000	34.38	20	1.5417
.0320	50	.0320	.9995	.0320	31.24	1.001	31.36	10	1.5388
.0349	**2°00′**	.0349	.9994	.0349	28.64	1.001	28.65	**88°00′**	1.5359
.0378	10	.0378	.9993	.0378	26.43	1.001	26.45	50	1.5330
.0407	20	.0407	.9992	.0407	24.54	1.001	24.56	40	1.5301
.0436	30	.0436	.9990	.0437	22.90	1.001	22.93	30	1.5272
.0465	40	.0465	.9989	.0466	21.47	1.001	21.49	20	1.5243
.0495	50	.0494	.9988	.0495	20.21	1.001	20.23	10	1.5213
.0524	**3°00′**	.0523	.9986	.0524	19.08	1.001	19.11	**87°00′**	1.5184
.0553	10	.0552	.9985	.0553	18.07	1.002	18.10	50	1.5155
.0582	20	.0581	.9983	.0582	17.17	1.002	17.20	40	1.5126
.0611	30	.0610	.9981	.0612	16.35	1.002	16.38	30	1.5097
.0640	40	.0640	.9980	.0641	15.60	1.002	15.64	20	1.5068
.0669	50	.0669	.9978	.0670	14.92	1.002	14.96	10	1.5039
.0698	**4°00′**	.0698	.9976	.0699	14.30	1.002	14.34	**86°00′**	1.5010
.0727	10	.0727	.9974	.0729	13.73	1.003	13.76	50	1.4981
.0756	20	.0756	.9971	.0758	13.20	1.003	13.23	40	1.4952
.0785	30	.0785	.9969	.0787	12.71	1.003	12.75	30	1.4923
.0814	40	.0814	.9967	.0816	12.25	1.003	12.29	20	1.4893
.0844	50	.0843	.9964	.0846	11.83	1.004	11.87	10	1.4864
.0873	**5°00′**	.0872	.9962	.0875	11.43	1.004	11.47	**85°00′**	1.4835
.0902	10	.0901	.9959	.0904	11.06	1.004	11.10	50	1.4806
.0931	20	.0929	.9957	.0934	10.71	1.004	10.76	40	1.4777
.0960	30	.0958	.9954	.0963	10.39	1.005	10.43	30	1.4748
.0989	40	.0987	.9951	.0992	10.08	1.005	10.13	20	1.4719
.1018	50	.1016	.9948	.1022	9.788	1.005	9.839	10	1.4690
		cos θ	sin θ	cot θ	tan θ	csc θ	sec θ	degrees θ	radians

Table 5 *(continued)*

radians	degrees θ	sin θ	cos θ	tan θ	cot θ	sec θ	csc θ		
.1047	**6°00′**	.1045	.9945	.1051	9.514	1.006	9.567	**84°00′**	1.4661
.1076	10	.1074	.9942	.1080	9.255	1.006	9.309	50	1.4632
.1105	20	.1103	.9939	.1110	9.010	1.006	9.065	40	1.4603
.1134	30	.1132	.9936	.1139	8.777	1.006	8.834	30	1.4573
.1164	40	.1161	.9932	.1169	8.556	1.007	8.614	20	1.4544
.1193	50	.1190	.9929	.1198	8.345	1.007	8.405	10	1.4515
.1222	**7°00′**	.1219	.9925	.1228	8.144	1.008	8.206	**83°00′**	1.4486
.1251	10	.1248	.9922	.1257	7.953	1.008	8.016	50	1.4457
.1280	20	.1276	.9918	.1287	7.770	1.008	7.834	40	1.4428
.1309	30	.1305	.9914	.1317	7.596	1.009	7.661	30	1.4399
.1338	40	.1334	.9911	.1346	7.429	1.009	7.496	20	1.4370
.1367	50	.1363	.9907	.1376	7.269	1.009	7.337	10	1.4341
.1396	**8°00′**	.1392	.9903	.1405	7.115	1.010	7.185	**82°00′**	1.4312
.1425	10	.1421	.9899	.1435	6.968	1.010	7.040	50	1.4283
.1454	20	.1449	.9894	.1465	6.827	1.011	6.900	40	1.4254
.1484	30	.1478	.9890	.1495	6.691	1.011	6.765	30	1.4224
.1513	40	.1507	.9886	.1524	6.561	1.012	6.636	20	1.4195
.1542	50	.1536	.9881	.1554	6.435	1.012	6.512	10	1.4166
.1571	**9°00′**	.1564	.9877	.1584	6.314	1.012	6.392	**81°00′**	1.4137
.1600	10	.1593	.9872	.1614	6.197	1.013	6.277	50	1.4108
.1629	20	.1622	.9868	.1644	6.084	1.013	6.166	40	1.4079
.1658	30	.1650	.9863	.1673	5.976	1.014	6.059	30	1.4050
.1687	40	.1679	.9858	.1703	5.871	1.014	5.955	20	1.4021
.1716	50	.1708	.9853	.1733	5.769	1.015	5.855	10	1.3992
.1745	**10°00′**	.1736	.9848	.1763	5.671	1.015	5.759	**80°00′**	1.3963
.1774	10	.1765	.9843	.1793	5.576	1.016	5.665	50	1.3934
.1804	20	.1794	.9838	.1823	5.485	1.016	5.575	40	1.3904
.1833	30	.1822	.9833	.1853	5.396	1.017	5.487	30	1.3875
.1862	40	.1851	.9827	.1883	5.309	1.018	5.403	20	1.3846
.1891	50	.1880	.9822	.1914	5.226	1.018	5.320	10	1.3817
.1920	**11°00′**	.1908	.9816	.1944	5.145	1.019	5.241	**79°00′**	1.3788
.1949	10	.1937	.9811	.1974	5.066	1.019	5.164	50	1.3759
.1978	20	.1965	.9805	.2004	4.989	1.020	5.089	40	1.3730
.2007	30	.1994	.9799	.2035	4.915	1.020	5.016	30	1.3701
.2036	40	.2022	.9793	.2065	4.843	1.021	4.945	20	1.3672
.2065	50	.2051	.9787	.2095	4.773	1.022	4.876	10	1.3643
		cos θ	sin θ	cot θ	tan θ	csc θ	sec θ	degrees θ	radians

Table 5 *(continued)*

radians	degrees θ	sin θ	cos θ	tan θ	cot θ	sec θ	csc θ		
.2094	**12°00′**	.2079	.9781	.2126	4.705	1.022	4.810	**78°00′**	1.3614
.2123	10	.2108	.9775	.2156	4.638	1.023	4.745	50	1.3584
.2153	20	.2136	.9769	.2186	4.574	1.024	4.682	40	1.3555
.2182	30	.2164	.9763	.2217	4.511	1.024	4.620	30	1.3526
.2211	40	.2193	.9757	.2247	4.449	1.025	4.560	20	1.3497
.2240	50	.2221	.9750	.2278	4.390	1.026	4.502	10	1.3468
.2269	**13°00′**	.2250	.9744	.2309	4.331	1.026	4.445	**77°00′**	1.3439
.2298	10	.2278	.9737	.2339	4.275	1.027	4.390	50	1.3410
.2327	20	.2306	.9730	.2370	4.219	1.028	4.336	40	1.3381
.2356	30	.2334	.9724	.2401	4.165	1.028	4.284	30	1.3352
.2385	40	.2363	.9717	.2432	4.113	1.029	4.232	20	1.3323
.2414	50	.2391	.9710	.2462	4.061	1.030	4.182	10	1.3294
.2443	**14°00′**	.2419	.9703	.2493	4.011	1.031	4.134	**76°00′**	1.3265
.2473	10	.2447	.9696	.2524	3.962	1.031	4.086	50	1.3235
.2502	20	.2476	.9689	.2555	3.914	1.032	4.039	40	1.3206
.2531	30	.2504	.9681	.2586	3.867	1.033	3.994	30	1.3177
.2560	40	.2532	.9674	.2617	3.821	1.034	3.950	20	1.3148
.2589	50	.2560	.9667	.2648	3.776	1.034	3.906	10	1.3119
.2618	**15°00′**	.2588	.9659	.2679	3.732	1.035	3.864	**75°00′**	1.3090
.2647	10	.2616	.9652	.2711	3.689	1.036	3.822	50	1.3061
.2676	20	.2644	.9644	.2742	3.647	1.037	3.782	40	1.3032
.2705	30	.2672	.9636	.2773	3.606	1.038	3.742	30	1.3003
.2734	40	.2700	.9628	.2805	3.566	1.039	3.703	20	1.2974
.2763	50	.2728	.9621	.2836	3.526	1.039	3.665	10	1.2945
.2793	**16°00′**	.2756	.9613	.2867	3.487	1.040	3.628	**74°00′**	1.2915
.2822	10	.2784	.9605	.2899	3.450	1.041	3.592	50	1.2886
.2851	20	.2812	.9596	.2931	3.412	1.042	3.556	40	1.2857
.2880	30	.2840	.9588	.2962	3.376	1.043	3.521	30	1.2828
.2909	40	.2868	.9580	.2994	3.340	1.044	3.487	20	1.2799
.2938	50	.2896	.9572	.3026	3.305	1.045	3.453	10	1.2770
.2967	**17°00′**	.2924	.9563	.3057	3.271	1.046	3.420	**73°00′**	1.2741
.2996	10	.2952	.9555	.3089	3.237	1.047	3.388	50	1.2712
.3025	20	.2979	.9546	.3121	3.204	1.048	3.356	40	1.2683
.3054	30	.3007	.9537	.3153	3.172	1.049	3.326	30	1.2654
.3083	40	.3035	.9528	.3185	3.140	1.049	3.295	20	1.2625
.3113	50	.3062	.9520	.3217	3.108	1.050	3.265	10	1.2595
.3142	**18°00′**	.3090	.9511	.3249	3.078	1.051	3.236	**72°00′**	1.2566
.3171	10	.3118	.9502	.3281	3.047	1.052	3.207	50	1.2537
.3200	20	.3145	.9492	.3314	3.018	1.053	3.179	40	1.2508
.3229	30	.3173	.9483	.3346	2.989	1.054	3.152	30	1.2479
.3258	40	.3201	.9474	.3378	2.960	1.056	3.124	20	1.2450
.3287	50	.3228	.9465	.3411	2.932	1.057	3.098	10	1.2421
		cos θ	sin θ	cot θ	tan θ	csc θ	sec θ	degrees θ	radians

Table 5 *(continued)*

radians	degrees θ	$\sin\theta$	$\cos\theta$	$\tan\theta$	$\cot\theta$	$\sec\theta$	$\csc\theta$		
.3316	**19°00′**	.3256	.9455	.3443	2.904	1.058	3.072	**71°00′**	1.2392
.3345	10	.3283	.9446	.3476	2.877	1.059	3.046	50	1.2363
.3374	20	.3311	.9436	.3508	2.850	1.060	3.021	40	1.2334
.3403	30	.3338	.9426	.3541	2.824	1.061	2.996	30	1.2305
.3432	40	.3365	.9417	.3574	2.798	1.062	2.971	20	1.2275
.3462	50	.3393	.9407	.3607	2.773	1.063	2.947	10	1.2246
.3491	**20°00′**	.3420	.9397	.3640	2.747	1.064	2.924	**70°00′**	1.2217
.3520	10	.3448	.9387	.3673	2.723	1.065	2.901	50	1.2188
.3549	20	.3475	.9377	.3706	2.699	1.066	2.878	40	1.2159
.3578	30	.3502	.9367	.3739	2.675	1.068	2.855	30	1.2130
.3607	40	.3529	.9356	.3772	2.651	1.069	2.833	20	1.2101
.3636	50	.3557	.9346	.3805	2.628	1.070	2.812	10	1.2072
.3665	**21°00′**	.3584	.9336	.3839	2.605	1.071	2.790	**69°00′**	1.2043
.3694	10	.3611	.9325	.3872	2.583	1.072	2.769	50	1.2014
.3723	20	.3638	.9315	.3906	2.560	1.074	2.749	40	1.1985
.3752	30	.3665	.9304	.3939	2.539	1.075	2.729	30	1.1956
.3782	40	.3692	.9293	.3973	2.517	1.076	2.709	20	1.1926
.3811	50	.3719	.9283	.4006	2.496	1.077	2.689	10	1.1897
.3840	**22°00′**	.3746	.9272	.4040	2.475	1.079	2.669	**68°00′**	1.1868
.3869	10	.3773	.9261	.4074	2.455	1.080	2.650	50	1.1839
.3898	20	.3800	.9250	.4108	2.434	1.081	2.632	40	1.1810
.3927	30	.3827	.9239	.4142	2.414	1.082	2.613	30	1.1781
.3956	40	.3854	.9228	.4176	2.394	1.084	2.595	20	1.1752
.3985	50	.3881	.9216	.4210	2.375	1.085	2.577	10	1.1723
.4014	**23°00′**	.3907	.9205	.4245	2.356	1.086	2.559	**67°00′**	1.1694
.4043	10	.3934	.9194	.4279	2.337	1.088	2.542	50	1.1665
.4072	20	.3961	.9182	.4314	2.318	1.089	2.525	40	1.1636
.4102	30	.3987	.9771	.4348	2.300	1.090	2.508	30	1.1606
.4131	40	.4014	.9159	.4383	2.282	1.092	2.491	20	1.1577
.4160	50	.4041	.9147	.4417	2.264	1.093	2.475	10	1.1548
.4189	**24°00′**	.4067	.9135	.4452	2.246	1.095	2.459	**66°00′**	1.1519
.4218	10	.4094	.9124	.4487	2.229	1.096	2.443	50	1.1490
.4247	20	.4120	.9112	.4522	2.211	1.097	2.427	40	1.1461
.4276	30	.4147	.9100	.4557	2.194	1.099	2.411	30	1.1432
.4305	40	.4173	.9088	.4592	2.177	1.100	2.396	20	1.1403
.4334	50	.4200	.9075	.4628	2.161	1.102	2.381	10	1.1374
.4363	**25°00′**	.4226	.9063	.4663	2.145	1.103	2.366	**65°00′**	1.1345
.4392	10	.4253	.9051	.4699	2.128	1.105	2.352	50	1.1316
.4422	20	.4279	.9038	.4734	2.112	1.106	2.337	40	1.1286
.4451	30	.4305	.9026	.4770	2.097	1.108	2.323	30	1.1257
.4480	40	.4331	.9013	.4806	2.081	1.109	2.309	20	1.1228
.4509	50	.4358	.9001	.4841	2.066	1.111	2.295	10	1.1199
		$\cos\theta$	$\sin\theta$	$\cot\theta$	$\tan\theta$	$\csc\theta$	$\sec\theta$	degrees θ	radians

Table 5 *(continued)*

radians	degrees θ	sin θ	cos θ	tan θ	cot θ	sec θ	csc θ		
.4538	**26°00′**	.4384	.8988	.4877	2.050	1.113	2.281	**64°00′**	1.1170
.4567	10	.4410	.8975	.4913	2.035	1.114	2.268	50	1.1141
.4596	20	.4436	.8962	.4950	2.020	1.116	2.254	40	1.1112
.4625	30	.4462	.8949	.4986	2.006	1.117	2.241	30	1.1083
.4654	40	.4488	.8936	.5022	1.991	1.119	2.228	20	1.1054
.4683	50	.4514	.8923	.5059	1.977	1.121	2.215	10	1.1025
.4712	**27°00′**	.4540	.8910	.5095	1.963	1.122	2.203	**63°00′**	1.0996
.4741	10	.4566	.8897	.5132	1.949	1.124	2.190	50	1.0966
.4771	20	.4592	.8884	.5169	1.935	1.126	2.178	40	1.0937
.4800	30	.4617	.8870	.5206	1.921	1.127	2.166	30	1.0908
.4829	40	.4643	.8857	.5243	1.907	1.129	2.154	20	1.0879
.4858	50	.4669	.8843	.5280	1.894	1.131	2.142	10	1.0850
.4887	**28°00′**	.4695	.8829	.5317	1.881	1.133	2.130	**62°00′**	1.0821
.4961	10	.4720	.8816	.5354	1.868	1.134	2.118	50	1.0792
.4945	20	.4746	.8802	.5392	1.855	1.136	2.107	40	1.0763
.4974	30	.4772	.8788	.5430	1.842	1.138	2.096	30	1.0734
.5003	40	.4797	.8774	.5467	1.829	1.140	2.085	20	1.0705
.5032	50	.4823	.8760	.5505	1.816	1.142	2.074	10	1.0676
.5061	**29°00′**	.4848	.8746	.5543	1.804	1.143	2.063	**61°00′**	1.0647
.5091	10	.4874	.8732	.5581	1.792	1.145	2.052	50	1.0617
.5120	20	.4899	.8718	.5619	1.780	1.147	2.041	40	1.0588
.5149	30	.4924	.8704	.5658	1.767	1.149	2.031	30	1.0559
.5178	40	.4950	.8689	.5696	1.756	1.151	2.020	20	1.0530
.5207	50	.4975	.8675	.5735	1.744	1.153	2.010	10	1.0501
.5236	**30°00′**	.5000	.8660	.5774	1.732	1.155	2.000	**60°00′**	1.0472
.5265	10	.5025	.8646	.5812	1.720	1.157	1.990	50	1.0443
.5294	20	.5050	.8631	.5851	1.709	1.159	1.980	40	1.0414
.5323	30	.5075	.8616	.5890	1.698	1.161	1.970	30	1.0385
.5352	40	.5100	.8601	.5930	1.686	1.163	1.961	20	1.0356
.5381	50	.5125	.8587	.5969	1.675	1.165	1.951	10	1.0327
.5411	**31°00′**	.5150	.8572	.6009	1.664	1.167	1.942	**59°00′**	1.0297
.5440	10	.5175	.8557	.6048	1.653	1.169	1.932	50	1.0268
.5469	20	.5200	.8542	.6088	1.643	1.171	1.923	40	1.0239
.5498	30	.5225	.8526	.6128	1.632	1.173	1.914	30	1.0210
.5527	40	.5250	.8511	.6168	1.621	1.175	1.905	20	1.0181
.5556	50	.5275	.8496	.6208	1.611	1.177	1.896	10	1.0152
.5585	**32°00′**	.5299	.8480	.6249	1.600	1.179	1.887	**58°00′**	1.0123
.5614	10	.5324	.8465	.6289	1.590	1.181	1.878	50	1.0094
.5643	20	.5348	.8450	.6330	1.580	1.184	1.870	40	1.0065
.5672	30	.5373	.8434	.6371	1.570	1.186	1.861	30	1.0036
.5701	40	.5398	.8418	.6412	1.560	1.188	1.853	20	1.0007
.5730	50	.5422	.8403	.6453	1.550	1.190	1.844	10	.9977
		cos θ	sin θ	cot θ	tan θ	csc θ	sec θ	degrees θ	radians

Table 5 *(continued)*

radians	degrees θ	sin θ	cos θ	tan θ	cot θ	sec θ	csc θ		
.5760	**33°00′**	.5446	.8387	.6494	1.540	1.192	1.836	**57°00′**	.9948
.5789	10	.5471	.8371	.6536	1.530	1.195	1.828	50	.9919
.5818	20	.5495	.8355	.6577	1.520	1.197	1.820	40	.9890
.5847	30	.5519	.8339	.6619	1.511	1.199	1.812	30	.9861
.5876	40	.5544	.8323	.6661	1.501	1.202	1.804	20	.9832
.5905	50	.5568	.8307	.6703	1.492	1.204	1.796	10	.9803
.5934	**34°00′**	.5592	.8290	.6745	1.483	1.206	1.788	**56°00′**	.9774
.5963	10	.5616	.8274	.6787	1.473	1.209	1.781	50	.9745
.5992	20	.5640	.8258	.6830	1.464	1.211	1.773	40	.9716
.6021	30	.5664	.8241	.6873	1.455	1.213	1.766	30	.9687
.6050	40	.5688	.8225	.6916	1.446	1.216	1.758	20	.9657
.6080	50	.5712	.8208	.6959	1.437	1.218	1.751	10	.9628
.6109	**35°00′**	.5736	.8192	.7002	1.428	1.221	1.743	**55°00′**	.9599
.6138	10	.5760	.8175	.7046	1.419	1.223	1.736	50	.9570
.6167	20	.5783	.8158	.7089	1.411	1.226	1.729	40	.9541
.6196	30	.5807	.8141	.7133	1.402	1.228	1.722	30	.9512
.6225	40	.5831	.8124	.7177	1.393	1.231	1.715	20	.9483
.6254	50	.5854	.8107	.7221	1.385	1.233	1.708	10	.9454
.6283	**36°00′**	.5878	.8090	.7265	1.376	1.236	1.701	**54°00′**	.9425
.6312	10	.5901	.8073	.7310	1.368	1.239	1.695	50	.9396
.6341	20	.5925	.8056	.7355	1.360	1.241	1.688	40	.9367
.6370	30	.5948	.8039	.7400	1.351	1.244	1.681	30	.9338
.6400	40	.5972	.8021	.7445	1.343	1.247	1.675	20	.9308
.6429	50	.5995	.8004	.7490	1.335	1.249	1.668	10	.9279
.6458	**37°00′**	.6018	.7986	.7536	1.327	1.252	1.662	**53°00′**	.9250
.6487	10	.6041	.7969	.7581	1.319	1.255	1.655	50	.9221
.6516	20	.6065	.7951	.7627	1.311	1.258	1.649	40	.9192
.6545	30	.6088	.7934	7673	1.303	1.260	1.643	30	.9163
.6574	40	.6111	.7916	.7720	1.295	1.263	1.636	20	.9134
.6603	50	.6134	.7898	.7766	1.288	1.266	1.630	10	.9105
.6632	**38°00′**	.6157	.7880	.7813	1.280	1.269	1.624	**52°00′**	.9076
.6661	10	.6180	.7862	.7860	1.272	1.272	1.618	50	.9047
.6690	20	.6202	.7844	.7907	1.265	1.275	1.612	40	.9018
.6720	30	.6225	.7826	.7954	1.257	1.278	1.606	30	.8988
.6749	40	.6248	.7808	.8002	1.250	1.281	1.601	20	.8959
.6778	50	.6271	.7790	.8050	1.242	1.284	1.595	10	.8930
.6807	**39°00′**	.6293	.7771	.8098	1.235	1.287	1.589	**51°00′**	.8901
.6836	10	.6316	.7753	.8146	1.228	1.290	1.583	50	.8872
.6865	20	.6338	.7735	.8195	1.220	1.293	1.578	40	.8843
.6894	30	.6361	.7716	.8243	1.213	1.296	1.572	30	.8814
.6923	40	.6383	.7698	.8292	1.206	1.299	1.567	20	.8785
.6952	50	.6406	.7679	.8342	1.199	1.302	1.561	10	.8756
		cos θ	sin θ	cot θ	tan θ	csc θ	sec θ	degrees θ	radians

Table 5 *(continued)*

radians	degrees θ	sin θ	cos θ	tan θ	cot θ	sec θ	csc θ		
.6981	**40°00′**	.6428	.7660	.8391	1.192	1.305	1.556	**50°00′**	.8727
.7010	10	.6450	.7642	.8441	1.185	1.309	1.550	50	.8698
.7039	20	.6472	.7623	.8491	1.178	1.312	1.545	40	.8668
.7069	30	.6494	.7604	.8541	1.171	1.315	1.540	30	.8639
.7098	40	.6417	.7585	.8591	1.164	1.318	1.535	20	.8610
.7127	50	.6539	.7566	.8642	1.157	1.322	1.529	10	.8581
.7156	**41°00′**	.6561	.7547	.8693	1.150	1.325	1.524	**49°00′**	.8552
.7185	10	.6583	.7528	.8744	1.144	1.328	1.519	50	.8523
.7214	20	.6604	.7509	.8796	1.137	1.332	1.514	40	.8494
.7243	30	.6626	.7490	.8847	1.130	1.335	1.509	30	.8465
.7272	40	.6648	.7470	.8899	1.124	1.339	1.504	20	.8436
.7301	50	.6670	.7451	.8952	1.117	1.342	1.499	10	.8407
.7330	**42°00′**	.6691	.7431	.9004	1.111	1.346	1.494	**48°00′**	.8378
.7359	10	.6713	.7412	.9057	1.104	1.349	1.490	50	.8348
.7389	20	.6734	.7392	.9110	1.098	1.353	1.485	40	.8319
.7418	30	.6756	.7373	.9163	1.091	1.356	1.480	30	.8290
.7447	40	.6777	.7353	.9217	1.085	1.360	1.476	20	.8261
.7476	50	.6799	.7333	.9271	1.079	1.364	1.471	10	.8232
.7505	**43°00′**	.6820	.7314	.9325	1.072	1.367	1.466	**47°00′**	.8203
.7534	10	.6841	.7294	.9380	1.066	1.371	1.462	50	.8174
.7563	20	.6862	.7274	.9435	1.060	1.375	1.457	40	.8145
.7592	30	.6884	.7254	.9490	1.054	1.379	1.453	30	.8116
.7621	40	.6905	.7234	.9545	1.048	1.382	1.448	20	.8087
.7650	50	.6926	.7214	.9601	1.042	1.386	1.444	10	.8058
.7679	**44°00′**	.6947	.7193	.9657	1.036	1.390	1.440	**46°00′**	.8029
.7709	10	.6967	.7173	.9713	1.030	1.394	1.435	50	.7999
.7738	20	.6988	.7153	.9770	1.024	1.398	1.431	40	.7970
.7767	30	.7009	.7133	.9827	1.018	1.402	1.427	30	.7941
.7796	40	.7030	.7112	.9884	1.012	1.406	1.423	20	.7912
.7825	50	.7050	.7092	.9942	1.006	1.410	1.418	10	.7883
.7854	**45°00′**	.7071	.7071	1.0000	1.0000	1.414	1.414	**45°00′**	.7854
		cos θ	sin θ	cot θ	tan θ	csc θ	sec θ	degrees θ	radians

Appendix B *Answers to Odd-Numbered Problems*

Section 1.1

1. $3x - 5 = 7x + 12 - 17 - 4x$
$3x - 5 = 3x - 5$

3. $(x - 1)(x + 3) + 3 = x^2 + 2x$
$x^2 + 2x - 3 + 3 = x^2 + 2x$
$x^2 + 2x = x^2 + 2x$

5. $x = -\frac{14}{9}$ **7.** $y = \frac{8}{19}$ **9.** $w = \frac{4}{3}$

11. $x = -\frac{27}{59}$ **13.** $y = 2, 4$ **15.** $z = -\frac{1}{4}, 7$ **17.** $x = \frac{5}{11}$ **19.** No solution **21.** $w = -\frac{11}{2}$

23. $\frac{11''}{4}$ and $\frac{77''}{4}$ **25.** 46, 48, 50 **27.** 800 square feet **29.** 12 15-point problems; 2 10-point problems

31. 390 miles, 6 hours **33.** 50° **35.** $10\frac{1}{2}$ liters **37.** $\frac{69}{7}$ miles **39.** $133\frac{1}{3}$ lbs. of grade A; $66\frac{2}{3}$ lbs. of grade B

41. $2\frac{2}{5}$ hours **43.** 66 mph **45.** $1\frac{19}{80}$ minutes **47.** 1st pump, 4 hours; 2nd pump, 2 hours

49. (a) $A(x) + C(x) = B(x) + C(x)$ (b) $m = n$ (c) $A(x) = B(x)$

Section 1.2

1. $\frac{160{,}000}{9}$ **3.** 1 **5.** $\frac{1}{x}$ **7.** xy^2 **9.** $\frac{4y^6}{9x^2}$ **11.** $\frac{y^6}{x^9}$ **13.** $\frac{9}{(2x + y)^2}$ **15.** x^2y^2z **17.** 2 **19.** -2

21. 3 **23.** 8 **25.** $\frac{1}{8}$ **27.** 81 **29.** 27 **31.** 0.001 **33.** $\frac{99}{25}$ **35.** 27,000 **37.** 64 **39.** $\frac{1}{16}$

41. 25 **43.** $\frac{1}{2}y^{1/12}$ **45.** $y^{11/24}$ **47.** $x^{1/2}$ **49.** $(x + 2)^{3/2}$ **51.** $(x + y)^{5/6}$ **53.** $(x + y) + (x + y)^{7/6}$

55. $xy^2 - x^2y^3$ **57.** 4 **59.** 6 **61.** $\frac{16}{9}$ **63.** $\frac{1}{25}$ **65.** $2xy\sqrt[3]{2x}$ **67.** $3xyz^2\sqrt{2y}$ **69.** $9w\sqrt[3]{v}$

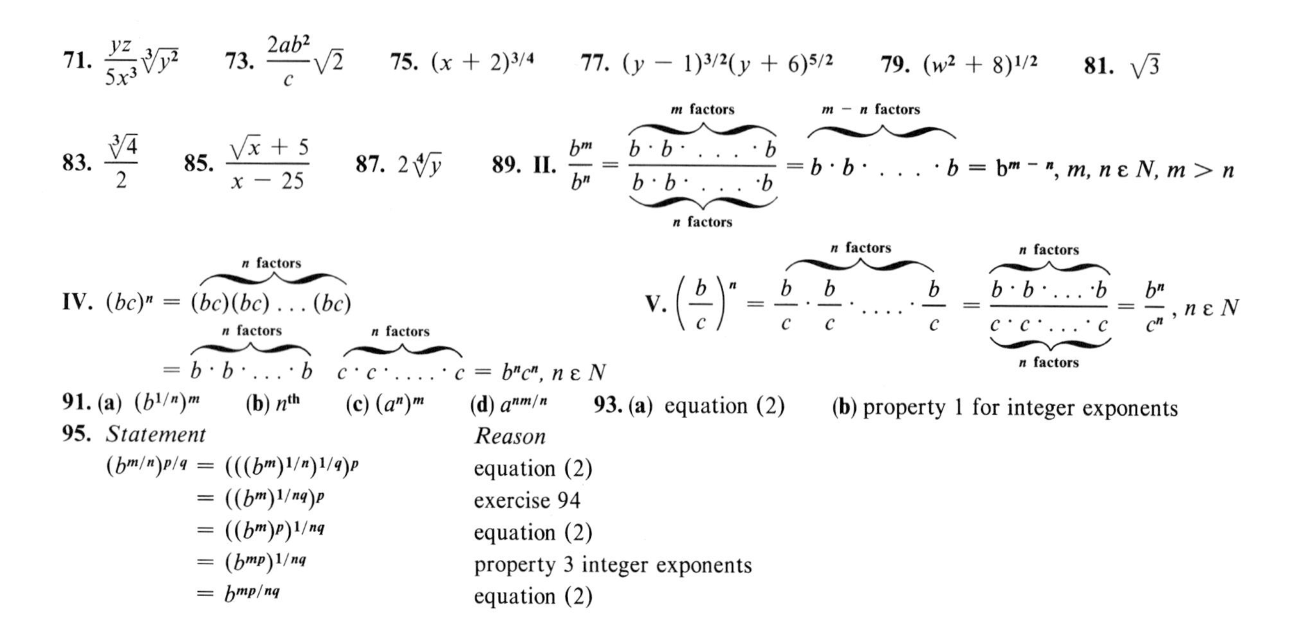

71. $\frac{yz}{5x^3}\sqrt[3]{y^2}$ **73.** $\frac{2ab^2}{c}\sqrt{2}$ **75.** $(x+2)^{3/4}$ **77.** $(y-1)^{3/2}(y+6)^{5/2}$ **79.** $(w^2+8)^{1/2}$ **81.** $\sqrt{3}$

83. $\frac{\sqrt[3]{4}}{2}$ **85.** $\frac{\sqrt{x}+5}{x-25}$ **87.** $2\sqrt[4]{y}$ **89. II.** $\frac{b^m}{b^n} = \frac{\overbrace{b\cdot b\cdot\ \ldots\ \cdot b}^{m\text{ factors}}}{\underbrace{b\cdot b\cdot\ \ldots\ \cdot b}_{n\text{ factors}}} = \overbrace{b\cdot b\cdot\ \ldots\ \cdot b}^{m-n\text{ factors}} = b^{m-n},\ m, n\ \varepsilon\ N,\ m > n$

IV. $(bc)^n = \overbrace{(bc)(bc)\ldots(bc)}^{n\text{ factors}}$

$= \overbrace{b\cdot b\cdot\ldots\cdot b}^{n\text{ factors}}\ \overbrace{c\cdot c\cdot\ldots\cdot c}^{n\text{ factors}} = b^nc^n,\ n\ \varepsilon\ N$

V. $\left(\frac{b}{c}\right)^n = \overbrace{\frac{b}{c}\cdot\frac{b}{c}\cdot\ldots\cdot\frac{b}{c}}^{n\text{ factors}} = \frac{\overbrace{b\cdot b\cdot\ldots\cdot b}^{n\text{ factors}}}{\underbrace{c\cdot c\cdot\ldots\cdot c}_{n\text{ factors}}} = \frac{b^n}{c^n},\ n\ \varepsilon\ N$

91. **(a)** $(b^{1/n})^m$ **(b)** n^{th} **(c)** $(a^n)^m$ **(d)** $a^{nm/n}$ **93.** **(a)** equation (2) **(b)** property 1 for integer exponents

95.

Statement	*Reason*
$(b^{m/n})^{p/q} = (((b^m)^{1/n})^{1/q})^p$	equation (2)
$= ((b^m)^{1/nq})^p$	exercise 94
$= ((b^m)^p)^{1/nq}$	equation (2)
$= (b^{mp})^{1/nq}$	property 3 integer exponents
$= b^{mp/nq}$	equation (2)

Section 1.3

1. $8x(x+3)$ **3.** $9y(y+3)^2$ **5.** $(x+2)(x+3)$ **7.** $(4x+5)(2x+7)$ **9.** $(11w-4)(w+3)$
11. $(5-4xy)^2$ **13.** $(x-3)(x+3)$ **15.** $(5xz-3y)(5xz+3y)$ **17.** $(x-2)(x+2)(x^2+4)$
19. $(2a-3b)(2a+3b)(4a^2+9b^2)$ **21.** $(xy-1)(xy+1)$ **23.** $(9w+5)(8w-3)$ **25.** $a(y^2+3y-5)$
27. $-2(3y-7x)(y+4x)$ **29.** $9y(x+1)(x+4)$ **31.** $(x+3)(x^2-3x+9)$
33. $(2x-ay)(4x^2+20xy+a^2y^2)$ **35.** $x(x^2+6x+12)$ **37.** $(x^2+3)(x^4-3x^2+9)$ **39.** $(x+a)(y+b)$
41. $(x+b)(bx+1)$ **43.** $(x-1)(x+1)(x+1)$ **45.** $(2x^2-1)(4-5x)$ **47.** $\left(y+\frac{1}{2}\right)\left(y+\frac{1}{2}\right)$
49. Not factorable **51.** $(x-\sqrt{3})(x+\sqrt{3})$ **53.** $(y+2^{1/3})(y^2-2^{1/3}y+2^{2/3})$ **55.** $(w^{1/2}+1)(w^{1/2}+1)$
57. $(4y^{2/5}+1)(y^{2/3}+3)$ **59.** $(x^{1/2}-1)(x+x^{1/2}+1)$ **61.** $(\sqrt{x}-2)(x+2\sqrt{x}+4)$
63. $x^{-5}y^{-4}(x-y)(x+y)$ **65.** $x^{-5}y^{-4}(y+x)(y^2-xy+y^2)$ **67.** $x^{-5}y^{-1}(x+2y)(x+y)$
69. $(y^n-10)(y^n+10)$ **71.** $y^n(4x+y)(x+y)$ **73.** $(2y^{2n}+3)(y^n-3)(y^n+3)$
75. $(a+1-x)(a+1+x)$ **77.** $(2x-3y-1)(2x+3y+1)$ **79.** $(2w-y+5)(2w+y+1)$
81. $(3x-5)^{2n-1}(y-1)^3(3x+y-6)$

Section 1.4

1. $a=3, b=7$ **3.** $a=\frac{1}{4}, b=-3$ **5.** $8+i$ **7.** $17+7i$ **9.** $-18-12i$ **11.** $2+3i$ **13.** $-1+5i$

15. $17-7i$ **17.** $2i$ **19.** $13i$ **21.** $4\sqrt{2}$ **23.** 3 **25.** $\frac{\sqrt{5}i}{20}$ **27.** -25 **29.** $1-i$ **31.** $2i$

33. $2+5i$ **35.** $6-i$ **37.** $\frac{8}{13}-\frac{1}{13}i$ **39.** $-8i$ **41.** $-\frac{6}{5}+\frac{11}{5}i$ **43.** $\frac{2+\sqrt{6}}{6}+\frac{\sqrt{2}-2\sqrt{3}}{6}i$

45. $\frac{-1}{3\sqrt{2}}i$ **47.** $\frac{2+2\sqrt{2}}{7}+\frac{\sqrt{3}+\sqrt{6}}{7}i$ **49.** $i^5=i$, $i^6=-1$, $i^7=-i$, and $i^8=1$. The pattern is $i, -1, -i, 1$, where $i^n=i^{n+4}$. **51.** $12-16i$ **53.** $52-47i$ **55.** Let $z_1=a_1+b_1i$, $z_2=a_2+b_2i$.

$$\begin{aligned}\overline{z_1+z_2} &= \overline{(a_1+b_1i)+(a_2+b_2i)} \\ &= \overline{(a_1+a_2)+(b_1+b_2)i} \\ &= (a_1+a_2)-(b_1+b_2)i \\ &= (a_1-b_1i)+(a_2-b_2i) \\ &= \bar{z}_1+\bar{z}_2\end{aligned}$$

57.
$$\begin{aligned}\overline{\left(\frac{z_1}{z_2}\right)} &= \overline{\left(\frac{a_1+b_1i}{a_2+b_2i}\right)} \\ &= \overline{\left(\frac{a_1+b_1i}{a_2+b_2i}\right)\cdot\left(\frac{a_2-b_2i}{a_2-b_2i}\right)} \\ &= \overline{\frac{(a_1a_2+b_1b_2)+(a_2b_1-a_1b_2)i}{a_2^2+b_2^2}} \\ &= \frac{a_1a_2+b_1b_2}{a_2^2+b_2^2}-\frac{a_2b_1-a_1b_2}{a_2^2+b_2^2}i \\ &= \frac{a_1a_2+b_1b_2}{a_2^2+b_2^2}+\frac{a_1b_2-a_2b_1}{a_2^2+b_2^2}i \\ &= \left(\frac{a_1-b_1i}{a_2-b_2i}\right)\left(\frac{a_2+b_2i}{a_2+b_2i}\right)=\frac{a_1-b_1i}{a_2-b_2i}=\frac{\bar{z}_1}{\bar{z}_2}\end{aligned}$$

Section 1.5

1. $x^2+3x+2=0$, $x=-2, x=-1$ **3.** $2x^2+3x-5=0$, $x=-\frac{5}{2}, x=1$ **5.** $2x^2-7x-4=0$, $x=-\frac{1}{2}, x=4$ **7.** $x^2+1=0$, $x=i, x=-i$

9. $x=-2\pm\sqrt{13}$ **11.** $x=-1\pm\sqrt{\frac{7}{3}}$ **13.** $\frac{1\pm\sqrt{73}}{6}$ **15.** $x=-2, -1$ **17.** $-2\pm\sqrt{13}$ **19.** $\pm\frac{5}{3}i$

21. $3\pm2\sqrt{6}$ **23.** $-\frac{7}{5}\pm\frac{1}{5}i$ **25.** $\frac{3\pm\sqrt{7}}{2}$ **27.** $\frac{-3\pm\sqrt{7}i}{2}$ **29.** $-3, \frac{1}{3}$ **31.** $x=-2\pm2\sqrt{3}$

33. $x=-\frac{1}{2}, x=2$ **35.** $x=\frac{3}{2}$ **37.** $x^2-3x+1=0$ **39.** $x^2-10x+25=0$

41. $x^2-4x+5=0$ **43.** $k=\frac{9}{20}$ **45.** Two real unequal roots **47.** $x=1, x=-2, x=3$

49. $x=-1, x=-2$ **51.** $x=\pm3, x=\pm2$ **53.** 3 or 1 **55.** 80 ft. by 200 ft. **57.** 550 mph

59. Car 1: $\frac{10\pm10\sqrt{301}}{3}\approx61.16$ mph; Car 2 ≈81.16 mph **61.** Either $17 or $18

63. Machine Y copies $\approx$ 42.29 copies/min. Machine X copies $\approx$ 47.29 copies/min.

65. No, not real. The discriminant of the quadratic equation is negative.

67. $r_1 + r_2 = \dfrac{-b - \sqrt{b^2 - 4ac}}{2a} + \dfrac{-b + \sqrt{b^2 - 4ac}}{2a} = \dfrac{-2b}{2a} = -\dfrac{b}{a}.$

$r_1 r_2 = \dfrac{-b - \sqrt{b^2 - 4ac}}{2a} \cdot \dfrac{-b + \sqrt{b^2 - 4ac}}{2a} = \dfrac{b^2 - (b^2 - 4ac)}{4a^2} = \dfrac{4ac}{4a^2} = \dfrac{c}{a}$

69. If r_3 is a root of $ax^2 + bx + c = 0$, then it is a root of $(x - r_1)(x - r_2) = 0$. It follows that $(r_3 - r_1)(r_3 - r_2) = 0$. But this is a contradiction. Since r_3 is unequal to either r_1 or r_2, $r_3 - r_1 \neq 0$ and $r_3 - r_2 \neq 0$. Therefore $(r_3 - r_1)(r_3 - r_2) \neq 0$ and r_3 is not a root.

Section 1.6

1. $\{x|x < 5\}$ $(-\infty, 5)$ **3.** $\{x|x \geq 4\}$ $[4, \infty)$ **5.** $\left\{x|x \geq \frac{29}{5}\right\}$ $\left[\frac{29}{5}, \infty\right)$ **7.** $\{x|-2 \leq x \leq 6\}$ $[-2, 6]$

9. $\left\{x|-\frac{5}{4} \leq x \leq \frac{3}{2}\right\}$ $\left[-\frac{5}{4}, \frac{3}{2}\right]$ **11.** $\{x|-24 < x < 12\}$ $(-24, 12)$ **13.** $\{x|-3 \leq x \leq 1\}$ $[-3, 1]$

15. $\left\{x|x < \frac{1}{2}\right\} \cup \{x|x > 2\}$; $\left(-\infty, \frac{1}{2}\right) \cup (2, \infty)$ **17.** $\{x|-1 - \sqrt{3} < x < -1 + \sqrt{3}\}$ $(-1 - \sqrt{3}, -1 + \sqrt{3})$

19. $\{x|-2 < x < 1\} \cup \{x|x > 3\}$ $(-2, 1) \cup (3, \infty)$ **21.** $\left\{x|-5 < x < -\frac{5}{7}\right\} \cup \{x|x > 0\}$ $\left(-5, -\frac{5}{7}\right) \cup (0, \infty)$

23. $\left\{x|\frac{2}{3} < x < \frac{13}{12}\right\}$ $\left(\frac{2}{3}, \frac{13}{12}\right)$

25. $\left\{x|x < \frac{-2 - \sqrt{10}}{3}\right\} \cup \left\{x|x > \frac{-2 + \sqrt{10}}{3}\right\}$ $\left(-\infty, \frac{-2 - \sqrt{10}}{3}\right) \cup \left(\frac{-2 + \sqrt{10}}{3}, \infty\right)$

27. $\{x|x < -3\} \cup \{x|0 < x < 1\}$ $(-\infty, -3) \cup (0, 1)$ **29.** $\{x|x \text{ is real}\}$ $(-\infty, \infty)$

31. $\left\{x|-4 < x \leq -\frac{5}{2}\right\} \cup \{x|0 \leq x < 2\}$ $\left(-4, -\frac{5}{2}\right] \cup [0, 2)$

33. $\{x|x < -3\} \cup \left\{x|-\frac{1}{2} < x < 2\right\}$ $(-\infty, -3) \cup \left(-\frac{1}{2}, 2\right)$ **35.** Between $\frac{75}{7}$ and $\frac{100}{7}$ cc of solution

37. Student must score at least 89. No, the student can not score high enough to earn an A.

39. Use $8\frac{1}{3}$ gallons of unleaded and $6\frac{2}{3}$ gallons of super-unleaded for best performance at the least cost.

41. Because there is no last real number to the right of 2. **43.** Pipe B must fill the pool in less than 3 hours.

45. $(x - y)^2 \geq 0 \rightarrow x^2 - 2xy + y^2 \geq 0$

$\rightarrow (x^2 + 2xy + y^2) - 4xy \geq 0$

$\rightarrow (x + y)^2 - 4xy \geq 0$

The last statement and given that $x + y > 0$, it follows that $\dfrac{(x + y)^2 - 4xy}{x + y} \geq 0$. Thus $x + y - \dfrac{4xy}{x + y} \geq 0$ and

$x + y \geq \dfrac{4xy}{x + y}.$

47. The product $(x - 1)(x + 2)$ can be positive or negative; both possibilities must be considered when solving the inequality. The solution set is the union of the solution sets of each of the following pairs of inequalities: $2(x + 2) < 5(x - 1)$ and $(x - 1)(x + 2) > 0$ and $2(x + 2) > 5(x - 1)$ and $(x - 1)(x + 2) < 0$.

49. **(a)** $<$ **(b)** $<$ **(c)** $B(r) + C(r)$ **(d)** $n + k$ **(e)** $<$ **(f)** $B(p)$ **(g)** $A(x) < B(x)$ **(h)** equivalent

Section 1.7

1. $x = 7$ **3.** $x = 2$ **5.** $x = -13$ **7.** No solution **9.** $x = \frac{1}{4}$ **11.** $x = -\frac{1}{2}$ **13.** $x = 27, x = 1$

15. $x = 625$ **17.** $x = 4, x = \frac{3}{5}$ **19.** $x = \pm\frac{1}{3}, x = \pm 2$ **21.** $x = \sqrt[3]{7}, x = 1$

23. $x = \pm\frac{2\sqrt{2}i}{\sqrt{3}}, x = \pm\sqrt{\frac{5}{2}}$ **25.** $x = \frac{1}{2}$ **27.** $x = 10, x = 1$ **29.** $x = 1$ **31.** $x = -1$

33. $x = 26, x = -2$ **35.** $x = 4$ **37.** $x = 4$ **39.** $s = \frac{gt^2}{2}$ **41.** $C = \frac{1}{4L\pi^2 f^2}$

43. $h = 490.5$ ft.

Section 1.8

1. $x = 4$ or $x = -8$; $-8 < x < 4$; $x > 4$ or $x < -8$

3. $x = \frac{11}{3}$ or $x = -\frac{7}{3}$; $-\frac{7}{3} < x < \frac{11}{3}$; $x > \frac{11}{3}$ or $x < -\frac{7}{3}$

5. $x = \frac{1}{3}$ or $x = 5$; $5 > x > \frac{1}{3}$; $x < \frac{1}{3}$ or $x > 5$

7. $x = \frac{8}{9}$ or $x = \frac{16}{9}$; $\frac{16}{9} > x > \frac{8}{9}$; $x < \frac{8}{9}$ or $x > \frac{16}{9}$

9. $x = \frac{5 \pm \sqrt{29}}{2}$ or $x = \frac{5 \pm \sqrt{5}}{2}$ **11.** $-\frac{12}{5} \leq x \leq \frac{6}{5}$ **13.** $x \leq \frac{2}{9}$ or $x \geq \frac{2}{3}$ **15.** $-\frac{5}{3} < x < 7$

17. $x = \frac{-3 \pm \sqrt{17}}{2}$ or $x = \frac{-15 \pm \sqrt{105}}{10}$ **19.** $11 > x > -10$ **21.** $x > 4$ or $x < -\frac{28}{3}$ **23.** $-\frac{33}{2} \leq x \leq 11$

25. $|x - 5| < 4$ **27.** $|x + 5| \geq 3$ **29.** $\left|x + \frac{5}{2}\right| < \frac{11}{2}$ **31.** $|x - 2.45| \geq 1.87$ **33.** No solution in the reals.

35. $x = -\frac{17}{4}$ or $x = -\frac{13}{6}$ **37.** $\left(-\infty, -\frac{2}{7}\right] \cup \left[\frac{8}{5}, \infty\right)$ **39.** $\left[-\frac{20}{7}, -\frac{3}{2}\right) \cup \left(-\frac{3}{2}, -\frac{16}{17}\right]$

41. $\left(-\infty, \frac{44}{23}\right) \cup \left(\frac{52}{17}, \infty\right)$ **43.** $[-1 - \sqrt{6}, -1 - \sqrt{2}] \cup [-1 + \sqrt{2}, -1 + \sqrt{6}]$ **45.** $(2, \infty)$

47. $\left(-\infty, -\frac{5}{2}\right) \cup (0, \infty)$ **49.** $x = 2, x = -\frac{4}{3}$ **51.** $\left(-\frac{3}{2}, \infty\right)$ **53.** $\left(-3, -\frac{1}{7}\right) \cup \left(\frac{11}{3}, \infty\right)$

55. $\left[-\sqrt{\frac{51}{10}}, -\frac{7}{\sqrt{10}}\right] \cup \left[\frac{7}{\sqrt{10}}, \sqrt{\frac{51}{10}}\right]$ **57.** 4 or 10

59. $a = a - b + b$. Then $|a| = |a - b + b| \leq |a - b| + |b|$ by the Triangle Ineq. Therefore, $|a| - |b| \leq |a - b|$, i.e. $|a - b| \geq |a| - |b|$. **61.** **(a)** suppose $a < b$. Then $a - b < 0$ and $b - a > 0$. $|a - b| = -(a - b) = b - a = |b - a|$ **(b)** suppose $a = b$. Then $a - b = b - a = 0$. $|a - b| = 0 = |b - a|$ **(c)** suppose $a > b$. Then $a - b > 0$ and $b - a < 0$. $|a - b| = a - b = -(b - a) = |b - a|$

Section 1.9

1. x can differ from 4 by less than $\frac{1}{12}$ of a unit. **3.** x can be in error by no more than ± 0.025.

5. $|(3x + 7) - 19| < 1 \rightarrow$

$$|3x - 12| < 1$$
$$3|x - 4| < 1$$
$$|x - 4| < \frac{1}{3}$$
$$-\frac{1}{3} < x - 4 < \frac{1}{3}$$
$$\frac{11}{3} < x < \frac{13}{3}$$

7. Error in measurement of a side must be $< \left|\frac{1}{81}\right|$.

9. x can differ from 5 by less than 0.0125 of a unit.

Chapter 1 Review Exercises

1. $\{x | -2 < x < 6, x \varepsilon J\}$ **3.** $3 + 4i$ **5.** $-78 - 40i$ **7.** $(2x + 3)(x + 2)$ **9.** $(y + 3)(y^2 - 3y + 9)$ **11.** $x^3(x^2 - 3x + 1)$ **13.** $(x + z)(x + y)$ **15.** -4 **17.** 27 **19.** $\frac{33}{256}$ **21.** $\frac{6{,}064}{3{,}087}$ **23.** $(y^2 - 5)(y^2 + 5)$ **25.** $[x - y + 2][x + y - 2]$ **27.** $(5x^{2n} - 1)(x^{2n} - 5)$ **29.** $(2y^{1/2} - 5)(5y^{1/2} + 1)$ **31.** $4 + i$ **33.** $-5i$ **35.** $\frac{9}{50} + \frac{37}{50}i$ **37.** $-\frac{6}{25} + \frac{8}{25}i$ **39.** $\frac{9z^6}{25x^2y^{16}}$ **41.** $\frac{1}{x^{7/18}y^{5/18}}$ **43.** $(9y - 4)(3y^2 + 8)$ **45.** $x^{-1}y^{-3}(x + y)(x^2 - xy + y^2)$ **47.** $\frac{52}{17}$ **49.** $x = 2, x = 4$ **51.** $x = -1$ **53.** $x = -3, x = 2$ **55.** $x = \frac{5}{2}, x = \frac{2}{3}$ **57.** $-3 \pm \sqrt{11}$ **59.** $x = -1$ or $x = -4$ **61.** $\frac{4 \pm \sqrt{85}}{3}$ **63.** $x < \frac{7}{8}$ **65.** $4 \geq x \geq -\frac{3}{2}$ **67.** $\frac{15}{4} \leq x \leq \frac{51}{4}$ **69.** $(-1, 2) \cup (11, \infty)$ **71.** $\left[-5, -\frac{1}{2}\right]$ **73.** $(-\infty, -4 - \sqrt{13}) \cup (-4 + \sqrt{13}, +\infty)$ **75.** $(-2, 0] \cup \left[\frac{2}{5}, 4\right)$ **77.** $x = 4$ **79.** $x = -125, x = 8$ **81.** No solution in R. **83.** No solution

85. $-3 \leq x \leq -\frac{2}{3}$ **87.** $x = \frac{-3 \pm \sqrt{33}}{2}$ **89.** $x < -\frac{27}{12}$ or $x > \frac{37}{12}$ **91.** $(-\infty, -6) \cup \left(-6, -\frac{9}{4}\right]$

93. $x = 3, x = -\frac{3}{13}$ **95.** $\left(-\frac{11}{5}, 7\right)$ **97.** $x = -3 \pm 2i$ **99.** $x = \frac{3 \pm \sqrt{57}}{4}$

101. Two unequal complex roots **103.** $x^2 + 5x - 6 = 0$ **105.** $x^2 - 2x + 2 = 0$ **107.** $k = -\frac{25}{16}$ **109.** No

111. 42 lbs. of walnuts; 98 lbs. of almonds **113.** 70 mph **115.** $5 \pm \sqrt{34} \approx 10.83$ mph **117.** $7,500
119. $w > -1 + \sqrt{41}$ **121.** The magnitude of the error must be less than 0.005 unit.

Chapter 1 Test 1

1. **(a)** $-125\sqrt[3]{5}$ **(b)** $\frac{3}{20}$ **2.** $11 - 2i, \frac{11}{58} - \frac{71}{58}i$ **3.** **(a)** $(x + 5)(x - 2)$ **(b)** $(3x - 2)(9x^2 + 6x + 4)$

4. $(3w + 2z)(2x + y)$ **5.** $\frac{1 \pm \sqrt{7}}{6}$ **6.** $x < \frac{4}{3}$ **7.** $x = 8$ or $x = 27$ **8.** $\left(-\infty, -\frac{5}{2}\right] \cup [6, \infty)$

9. $\frac{1}{2} \leq x \leq 4$ **10.** $x = 9$ **11.** Integers are 123, 125, 127.

Chapter 1 Test 2

1. **(a)** $\frac{243}{64}$ **(b)** $\frac{x^2z^{1/2}}{y^4}$ **2.** $z_1 = a + bi, z_2 = a - bi.\ z_1 \cdot z_2 = a^2 + b^2 - abi + abi = a^2 + b^2$, a real number
3. **(a)** $3x(10x^2 - 16x - 3)$ **(b)** $(x^{1/4} + 4)(x^{1/2} - 4x^{1/4} + 16)$ **4.** $(3x - 2y - 3)(3x + 2y + 3)$
5. $\frac{-4 \pm \sqrt{10}}{2}$ **6.** No real solution **7.** $\left(-\infty, \frac{3}{4}\right] \cup \left[\frac{7}{3}, \infty\right)$ **8.** $-3 \leq x \leq -\frac{3}{7}$

9. $\left(-\infty, -\frac{11}{5}\right) \cup \left(-\frac{5}{3}, 0\right) \cup (1, \infty)$ **10.** $k = \pm 8$ **11.** Speed $= \frac{500 + 20\sqrt{634}}{2} \approx 501.8$ mph

Section 2.1

1. A function (one y for each x). **3.** Not a function (two different ys for $x = -8$).

5. Not a function (more than one y for $x = 1$). **7.** Domain: reals; range: reals **9.** Domain: R; range: $\left[-\frac{5}{4}, \infty\right)$

11. Domain: R; range: $\left[-\frac{1}{8}, \infty\right)$ **13.** Domain: $[0, \infty)$; range: $[0, \infty)$
15. Domain: $(-\infty, -2] \cup [0, \infty)$; range: $[0, \infty)$ **17.** Domain: reals except 2; range: reals except 2
19. Domain: all reals except 0,5; range: all reals except 0 **21.** 6; 14; $-4x + 6$; $8x + 2$
23. $\sqrt{5}$; $\sqrt{2}$; $\sqrt{2 - x}$; $\sqrt{15 - 3x}$

25. $\sqrt{9x^2 + 18x + 8}$; $\sqrt{x^2 + (2a + 2)x + a^2 + 2a}$ **27.** $-\frac{5}{2}$; -1; $\frac{10x + 1}{5x - 4}$; $\frac{2 + 5x}{1 - 2x}$

29. No **31.** Yes **33.** No **35.** Increasing: $(-1, 2)$; decreasing: $(2, 4)$; constant: $(-4, -1)$
37. Increasing: $(2, 4)$; decreasing: $(-3, -1)(4, 5)$; constant: $(-1, 2)$ **39.** Increasing: $(-4, -2)$; decreasing: $(0, 2)$; constant: nowhere **41.** Yes, for each value of the independent variable (1 through 5) there is associated exactly one value of the dependent variable (closing price). **43.** Not a function if you have favorable encounters with more than one person or unfavorable encounters with more than one person. A function if you have only one favorable and/or one unfavorable encounter.

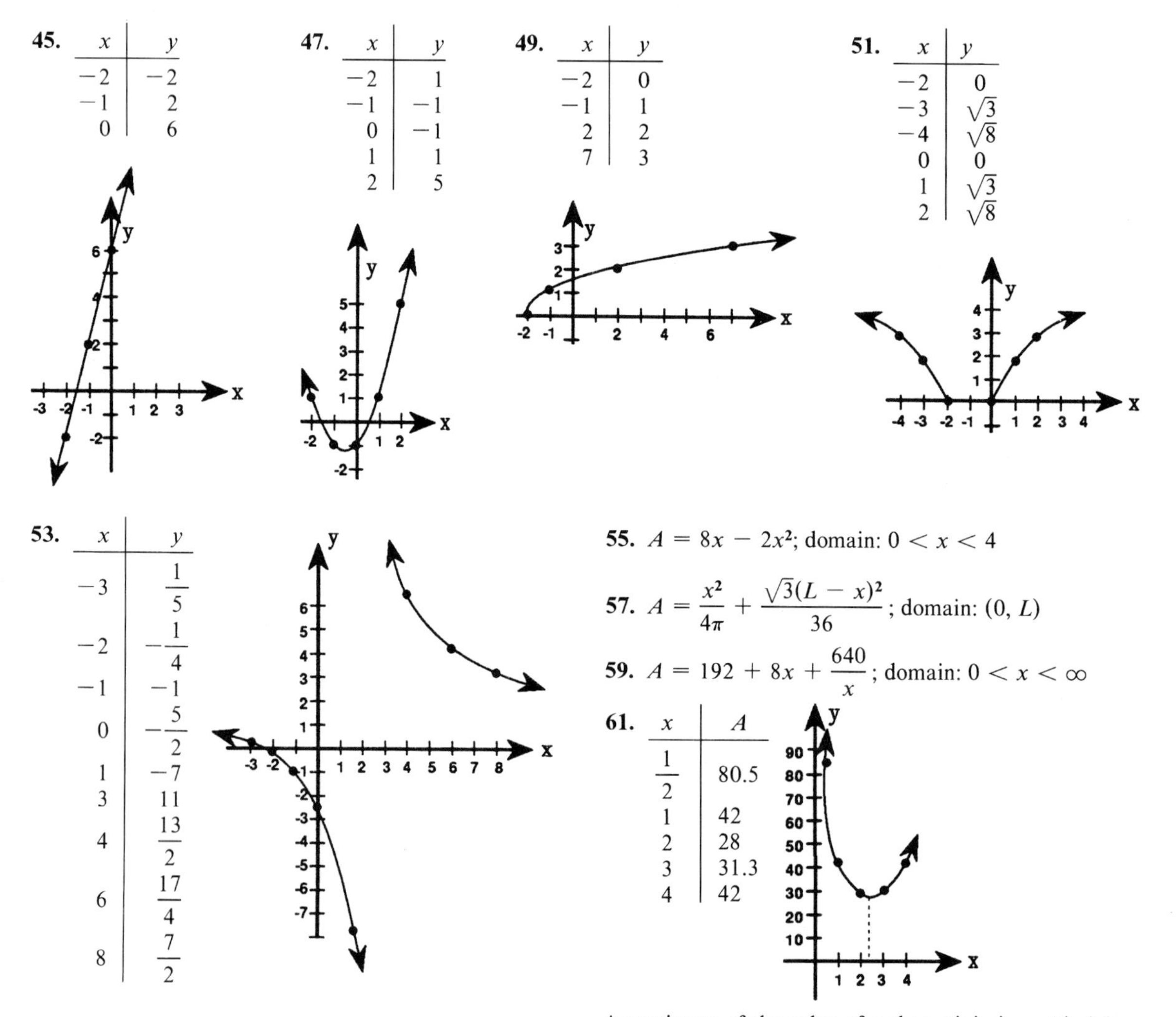

45.

x	y
-2	-2
-1	2
0	6

47.

x	y
-2	1
-1	-1
0	-1
1	1
2	5

49.

x	y
-2	0
-1	1
2	2
7	3

51.

x	y
-2	0
-3	$\sqrt{3}$
-4	$\sqrt{8}$
0	0
1	$\sqrt{3}$
2	$\sqrt{8}$

53.

x	y
-3	$\frac{1}{5}$
-2	$-\frac{1}{4}$
-1	-1
0	$-\frac{5}{2}$
1	-7
3	11
4	$\frac{13}{2}$
6	$\frac{17}{4}$
8	$\frac{7}{2}$

55. $A = 8x - 2x^2$; domain: $0 < x < 4$

57. $A = \frac{x^2}{4\pi} + \frac{\sqrt{3}(L - x)^2}{36}$; domain: $(0, L)$

59. $A = 192 + 8x + \frac{640}{x}$; domain: $0 < x < \infty$

61.

x	A
$\frac{1}{2}$	80.5
1	42
2	28
3	31.3
4	42

An estimate of the value of x that minimizes A is 2.2.

Section 2.2

1. $y - 3 = 2(x - 1)$, $y = 2x + 1$, $-2x + y - 1 = 0$

3. $y - 2 = -\frac{3}{5}(x + 3)$, $y = -\frac{3}{5}x + \frac{1}{5}$, $3x + 5y - 1 = 0$

5. $y - 0 = -\frac{5}{2}(x - 0)$, $y = -\frac{5}{2}x$, $5x + 2y = 0$

7. $y - 4 = -\frac{8}{3}(x - 5)$, $y = -\frac{8}{3}x + \frac{52}{3}$, $8x + 3y - 52 = 0$

9. $y = 2$ **11.** $x = 6$ **13.** $y = 0$

15. $x = 0$ **17.** $x = 3$ **19.** $y = 2$ **21.** $y - 0 = -2(x - 0)$, or $y = -2x$; $y - 0 = \frac{1}{2}(x - 0)$, or $y = \frac{1}{2}x$

23. $y + 2 = 5(x + 5)$; $y + 2 = -\frac{1}{5}(x + 5)$ **25.** $y - 3 = \frac{3}{2}(x - 4)$; $y - 3 = -\frac{2}{3}(x - 4)$

27. $y - 0 = -\frac{2}{3}(x - 7)$; $y - 0 = \frac{3}{2}(x - 7)$ **29.** Perpendicular **31.** Perpendicular **33.** Neither

35. Perpendicular **37.** (a) $2\sqrt{5}$ (b) $(2, 7)$ (c) $y - 7 = \frac{1}{2}(x - 2)$ **39.** (a) $\sqrt{170}$ (b) $\left(\frac{1}{2}, \frac{1}{2}\right)$

(c) $y - \frac{1}{2} = -\frac{11}{7}\left(x - \frac{1}{2}\right)$ **41.** (a) 5 (b) $\left(\frac{5}{2}, 0\right)$ (c) $x = \frac{5}{2}$ **43.** $\sqrt{(2 - 3)^2 + (4 - 6)^2} = \sqrt{1 + 4} = \sqrt{5}$; $\sqrt{(6 - 2)^2 + (2 - 4)^2} = \sqrt{16 + 4} = \sqrt{20}$; $\sqrt{(6 - 3)^2 + (2 - 6)^2} = \sqrt{9 + 16} = \sqrt{25}$; $(5)^2 + (\sqrt{20})^2 = (\sqrt{25})^2$

45.

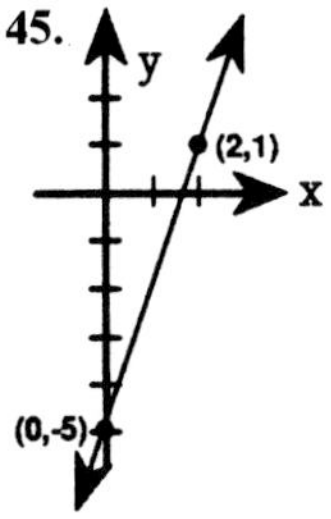

47.

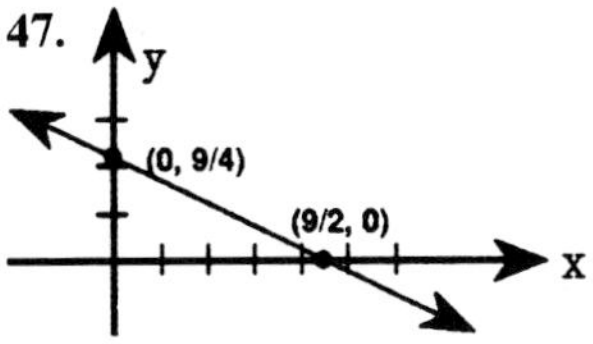

49.

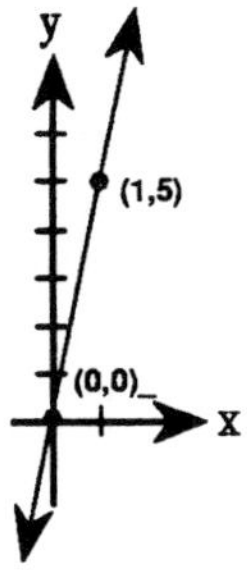

51.

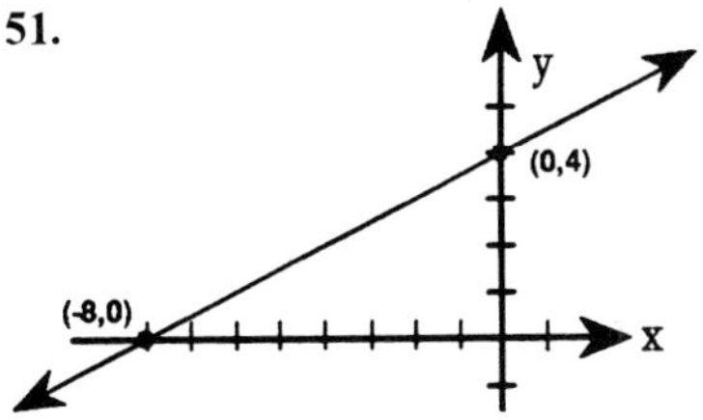

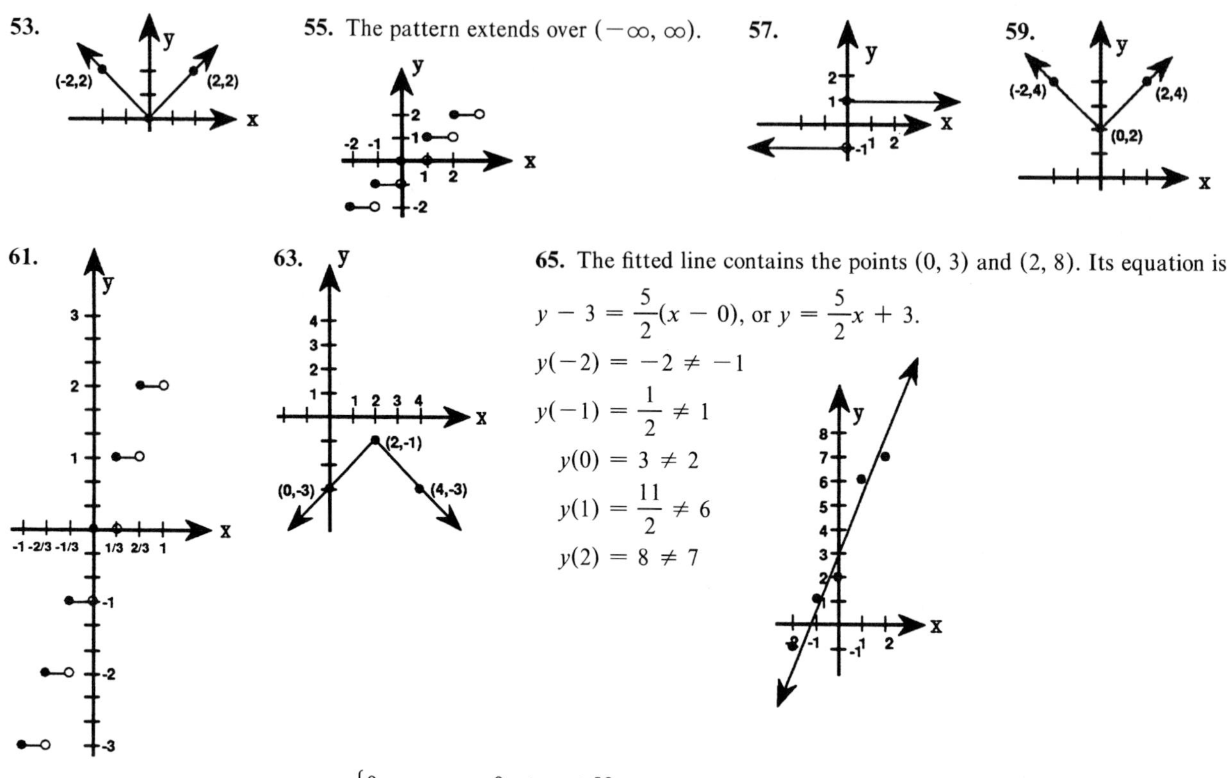

55. The pattern extends over $(-\infty, \infty)$.

65. The fitted line contains the points (0, 3) and (2, 8). Its equation is

$y - 3 = \frac{5}{2}(x - 0)$, or $y = \frac{5}{2}x + 3$.

$y(-2) = -2 \neq -1$

$y(-1) = \frac{1}{2} \neq 1$

$y(0) = 3 \neq 2$

$y(1) = \frac{11}{2} \neq 6$

$y(2) = 8 \neq 7$

67. Let x = lbs. of baggage. $C = \begin{cases} 0 & 0 \leq x \leq 50 \\ 3(x - 50) & x > 50 \end{cases}$ If $\Delta x = 1$, then $\Delta C = \$3$.

69. $V = 10{,}000t + 10{,}000$. The slope is the amount the computer depreciates each year.

71. Angles C and Q are right angles. They are formed by lines that are parallel to the coordinate axes. $\angle ABC = \angle PQR$. They are corresponding angles formed by the transversal $\overline{BR}$ and parallel lines ℓ_1 and ℓ_2. Angles BAC and QPR are equal. If two angles of one triangle equal a corresponding pair of angles in a second triangle, then the third angles in each triangle are equal. The x-coordinates at A and P are the same, and, the x-coordinates at C and R are the same. Thus, the length of the horizontal line segments $\overline{AC}$ and $\overline{PR}$ are the same. Therefore, ΔABC is congruent to ΔPQR by angle-side-angle.

73. We are given $m_1 = -\frac{1}{m_2}$. Therefore $\frac{h}{a} = \frac{-1}{-\frac{h}{b}}$ or $\frac{h}{a} = \frac{b}{h}$, or $h^2 = ab$. Since ΔABD and ΔCDB are right triangles, $\overline{AB}^2 = a^2 + h^2$ and $\overline{BC}^2 = h^2 + b^2$. Now $\overline{AB}^2 + \overline{BC}^2 = (a^2 + h^2) + (h^2 + b^2) = a^2 + b^2 + 2h^2 = a^2 + b^2 + 2(ab) = (a + b)^2 = \overline{AC}^2$. Since the sides of ΔABC satisfy the Pythagorean Theorem, ΔABC is a right triangle and $\overline{AB}$ is perpendicular to $\overline{BC}$.

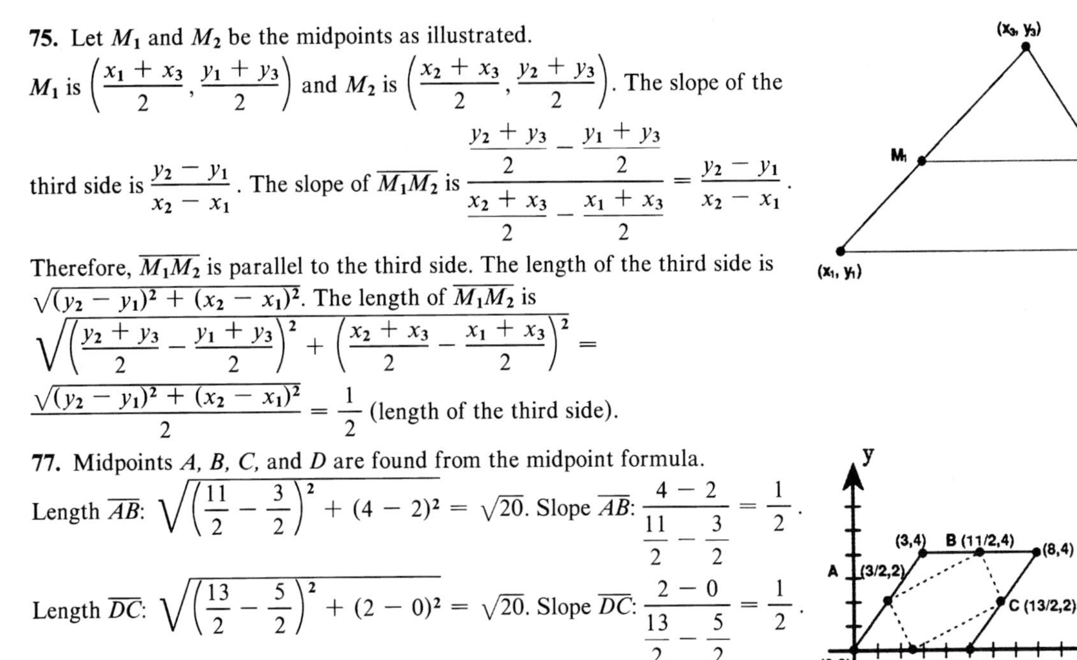

75. Let M_1 and M_2 be the midpoints as illustrated.

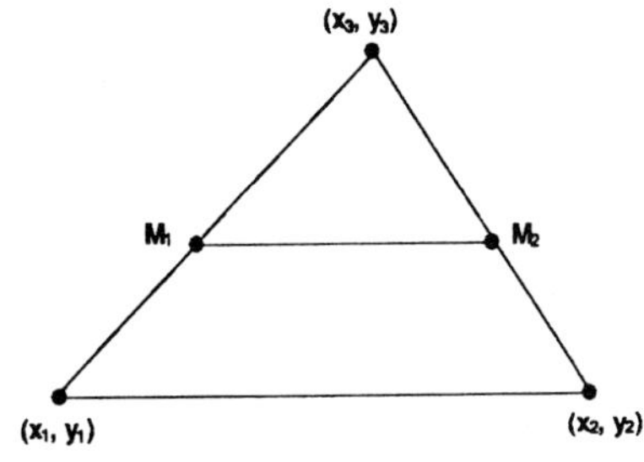

M_1 is $\left(\frac{x_1 + x_3}{2}, \frac{y_1 + y_3}{2}\right)$ and M_2 is $\left(\frac{x_2 + x_3}{2}, \frac{y_2 + y_3}{2}\right)$. The slope of the third side is $\frac{y_2 - y_1}{x_2 - x_1}$. The slope of $\overline{M_1M_2}$ is $\dfrac{\frac{y_2 + y_3}{2} - \frac{y_1 + y_3}{2}}{\frac{x_2 + x_3}{2} - \frac{x_1 + x_3}{2}} = \frac{y_2 - y_1}{x_2 - x_1}$.

Therefore, $\overline{M_1M_2}$ is parallel to the third side. The length of the third side is $\sqrt{(y_2 - y_1)^2 + (x_2 - x_1)^2}$. The length of $\overline{M_1M_2}$ is

$$\sqrt{\left(\frac{y_2 + y_3}{2} - \frac{y_1 + y_3}{2}\right)^2 + \left(\frac{x_2 + x_3}{2} - \frac{x_1 + x_3}{2}\right)^2} =$$

$$\frac{\sqrt{(y_2 - y_1)^2 + (x_2 - x_1)^2}}{2} = \frac{1}{2} \text{ (length of the third side).}$$

77. Midpoints A, B, C, and D are found from the midpoint formula.

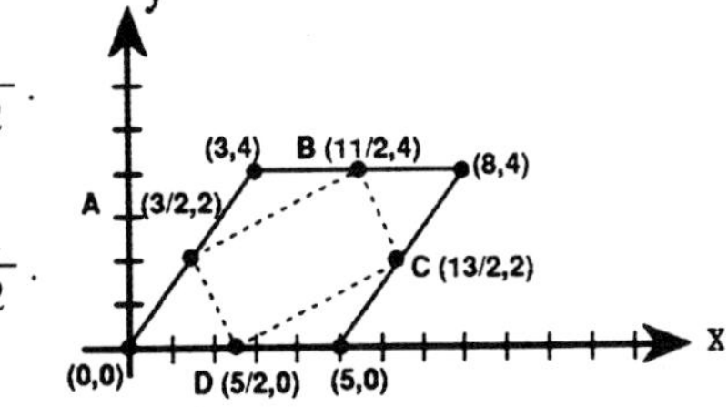

Length $\overline{AB}$: $\sqrt{\left(\frac{11}{2} - \frac{3}{2}\right)^2 + (4 - 2)^2} = \sqrt{20}$. Slope $\overline{AB}$: $\dfrac{4 - 2}{\frac{11}{2} - \frac{3}{2}} = \frac{1}{2}$.

Length $\overline{DC}$: $\sqrt{\left(\frac{13}{2} - \frac{5}{2}\right)^2 + (2 - 0)^2} = \sqrt{20}$. Slope $\overline{DC}$: $\dfrac{2 - 0}{\frac{13}{2} - \frac{5}{2}} = \frac{1}{2}$.

If a pair of opposite sides of a quadrilateral are equal and parallel, the quadrilateral is a parallelogram.

Slope $BC = \dfrac{4 - 2}{\frac{11}{2} - \frac{13}{2}} = -2$. $\overline{AB}$ is perpendicular to BC because their slopes are negative reciprocals. Thus $ABCD$ is a rectangle. A parallelogram with a right angle is a rectangle.

Section 2.3

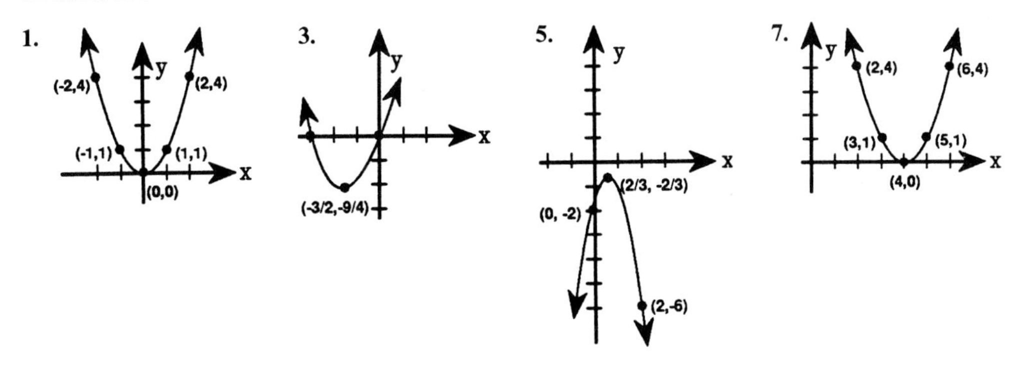

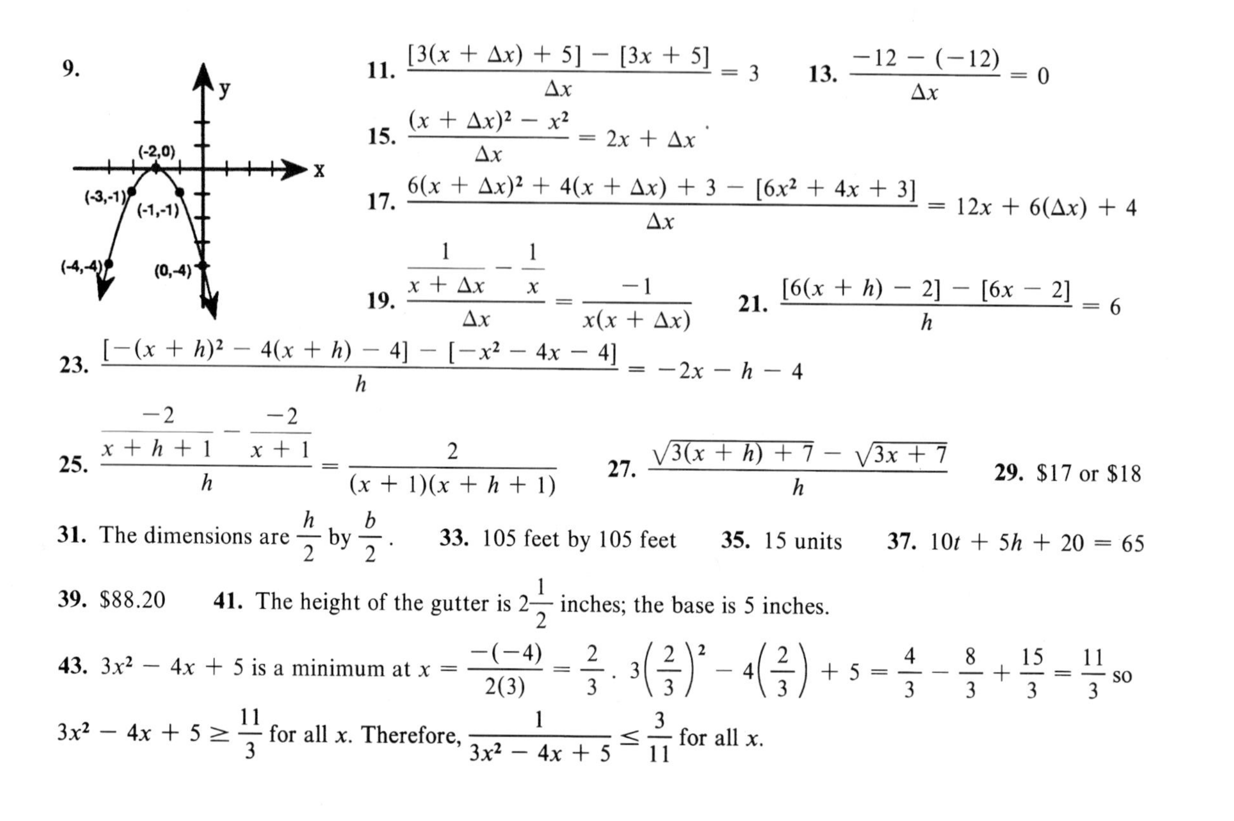

9. (graph)

11. $\dfrac{[3(x + \Delta x) + 5] - [3x + 5]}{\Delta x} = 3$ **13.** $\dfrac{-12 - (-12)}{\Delta x} = 0$

15. $\dfrac{(x + \Delta x)^2 - x^2}{\Delta x} = 2x + \Delta x$

17. $\dfrac{6(x + \Delta x)^2 + 4(x + \Delta x) + 3 - [6x^2 + 4x + 3]}{\Delta x} = 12x + 6(\Delta x) + 4$

19. $\dfrac{\dfrac{1}{x + \Delta x} - \dfrac{1}{x}}{\Delta x} = \dfrac{-1}{x(x + \Delta x)}$ **21.** $\dfrac{[6(x + h) - 2] - [6x - 2]}{h} = 6$

23. $\dfrac{[-(x + h)^2 - 4(x + h) - 4] - [-x^2 - 4x - 4]}{h} = -2x - h - 4$

25. $\dfrac{\dfrac{-2}{x + h + 1} - \dfrac{-2}{x + 1}}{h} = \dfrac{2}{(x + 1)(x + h + 1)}$ **27.** $\dfrac{\sqrt{3(x + h) + 7} - \sqrt{3x + 7}}{h}$ **29.** \$17 or \$18

31. The dimensions are $\dfrac{h}{2}$ by $\dfrac{b}{2}$. **33.** 105 feet by 105 feet **35.** 15 units **37.** $10t + 5h + 20 = 65$

39. \$88.20 **41.** The height of the gutter is $2\frac{1}{2}$ inches; the base is 5 inches.

43. $3x^2 - 4x + 5$ is a minimum at $x = \dfrac{-(-4)}{2(3)} = \dfrac{2}{3}$. $3\left(\dfrac{2}{3}\right)^2 - 4\left(\dfrac{2}{3}\right) + 5 = \dfrac{4}{3} - \dfrac{8}{3} + \dfrac{15}{3} = \dfrac{11}{3}$ so $3x^2 - 4x + 5 \geq \dfrac{11}{3}$ for all x. Therefore, $\dfrac{1}{3x^2 - 4x + 5} \leq \dfrac{3}{11}$ for all x.

Section 2.4

1. (**a**) $(-2, 5)$
(**b**) $(2, -5)$
(**c**) $(-2, -5)$

3. (**a**) $(1, -5)$
(**b**) $(-1, 5)$
(**c**) $(1, 5)$

5. (**a**) $(6, 8)$
(**b**) $(-6, -8)$
(**c**) $(6, -8)$

7. (**a**) $(0, -2)$
(**b**) $(0, 2)$
(**c**) $(0, 2)$

9. (**a**) x-int: ± 1; y-int: -1
(**b**) domain: all reals; range: $y \geq -1$
(**c**) sym. to y-axis

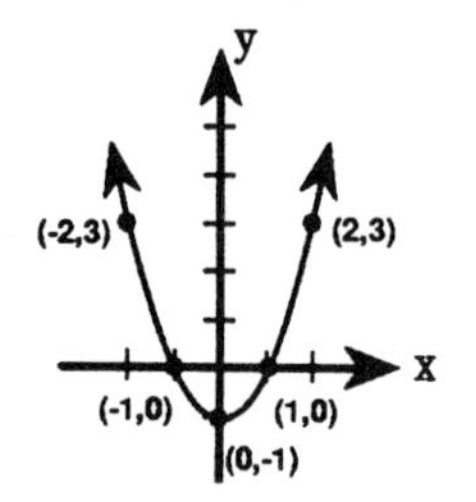

11. **(a)** x-int: 3, -1; y-int: -3
(b) domain: Reals; range: $y \geq -4$
(c) no symmetry

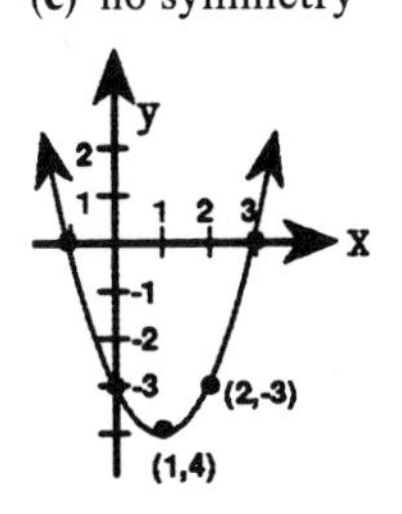

13. **(a)** x- and y-int: 0
(b) domain: $x \geq 0$; range: $y \geq 0$
(c) no symmetry

15. **(a)** x- and y-int: 0
(b) domain: $x \geq 0$; range: $y \geq 0$
(c) no symmetry

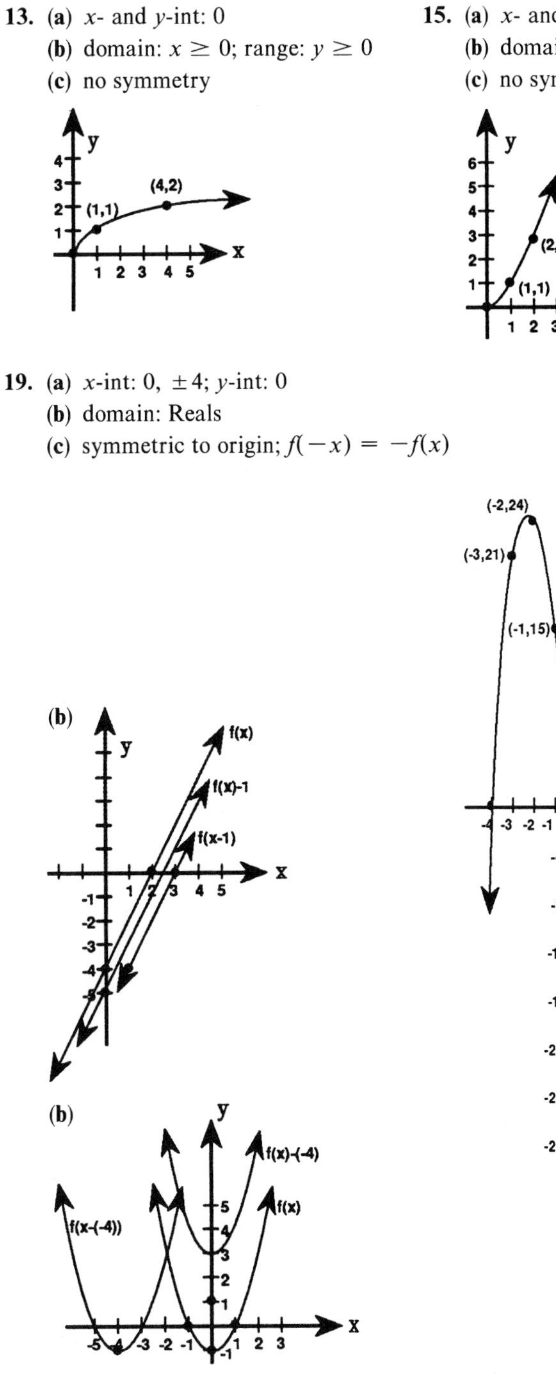

17. **(a)** x-int: ± 3; y-int: 3
(b) domain: $[-3, 3]$; range: $[0, 3]$
(c) symmetric to y-axis

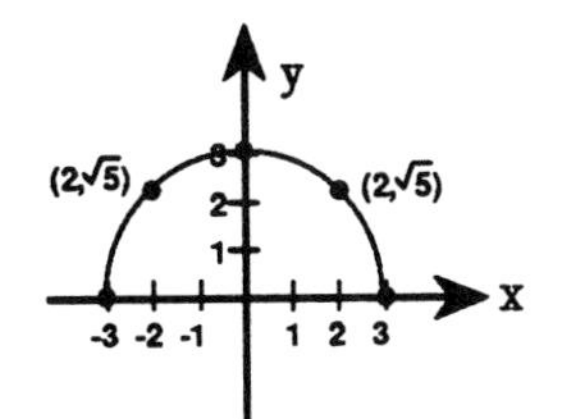

19. **(a)** x-int: 0, ± 4; y-int: 0
(b) domain: Reals
(c) symmetric to origin; $f(-x) = -f(x)$

21. **(a)**

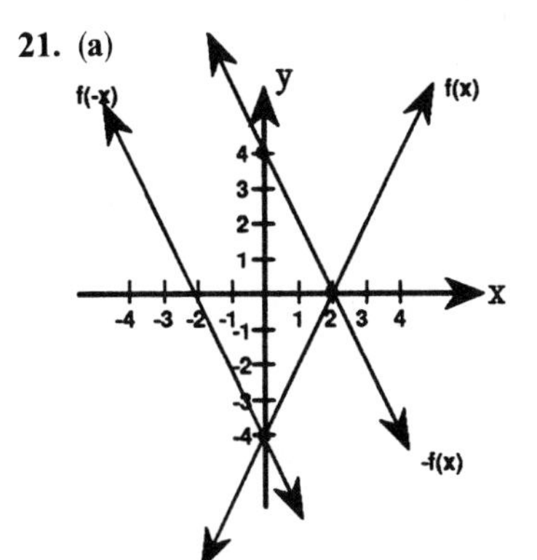

(b)

23. **(a)**

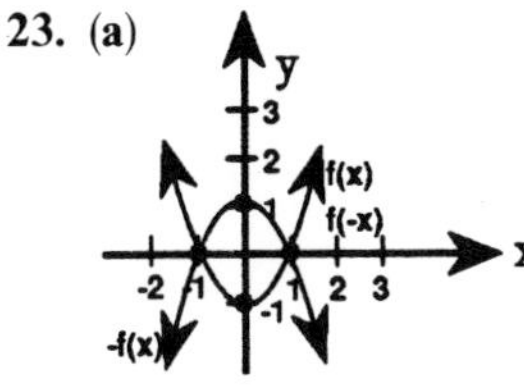

(b)

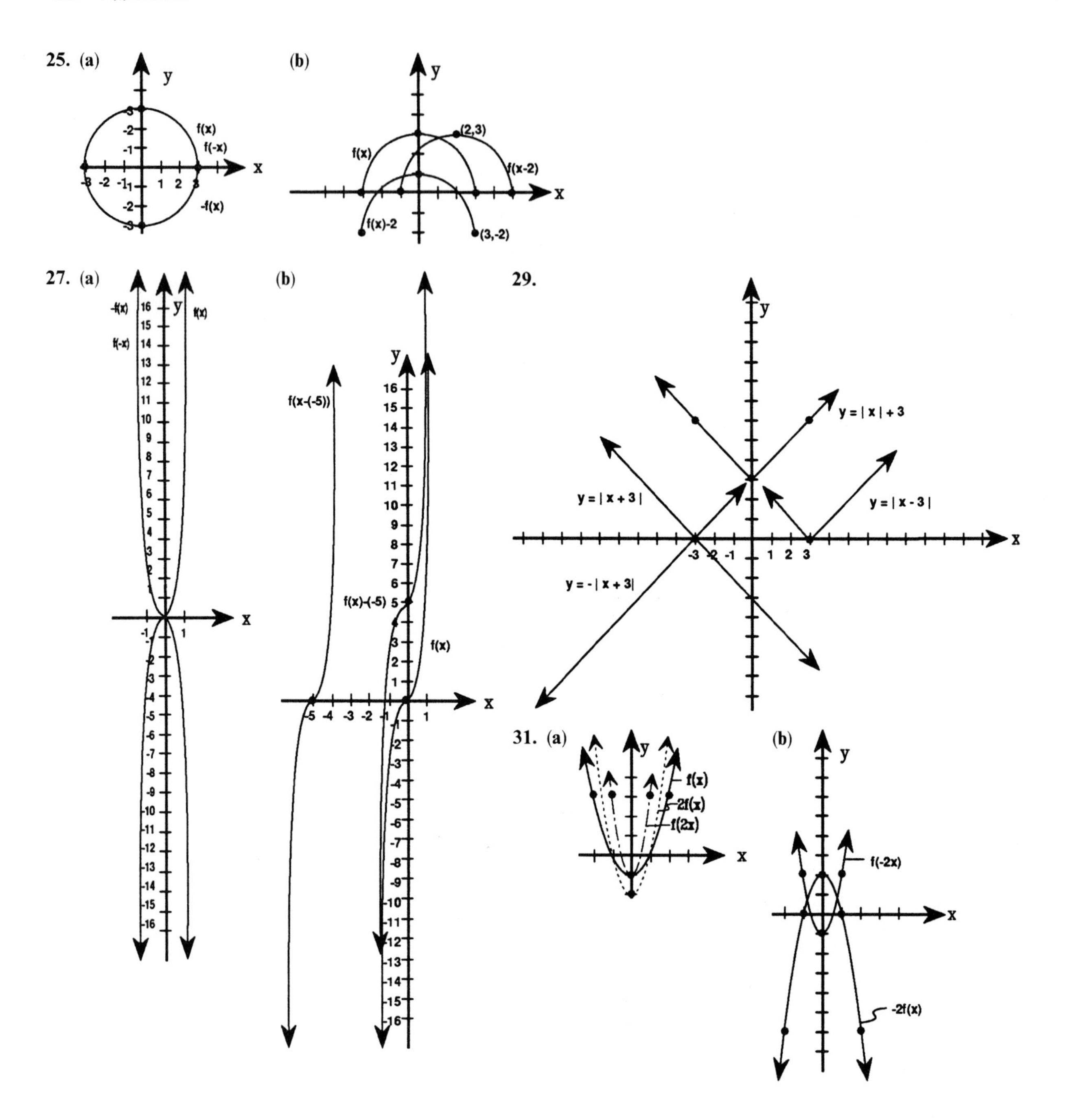
25. (a)
y
x
f(x)
f(-x)
-f(x)
(b)
y
x
f(x)
(2,3)
f(x-2)
f(x)-2
(3,-2)
27. (a)
-f(x)
f(-x)
f(x)
y
x
(b)
y
x
f(x-(-5))
f(x)-(-5)
f(x)
29.
y
x
y = | x | + 3
y = | x + 3 |
y = | x - 3 |
y = - | x + 3|
31. (a)
y
x
f(x)
2f(x)
f(2x)
(b)
y
x
f(-2x)
-2f(x)

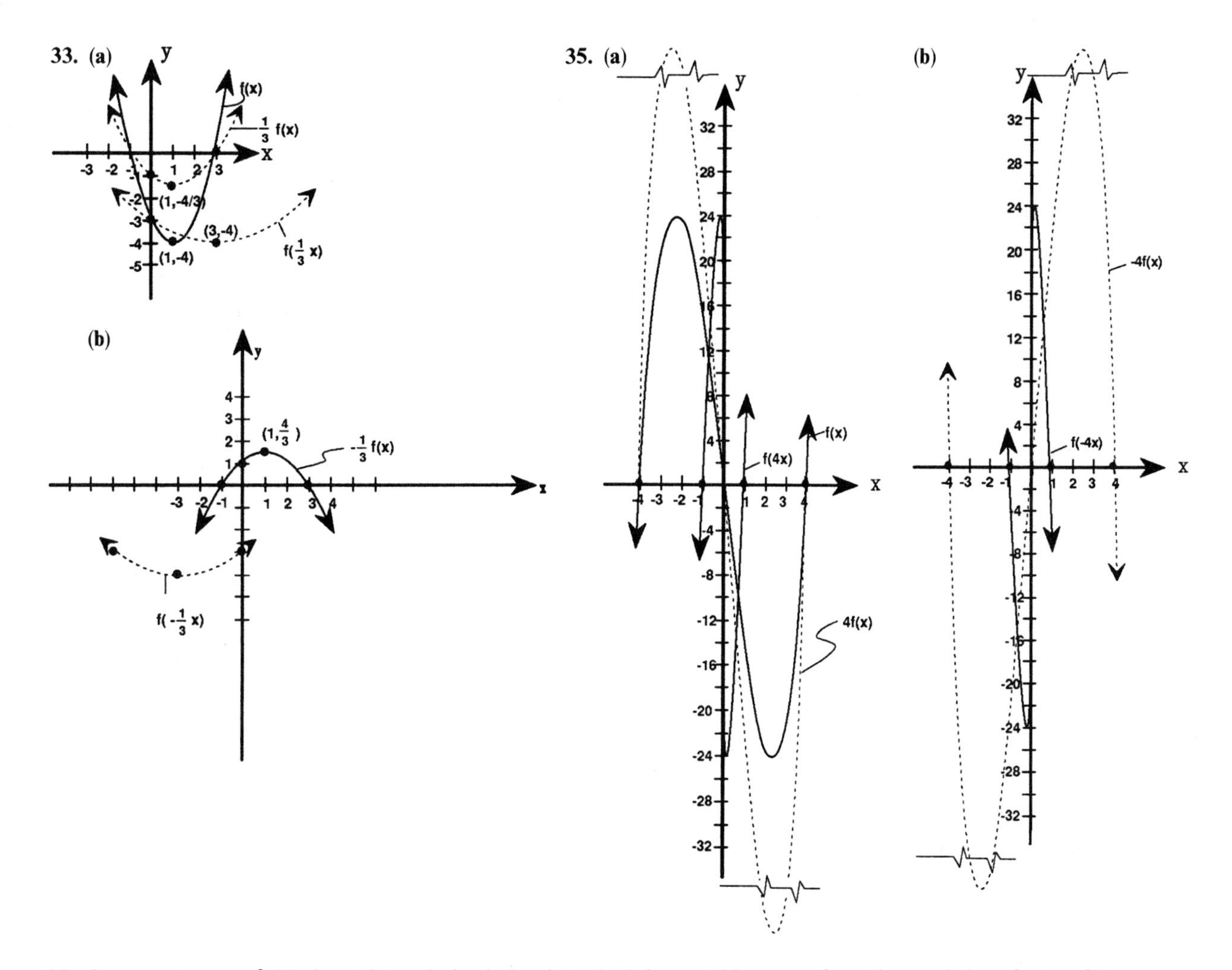

37. Compare to $y = x^2$. Horizontal translation is 1 unit to the left, stretching away from the x-axis by a factor of 3.

39. Compare to $y = |x|$. Horizontal translation is 1 unit to the right with contraction toward x-axis by a factor of $\frac{1}{2}$; reflection is in the x-axis and vertical translation 5 units up.

41. Compare to $y = x^3$. Contraction is toward the x-axis by a factor of $\frac{1}{8}$; reflection in the x-axis.

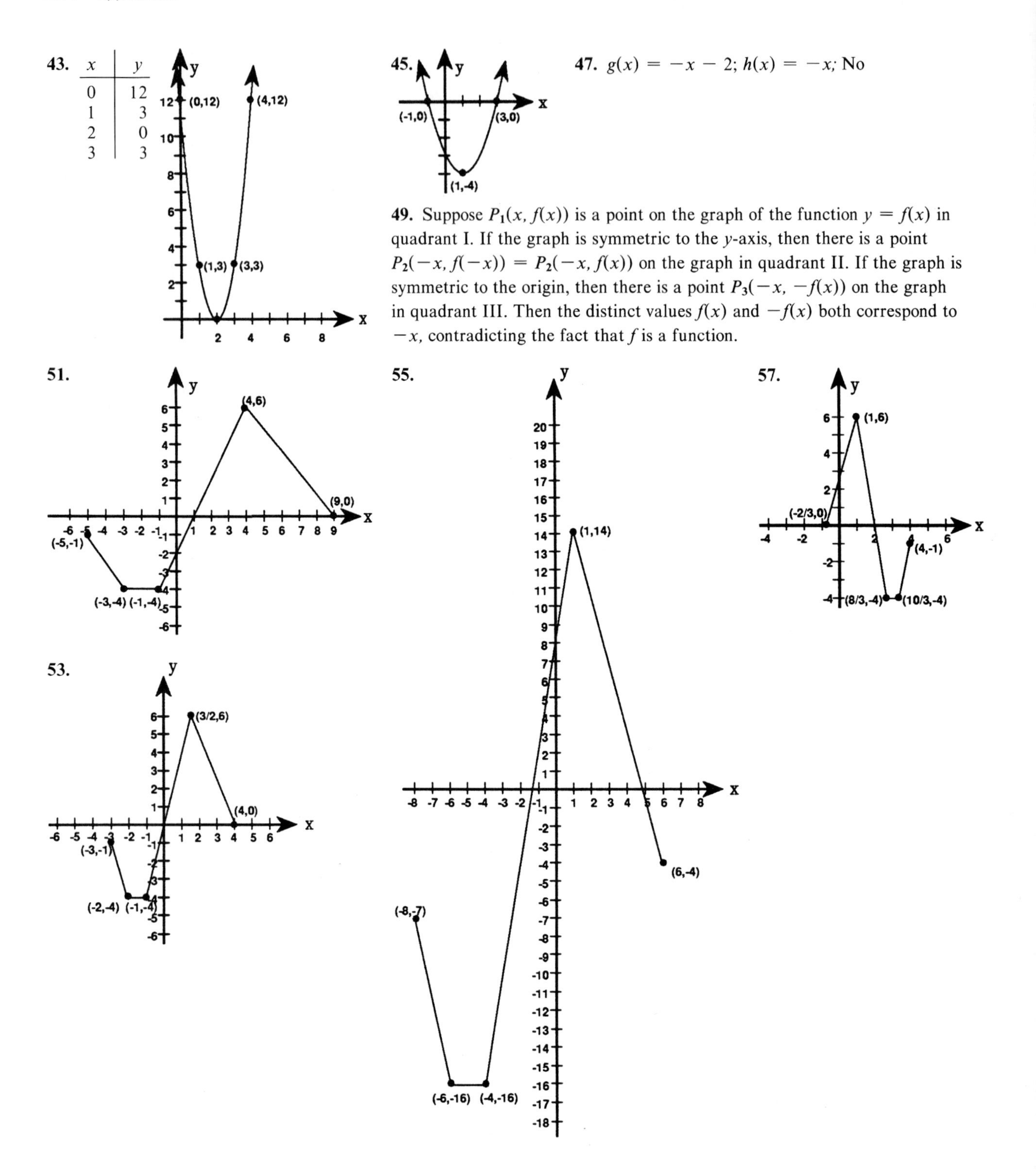

43.

x	y
0	12
1	3
2	0
3	3

45.

47. $g(x) = -x - 2$; $h(x) = -x$; No

49. Suppose $P_1(x, f(x))$ is a point on the graph of the function $y = f(x)$ in quadrant I. If the graph is symmetric to the y-axis, then there is a point $P_2(-x, f(-x)) = P_2(-x, f(x))$ on the graph in quadrant II. If the graph is symmetric to the origin, then there is a point $P_3(-x, -f(x))$ on the graph in quadrant III. Then the distinct values $f(x)$ and $-f(x)$ both correspond to $-x$, contradicting the fact that f is a function.

51.

53.

55.

57.

Section 2.5

1. $(f+g)(x) = 3x - 2$ domain = Reals
$(f-g)(x) = -x - 4$ domain = R
$(fg)(x) = 2x^2 + x - 6x - 3$
$= 2x^2 - 5x - 3$ domain = R
$\left(\frac{f}{g}\right)(x) = \frac{x-3}{2x+1}$ domain $= \left\{x \middle| x \text{ is real}, x \neq -\frac{1}{2}\right\}$

3. $(f+g)(x) = 2x^2 + 4x + 2$ domain = R
$(f-g)(x) = -4x - 4$ domain = R
$(fg)(x) = x^4 + 4x^3 + 3x^2 - x^2 - 4x - 3$
$= x^4 + 4x^3 + 2x^2 - 4x - 3$ domain = R
$\left(\frac{f}{g}\right)(x) = \frac{x^2-1}{x^2+4x+3} = \frac{(x-1)(x-1)}{(x+3)(x+1)}$
$= \frac{x-1}{x+3}$ domain $= \{x|x \text{ is real}, x \neq -3, x \neq -1\}$

5. $(f+g)(x) = \sqrt{x} + |x|$ domain: $[0, \infty)$
$(f-g)(x) = \sqrt{x} - |x|$ domain: $[0, \infty)$
$(fg)(x) = \sqrt{x}\,(|x|)$ domain: $[0, \infty)$
$\left(\frac{f}{g}\right)(x) = \frac{\sqrt{x}}{|x|}$ domain: $(0, \infty)$

7. $(f+g)(x) = \sqrt{16 - x^2} + \sqrt{x^2 - 1}$ domain: $[-4, -1] \cup [1, 4]$
$(f-g)(x) = \sqrt{16 - x^2} - \sqrt{x^2 - 1}$ domain: same
$(fg)(x) = \sqrt{16 - x^2}\sqrt{x^2 - 1}$ domain: same
$\left(\frac{f}{g}\right)(x) = \frac{\sqrt{16 - x^2}}{\sqrt{x^2 - 1}}$ domain: $[-4, -1) \cup (1, 4]$

9. $(f+g)(x) = \begin{cases} x + 1, x < 0 \\ x^2 + x + 1, x > 1 \end{cases}$

$(f-g)(x) = \begin{cases} 1 - x, x < 0 \\ -x^2 + x + 1, x > 1 \end{cases}$

$(fg)(x) = \begin{cases} x, x < 0 \\ x^3 + x^2, x > 1 \end{cases}$

$\left(\frac{f}{g}\right)(x) = \begin{cases} \frac{1}{x}, x < 0 \\ \frac{x+1}{x^2}, x > 1 \end{cases}$

11. $2x - 2$; domain: Reals **13.** $\sqrt{|x|}$; domain: Reals **15.** $(3x^2 - 2x + 1)^5$; domain: Reals
17. $\sqrt{17 - x^2}$; domain: $[-\sqrt{17}, -1] \cup [1, \sqrt{17}]$ **19.** $(f \circ g)(x) = 2x - 2$; $(g \circ f)(x) = 2x - 5$
21. $(f \circ g)(x) = \sqrt{|x|}$ domain: Reals; $(g \circ f)(x) = \sqrt{x}$ domain: Reals ≥ 0 **23.** $g(x) = x - 3, f(x) = x^2$
25. $g(x) = 2x, f(x) = \sqrt[3]{x}$ **27.** $g(x) = x^3, f(x) = |x|$ **29.** **(a)** $f^{-1}(x) = \frac{x}{2}$
(b) $f(f^{-1}(x)) = f\left(\frac{x}{2}\right) = 2\left(\frac{x}{2}\right) = x; f^{-1}(f(x)) = f^{-1}(2x) = \frac{2x}{2} = x$
(c) f: domain: $[2, \infty]$ range: $[4, \infty]$; f^{-1}: domain: $[4, \infty]$ range: $[2, \infty]$. (They are reversed.) **(d)** $\frac{1}{f} = \frac{1}{2x}$; No

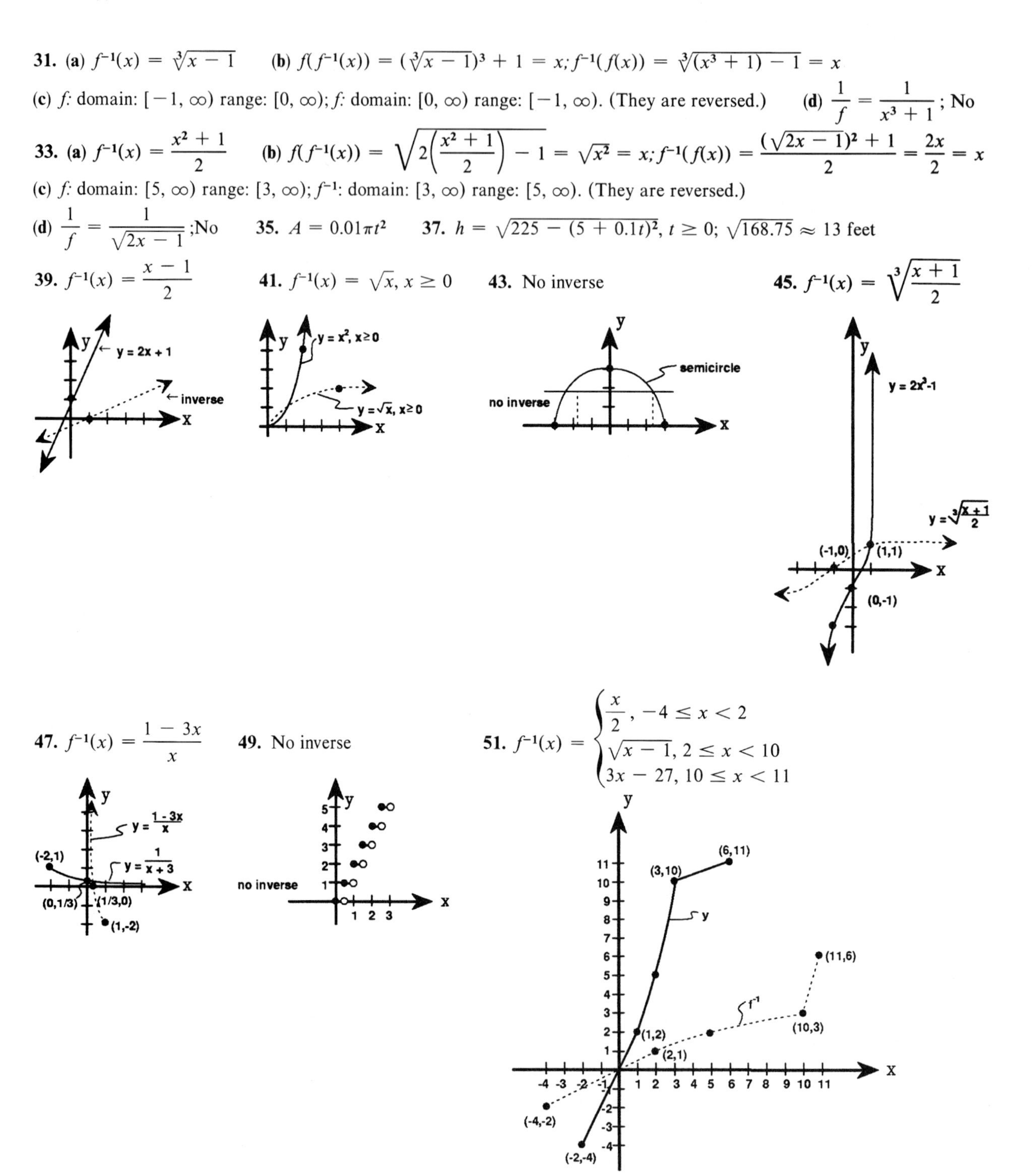

31. **(a)** $f^{-1}(x) = \sqrt[3]{x-1}$ **(b)** $f(f^{-1}(x)) = (\sqrt[3]{x-1})^3 + 1 = x; f^{-1}(f(x)) = \sqrt[3]{(x^3+1)-1} = x$

(c) f: domain: $[-1, \infty)$ range: $[0, \infty)$; f: domain: $[0, \infty)$ range: $[-1, \infty)$. (They are reversed.) **(d)** $\frac{1}{f} = \frac{1}{x^3+1}$; No

33. **(a)** $f^{-1}(x) = \frac{x^2+1}{2}$ **(b)** $f(f^{-1}(x)) = \sqrt{2\left(\frac{x^2+1}{2}\right) - 1} = \sqrt{x^2} = x; f^{-1}(f(x)) = \frac{(\sqrt{2x-1})^2+1}{2} = \frac{2x}{2} = x$

(c) f: domain: $[5, \infty)$ range: $[3, \infty)$; f^{-1}: domain: $[3, \infty)$ range: $[5, \infty)$. (They are reversed.)

(d) $\frac{1}{f} = \frac{1}{\sqrt{2x-1}}$;No **35.** $A = 0.01\pi t^2$ **37.** $h = \sqrt{225 - (5 + 0.1t)^2}$, $t \geq 0$; $\sqrt{168.75} \approx 13$ feet

39. $f^{-1}(x) = \frac{x-1}{2}$ **41.** $f^{-1}(x) = \sqrt{x}$, $x \geq 0$ **43.** No inverse **45.** $f^{-1}(x) = \sqrt[3]{\frac{x+1}{2}}$

47. $f^{-1}(x) = \frac{1-3x}{x}$ **49.** No inverse **51.** $f^{-1}(x) = \begin{cases} \frac{x}{2}, & -4 \leq x < 2 \\ \sqrt{x-1}, & 2 \leq x < 10 \\ 3x - 27, & 10 \leq x < 11 \end{cases}$

53. $=\dfrac{x^2}{25}+\dfrac{2x}{5}-1$ **55.** $\dfrac{2x^3}{125}-\dfrac{2x}{5}$ **57.** $\dfrac{2x}{20x^2-5}$ **59.** $\dfrac{2x^2}{25}-2$ **61.** **(a)** $y = x$

(b) Suppose $y = f(x)$ is a strictly increasing function. Then f is either one-to-one or it is not one-to-one. Suppose f is not one-to-one. There must exist $x_1 \neq x_2$ in the domain such that $f(x_1) = f(x_2)$. Without loss of generality, let $x < x_2$. Then $x_1 < x_2$ with $f(x_1) = f(x_2)$ contradicts f being strictly increasing. Therefore f is one-to-one.

63. **(a)** $(f+g)(-x) = f(-x) + g(-x) = f(x) + g(x) = (f+g)x$

(b) $(fg)(-x) = [f(-x)][g(-x)] = f(x)g(x) = (fg)(x)$

65. Since f has an inverse, $f^{-1}(f(g(x))) = g(x)$. Since g has an inverse, $g^{-1}(f^{-1}(f(g(x))) = g^{-1}(g(x)) = x$. But $(f \circ g)^{-1}[(f \circ g)(x)] = x$. Therefore $(f \circ g)^{-1}(x) = g^{-1}(f^{-1}(x))$.

67. **(a)** $f(g(x)) = 9x^2 + 12x - 1,\ x \geq -\dfrac{2}{3}$ **(b)**

$$\begin{cases} x = 9y^2 + 12y - 1 \\ 0 = 9y^2 + 12y - (x+1) \\ y = \dfrac{-12 \pm \sqrt{180 + 36x}}{18} = \dfrac{-2 \pm \sqrt{5+x}}{3},\ x \geq -5 \\ (f \circ g)^{-1}(x) = \dfrac{-2 + \sqrt{5+x}}{3},\ x \geq -5 \end{cases}$$

(c) $f^{-1}(x) = \sqrt{x+5},\ x \geq -5,\ g^{-1}(x) = \dfrac{x-2}{3};\ g^{-1}(f^{-1}(x)) = \dfrac{\sqrt{x+5}-2}{3},\ x \geq -5$

Chapter 2 Review Exercises

1. Is a function. Inverse function: $\{(2, 5), (6, 2), (4, 6), (3, 4)\}$ **3.** Not a function **5.** **(a)** $5\sqrt{5}$ **(b)** $\left(-\dfrac{1}{2}, -1\right)$

(c)
$$y - 4 = -2(x+3)$$
$$y = -2x - 2$$
$$2x + y + 2 = 0$$

7. **(a)** $13\sqrt{2}$ **(b)** $\left(-\dfrac{9}{2}, \dfrac{7}{2}\right)$ **(c)**
$$y + 5 = \frac{17}{7}(x+8)$$
$$y = \frac{17}{7}x + \frac{101}{7}$$
$$17x - 7y + 101 = 0$$

9. $y = -3$ **11.** $4x - 6 + 2h$

13. (-5,4) (-1,4) (-3,0)

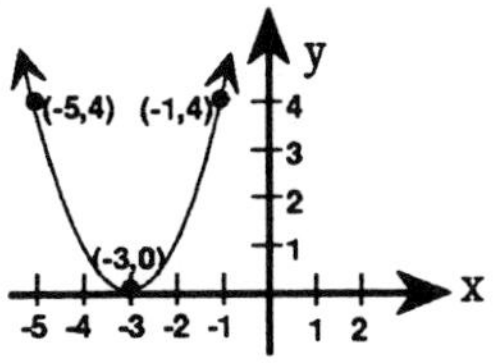

15. (-1/2, 21/4) (0,5) (-2,3) (1,3)

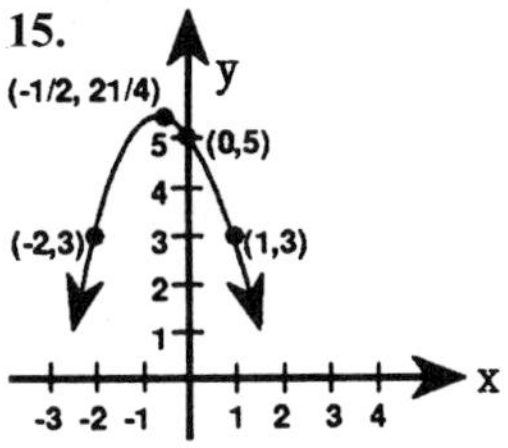

17. $(-\frac{1}{\sqrt{3}}, 0)$ $(\frac{1}{\sqrt{3}}, 0)$ (-1,-2) (1,-2)

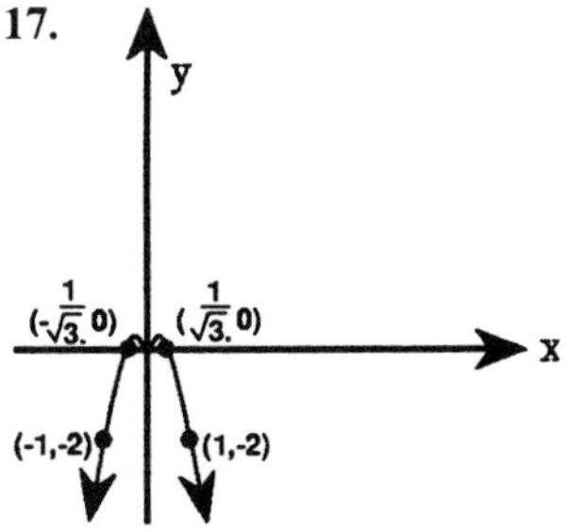

19. (0,4) (-2,0) (2,0)

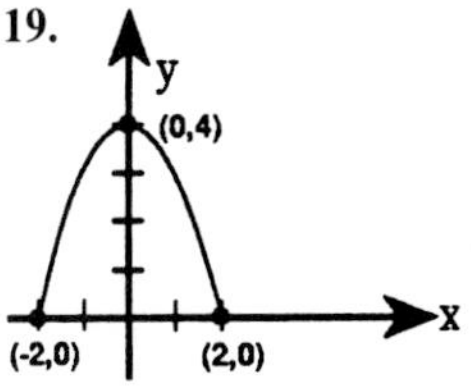

21. **(a)** $(f+g)(x) = 10x - 4$ domain: R **(b)** $(f-g)(x) = -6x - 6$ domain: R

(c) $(fg)(x) = 16x^2 - 38x - 5$ domain: R **(d)** $\left(\dfrac{f}{g}\right)(x) = \dfrac{2x-5}{8x+1}$, domain: all reals except $-\dfrac{1}{8}$

(e) $(f \circ g)(x) = 16x - 3$ domain: reals **(f)** $(g \circ f)(x) = 16x - 39$ domain: reals

23. (**a**) $(f+g)(x) = \sqrt{4-x^2} + (x-2)$ domain: $[-2, 2]$ (**b**) $(f-g)(x) = \sqrt{4-x^2} - x + 2$ domain: same (**c**) $(fg)(x) = \sqrt{4-x^2}(x-2)$ domain: same (**d**) $\left(\dfrac{f}{g}\right)(x) = \dfrac{\sqrt{4-x^2}}{x-2}$ domain: $[-2, 2]$ (**e**) $(f \circ g)(x) = \sqrt{4x-x^2}$ domain: $[0, 4]$ (**f**) $(g \circ f)(x) = \sqrt{4-x^2} - 2$ domain: $[-2, 2]$ **25.** (**a**) $y = 3x$ (**b**) $y = -\dfrac{1}{3}x$ **27.** (**a**) $y + 7 = -2(x+3)$ (**b**) $y + 7 = \dfrac{1}{2}(x+3)$

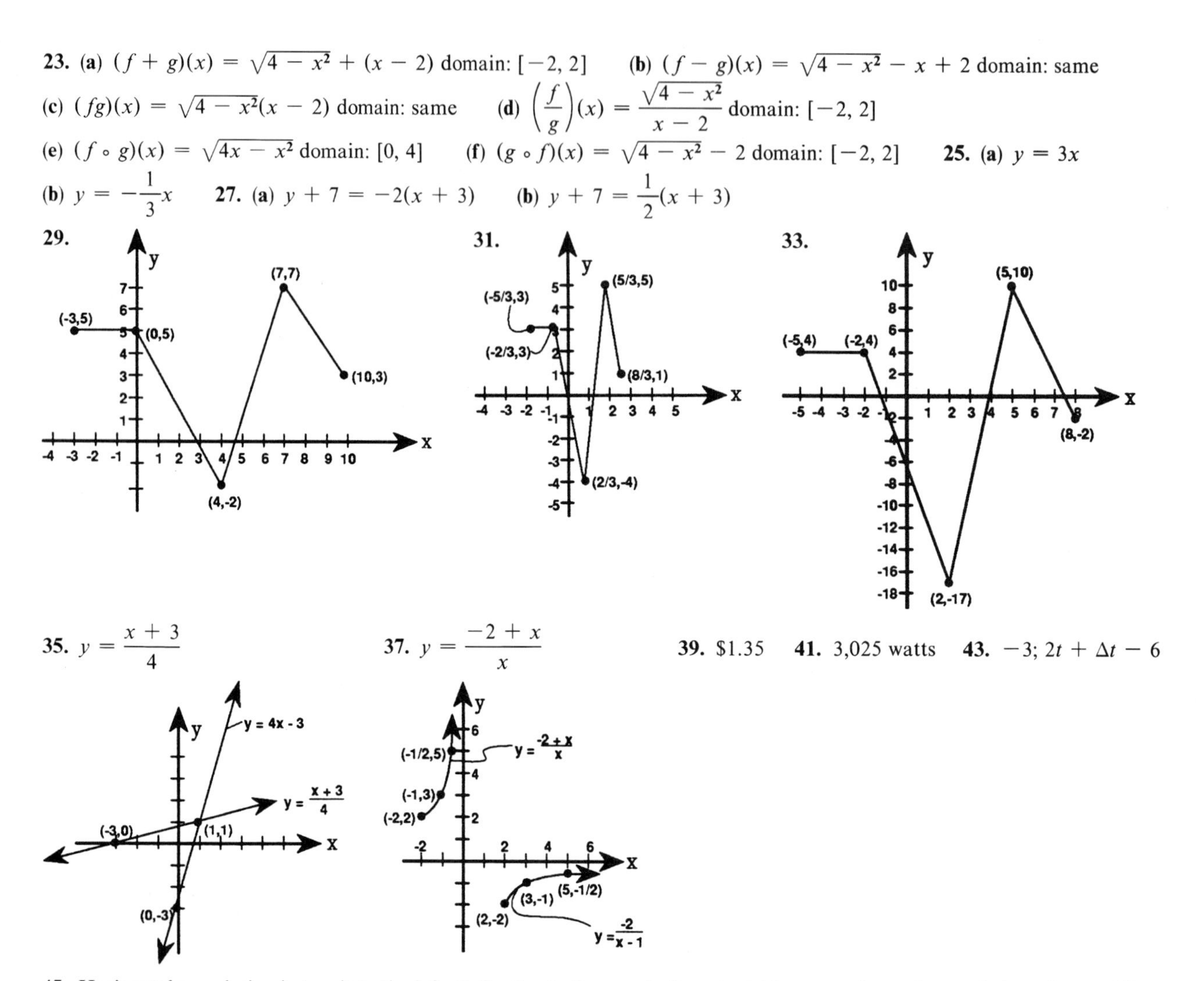

35. $y = \dfrac{x+3}{4}$ **37.** $y = \dfrac{-2+x}{x}$ **39.** \$1.35 **41.** 3,025 watts **43.** -3; $2t + \Delta t - 6$

45. Horizontal translation is 1 unit to the left. Reflection in the x-axis then stretching away from the x-axis by a factor of 3. Vertical translations up 5 units. **47.** The horizontal translation is 3 units to the right, stretching away from the x-axis by a factor of 4; the vertical translation is up 2 units.

Chapter 2 Test 1

1. Domain: $[-3, 3]$; range: $[0, 3]$ **2.** $y - 1 = -\dfrac{7}{10}(x-4)$ **3.** $y - 0 = -\dfrac{2}{3}(x-5)$

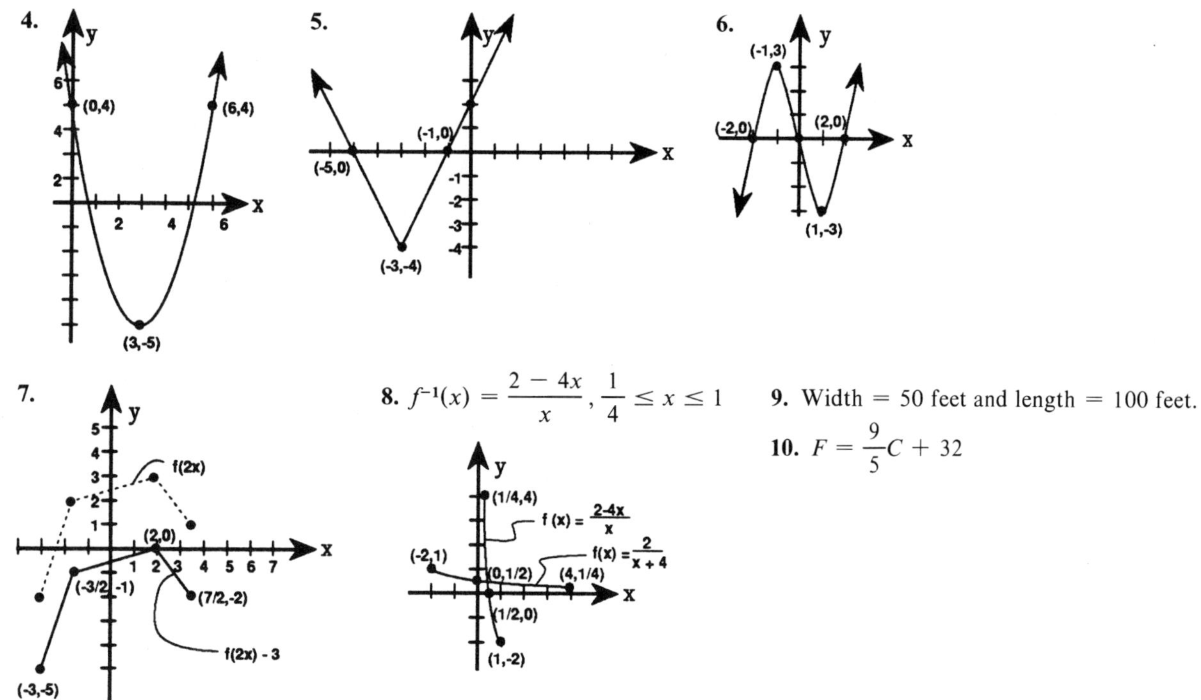

8. $f^{-1}(x) = \dfrac{2 - 4x}{x}, \ \dfrac{1}{4} \leq x \leq 1$

9. Width = 50 feet and length = 100 feet.

10. $F = \dfrac{9}{5}C + 32$

Chapter 2 Test 2

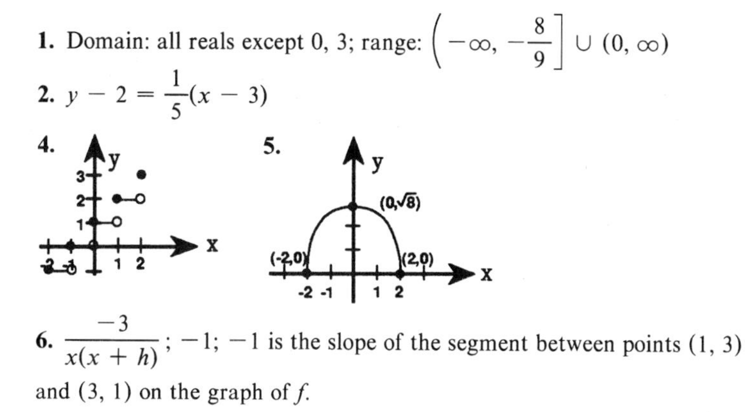

1. Domain: all reals except 0, 3; range: $\left(-\infty, -\dfrac{8}{9}\right] \cup (0, \infty)$

2. $y - 2 = \dfrac{1}{5}(x - 3)$

4.

5.

6. $\dfrac{-3}{x(x + h)}$; -1; -1 is the slope of the segment between points (1, 3) and (3, 1) on the graph of f.

3.

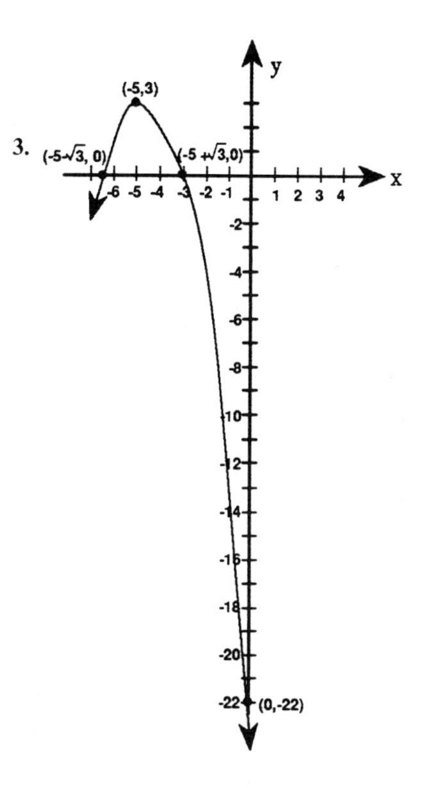

7. Start with $y = x^2$, and translate horizontally to the left 2 units. Then reflect in the x-axis. Then stretch away from the x-axis by a factor of 3. Then translate vertically down 1 unit. **8.** $f^{-1}(x) = \begin{cases} 1 - \sqrt{x - 4}, & 4 \leq x \leq 13 \\ \dfrac{7 - x}{3}, & -5 \leq x \leq 4 \end{cases}$

9. $V = -5{,}000t + 75{,}000$ **10.** $715

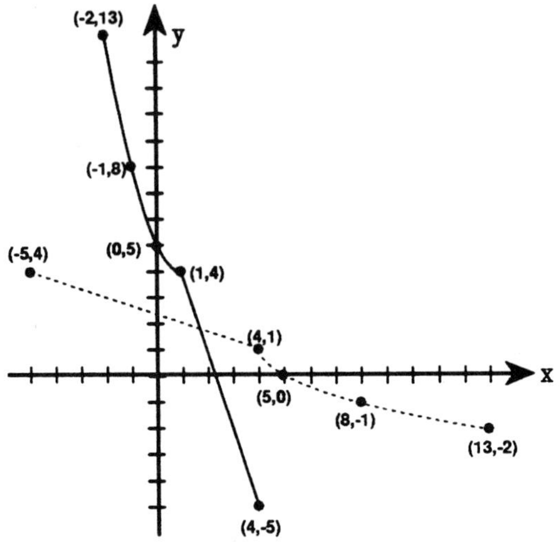

Section 3.1

1. (4, 4) **3.** Coincident lines $\{(x, y) | x - 2y = 11\}$ **5.** $\left(\frac{5}{16}, \frac{51}{16}\right)$ **7.** $\left(-\frac{23}{97}, \frac{25}{97}\right)$ **9.** $\left(\frac{52}{19}, \frac{36}{19}\right)$ **11.** $\emptyset$ **13.** $(-3, 0)$ **15.** $\emptyset$ **17.** $\left(\frac{3{,}235}{1{,}261}, \frac{134}{97}\right)$ **19.** $\left(\frac{19}{26}, \frac{58}{65}\right)$ **21.** $\left(\frac{1}{4}, \frac{1}{2}\right)$ **23.** $\left(\frac{17}{8}, \frac{34}{7}\right)$ **25.** $y = 2x + b$ **27.** $y - 1 = m(x - 5)$ **29.** $\frac{15}{26}\sqrt{26}$ **31.** $\frac{67}{15}$ **33.** $\frac{15}{116}\sqrt{58}$ **35.** **(a)** $s = 5, t = 2$ **(b)** $s = 1, t = -3$ **(c)** $s = 9, t = 7$ **37.** **(a)** $s = 5, t = 2$ **(b)** $s = 8, t = -5$ **(c)** $s = 37, t = -18$ **39.** **(a)** $s = 14, t = 3$ **(b)** $s = 3, t = -5$ **(c)** $s = 31, t = 1$ **41.** 24 feet by 30 feet **43.** $7,200 at 12% and $10,800 at 9% **45.** $a = -6, b = -18$ **47.** 40 ounces of $\frac{1}{5}$ gold alloy; 20 ounces of $\frac{1}{2}$ gold alloy **49.** Plane's airspeed = $506\frac{2}{3}$ mph; wind velocity = $26\frac{2}{3}$ mph **51.** $p = \$50; S = D = 1{,}075$ **53.** 84 model A homes; 36 model B homes **55.** $y = -\frac{3}{8}x + \frac{23}{4}$

57. $ax + by + c = 0$

$\perp: bx - ay + (ay_0 - bx_0) = 0$

$a\ell_1 + b\ell_2: (a^2 + b^2)x + (ac + aby_0 - b^2x_0) = 0$

$$x_1 = -\frac{ac + aby_0 - b^2x_0}{a^2 + b^2}$$

$b\ell_1 - a\ell_2: (b^2 + a^2)y + (bc - a^2y_0 + abx_0) = 0$

$$y_1 = \frac{bc + abx_0 - a^2y_0}{a^2 + b^2}$$

$$\text{distance} = \sqrt{\left(\frac{b^2x_0 - aby_0 - ac}{a^2 + b^2} - x_0\right)^2 + \left(\frac{a^2y_0 - bc - abx_0}{a^2 + b^2} - y_0\right)^2}$$

$$= \sqrt{\frac{(b^2x_0 - aby_0 - ac - a^2x_0 - b^2x_0)^2 + (a^2y_0 - bc - abx_0 - a^2y_0 - b^2y_0 - by_0)^2}{(a^2 + b^2)^2}}$$

$$= \sqrt{\frac{[a(-ax_0 - by_0 - c)]^2 + [b(-ax_0 - by_0 - c)]^2}{(a^2 + b^2)^2}}$$

$$= \sqrt{\frac{a^2(ax_0 + by_0 + c)^2 + b^2(ax_0 + by_0 + c)^2}{(a^2 + b^2)^2}}$$

$$= \sqrt{\frac{(a^2 + b^2)(ax_0 + by_0 + c)^2}{(a^2 + b^2)^2}}$$

$$= \frac{|a + by_0 + c|}{\sqrt{a^2 + b^2}}, \text{ because } \sqrt{w^2} = |w|$$

Section 3.2

1. $(-6, -11, 31)$ **3.** $(2, 3, 1)$ **5.** $(2, 1, -3)$ **7.** $(2, -1, 1)$ **9.** $(1, 5, -3)$ **11.** $\left(\frac{1}{2}, \frac{1}{4}, \frac{3}{4}\right)$

13. No solution **15.** $(-5, -1, 5)$ **17.** $(1, 5, -3)$ **19.** $\left(\frac{7}{17}, -\frac{73}{17}, -\frac{2}{17}\right)$ **21.** $(60, 20, 100)$

23. $(46, 17, 68)$ **25.** $\frac{20}{11}$ lbs. of cashews; $\frac{40}{11}$ lbs. of peanuts; $\frac{50}{11}$ lbs. of almonds. **27.** $\frac{30}{11}$ hours **29.** 3 type A cargo planes; 6 type B cargo planes; 11 type C cargo planes **31.** $y = -\frac{1}{15}x^2 - \frac{7}{15}x + 2$ **33.** $y = -\frac{3}{16}x^2$

35. and 37. Choose 3 of the 5 data points and proceed as in Exercises 31 and 33 to find the equation of a parabola ($y = ax^2 + bx + c$) that passes through the chosen 3 points. You may in fact choose 3 points that are close to the data points. There are an infinite number of possible parabolas that can be generated. **39. (a)** y-axis **(b)** x-z plane

Section 3.3

1. (1, 2, 0) **3.** (1, −3, 2) **5.** (4, −2, 3) **7.** (−6, −11, 31) **9.** (2, 3, 1) **11.** (2, 1, −3) **13.** (2, −1, 1) **15.** (1, 5, −3) **17.** $\left(\frac{1}{2}, \frac{1}{4}, \frac{3}{4}\right)$ **19.** No solution. The entry in each column of the last row, except the last column, is 0. **21.** (−5, −1, 5) **23.** (1, 5, −3) **25.** $\left(\frac{7}{17}, -\frac{73}{17}, -\frac{2}{17}\right)$ **27.** (1, 2, 0)
29. Inconsistent. Each column in the last row, except the last column, has a 0 entry. **31.** (4, −2, 3)
33. Inconsistent. Each column of the last row, except the last column, has a 0 entry. **35.** (0, 2, −2, 2)
37. Inconsistent. Each column in the last row, except the last column, has a 0 entry. **39.** If (x_0, y_0, z_0) is a solution of $ax + by + cz = d$, then $ax_0 + by_0 + cz_0 = d$ and $k(ax_0 + by_0 + cz_0) = kd$. Therefore, (x_0, y_0, z_0) is a solution of $k(ax + by + cz) = kd$. If (x_1, y_1, z_1) is a solution of $k(ax + by + cz) = kd, k \neq 0$, then $\frac{1}{k}(k(ax_1 + by_1 + cz_1)) = \frac{1}{k}(kd)$ and $ax_1 + by_1 + cz_1 = d$. Therefore (x_1, y_1, z_1) is a solution of $ax + by + cz = d$. Thus, every solution of $ax + by + cz = d$ is also a solution of $k(ax + by + cz) = kd$ and vice versa. The equations are equivalent because their solution sets are equal.

Section 3.4

1. $\left(\frac{7}{8} - \frac{5}{4}z, \frac{11}{8} + \frac{3}{4}z, z\right)$; $\left(\frac{7}{8}, \frac{11}{8}, 0\right)$, $\left(\frac{17}{8}, \frac{5}{8}, 1\right)$, $\left(-\frac{3}{8}, \frac{17}{8}, 1\right)$ **3.** Inconsistent
5. $(1 + 2z, 2 + 2z, z, z)$; $(-1, 0, -1, -1)$, $(1, 2, 0, 0)$, and $(3, 4, 1, 1)$
7. $\left(-9 + x - 18z, x, \frac{4}{3} + \frac{8}{3}z, z\right)$; $\left(-9, 0, \frac{4}{3}, 0\right)$, $(-27, 0, 4, 1)$, $\left(-8, 1, \frac{4}{3}, 0\right)$
9. $\left(-\frac{3}{7}z, \frac{2}{7}z, \frac{3}{7}z, z\right)$; $\left(-\frac{11}{7}, -\frac{2}{7}, -\frac{2}{7}, -1\right)$, $(0, 0, 0, 0)$, $\left(\frac{11}{7}, \frac{2}{7}, \frac{2}{7}, 1\right)$ **11.** $\left(\frac{20}{3}, \frac{5}{3}, \frac{10}{3}\right)$
13. $\left(\frac{4}{13} + \frac{11}{13}y - \frac{12}{13}z, -\frac{2}{13} + \frac{1}{13}y - \frac{7}{13}z, y, z\right)$; $\left(\frac{4}{13}, -\frac{2}{13}, 0, 0\right)$, $\left(\frac{15}{13}, -\frac{1}{13}, 1, 0\right)$, $\left(\frac{8}{13}, -\frac{9}{13}, 0, 1\right)$
15. Current through 7 ohms is 1.5 amps. Current through 6 ohms is 0.75 amps. Current through 12 ohms is 0.75 amps.
17. Current through 50 ohms is $\frac{5}{3}$ amps. Current through 80 ohms is $\frac{1}{3}$ amps. Current through 20 ohms is $\frac{4}{3}$ amps.
19. Minimum traffic flow along *CD* is 0. Maximum traffic flow along *CB* is 750.
21. $x_1 = 350$
$x_2 = 450$
$x_3 = 700$
$x_4 = 550$
$x_5 = 650$
$x_6 = x_7 = 500$
23. There are 9 different possible coin collections. Let z assume the whole number values 0 through 8 and find the corresponding values for x and y. **25.** $2a + 11b = 16c$

Chapter 3 Review Exercises

1. (4, 2) **3.** No solution **5.** $\left(\frac{10}{7}, \frac{20}{21}\right)$ **7.** (1, 2, 3) **9.** $\left(\frac{13}{7} - \frac{6}{7}z, \frac{16}{7}z - \frac{9}{7}, z\right)$; infinite set of solutions.
11. $(4, 2 - 2z, z)$; infinite set of solutions. **13.** $\left(\frac{25}{4}, -\frac{41}{4}, -\frac{9}{2}\right)$
15. $(10 - 3z, 5z - 7, z)$; infinite set of solutions. **17.** $\left(2 + 6z, \frac{7}{6} - \frac{5}{2}z, z\right)$; infinite solution set.
19. $\left(-\frac{244}{51}, -\frac{367}{153}, -\frac{1{,}216}{459}, \frac{1{,}682}{459}\right)$ **21.** $\left(\frac{1}{2} + \frac{1}{3}z, \frac{13}{4} - \frac{13}{6}z, \frac{11}{4} - \frac{3}{2}z, z\right)$; an infinite number of solutions.
23. (3, 2, 1) **25.** $\frac{23}{\sqrt{45}}$ **27.** $y = -\frac{7}{24}x^2 + \frac{61}{12}x - 25$ **29.** $\frac{59}{3}$ cm, $\frac{59}{3}$ cm, $\frac{44}{3}$ cm **31.** $a = \frac{13}{2}, b = \frac{26}{7}$
33. 31 of brand A cameras; 22 of brand B cameras **35.** The first number is 273, the second number is 86, and the third number is 154. **37.** 10 tons of fertilizer A; $\frac{20}{3}$ tons of fertilizer B; $\frac{40}{3}$ tons of fertilizer C
39. $\frac{75}{11}$ pounds of pasta A; $\frac{100}{11}$ pounds of pasta B; $\frac{100}{11}$ pounds of pasta C
41. Current through 15 ohms $= \frac{23}{93}$ amperes. Current through 18 ohms $= \frac{110}{93}$ amperes. Current through 20 ohms $= \frac{87}{93}$ amperes. **43.** Maximum traffic flow is 1,260; minimum traffic flow is 85. **45.** $\frac{31}{18} = C$

Chapter 3 Test 1

1. $\left(\frac{44}{49}, \frac{46}{49}\right)$ **2.** $\left(\frac{31}{35}, -\frac{11}{35}, \frac{79}{35}\right)$ **3.** $\left(\frac{173}{182}, \frac{385}{182}, \frac{149}{91}\right)$ **4.** $\left(\frac{197}{26}, -\frac{73}{26}, \frac{159}{26}\right)$ **5.** No solution
6. $\left(\frac{120 - 7z}{49}, \frac{56z - 57}{98}, z\right)$; infinite solution set **7.** Length $= \frac{313}{4}$ feet; width $= \frac{107}{4}$ feet **8.** 18 ounces of 25% alloy and 27 ounces of 40% alloy **9.** 4 nickels; 25 dimes; 6 quarters **10.** $y = -\frac{2}{5}x^2 + \frac{7}{10}x + \frac{41}{5}$

Chapter 3 Test 2

1. $\left(\frac{294}{87}, \frac{123}{87}\right)$ **2.** $\left(\frac{61}{20}, \frac{67}{20}, \frac{19}{10}\right)$ **3.** $\left(\frac{123}{95}, \frac{863}{95}, \frac{568}{95}\right)$ **4.** $\left(\frac{407}{153}, -\frac{35}{51}, \frac{13}{17}\right)$ **5.** $\left(\frac{183}{104}, \frac{46}{13}, \frac{93}{208}, \frac{389}{208}\right)$
6. $\left(\frac{17z + 1}{3}, \frac{10z - 4}{3}, \frac{14 - 11z}{3}, z\right)$; an infinite number of solutions. **7.** Average airspeed is $\frac{11{,}180}{23}$ mph. Average wind speed is $\frac{780}{23}$ mph. **8.** 10 pounds of food A; 25 pounds of food B; 5 pounds of food C

9.

Nickels	Dimes	Quarters	Value
x	y	z	
6	12	3	\$2.25
5	10	6	2.75
4	8	9	3.25
3	6	12	3.75
2	4	15	4.25
1	2	18	4.75

10. $13 = C$

Section 4.1

1. **(a)** $4x - 14$ **(b)** $-5x + 2$ **(c)** $3x^2 - 26x + 48$
3. **(a)** $2x^2 + 10x - 2$ **(b)** $2x^2 - 2x - 4$ **(c)** $12x^3 + 26x^2 - 14x - 3$
5. **(a)** $-5x^2 - 5x + 12$ **(b)** $5x^2 - 11x - 6$ **(c)** $40x^3 - 39x^2 - 63x + 27$
7. **(a)** $4x^2 + 2x + 3$ **(b)** $-2x^2 - 2x + 5$ **(c)** $3x^4 + 2x^3 + 11x^2 + 8x - 4$
9. **(a)** $6x^2 - 7x + 5$ **(b)** $8x^2 + 3x - 11$ **(c)** $-7x^4 - 33x^3 + 69x^2 - x - 24$
11. **(a)** $x^3 + 8x^2 - 7x + 9$ **(b)** $x^3 + 6x^2 - 3x - 1$ **(c)** $x^5 + 5x^4 - 14x^3 + 49x^2 + 20$
13. **(a)** $14x^3 - 8x^2 - 5x + 14$ **(b)** $-6x^3 + 8x^2 - 5x + 2$ **(c)** $40x^6 - 32x^5 - 50x^4 + 144x^3 - 64x^2 - 30x + 4$
15. **(a)** $14x^3 - 4x^2 + 13x - 7$ **(b)** $-10x^3 - 6x^2 + 3x + 13$ **(c)** $24x^6 - 58x^5 + 101x^4 - x^3 + 93x^2 - 65x - 30$
17. $6x^2 - x - 2 = (2x + 1)(3x - 2) + 0$ **19.** $8x^3 - 1 = (2x - 1)(4x^2 + 2x + 1) + 0$
21. $8x^4 + 12x^3 - 6x^2 + 8x + 3 = (4x^2 + 6x - 1)(2x^2 - 1) + (14x + 2)$
23. $6x^4 - 2x^2 + 5 = (3x^2 + 4x + 3)\left(2x^2 - \frac{8}{3}x + \frac{8}{9}\right) + \left(\frac{40}{9}x + \frac{21}{9}\right)$
25. $x^3 + 5x^2 - 7x + 2 = (x + 3)(x^2 + 2x - 13) + 41$ **27.** $6x^3 - 10x = (x - 1)(6x^2 + 6x - 4) + (-4)$
29. $8x^4 - 32x^3 + 27x - 108 = (x - 4)(8x^3 + 27) + 0$
31. $a_nb_mx^{n+m}$, $(a_nb_{m-1} + a_{m-1}b_m)x^{n+m-1}$, $(a_nb_{m-2} + a_{n-1}b_{m-1} + a_{n-2}b_m)x^{n+m-2}$

33.
$$\begin{array}{c|cccc} & \frac{3}{2} & -3 & -4 & -2 \\ \hline \frac{1}{2} & \frac{3}{2} & -\frac{9}{4} & -\frac{41}{8} & -\frac{9}{16} \end{array}$$
$$3x^3 - 6x^2 - 8x + 4 = (2x - 1)\left(\frac{3}{2}x^2 - \frac{9}{4}x - \frac{41}{8}\right) + \left(-\frac{9}{8}\right)$$

35.
$$\begin{array}{c|cccccc} & \overbrace{1 \quad 0 \quad 0 \quad 0 \ldots 0}^{n \text{ positions}} & -1 \\ \hline 1 & 1 \quad 1 \quad 1 \quad 1 \ldots 1 & 0 \end{array}$$

Section 4.2

1. $F(3) = -46$ **3.** $F(-2) = 4$ **5.** $F(2) = -12$ **7.** $F(-3) = -134$ **9.**
$$\begin{array}{c|cccc} & 2 & 1 & -5 & 2 \\ \hline 1 & 2 & 3 & -2 & 0 \end{array}$$

11.

	3	8	−7	−12
$\frac{4}{3}$	3	12	9	0

13.

	2	−5	−1	−5	−3
3	2	1	2	1	0

15.

	3	−11	9	13	−10
$\frac{2}{3}$	3	−9	3	15	0

17. $(x - 1)(x - 2)(x + 2)$ **19.** $(x - 1)(x + 2)(x + 2)$ **21.** $(x + 3)(x + 3)(x + 3)$
23. $(x - 2)(x - 2)(x + 3)(x + 3)$ **25.** $(x - 1)(x + 2)(x + 5)(x + 5)$

27.

	2	7	2	−6	
−1	2	5	−3	−3	← root
−2	2	3	−4	2	

29.

	6	−11	15	−22	6	
1	6	−5	10	−12	−6	← root
2	6	1	17	12	30	

31. From the Division Algorithm, $F(4) = (x - c)\,Q(x) + R$. Since $x - c$ is a factor of $F(x)$, $R = 0$. Thus, $F(c) = (c - c)\,Q(c) + 0 = 0$.

Section 4.3

1. $x^3 + 6x^2 - x - 30 = 0$ **3.** $x^4 - 2x^3 - 3x^2 + 4x + 4 = 0$ **5.** $x^4 + 2x^3 - 2x^2 + 8 = 0$ **7.** $\{1, 2, -2\}$
9. $\{-3, -2, -1\}$ **11.** $\left\{\frac{-1 - \sqrt{7}}{4}, \frac{1}{2}, \frac{-1 + \sqrt{7}}{4}\right\}$ **13.** $\left\{-\frac{2}{3}, -2i, 2i\right\}$ **15.** $\{-3, -2, 1, 2\}$
17. $\left\{-1, \frac{2}{3}, 2 - i, 2 + i\right\}$ **19.** $\left\{-\frac{3}{4}, -\frac{1}{3}, \frac{1}{2}\right\}$ **21.** $\left\{-\frac{2}{3}, \frac{1}{2}, 1 - \sqrt{3}i, 1 + \sqrt{3}i\right\}$
23. The only rational zero is 1.
25. By Descartes' Rule of Signs, the equation has 1 positive root and 3 or 1 negative roots. Since $p \in \{\pm 1, \pm 3\}$ and $q \in \{\pm 1, \pm 2\}$ the only possible rational roots are ± 1, $\pm\frac{1}{2}$, ± 3, or $\pm\frac{3}{2}$.

	2	−3	−8	−5	−3
$\frac{1}{2}$	2	−2	−9	$-\frac{19}{2}$	$-\frac{31}{4}$
1	2	−1	−9	−14	−17
$\frac{3}{2}$	2	0	−8	−17	$-\frac{57}{2}$
3	2	3	1	−2	−9
$-\frac{1}{2}$	2	−4	−6	−2	−2
−1	2	−5	−3	−2	−1
$-\frac{3}{2}$	2	−6	1	$-\frac{13}{2}$	$\frac{27}{4}$
−3	2	−9	19	−62	183

Not one of the possible rational roots yields a zero remainder. Hence, the equation has no rational roots.
27. The square root of 3 is a root of the equation $x^2 - 3 = 0$. But the only possible rational roots of the equation are ± 1 and ± 3. Therefore $\sqrt{3}$ is not a rational number.
29. Since $B(x)$ is a factor of $F(x)$, there is a polynomial $Q(x)$ such that, $F(x) = B(x) \cdot Q(x)$. Also, since $A(x)$ is a factor of $B(x)$, there is a polynomial $P(x)$ such that, $B(x) = A(x) \cdot P(x)$. Now we have, $F(x) = A(x) \cdot P(x) \cdot Q(x)$. Therefore, A is a factor of F.

31. **(a)** The leading coefficient of $F(x)$ is the product of the leading coefficients $(x - c)$ and $Q(x)$. Since the leading coefficient of $x - c$ is 1, the leading coefficient of $Q(x)$ must be positive.
(b) From **(a)**, the leading coefficient of $Q(x)$ is positive. Since the coefficients alternate in sign, the even subscripted coefficients will be positive and R will be negative if n is even. If n is odd, the odd subscripted coefficient will be positive, b_0 will be negative and R will be positive.
(c) $Q(p) = b_n p^n + b_{n-1} p^{n-1} + \cdots + b_1 p + b_0$. If n is even $b_n > 0$ and $p^n > 0$, $b_{n-1} < 0$ and $p^{n-1} < 0, \ldots, b_1 < 0$ and $p < 0$, and $b_0 > 0$. Hence $Q(p) > 0$. Because $R < 0$ when n is even and because $p - c < 0$, then $F(p) = (p - c)\,Q(p) + R < 0$.
(d) If n is odd, $b_n > 0$ and $p^n < 0$ (because $p < 0$), $b_{n-1} < 0$ and $p^{n-1} > 0 \ldots, b_1 > 0$ and $p < 0$, and $b_0 < 0$. Hence $Q(p) < 0$. Because $R > 0$ when n is odd and because $p - c < 0$ then $F(p) = (p - c)\,Q(p) + R > 0$.
(e) From **(c)** and **(d)**, $F(p) \neq 0$ if $p < c$. Hence, any real number less than c cannot be a zero of F.

Section 4.4

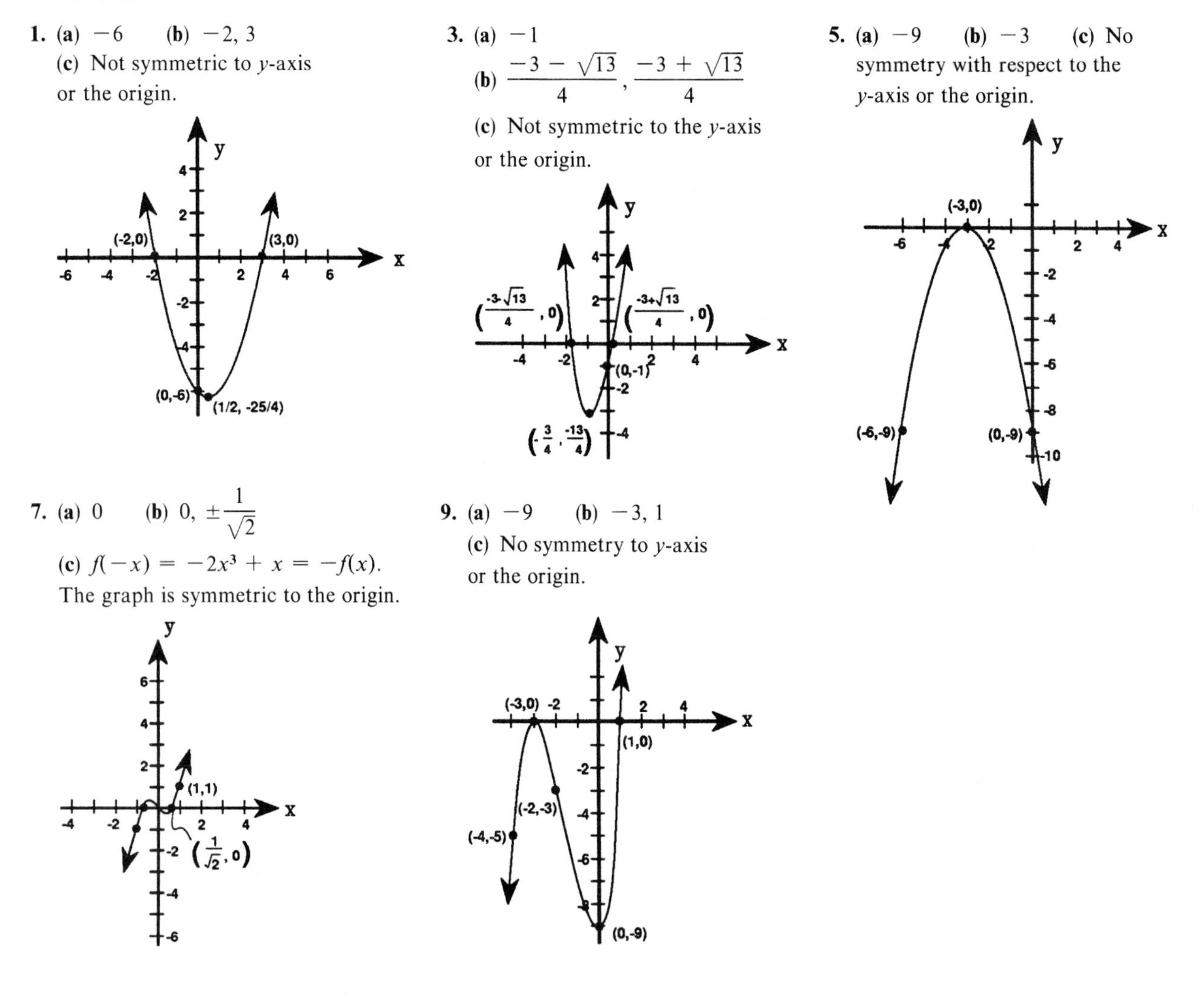

1. **(a)** -6 **(b)** $-2, 3$
(c) Not symmetric to y-axis or the origin.

3. **(a)** -1
(b) $\dfrac{-3 - \sqrt{13}}{4}, \dfrac{-3 + \sqrt{13}}{4}$
(c) Not symmetric to the y-axis or the origin.

5. **(a)** -9 **(b)** -3 **(c)** No symmetry with respect to the y-axis or the origin.

7. **(a)** 0 **(b)** $0, \pm\dfrac{1}{\sqrt{2}}$
(c) $f(-x) = -2x^3 + x = -f(x)$.
The graph is symmetric to the origin.

9. **(a)** -9 **(b)** $-3, 1$
(c) No symmetry to y-axis or the origin.

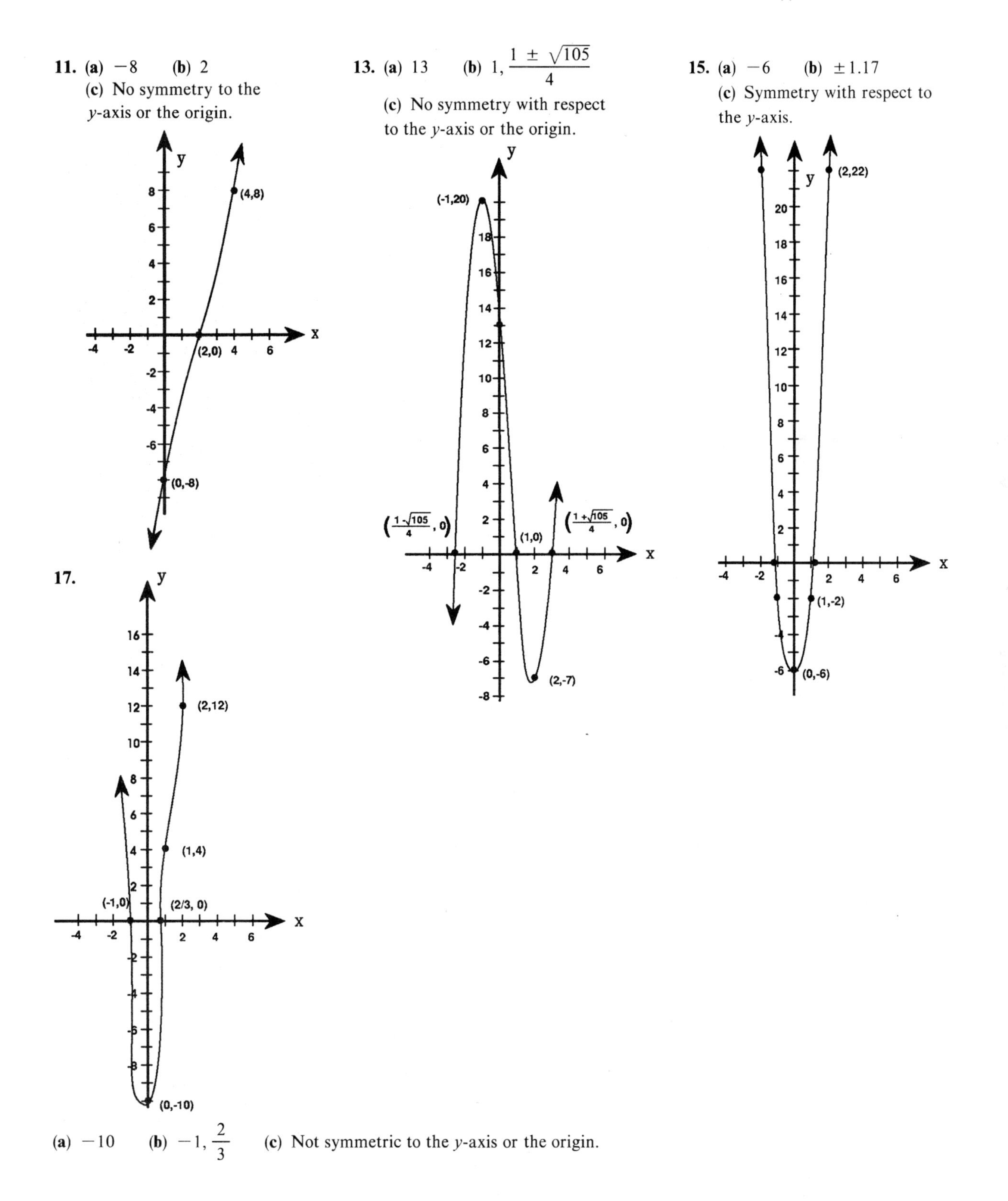

11. **(a)** -8 **(b)** 2
(c) No symmetry to the y-axis or the origin.

13. **(a)** 13 **(b)** $1, \frac{1 \pm \sqrt{105}}{4}$
(c) No symmetry with respect to the y-axis or the origin.

15. **(a)** -6 **(b)** ± 1.17
(c) Symmetry with respect to the y-axis.

17.

(a) -10 **(b)** $-1, \frac{2}{3}$ **(c)** Not symmetric to the y-axis or the origin.

19. **(a)** 9 **(b)** $-3, 1$ **(c)** Not symmetric to the y-axis or the origin.

21. **(a)** 0 **(b)** $0, -2$ **(c)** Not symmetric to the y-axis or the origin.

23. **(a)** 0 **(b)** $-4, -1, 0, 3$ **(c)** Not symmetric to the y-axis or the origin.

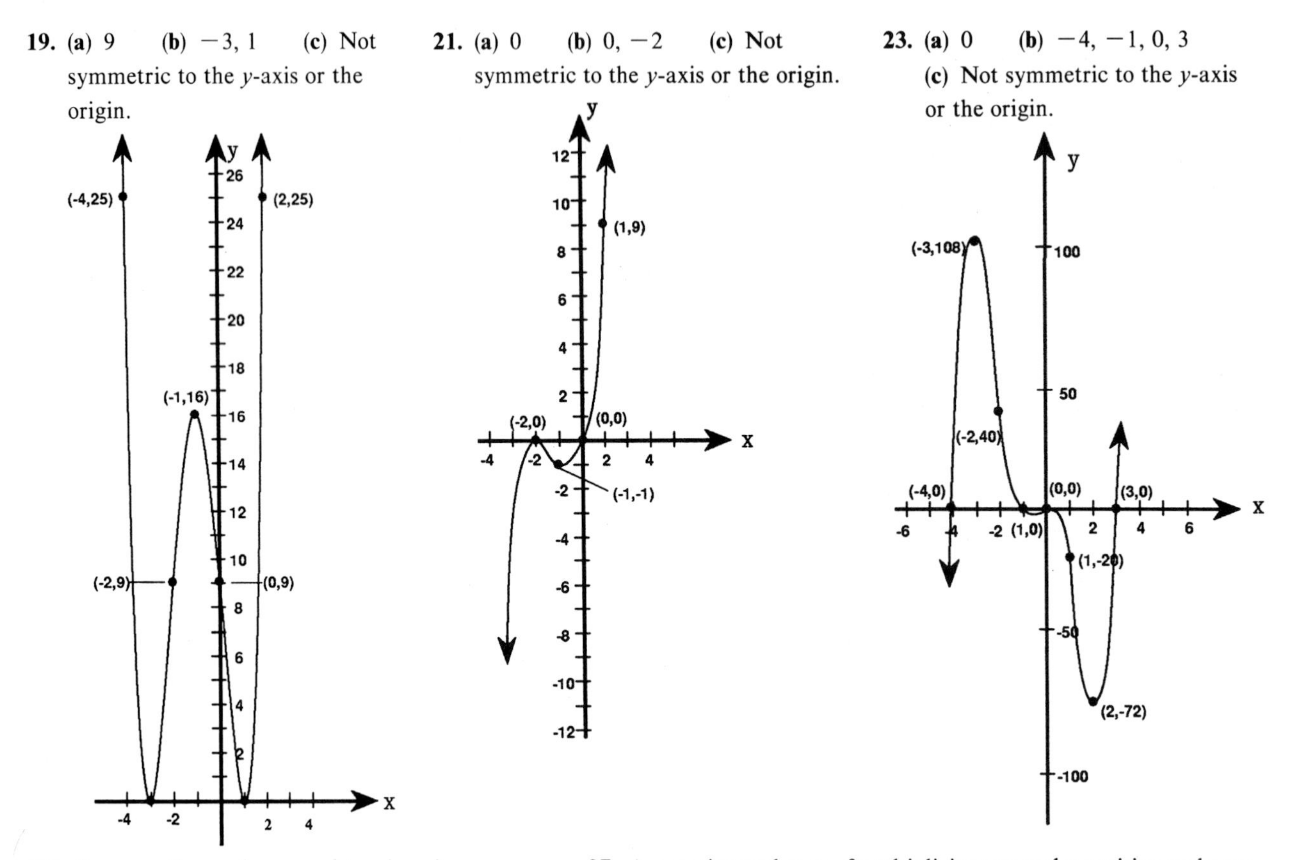

25. One negative real zero and two imaginary zeros **27.** A negative real zero of multiplicity two and a positive real zero of multiplicity two **29.** $F(x) = (x + 1)(x - 2)^2$ **31.** $F(x) = (x - 1)^2$ **33.** $c = 9$
35. $F(x) = (x + 2)(x - 1)(x - 3)$. The zeros of $F(x)$ are -2, 1, and 3. $F(-x) = (-x + 2)(-x - 1)(-x - 3)$. The zeros of $F(-x)$ are 2, -1, and -3. **37.** $y = x^3 + 3x^2 + 6x - 10$
39. $V = 4x^3 - 66x^2 + 270x$, $0 < x < 7.5$. The length of the side of the square cut from each corner is either 5″ or 0.95″. A maximum value of about 310 cubic inches results from a side length of about 3.25″.

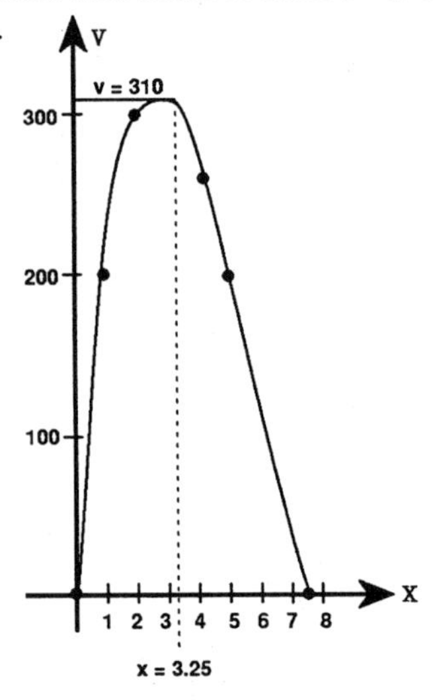

41. The radius is 6″ and the height is 8″. $V = \pi r^2\left[20 - \frac{2}{3}r\right]$, $0 < r < 10$.

43. The population increases over the time interval $\left[0, \frac{49}{3}\right]$.

45. $r = 3$, $h = 5$ or $r = 2.65$, $h = 6.41$. $V = \pi(24r - r^3)$. The radius that produces the maximum volume is about 2.8 feet. The maximum volume is about 143 cubic feet.

47. By the Fundamental Theorem of Algebra, the second degree polynomial $F(x) - G(x)$ must have exactly two zeros in the set of complex numbers. Since $F(x) - G(x)$ is a second degree polynomial that has at least three zeros (x_1, x_2, and x_3), it must be identically zero. That is, $a_2 = b_2$, $a_1 = b_1$, and $a_0 = b_0$.

Section 4.5

1. $x = 0.35$ **3.** $x = -1.35$ **5.** $x = -2.78$ **7.** 199, −10.583 **9.** 2,011, 195.9601 **11.** 3,010, 5,277.3541 **13.** 0.35 **15.** −0.32 **17.**

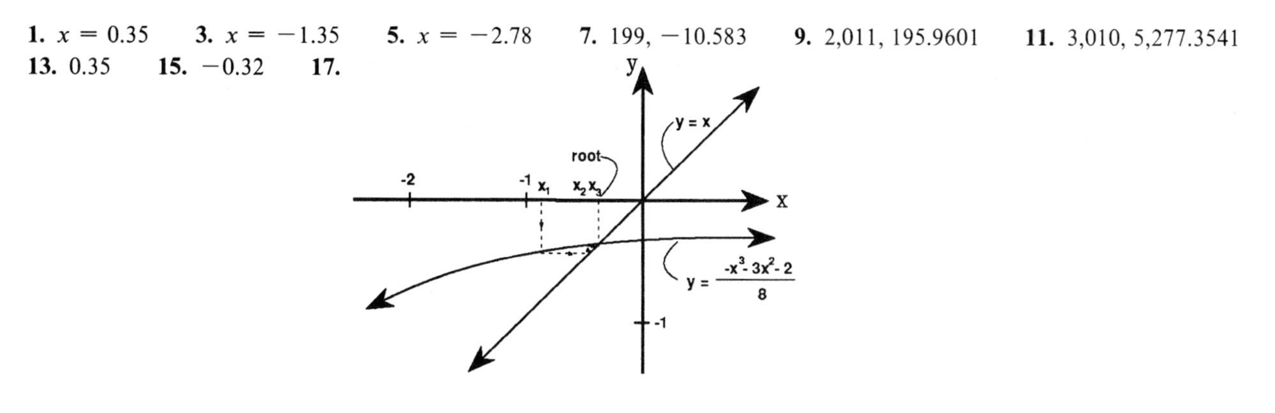

Chapter 4 Review Exercises

1. **(a)** $-x^2 + x + 8$ **(b)** $5x^2 - 11x - 8$ **(c)** $-6x^4 + 27x^3 - 14x^2 - 40x$ **3.** **(a)** $9x^3 + 3x^2 - 15x + 21$ **(b)** $-3x^3 + 7x^2 - 5x + 3$ **(c)** $18x^6 + 24x^5 - 85x^4 + 94x^3 + 71x^2 - 150x + 108$ **5.** **(a)** $x^4 - 7x^3 + 6x^2 + 6$ **(b)** $x^4 + 7x^3 + 24x^2 - 14$ **(c)** $-28x^7 - 9x^6 - 105x^5 - 125x^4 + 28x^3 + 186x^2 - 40$

7. $5x^3 - 8x^2 + 6x - 9 = (x^2 + 3x)(5x - 23) + (75x + 9)$

9. $20x^4 - 5x^3 + 9x - 4 = (4x^2 + 5x - 7)\left(5x^2 - \frac{15}{2}x + \frac{145}{8}\right) + \left(\frac{1{,}073}{8}x + \frac{983}{8}\right)$

11. $5x^3 + 7x^2 - 11x + 4 = (x + 2)(5x^2 - 3x - 5) + 14$ 13. $4x^3 - 15x^2 + 10x - 3 = (x - 3)(4x^2 - 3x + 1)$

15. $x^4 + 2x - 9 = (x - 5)(x^3 + 5x^2 + 25x + 127) + 626$ 17. $F(-2) = 44$ 19. $F(2) = 48$

21. $\begin{array}{r|rrrr} & 3 & 3 & -2 & -4 \\ \hline 1 & 3 & 6 & 4 & 0 \end{array}$

23. $\begin{array}{r|rrrrr} & 6 & 5 & -14 & 3 & 1 \\ \hline \frac{1}{2} & 6 & 8 & -10 & -2 & 0 \end{array}$

25. $2x^3 - 5x^2 - x + 6 = 2(x - 2)\left(x - \frac{3}{2}\right)(x + 1)$

27. $4x^4 + 16x^3 + 17x^2 + 4x + 4 = 4(x + 2)(x + 2)\left(x + \frac{1}{2}i\right)\left(x - \frac{1}{2}i\right)$ 29. $\left\{-3, \frac{2}{3}\right\}$ 31. $\{-2, 4\}$

33. $\{0\}$

35.

37.

39. Symmetric to the origin because $f(-x) = -f(x)$.

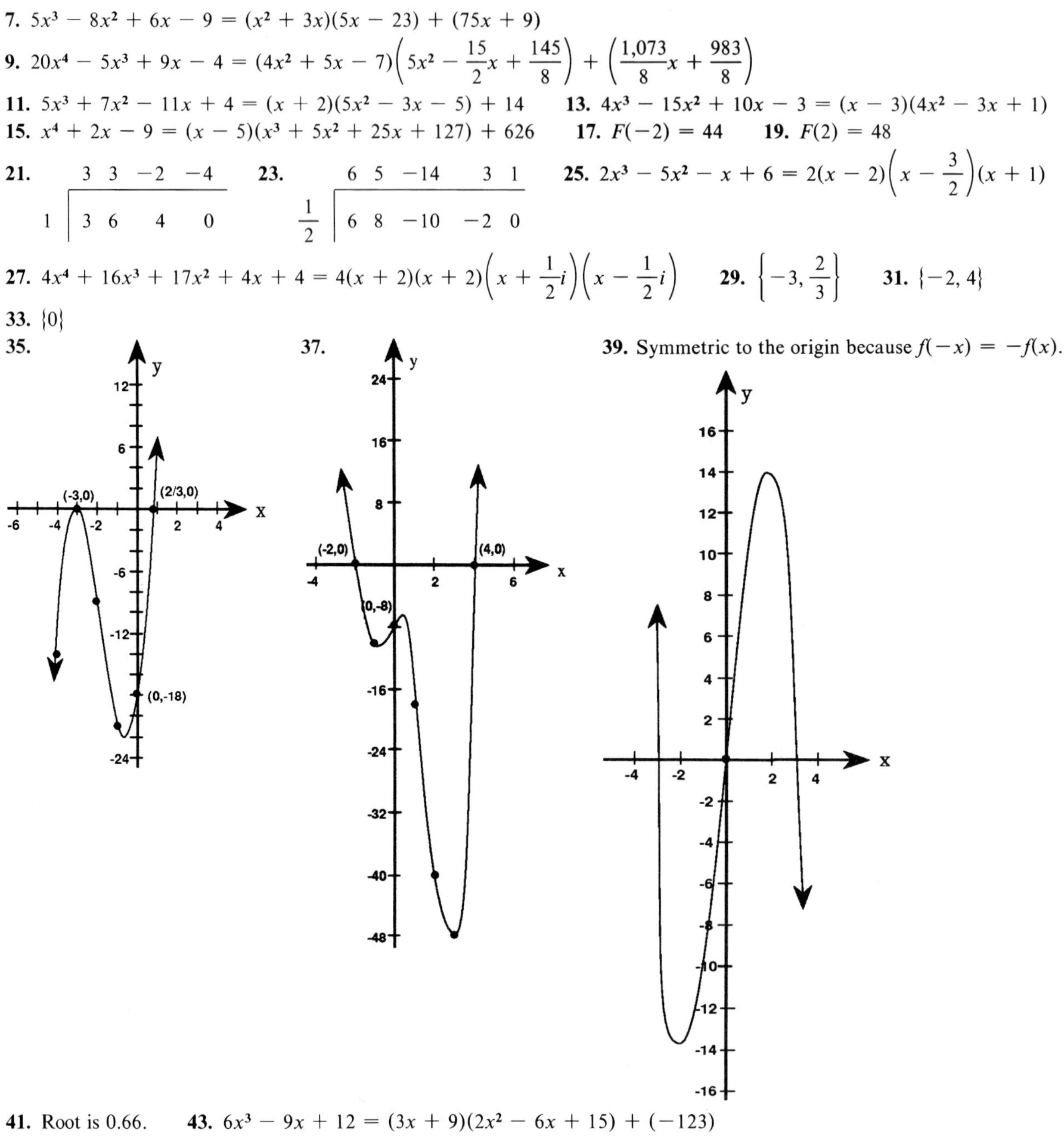

41. Root is 0.66. 43. $6x^3 - 9x + 12 = (3x + 9)(2x^2 - 6x + 15) + (-123)$

45. $n - 1$ zeros

	1	0	0	0	. . .	0	−1
−1	1	−1	1	−1	. . .	−1	1

Counting from left to right the negative ones fall under the odd-numbered zeros. Therefore the last zero before −1 must be odd. Since $n - 1$ must be odd, n must be even.

47. The possible rational roots are $\pm\frac{1}{2}$, ± 1, $\pm\frac{5}{2}$, or ± 5. There are three or one possible positive roots. There is only one negative root.

	2	6	−7	9	−5	
$\frac{1}{2}$	2	7	$-\frac{7}{2}$	$\frac{29}{4}$	$-\frac{11}{8}$	← root
1	2	8	1	10	5	← upper bound
$-\frac{1}{2}$	2	5	$-\frac{19}{2}$	$\frac{55}{4}$	$-\frac{95}{8}$	
−1	2	4	−11	20	−25	
$-\frac{5}{2}$	2	1	$-\frac{19}{2}$	$\frac{131}{4}$	$-\frac{695}{8}$	← root
−5	2	−4	13	−56	275	← lower bound

There is a positive irrational root and a negative irrational root, and there are two complex roots.

49. A root of the equation $x^3 - 9 = 0$ is $\sqrt[3]{9}$. But, the only possible positive rational roots of the equation are 1, 3, or 9. Therefore $\sqrt[3]{9}$ is not a rational number.

51. One negative real root and two positive real roots

53. $y = (x - 3)^2\left(x^2 + \frac{4}{9}\right)$

55. **57.** **59.**

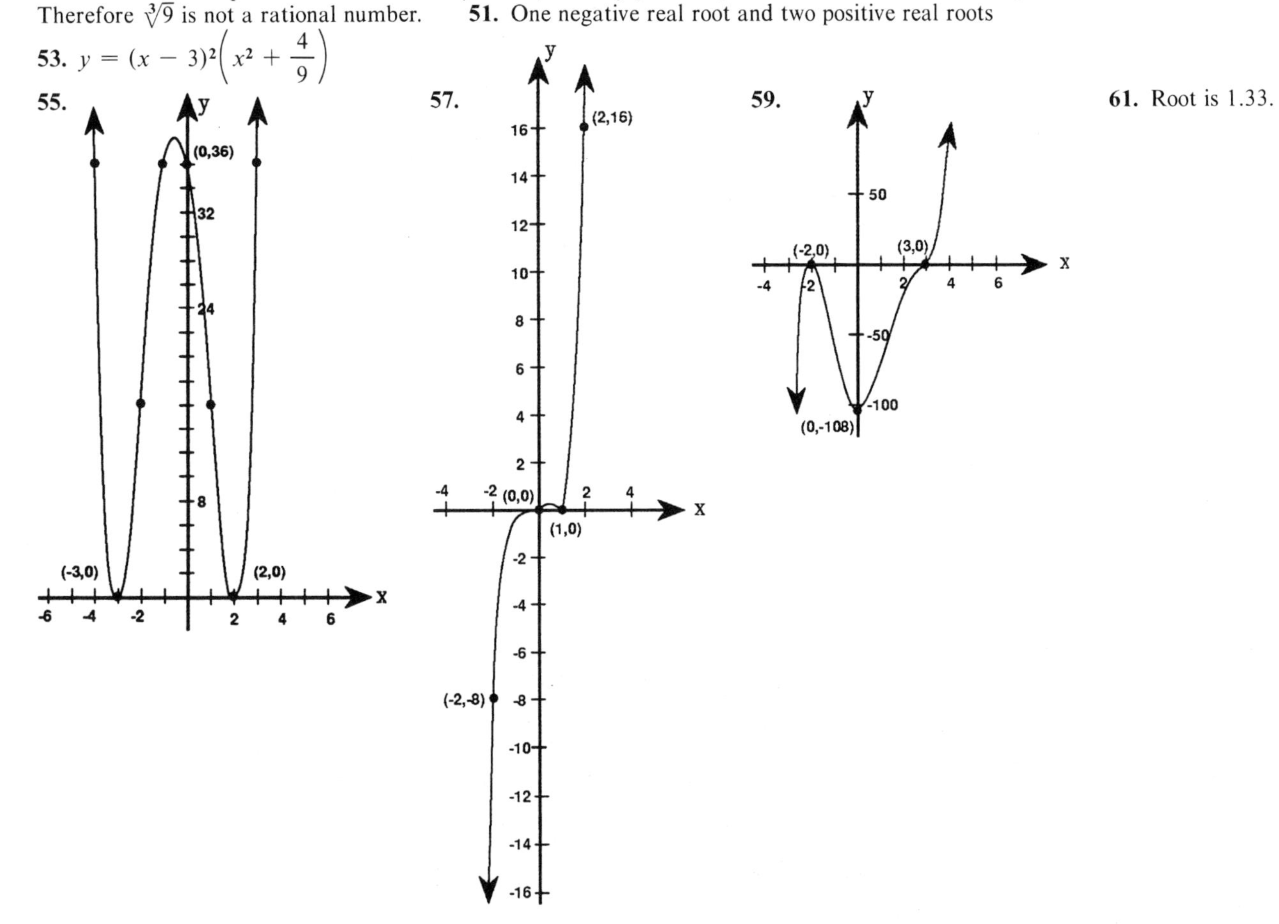

61. Root is 1.33.

Chapter 4 Test 1

1. $-3x^3 + 4x^2 + 15x - 9$, $-10x^6 + 8x^5 - 12x^4 + 55x^3 + 26x^2 - 75x - 14$

2. $2x^4 - 10x^3 + 5x^2 + 8x - 1 = (x + 3)(2x^3 - 16x^2 + 53x - 151) + 452$ **3.** $F(-3) = -63$

4. $\{-2, 1, 6\}$ **5.** $\{-3\}$ **6.** **7.**

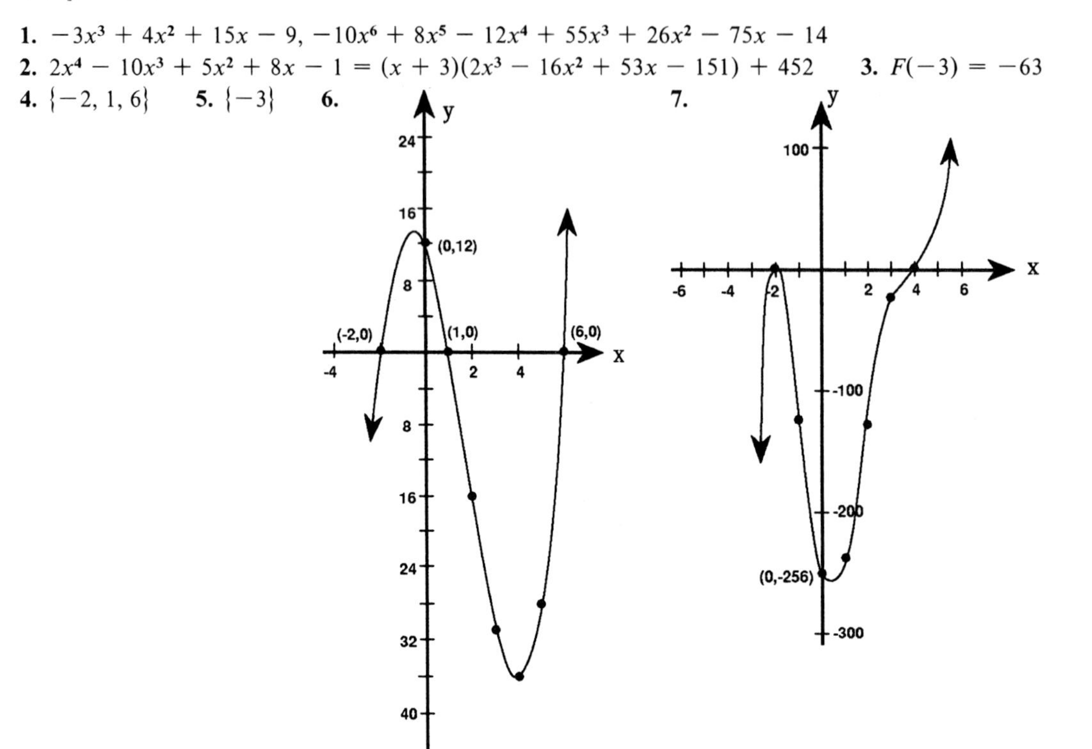

8. The possible rational roots are $\pm\frac{1}{3}$, ± 1, $\pm\frac{5}{3}$, ± 5. There is one negative root and there may be two or zero positive roots.

	3	7	-6	5	
$\frac{1}{3}$	3	8	$-\frac{10}{3}$	$\frac{35}{9}$	
1	3	10	4	9	← upper bound
$-\frac{1}{3}$	3	6	-8	$\frac{23}{3}$	
-1	3	4	-10	15	
$-\frac{5}{3}$	3	2	$-\frac{28}{3}$	$\frac{185}{9}$	← root
-5	3	-8	34	-165	

There is one negative irrational root and a pair of complex conjugate roots.

Chapter 4 Test 2

1. $\left\{-9, \frac{2}{3}, 4\right\}$ **2.** $\{5\}$

3. 4.

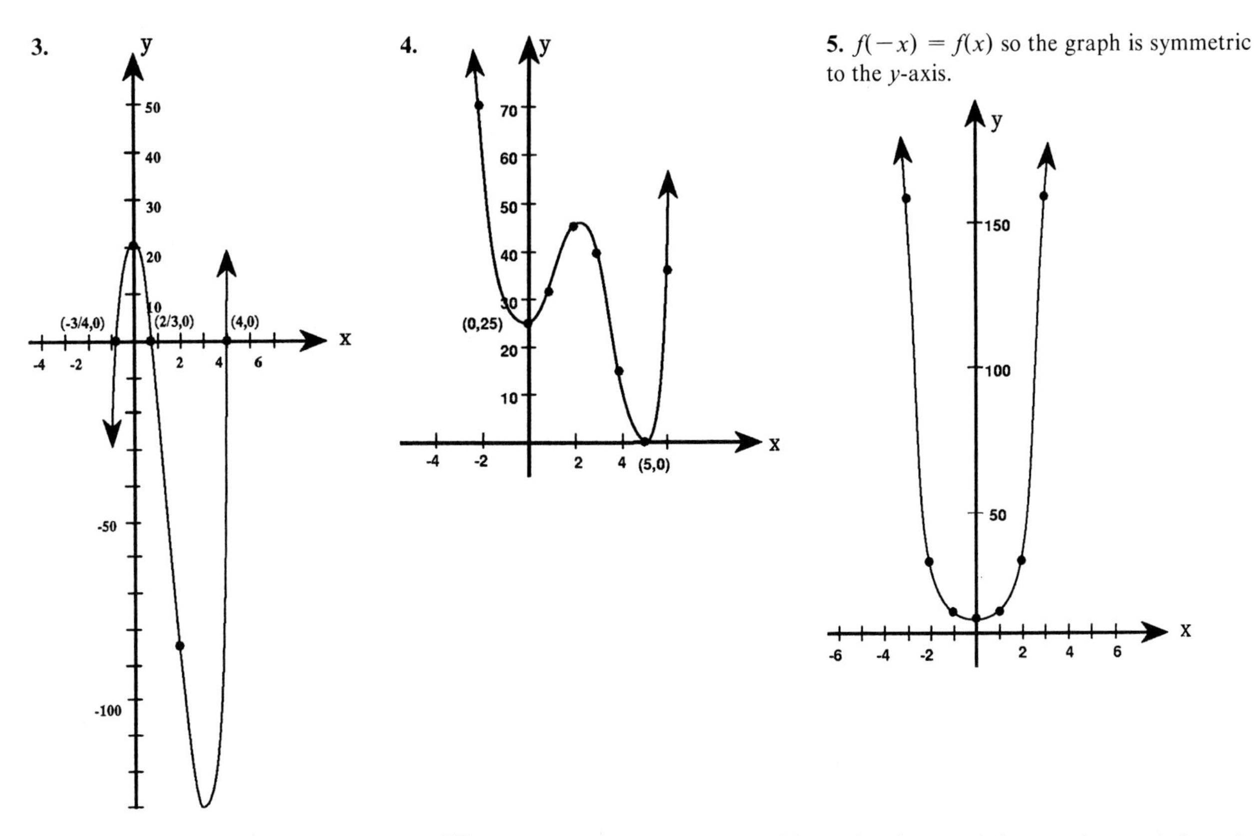

5. $f(-x) = f(x)$ so the graph is symmetric to the y-axis.

6. A root of the equation $x^3 - 4 = 0$ is $\sqrt[3]{4}$. However, the only possible positive rational roots of the equation are 1, 2, or 4. Therefore $\sqrt[3]{4}$ is not a rational number. 7. One positive zero and a pair of complex conjugate zeros
8. $y = (x + 3)(x)(x - 4)^2$

Section 5.1

1. Reals except ± 1, $R(x) = \dfrac{1}{x - 1}, x \neq \pm 1$ 3. Reals except -2, $S(x) = \dfrac{x - 3}{x + 2}, x \neq -2$

5. Reals except 0, $-\dfrac{5}{3}$, $T(x) = \dfrac{x - 2}{2x}, x \neq 0, -\dfrac{5}{3}$

7. $y = \dfrac{1}{x}$

9. $y = \dfrac{2}{(x+2)^2}$, (0,1/2), $x = 2$

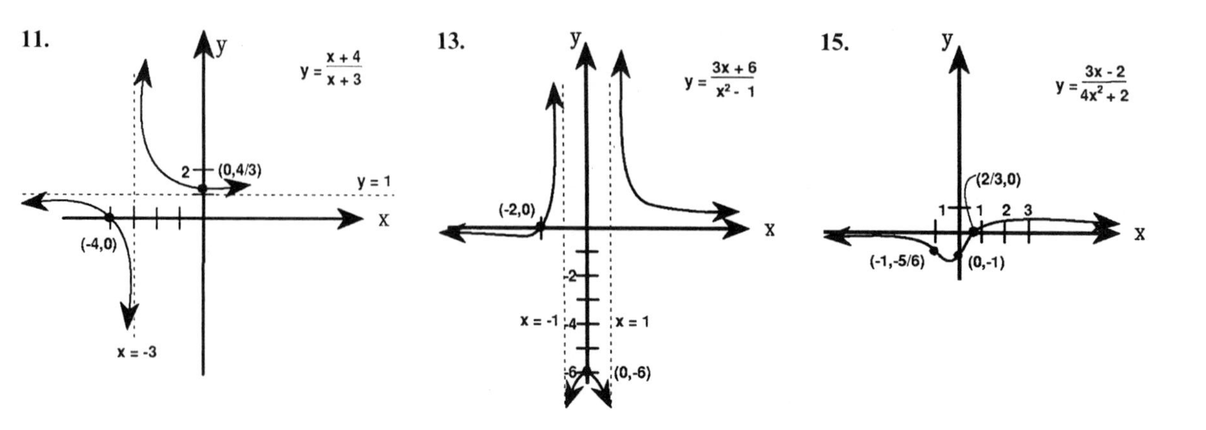

11.
y = (x + 4)/(x + 3)
(0,4/3)
y = 1
(-4,0)
x = -3
13.
y = (3x + 6)/(x² - 1)
(-2,0)
x = -1
x = 1
(0,-6)
15.
y = (3x - 2)/(4x² + 2)
(2/3,0)
1 2 3
(-1,-5/6)
(0,-1)

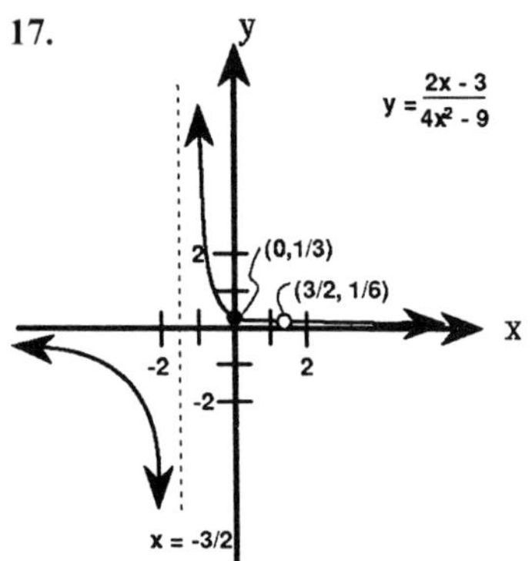

17.
y = (2x - 3)/(4x² - 9)
(0,1/3)
(3/2, 1/6)
-2
2
x = -3/2

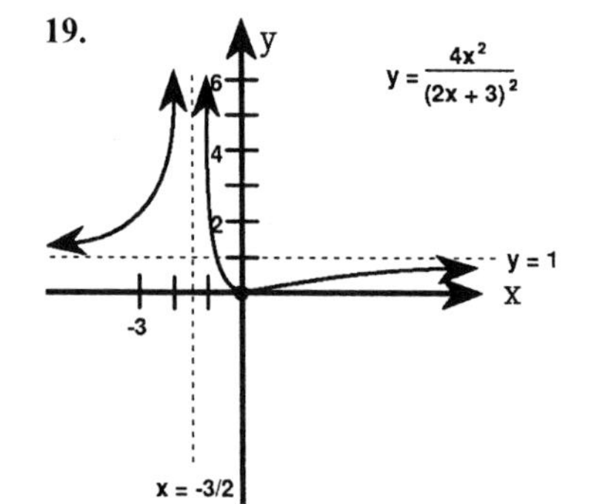

19.
y = 4x²/(2x + 3)²
y = 1
-3
x = -3/2

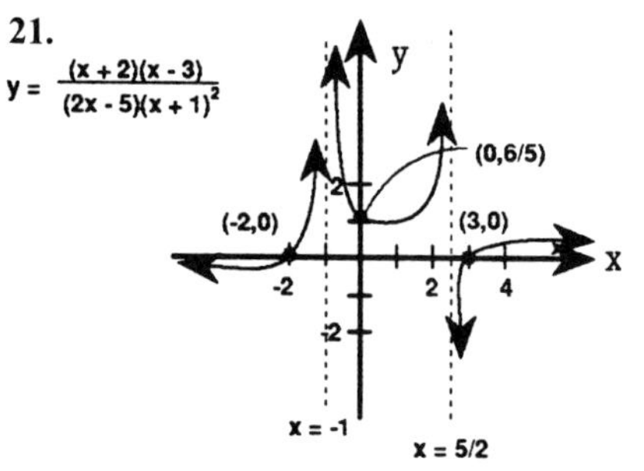

21.
y = (x + 2)(x - 3)/((2x - 5)(x + 1)²)
(0,6/5)
(-2,0)
(3,0)
x = -1
x = 5/2

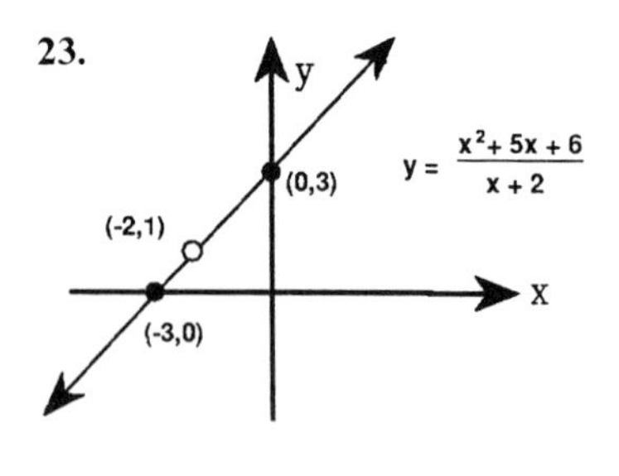

23.
y = (x² + 5x + 6)/(x + 2)
(0,3)
(-2,1)
(-3,0)

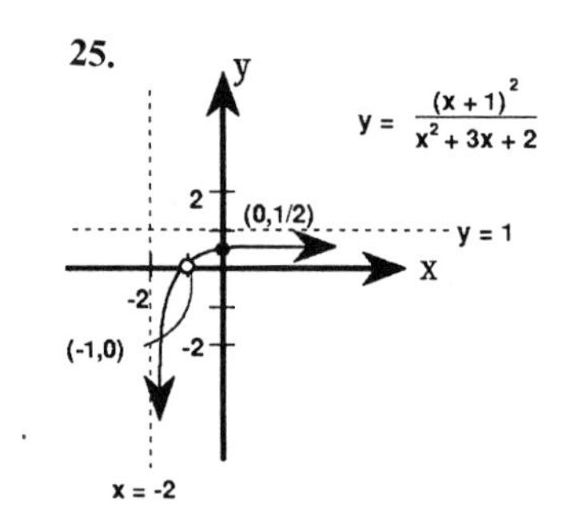

25.
y = (x + 1)²/(x² + 3x + 2)
(0,1/2)
y = 1
(-1,0)
x = -2

27.

29. (**a**) (**b**)

(**c**) same as (**b**) (**d**) Shift 1 unit left and 2 units up

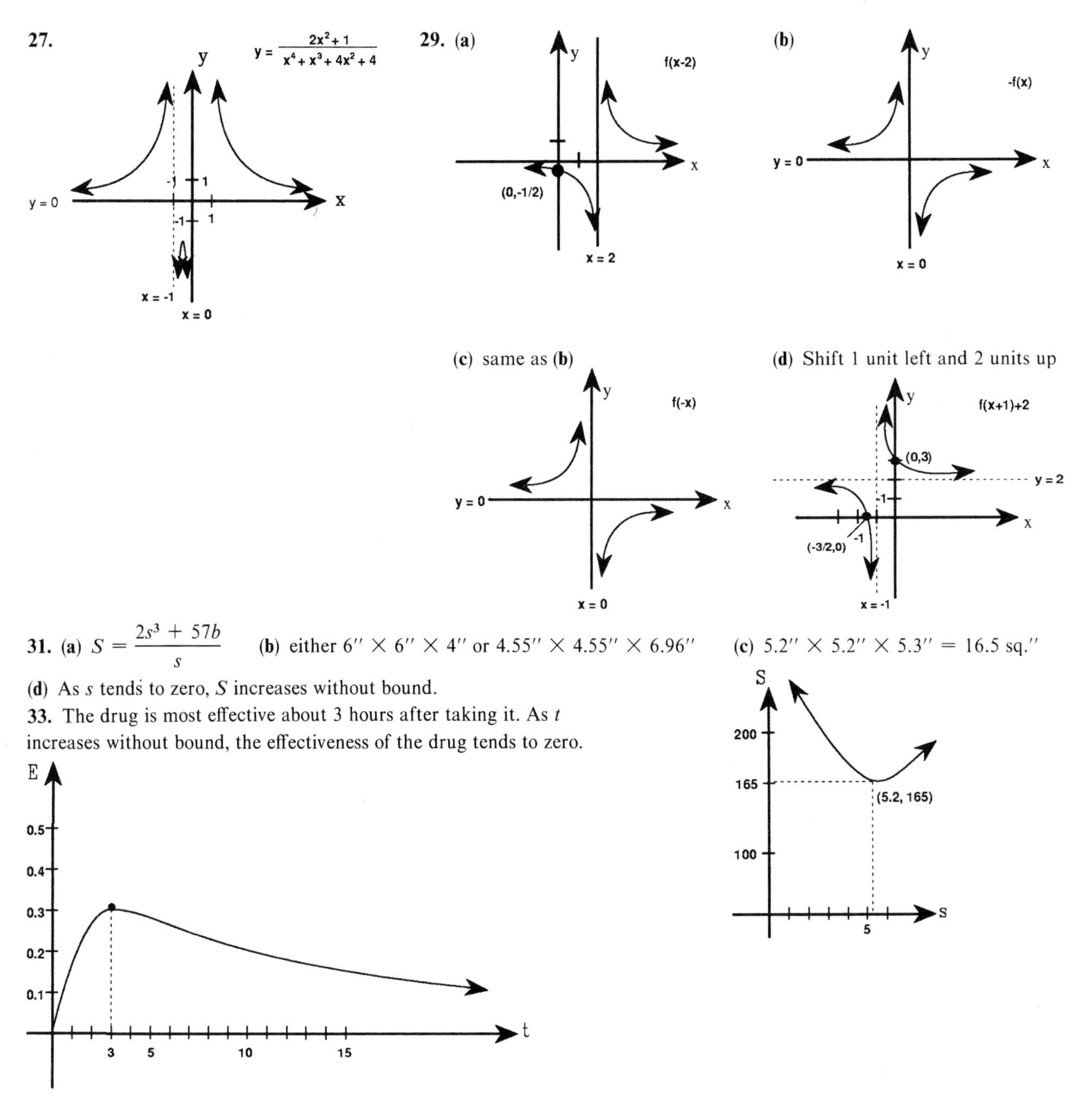

31. (**a**) $S = \dfrac{2s^3 + 57b}{s}$ (**b**) either 6″ × 6″ × 4″ or 4.55″ × 4.55″ × 6.96″ (**c**) 5.2″ × 5.2″ × 5.3″ = 16.5 sq.″

(**d**) As s tends to zero, S increases without bound.

33. The drug is most effective about 3 hours after taking it. As t increases without bound, the effectiveness of the drug tends to zero.

35. $c = 504$ **37.** $c = 3,405$ **39.**

41.

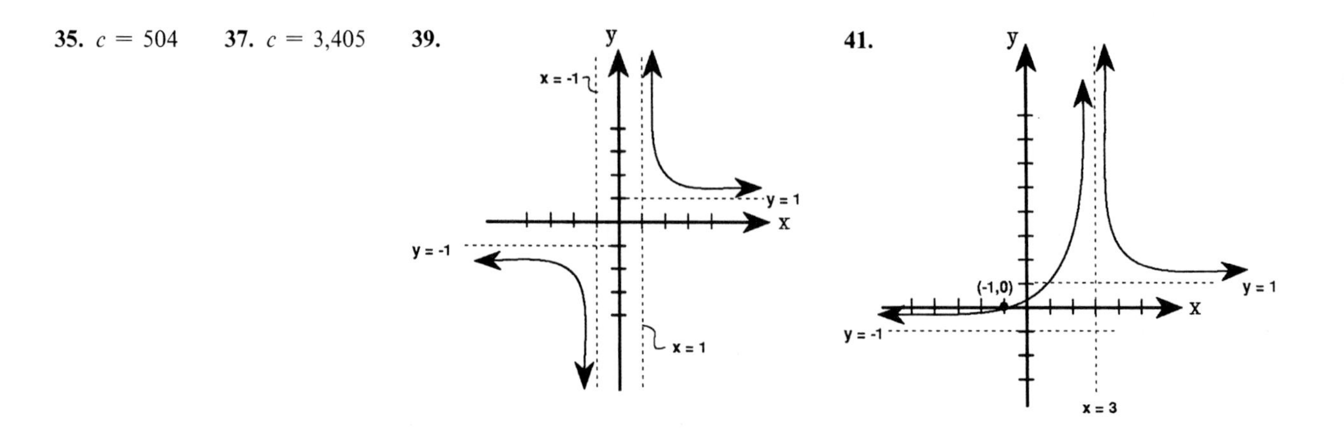

Section 5.2

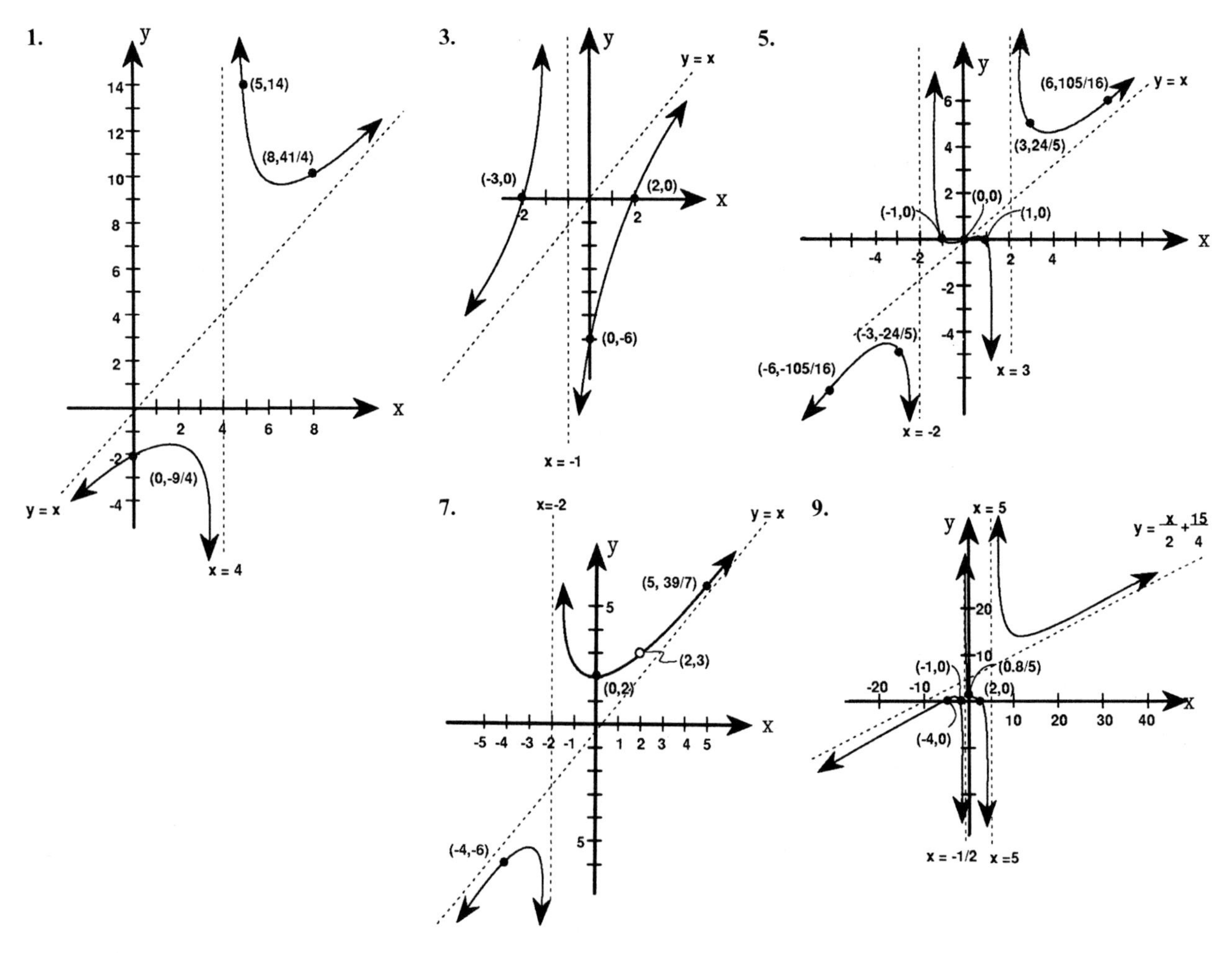

11.

$$\begin{array}{r|rrrr} & 1 & 0 & 0 & -1 \\ 2 & 1 & 2 & 4 & 7 \end{array}$$

So, $y = f(x) = \dfrac{x^3 - 1}{x - 2} = x^2 + 2x + 4 + \dfrac{7}{x - 2}$. For large $|x|$, $\dfrac{7}{x - 2} \approx 0$ and $f(x) \approx x^2 + 2x + 1$.

Section 5.3

1. $\dfrac{\frac{19}{17}}{(2x - 5)} + \dfrac{\frac{14}{17}}{(3x + 1)}$ **3.** $\dfrac{\frac{7}{3}}{x - 2} + \dfrac{\frac{2}{3}}{x + 1}$ **5.** $\dfrac{3}{x + 1} + \dfrac{2}{(x + 1)^2}$ **7.** $\dfrac{3}{x} - \dfrac{2}{x^2} - \dfrac{2}{x - 1}$

9. $\dfrac{4}{x - 2} + \dfrac{x + 8}{x^2 + 4}$ **11.** $\dfrac{-\frac{1}{4}}{x + 1} + \dfrac{\frac{1}{4}x + \frac{1}{4}}{x^2 - 6x + 13}$ **13.** $\dfrac{-\frac{1}{2}}{x + 1} + \dfrac{2}{x + 2} + \dfrac{-\frac{3}{2}}{x + 3}$

15. $\dfrac{\frac{55}{27}}{x + 2} + \dfrac{-\frac{28}{9}}{(x + 2)^2} + \dfrac{\frac{26}{27}}{x - 1} + \dfrac{\frac{2}{9}}{(x - 1)^2}$

17. $\dfrac{3x^2 - 5x + 12}{(x - 1)^2(x + 4)} = \dfrac{Fx + G}{(x - 1)^2} + \dfrac{H}{x + 4}$

$$\begin{aligned} 3x^2 - 5x + 12 &= (Fx + G)(x + 4) + H(x - 1)^2 \\ &= Fx^2 + 4Fx + Gx + 4G + Hx^2 - 2Hx + H \\ &= (F + H)x^2 + (4F + G - 2H)x + (4G + H) \end{aligned}$$

So, $\left.\begin{aligned} 3 &= F + H \\ -5 &= 4F + G - 2H \\ 12 &= 4G + H \end{aligned}\right\}$ $\begin{aligned} 2R_1 + R_2 \quad 1 &= 6F + G \\ 2R_3 + R_2 \quad 19 &= 4F + 9G \end{aligned}\Big\}$ $-9R_1 + R_2$ $\quad 10 = -50F$, $\quad F = -\dfrac{1}{5}$

$H = 3 + \dfrac{1}{5} = \dfrac{16}{5}$ $\qquad G = \dfrac{12 - \frac{16}{5}}{4} = \dfrac{11}{5}$

$$\frac{3x^2 - 5x + 12}{(x - 1)^2(x + 4)} = \frac{-\frac{1}{5}x + \frac{11}{5}}{(x - 1)^2} + \frac{\frac{16}{5}}{x + 4}$$

But, $\dfrac{-\frac{1}{5}x + \frac{11}{5}}{(x - 1)^2} = \dfrac{-\frac{1}{5}(x - 1) + 2}{(x - 1)} = \dfrac{2}{(x - 1)^2} - \dfrac{\frac{1}{5}}{x - 1}$.

Thus, the result obtained is the same as in example 1 of this section.

19. **(a)** Since $P(0) = a_0$, $Q(0) = b_0$ and $P(x) = Q(x)$ for all x, it follows that $a_0 = b_0$ and $a_0 - b_0 = 0$.
(b) $(P - Q)(x) = 0$ for all x. **(c)** A jth degree polynomial must have exactly j complex zeros.

Chapter 5 Review Exercises

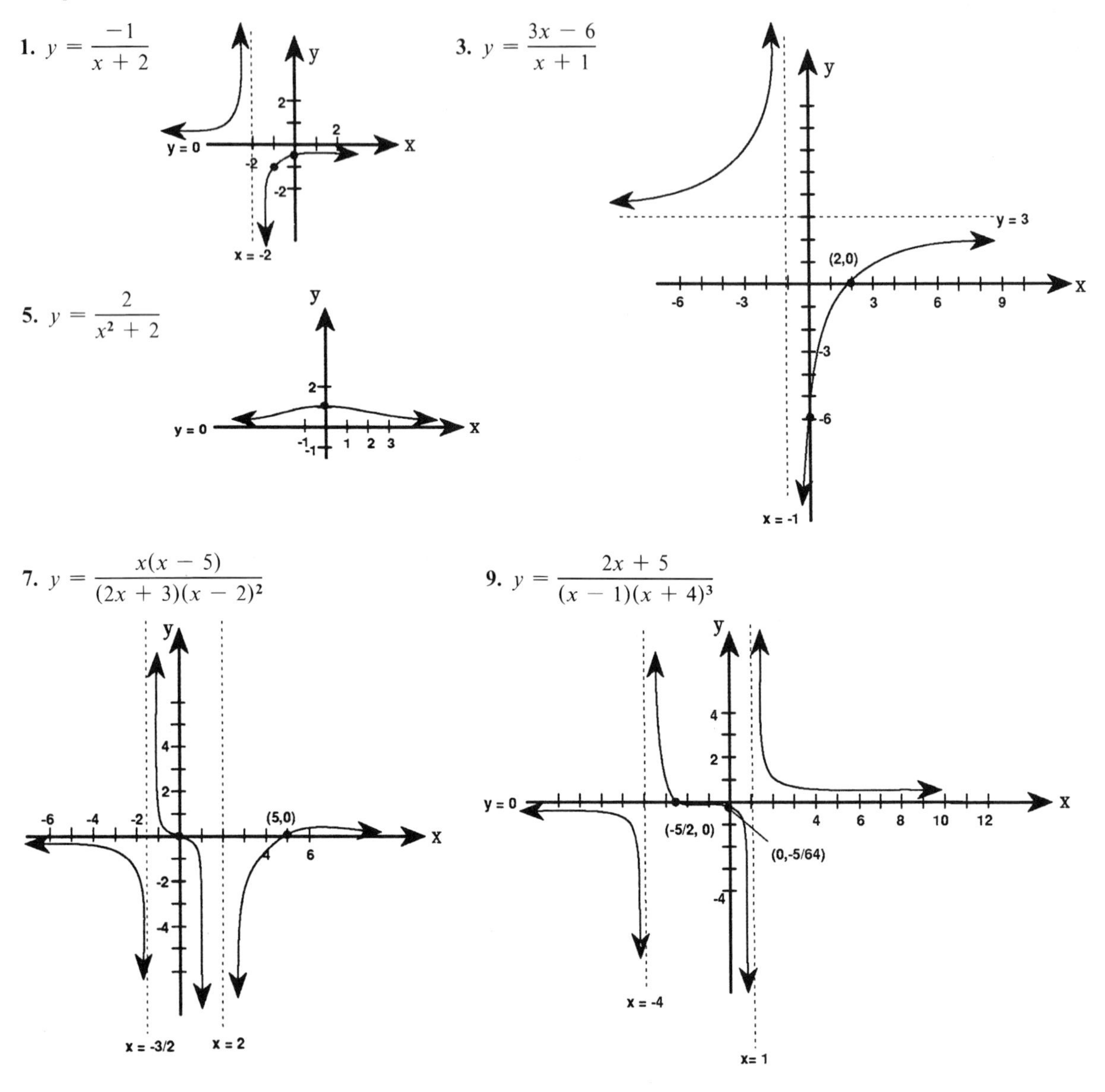

11. **13.**

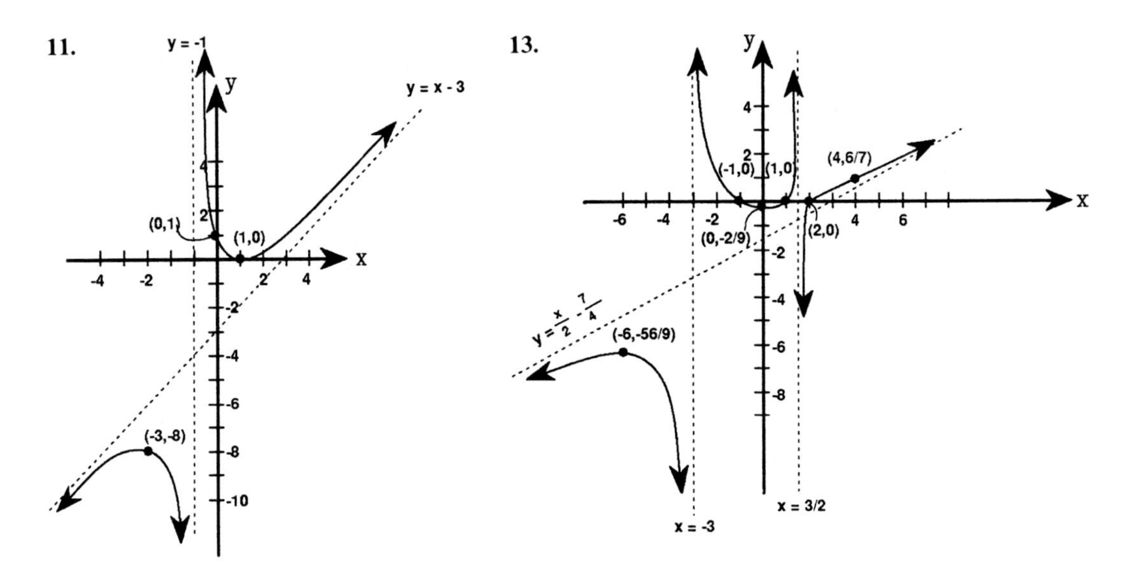

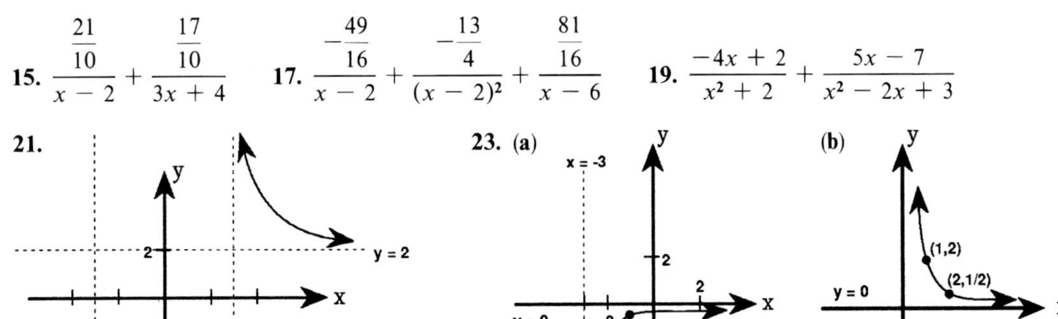

15. $\dfrac{\frac{21}{10}}{x-2} + \dfrac{\frac{17}{10}}{3x+4}$ **17.** $\dfrac{-\frac{49}{16}}{x-2} + \dfrac{-\frac{13}{4}}{(x-2)^2} + \dfrac{\frac{81}{16}}{x-6}$ **19.** $\dfrac{-4x+2}{x^2+2} + \dfrac{5x-7}{x^2-2x+3}$

21.

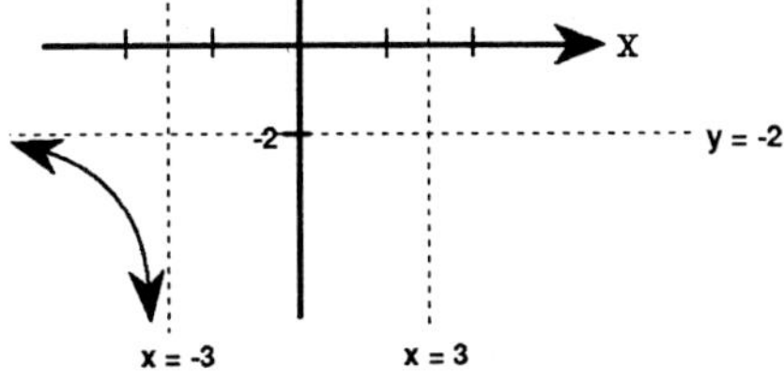

23. **(a)** **(b)**

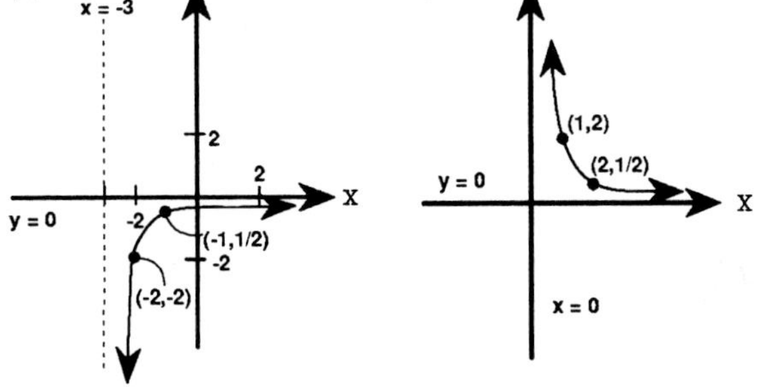

(c) **(d)**

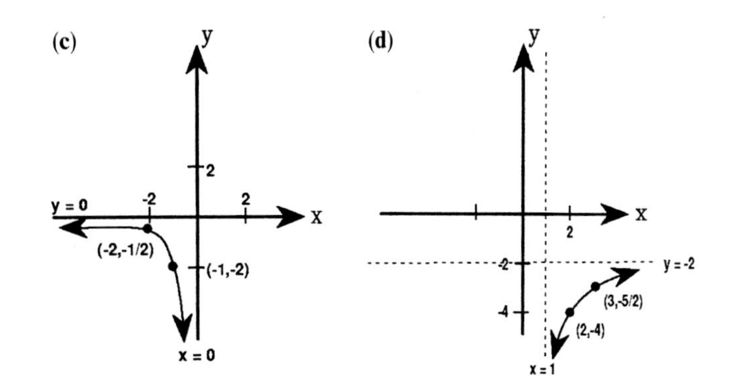

25. (**a**) $4\left(\frac{y^3 + 90}{y}\right)$ (**b**) 10″ wide, 5″ deep, $\frac{12}{5}$″ high or $(\sqrt{97} - 5)$″ wide, $\frac{\sqrt{97} - 5}{2}$″ deep, $\frac{240}{(\sqrt{97} - 5)^2}$″ high

(**c**)

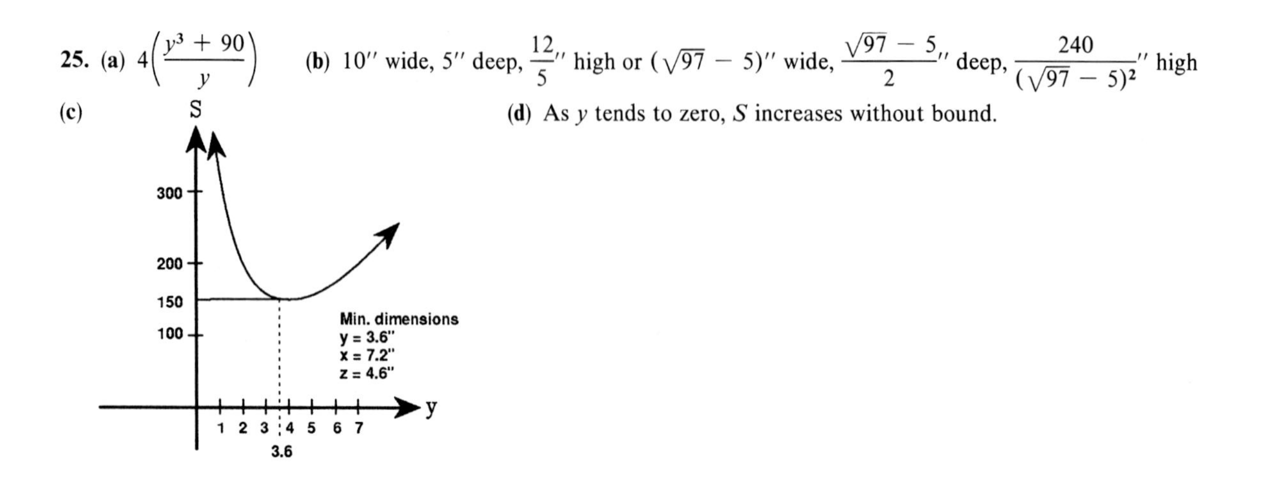

(**d**) As y tends to zero, S increases without bound.

Chapter 5 Test 1

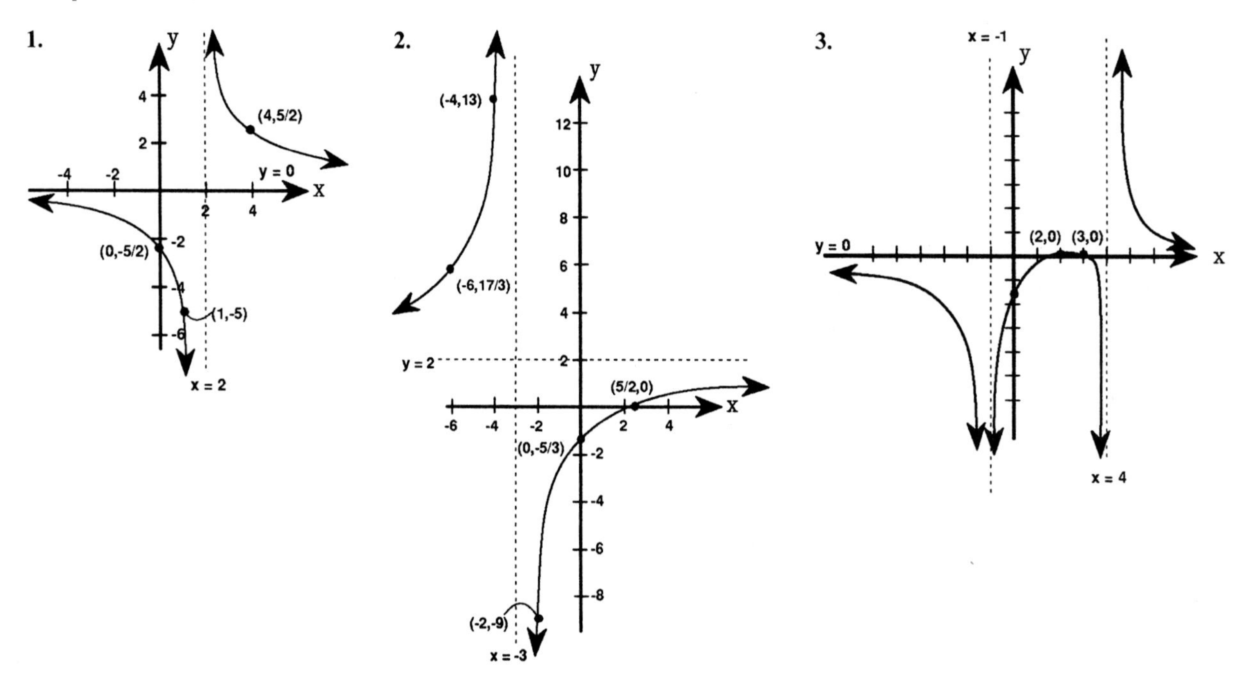

4.

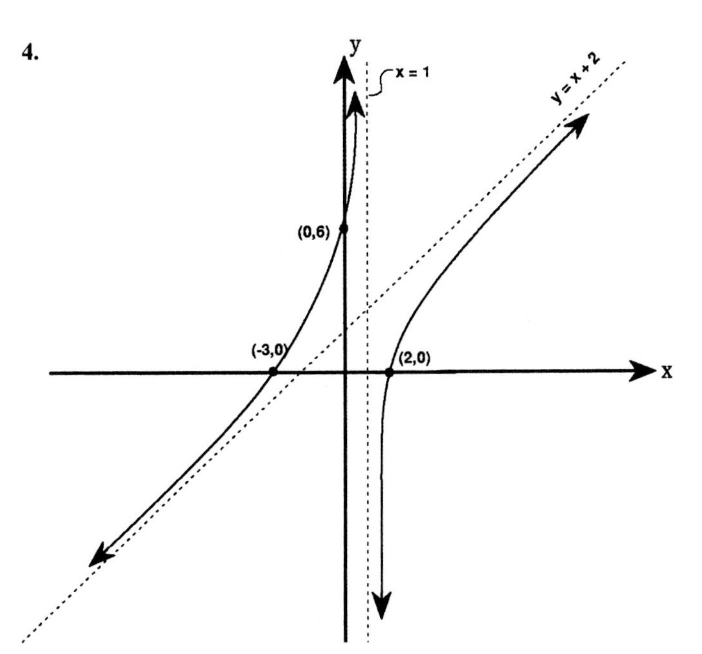

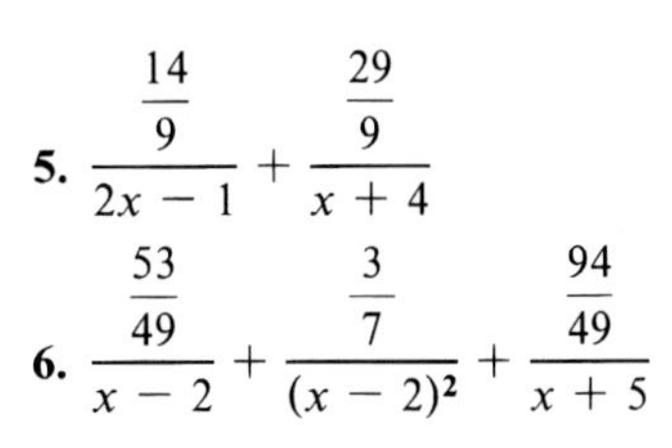

5. $\dfrac{\frac{14}{9}}{2x - 1} + \dfrac{\frac{29}{9}}{x + 4}$

6. $\dfrac{\frac{53}{49}}{x - 2} + \dfrac{\frac{3}{7}}{(x - 2)^2} + \dfrac{\frac{94}{49}}{x + 5}$

Chapter 5 Test 2

1. 2.

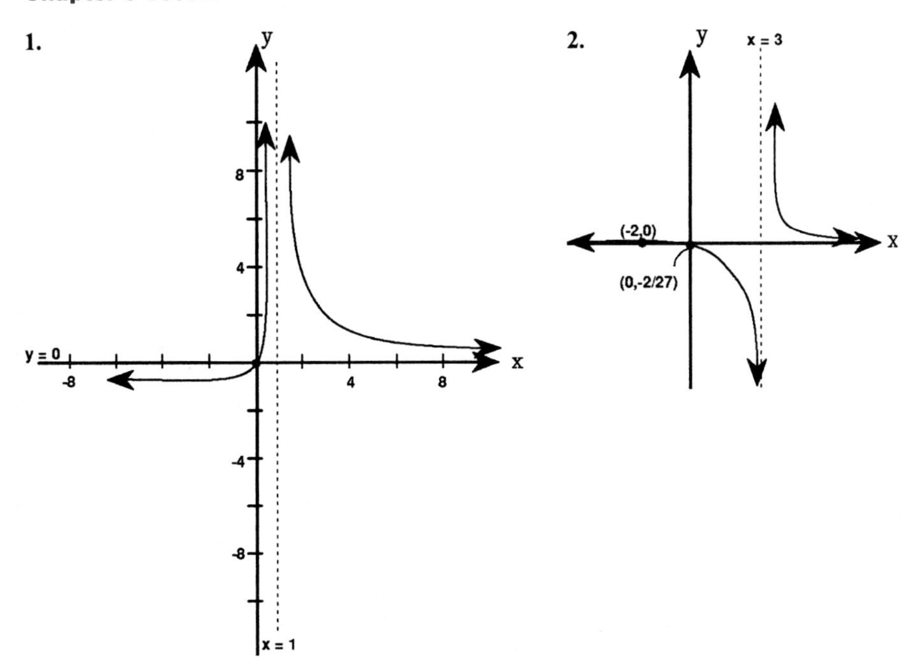

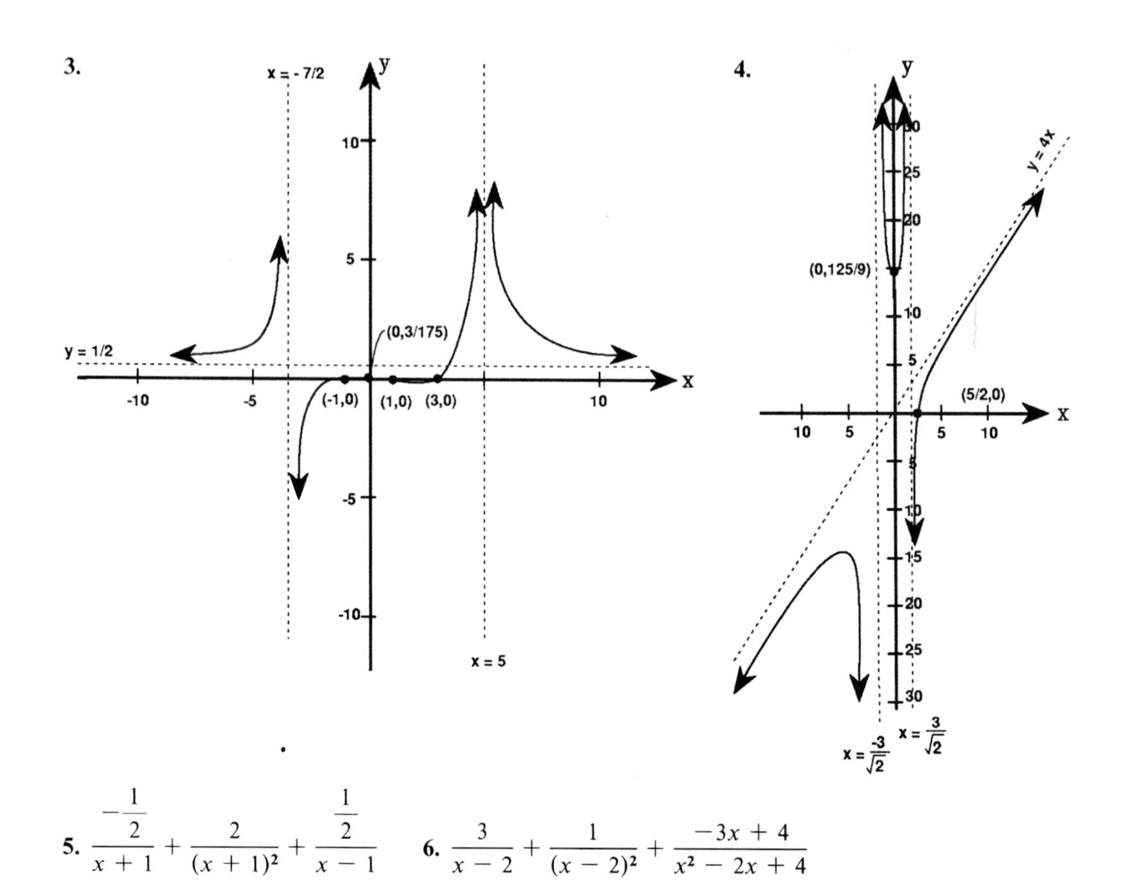

5. $\dfrac{-\frac{1}{2}}{x+1} + \dfrac{2}{(x+1)^2} + \dfrac{\frac{1}{2}}{x-1}$ 6. $\dfrac{3}{x-2} + \dfrac{1}{(x-2)^2} + \dfrac{-3x+4}{x^2-2x+4}$

Section 6.1

1. Decreasing on reals 3. Increasing on reals 5. Neither 7. Decreasing $(\infty, x]$, (x_2, x_3); increasing (x_1, x_2), (x_3, ∞)

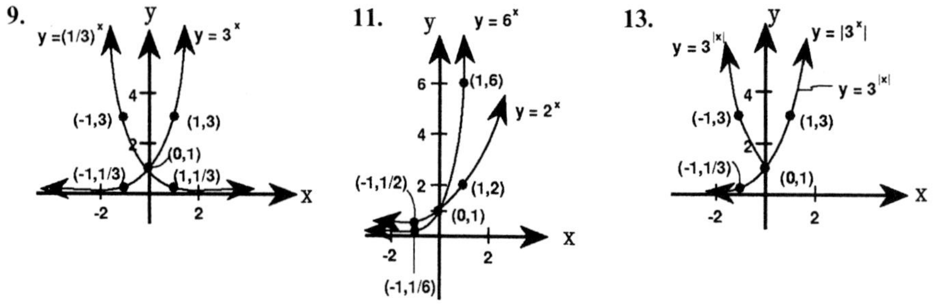

15. (a) 4,000 bacteria (b) 32,000 bacteria (c) 101,594 bacteria 17. (a) 27,104 (b) 36,731 (c) 49,777

19. $3^{\sqrt{2}+\sqrt{3}}$ 21. $5^{1+\sqrt{7}-\sqrt{5}}$ 23. $(x+1)^{\sqrt{10}}$ 25. $x^{2\sqrt{3}-4}y^{\sqrt{3}-2}$ 27. $\dfrac{x^{\pi}}{y^{\pi}}$

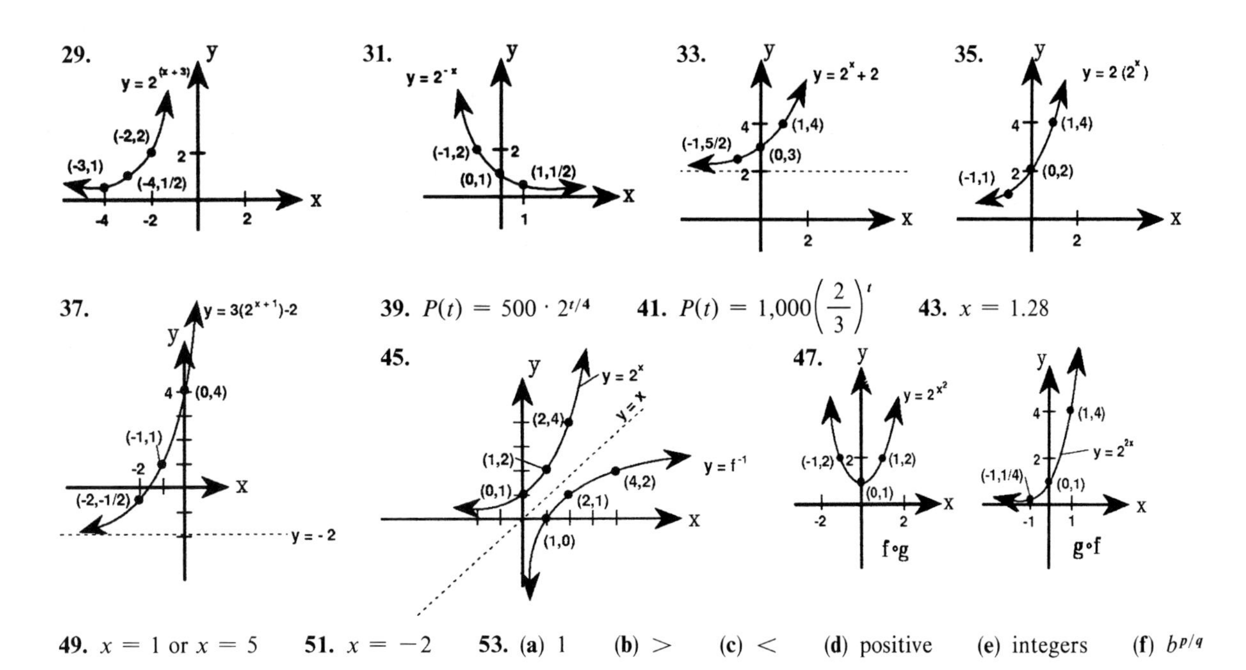

39. $P(t) = 500 \cdot 2^{t/4}$ **41.** $P(t) = 1{,}000\left(\frac{2}{3}\right)^t$ **43.** $x = 1.28$

49. $x = 1$ or $x = 5$ **51.** $x = -2$ **53.** **(a)** 1 **(b)** $>$ **(c)** $<$ **(d)** positive **(e)** integers **(f)** $b^{p/q}$

Section 6.2

1. **(a)** 232,085,893 **(b)** 256,494,579 **(c)** 346,231,467 **3.** **(a)** 3,664,208 **(b)** 4,475,474 **(c)** 8,154,845
5. **(a)** 0.0978 g **(b)** $8.006 \cdot 10^{-29}$ g **(c)** 0 g **7.** **(a)** 96.5895 g **(b)** 70.6805 g **(c)** 17.6400 g
9. \$4,875.44 **11.** \$10,892.55 **13.** \$1,822.12 **15.** **(a)** 0.1903 amps **(b)** 2.0 amps **(c)** 2.0 amps **17.** e

19. (-1,e) (0,1) (1,e⁻¹) **21.** (1,e²) (1/2,e) (0,1) (-1,e⁻²) **23.** (3,e) (1,e⁻¹) (2,1) (0,e⁻²) **25.** (-1,3.1) (1,3.1)

27. **29.** i = 2 (0.05,1.99) (0.01,1.26) t (seconds)

31.

33. $x = 5$ or $x = -4$ **35.** $x = -4$ **37.** 8.30% **39.** \$7,417.40

41. $x = 0.567$ **43.** f is called a damping function because as t increases, f approaches 0 as $e^{-0.1t}$ has the effect of damping out the linear factor $6 - 2t$.

45.

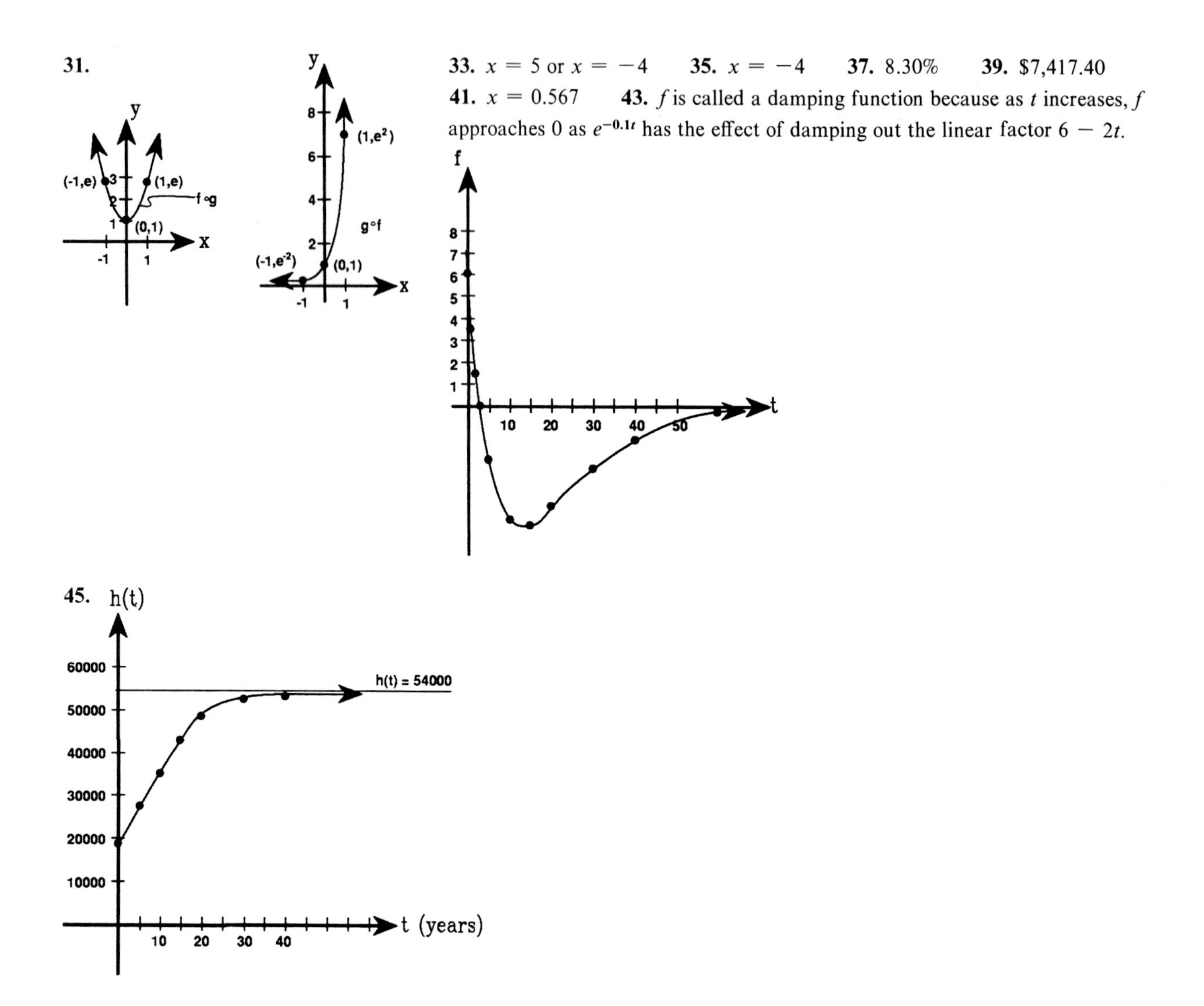

Section 6.3

1. 4 **3.** $\frac{1}{4}$ **5.** 4 **7.** 2 **9.** 1 **11.** $\frac{1}{3}$ **13.** $x = 32$ **15.** $b = 7$ **17.** $b = 10$ **19.** $6 < y < 7$

21. $3 < y < 4$ **23.** $4 < y < 5$ **25.** 6.4918531 **27.** 3.091667 **29.** 4.4231754 **31.** 9 **33.** -1

35. $\frac{6}{5}$

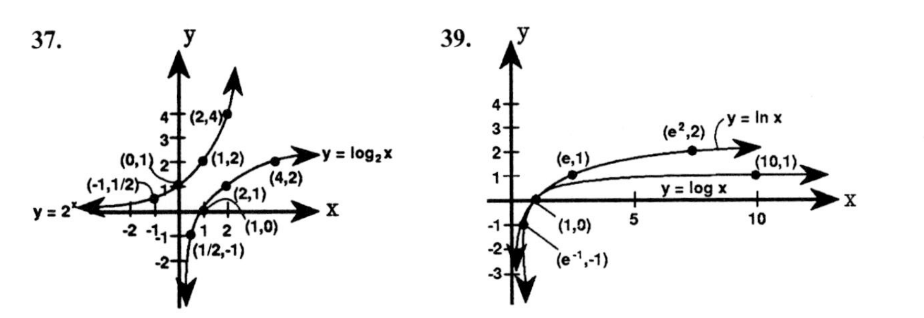

41. $\log_b x + \log_b y + \log_b z$ **43.** $\log_b x + \log_b y - \log_b z$ **45.** $\frac{1}{3}\log_b x - \frac{2}{3}\log_b y$ **47.** $\log_b\left(\frac{xz^6}{y}\right)$

49. $\log_b\left(\frac{x^{2/3}}{y^3z^5}\right)$ **51.** $\log_b\left[\frac{(z-x)^3}{(x-y)^3(y-z)^6}\right]$

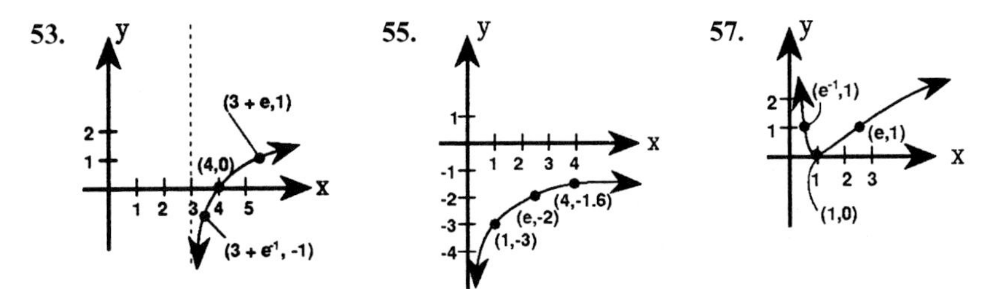

59. $\log_9 x = \frac{\log x}{\log 9} = \frac{\log x}{\log 3^2} = \frac{\log x}{2\log 3} = \frac{1}{2}\log_3 x$ **61.** $x = 125$ **63.** Let $m = \log_b b$, then $b^m = b$ and $m = 1$.
65. $f(x) = \log_b x$ and $g(x) = x^2 - 4$ **67.** $f(x) = e^x$ and $g(x) = \log_b 2x$ **69.** x^2 **71.** $3x$ **73.** $y = 10^4x$
75. $y = 1{,}000x$

Section 6.4

1. $x = 97$ **3.** $x = \frac{e^{3.8}+6}{2} \approx 25.35$ **5.** $x = 3$ **7.** $\emptyset$ **9.** $x = \frac{5+3\sqrt{19}}{6}$ **11.** $x = \frac{\log 28}{\log 5} - 2 \approx 0.07$

13. $x = \pm\sqrt{\frac{\log 9}{\log 2}} \approx \pm 1.78$ **15.** $\emptyset$ **17.** $x = 2 - \ln 2 \approx 1.31$ **19.** $x = \frac{\log \frac{76}{126^2}}{\log \frac{126}{76}} \approx -10.57$

21. Freeway traffic is almost a third of a million times louder than suburban street traffic. **23.** 2,378 years old
25. 8.67 years **27.** 15,625; 2,011 **29.** 69.66 years **31.** 0.0092 sec. **33.** About 63 times stronger
35. \$73.74 **37.** 8.65 hours **39.** 13.84 minutes **41.** About 7.2 days **43.** $\{1, 100\}$ **45.** $x = 10{,}000{,}000{,}000$
47. $x = 3.146$ **49.** 0.0028879 ($x = 0.5$); 0.0516152 ($x = 1$); 0.2941891 ($x = 1.5$); 1.0557228 ($x = 2$)

Chapter 6 Review Exercises

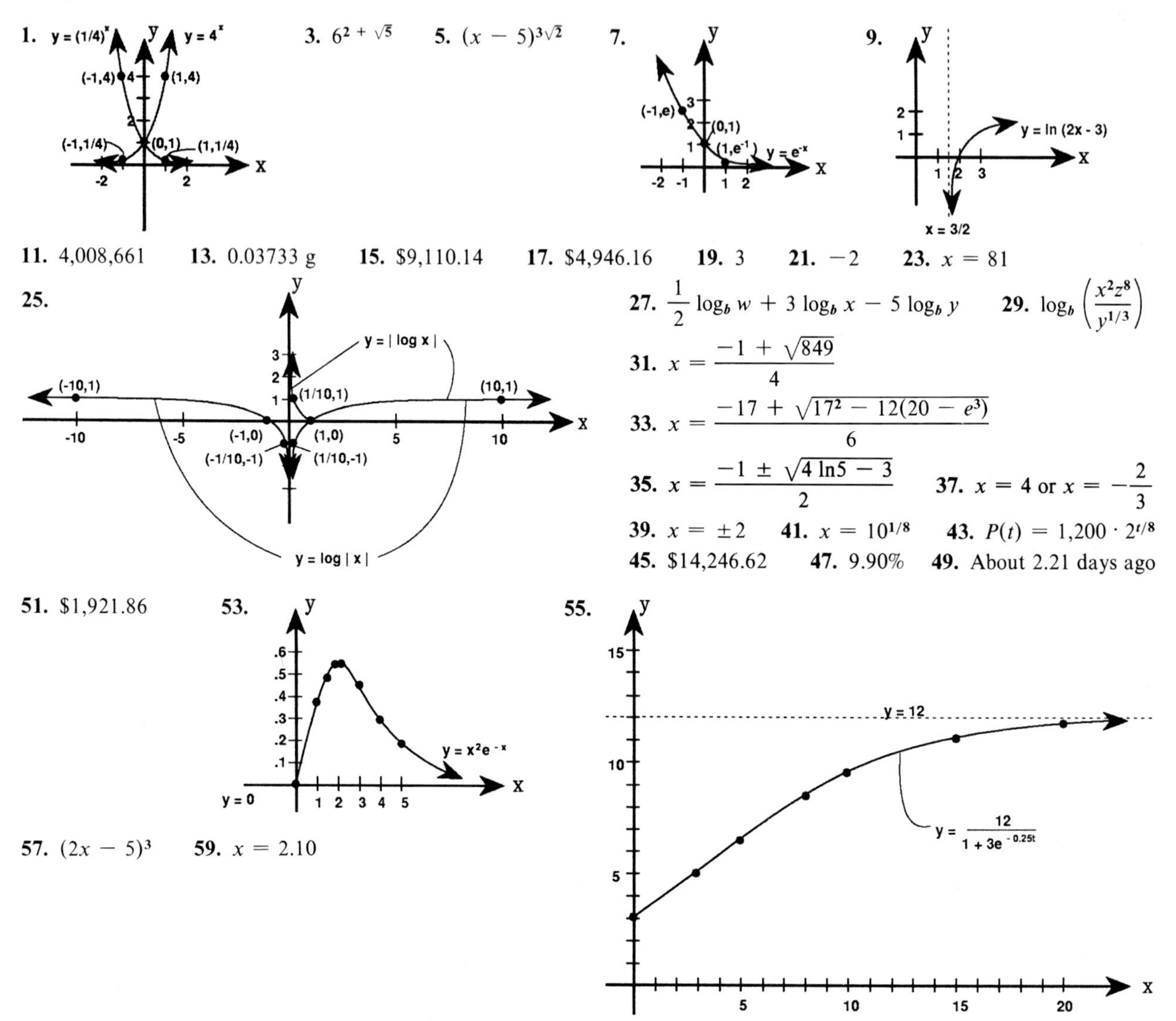

3. $6^{2+\sqrt{5}}$ **5.** $(x - 5)^{3\sqrt{2}}$

11. 4,008,661 **13.** 0.03733 g **15.** \$9,110.14 **17.** \$4,946.16 **19.** 3 **21.** -2 **23.** $x = 81$

27. $\frac{1}{2}\log_b w + 3\log_b x - 5\log_b y$ **29.** $\log_b\left(\frac{x^2z^8}{y^{1/3}}\right)$

31. $x = \frac{-1 + \sqrt{849}}{4}$

33. $x = \frac{-17 + \sqrt{17^2 - 12(20 - e^3)}}{6}$

35. $x = \frac{-1 \pm \sqrt{4\ln 5 - 3}}{2}$ **37.** $x = 4$ or $x = -\frac{2}{3}$

39. $x = \pm 2$ **41.** $x = 10^{1/8}$ **43.** $P(t) = 1{,}200 \cdot 2^{t/8}$

45. \$14,246.62 **47.** 9.90% **49.** About 2.21 days ago

51. \$1,921.86

57. $(2x - 5)^3$ **59.** $x = 2.10$

Chapter 6 Test 1

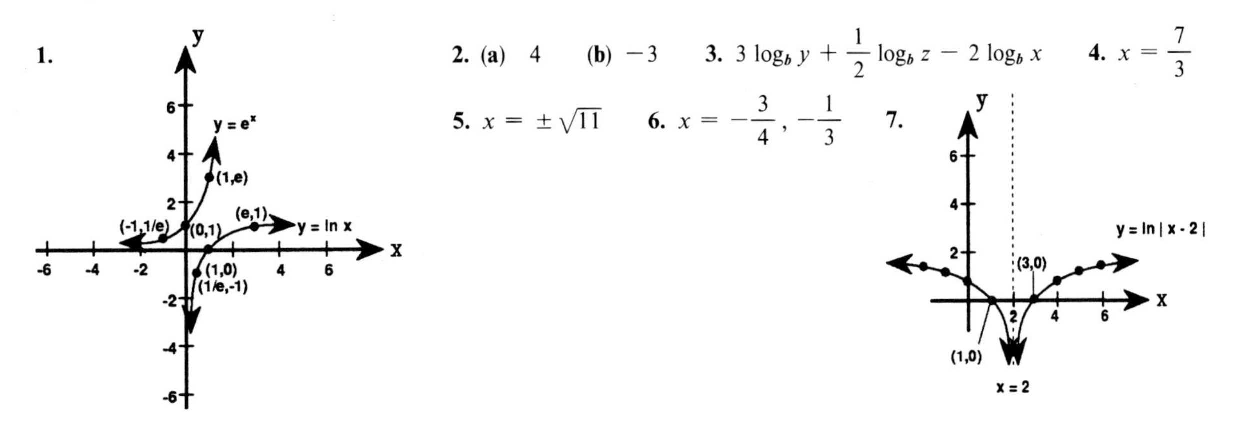

1. (graph) **2.** **(a)** 4 **(b)** -3 **3.** $3\log_b y + \frac{1}{2}\log_b z - 2\log_b x$ **4.** $x = \frac{7}{3}$

5. $x = \pm\sqrt{11}$ **6.** $x = -\frac{3}{4}, -\frac{1}{3}$ **7.** (graph)

8. About 23.45 hours **9.** 13.94% **10.** 1.31%

Chapter 6 Test 2

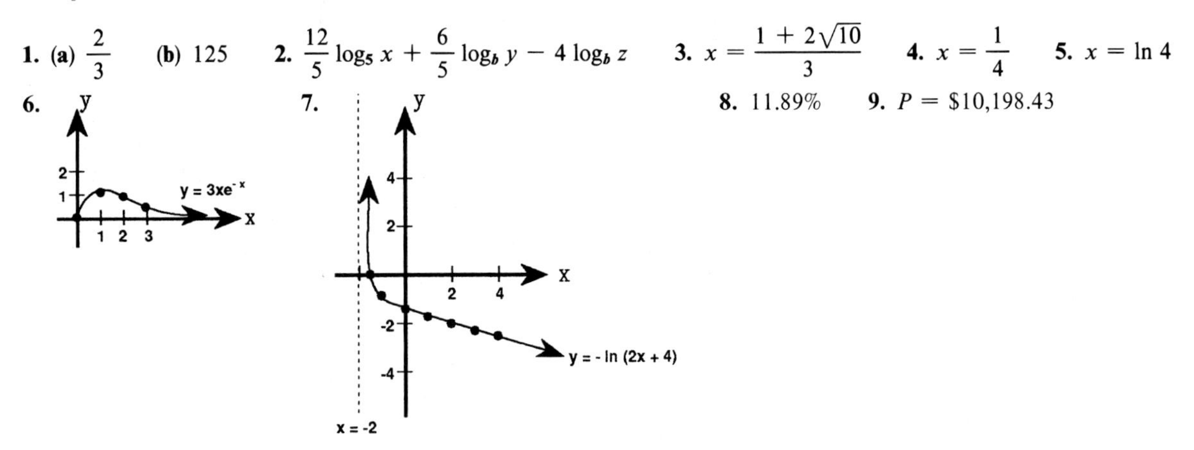

1. **(a)** $\frac{2}{3}$ **(b)** 125 **2.** $\frac{12}{5}\log_5 x + \frac{6}{5}\log_b y - 4\log_b z$ **3.** $x = \frac{1 + 2\sqrt{10}}{3}$ **4.** $x = \frac{1}{4}$ **5.** $x = \ln 4$

6. (graph) **7.** (graph) **8.** 11.89% **9.** $P = \$10{,}198.43$

10. 50 people originally; 5,000 people; 17.86 days

Section 7.1

1. $\frac{\sqrt{3}}{2}$ **3.** **(a)** $-\frac{1}{2}$ **(b)** $-\frac{1}{\sqrt{2}}$ **(c)** $\frac{-\sqrt{3}}{2}$ **(d)** -1 **5.** **(a)** 1 **(b)** $\frac{\sqrt{3}}{2}$ **(c)** $\frac{1}{\sqrt{2}}$ **(d)** $\frac{1}{2}$ **(e)** 0

7. **(a)** 1 **(b)** $\frac{\sqrt{3}}{2}$ **(c)** $\frac{1}{\sqrt{2}}$ **(d)** $\frac{1}{2}$ **(e)** 0 **9.** **(a)** $-\frac{\sqrt{3}}{2}$ **(b)** $-\frac{1}{\sqrt{2}}$ **(c)** $-\frac{1}{2}$ **(d)** 0

11. **(a)** 0 **(b)** $-\frac{1}{2}$ **(c)** $-\frac{1}{\sqrt{2}}$ **(d)** $-\frac{\sqrt{3}}{2}$ **(e)** -1 **13.** **(a)** $5\sqrt{3}$ cm **(b)** 0 cm

(c) $\frac{-10}{\sqrt{2}}$ cm **(d)** 0 cm

15. $f(x) = x - n,\ n - 1 \le x < n,\ n = 1, 2, 3, 4, 5, 6$

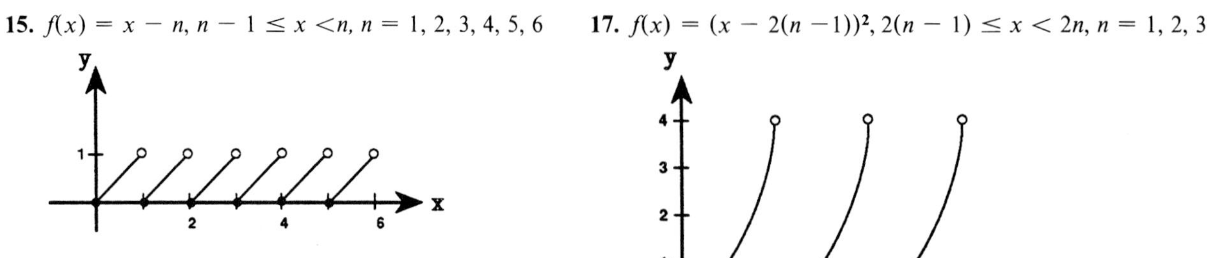

17. $f(x) = (x - 2(n - 1))^2,\ 2(n - 1) \le x < 2n,\ n = 1, 2, 3$

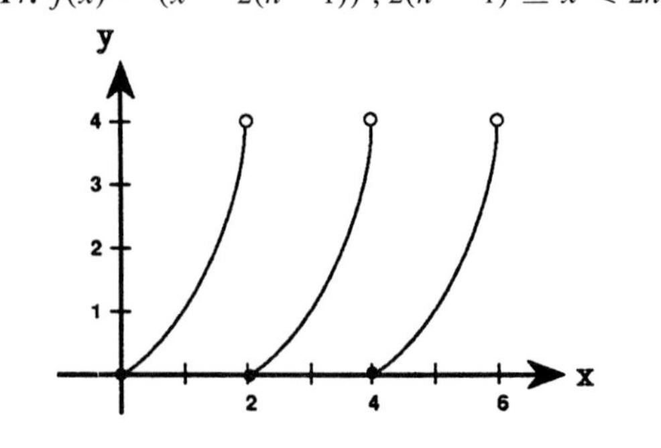

19. Periodic with period 2

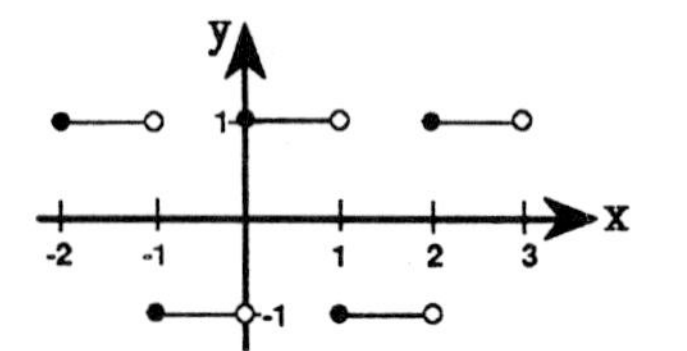

21. Not periodic

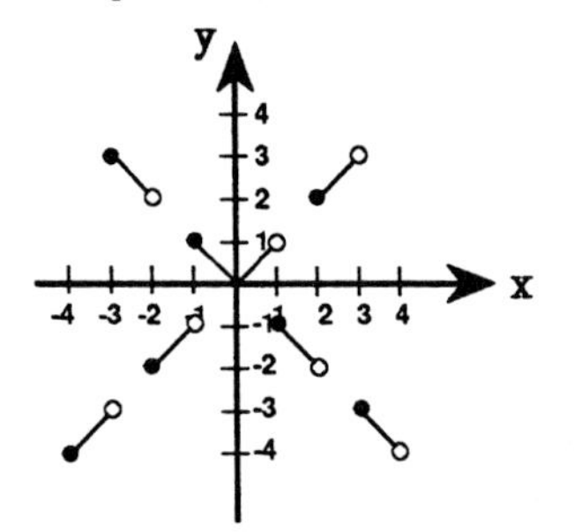

23. Periodic with period 1

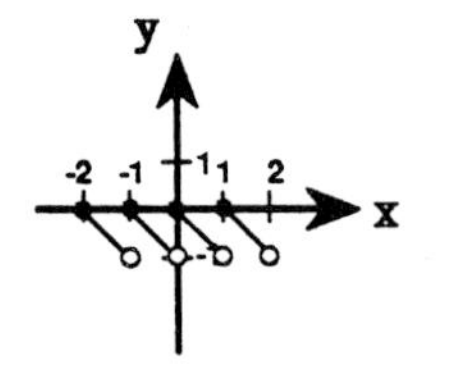

25. No. A periodic function cannot be a 1:1 function. There is more than one domain element associated with each range element.

27.
$$
\begin{aligned}
(f + g)(x + p) &= f(x + p) + g(x + p) \\
&= f(x) + g(x) \\
&= (f + g)(x)
\end{aligned}
$$

Section 7.2

1. $\dfrac{11\pi}{36}$ **3.** $\dfrac{13\pi}{18}$ **5.** $\dfrac{103\pi}{45}$ **7.** $\dfrac{\pi}{5}$ **9.** $135°$ **11.** $300°$ **13.** $\dfrac{360}{\pi} \approx 114.6°$ **15.** $\dfrac{1{,}404}{\pi} \approx 446.9°$

17. $\dfrac{4\pi}{3}$ inches **19.** $\dfrac{8\pi}{9}$ m **21.** 53 cm **23.** 340 ft.

25. II **27.** II **29.** IV

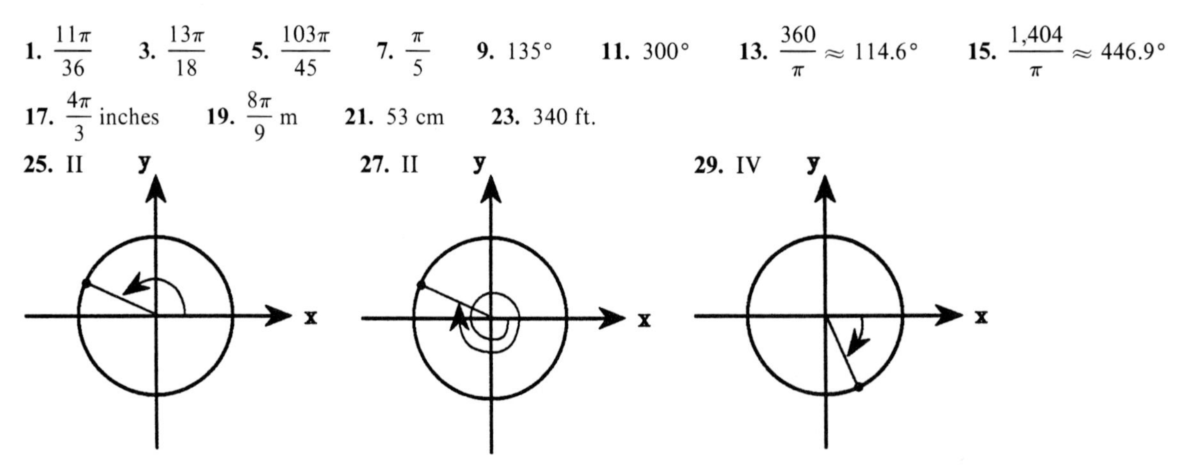

31. I

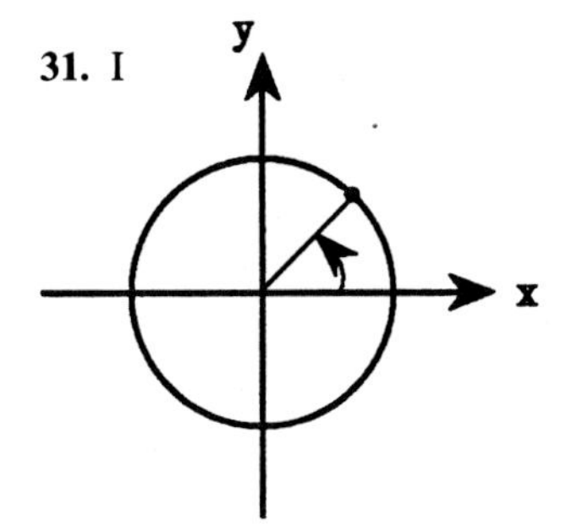

33. $\dfrac{A}{\pi r^2} = \dfrac{\theta}{2\pi}$ so, $A = \dfrac{r^2\theta}{2}$ **35.** 215.45 cm **37.** 57.15 cm **39.** 0.5

41. 64π radians/minute

Section 7.3

	$\sin\theta$	$\cos\theta$	$\tan\theta$	$\csc\theta$	$\sec\theta$	$\cot\theta$
1.	$\frac{3}{4}$	$\frac{\sqrt{7}}{4}$	$\frac{3}{\sqrt{7}}$	$\frac{4}{3}$	$\frac{4}{\sqrt{7}}$	$\frac{\sqrt{7}}{3}$
3.	$\frac{1}{\sqrt{5}}$	$\frac{2}{\sqrt{5}}$	$\frac{1}{2}$	$\sqrt{5}$	$\frac{\sqrt{5}}{2}$	2
5.	$\frac{5}{13}$	$\frac{12}{13}$	$\frac{5}{12}$	$\frac{13}{5}$	$\frac{13}{12}$	$\frac{12}{5}$
7.	$\frac{2}{5}$	$\frac{\sqrt{21}}{5}$	$\frac{2}{\sqrt{21}}$	$\frac{5}{2}$	$\frac{5}{\sqrt{21}}$	$\frac{\sqrt{21}}{2}$
9.	$\frac{1}{\sqrt{10}}$	$\frac{3}{\sqrt{10}}$	$\frac{1}{3}$	$\sqrt{10}$	$\frac{\sqrt{10}}{3}$	3
11.	$\frac{\sqrt{7}}{4}$	$\frac{3}{4}$	$\frac{\sqrt{7}}{3}$	$\frac{4}{\sqrt{7}}$	$\frac{4}{3}$	$\frac{3}{\sqrt{7}}$

13. $\frac{1}{\sqrt{2}}$ **15.** $-\sqrt{3}$ **17.** $-\frac{1}{2}$ **19.** $-\frac{2}{\sqrt{3}}$ **21.** -1 **23.** $-\frac{2}{\sqrt{3}}$ **25.** II **27.** I **29.** III **31.** $\frac{\pi}{4}$

33. $\frac{\pi}{3}$ **35.** $\frac{\pi}{2}$ **37.** $\frac{11\pi}{6}$

	$\sin\theta$	$\cos\theta$	$\tan\theta$	$\csc\theta$	$\sec\theta$	$\cot\theta$
39.	$\frac{3}{10}$	$-\frac{\sqrt{91}}{10}$	$-\frac{3}{\sqrt{91}}$	$\frac{10}{3}$	$-\frac{10}{\sqrt{91}}$	$-\frac{\sqrt{91}}{3}$
41.	$\frac{2}{\sqrt{5}}$	$-\frac{1}{\sqrt{5}}$	-2	$\frac{\sqrt{5}}{2}$	$-\sqrt{5}$	$-\frac{1}{2}$
43.	$\frac{\sqrt{7}}{4}$	$-\frac{3}{4}$	$-\frac{\sqrt{7}}{3}$	$\frac{4}{\sqrt{7}}$	$-\frac{4}{3}$	$-\frac{3}{\sqrt{7}}$
45.	$-\frac{\sqrt{3}}{2}$	$-\frac{1}{2}$	$\sqrt{3}$	$-\frac{2}{\sqrt{3}}$	-2	$\frac{1}{\sqrt{3}}$
47.	$-\frac{5}{\sqrt{26}}$	$-\frac{1}{\sqrt{26}}$	5	$-\frac{\sqrt{26}}{5}$	$-\sqrt{26}$	$\frac{1}{5}$
49.	$-\frac{12}{13}$	$-\frac{5}{13}$	$\frac{12}{5}$	$-\frac{13}{12}$	$-\frac{13}{5}$	$\frac{5}{12}$

51. Drop a perpendicular from P to the x-axis to obtain a point A on the x-axis, and drop a perpendicular from Q to the x-axis to obtain point B. Triangles PAO and QBO are congruent by a.s.a. Angle POA = angle $QOB = \pi - \theta$. Angle A and angle B are equal because they are right angles. Hence angle BQO = angle APO. Finally $OQ = OP$ because they are radii in the same circle. Now the length of AP equals the length of BQ, and the length of OA equals the length of OB. Thus Q is a reflection of P in the y-axis.

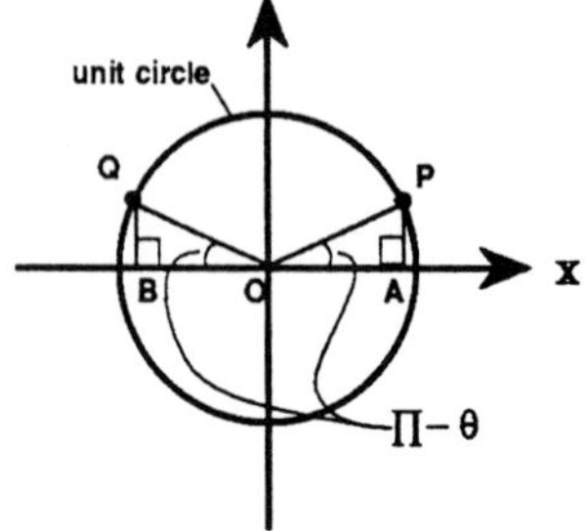

53. Triangles PAO and QAO are congruent. Therefore Q is a reflection of P in the x-axis. If (x, y) are the coordinates of P, the $(x, -y)$ are the coordinates of Q. Thus, $\sin\theta = -y = -\sin(2\pi - \theta)$, $\cos\theta = x = \cos(2\pi - \theta)$, and $\tan\theta = -\dfrac{y}{x} = -\tan(2\pi - \theta)$.

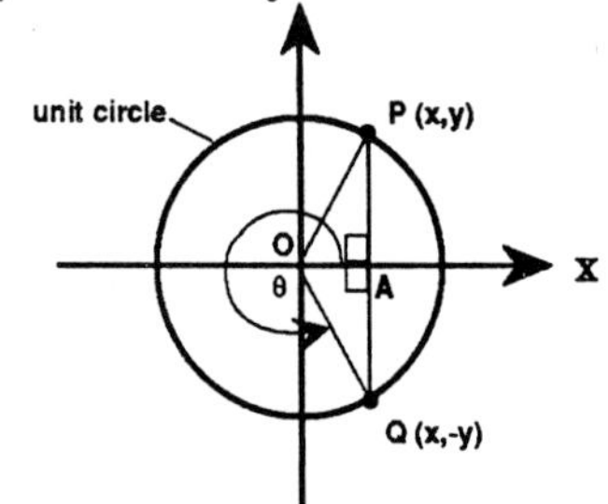

Section 7.4

1. $\csc\theta = \dfrac{3}{2}$ **3.** $\cot\theta = \dfrac{1}{4}$ **5.** $\cos\theta = \dfrac{1}{2}$ **7.** $-\dfrac{\sqrt{5}}{3}$ **9.** $-\dfrac{3\sqrt{7}}{8}$

	$\sin\theta$	$\cos\theta$	$\tan\theta$	$\csc\theta$	$\sec\theta$	$\cot\theta$
11.	$\frac{1}{4}$	$\frac{\sqrt{15}}{4}$	$\frac{1}{\sqrt{15}}$	4	$\frac{4}{\sqrt{15}}$	$\sqrt{15}$
13.	$-\frac{\sqrt{5}}{3}$	$\frac{2}{3}$	$-\frac{\sqrt{5}}{2}$	$-\frac{3}{\sqrt{5}}$	$\frac{3}{2}$	$-\frac{2}{\sqrt{5}}$
15.	$\frac{2}{\sqrt{5}}$	$-\frac{1}{\sqrt{5}}$	-2	$\frac{\sqrt{5}}{2}$	$-\sqrt{5}$	$-\frac{1}{2}$

17. Points $P(-x, -y)$ and $Q(-x, y)$ are reflections of each other with respect to the x-axis because triangles OQA and OPA are congruent. Hence, $\sin(-\theta) = -y = -\sin\theta$, $\cos(-\theta) = -x = \cos\theta$, and $\tan(-\theta) = \dfrac{-y}{-x} = -\left(\dfrac{y}{-x}\right) = -\tan\theta$.

19. $\sin(\beta + 6\pi) = \sin(\beta + 3 \cdot 2\pi) = \sin\beta = y$ **21.** $\dfrac{1}{\sqrt{2}}$ **23.** -1 **25.** 1

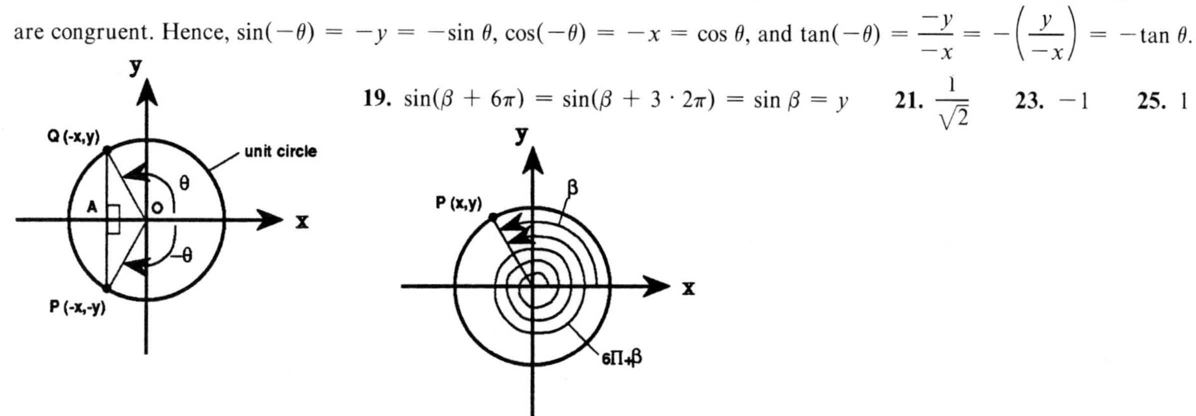

27. $\sqrt{2}$ **29.** $\sqrt{3}$ **31.** 1 **33.** $-0.0731, -0.0699$ **35.** 1.3589, 1.3409 **37.** 0.9310, 0.9287

39. 0.0060, 0.0108 **41.** -1.0881, 1.0903 **43.** $-1.5435, -1.5523$ **45.** $\tan\theta = \dfrac{y}{x} = \cot\left(\dfrac{\pi}{2} - \theta\right)$

47. Triangles OPA and OQB are congruent. Therefore $\overline{PA} = \overline{QB}$ and $\overline{OA} = \overline{OB}$. But the negative of $\overline{PA}$ is the x-coordinate of P and the negative of $\overline{QB}$ is the y-coordinate of Q. Thus, the x-coordinate of P equals the y-coordinate of Q. Similarly, the y-coordinate of P equals the x-coordinate of Q (see figure). Now $\sin\theta = y = \cos\left(\dfrac{\pi}{2} - \theta\right)$.

49. $\cot\beta = \cot\left(\dfrac{\pi}{2} - \left(\dfrac{\pi}{2} - \beta\right)\right) = \tan\left(\dfrac{\pi}{2} - \beta\right)$

Section 7.5

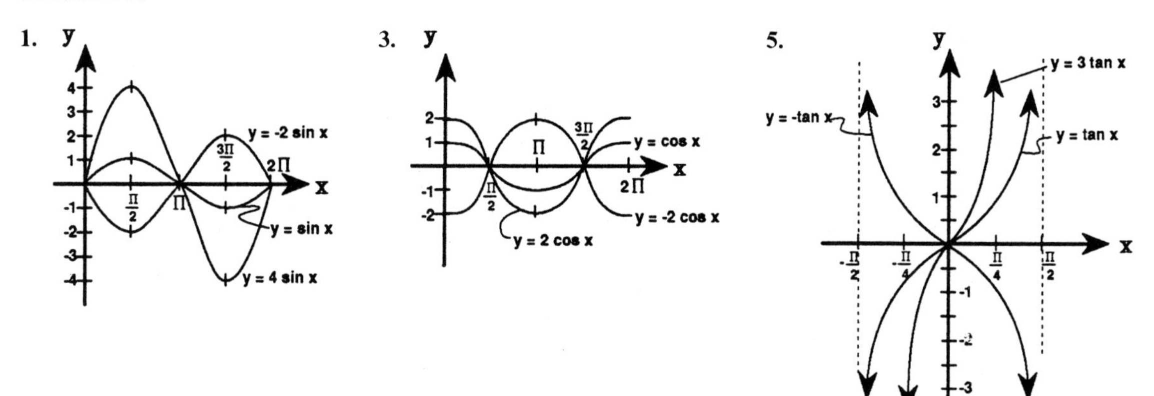

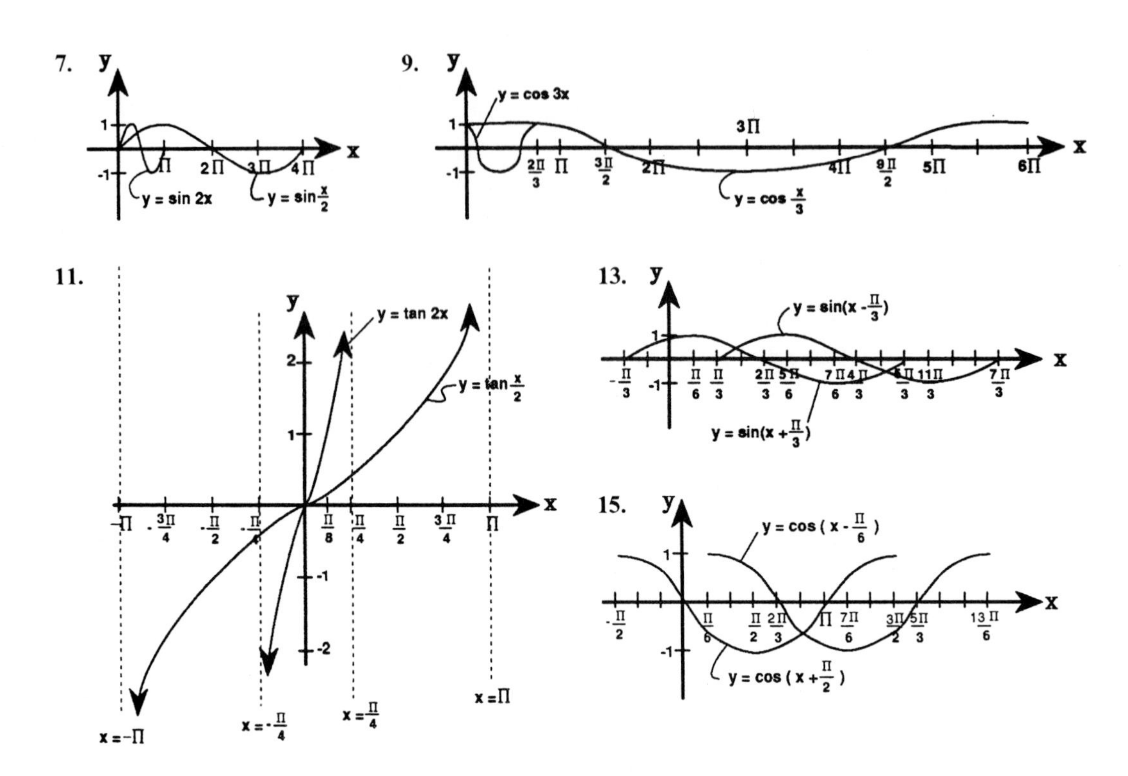
7.
y = sin 2x
y = sin x/2
9.
y = cos 3x
3Π
y = cos x/3
11.
y = tan 2x
y = tan x/2
x = −Π
x = −Π/4
x = Π/4
x = Π
13.
y = sin(x − Π/3)
y = sin(x + Π/3)
15.
y = cos (x − Π/6)
y = cos (x + Π/2)

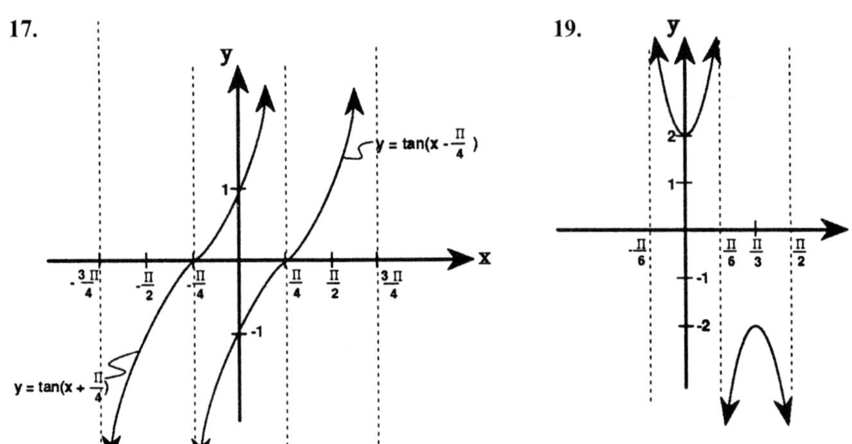
17.
y = tan(x − Π/4)
y = tan(x + Π/4)
19.

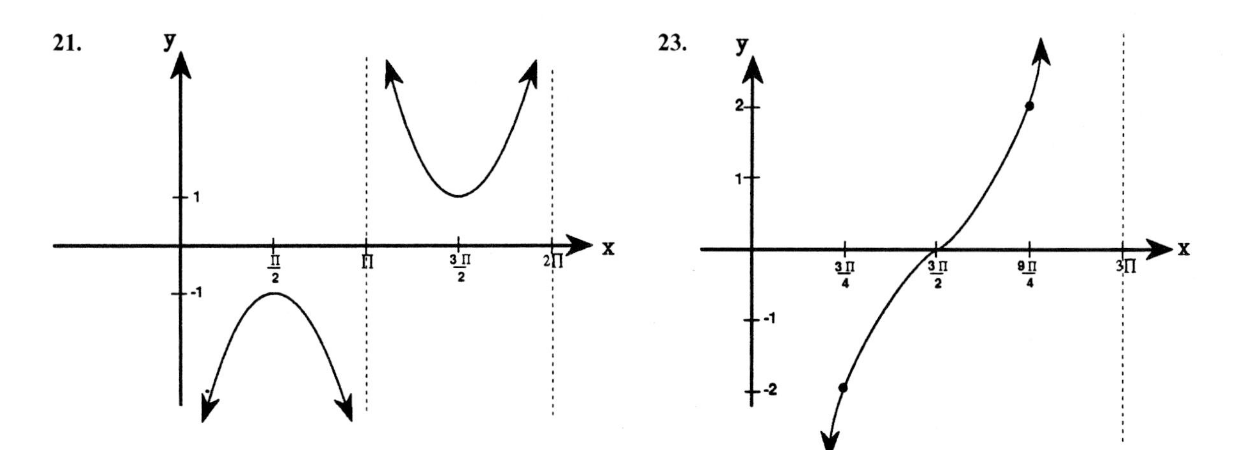

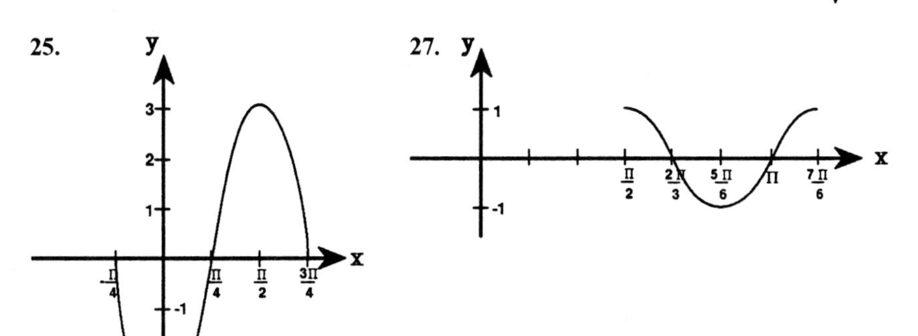

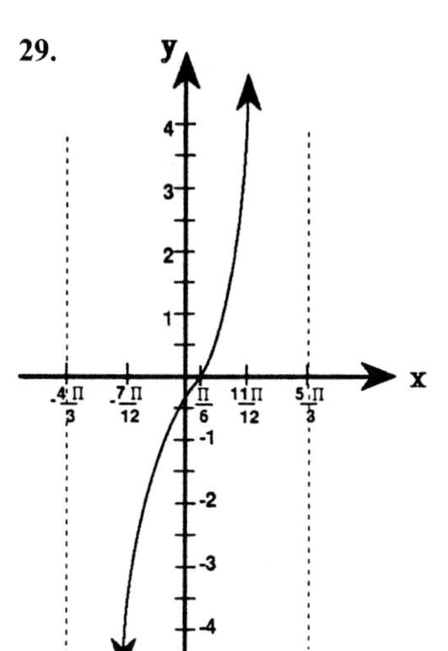

31. $y = 2 \sin 4\left(x - \frac{\pi}{3}\right)$ **33.** $y = \frac{1}{2} \cos 2\left(x + \frac{\pi}{4}\right)$

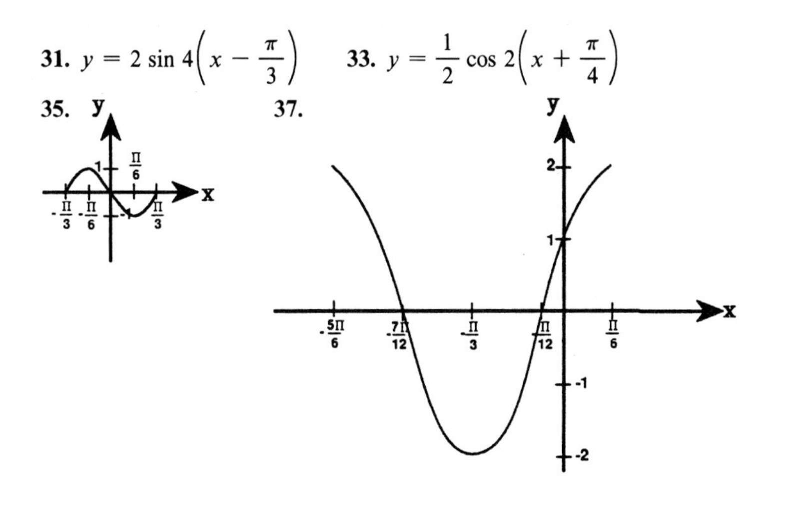

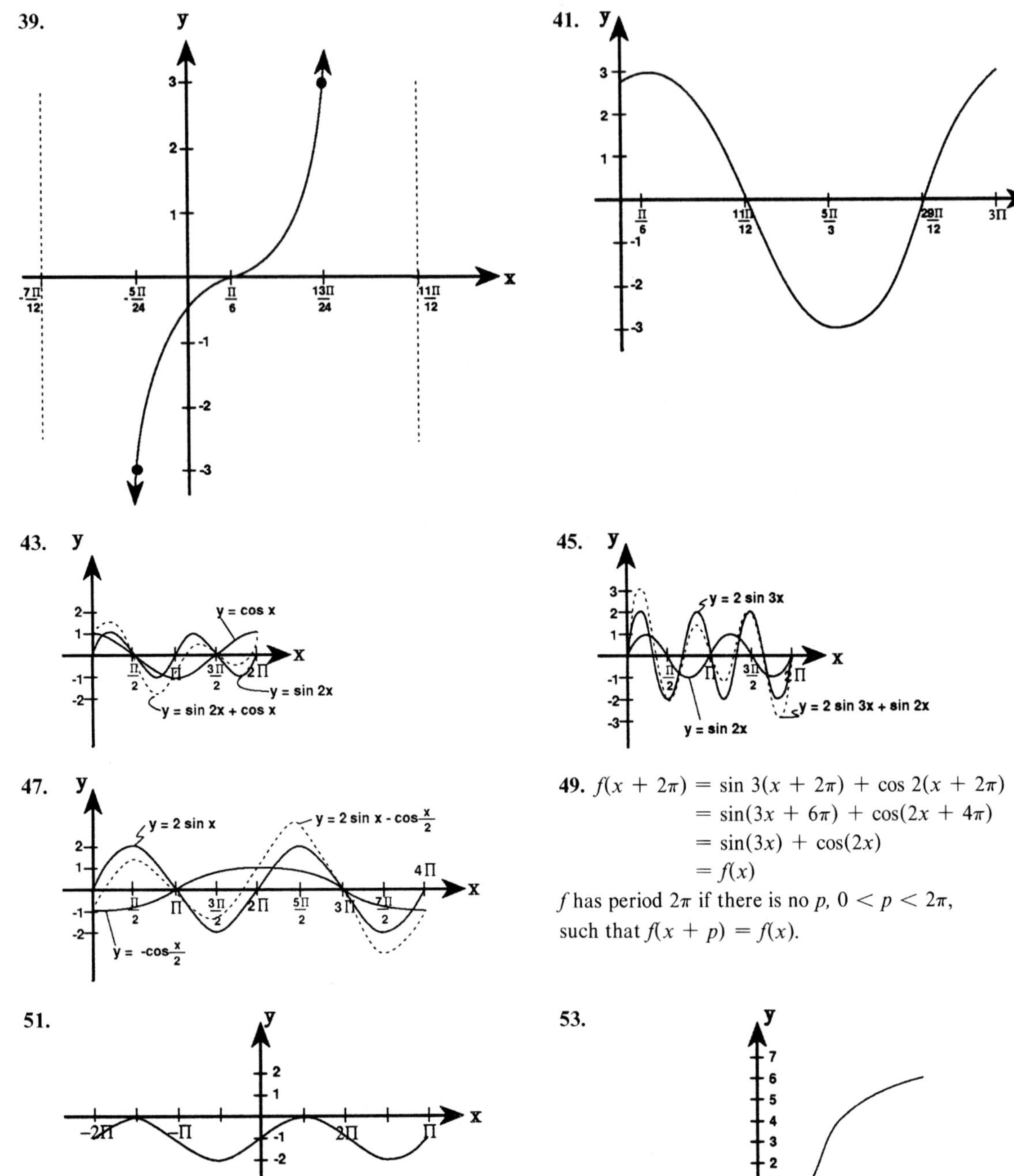

49. $f(x + 2\pi) = \sin 3(x + 2\pi) + \cos 2(x + 2\pi)$
$= \sin(3x + 6\pi) + \cos(2x + 4\pi)$
$= \sin(3x) + \cos(2x)$
$= f(x)$

f has period 2π if there is no p, $0 < p < 2\pi$, such that $f(x + p) = f(x)$.

53.

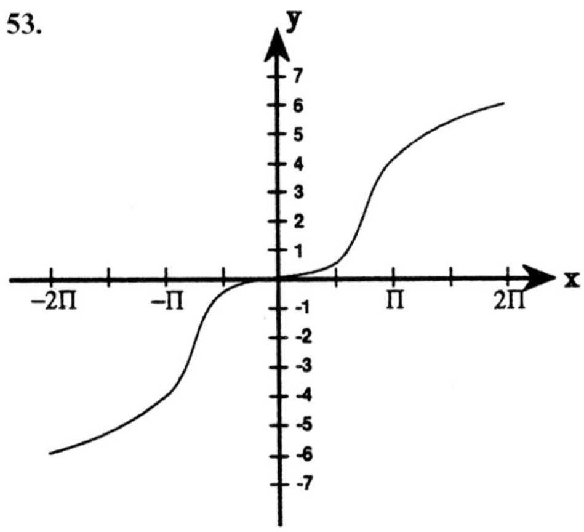

55.

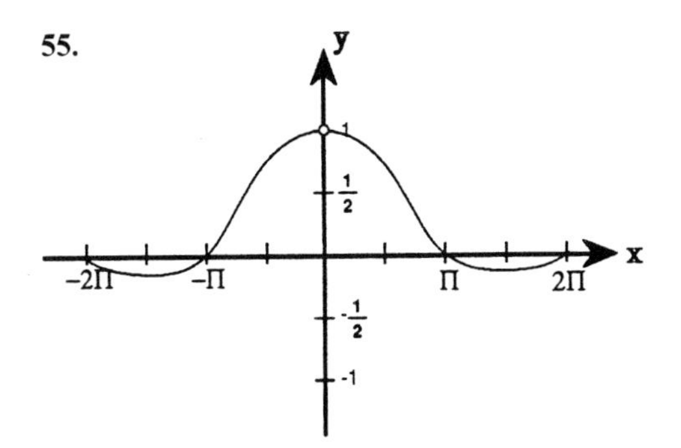

Section 7.6

1. π **3.** $\frac{\pi}{3}$ **5.** $\frac{\pi}{3}$ **7.** $-\frac{\pi}{3}$ **9.** $-\frac{\pi}{3}$ **11.** $\frac{\pi}{2}$ **13.** 0.930 **15.** -1.223 **17.** 2.450 **19.** 1.490

21. 0.193 **23.** $\frac{\pi}{6}$ **25.** $\frac{1}{2}$ **27.** $-\frac{3}{2}$ **29.** $\frac{5}{\sqrt{26}}$ **31.** $\frac{12}{13}$ **33.** $\frac{\sqrt{35}}{6}$ **35.** $\frac{3\sqrt{7}}{8}$

37. **39.** **41.** **43.**

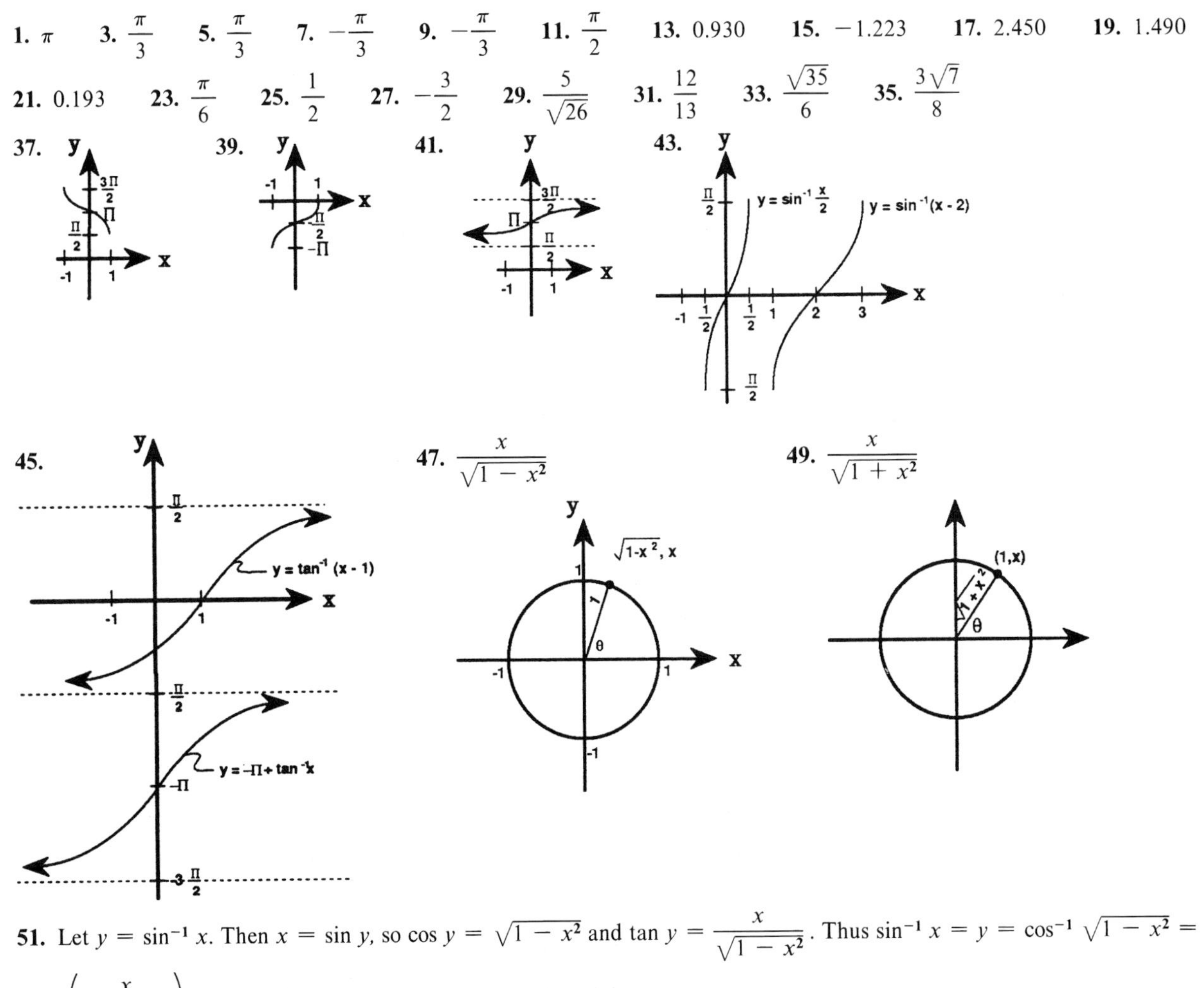

45.

47. $\frac{x}{\sqrt{1 - x^2}}$ **49.** $\frac{x}{\sqrt{1 + x^2}}$

51. Let $y = \sin^{-1} x$. Then $x = \sin y$, so $\cos y = \sqrt{1 - x^2}$ and $\tan y = \frac{x}{\sqrt{1 - x^2}}$. Thus $\sin^{-1} x = y = \cos^{-1} \sqrt{1 - x^2} =$

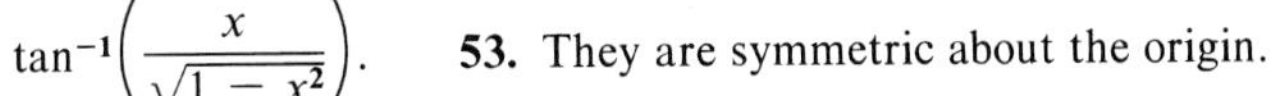

$\tan^{-1}\left(\frac{x}{\sqrt{1 - x^2}}\right)$. **53.** They are symmetric about the origin.

Chapter 7 Review Exercises

1. $\frac{4\pi}{15}$ **3.** $120°$ **5.** 10.56 cm **7.** 1 **9.** $-\sqrt{3}$ **11.** $-\frac{1}{\sqrt{2}}$ **13.** $-\frac{\sqrt{3}}{2}$ **15.** $\frac{\pi}{4}$ **17.** $\frac{\pi}{6}$ **19.** $\frac{3\pi}{4}$

21. $\frac{\pi}{3}$ **23.** $\frac{\pi}{4}$ **25.** $-\frac{1}{5}$ **27.** III **29.** III

	sin θ	cos θ	tan θ	csc θ	sec θ	cot θ
31.	$\frac{3}{5}$	$\frac{4}{5}$	$\frac{3}{4}$	$\frac{5}{3}$	$\frac{5}{4}$	$\frac{4}{3}$
33.	$-\frac{15}{17}$	$-\frac{8}{17}$	$\frac{15}{8}$	$-\frac{17}{15}$	$-\frac{17}{8}$	$\frac{8}{15}$
35.	$-\frac{8}{17}$	$\frac{15}{17}$	$-\frac{8}{15}$	$-\frac{17}{8}$	$\frac{17}{15}$	$-\frac{15}{8}$
37.	$-\frac{\sqrt{5}}{3}$	$-\frac{2}{3}$	$\frac{\sqrt{5}}{2}$	$-\frac{3}{\sqrt{5}}$	$-\frac{3}{2}$	$\frac{2}{\sqrt{5}}$

39. −0.781 **41.** 0.170 **43.** 0.315 **45.** 3.690 **47.** 0.804 **49.** 1.855

51. **53.** **55.** **57.** **59.**

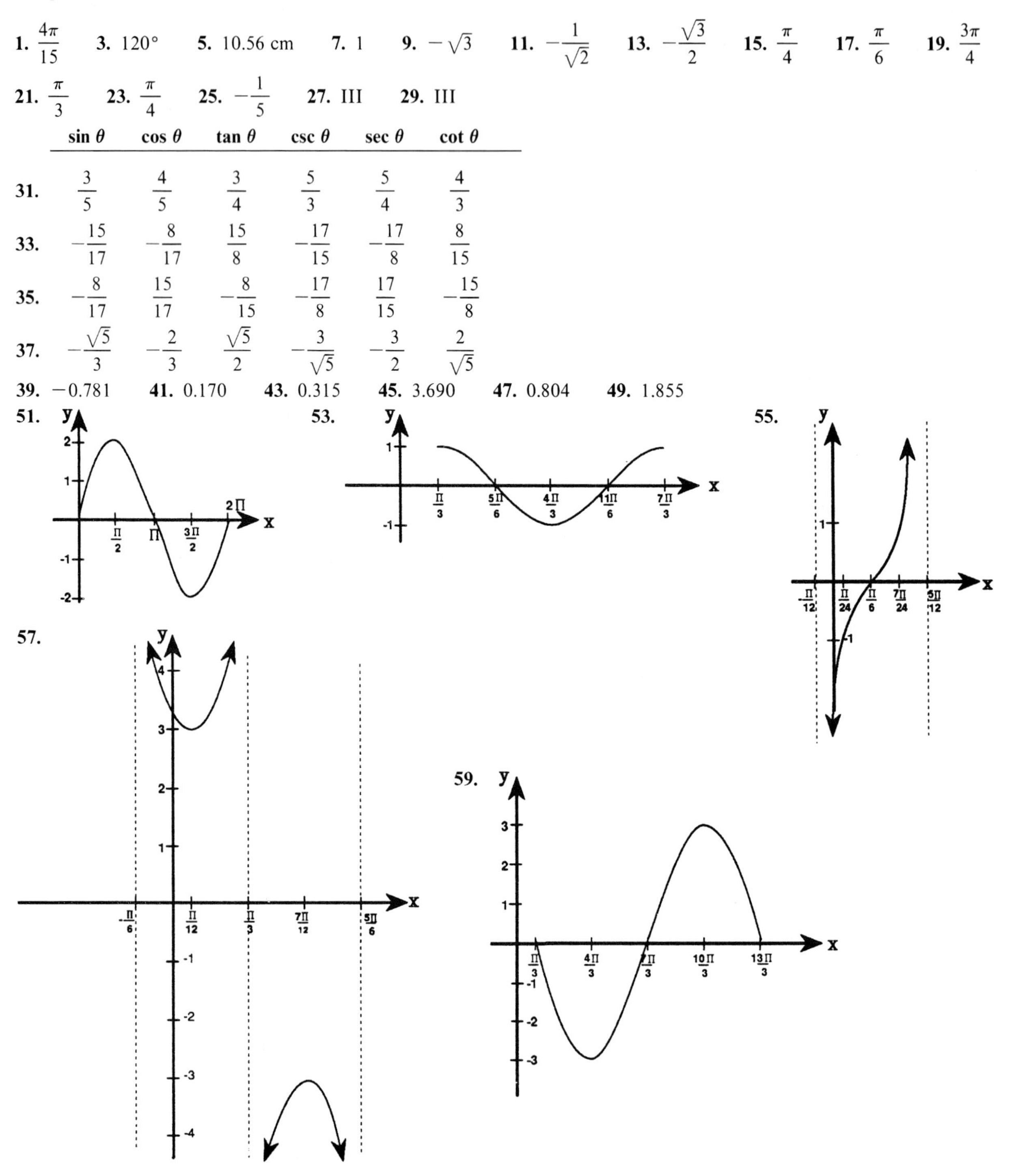

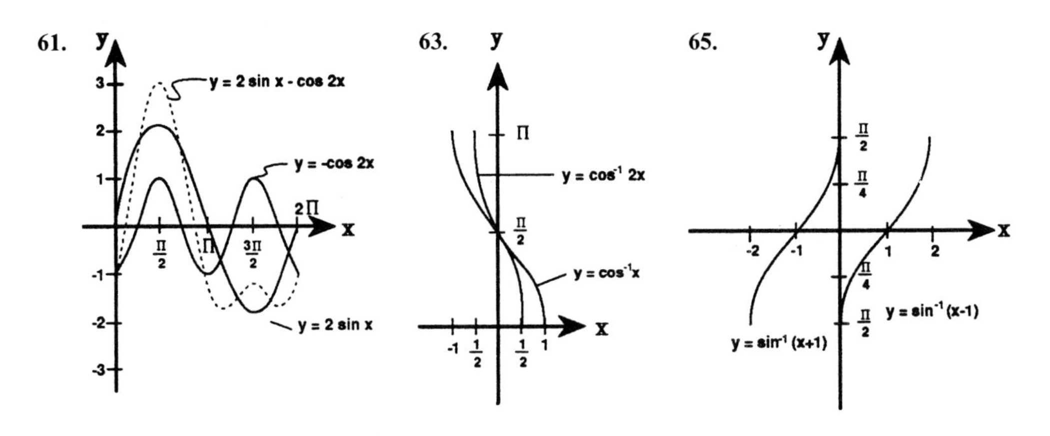

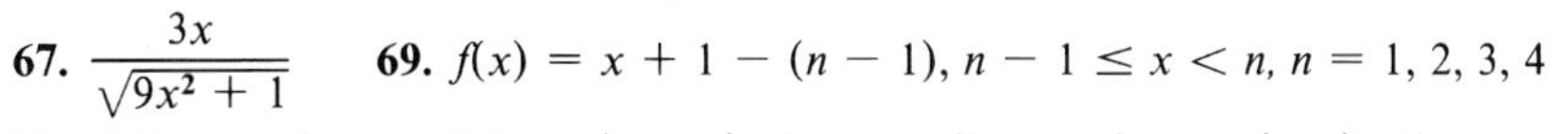

67. $\dfrac{3x}{\sqrt{9x^2 + 1}}$ 69. $f(x) = x + 1 - (n - 1), n - 1 \leq x < n, n = 1, 2, 3, 4$

71. If the coordinates of P are $(-x, y)$, then coordinates of Q are (x, y). Thus, $\cos \theta = -x = -\cos(\pi - \theta)$.

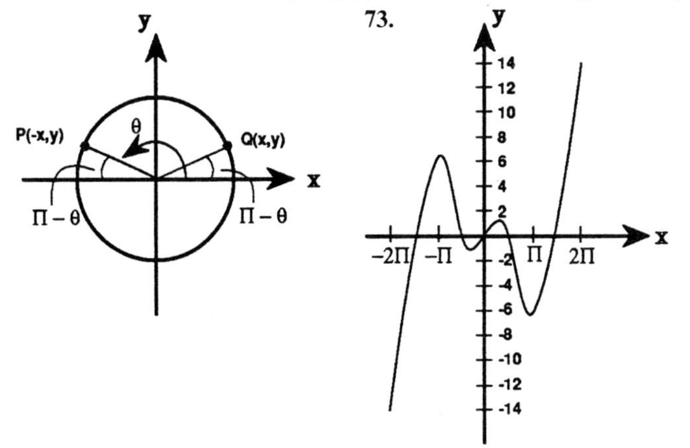

Chapter 7 Test 1

1. **(a)** $\dfrac{\pi}{3}$ **(b)** $150°$ **2.** **(a)** 3.15 cm **(b)** $A = \dfrac{r^2\theta}{2}$ **3.** **(a)** III **(b)** II **4.** **(a)** $-\dfrac{1}{2}$ **(b)** $-\sqrt{3}$

5. **(a)** $\dfrac{\pi}{3}$ **(b)** $-\dfrac{\pi}{3}$ **6.** **(a)** $\dfrac{1}{3}$ **(b)** $-\dfrac{\pi}{4}$

7. $\sin \theta = -\dfrac{15}{17}$, $\tan \theta = -\dfrac{15}{8}$, $\csc \theta = -\dfrac{17}{15}$, $\sec \theta = \dfrac{17}{8}$, $\cot \theta = -\dfrac{8}{15}$

8. **9.** **10.** $\sqrt{1 - 4x^2}$

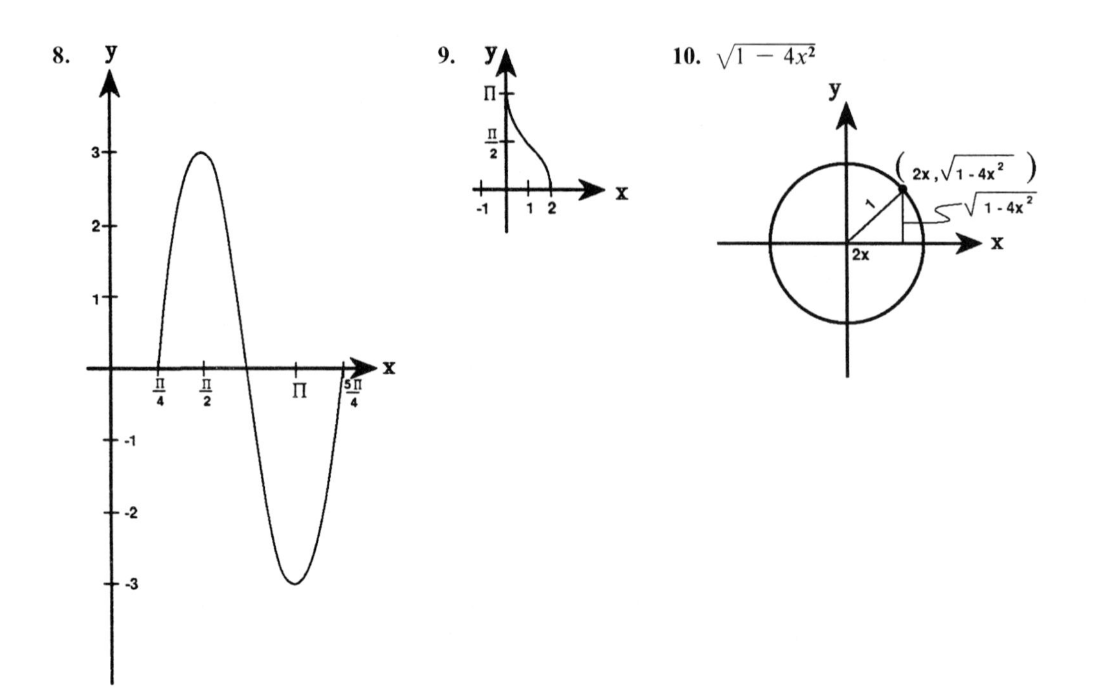

Chapter 7 Test 2

1. **(a)** $\frac{62\pi}{45}$ **(b)** $-\frac{468°}{\pi}$ **2.** **(a)** $\frac{5}{2}$ **(b)** 20 in.^{2} **3.** **(a)** $-\pi$ **(b)** $\frac{2\pi}{3}$ **4.** **(a)** $\frac{1}{5}$ **(b)** $2\pi - 6.4$

5. **(a)** 2.191 **(b)** -0.357 **6.** $\frac{x+2}{\sqrt{1-(x+2)^2}}$

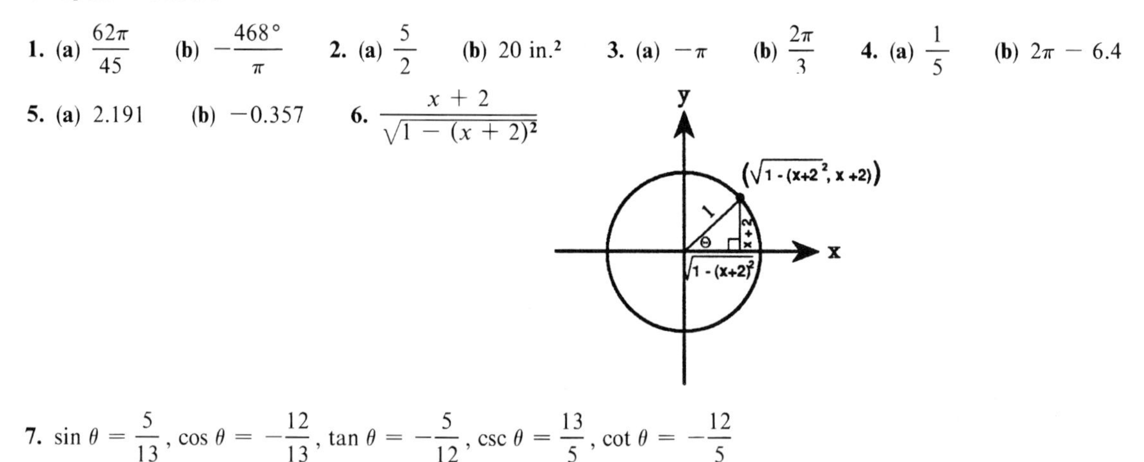

7. $\sin\theta = \frac{5}{13}$, $\cos\theta = -\frac{12}{13}$, $\tan\theta = -\frac{5}{12}$, $\csc\theta = \frac{13}{5}$, $\cot\theta = -\frac{12}{5}$

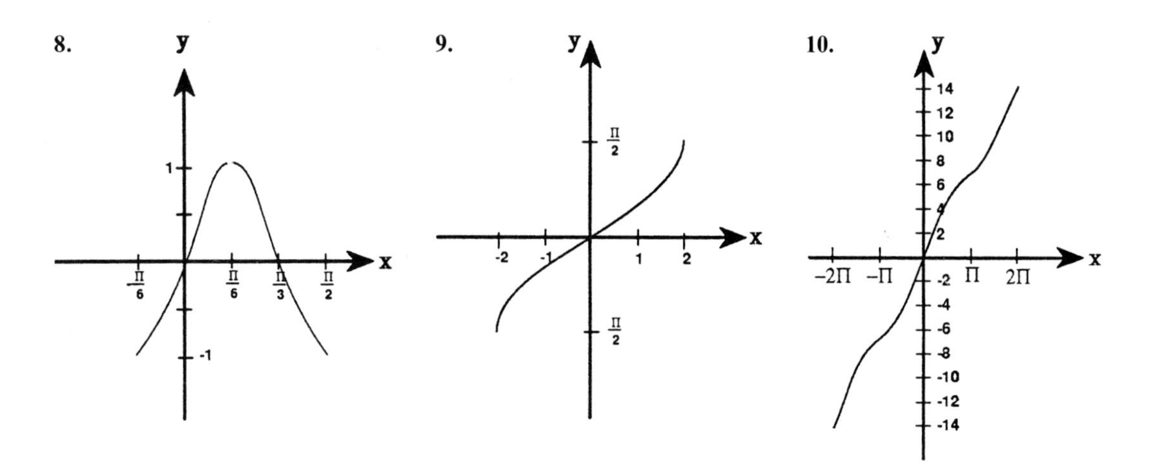

Section 8.1

1. $\tan^2 x + 1 = \dfrac{\sin^2 x}{\cos^2 x} + 1 = \dfrac{\sin^2 x + \cos^2 x}{\cos^2 x} = \dfrac{1}{\cos^2 x} = \sec^2 x$

3. $\sin(x - y) = \sin(x + (-y)) = \sin x \cos(-y) + \sin(-y) \cos x$
$= \sin x \cos y - \cos x \sin y$

5. $\tan (x + y) = \dfrac{\sin(x + y)}{\cos(x + y)} = \dfrac{\sin x \cos y + \cos x \sin y}{\cos x \cos y - \sin x \sin y}$
$= \dfrac{\dfrac{\sin x \cos y + \cos x \sin y}{\cos x \cos y}}{\dfrac{\cos x \cos y - \sin x \sin y}{\cos x \cos y}}$
$= \dfrac{\tan x + \tan y}{1 - \tan x \tan y}$

7. $\cos x \cos y = \dfrac{1}{2}((\cos x \cos y - \sin x \sin y) + (\cos x \cos y + \sin x \sin y))$
$= \dfrac{1}{2}(\cos (x + y) + \cos(x - y))$

9. $\sin x \sec x = (\sin x)\left(\dfrac{1}{\cos x}\right) = \tan x$

11. $\dfrac{\cos x}{1 - \sin x} = \dfrac{\cos x}{1 - \sin x} \cdot \dfrac{1 + \sin x}{1 + \sin x} = \dfrac{(\cos x)(1 + \sin x)}{1 - \sin^2 x}$
$= \dfrac{(\cos x)(1 + \sin x)}{\cos^2 x} = \dfrac{1 + \sin x}{\cos x}$

13. $\dfrac{\sec^2 x - 1}{\sec^2 x} = \dfrac{\tan^2 x}{\sec^2 x} = \dfrac{\dfrac{\sin^2 x}{\cos^2 x}}{\dfrac{1}{\cos^2 x}} = \sin^2 x$

15. $$\begin{aligned}(\csc^2 x - 1)(1 - \cos^2 x) &= (\cot^2 x)(\sin^2 x)\\ &= \frac{\cos^2 x}{\sin^2 x}\sin^2 x\\ &= \cos^2 x\\ &= \frac{1}{\sec^2 x}\end{aligned}$$

17. $$\begin{aligned}(\cot x - \csc x)^2 &= \cot^2 x - 2\cot x\csc x + \csc^2 x\\ &= \frac{\cos^2 x - 2\cos x + 1}{\sin^2 x}\\ &= \frac{(\cos x - 1)^2}{(\cos x - 1)(\cos x + 1)} = \frac{\cos x - 1}{\cos x + 1}\end{aligned}$$

19. $$\begin{aligned}\sin(x - \pi) &= \sin x\cos\pi - \sin\pi\cos x\\ &= (\sin x)(-1) - (0)\cos x\\ &= -\sin x\end{aligned}$$

21. $\cos(x + y) - \cos(x - y) = \cos x\cos y - \sin x\sin y - (\cos x\cos y + \sin x\sin y) = -2\sin x\sin y$

23. $$\begin{aligned}\cos 3x &= \cos(5x - 2x)\\ &= \cos 2x\cos 5x + \sin 2x\sin 5x\end{aligned}$$

25. $$\begin{aligned}\frac{\sin(x + y)}{\sin(x - y)} &= \frac{\sin x\cos y + \cos x\sin y}{\sin x\cos y - \cos x\sin y}\\ &= \frac{\dfrac{\sin x\cos y + \cos x\sin y}{\cos x\cos y}}{\dfrac{\sin x\cos y - \cos x\sin y}{\cos x\cos y}}\\ &= \frac{\tan x + \tan y}{\tan x - \tan y}\end{aligned}$$

27. $$\begin{aligned}\cos\frac{5x}{6} &= \cos\left(\frac{x}{2} + \frac{x}{3}\right)\\ &= \cos\frac{x}{2}\cos\frac{x}{3} - \sin\frac{x}{2}\sin\frac{x}{3}\end{aligned}$$

29. $$\begin{aligned}\csc(x - y) &= \frac{1}{\sin(x - y)} = \frac{1}{\sin x\cos y - \sin y\cos x}\\ &= \frac{\dfrac{1}{\sin x\sin y}}{\dfrac{\sin x\cos y - \sin y\cos x}{\sin x\sin y}}\\ &= \frac{\csc x\csc y}{\cot y - \cot x}\end{aligned}$$

31. $\sin(x + y) = \dfrac{-2 + 2\sqrt{30}}{15}$, $\cos(x + y) = \dfrac{-\sqrt{5} - 4\sqrt{6}}{15}$, $x + y = 8.785$

33. $$\begin{aligned}\sin(x + y) &= \frac{-15 + \sqrt{273}}{32}\\ \cos(x + y) &= \frac{-5\sqrt{7} - 3\sqrt{39}}{32}\\ x + y &= 3.094\end{aligned}$$

35. $\sin(x+y+z) = \sin(x+y)\cos z + \cos(x+y)\sin z$
$= (\sin x\cos y + \sin y\cos x)\cos z + (\cos x\cos y - \sin x\sin y)\sin z$
$= \sin x\cos y\cos z - \sin x\sin y\sin z + \cos x\sin y\cos z + \cos x\cos y\sin z$

37. $$\frac{\sin x+\cos x-1}{\sin x-\cos x+1} = \frac{\sin x+\cos x-1}{\sin x-\cos x+1}\cdot\frac{(1-\sin x)}{(1-\sin x)} = \frac{(\sin x+\cos x-1)(1-\sin x)}{-\cos x+1-\sin^2 x+\sin x\cos x} = \frac{(\sin x+\cos x-1)(1-\sin x)}{\cos x(-1+\cos x+\sin x)} = \frac{1-\sin x}{\cos x}$$

39. Not an identity—fails when $x = 0$.

41. $$\frac{\tan(x+y)-\tan z}{1+\tan(x+y)\tan z} = \tan((x+y)-z) = \tan(x+(y-z)) = \frac{\tan x+\tan(y-z)}{1-\tan x\tan(y-z)}$$

43. Not an identity—fails when $x = 0$ and $y = \dfrac{\pi}{4}$.

45. $w = x + y$ $\quad \sin w + \sin z = \sin(x+y) + \sin(x-y)$
$\underline{z = x - y}$
$w + z = 2x$ $\quad = 2\sin x\cos y$
$w - z + 2y$ $\quad = 2\sin\left(\dfrac{w+z}{2}\right)\cos\left(\dfrac{w-z}{2}\right)$

47. $$\frac{\sin 5x+\sin 9x}{\cos 5x-\cos 9x} = \frac{2\sin(7x)\cos(-2x)}{-2\sin(7x)\sin(-2x)} = \frac{2\sin(7x)\cos(2x)}{2\sin(7x)\sin(2x)} = \cot 2x$$

49. $\cos(x-y) + \cos(x+y) = (\cos x\cos y + \sin x\sin y) + (\cos x\cos y - \sin x\sin y)$
$= 2\cos x\cos y$

Let $w = x + y$, $z = x - y$, then $x = \dfrac{w+z}{2}$ and $y = \dfrac{w-z}{2}$.

Now $\cos w + \cos z = 2\cos\left(\dfrac{w+z}{2}\right)\cos\left(\dfrac{w-z}{2}\right)$. Using this result and Exercise 45, we can write $\dfrac{\sin 3x+\sin x}{\cos 3x+\cos x}$ $= \dfrac{2\sin 2x\cos x}{2\cos 2x\cos x} = \tan 2x.$

51. $\cos(x+y)\cos(x-y) = (\cos x\cos y - \sin x\sin y)(\cos x\cos y + \sin x\sin y)$
$= \cos^2 x\cos^2 y - \sin^2 x\sin^2 y$
$= \cos^2 x(1-\sin^2 y) - (1-\cos^2 x)\sin^2 y$
$= \cos^2 x - \sin^2 y$

Section 8.2

1. $\sin(2x) = \sin(x+x) = \sin x\cos x + \cos x\sin x = 2\sin x\cos x$

3. $\cos(2x) = \cos^2 x - \sin^2 x = \cos^2 x - (1-\cos^2 x) = 2\cos^2 x - 1$

5. From Exercise 4, $\cos x = 1 - 2\sin^2 \frac{x}{2}$

$$\sin^2 \frac{x}{2} = \frac{1}{2}(1 - \cos x)$$

$$\sin \frac{x}{2} = \pm\sqrt{\frac{1 - \cos x}{2}}$$

7. $$\tan 2x = \frac{\sin 2x}{\cos 2x} = \frac{2\sin x \cos x}{\cos^2 x - \sin^2 x} = \frac{\dfrac{2\sin x \cos x}{\cos^2 x}}{\dfrac{\cos^2 x - \sin^2 x}{\cos^2 x}} = \frac{2\tan x}{1 - \tan^2 x}$$

9. $$\begin{aligned}\sin 3x &= \sin x \cos 2x + \sin 2x \cos x \\ &= \sin x(1 - 2\sin^2 x) + 2\sin x \cos^2 x \\ &= \sin x - 2\sin^3 x + 2\sin x - 2\sin^3 x \\ &= 3\sin x - 4\sin^3 x\end{aligned}$$

11. $$\begin{aligned}\sin^4 x = (\sin^2 x)^2 &= \left(\frac{(1 - \cos 2x)}{2}\right)^2 \\ &= \frac{1}{4} - \frac{1}{2}\cos 2x + \frac{1}{4}\cos^2 2x \\ &= \frac{1}{4} - \frac{1}{2}\cos 2x + \frac{1}{4}\left(\frac{1 + \cos 4x}{2}\right) \\ &= \frac{3}{8} - \frac{1}{2}\cos 2x + \frac{1}{8}\cos 4x\end{aligned}$$

13. $$\begin{aligned}2\csc 2x = \frac{2}{\sin 2x} = \frac{2}{2\sin x \cos x} &= \frac{\sin^2 x + \cos^2 x}{\sin x \cos x} \\ &= \tan x + \cot x\end{aligned}$$

15. $$\begin{aligned}\cot^2 \frac{x}{2} &= \csc^2 \frac{x}{2} - 1 = \left(\frac{1}{\sin^2 \frac{x}{2}}\right) - 1 = \left(\frac{2}{1 - \cos x}\right) - 1 \\ &= \frac{1 + \cos x}{1 - \cos x} = \frac{\dfrac{1 + \cos x}{\cos x}}{\dfrac{1 - \cos x}{\cos x}} = \frac{\sec x + 1}{\sec x - 1}\end{aligned}$$

17. $$\sec 2x = \frac{1}{\cos 2x} = \frac{1}{2\cos^2 x - 1} = \frac{\dfrac{1}{\cos^2 x}}{\dfrac{2\cos^2 x - 1}{\cos^2 x}} = \frac{\sec^2 x}{2 - \sec^2 x}$$

19. $\cos 4x = 2\cos^2 2x - 1 = 2(2\cos^2 x - 1)^2 - 1 = 8\cos^4 x - 8\cos^2 x + 1$

21. $\sin 6x \tan 3x = 2\sin 3x \cos 3x \tan 3x = 2\sin^2 3x$

23. $2\sin^2 \frac{x}{2} \tan x = (1 - \cos x)\tan x = \tan x - \sin x$

25. $$\tan \frac{x}{2} = \frac{\sin \frac{x}{2}}{\cos \frac{x}{2}} \cdot \frac{2\cos \frac{x}{2}}{2\cos \frac{x}{2}} = \frac{\sin x}{2\cos^2 \frac{x}{2}} = \frac{\sin x}{1 + \cos x}$$

27. $\sec 2x = \dfrac{1}{\cos 2x} = \dfrac{1}{2\cos^2 x - 1} = \dfrac{1}{\dfrac{2}{\sec^2 x} - 1} = \dfrac{\sec^2 x}{2 - \sec^2 x}$

29.
$$\tan 3x = \tan(2x + x) = \frac{\tan 2x + \tan x}{1 - \tan 2x \tan x} = \frac{\dfrac{2\tan x}{1 - \tan^2 x} + \tan x}{1 - \dfrac{2\tan x}{1 - \tan^2 x}\tan x} = \frac{2\tan x + \tan x\,(1 - \tan^2 x)}{1 - \tan^2 x - 2\tan^2 x} = \frac{3\tan x - \tan^3 x}{1 - 3\tan^2 x}$$

31.
$$\begin{aligned}\cos 5x &= \cos x \cos 4x - \sin x \sin 4x = \cos x(2\cos^2 2x - 1) - 2\sin x \sin 2x \cos 2x\\ &= 2\cos x(2\cos^2 2x - 1)^2 - \cos x - 2\sin x(2\sin x\cos x)(2\cos^2 x - 1)\\ &= 8\cos^5 x - 8\cos^3 x + \cos x - 4(1 - \cos^2 x)(2\cos^3 x - \cos x)\\ &= 16\cos^5 x - 20\cos^3 x + 5\cos x\end{aligned}$$

33. $\dfrac{\sqrt{2\sqrt{2} + \sqrt{3} - 1}}{2\sqrt[4]{2}}$

35.
$$\frac{1}{2\sqrt{2\sqrt{2\sqrt{2\sqrt{2}}}}}\left[4\sqrt{2\sqrt{2\sqrt{2}}} + (1 - \sqrt{3})\sqrt{2\sqrt{2\sqrt{2}} + \sqrt{3\sqrt{2} - 3} + \sqrt{\sqrt{2} - 1}}\right.$$
$$\left. - (1 + \sqrt{3})\sqrt{2\sqrt{2\sqrt{2}} - \sqrt{3\sqrt{2} - 3} - \sqrt{\sqrt{2} - 1}}\right]^{1/2}$$

37. $\dfrac{\sqrt{35}}{18}$ **39.** $-\sqrt{0.3}$ **41.** $\sin x = \sqrt{\dfrac{\sqrt{5} - 1}{2\sqrt{5}}}$, $\cos x = -\sqrt{\dfrac{\sqrt{5} + 1}{2 + \sqrt{5}}}$, $\tan x = -\sqrt{\dfrac{\sqrt{5} - 1}{\sqrt{5} + 1}}$

43. $\dfrac{\sqrt{273} - 3}{20}$ **45.** $\dfrac{10 - \sqrt{195}}{24}$ **47.** $\dfrac{4\sqrt{6}}{25}$ **49.** $-\dfrac{31}{50}$ **51.** Not an identity—fails for $x = \dfrac{\pi}{4}$.

53.
$$\frac{1 - \sin 2x}{\cos 2x} = \frac{1 - 2\sin x\cos x}{\cos^2 x - \sin^2 x} = \frac{\cos^2 x + \sin^2 x - 2\sin x\cos x}{\cos^2 x - \sin^2 x} = \frac{(\cos x - \sin x)^2}{(\cos x - \sin x)(\cos x + \sin x)} = \frac{\cos x - \sin x}{\cos x + \sin x} = \frac{1 - \tan x}{1 + \tan x}$$

55. $1 - 2x^2$ **57.** $\sqrt{\dfrac{1 - x}{2}}$

59. Let $y = \sin^{-1} x$. Then $x = \sin y = \cos\left(\dfrac{\pi}{2} - y\right)$ so $\cos^{-1} x = \dfrac{\pi}{2} - y$. Thus $\sin^{-1} x + \cos^{-1} x = y + \dfrac{\pi}{2} - y = \dfrac{\pi}{2}$

61. $\sec^2 \frac{x}{2} = \tan^2 \frac{x}{2} + 1 = z^2 + 1$

$\cos^2 \frac{x}{2} = \frac{1}{z^2 + 1}$. $\cos x = 2\cos^2 \frac{x}{2} - 1 = \frac{2}{z^2 + 1} - 1 = \frac{1 - z^2}{1 + z^2}$

$$\begin{aligned} \sin x &= 2 \sin \frac{x}{2} \cos \frac{x}{2} \\ &= 2\frac{\sin \frac{x}{2}}{\cos \frac{x}{2}} \cdot \cos^2 \frac{x}{2} \\ &= 2 \tan \frac{x}{2} \cos^2 \frac{x}{2} \\ &= 2 \cdot z \cdot \frac{1}{z^2 + 1} \\ &= \frac{2z}{1 + z^2} \end{aligned}$$

Section 8.3

1. $x = \frac{\pi}{3}, \frac{2\pi}{3}$ **3.** $x = \frac{\pi}{18}, \frac{5\pi}{18}, \frac{13\pi}{18}, \frac{17\pi}{18}, \frac{25\pi}{18}, \frac{29\pi}{18}$ **5.** $x = \frac{\pi}{3}, \frac{4\pi}{3}$ **7.** $x = \frac{3\pi}{2}$

9. $\left\{\frac{\pi}{2}, 2.034, \frac{3\pi}{2}, 5.176\right\}$ **11.** $\left\{0, \frac{2\pi}{3}, \frac{4\pi}{3}\right\}$ **13.** $\{1.122, 2.446, 3.837, 5.162\}$ **15.** $\{0.876, 1.97, 4.31, 5.41\}$

17. $\{0.955, 2.19, 4.10, 5.33\}$ **19.** $\{0, \pi\}$ **21.** $\{1.34, 2.39, 3.90, 4.94\}$ **23.** $\left\{\frac{\pi}{4}, \frac{3\pi}{4}, \frac{5\pi}{4}, \frac{7\pi}{4}\right\}$

25. $\left\{0.365, \frac{\pi}{4}, 1.205, 3.505, \frac{5\pi}{4}, 4.345\right\}$ **27.** $\left\{\frac{\pi}{3}, \pi, \frac{5\pi}{3}\right\}$ **29.** $\emptyset$ **31.** $x = \frac{1}{\sqrt{5}}$ **33.** $x = \frac{9}{5}$

35. $\left\{\frac{\pi}{3}, 2.47, 3.82, \frac{5\pi}{3}\right\}$ **37.** $\{1.76, 2.82, 4.90, 5.96\}$ **39.** $\left\{0, \frac{\pi}{4}, 1.32, \frac{\pi}{2}, \frac{3\pi}{4}, \pi, \frac{5\pi}{4}, \frac{3\pi}{2}, 4.96, \frac{7\pi}{4}\right\}$

41. $x = \sin\left(\frac{2\pi}{3} - \tan^{-1} x\right)$

$x = \sin \frac{2\pi}{3} \cos(\tan^{-1} x) - \cos \frac{2\pi}{3} \sin(\tan^{-1} x)$

$x = \frac{\sqrt{3}}{2} \cdot \frac{1}{\sqrt{1 + x^2}} - \left(-\frac{1}{2}\right) \cdot \frac{x}{1 + x^2}$

$x = \frac{\sqrt{3} + x}{2\sqrt{1 + x^2}}$

$x^2 \cdot 4(1 + x^2) = 3 + 2\sqrt{3}x + x^2$

$4x^4 + 3x^2 - 2\sqrt{3}x - 3 = 0$

	4	0	3	$-2\sqrt{3}$	-3	
0.9	4	3.6	6.24	≈ 2.152	≈ -1.063	← root
1	4	4	7	≈ 3.586	≈ 0.536	

43. $x = 3.892$

45. $|\tan x + \cot x| = \left|\dfrac{\sin x}{\cos x} + \dfrac{\cos x}{\sin x}\right|$

$$A = \frac{1}{|\sin x \cos x|}$$

$$A = \frac{1}{\left|\dfrac{1}{2}\sin 2x\right|}$$

$$A = \frac{2}{|\sin 2x|}$$

$$A = 2|\csc 2x|,\ |\csc \theta| \geq 1$$

$$A \geq 2$$

47. $A = \dfrac{1}{9}\left(\sin\theta + \dfrac{\sin 2\theta}{2}\right)$. Estimate of θ for maximum area is 0.35π.

A
0.2
0.1
0
$\frac{\pi}{6}$ $\frac{\pi}{4}$ $\frac{\pi}{3}$ $\frac{\pi}{2}$
θ

Section 8.4

1. 1.434 **3.** 0.824 **5.** 0.733 **7.** 0.622 **9.** −0.719 **11.** 0.40

Chapter 8 Review Exercises

1. $\sec x \cot x = \dfrac{1}{\cos x}\cdot\dfrac{\cos x}{\sin x} = \dfrac{1}{\sin x} = \csc x$ **3.** $\dfrac{\tan x}{1+\tan^2 x} = \dfrac{\tan x}{\sec^2 x} = \left(\dfrac{\sin x}{\cos x}\right)(\cos^2 x) = \sin x \cos x$

5. $\sin 7x = \sin(2x + 5x)$
$= \sin 2x \cos 5x + \cos 2x \sin 5x$

7. $\cos(x+y) + \cos(x-y) = (\cos x \cos y - \sin x \sin y) + (\cos x \cos y + \sin x \sin y)$
$= 2\cos x \cos y$

9. $\dfrac{\sin 2x}{1+\cos 2x} = \dfrac{2\cos x \sin x}{1 + (2\cos^2 x - 1)} = \dfrac{2\cos x \sin x}{2\cos^2 x} = \tan x$

11. $(\tan x - \cot x)\tan 2x = \dfrac{(\tan x - \cot x)(2\tan x)}{1-\tan^2 x} = \dfrac{-2(1-\tan^2 x)}{1-\tan^2 x} = -2$

13. Using the result from exercise 12: $\dfrac{\sin 3x + \sin x}{\cos 3x + \cos x} = \dfrac{\sin 3x - \sin(-x)}{\cos 3x + \cos(-x)} = \tan\left[\dfrac{3x-(-x)}{2}\right] = \tan 2x$

15. $\dfrac{6-\sqrt{35}}{12}$, $x + y \approx 3.13$ **17.** $\dfrac{\sqrt{15}}{8}$ **19.** 0.570 **21.** $\dfrac{33}{65}$ **23.** $-\dfrac{\sqrt{35}}{17}$ **25.** $\left\{\dfrac{\pi}{6}, \dfrac{5\pi}{6}, \dfrac{7\pi}{6}, \dfrac{11\pi}{6}\right\}$

27. $\left\{\dfrac{7\pi}{6}, \dfrac{3\pi}{2}, \dfrac{11\pi}{6}\right\}$

29. $\left\{\frac{3\pi}{4}, \frac{7\pi}{4}\right\}$ **31.** $\left\{\frac{\pi}{2}, 4.07\right\}$ **33.** $x \approx 0.403$

35. Not an identity—fails for $x = y = \frac{\pi}{6}$. **37.** $\csc 2x = \frac{1}{\sin 2x} = \frac{1}{2\cos x \sin x} = \frac{\frac{1}{\cos^2 x}}{\frac{2\sin x}{\cos x}}$

$$= \frac{\sec^2 x}{2\tan x} = \frac{1 + \tan^2 x}{2\tan x}$$

39.
$$\frac{\sin 3x}{\sin x} - \frac{\sin 4x}{\sin 2x} = \frac{\sin 3x}{\sin x} - \frac{2\cos 2x \sin 2x}{\sin 2x}$$
$$= \frac{\sin x \cos 2x + \cos x \sin 2x}{\sin x} - 2\cos 2x$$
$$= \frac{-\sin x \cos 2x + \cos x \sin 2x}{\sin x}$$
$$= \frac{\sin(2x - x)}{\sin x} = 1$$

41. $\left\{0, \frac{\pi}{9}, \frac{\pi}{3}, \frac{5\pi}{9}, \frac{2\pi}{3}, \frac{7\pi}{9}, \pi, \frac{11\pi}{9}, \frac{4\pi}{3}, \frac{13\pi}{9}, \frac{5\pi}{3}, \frac{17\pi}{9}\right\}$ **43.** $\left\{\frac{2\pi}{3}, \frac{4\pi}{3}\right\}$

45. 1.31 **47.** **(a)** $C = 5k - 2k \cot\theta + 4k \csc\theta$ **(b)** S is approximately 4.07 miles from B.
(c) Minimum cost for laying the cable is about $8.5k$. Position of S for minimum cost is about 0.37 miles from B.

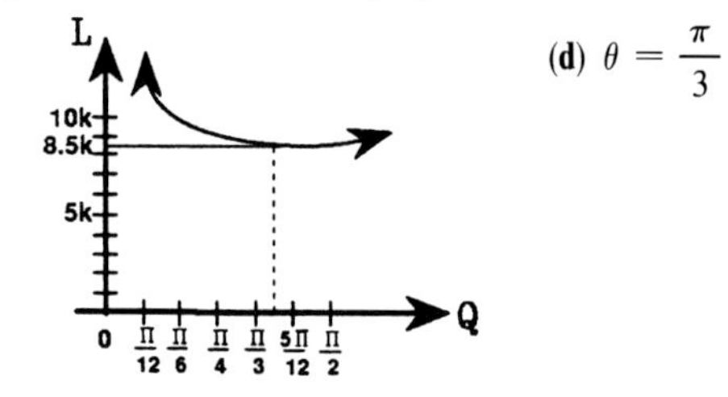

(d) $\theta = \frac{\pi}{3}$

Chapter 8 Test 1

1. $\cos\left(2x - \frac{\pi}{2}\right) = \cos 2x \cos\frac{\pi}{2} + \sin 2x \sin\frac{\pi}{2} = \sin 2x$ **2.** $\frac{\sin 2x \cos x}{\cot^2 x} = \frac{2\cos^2 x \sin x}{\frac{\cos^2 x}{\sin^2 x}} = 2\sin^3 x$

3. $\tan 2x = \frac{\sin 2x}{\cos 2x} = \frac{2\sin x \cos x}{\cos^2 x - \sin^2 x} = \frac{\frac{2\sin x}{\cos x}}{1 - \frac{\sin^2 x}{\cos^2 x}} = \frac{2\tan x}{1 - \tan^2 x}$ **4.** $\frac{\sin x}{1 + \cos x} = \frac{2\sin\frac{x}{2}\cos\frac{x}{2}}{1 + 2\sin^2\frac{x}{2} - 1} = \cot\frac{x}{2}$

5. $\frac{-4\sqrt{5}}{9}$ **6.** $\frac{33}{65}$ **7.** $x = \frac{\pi}{2}, \frac{3\pi}{2}$ **8.** $x = \frac{\pi}{4}$ **9.** $\{0.31, 1.79, 2.40, 3.88, 2.40, 5.98\}$

10. $\sin\frac{\pi}{6} + \sin 2\left(\frac{\pi}{6}\right) = \frac{1 + \sqrt{3}}{2} \neq \sin\frac{3\pi}{6} = 1$. The equation is not an identity.

Chapter 8 Test 2

1. $$\frac{(\cos^2 x - \sin^2 x)^2}{\cos^4 x - \sin^4 x} = \frac{(\cos^2 x - \sin^2 x)^2}{(\cos^2 x - \sin^2 x)(\cos^2 x + \sin^2 x)}$$
$$= \cos^2 x - \sin^2 x$$
$$= 2\cos^2 x - 1$$

2. $$\cot(x + y) = \frac{\cos(x + y)}{\sin(x + y)}$$
$$= \frac{\cos x \cos y - \sin x \sin y}{\sin x \cos y + \sin y \cos x}$$
$$= \frac{\dfrac{\cos x \cos y}{\sin x \sin y} - \dfrac{\sin x \sin y}{\sin x \sin y}}{\dfrac{\sin x \cos y}{\sin x \sin y} + \dfrac{\sin y \cos x}{\sin x \sin y}}$$
$$= \frac{\cot x \cot y - 1}{\cot x + \cot y}$$

3. $$\frac{\sin x + \sin 3x}{\cos x - \cos 3x} = \frac{\sin\left[\dfrac{x + 3x}{2} + \dfrac{x - 3x}{2}\right] + \sin\left[\dfrac{x + 3x}{2} - \dfrac{x - 3x}{2}\right]}{\cos\left[\dfrac{x + 3x}{2} + \dfrac{x - 3x}{2}\right] - \cos\left[\dfrac{x + 3x}{2} - \dfrac{x - 3x}{2}\right]}$$
$$= \frac{2\sin\left[\dfrac{x + 3x}{2}\right]\cos\left[\dfrac{x - 3x}{2}\right]}{-2\sin\left[\dfrac{x + 3x}{2}\right]\sin\left[\dfrac{x - 3x}{2}\right]}$$
$$= -\cot(-x) = \cot x$$

4. $$\frac{1 + \csc 2x}{\cot x} = \frac{1 + \dfrac{1}{\sin 2x}}{\cot}$$
$$= \frac{\sin 2x + 1}{\sin 2x \cot x}$$
$$= \frac{2\sin x \cos x + 1}{2\sin x \cos x \dfrac{\cos x}{\sin x}}$$
$$= \frac{2\sin x \cos x + 1}{2\cos^2 x}$$
$$= \tan x + \frac{1}{2}\sec^2 x$$
$$= \frac{2\tan x + (\tan^2 x + 1)}{2}$$
$$= \frac{(\tan x + 1)^2}{2}$$

5. $-\dfrac{56}{65}$, $x + y \approx 11.53$ 6. $\dfrac{240}{161}$ 7. $\left\{0, \dfrac{\pi}{4}, \dfrac{3\pi}{4}, \pi, \dfrac{5\pi}{4}, \dfrac{7\pi}{4}\right\}$ 8. $\{1.17, 2.27\}$ 9. $x = \dfrac{1}{\sqrt{2}}$

10. $\cot\frac{\pi}{4} = 1 \neq \dfrac{1+\cos\frac{\pi}{4}}{\sin\left(2\cdot\frac{\pi}{4}\right)} = 1 + \frac{1}{\sqrt{2}}$. The equation is not an identity.

Section 9.1

1. 3,000 rpm **3.** $\frac{1{,}250}{3\pi}$ rpm **5.** 4,000 rpm or $8{,}000\pi$ rad/min **7.** $\frac{9\pi}{5}$ ft/sec **9.** $\frac{10\pi}{3}$ ft/sec **11.** $\frac{3{,}360}{\pi}$ rpm

13. $\frac{640}{3}$ rpm **15.** 10 cm and 25 cm **17.** $\frac{15{,}000}{11}$ mph **19.** 2,500 ft/min

Section 9.2

1. $B = 68°$, $a = 5.1$, $c = 13.5$ **3.** $c = 20.3$, $A = 42°$, $B = 48°$ **5.** $A = 44°$, $a = 83.4$, $b = 86.3$

7. $\sec A = \frac{\text{hyp}}{\text{adj}}$, $\csc A = \frac{\text{hyp}}{\text{opp}}$, $\cot A = \frac{\text{adj}}{\text{opp}}$ **9.** 2,257′ **11.** 625′ or 196′ **13.** 2,796 mi **15.** 62.79″

17. 9:30 A.M. **19.** 557.4 ft., N 31°-0′ E **21.** $\overline{AB} = 3.35''$, $\overline{AD} = 2.53''$ **23.** 102.8″

25. $h = x(\tan\theta - \tan\alpha)$ **27.** 183.7 ft/sec

29. $\theta = \tan^{-1}\frac{4}{x} - \tan^{-1}\frac{1}{x}$. θ tends to 0 as x tends to 0 or as x tends to ∞. θ is maximized near $x = 2$.

31. 4.4°

33. Since $\alpha = \beta$, $\frac{a+b}{x} = \tan 2\alpha$

$$
\begin{aligned}
a + b &= x\left(\frac{2\tan\alpha}{1-\tan^2\alpha}\right)\\
b &= x\left(\frac{\frac{2a}{x}}{1-\frac{a^2}{x^2}}\right) - a\\
b &= \frac{2ax^2}{x^2-a^2} - a\\
b &= \frac{ax^2+a^3}{x^2-a^2}\\
bx^2 - a^2b &= ax^2 + a^3\\
x^2(b-a) &= a^2(b+a)\\
x &= a\sqrt{\frac{b+a}{b-a}}
\end{aligned}
$$

Section 9.3

1. $B = 62°$, $a = 11.5$, $c = 16.7$ **3.** $b = 20.3$, $A = 35°$, $C = 69°$ **5.** $A = 40°$, $b = 5.7$, $c = 12.3$
7. $c = 12.8$, $A = 24.4°$, $B = 136.0°$ **9.** $C = 1.06$, $a = 50.3$, $b = 70.7$ **11.** $c = 11.6$, $A = 1.13$, $B = 1.05$
13. $B = 0.33$, $C = 1.49$, $c = 195.6$ **15.** $A = 85°$, $C = 60°$, $a = 417.5$; or $A = 25°$, $C = 120°$, $a = 175.2$
17. No solution **19.** $A = 33.8°$, $B = 117.5°$, $C = 28.7°$ **21.** 92 km/hr and 82 km/hr **23.** $\theta = 23°$
25. $\overline{PQ} = 32.5$ mi **27.** $x = \dfrac{84}{37}$ **29.** 1:06.08 P.M. **31.** Angle $BCA = 52°$, angle $BAC = 46°$, angle $ABC = 82°$

33. $h = c \sin A$, so the area is $\frac{1}{2}bh = \frac{1}{2}bc \sin A$.

35.
$$\frac{1}{2}bc(1 + \cos A) = \frac{1}{2}bc\left(1 + \frac{b^2 + c^2 - a^2}{2bc}\right) = \frac{1}{2}bc\left(\frac{2bc + b^2 + c^2 - a^2}{2bc}\right) = \frac{1}{4}(b + c - a)(b + c + a).$$
$$\frac{1}{2}bc(1 - \cos A) = \frac{1}{2}bc\left(1 - \frac{b^2 + c^2 - a^2}{2bc}\right) = \frac{1}{2}bc\left(\frac{2bc - b^2 - c^2 + a^2}{2bc}\right) = \frac{1}{4}(a - b + c)(a + b - c).$$

37. $\cos A + \cos B + \cos C$
$$= \frac{b^2 + c^2 - a^2}{2bc} + \frac{a^2 + c^2 - b^2}{2ac} + \frac{a^2 + b^2 - c^2}{2ab}$$
$$= \frac{ab^2 + ac^2 - a^3 + a^2b + bc^2 - b^3 + a^2c + b^2c - c^3}{2abc}$$
$$= \frac{(a - b + c)(a + b - c)(-a + b + c)}{2abc} + 1$$
$$= 1 + \sqrt{\frac{(a - b + c)^2(a + b - c)^2(-a + b + c)^2}{4a^2b^2c^2}}$$
$$= 1 + 4\sqrt{\frac{(a - b + c)(a + b - c)}{4bc}} \cdot \sqrt{\frac{(-a + b + c)(a + b - c)}{4ac}} \cdot \sqrt{\frac{(-a + b + c)(a - b + c)}{4ab}}$$
$$= 1 + 4\sqrt{\frac{1 - \cos A}{2}}\sqrt{\frac{1 - \cos B}{2}}\sqrt{\frac{1 - \cos C}{2}}$$
$$= 1 + 4 \sin \frac{A}{2} \sin \frac{B}{2} \sin \frac{C}{2}$$

39. $\cot A = \dfrac{\cos A}{\sin A} = \dfrac{bc \cos A}{2a} = \dfrac{-a^2 + b^2 + c^2 + 2bc}{4a} - \dfrac{2bc}{4a} = \dfrac{-a^2 + b^2 + c^2}{4a}$;

hence, $\cot A \cot B + \cot A \cot C + \cot B \cot C$
$$= \frac{1}{(4a)^2}[(-a^2 + b^2 + c^2)(a^2 - b^2 + c^2) + (-a^2 + b^2 + c^2)(a^2 + b^2 - c^2) + (a^2 - b^2 + c^2)(a^2 + b^2 - c^2)]$$
$$= \frac{4^2}{(4a)^2}\left[\frac{(a + b + c)(-a + b + c)(a - b + c)(a + b - c)}{4^2}\right]$$
$$= \frac{4^2}{(4a)^2} \cdot a^2 = 1$$

Section 9.4

1. **(a)** **(b)** $(4\sqrt{2}, 4\sqrt{2})$ **3.** **(a)** **(b)** $(-3\sqrt{3}, 3)$

5. **(a)** **(b)** $(50, -50\sqrt{3})$ **7.** **(a)** $(5, 8)$ **(b)** $\sqrt{89}$ **9.** **(a)** $(-3, -8)$ **(b)** $\sqrt{73}$
11. **(a)** $(-13, 8)$ **(b)** $\sqrt{233}$ **13.** $\mathbf{A} + \mathbf{B} = (8, 2)$

15. $\mathbf{A} + \mathbf{B} = (-2, -5)$ **17.** $\mathbf{A} + \mathbf{B} = (-5, 1)$

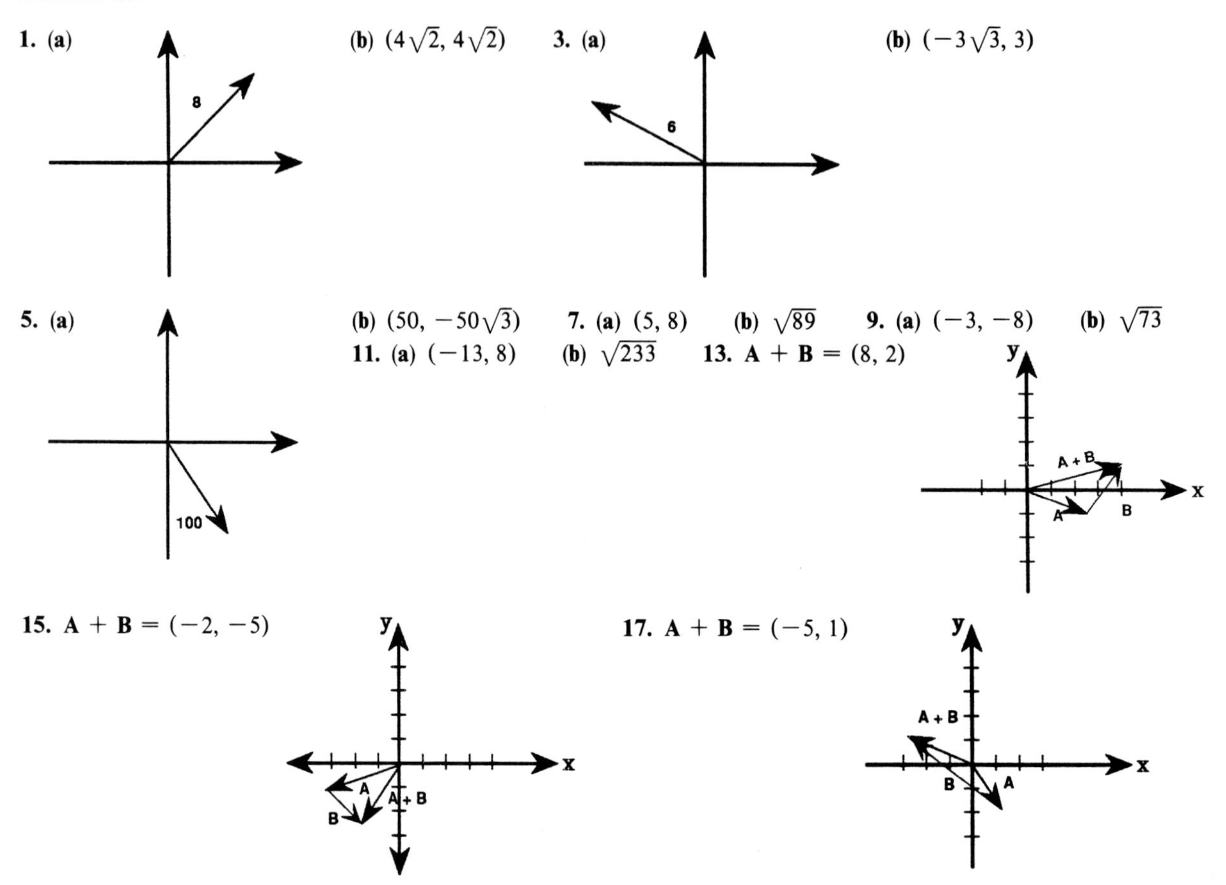

19. S 84° E **21.** The ground speed is 483.6 mph and the heading is 43°. **23.** The swimmer should head N 60° E, and his actual speed is $2\sqrt{3}$ mph. **25.** Heading 14.4°, flying time 0.95 hr. **27.** Current's speed is 3.1 mph. Current's direction is S 64° E. **29.** 11,472 lbs. tension; 10,000 lbs. compression **31.** 652 lbs.; 737 lbs. **33.** 235 lbs. **35.** $\theta = 80.4°$ **37.** $||\mathbf{T}|| = 1{,}649$ lbs.; $||\mathbf{C}|| = 566$ lbs.

Section 9.5

1. $2(\sin 8t)(\cos 2t)$ **3.** $20(\sin 26\pi t)(\cos 2\pi t)$ **5.** $2(\sin 110\pi t)(\cos 10\pi t)$

7. 2 beats per second

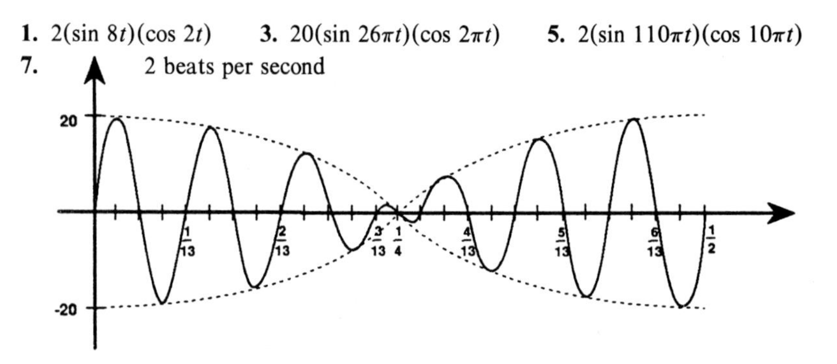

9. $y = A(1 + C\sin(2\pi \cdot 500)t)\sin(2\pi \cdot 100{,}000))$. The audio carrying signals are $\frac{C}{2}\cos 2\pi(100{,}000 - 400)t$ and $\frac{C}{2}\cos 2\pi(100{,}000 + 400)t$. The frequencies are 100,400 and 99,600. **11.** $y = 5\sin(2\pi \cdot 3{,}000t)$, where t is in minutes. **13.** $y = 10\sin(2\pi \cdot 25t)$, where t is in seconds.

Chapter 9 Review Exercises

1. 6,000 rpm **3.** $B = 60°$, $a = \frac{25}{\sqrt{3}}$, $c = \frac{50}{\sqrt{3}}$ **5.** $A = 60°$, $B = 30°$, $b = 868$ **7.** $B = 75°$, $a = 8.8$, $c = 10.8$ **9.** $a = 135.8$, $B = 46°$, $C = 51°$ **11.** $A = 125°$, $B = 14°$, $C = 41°$ **13.** $(-133, 43)$ **15.** $\mathbf{A} + \mathbf{B} = (5, 8)$

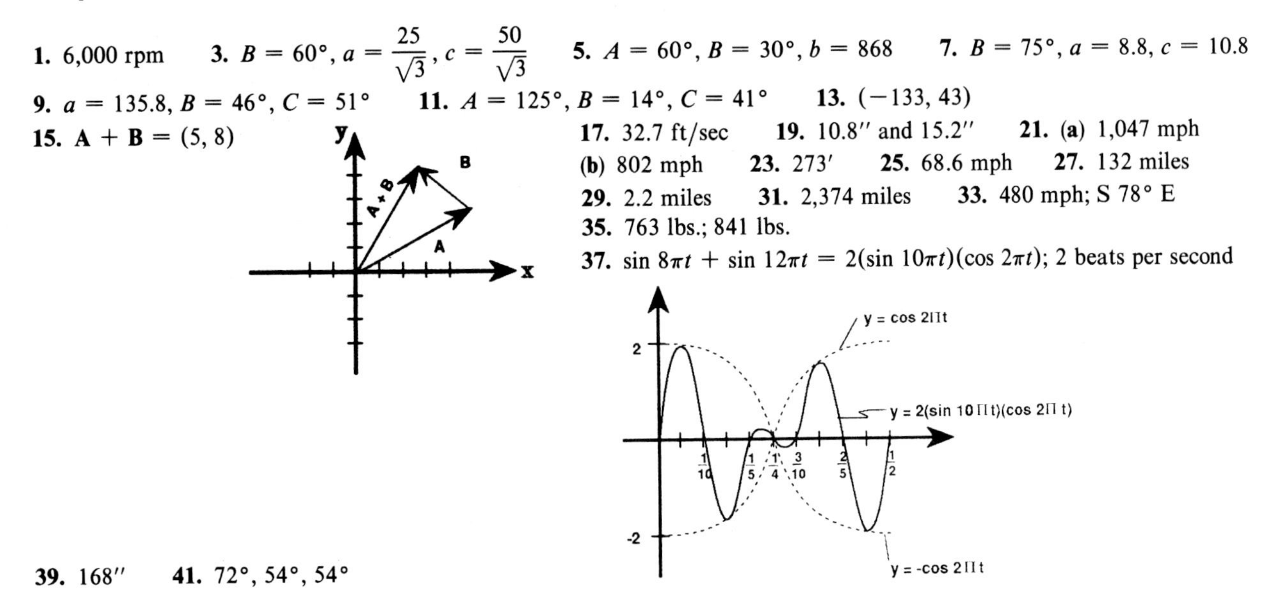

17. 32.7 ft/sec **19.** 10.8″ and 15.2″ **21.** (a) 1,047 mph (b) 802 mph **23.** 273′ **25.** 68.6 mph **27.** 132 miles **29.** 2.2 miles **31.** 2,374 miles **33.** 480 mph; S 78° E **35.** 763 lbs.; 841 lbs. **37.** $\sin 8\pi t + \sin 12\pi t = 2(\sin 10\pi t)(\cos 2\pi t)$; 2 beats per second

39. 168″ **41.** 72°, 54°, 54°

Chapter 9 Test 1

1. 54°; 144 cm; 198 cm **2.** $A = 49°$, $a = 15.5$, $c = 8.0$ **3.** $b = 16.5$, $A = 60°$, $C = 48°$ **4.** $3{,}200\pi$ in/min. **5.** 1,825 ft. **6.** 363 mi. **7.** 772 ft. **8.** 153° **9.** Boat heading 78.5°; trip time 0.026 hrs. **10.** $\theta = 68°$

Chapter 9 Test 2

1. 23°, 67°, 13 units **2.** $B = 119°$, $C = 7°$, $c = 16.1$ **3.** $A = 16.2$, $B = 115.2°$, $C = 48.6°$ **4.** 907 mph **5.** 118′ **6.** 9:42 A.M. **7.** 252 mph **8.** 36.6°, 55.3°, 88.1°
9. The ground speed is 545 mph and the bearing is N 86.3° E.
10. 2,120 lb. tension in the 26° cable; 2,000 lb. tension in the 32° cable

Section 10.1

1. $\left(3\sqrt{2}, \frac{\pi}{4}\right); \left(3\sqrt{2}, \frac{9\pi}{4}\right); \left(-3\sqrt{2}, -\frac{3\pi}{4}\right); \left(-3\sqrt{2}, \frac{5\pi}{4}\right)$ **3.** $\left(-2, \frac{\pi}{6}\right); \left(-2, -\frac{11\pi}{6}\right); \left(2, \frac{7\pi}{6}\right); \left(2, -\frac{5\pi}{6}\right)$

5. $(5, \pi); (5, -\pi); (-5, 0); (-5, 2\pi)$ **7.** $\left(\frac{\sqrt{3}}{2}, \frac{1}{2}\right)$ **9.** $(1, -\sqrt{3})$ **11.** $(4\sqrt{2}, -4\sqrt{2})$ **13.** $\theta = \frac{\pi}{4}$

15. $r = 3$ **17.** $\theta = \tan^{-1}(-3)$ **19.** $r = 2\sqrt{3}$ **21.** $y = \frac{\sqrt{3}}{3}x$ **23.** $x^2 + y^2 = 9$ **25.** $y = -x$

27. $x^2 + y^2 = 1$ **29.** $r = 2\cos\theta$ **31.** $r = \frac{5}{\cos\theta + 2\sin\theta}$ **33.** $r = 2\sec\theta$ **35.** $r = 6\cos\theta - 4\sin\theta$

37. $r = 2\tan\theta\sec\theta$ **39.** $x = 5$ **41.** $x^2 + y^2 = 4y$ **43.** $x^2 + y^2 = 4x$ **45.** $y = \sqrt{3}x - 2$

47. $3x^2 + 4y^2 - 12x = 36$ **49.** $4x^2 - 5y^2 - 36y = 36$ **51.** $y = \frac{1}{2x}$

53. $d = \sqrt{r_1^2 + r_2^2 - 2r_1r_2\cos(\theta_1 - \theta_2)}$, $\sqrt{10 - 6\cos\frac{7\pi}{12}}$ **55.** $r = \frac{-6}{1 - \cos\theta}$

Section 10.2

1.

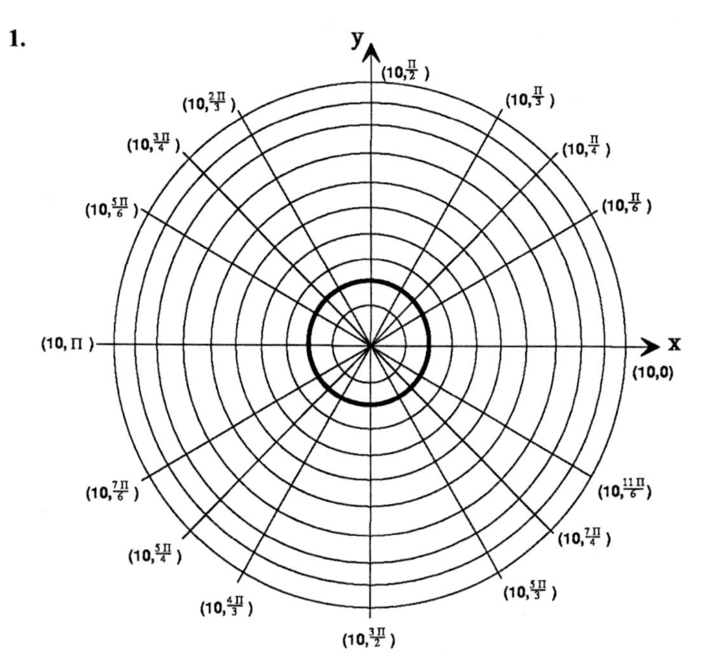

3.

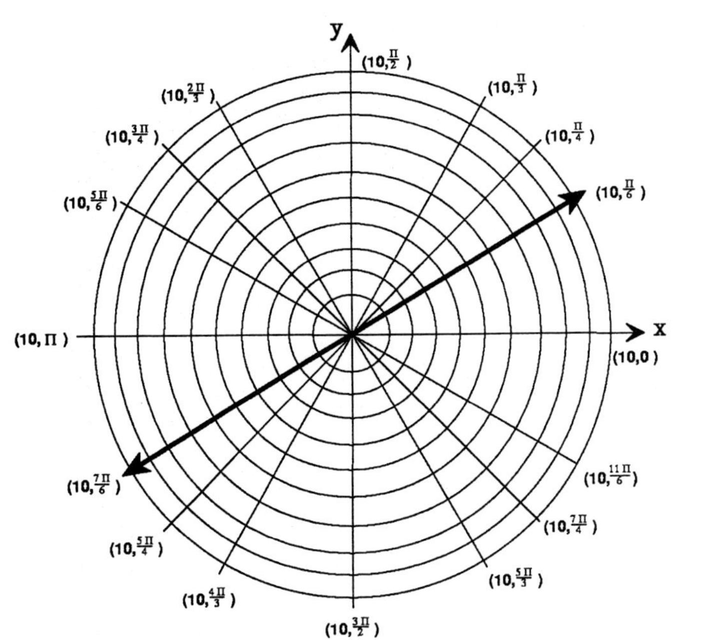

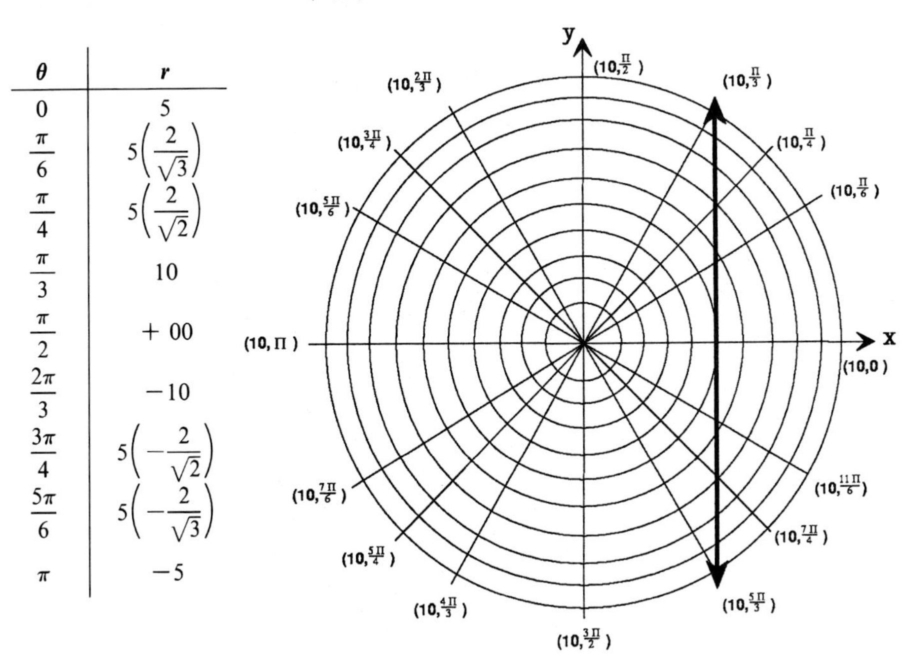

θ	r
0	5
$\frac{\pi}{6}$	$5\left(\frac{2}{\sqrt{3}}\right)$
$\frac{\pi}{4}$	$5\left(\frac{2}{\sqrt{2}}\right)$
$\frac{\pi}{3}$	10
$\frac{\pi}{2}$	$+\ 00$
$\frac{2\pi}{3}$	-10
$\frac{3\pi}{4}$	$5\left(-\frac{2}{\sqrt{2}}\right)$
$\frac{5\pi}{6}$	$5\left(-\frac{2}{\sqrt{3}}\right)$
π	-5

7.

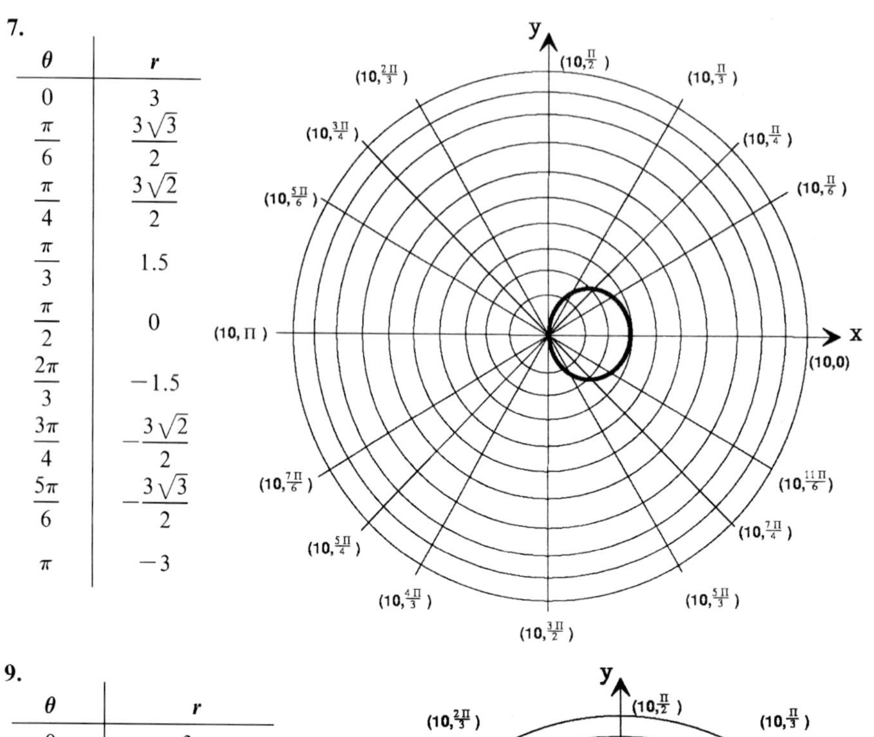

θ	r
0	3
$\frac{\pi}{6}$	$\frac{3\sqrt{3}}{2}$
$\frac{\pi}{4}$	$\frac{3\sqrt{2}}{2}$
$\frac{\pi}{3}$	1.5
$\frac{\pi}{2}$	0
$\frac{2\pi}{3}$	-1.5
$\frac{3\pi}{4}$	$-\frac{3\sqrt{2}}{2}$
$\frac{5\pi}{6}$	$-\frac{3\sqrt{3}}{2}$
π	-3

9.

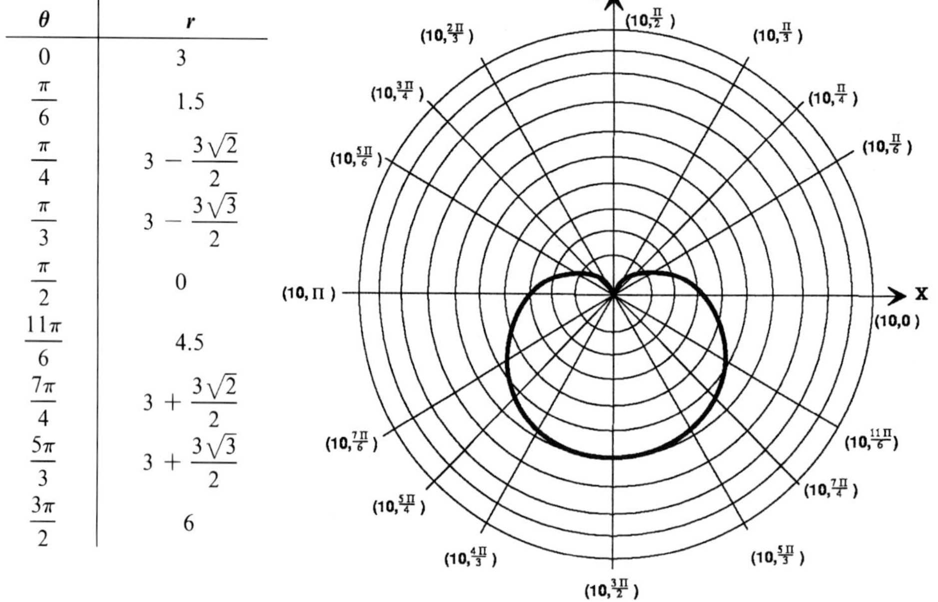

θ	r
0	3
$\frac{\pi}{6}$	1.5
$\frac{\pi}{4}$	$3-\frac{3\sqrt{2}}{2}$
$\frac{\pi}{3}$	$3-\frac{3\sqrt{3}}{2}$
$\frac{\pi}{2}$	0
$\frac{11\pi}{6}$	4.5
$\frac{7\pi}{4}$	$3+\frac{3\sqrt{2}}{2}$
$\frac{5\pi}{3}$	$3+\frac{3\sqrt{3}}{2}$
$\frac{3\pi}{2}$	6

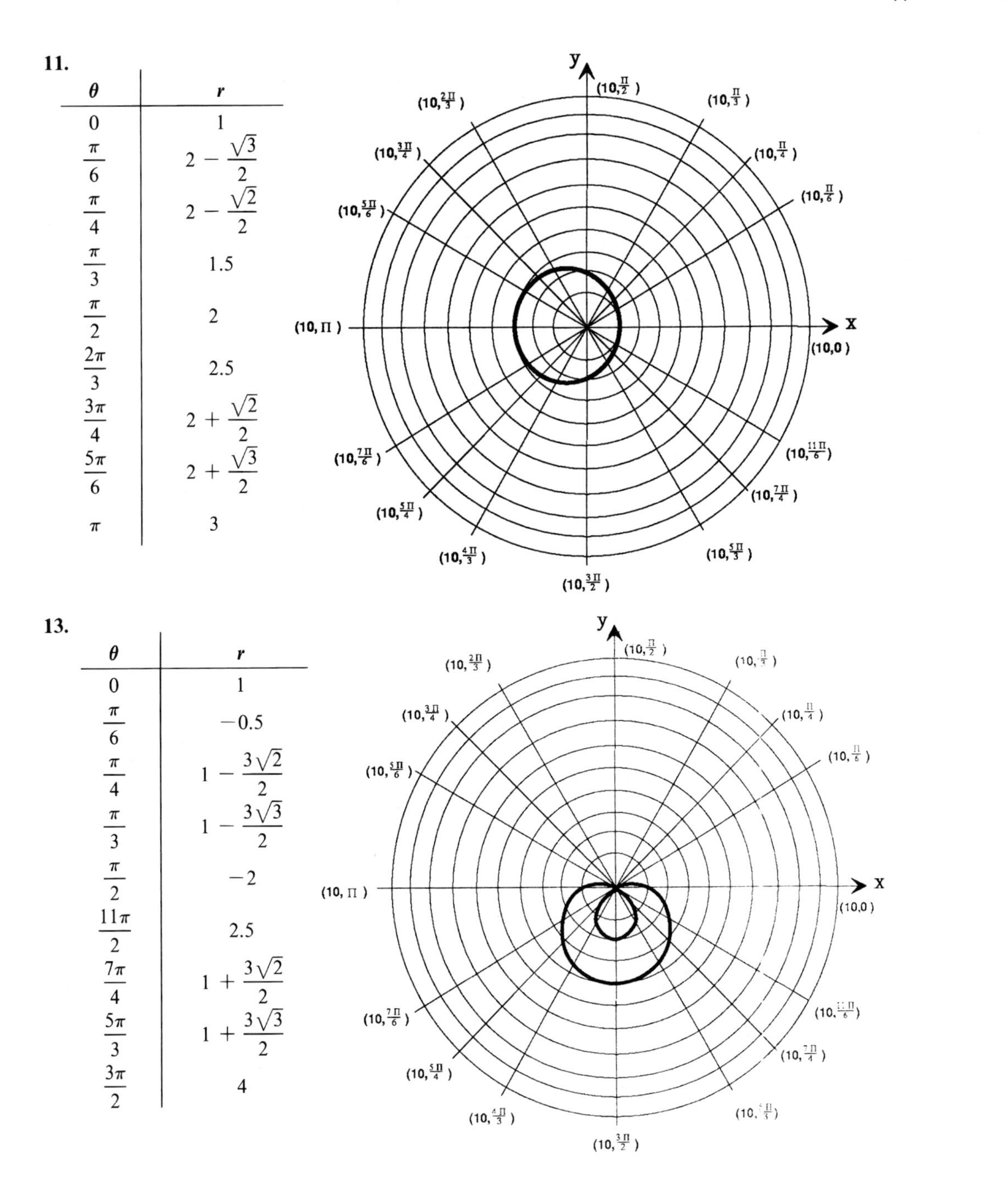

11.

θ	r
0	1
$\frac{\pi}{6}$	$2 - \frac{\sqrt{3}}{2}$
$\frac{\pi}{4}$	$2 - \frac{\sqrt{2}}{2}$
$\frac{\pi}{3}$	1.5
$\frac{\pi}{2}$	2
$\frac{2\pi}{3}$	2.5
$\frac{3\pi}{4}$	$2 + \frac{\sqrt{2}}{2}$
$\frac{5\pi}{6}$	$2 + \frac{\sqrt{3}}{2}$
π	3

13.

θ	r
0	1
$\frac{\pi}{6}$	-0.5
$\frac{\pi}{4}$	$1 - \frac{3\sqrt{2}}{2}$
$\frac{\pi}{3}$	$1 - \frac{3\sqrt{3}}{2}$
$\frac{\pi}{2}$	-2
$\frac{11\pi}{2}$	2.5
$\frac{7\pi}{4}$	$1 + \frac{3\sqrt{2}}{2}$
$\frac{5\pi}{3}$	$1 + \frac{3\sqrt{3}}{2}$
$\frac{3\pi}{2}$	4

15.

θ	r
0 & $\frac{2\pi}{3}$	1
$\frac{\pi}{12}$ & $\frac{7\pi}{12}$ & $\frac{3\pi}{4}$	$\frac{\sqrt{2}}{2}$
$\frac{\pi}{6}$ & $\frac{\pi}{2}$ & $\frac{5\pi}{6}$	0
$\frac{\pi}{4}$ & $\frac{5\pi}{12}$ & $\frac{11\pi}{12}$	$-\frac{\sqrt{2}}{2}$
$\frac{\pi}{3}$ & π	-1

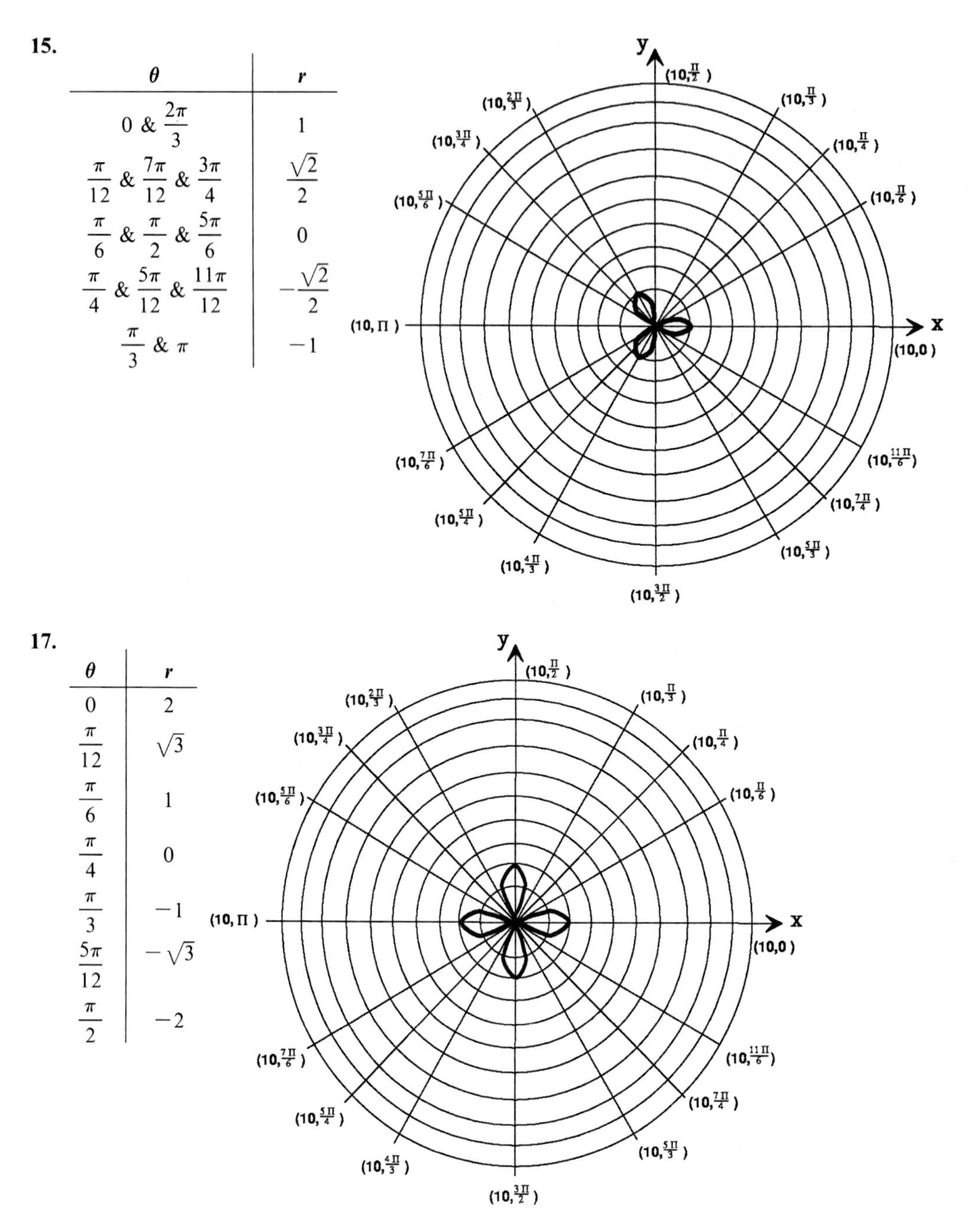

17.

θ	r
0	2
$\frac{\pi}{12}$	$\sqrt{3}$
$\frac{\pi}{6}$	1
$\frac{\pi}{4}$	0
$\frac{\pi}{3}$	-1
$\frac{5\pi}{12}$	$-\sqrt{3}$
$\frac{\pi}{2}$	-2

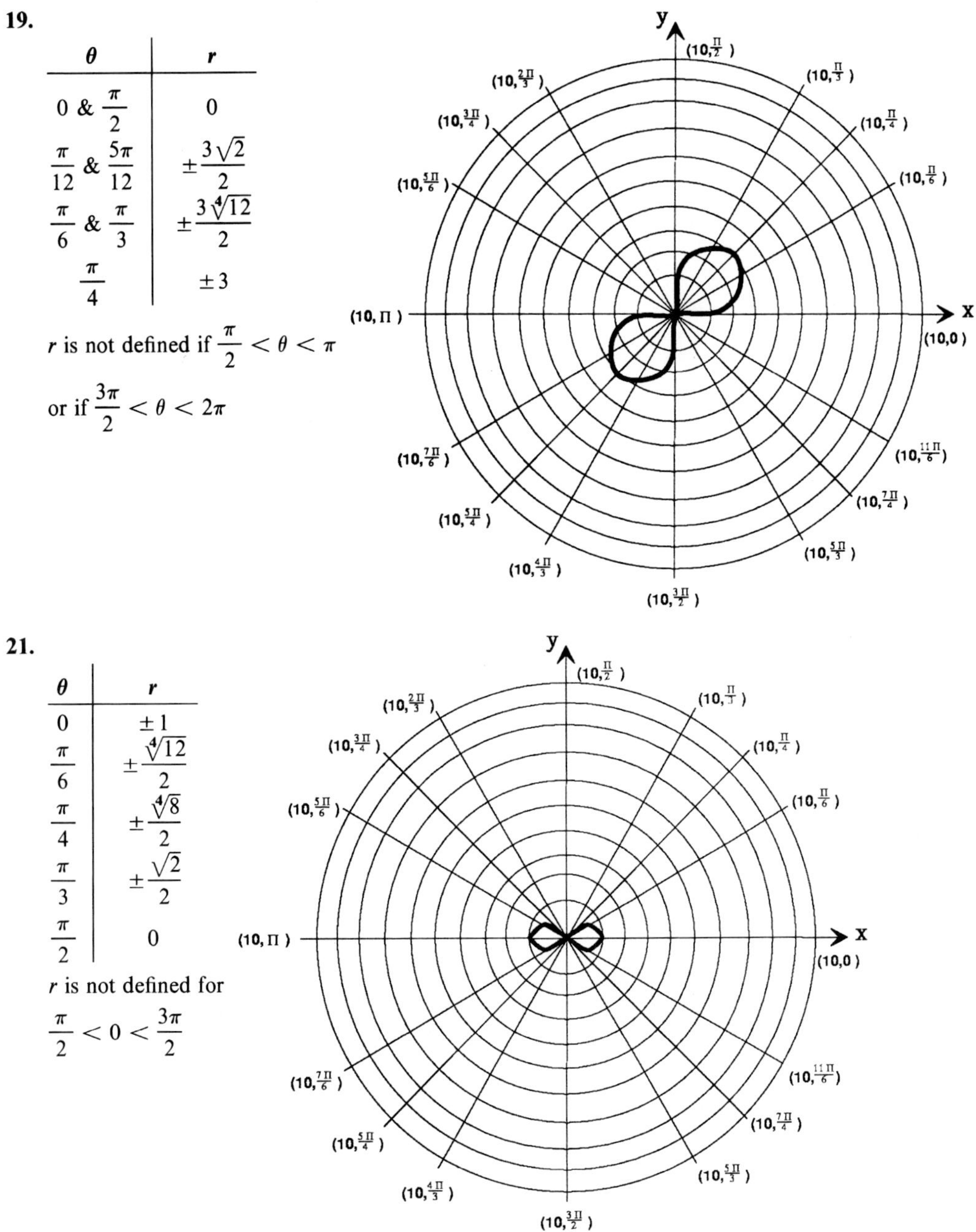

19.

θ	r
0 & $\frac{\pi}{2}$	0
$\frac{\pi}{12}$ & $\frac{5\pi}{12}$	$\pm\frac{3\sqrt{2}}{2}$
$\frac{\pi}{6}$ & $\frac{\pi}{3}$	$\pm\frac{3\sqrt[4]{12}}{2}$
$\frac{\pi}{4}$	± 3

r is not defined if $\frac{\pi}{2} < \theta < \pi$

or if $\frac{3\pi}{2} < \theta < 2\pi$

21.

θ	r
0	± 1
$\frac{\pi}{6}$	$\pm\frac{\sqrt[4]{12}}{2}$
$\frac{\pi}{4}$	$\pm\frac{\sqrt[4]{8}}{2}$
$\frac{\pi}{3}$	$\pm\frac{\sqrt{2}}{2}$
$\frac{\pi}{2}$	0

r is not defined for

$\frac{\pi}{2} < 0 < \frac{3\pi}{2}$

23.

θ	r
0	0
$\frac{\pi}{6}$	$\frac{\pi}{6}$
$\frac{\pi}{3}$	$\frac{\pi}{3}$

etc.

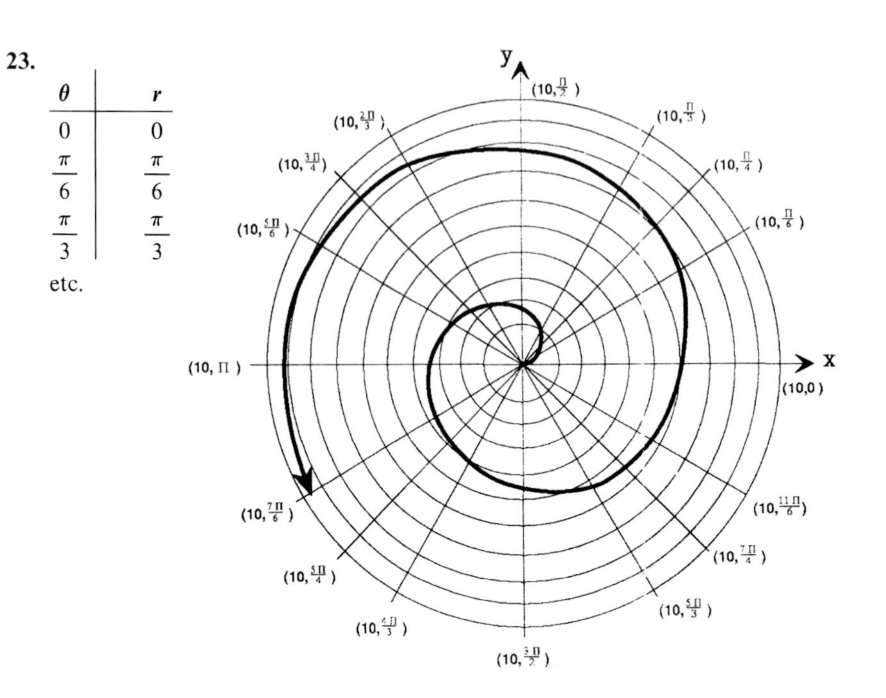

25. Equation $-r = 2 \cos 4(\theta + \pi)$ is equivalent to $r = -2 \cos 4\theta$. Thus, if (r, θ) is a solution of $r = 2 \cos 4\theta$, then $(-r, \theta + \pi)$ is a solution of $-r = 2 \cos 4(\theta + \pi)$. But (r, θ) and $(-r, \theta + \pi)$ are polar coordinate descriptions of the same point. Therefore, the graph of $r = 2 \cos 4\theta$ and $-r = 2 \cos 4(\theta + \pi)$ are the same.

27. x-axis: If the same or an equivalent equation is obtained when θ is replaced by $\theta + \pi$ and r by $-r$. y-axis: If the same or an equivalent equation is obtained when θ is replaced by $2\pi - \theta$ and r by $-r$. origin: If the same or an equivalent equation is obtained when θ is replaced by $-\pi + \theta$.

29. Graph is symmetric to y-axis because $r = \sin(\pi - \theta) \cos^2(\pi - \theta)$ is equivalent to $r = \sin\theta \cos^2\theta$.

θ	r
$0; \frac{\pi}{2}$	0
$\frac{\pi}{6}$	$\frac{3}{8}$
$\frac{\pi}{4}$	$\frac{\sqrt{2}}{4}$
$\frac{\pi}{3}$	$\frac{\sqrt{3}}{8}$
$\frac{11\pi}{6}$	$-\frac{3}{8}$
$\frac{7\pi}{4}$	$-\frac{\sqrt{2}}{4}$
$\frac{5\pi}{3}$	$-\frac{\sqrt{3}}{8}$

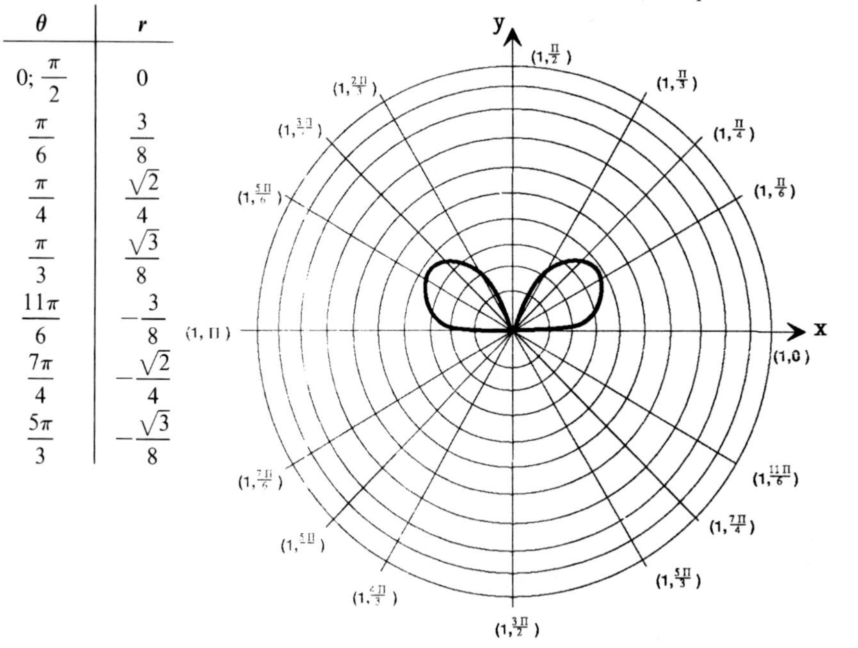

31. Symmetric to the x-axis because $\sin(-\theta)\tan(-\theta) = \sin\theta\tan\theta$.

θ	r
0 & π	0
$\frac{\pi}{6}$	$\frac{\sqrt{3}}{6}$
$\frac{\pi}{4}$	$\frac{\sqrt{2}}{2}$
$\frac{\pi}{3}$	1.5
$\frac{2\pi}{3}$	-1.5
$\frac{3\pi}{4}$	$-\frac{\sqrt{2}}{2}$
$\frac{5\pi}{6}$	$-\frac{\sqrt{3}}{6}$

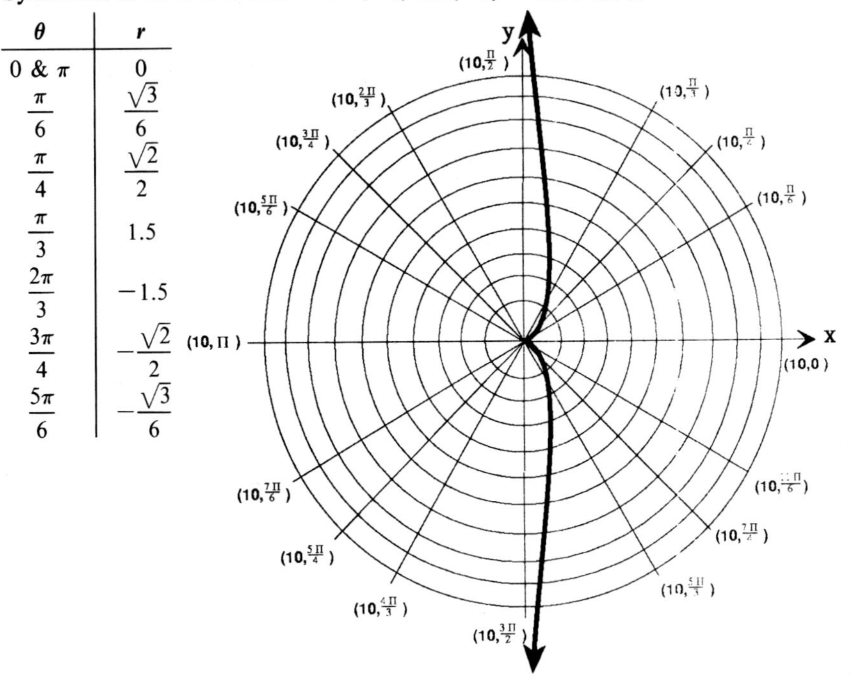

Section 10.3

1. (a)

t	$x = t - 2$	$y = 2t + 1$
2	0	5
1	−1	3
0	−2	1
−1	−3	−1
−2	−4	−3

(b) $y = 2x + 5$

3. (a)

t	$x = t - 1$	$y = 2t + 3$
2	1	7
1	0	5
0	−1	3
−1	−2	1
−2	−3	−1

(b) $y = 2x + 5$

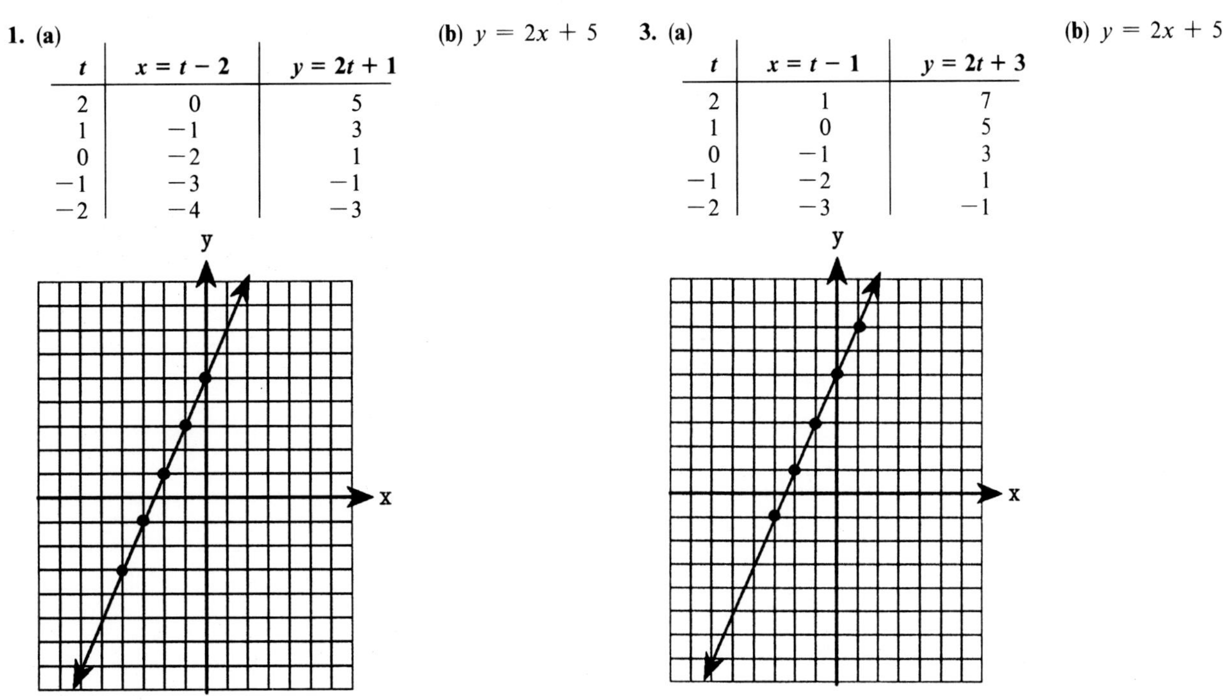

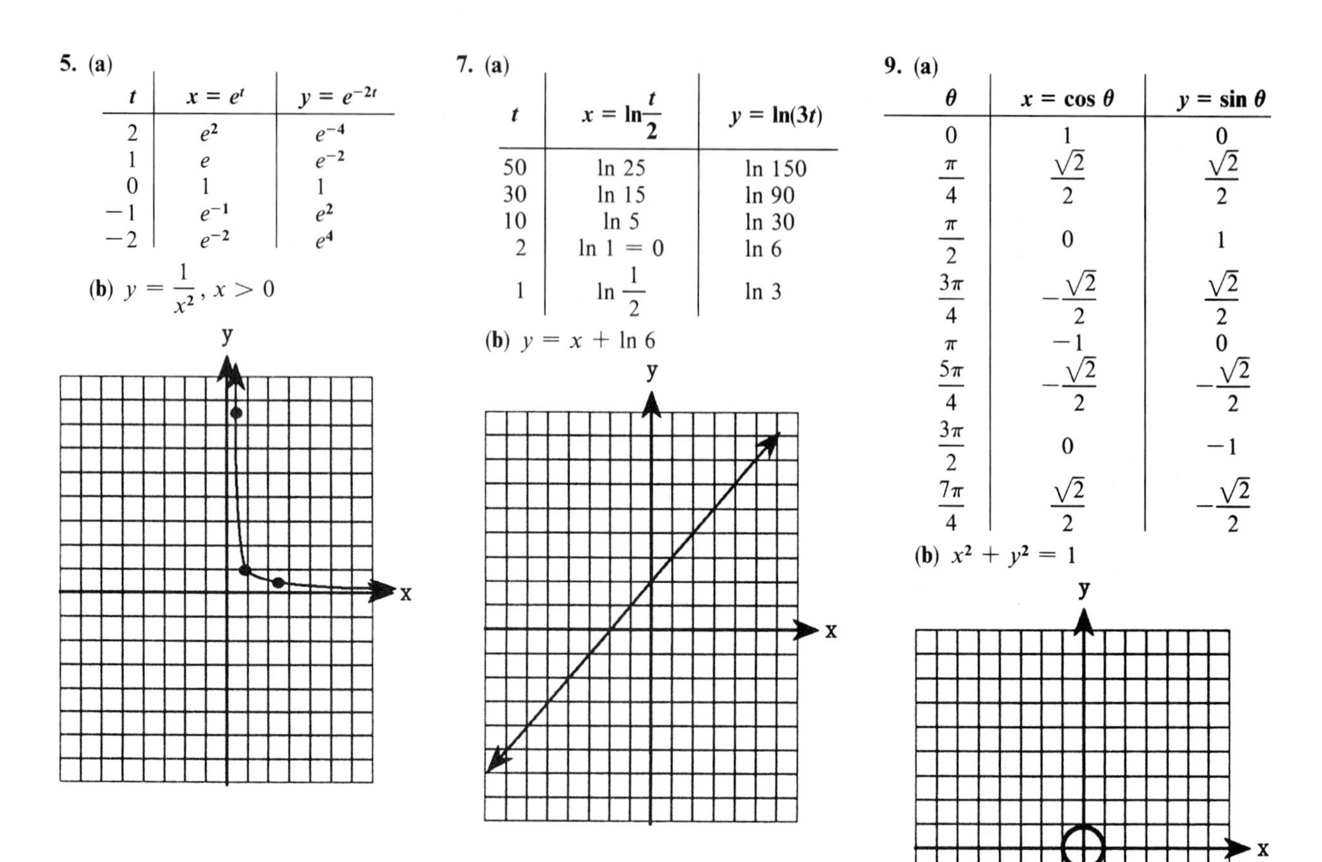

5. (a)

t	$x = e^t$	$y = e^{-2t}$
2	e^2	e^{-4}
1	e	e^{-2}
0	1	1
-1	e^{-1}	e^2
-2	e^{-2}	e^4

(b) $y = \frac{1}{x^2}, x > 0$

7. (a)

t	$x = \ln\frac{t}{2}$	$y = \ln(3t)$
50	ln 25	ln 150
30	ln 15	ln 90
10	ln 5	ln 30
2	$\ln 1 = 0$	ln 6
1	$\ln\frac{1}{2}$	ln 3

(b) $y = x + \ln 6$

9. (a)

θ	$x = \cos\theta$	$y = \sin\theta$
0	1	0
$\frac{\pi}{4}$	$\frac{\sqrt{2}}{2}$	$\frac{\sqrt{2}}{2}$
$\frac{\pi}{2}$	0	1
$\frac{3\pi}{4}$	$-\frac{\sqrt{2}}{2}$	$\frac{\sqrt{2}}{2}$
π	-1	0
$\frac{5\pi}{4}$	$-\frac{\sqrt{2}}{2}$	$-\frac{\sqrt{2}}{2}$
$\frac{3\pi}{2}$	0	-1
$\frac{7\pi}{4}$	$\frac{\sqrt{2}}{2}$	$-\frac{\sqrt{2}}{2}$

(b) $x^2 + y^2 = 1$

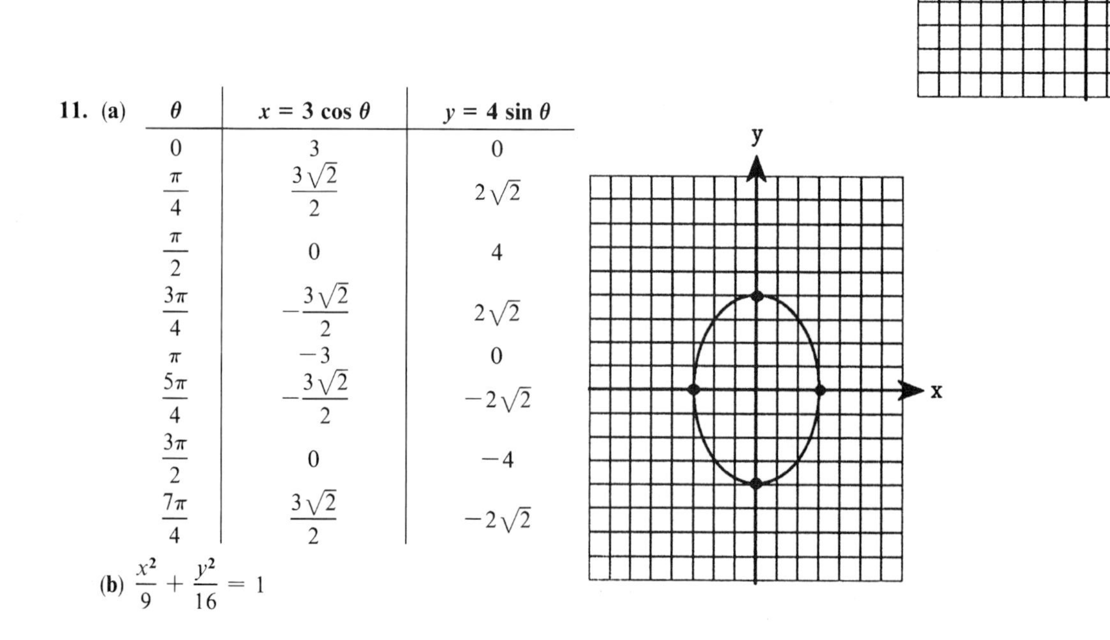

11. (a)

θ	$x = 3\cos\theta$	$y = 4\sin\theta$
0	3	0
$\frac{\pi}{4}$	$\frac{3\sqrt{2}}{2}$	$2\sqrt{2}$
$\frac{\pi}{2}$	0	4
$\frac{3\pi}{4}$	$-\frac{3\sqrt{2}}{2}$	$2\sqrt{2}$
π	-3	0
$\frac{5\pi}{4}$	$-\frac{3\sqrt{2}}{2}$	$-2\sqrt{2}$
$\frac{3\pi}{2}$	0	-4
$\frac{7\pi}{4}$	$\frac{3\sqrt{2}}{2}$	$-2\sqrt{2}$

(b) $\frac{x^2}{9} + \frac{y^2}{16} = 1$

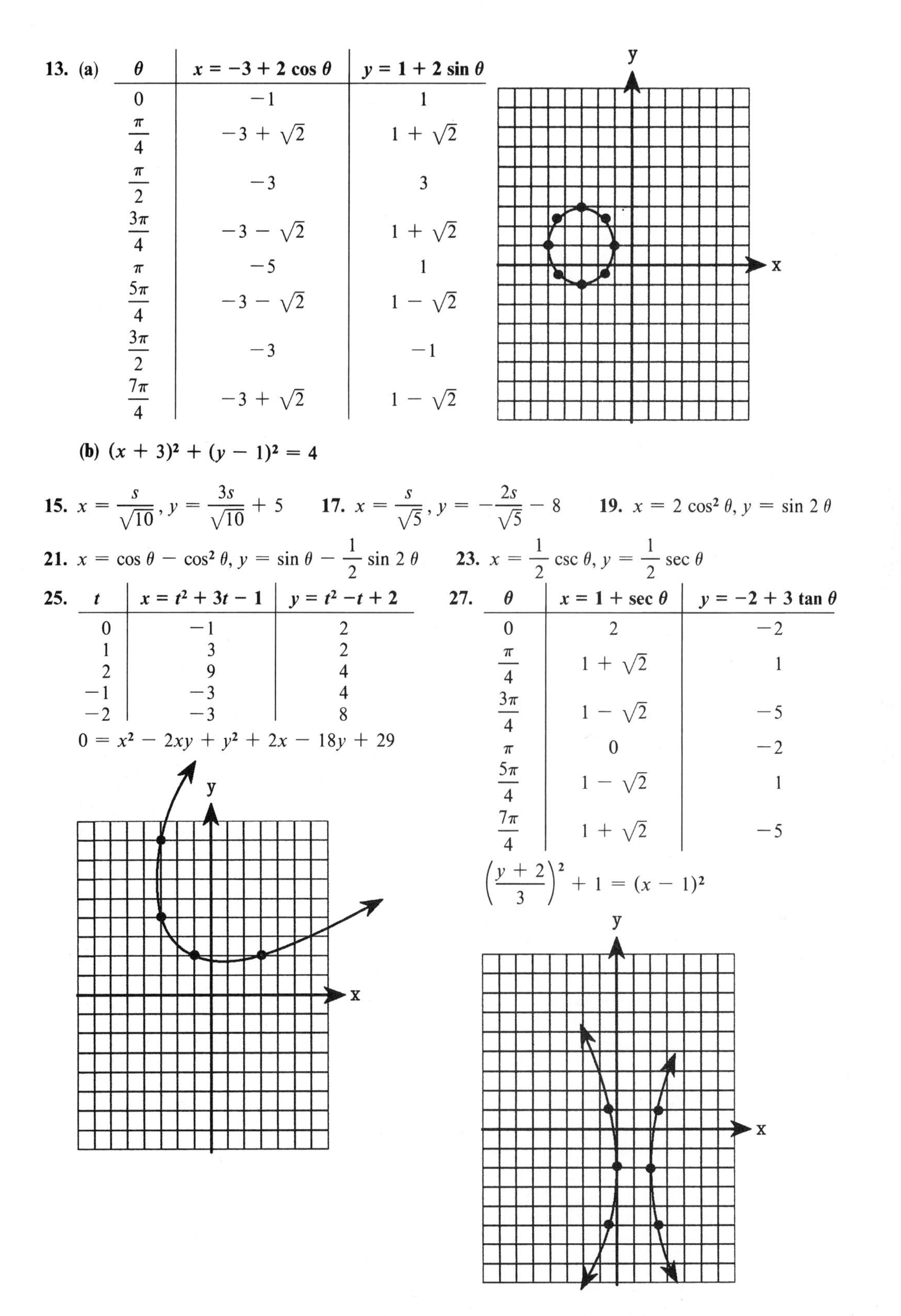

13. **(a)**

θ	$x = -3 + 2\cos\theta$	$y = 1 + 2\sin\theta$
0	-1	1
$\frac{\pi}{4}$	$-3 + \sqrt{2}$	$1 + \sqrt{2}$
$\frac{\pi}{2}$	-3	3
$\frac{3\pi}{4}$	$-3 - \sqrt{2}$	$1 + \sqrt{2}$
π	-5	1
$\frac{5\pi}{4}$	$-3 - \sqrt{2}$	$1 - \sqrt{2}$
$\frac{3\pi}{2}$	-3	-1
$\frac{7\pi}{4}$	$-3 + \sqrt{2}$	$1 - \sqrt{2}$

(b) $(x + 3)^2 + (y - 1)^2 = 4$

15. $x = \frac{s}{\sqrt{10}}, y = \frac{3s}{\sqrt{10}} + 5$ **17.** $x = \frac{s}{\sqrt{5}}, y = -\frac{2s}{\sqrt{5}} - 8$ **19.** $x = 2\cos^2\theta, y = \sin 2\theta$

21. $x = \cos\theta - \cos^2\theta, y = \sin\theta - \frac{1}{2}\sin 2\theta$ **23.** $x = \frac{1}{2}\csc\theta, y = \frac{1}{2}\sec\theta$

25.

t	$x = t^2 + 3t - 1$	$y = t^2 - t + 2$
0	-1	2
1	3	2
2	9	4
-1	-3	4
-2	-3	8

$0 = x^2 - 2xy + y^2 + 2x - 18y + 29$

27.

θ	$x = 1 + \sec\theta$	$y = -2 + 3\tan\theta$
0	2	-2
$\frac{\pi}{4}$	$1 + \sqrt{2}$	1
$\frac{3\pi}{4}$	$1 - \sqrt{2}$	-5
π	0	-2
$\frac{5\pi}{4}$	$1 - \sqrt{2}$	1
$\frac{7\pi}{4}$	$1 + \sqrt{2}$	-5

$\left(\frac{y + 2}{3}\right)^2 + 1 = (x - 1)^2$

29. $r(1 - \sin\theta) = 2$ is equivalent to $r = y + 2$. Squaring both sides of this last equation yields

$$\begin{aligned} r^2 &= y^2 + 4y + 4 \\ x^2 + y^2 &= y^2 + 4y + 4 \\ x^2 &= 4y + 4. \end{aligned}$$

Solving $x = 2(t + 1)$ for t gives us $t = \dfrac{x - 2}{2}$. Substituting this value of t in $y = t(t + 2)$ generates equations

$$\begin{aligned} y &= \left(\frac{x-2}{2}\right)\left(\frac{x-2}{2} + 2\right) \\ y &= \frac{x^2 - 4}{4} \\ x^2 &= 4y + 4. \end{aligned}$$

Since the polar coordinate equation and the pair of parametric equations generate the same rectangular coordinate equation, they determine the same curve.

31. $x = a(\cos\theta + \theta\sin\theta)$, $y = a(\sin\theta - \theta\cos\theta)$

Section 10.4

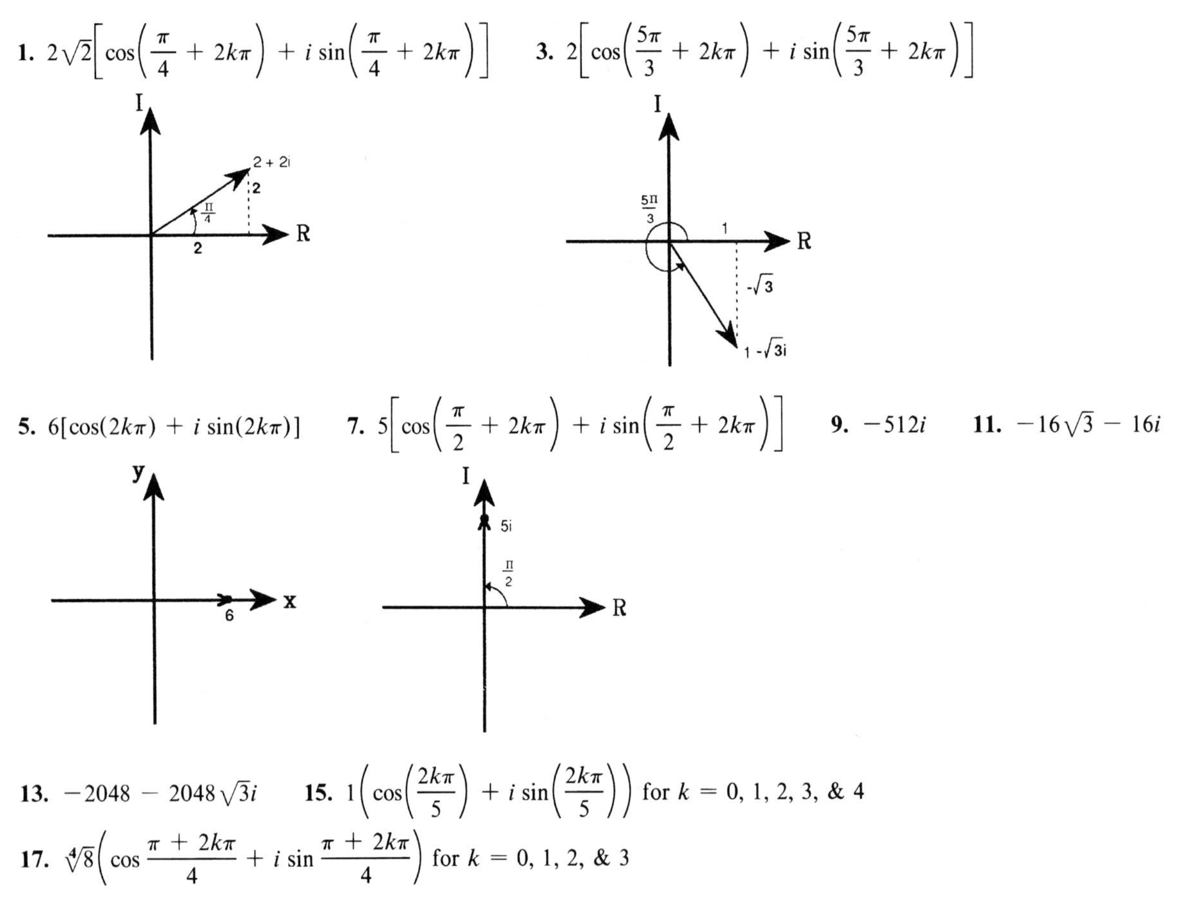

1. $2\sqrt{2}\left[\cos\left(\frac{\pi}{4} + 2k\pi\right) + i\sin\left(\frac{\pi}{4} + 2k\pi\right)\right]$

3. $2\left[\cos\left(\frac{5\pi}{3} + 2k\pi\right) + i\sin\left(\frac{5\pi}{3} + 2k\pi\right)\right]$

5. $6[\cos(2k\pi) + i\sin(2k\pi)]$

7. $5\left[\cos\left(\frac{\pi}{2} + 2k\pi\right) + i\sin\left(\frac{\pi}{2} + 2k\pi\right)\right]$

9. $-512i$

11. $-16\sqrt{3} - 16i$

13. $-2048 - 2048\sqrt{3}i$

15. $1\left(\cos\left(\frac{2k\pi}{5}\right) + i\sin\left(\frac{2k\pi}{5}\right)\right)$ for $k = 0, 1, 2, 3, \& 4$

17. $\sqrt[4]{8}\left(\cos\frac{\pi + 2k\pi}{4} + i\sin\frac{\pi + 2k\pi}{4}\right)$ for $k = 0, 1, 2, \& 3$

19. $2\left(\cos\dfrac{\pi + 2k\pi}{6} + i\sin\dfrac{\pi + 2k\pi}{6}\right)$ for $k = 0, 1, 2, 3, 4, \& \ 5$

21. $1\left(\cos\dfrac{\pi + 4k\pi}{8} + i\sin\dfrac{\pi + 4k\pi}{8}\right)$ for $k = 0, 1, 2, \& \ 3$

23. $\sqrt[3]{2}\left(\cos\dfrac{5\pi + 6k\pi}{9} + i\sin\dfrac{5\pi + 6k\pi}{9}\right)$ for $k = 0, 1, \& \ 2$

25. $2\left(\cos\dfrac{11\pi + 12k\pi}{30} + i\sin\dfrac{11\pi + 12k\pi}{30}\right)$ for $k = 0, 1, 2, 3, \& \ 4$

27.
$$\begin{aligned}\frac{r_1(\cos\theta_1 + i\sin\theta_1)}{r_2(\cos\theta_2 + i\sin\theta_2)} &= \frac{r_1(\cos\theta_1 + i\sin\theta_1)(\cos\theta_2 - i\sin\theta_2)}{r_2(\cos\theta_2 + i\sin\theta_2)(\cos\theta_2 - i\sin\theta_2)}\\ &= \frac{r_1}{r_2}\,\frac{(\cos\theta_1\cos\theta_2 + \sin\theta_1\sin\theta_2 + i(\sin\theta_1\cos_2 - \sin\theta_2\cos\theta_1)}{\cos^2\theta_2 + \sin^2\theta_2}\\ &= \frac{r_1}{r_2}(\cos(\theta_1 - \theta_2) + i\sin(\theta_1 - \theta_2))\end{aligned}$$

Chapter 10 Review Exercises

1. $\left(4\sqrt{2}, \dfrac{7\pi}{4}\right)$ and $\left(-4\sqrt{2}, \dfrac{3\pi}{4}\right)$ **3.** $x = \dfrac{3}{2}, y = \dfrac{3\sqrt{3}}{2}$ **5.** $2\left(\cos\dfrac{11\pi}{6} + i\sin\dfrac{11\pi}{6}\right)$ **7.** $-972 - 972i$

9. $\tan\theta = -5$ **11.** $r^2 = \dfrac{4}{1 + \sin 2\theta}$ **13.** $r = -6\cos\theta$ **15.** $r = 9\tan\theta\sec\theta$ **17.** $y = \dfrac{x}{\sqrt{3}}$ **19.** $y = -2$

21. $y^2 = 16 + 8x$ **23.** $x^2 - y^2 = \dfrac{5}{2}$

25.

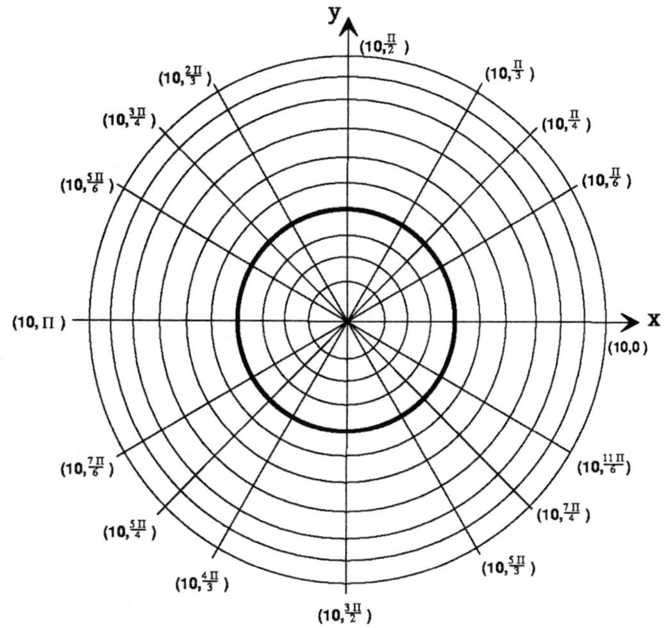

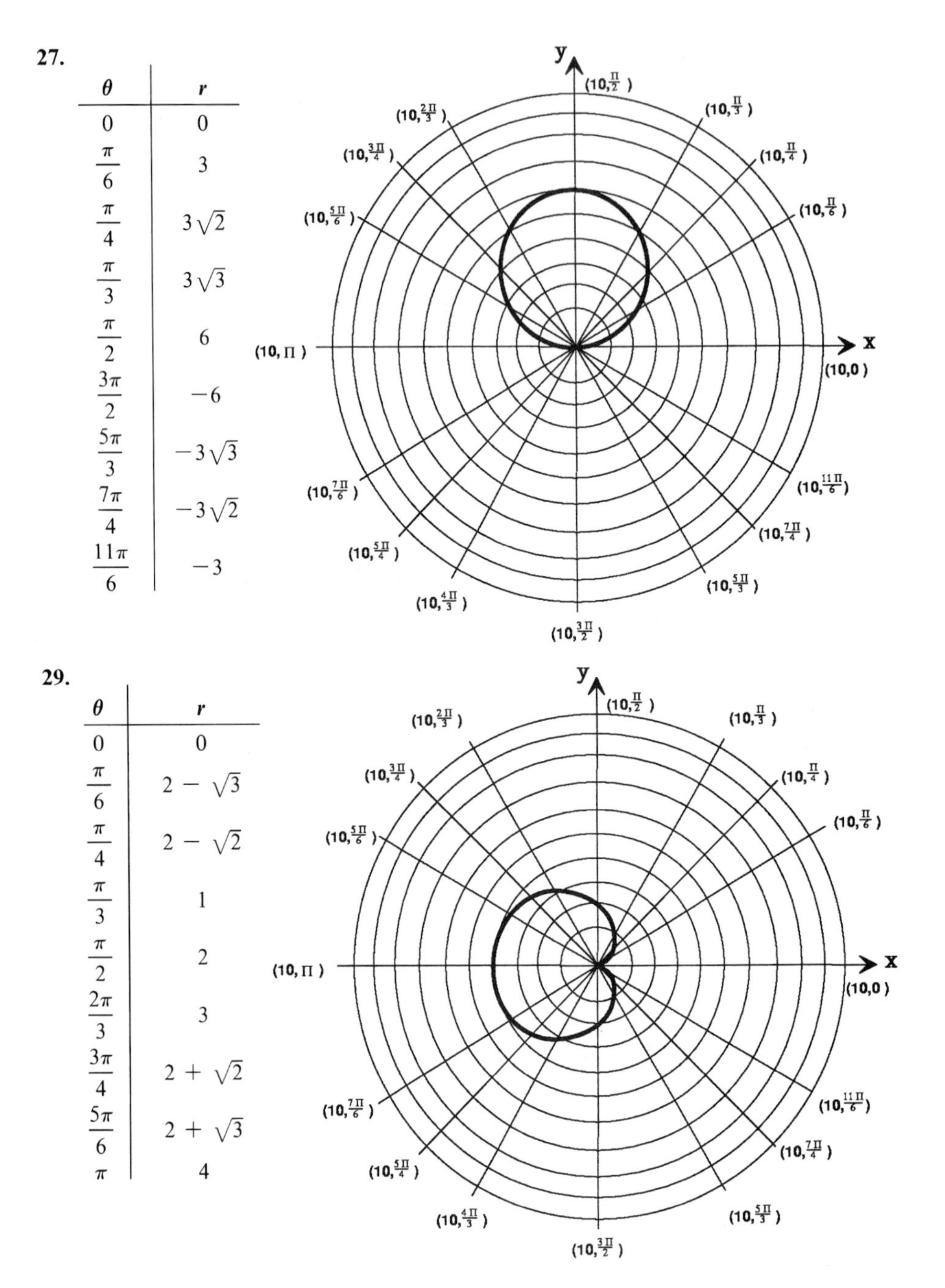

27.

θ	r
0	0
$\frac{\pi}{6}$	3
$\frac{\pi}{4}$	$3\sqrt{2}$
$\frac{\pi}{3}$	$3\sqrt{3}$
$\frac{\pi}{2}$	6
$\frac{3\pi}{2}$	-6
$\frac{5\pi}{3}$	$-3\sqrt{3}$
$\frac{7\pi}{4}$	$-3\sqrt{2}$
$\frac{11\pi}{6}$	-3

29.

θ	r
0	0
$\frac{\pi}{6}$	$2-\sqrt{3}$
$\frac{\pi}{4}$	$2-\sqrt{2}$
$\frac{\pi}{3}$	1
$\frac{\pi}{2}$	2
$\frac{2\pi}{3}$	3
$\frac{3\pi}{4}$	$2+\sqrt{2}$
$\frac{5\pi}{6}$	$2+\sqrt{3}$
π	4

31.

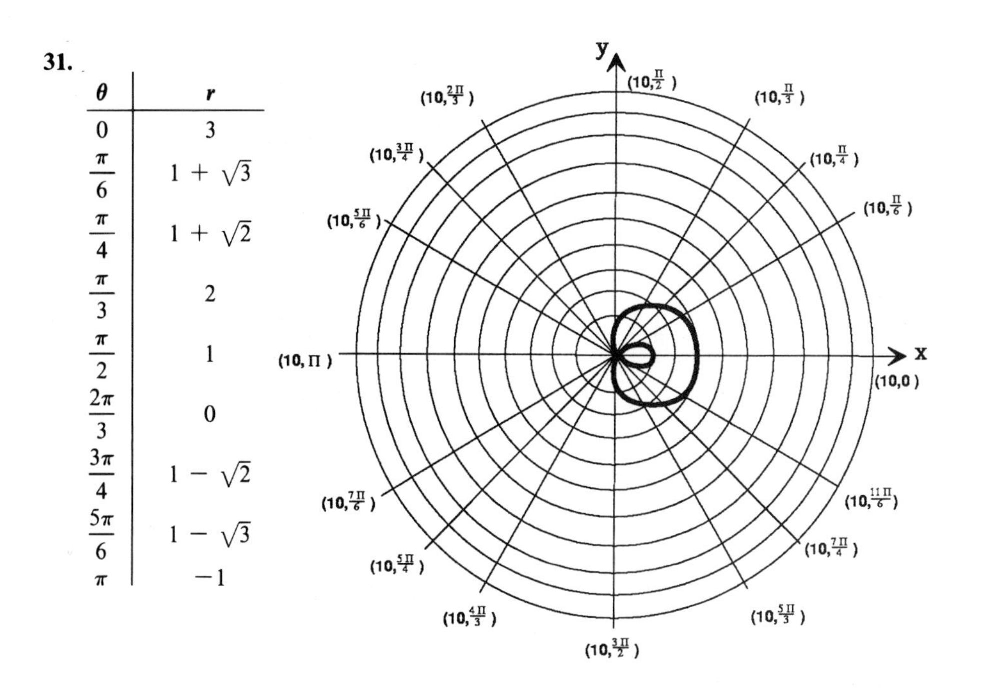

θ	r
0	3
$\frac{\pi}{6}$	$1+\sqrt{3}$
$\frac{\pi}{4}$	$1+\sqrt{2}$
$\frac{\pi}{3}$	2
$\frac{\pi}{2}$	1
$\frac{2\pi}{3}$	0
$\frac{3\pi}{4}$	$1-\sqrt{2}$
$\frac{5\pi}{6}$	$1-\sqrt{3}$
π	-1

33.

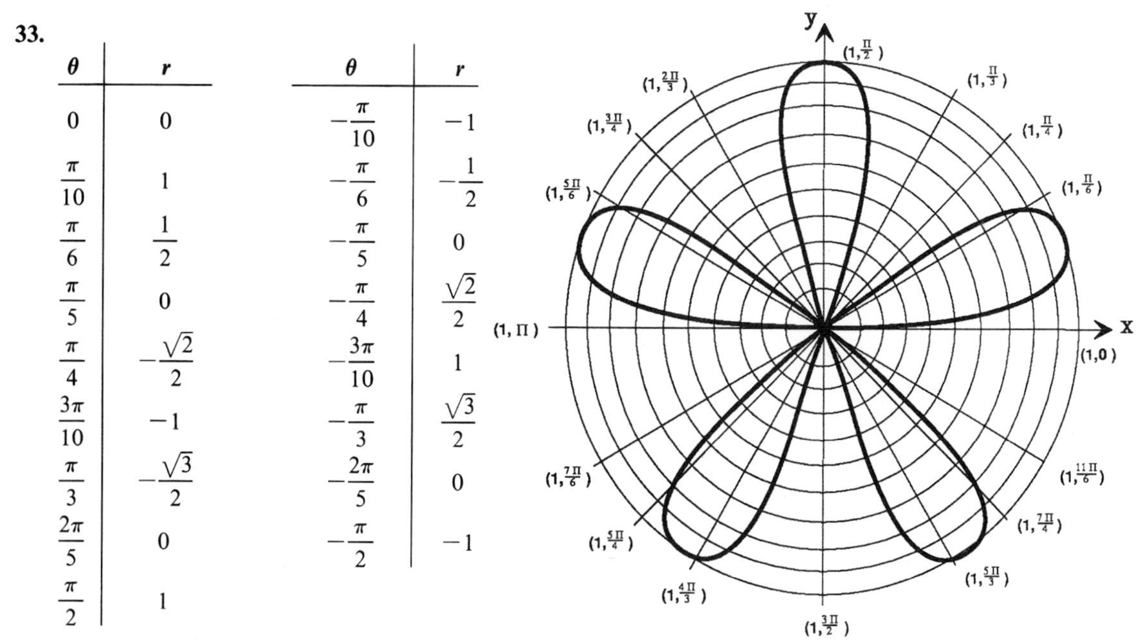

θ	r
0	0
$\frac{\pi}{10}$	1
$\frac{\pi}{6}$	$\frac{1}{2}$
$\frac{\pi}{5}$	0
$\frac{\pi}{4}$	$-\frac{\sqrt{2}}{2}$
$\frac{3\pi}{10}$	-1
$\frac{\pi}{3}$	$-\frac{\sqrt{3}}{2}$
$\frac{2\pi}{5}$	0
$\frac{\pi}{2}$	1

θ	r
$-\frac{\pi}{10}$	-1
$-\frac{\pi}{6}$	$-\frac{1}{2}$
$-\frac{\pi}{5}$	0
$-\frac{\pi}{4}$	$\frac{\sqrt{2}}{2}$
$-\frac{3\pi}{10}$	1
$-\frac{\pi}{3}$	$\frac{\sqrt{3}}{2}$
$-\frac{2\pi}{5}$	0
$-\frac{\pi}{2}$	-1

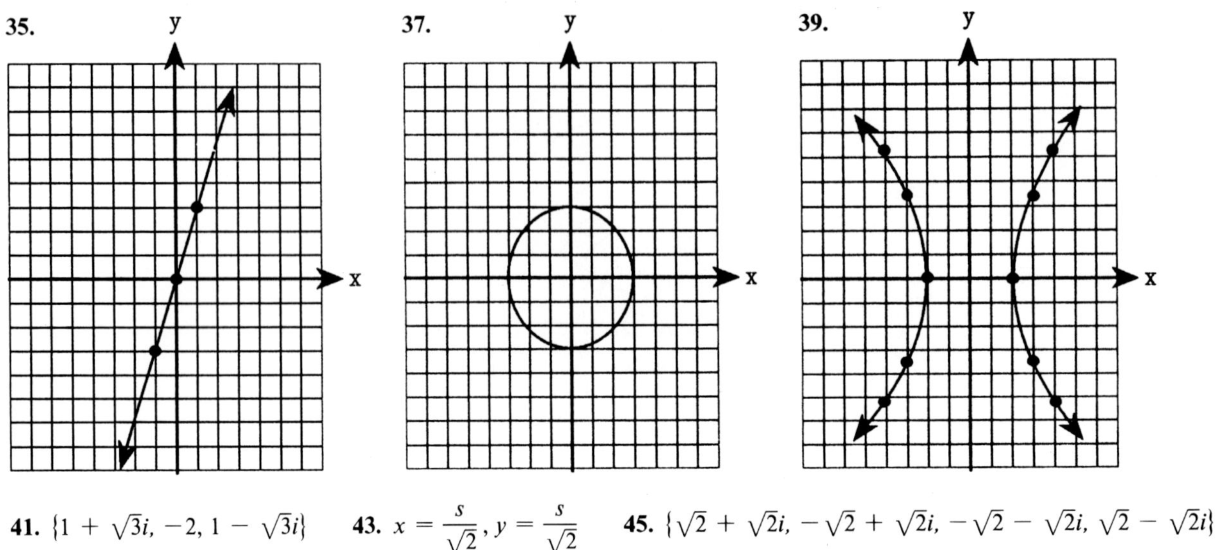

41. $\{1 + \sqrt{3}i, -2, 1 - \sqrt{3}i\}$ **43.** $x = \dfrac{s}{\sqrt{2}}, y = \dfrac{s}{\sqrt{2}}$ **45.** $\{\sqrt{2} + \sqrt{2}i, -\sqrt{2} + \sqrt{2}i, -\sqrt{2} - \sqrt{2}i, \sqrt{2} - \sqrt{2}i\}$

47. $x = 2 \sin \theta \cos^2 \theta, y = 2 \sin^2 \theta \cos \theta$ **49.** $0 = 4x^2 - 4xy + y^2 - 11x - 55y$

Chapter 10 Test 1

1. **(a)** $x = \dfrac{\sqrt{3}}{2}$ and $y = \dfrac{1}{2}$ **(b)** $\left(2\sqrt{3}, \dfrac{\pi}{3}\right)$ **2.** $r = \dfrac{1}{2} \tan \theta \sec \theta$ **3.** $x^2 + y^2 - 5y = 0$

4. $r = 3 \sec \theta = \dfrac{3}{\cos \theta}$

$r \cos \theta = 3$

$x = 3$

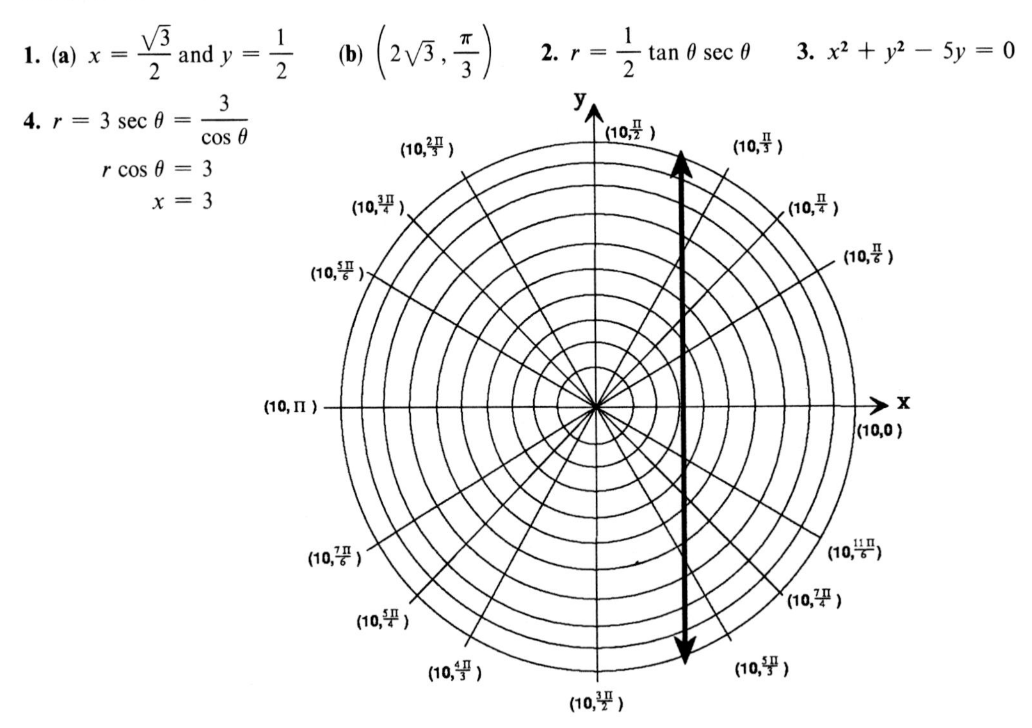

5.

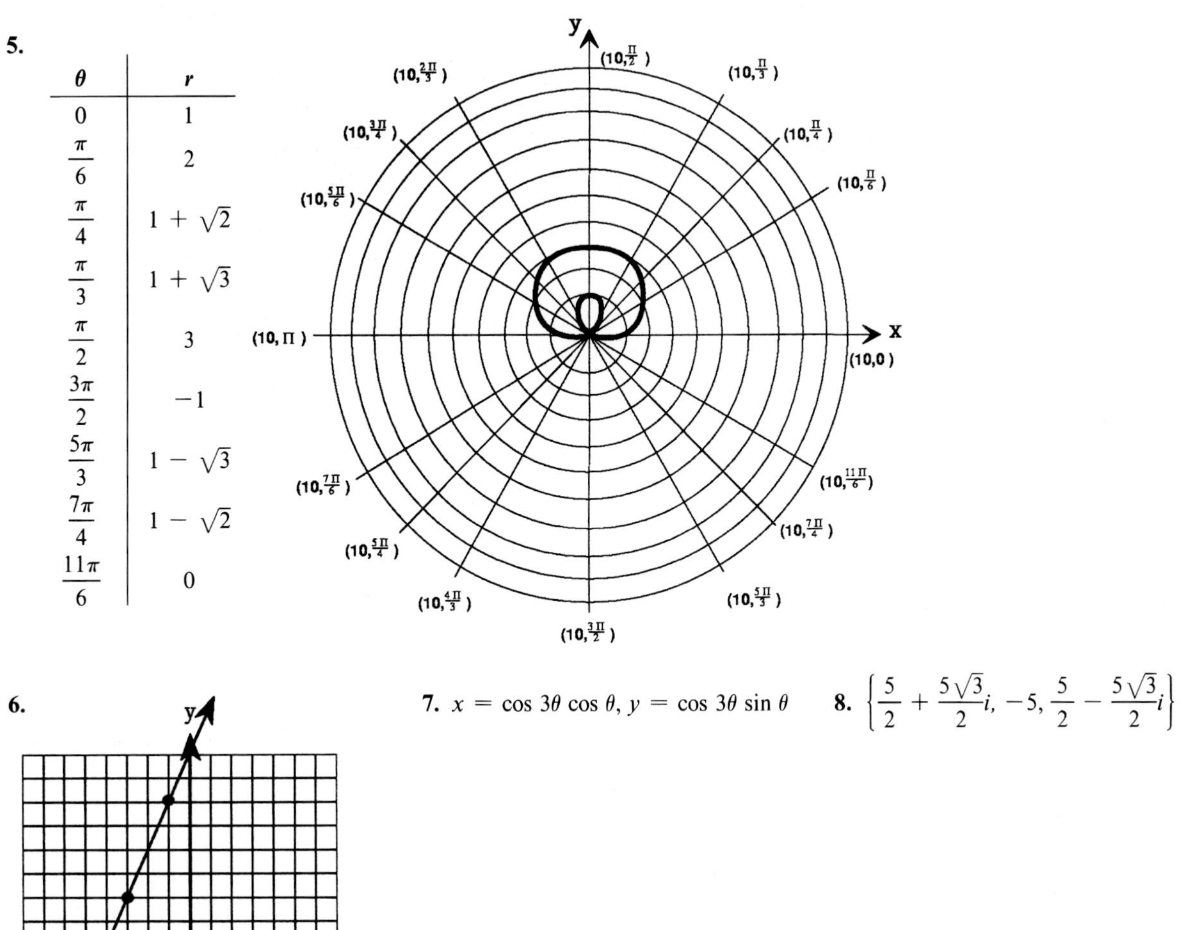

θ	r
0	1
$\frac{\pi}{6}$	2
$\frac{\pi}{4}$	$1 + \sqrt{2}$
$\frac{\pi}{3}$	$1 + \sqrt{3}$
$\frac{\pi}{2}$	3
$\frac{3\pi}{2}$	-1
$\frac{5\pi}{3}$	$1 - \sqrt{3}$
$\frac{7\pi}{4}$	$1 - \sqrt{2}$
$\frac{11\pi}{6}$	0

6.

7. $x = \cos 3\theta \cos \theta,\ y = \cos 3\theta \sin \theta$

8. $\left\{\frac{5}{2} + \frac{5\sqrt{3}}{2}i,\ -5,\ \frac{5}{2} - \frac{5\sqrt{3}}{2}i\right\}$

Chapter 10 Test 2

1. **(a)** $x = \frac{5\sqrt{2}}{2}$ and $y = -\frac{5\sqrt{2}}{2}$ **(b)** $\left(\sqrt{41}, \tan^{-1}\left(-\frac{5}{4}\right)\right)$ **2.** $r = 5 \sin \theta - 2 \cos \theta$

3. $0 = 8x^2 - y^2 + 12x + 4$

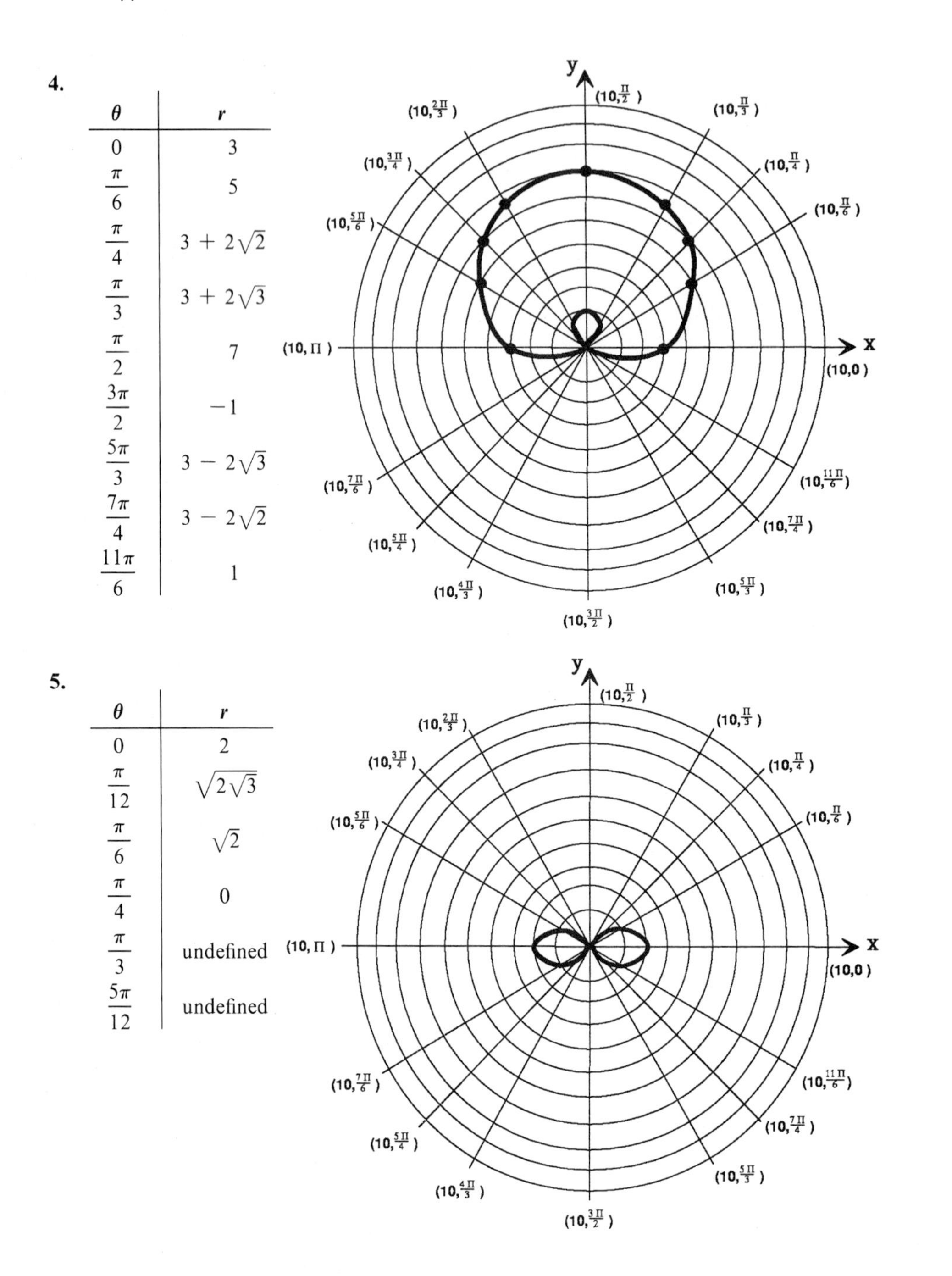

4.

θ	r
0	3
$\frac{\pi}{6}$	5
$\frac{\pi}{4}$	$3 + 2\sqrt{2}$
$\frac{\pi}{3}$	$3 + 2\sqrt{3}$
$\frac{\pi}{2}$	7
$\frac{3\pi}{2}$	-1
$\frac{5\pi}{3}$	$3 - 2\sqrt{3}$
$\frac{7\pi}{4}$	$3 - 2\sqrt{2}$
$\frac{11\pi}{6}$	1

5.

θ	r
0	2
$\frac{\pi}{12}$	$\sqrt{2\sqrt{3}}$
$\frac{\pi}{6}$	$\sqrt{2}$
$\frac{\pi}{4}$	0
$\frac{\pi}{3}$	undefined
$\frac{5\pi}{12}$	undefined

6.

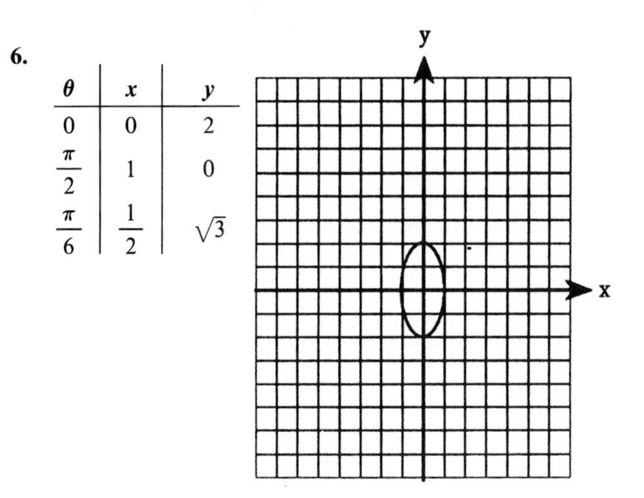

θ	x	y
0	0	2
$\frac{\pi}{2}$	1	0
$\frac{\pi}{6}$	$\frac{1}{2}$	$\sqrt{3}$

7. $x = \cos\theta - \sin 2\theta,\ y = \sin\theta - 2\sin^2\theta$

8. $2^{1/8}\left[\cos\dfrac{(3+8k)\pi}{16} + i\sin\dfrac{(3+8k)\pi}{16}\right],\ k = 0, 1, 2, 3$

Section 11.1

1. $(x-1)^2 + (y+5)^2 = 16$
$x^2 + y^2 - 2x + 10y + 10 = 0$
3. $(x+7)^2 + (y+3)^2 = 4$
$x^2 + y^2 + 14x + 6y + 54 = 0$
5. $C(-3, 4), r = 5$
7. $C(-2, -4), r = 25$ **9.** Not a circle; just the point (2, 5) **11.** $17x^2 + 17y^2 - 91x - 13y = 0$
13. $x^2 + y^2 - 6x - 2y - 7 = 0$ **15.** $x^2 + y^2 + 14x - 4y + 49 = 0$
17. $29x^2 + 29y^2 - 232x + 174y + 709 = 0$ **19.** $x^2 + y^2 - 4x - 21 = 0$ **21.** $x^2 + y^2 + 4x - 6y + 5 = 0$
23. $(x-0)^2 + (y-0)^2 = (2)^2$ **25.** $[x-(-1)]^2 + (y-2)^2 = (3)^2$ **27.** (3.67, 1.52), 1.52 or (6.82, 2.82), 2.82
29. As θ increases from 0 to 2π, the parametrization $x = 2\cos\theta$, $y = 2\sin\theta$ traces out the circle counterclockwise starting at (2, 0). The other parametrization traces the circle clockwise starting at (0, 2).
31.

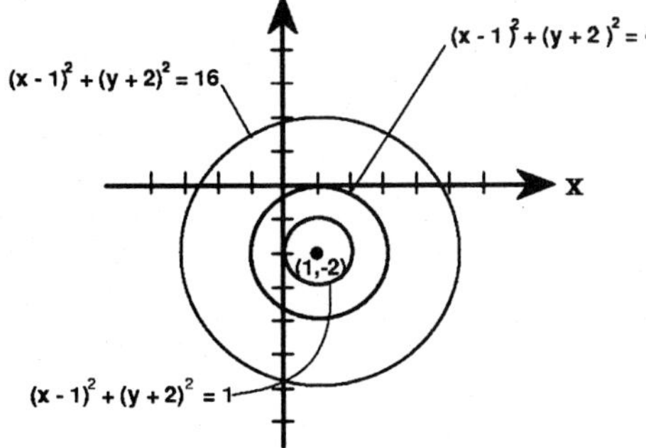

33. $(x-6)^2 + (y-2)^2 = R^2$
35. $(x-h)^2 + [(y-(5-h)]^2 = 4$
37. $(x-h)^2 + (y-2h)^2 = 5h^2$ **39.** $x^2 + y^2 - 6x - 4y = 0$

Section 11.2

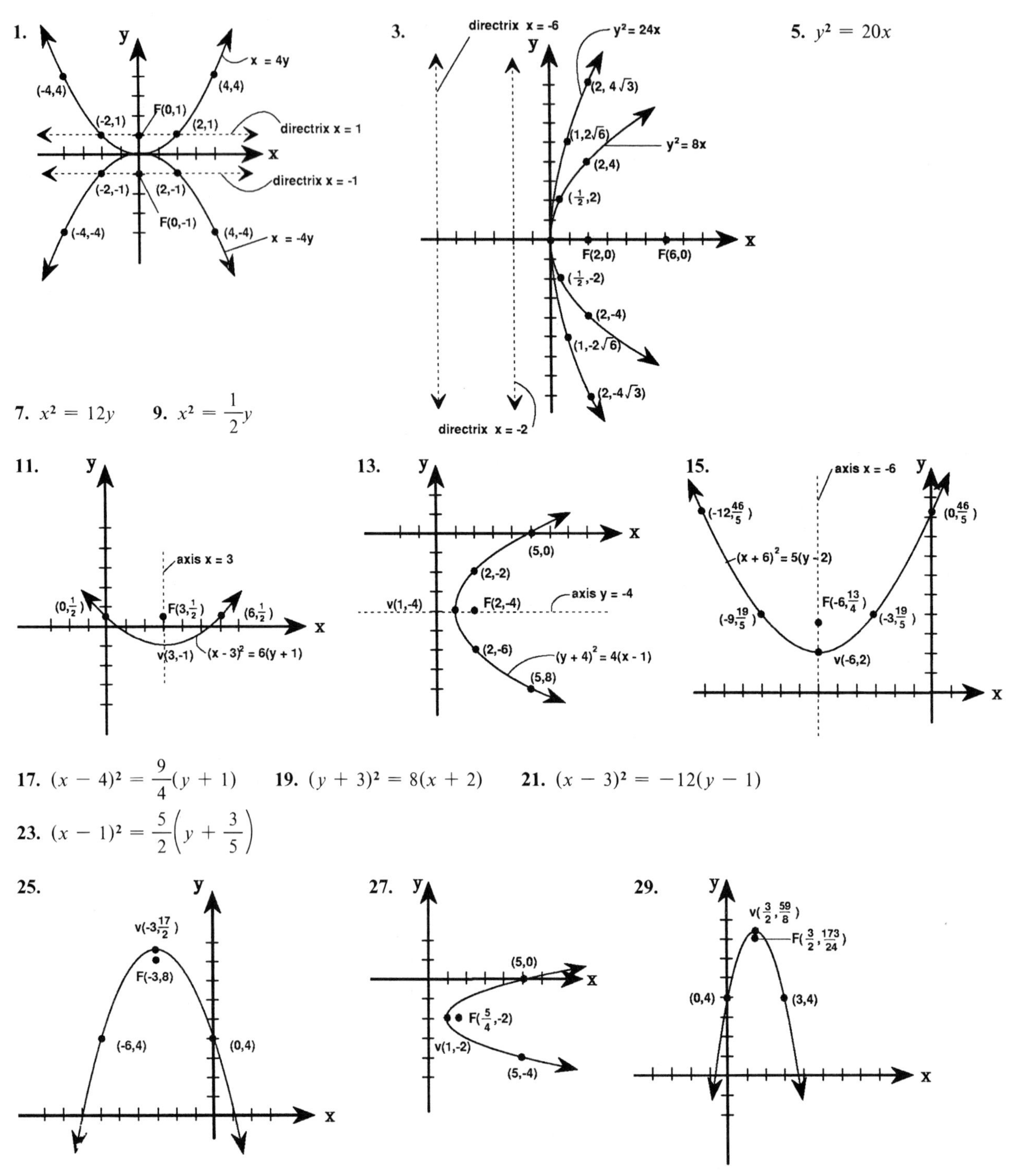

5. $y^2 = 20x$

7. $x^2 = 12y$ **9.** $x^2 = \frac{1}{2}y$

17. $(x - 4)^2 = \frac{9}{4}(y + 1)$ **19.** $(y + 3)^2 = 8(x + 2)$ **21.** $(x - 3)^2 = -12(y - 1)$

23. $(x - 1)^2 = \frac{5}{2}\left(y + \frac{3}{5}\right)$

31. $x^2 - 2xy + y^2 + 2x + 2y - 1 = 0$ **33.** $9x^2 + 6xy + y^2 - 42x - 94y + 49 = 0$ **35.** 50 feet

37. Let the coordinates of P_2 be (x, y) then $\overline{BR} = x$ and $\overline{RP_2} = y$. In the similar triangles BRP_2 and BEQ

$$\frac{x}{y} = \frac{\ell}{\left(\frac{n-2}{n}\right)h}, \tag{1}$$

where $h = \overline{EA} = n(b_i - b_{i-1})$, and $b_i - b_{i-1}$ is the length of one of the n subdivisions of $\overline{EA}$. Since $\overline{AQ}$ encompasses two subdivisions, $\overline{EQ} = \left(\frac{n-2}{n}\right)h$. Similarly, $x = \overline{BR} = \left(\frac{n-2}{n}\right)\ell$. Now, from (1),

$$\frac{x}{y\ell} = \frac{n}{(n-2)h}$$

$$\left(\frac{x}{y\ell}\right)x = \frac{n}{(n-2)h} \cdot \frac{n-2}{n}\ell$$

$$x^2 = \frac{\ell^2}{h}y.$$

The last equation relates the x- and y-coordinates of P_2, and shows that P_2 is a point on a parabola. But the relationship between the x and y coordinates of each of the points, $P_1, P_2, P_3, \ldots, P_n$ is exactly the same. Thus the points determine a parabola.

Section 11.3

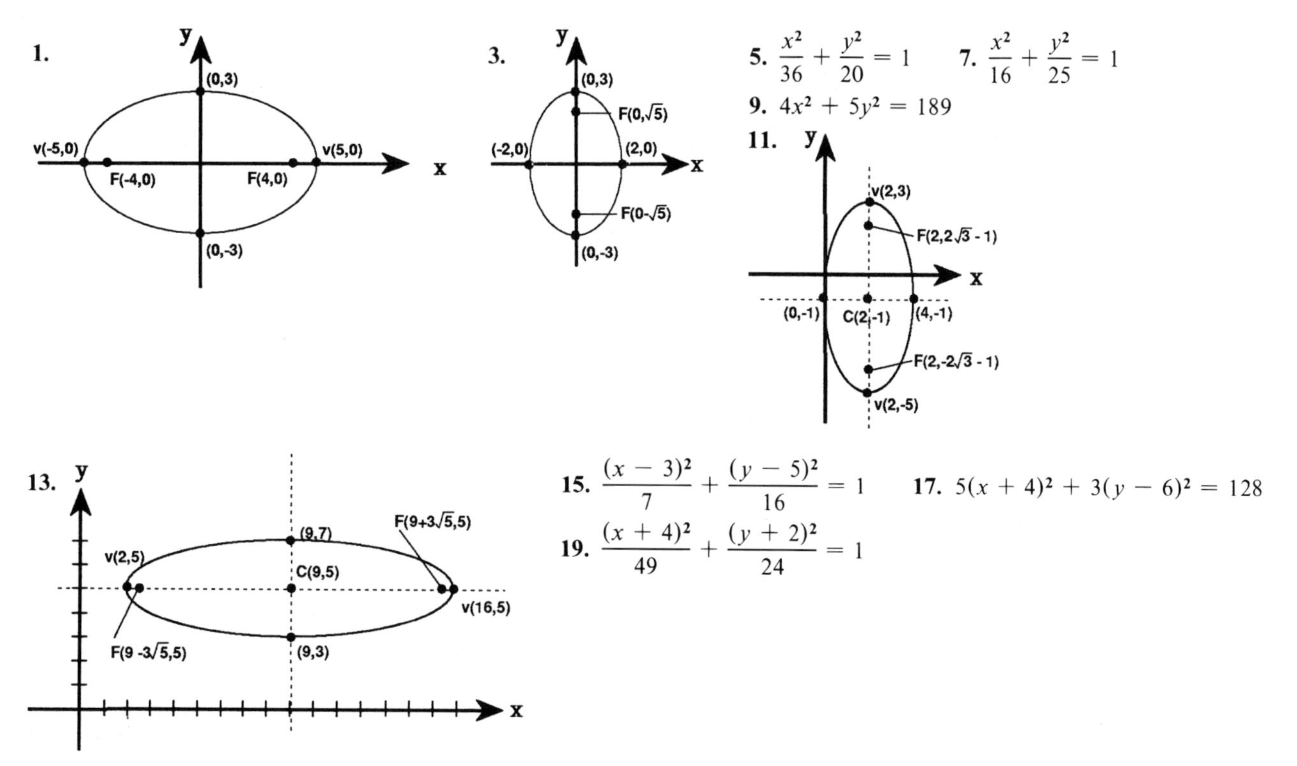

5. $\dfrac{x^2}{36} + \dfrac{y^2}{20} = 1$ **7.** $\dfrac{x^2}{16} + \dfrac{y^2}{25} = 1$

9. $4x^2 + 5y^2 = 189$

15. $\dfrac{(x-3)^2}{7} + \dfrac{(y-5)^2}{16} = 1$ **17.** $5(x+4)^2 + 3(y-6)^2 = 128$

19. $\dfrac{(x+4)^2}{49} + \dfrac{(y+2)^2}{24} = 1$

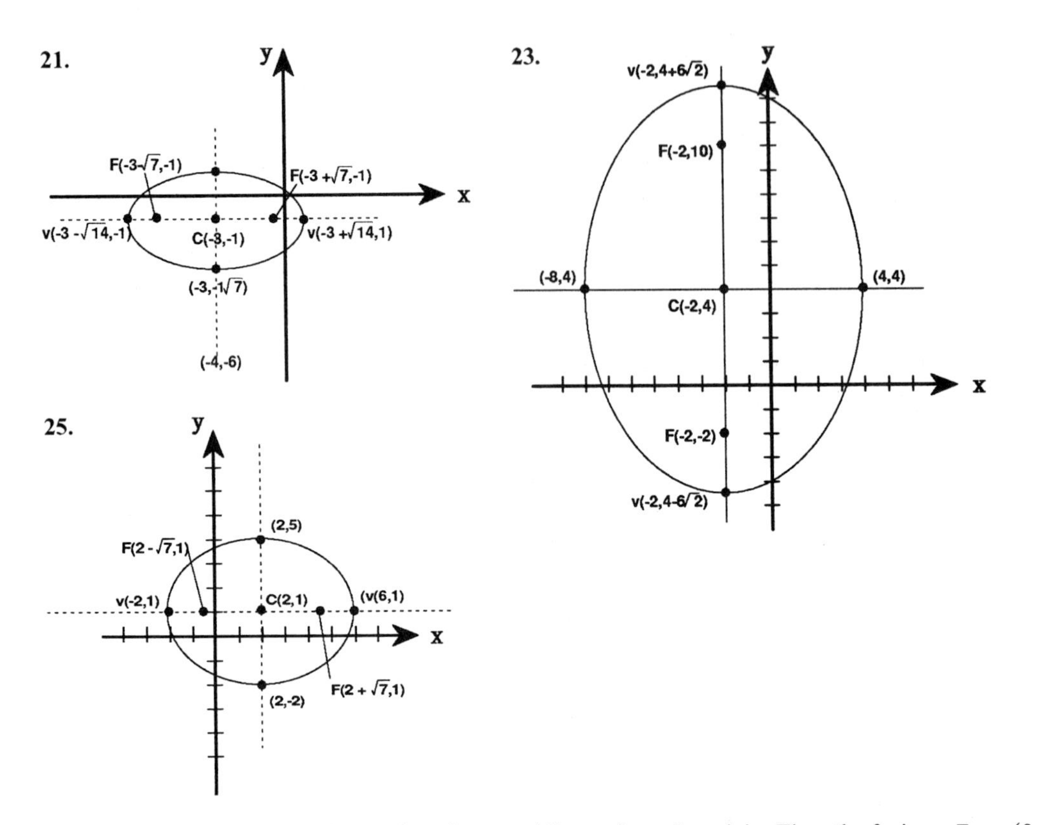

27. Assume the foci are on the y-axis and are equidistant from the origin. Then the foci are $F_1 = (0, c)$ and $F_2 = (0, -c)$. For any point $P = (x, y)$ on the ellipse, the sum of the distance from P to F_1 and from P to F_2 is some constant represented by $(2a)$. That is,

$$\begin{aligned}
\sqrt{(x-0)^2 + (y-c)^2} + \sqrt{(x-0)^2 + (y+c)^2} &= 2a \\
\sqrt{x^2 + y^2 - 2cy + c^2} &= 2a - \sqrt{x^2 + y^2 + 2cy + c^2} \\
x^2 + y^2 - 2cy + c^2 &= 4a^2 - 4a\sqrt{x^2 + y^2 + 2cy + c^2} + x^2 + y^2 + 2cy + c^2 \\
a\sqrt{x^2 + y^2 + 2cy + c^2} &= a^2 + cy \\
a^2x^2 + a^2y^2 + 2a^2cy + a^2c^2 &= a^4 + 2a^2cy + c^2y^2 \\
a^2x^2 + (a^2 - c^2)y^2 &= a^2(a^2 - c^2).
\end{aligned}$$

Letting $a^2 - c^2 = b^2$, and dividing both sides by a^2b^2 yields $\dfrac{x^2}{b^2} + \dfrac{y^2}{a^2} = 1$.

29. $\dfrac{(x - 120)^2}{200^2} + \dfrac{y^2}{160^2} = 1$ **31.** $\dfrac{x^2}{4} + \dfrac{y^2}{9} = 1$ **33.** $\dfrac{x^2}{81} + \dfrac{y^2}{16} = 1$

35. As θ assumes values from 0 to 2π, the parametrization in exercise 32 traces the ellipse in the counterclockwise direction starting at (6, 0) while the parametrization in exercise 34 traces the ellipse in the clockwise direction starting at (0, 4).

Section 11.4

1. **3.**

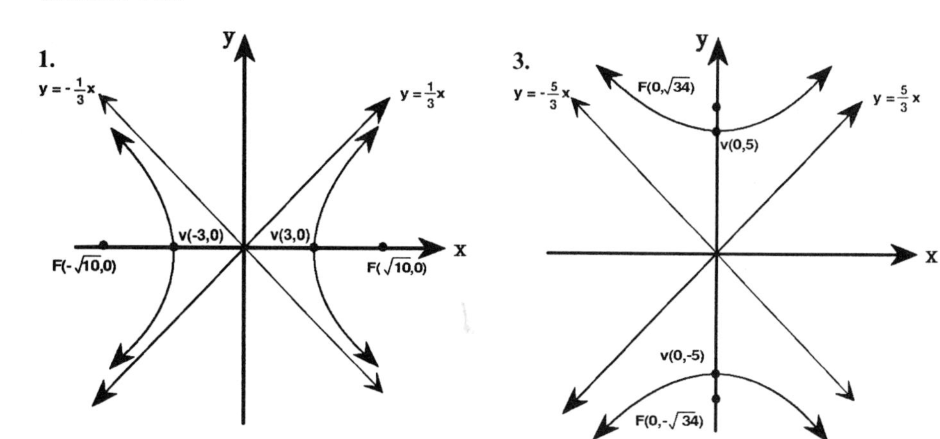

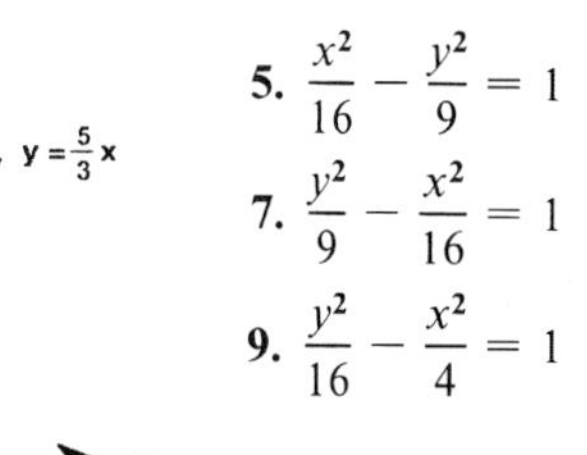

5. $\dfrac{x^2}{16} - \dfrac{y^2}{9} = 1$

7. $\dfrac{y^2}{9} - \dfrac{x^2}{16} = 1$

9. $\dfrac{y^2}{16} - \dfrac{x^2}{4} = 1$

11.

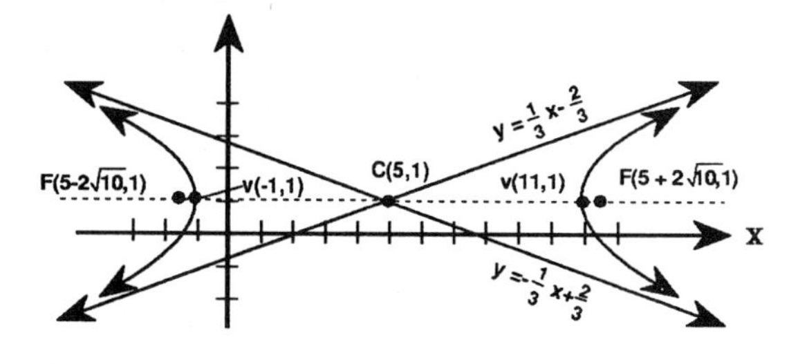

13.

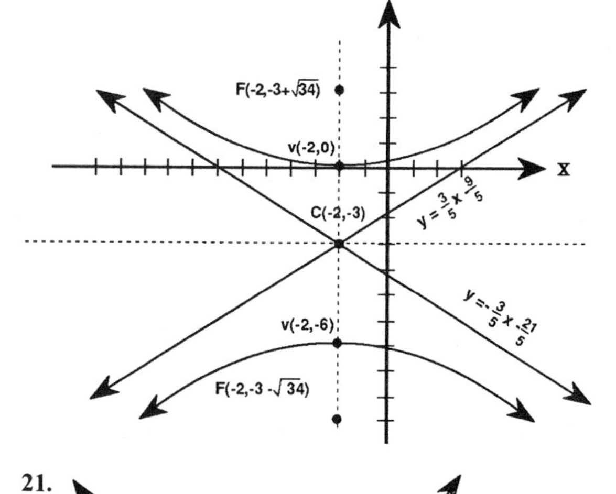

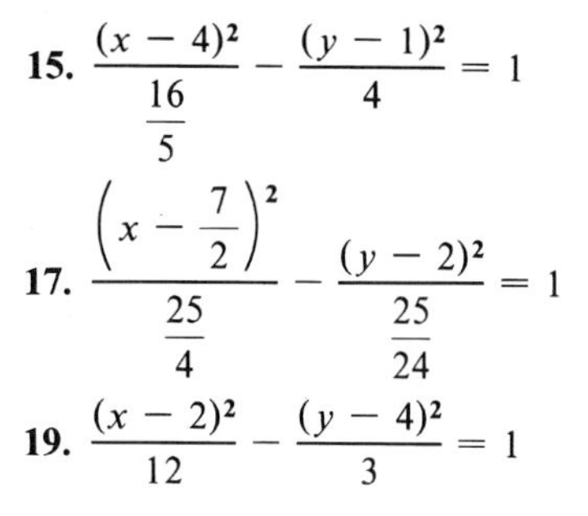

15. $\dfrac{(x-4)^2}{\frac{16}{5}} - \dfrac{(y-1)^2}{4} = 1$

17. $\dfrac{\left(x - \frac{7}{2}\right)^2}{\frac{25}{4}} - \dfrac{(y-2)^2}{\frac{25}{24}} = 1$

19. $\dfrac{(x-2)^2}{12} - \dfrac{(y-4)^2}{3} = 1$

21.

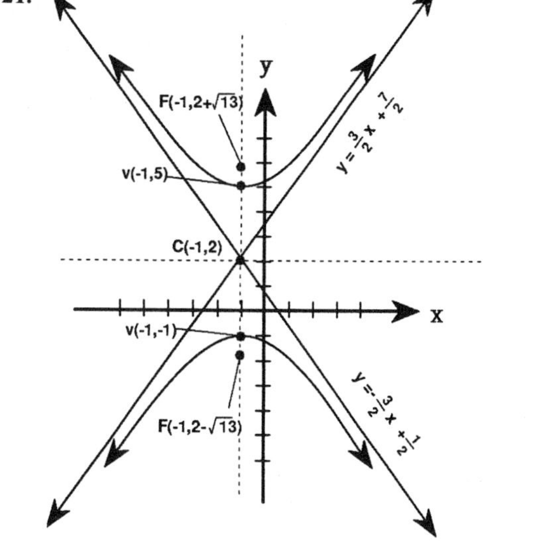

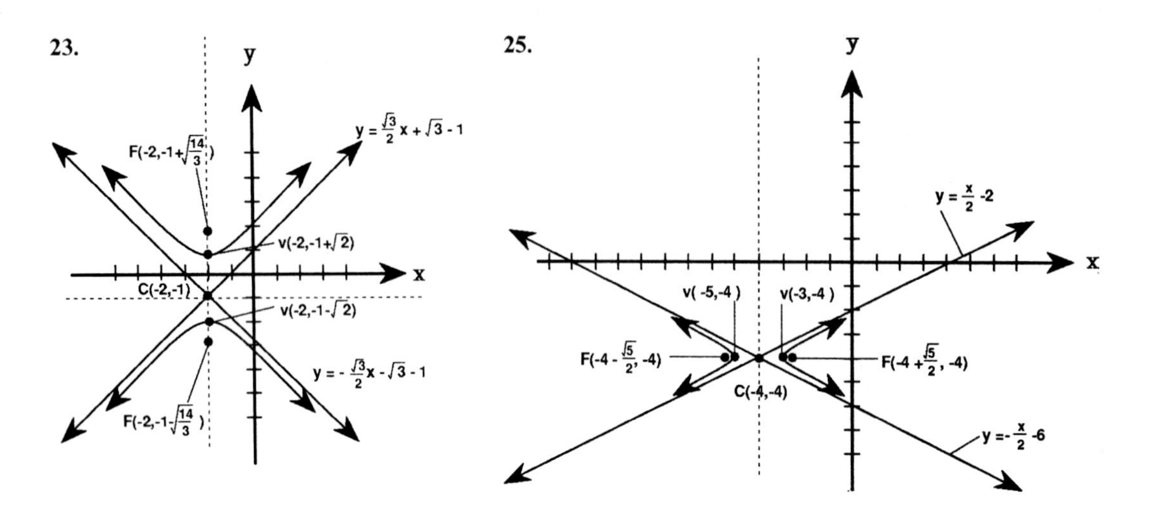

27. Ellipse **29.** Circle **31.** Parabola **33.** Hyperbola

35. $\left|\sqrt{(x-0)^2+(y+c)^2}-\sqrt{(x-0)^2+(y-c)^2}\right| = 2a$

$x^2 + y^2 + 2cy + c^2 - 2\sqrt{(x^2+y^2+2cy+c^2)(x^2+y^2-2cy+c^2)} + x^2 + y^2 - 2cy + c^2 = 4a^2$

$2x^2 + 2y^2 + 2c^2 - 4a^2 = 2\sqrt{(x^2+y^2+2cy+c^2)(x^2+y^2-2cy+c^2)}$

$(x^2 + y^2 + c^2 - 2a^2)^2 = (x^2+y^2+2cy+c^2)(x^2+y^2-2cy+c^2)$

$4c^2y^2 + 4a^4 - 4a^2x^2 - 4a^2y^2 - 4a^2c^2 = 0$. Dividing both sides by 4 leads to

$(c^2 - a^2)y^2 - a^2x^2 = a^2(c^2 - a^2)$. Letting $b^2 = c^2 - a^2$ and dividing a^2b^2 leads to

$$\frac{y^2}{a^2} - \frac{x^2}{b^2} = 1.$$

37. $\frac{x^2}{4} - \frac{y^2}{9} = 1$ **39.** $\frac{y^2}{9} - \frac{x^2}{4} = 1$ **41.** A point of intersection of the two circles is r_1 units from the center of one circle and r_2 units from the center of the other circle. Since $|r_1 - r_2| = 2a$, the point of intersection is on a hyperbola whose foci are the centers of the circle. **43.** (1,250, 2,050)

Section 11.5

1.

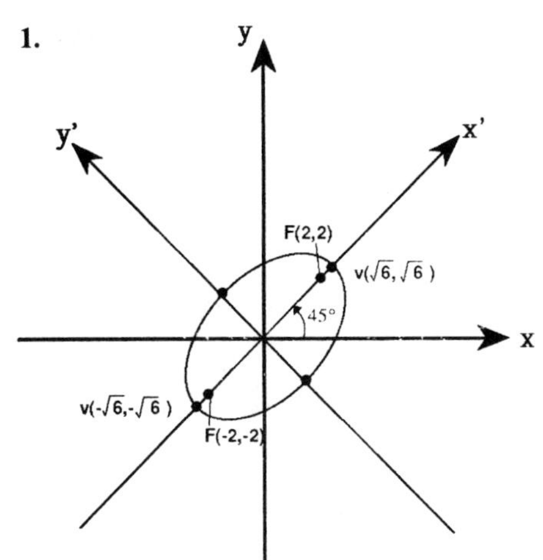

3.

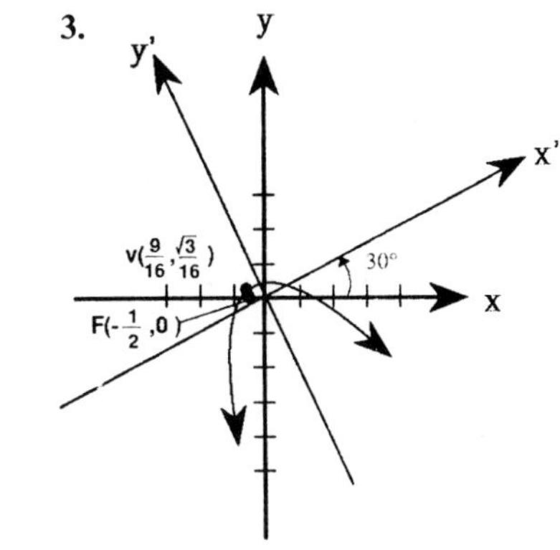

5.

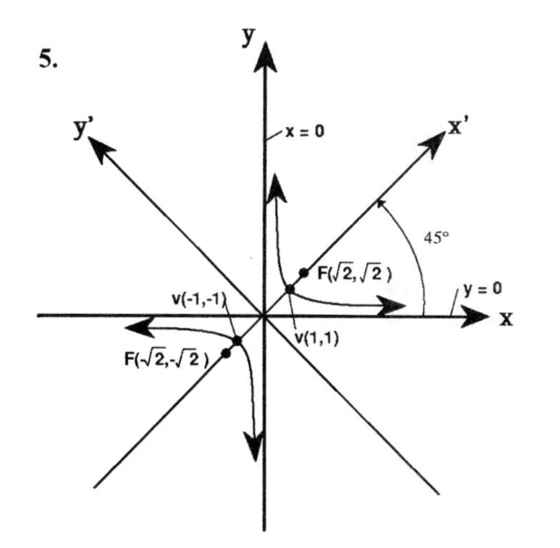

7. $\dfrac{x''^2}{12} + \dfrac{y''^2}{4} = 1$

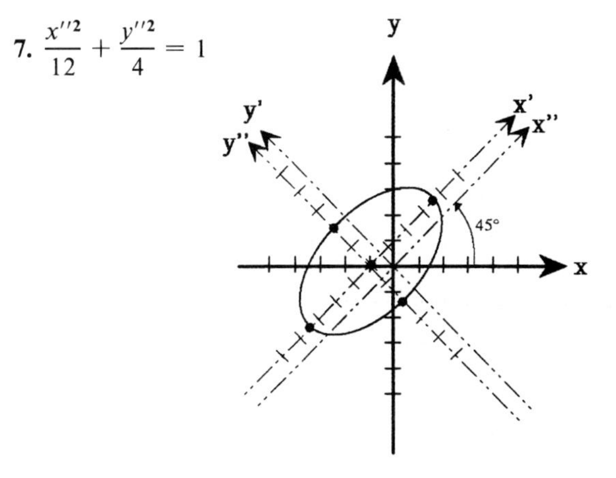

9. $x''^2 = 4y''$

11. $\dfrac{y''^2}{2} - \dfrac{x''^2}{3} = 1$

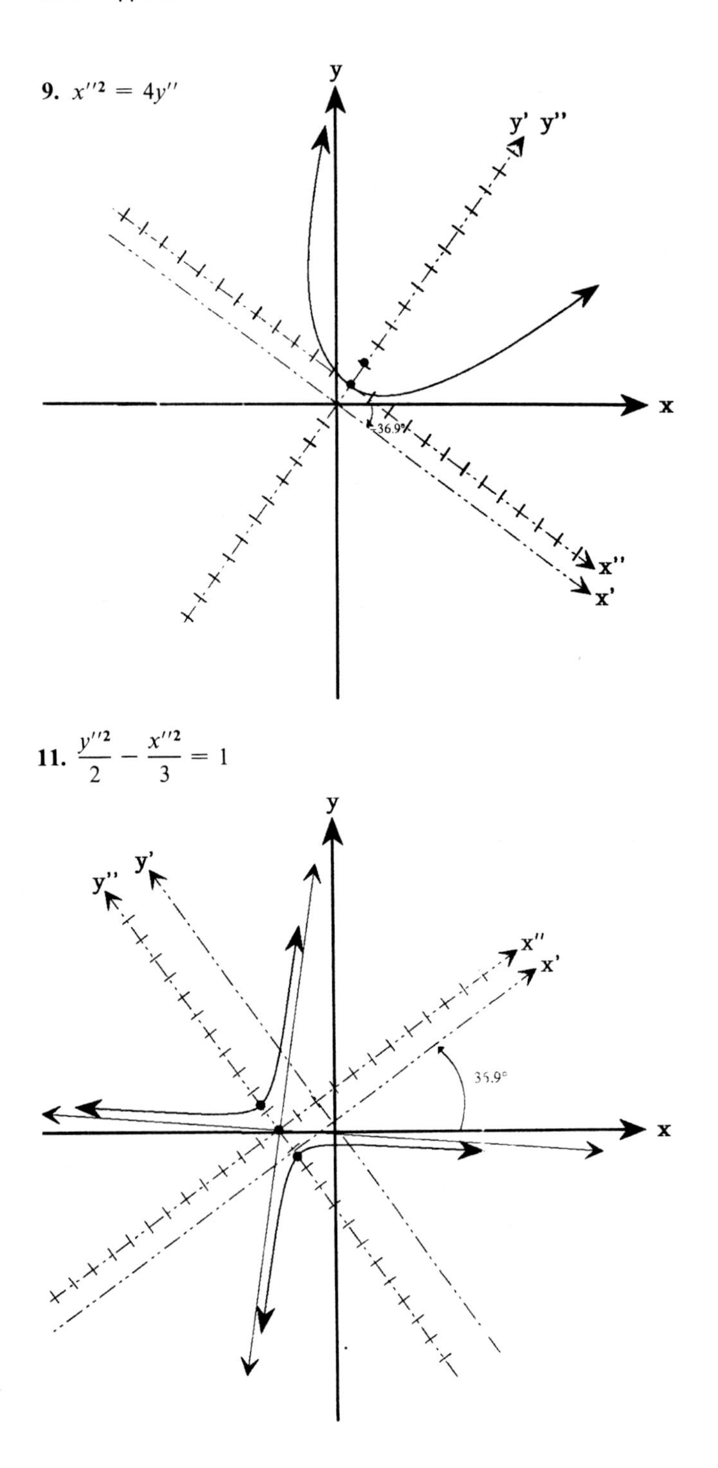

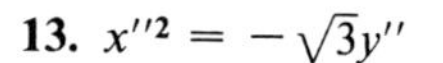

13. $x''^2 = -\sqrt{3}y''$

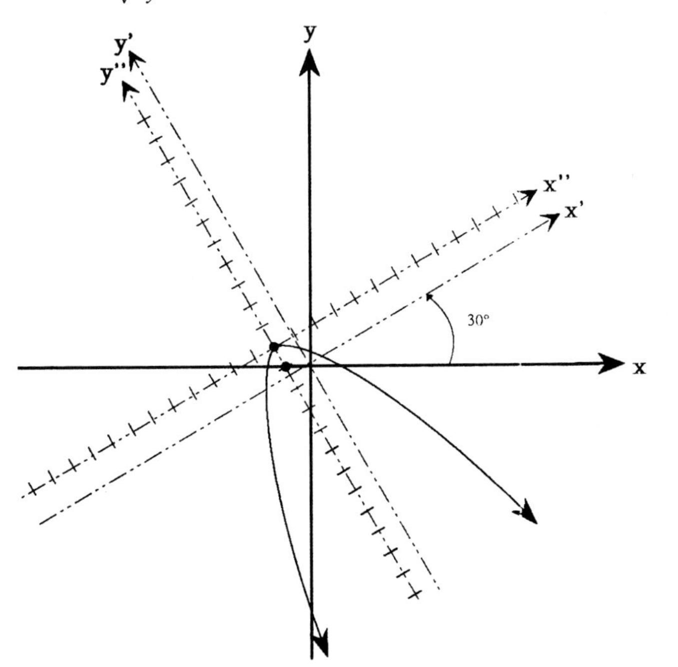

15. Vertices: $(\sqrt{6} - 1, \sqrt{6})$ and $(-\sqrt{6} - 1, -\sqrt{6})$; foci: $(1, 2)$ and $(-3, -2)$

17. Vertices: $\left(-2 - \frac{3}{5}\sqrt{2}, \frac{4}{5}\sqrt{2}\right)$ and $\left(-2 + \frac{3}{5}\sqrt{2}, -\frac{4}{5}\sqrt{2}\right)$; foci: $\left(-2 - \frac{3}{5}\sqrt{5}, \frac{4}{5}\sqrt{5}\right)$ and $\left(-2 + \frac{3}{5}\sqrt{5}, -\frac{4}{5}\sqrt{5}\right)$

19. We have $Ax^2 + Bxy + Cy^2 + Dx + Ey + F = 0$, $x = x' \cos\theta - y' \sin\theta$ and $y = y' \cos\theta + x' \sin\theta$. Then $A[(x')^2 \cos^2\theta - 2x'y' \cos\theta \sin\theta + (y')^2 \sin^2\theta] + B[x'y' \cos^2\theta + (x')^2 \sin\theta \cos\theta - (y')^2 \sin\theta \cos\theta - x'y' \sin^2\theta] + C[(y')^2 \cos^2\theta + 2x'y' \cos\theta \sin\theta + (x')^2 \sin^2\theta] + D[x' \cos\theta - y' \sin\theta] + E[y' \cos\theta + x' \sin\theta] + F = 0$. The coefficient A' of $(x')^2$ is $[A \cos^2\theta + B \sin\theta \cos\theta + C \sin^2\theta]$. The coefficient of B' of $x'y'$ is $[B \cos^2\theta - 2A \cos\theta \sin\theta - B \sin^2\theta + 2C \cos\theta \sin\theta] = [B(\cos^2\theta - \sin^2\theta) - 2(A - C) \sin\theta \cos\theta]$. The coefficient C' of $(y')^2$ is $[A \sin^2\theta - B \sin\theta \cos\theta + C \cos^2\theta]$.

Section 11.6

1. $\dfrac{x^2}{64} + \dfrac{y^2}{48} = 1$ 3. $\dfrac{x^2}{4} - \dfrac{y^2}{12} = 1$ 5. $\dfrac{x^2}{25} + \dfrac{y^2}{\frac{975}{64}} = 1$ 7. $\dfrac{y^2}{\frac{208}{9}} - \dfrac{x^2}{\frac{832}{81}} = 1$ 9. $(x - 2)^2 = 4y$

11. $\dfrac{\left(x + \frac{3}{5}\right)^2}{\frac{144}{25}} - \dfrac{y^2}{\frac{36}{5}} = 1$ 13. $r = \dfrac{2}{1 + \cos\theta}$ 15. $r = \dfrac{5}{2 - \cos\theta}$ 17. $r = \dfrac{4}{3 + 4\sin\theta}$

19. $y^2 = 12(x + 3)$ 21. $\dfrac{x^2}{\frac{1}{8}} + \dfrac{\left(y - \frac{1}{8}\right)}{\frac{9}{64}} = 1$ 23. $\dfrac{(x + 9)^2}{36} - \dfrac{y^2}{45} = 1$

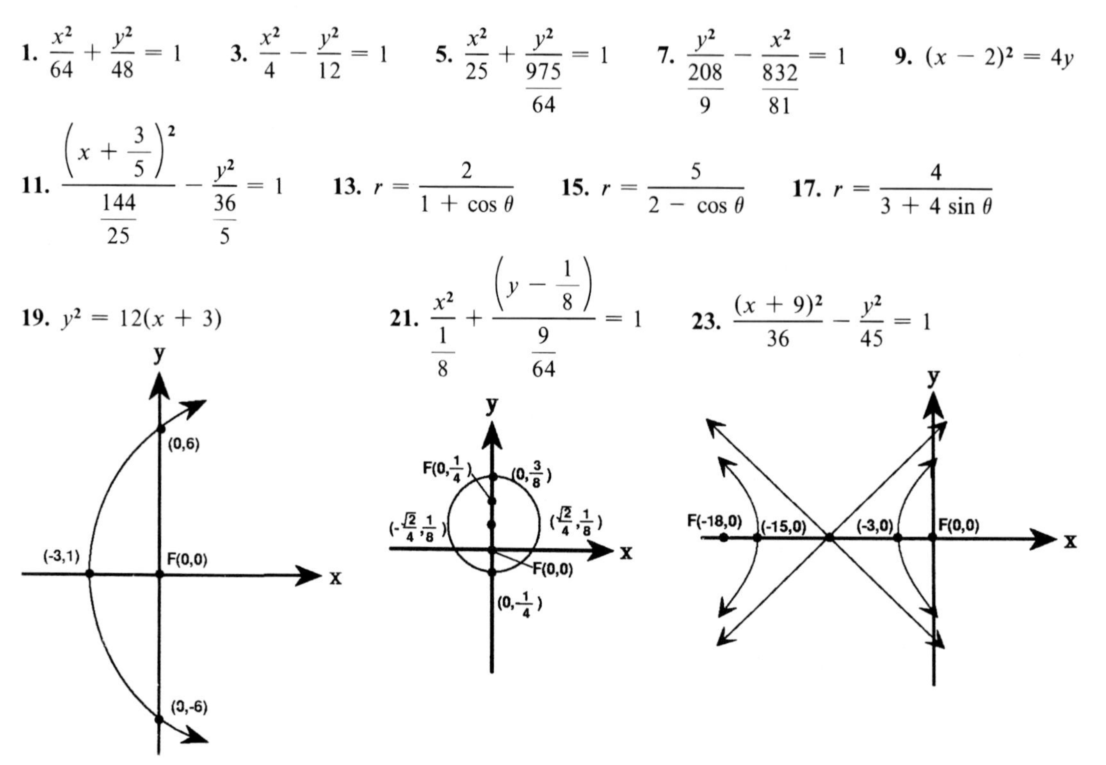

25. Say $\sqrt{(x - c)^2 + y^2} = 2a - \sqrt{(x - c)^2 + y^2}$. Then squaring both sides gives $(x - c)^2 + y^2 = 4a^2 - 4a\sqrt{(x - c)^2 + y^2} + (x - c)^2 + y^2$, which leads to $a\sqrt{(x - c)^2 + y^2} = a^2 + cx$.

$$\sqrt{(x + c)^2 + y^2} = a + \frac{c}{a}x = \frac{c}{a}\left(x - \left(-\frac{a^2}{c}\right)\right). \tag{1}$$

The last line reads, "The distance from a point (x, y) on the ellipse to its focus $(-c, 0)$ is equal to $\dfrac{c}{a}$ times the distance of the point from the line $x = \dfrac{a^2}{c}$. Call the focus $(-c, 0)$, F_2 and the line $x = -\dfrac{a^2}{c}$, D_2. Then equation (1) can be written $d(F_2) = ed(D_2)$.

27. With the horizontal directrix below the pole, the distance from point (r, θ) to the directrix equals $(d + r\sin\theta)$. The distance from (r, θ) to the focus at the pole equals r. Then,

$$e = \frac{\text{distance from point to focus}}{\text{distance from point to directrix}} = \frac{r}{d + r\sin\theta}.$$

Now

$$ed + er\sin\theta = r$$
$$ed = r(1 - e\sin\theta)$$
$$r = \frac{ed}{1 - e\sin\theta}$$

29. $41x^2 + 16xy + 29y^2 + 206x + 38y + 194 = 0$ 31. $4x^2 + 36xy + 31y^2 - 20x - 40y - 100 = 0$

33. $r = \dfrac{94{,}962 \times 10^5}{100 - 2\cos\theta}$

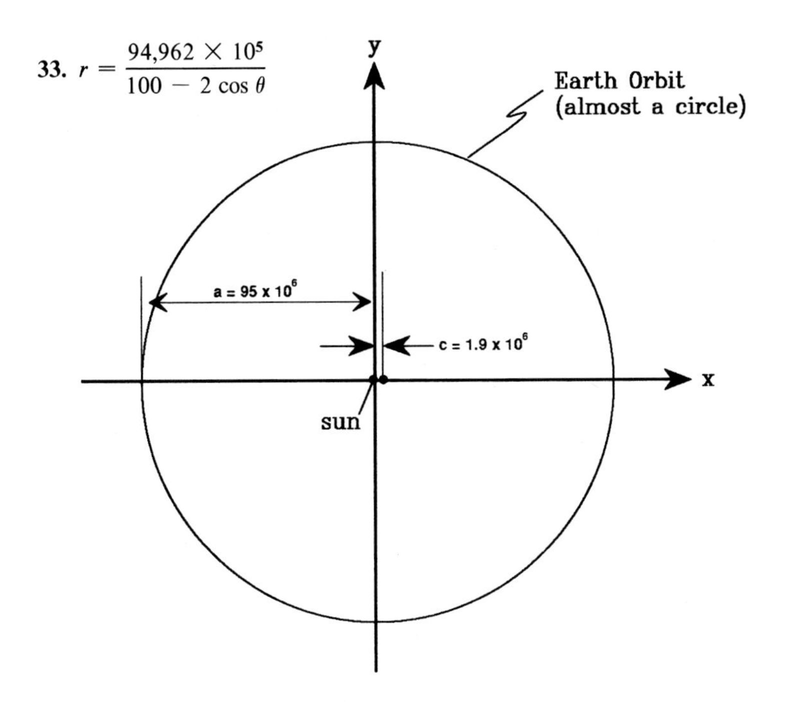

Section 11.7

1. 12.53° **3.** 31.33° **5.** 71.57° **7.** 34.5°, 40.2°, and 105.3° **9.** 90°, 45°, and 45°

11. $y - 2 = \left(\dfrac{3 + \sqrt{3}}{3 - \sqrt{3}}\right)(x - 2)$

$y - 2 = \left(\dfrac{3 - \sqrt{3}}{3 + \sqrt{3}}\right)(x - 2)$

13. $y = \left(\dfrac{4\sqrt{3} + 1}{4 - \sqrt{3}}\right)(x + 4)$

$y = \left(\dfrac{1 - 4\sqrt{3}}{4 + \sqrt{3}}\right)(x + 4)$

15. $y - 4 = 4(x - 2)$

17. $y - 5 = -2(x + 2)$ **19.** $y - 1 = -\dfrac{2}{5}(x + 1)$ **21.** $y - 2\sqrt{2} = -\dfrac{\sqrt{2}}{4}(x - 1)$ **23.** $y - 1 = -\dfrac{2}{3}x$

25. $y + 2 = (-6 - 2\sqrt{11})(x + 3), y + 2 = (-6 + 2\sqrt{11})(x + 3)$

27. $y - 2 = \dfrac{4}{5}(-1 - \sqrt{2})x, y - 2 = \dfrac{4}{5}(-1 + \sqrt{2})x$

29. $y + 3 = \left(\dfrac{-32 - 5\sqrt{55}}{9}\right)(x - 5), y + 3 = \left(\dfrac{-32 + 5\sqrt{55}}{9}\right)(x + 5)$

31. $y + \dfrac{5}{4} = -\left(x - \dfrac{5}{4}\right), y + \dfrac{5}{4} = x - \dfrac{5}{4}$

33. The family of lines through (x_0, y_0) is $y - y_0 = m(x - x_0)$. The member of the family that is tangent to the ellipse intersects the ellipse only at (x_0, y_0). We solve the equations $y = mx + (y_0 - mx_0)$ and $\frac{x^2}{a^2} + \frac{y^2}{b^2} = 1$ (or $b^2x^2 + a^2y^2 = a^2b^2$) simultaneously to produce $(a^2m^2 + b^2)x^2 + (2a^2m(y_0 - mx_0))x + (a^2(y_0 - mx_0)^2 - a^2b^2) = 0$. Because the equation has only the one solution x_0, its discriminant must be 0. Therefore,

$$\begin{aligned} x_0 &= -\frac{2a^2m(y_0 - mx_0)}{2(a^2m^2 + b^2)} \\ a^2m^2x_0 + b^2x_0 &= -a^2my_0 + a^2m^2x_0 \\ -\frac{b^2x_0}{a^2y_0} &= m. \end{aligned}$$

35. The family of lines through (x_0, y_0) is $y - y_0 = m(x - x_0)$. The member of the family that is tangent to the hyperbola intersects the hyperbola only at (x_0, y_0). We solve the equations $y = mx + (y_0 - mx_0)$ and $\frac{x^2}{a^2} - \frac{y^2}{b^2} = 1$ (or $b^2x^2 - a^2y^2 = a^2b^2$) simultaneously to produce $(b^2 - a^2m^2)x^2 - (2a^2m(y_0 - mx_0))x - (a^2b^2 + a^2(y_0 - mx_0)^2 = 0$. Because the equation has only the one solution x_0, its discriminant must be 0. Therefore,

$$\begin{aligned} x_0 &= \frac{2a^2m(y_0 - mx_0)}{2(b^2 - a^2m^2)} \\ b^2x_0 - a^2m^2x_0 &= a^2my_0 - a^2m^2x_0 \\ \frac{b^2x_0}{a^2y_0} &= m. \end{aligned}$$

Chapter 11 Review Exercises

1. $(x - 1)^2 + (y - 2)^2 = 4^2$ **3.** $(y - 0)^2 = -4(x - 0)^2$ **5.** $\frac{x^2}{16} + \frac{y^2}{12} = 1$ **7.** $\frac{y^2}{16} - \frac{x^2}{20} = 1$

9. **11.** **13.**

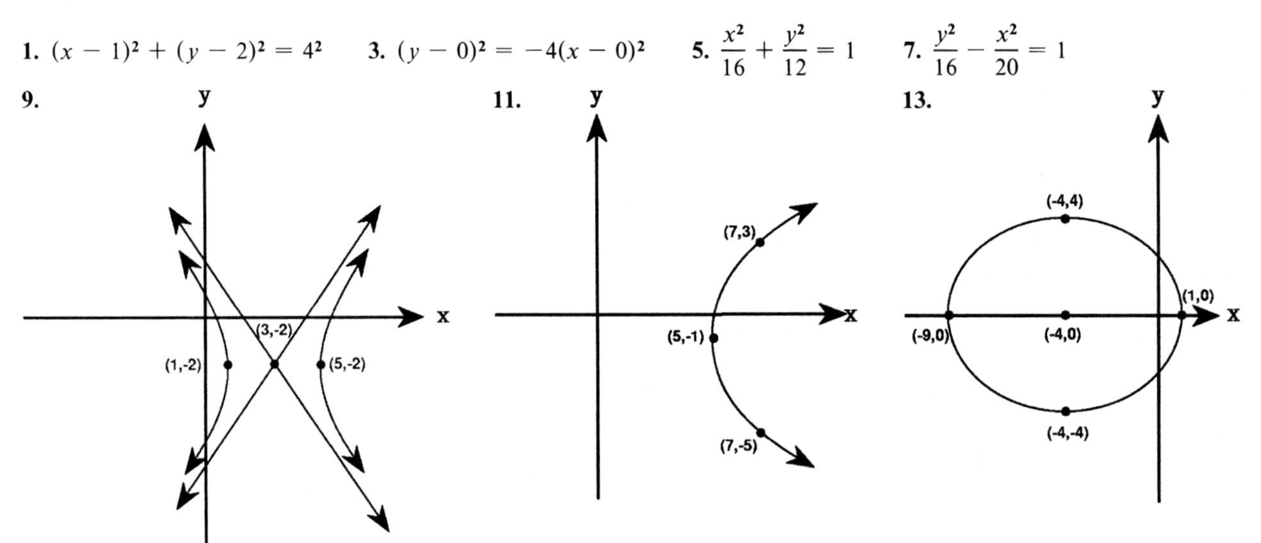

15. $x^2 + y^2 + 11x - 3y - 10 = 0$ **17.** $(y-3)^2 = \frac{9}{4}(x-1)$ **19.** $5(x-3)^2 + 7(x+1)^2 = 108$

21. $\frac{(x-3)^2}{25} - \frac{(y-2)^2}{\frac{50}{3}} = 1$

23. **25.** **27.**

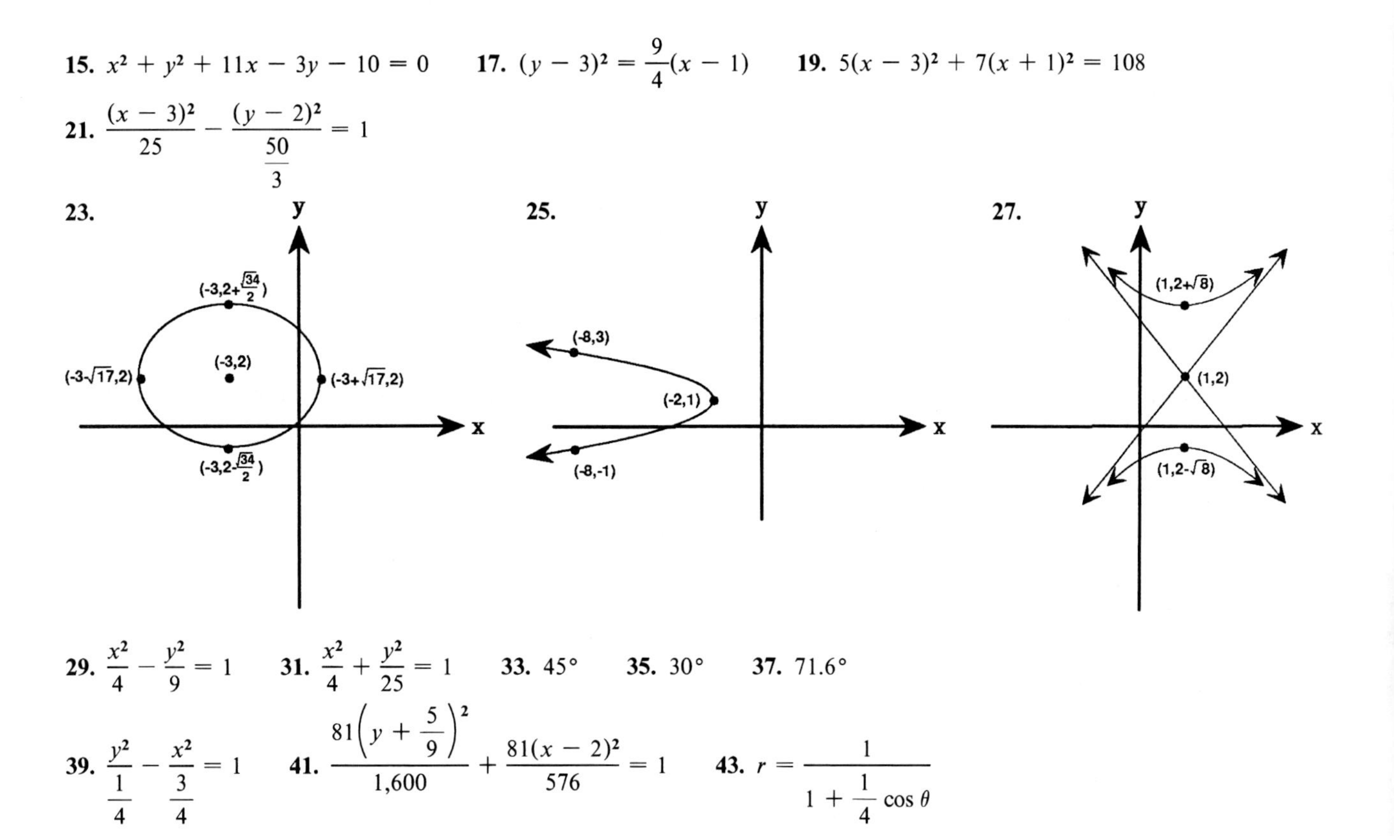

29. $\frac{x^2}{4} - \frac{y^2}{9} = 1$ **31.** $\frac{x^2}{4} + \frac{y^2}{25} = 1$ **33.** 45° **35.** 30° **37.** 71.6°

39. $\frac{y^2}{\frac{1}{4}} - \frac{x^2}{\frac{3}{4}} = 1$ **41.** $\frac{81\left(y + \frac{5}{9}\right)^2}{1{,}600} + \frac{81(x-2)^2}{576} = 1$ **43.** $r = \frac{1}{1 + \frac{1}{4}\cos\theta}$

45. $$\frac{\left(x - \frac{6}{7}\right)^2}{\frac{64}{49}} + \frac{y^2}{\frac{4}{7}} = 1$$

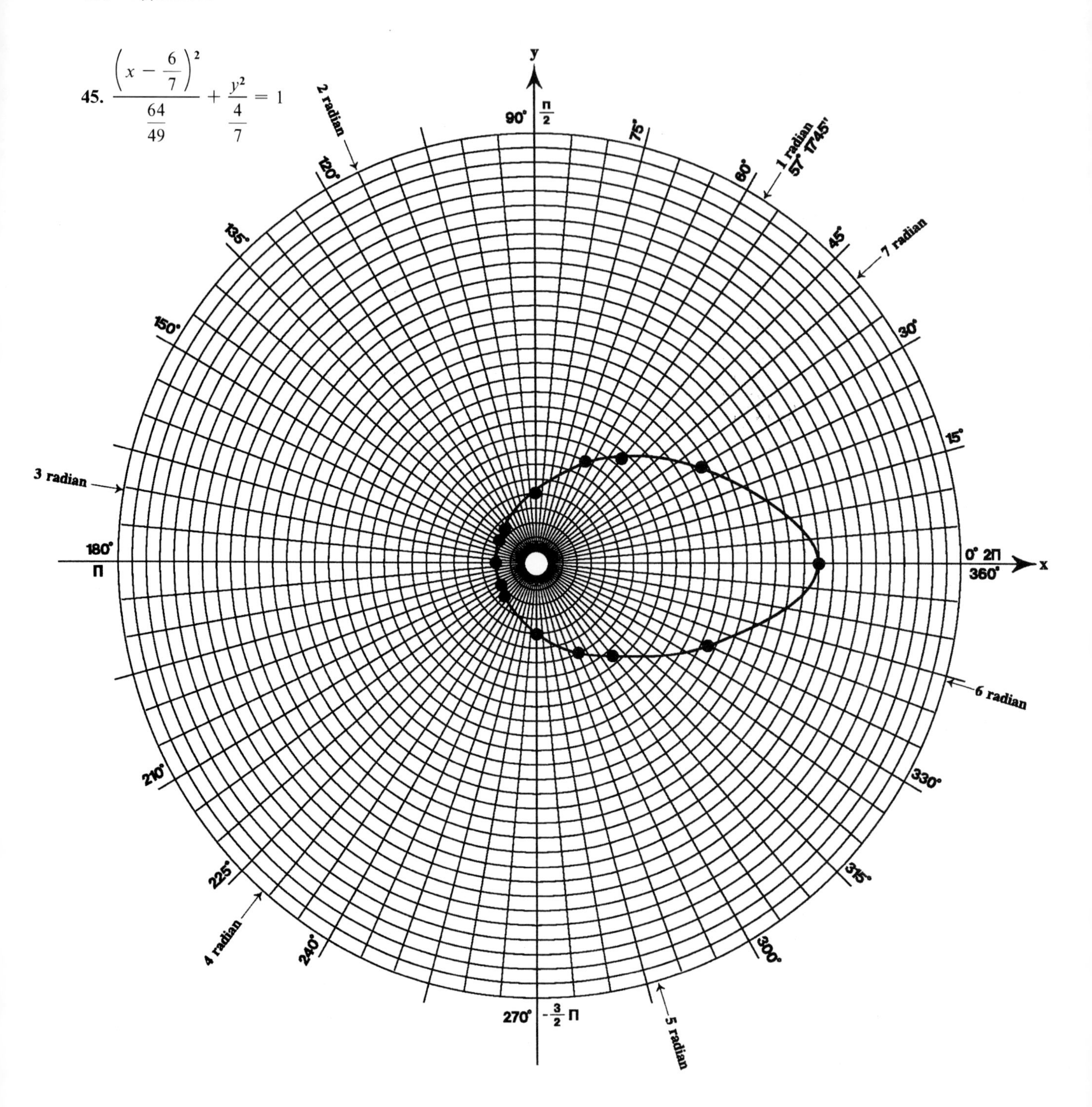

47. $y = x - 2$ **49.** (1.23, 5.21), 1.23 and (0.77, 3.26), 0.77

51.

	(x'', y'')	(x', y')	(x, y)
vertex	$(0, 0)$	$\left(-\frac{\sqrt{3}}{4}, \frac{3}{8}\right)$	$\left(-\frac{9}{16}, \frac{\sqrt{3}}{16}\right)$
focus	$\left(0, -\frac{1}{8}\right)$	$\left(-\frac{\sqrt{3}}{4}, \frac{1}{4}\right)$	$\left(-\frac{1}{2}, 0\right)$

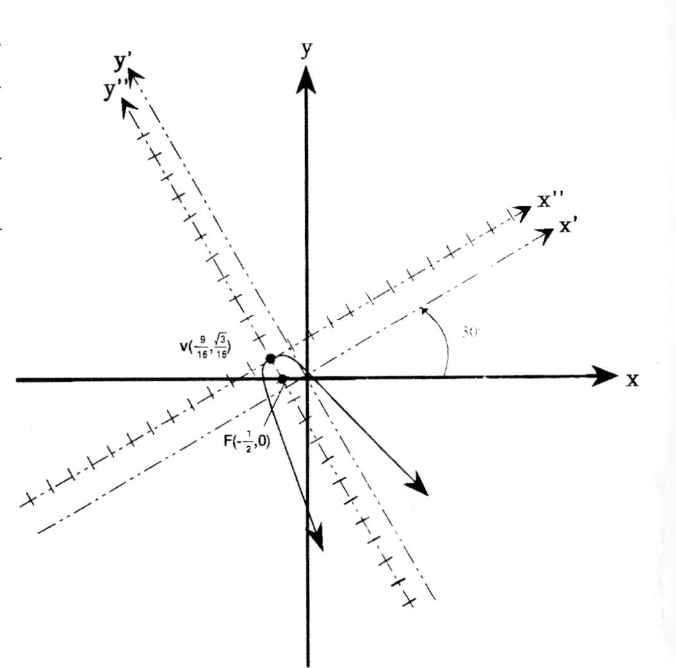

53. $y = (3 + 2\sqrt{3})(x - 2)$ and $y = (3 - 2\sqrt{3})(x - 2)$

Chapter 11 Test 1

1. $(x - 4)^2 + (y - 5)^2 = 4$ **2.** $(y - 4)^2 = -16(x - 1)$ **3.** $\frac{(x + 2)^2}{36} + \frac{(y + 2)^2}{20} = 1$

4. $\frac{(y - 5)^2}{9} - \frac{(x - 3)^2}{16} = 1$

5. **6.** **7.**

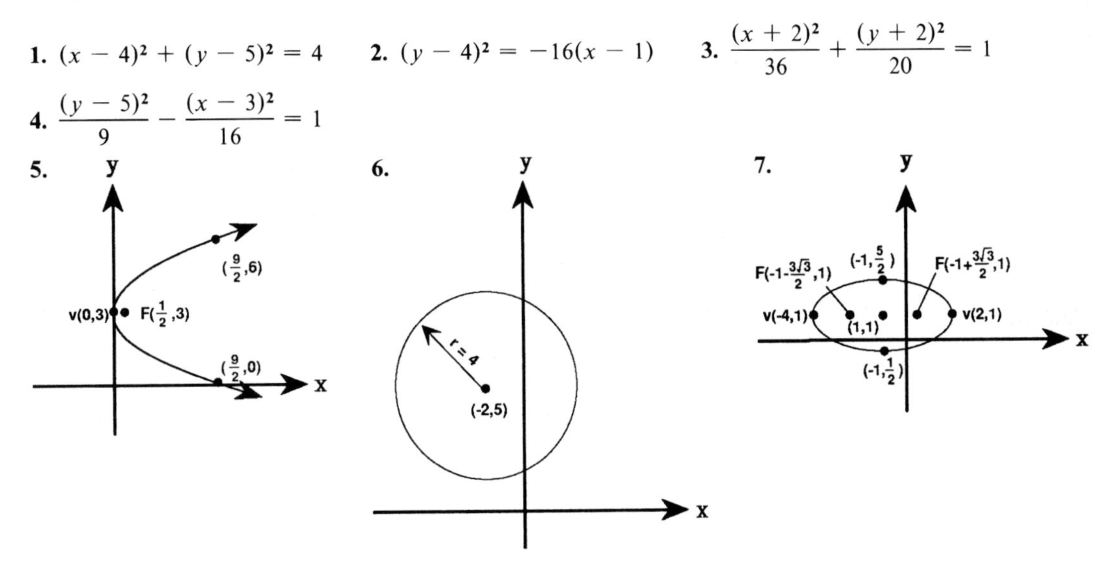

8.

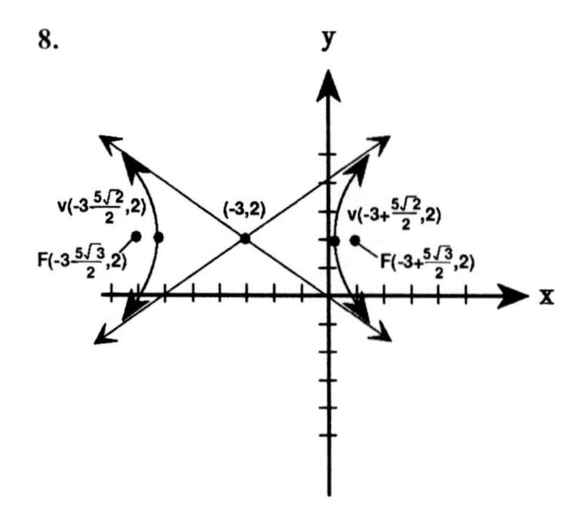

9. $4x^2 - 4xy + y^2 + 12x + 24y - 36 = 0$ **10.** $67.5°$

Chapter 11 Test 2

1. $(x - 7)^2 + (y - 4)^2 = 16$ **2.** $(x - 2)^2 = \frac{4}{5}(y + 5)$ **3.** $\frac{(y - 4)^2}{\frac{140}{3}} + \frac{(x + 5)^2}{\frac{140}{32}} = 1$

4. **5.** **6.**

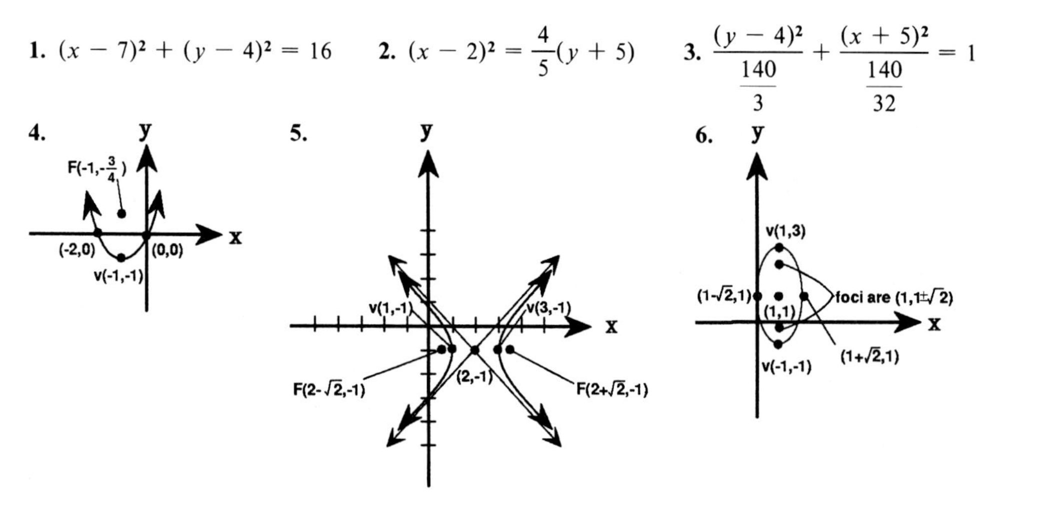

7. $1 = \frac{(x - 2)^2}{9} + \frac{(y + 1)^2}{4}$ **8.** $y - 4 = \left(2 - \frac{\sqrt{15}}{2}\right)(x - 4)$ and $y - 4 = \left(2 + \frac{\sqrt{15}}{2}\right)(x - 4)$ **9.** $71.56°$
10. $5x^2 + 9y^2 + 32x - 36y - 28 = 0$

Section 12.1

1. $x = 3$ and $y = 3\sqrt{3}$ **3.** $x = 130(\sqrt{6} - \sqrt{2})$ and $y = 130(\sqrt{6} + \sqrt{2})$ **5.** $x = \frac{15}{2}$ and $y = -\frac{15\sqrt{3}}{2}$

7. **(a)** $(2, 5)$ **(b)** $\sqrt{2^2 + 5^2} = \sqrt{29}$ **9.** **(a)** $(6, 8)$ **(b)** $\sqrt{6^2 + 8^2} = 10$

11. **(a)** $(-8, 15)$ **(b)** $\sqrt{(-8)^2 + (15)^2} = 17$ **13.** $x = \frac{20}{7}$ and $y = -\frac{12}{7}$ **15.** $x = -2$ and $y = -1$

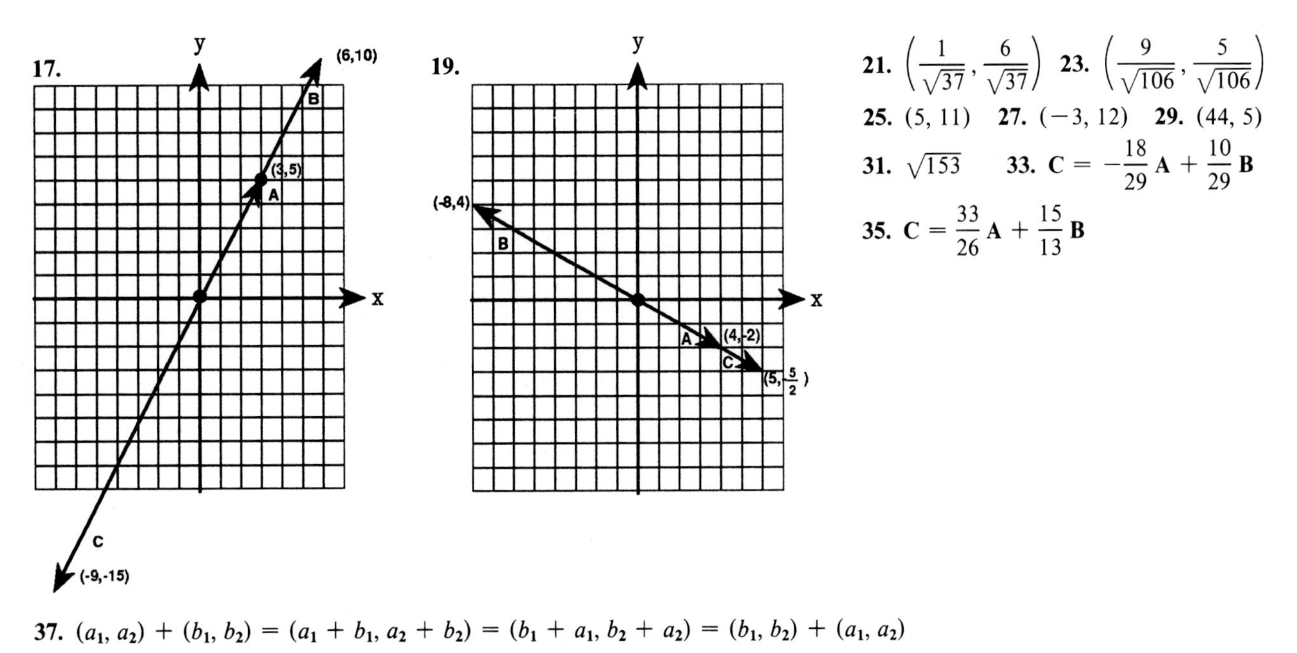

21. $\left(\frac{1}{\sqrt{37}}, \frac{6}{\sqrt{37}}\right)$ **23.** $\left(\frac{9}{\sqrt{106}}, \frac{5}{\sqrt{106}}\right)$

25. (5, 11) **27.** (−3, 12) **29.** (44, 5)

31. $\sqrt{153}$ **33.** $\mathbf{C} = -\frac{18}{29}\mathbf{A} + \frac{10}{29}\mathbf{B}$

35. $\mathbf{C} = \frac{33}{26}\mathbf{A} + \frac{15}{13}\mathbf{B}$

37. $(a_1, a_2) + (b_1, b_2) = (a_1 + b_1, a_2 + b_2) = (b_1 + a_1, b_2 + a_2) = (b_1, b_2) + (a_1, a_2)$

39. $\mathbf{C} = k\mathbf{A} + t\mathbf{B}$ if the system $\begin{cases} x = kx_1 + tx_2 \\ y = ky_1 + ty_2 \end{cases}$ has a solution. The system will have a solution if $\frac{x_1}{y_1} \neq \frac{x_2}{y_2}$. But $\frac{x_1}{y_1} \neq \frac{x_2}{y_2}$ if and only if **A** and **B** are not parallel vectors. **41.** $||\mathbf{P} - \mathbf{C}|| = \sqrt{(x-h)^2 + (y-k)^2}$, $||\mathbf{R}|| = r$. Squaring both sides of the latter equation gives $(x-h)^2 + (y-k)^2 = r^2$.

43. $||\mathrm{P} - \mathrm{F}_1|| + ||\mathrm{P} - \mathrm{F}_2|| = 2a$ gives $\sqrt{(x-c)^2 + (y+0)^2} + \sqrt{(x+c)^2 + (y-0)^2} = 2a$, which gives $(\sqrt{(x+c)^2 + y^2})^2 = (2a - \sqrt{(x-c)^2 + y^2})^2$, simplifying to $(a^2 - c^2)x^2 + a^2y^2 = a^2(a^2 - c^2)$. Dividing each side by $a^2(a^2 - c)$ and writing b^2 in place of $a^2 - c^2$ gives $\frac{x^2}{a^2} = \frac{y^2}{b^2} = 1$

45. Let M_1, M_2, M_3, and M_4 be the midpoints for sides $\overline{AB}$, $\overline{BC}$, $\overline{CD}$, and $\overline{DA}$ of the quadrilateral $ABCD$.

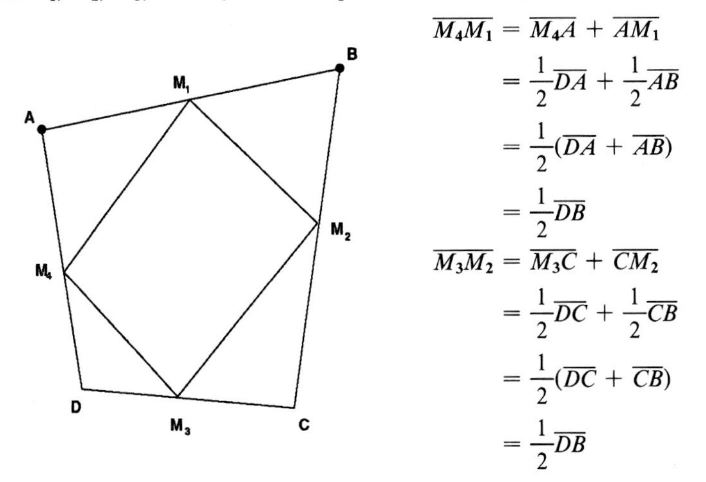

$$\begin{aligned}
\overline{M_4M_1} &= \overline{M_4A} + \overline{AM_1} \\
&= \frac{1}{2}\overline{DA} + \frac{1}{2}\overline{AB} \\
&= \frac{1}{2}(\overline{DA} + \overline{AB}) \\
&= \frac{1}{2}\overline{DB} \\
\overline{M_3M_2} &= \overline{M_3C} + \overline{CM_2} \\
&= \frac{1}{2}\overline{DC} + \frac{1}{2}\overline{CB} \\
&= \frac{1}{2}(\overline{DC} + \overline{CB}) \\
&= \frac{1}{2}\overline{DB}
\end{aligned}$$

Since opposite sides $\overline{M_4M_1}$ and $\overline{M_3M_2}$ of quadrilateral $M_1M_2M_3M_4$ are equal and parallel, the quadrilateral is a parallelogram.

Section 12.2

1. $20\sqrt{3}$ **3.** 50 **5.** $\left(\frac{2\sqrt{5}}{5}, -\frac{\sqrt{5}}{5}\right)$ **7.** $\left(\frac{2}{\sqrt{5}}, \frac{1}{\sqrt{5}}\right)$ **9.** $\left(\frac{9\sqrt{97}}{97}, -\frac{4\sqrt{97}}{97}\right)$ **11.** 15.3° **13.** 88.8°

15. 78.1° **17.** $\left(-\frac{\sqrt{10}}{10}, \frac{3\sqrt{10}}{10}\right)$ **19.** $\left(\frac{10\sqrt{109}}{109}, -\frac{3\sqrt{109}}{109}\right)$ **21.** $\left(-\frac{\sqrt{2}}{2}, \frac{\sqrt{2}}{2}\right)$ **23.** $\frac{13\sqrt{5}}{5}$ **25.** $\frac{65\sqrt{73}}{73}$

27. $3\sqrt{10}$ **29.** $\frac{9\sqrt{10}}{10}$ **31.** $(6, 3) \cdot (10, 0) = 60$ **33.** $(12, 5) \cdot (11, 12) = 192$ **35.** $\frac{11\sqrt{5}}{5}$ **37.** $\frac{65}{73}\sqrt{73}$

39. $4x - 3y - 9 = 0$ or $4x - 3y + 21 = 0$ **41.** $27x - 99y + 88 = 0$ and $77x + 21y + 68 = 0$

43. 37.5 square units **45.** Let $\mathbf{A} = (5, 5)$, $\mathbf{B} = (0, 1)$ and $\mathbf{C} = (1, 0)$. $\mathbf{A} \cdot \mathbf{B} = \mathbf{A} \cdot \mathbf{C} = 5$, but $\mathbf{B} \neq \mathbf{C}$.

47.
$$\begin{aligned} \|\mathbf{A} + \mathbf{B}\|^2 &= (\mathbf{A} + \mathbf{B}) \cdot (\mathbf{A} + \mathbf{B}) \\ &= \|\mathbf{A}\|^2 + 2\mathbf{A} \cdot \mathbf{B} + \|\mathbf{B}\|^2 && \text{(exercise 46)} \\ &\leq \|\mathbf{A}\|^2 + 2\|\mathbf{A}\|\,\|\mathbf{B}\| + \|\mathbf{B}\|^2 && \text{(Cauchy-Schwartz Inequality)} \\ &= (\|\mathbf{A}\| + \|\mathbf{B}\|)^2 \end{aligned}$$

Hence, $\|\mathbf{A} + \mathbf{B}\| \leq \|\mathbf{A}\| + \|\mathbf{B}\|$.

49. In triangle ABC,

$\|\overline{AC}\| = \|\overline{BC}\|$ and $\|\overline{AD}\| = \|\overline{DB}\|$. We want to show that $\overline{CD} \cdot \overline{AB} = 0$.

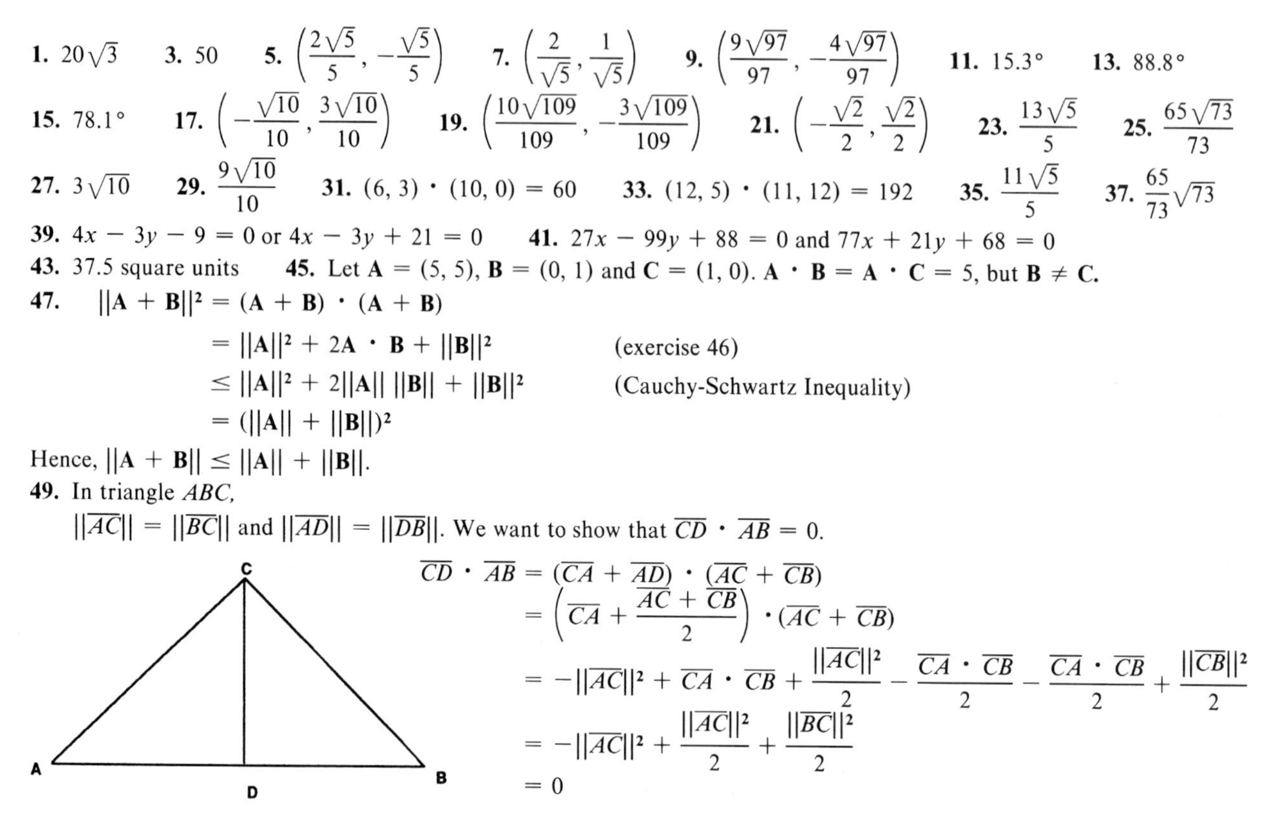

$$\begin{aligned} \overline{CD} \cdot \overline{AB} &= (\overline{CA} + \overline{AD}) \cdot (\overline{AC} + \overline{CB}) \\ &= \left(\overline{CA} + \frac{\overline{AC} + \overline{CB}}{2}\right) \cdot (\overline{AC} + \overline{CB}) \\ &= -\|\overline{AC}\|^2 + \overline{CA} \cdot \overline{CB} + \frac{\|\overline{AC}\|^2}{2} - \frac{\overline{CA} \cdot \overline{CB}}{2} - \frac{\overline{CA} \cdot \overline{CB}}{2} + \frac{\|\overline{CB}\|^2}{2} \\ &= -\|\overline{AC}\|^2 + \frac{\|\overline{AC}\|^2}{2} + \frac{\|\overline{BC}\|^2}{2} \\ &= 0 \end{aligned}$$

51. The perpendicular bisectors of AC and BC intersect at G. Let D be the midpoint of AB. We must show that $\overline{GD} \cdot \overline{AB} = 0$.

$$\begin{aligned} \overline{GD} \cdot \overline{AB} &= (\overline{GF} + \overline{FA} + \overline{AD}) \cdot (\overline{AC} + \overline{CB}) \\ &= \overline{GF} \cdot \overline{AC} + \overline{GF} \cdot \overline{CB} + \overline{FA} \cdot \overline{AC} + \overline{FA} \cdot \overline{CB} + \overline{AD} \cdot \overline{AC} + \overline{AD} \cdot \overline{CB} \\ &= 0 + \overline{GF} \cdot \overline{CB} - \frac{\|\overline{AC}\|^2}{2} - \frac{\overline{AC} \cdot \overline{CB}}{2} + \overline{AD} \cdot (\overline{AC} + \overline{CB}) \\ &= \overline{GF} \cdot \overline{CB} - \frac{\|\overline{AC}\|^2}{2} - \frac{\overline{AC} \cdot \overline{CB}}{2} + \frac{(\overline{AC} + \overline{CB})}{2}(\overline{AC} + \overline{CB}) \\ &= \overline{GF} \cdot \overline{CB} - \frac{\|\overline{AC}\|^2}{2} - \frac{\overline{AC} \cdot \overline{CB}}{2} + \frac{\|\overline{AC}\|^2}{2} + \overline{AC} \cdot \overline{CB} + \frac{\|\overline{CB}\|^2}{2} \\ &= \overline{GF} \cdot \overline{CB} + \frac{\overline{AC} \cdot \overline{CB}}{2} + \frac{\|\overline{CB}\|^2}{2} \\ &= \left(\overline{GF} + \frac{\overline{AC}}{2}\right) \cdot \overline{CB} + \frac{\|\overline{CB}\|^2}{2} \\ &= (\overline{GE} + \overline{EC}) \cdot \overline{CB} + \frac{\|\overline{CB}\|^2}{2} \\ &= \overline{GE} \cdot \overline{CB} - \frac{\|\overline{CB}\|^2}{2} + \frac{\|\overline{CB}\|^2}{2} \\ &= 0 \end{aligned}$$

C F E G A D B

Section 12.3

1. **(a)** $(x, y) = (0, 2) + t(1, 5)$ **(b)** $x = t$ and $y = 2 + 5t$ **3.** **(a)** $(x, y) = \left(0, -\frac{5}{2}\right) + t(3, -1)$
(b) $x = 3t$ and $y = -\frac{5}{2} - t$ **5.** **(a)** $(x, y) = (2, 5) + t(2, 1)$ **(b)** $x = 2 + 2t$ and $y = 5 + t$
7. **(a)** $(x, y) = (0, 0) + t(7, 10)$ **(b)** $x = 7t$ and $y = 10t$ **9.** $x + 3y = 5$ **11.** $2x + y = 2$
13. $\left(5 + \frac{18}{\sqrt{10}}, -3 + \frac{6}{\sqrt{10}}\right)$ $\left(5 - \frac{18}{\sqrt{10}}, -3 - \frac{6}{\sqrt{10}}\right)$ **15.** $\left(\frac{6}{\sqrt{109}}, 4 - \frac{20}{\sqrt{109}}\right)$ $\left(-\frac{6}{\sqrt{109}}, 4 + \frac{20}{\sqrt{109}}\right)$ **17.** $\left(\frac{13}{11}, -\frac{47}{11}\right)$ **19.** $\left(-\frac{39}{2}, -39\right)$ **21.** $(6, 4)$
23. $\left(\frac{33}{4}, -\frac{17}{4}\right)$ **25.** $\left(\frac{28}{5}, \frac{23}{5}\right)$ **27.** $\left(\frac{24}{7}, -\frac{50}{7}\right)$ **29.** $(x, y) = (1, 3) + t(6, 0)$
31. $(x, y) = (1, 2) + t(5, -3)$ **33.** $(1, -1)$ **35.** Collinear **37.** Not collinear
39. $\overline{AM} = \overline{MB}$ because M is the midpoint of $\overline{AB}$.

$$\begin{aligned}\overline{PA} + \overline{PB} &= (\overline{PM} + \overline{MA}) + (\overline{PM} + \overline{MB}) \\ &= 2\overline{PM} - \overline{AM} + \overline{MB} \\ &= 2\overline{PM}\end{aligned}$$

41. Add diagonal DB to the parallelogram in figure 12.22 and let Q be the point of intersection of AC and DB. Because the opposite sides of a parallelogram are equal and parallel, $\overline{AD} = \overline{BC}$ and $\overline{AB} = \overline{DC}$.

$$\begin{aligned}\overline{AC} &= \overline{AB} + \overline{BC} \\ \overline{AQ} &= k\overline{AC} \\ &= k\overline{AB} + k\overline{BC} \\ &= k\overline{AD} + k\overline{AB}\end{aligned}$$

Since $k + k = 1$, $k = \frac{1}{2}$ and $\overline{AQ} = \frac{1}{2}\overline{AC}$.

$$\begin{aligned}\overline{DB} &= \overline{DA} + \overline{AB} \\ \overline{DQ} &= k\overline{DB} \\ &= k\overline{DA} + k\overline{AB} \\ &= k\overline{DC} + k\overline{DA}\end{aligned}$$

Since $k + k = 1$, $k = \frac{1}{2}$ and $\overline{DQ} = \frac{1}{2}\overline{DB}$.

Section 12.4

1. $\sqrt{86}$ **3.** $5\sqrt{5}$ **5.** $x = 1 - t, y = 3 + 9t, z = 7 + 2t; \frac{x-1}{-1} = \frac{y-3}{9} = \frac{z-7}{2}$
7. $x = 2 + 9t, y = -10 - 12t, z = 1 + 9t; \frac{x-2}{9} = \frac{y+10}{-12} = \frac{z-1}{9}$
9. $x = 6 + 1t, y = -3 - 7t, z = 11 + 9t; \frac{x-6}{1} = \frac{y+3}{-7} = \frac{z-11}{9}$ **11.** $\frac{x-1}{3} = \frac{y-4}{-1} = \frac{z-1}{-3}$
13. $\frac{x+8}{-6} = \frac{y-12}{3} = \frac{z+4}{-9}$ **15.** $x = 1 + 2t, y = -3 + 4t, z = -2 + 9t$

17. $x = -5 + t, y = 5 + (-5)t, z = 6 + 2t$ **19.** $\frac{x-1}{1} + \frac{y-4}{-5} + \frac{z-1}{2}$ **21.** $\frac{x-2}{-46} = \frac{y-2}{-25} = \frac{z-2}{9}$
23. $27x + 6y - 13z = 0$ **25.** $-15x - 39y + 37z - 34 = 0$ **27.** $-11x + 27y + 4z + 59 = 0$
29. $x + y - z = 0$ **31.** $3x + 10y + 3z + 3 = 0$ **33.** x and y are negative, but z is positive. **35.** $z = 0$
37. A point moving on the line would experience no change in one of the coordinate directions.

$$\frac{x - x_0}{a} = \frac{y - y_0}{b}$$

39. Any line that is in a plane that is perpendicular to the given line and that also contains the point (3, 7, 1), is a solution of the problem. Since such a plane contains an infinity of lines, the problem has an infinite number of solutions.
41. Let the normal **N** have direction numbers A, B, and C. Now the plane has an equation of the form $Ax + By + Cz + D = 0$. Since (x_0, y_0, z_0) is a point in the plane, $D = -(Ax_0 + By_0 + Cz_0)$.

$$\frac{d}{\|\mathbf{V}\|} = |\cos\theta|$$

$$d = \frac{\|\mathbf{V}\|\,\|\mathbf{N}\|\,|\cos\theta|}{\|\mathbf{N}\|} = \frac{|\mathbf{V}\cdot\mathbf{N}|}{\|\mathbf{N}\|}$$

$$d = \frac{|A(x_1 - x_0) + B(y_1 - y_0) + C(z_1 - z_0)|}{\sqrt{A^2 + B^2 + C^2}}$$

$$d = \frac{|Ax_1 + By_1 + Cz_1 - (Ax_0 + By_0 + Cz_0)|}{\sqrt{A^2 + B^2 + C^2}}$$

$$d = \frac{|Ax_1 + By_1 + Cz_1 + D|}{\sqrt{A^2 + B^2 + C^2}}$$

43. $\frac{|D|}{\sqrt{A^2 + B^2 + C^2}}$

Chapter 12 Review Exercises

1. $\mathbf{V} = (7, -7)$, $\|\mathbf{V}\| = 7\sqrt{2}$ **3.** $\mathbf{V} = (2, 11, -11)$, $\|\mathbf{V}\| = \sqrt{246}$

5.

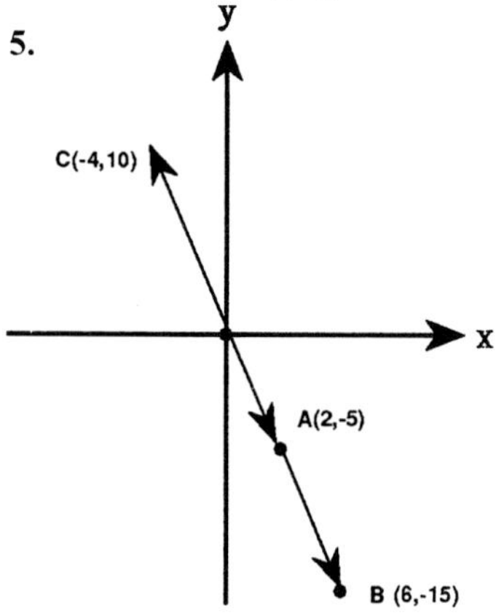

y
x
C(-4,10)
A(2,-5)
B (6,-15)

7. $\left(\frac{5}{13}, \frac{12}{13}\right)$ **9.** $\left(\frac{2}{\sqrt{38}}, -\frac{5}{\sqrt{38}}, \frac{3}{\sqrt{38}}\right)$ **11.** $(-19, 20)$ **13.** 4 **15.** 1

17. -480 **19.** $\left(\frac{1}{\sqrt{5}}, \frac{2}{\sqrt{5}}\right)$ **21.** $\left(\frac{2}{\sqrt{13}}, \frac{3}{\sqrt{13}}\right)$ **23.** 8.1° **25.** $\frac{5}{\sqrt{10}}$

27. $\frac{41}{\sqrt{29}}$ **29.** $P = (1, 3) + t(1, 7)$; $x = 1 + t$ and $y = 3 + 7t$

31. $P = t(2, 4)$; $x = 2t$ and $y = 4t$ **33.** $\left(\frac{29}{5}, \frac{17}{5}\right)$ and $\left(-\frac{19}{5}, -\frac{47}{5}\right)$

35. $\left(-\frac{47}{11}, \frac{5}{11}\right)$ **37.** $\left(\frac{1}{3}, \frac{8}{3}\right)$ **39.** $\frac{x-2}{-3} = \frac{y}{3} = \frac{z-6}{2}$;
$x = 2 - 3t, y = 3t, z = 6 + 2t$ **41.** $x = -2 + 5t, y = 1 + 4t, z = -5 + 2t$

43. $-34x + 13y + 50z + 95 = 0$ **45.** $\frac{8}{\sqrt{5}}$

47. $\left(\frac{2\sqrt{10}}{5} - 1\right)x + \left(\frac{\sqrt{10}}{5} + 1\right)y + 4 - \frac{\sqrt{10}}{5} = 0; \left(\frac{2\sqrt{10}}{5} + 1\right)x + \left(\frac{\sqrt{10}}{5} - 1\right)y - \left(4 + \frac{\sqrt{10}}{5}\right) = 0$

49. $(x, y) = (5, -3) + t(5, 2)$ **51.** $\frac{x - 6}{-14} = \frac{y + 1}{-11} = \frac{z + 4}{14}$

Chapter 12 Test 1

1.

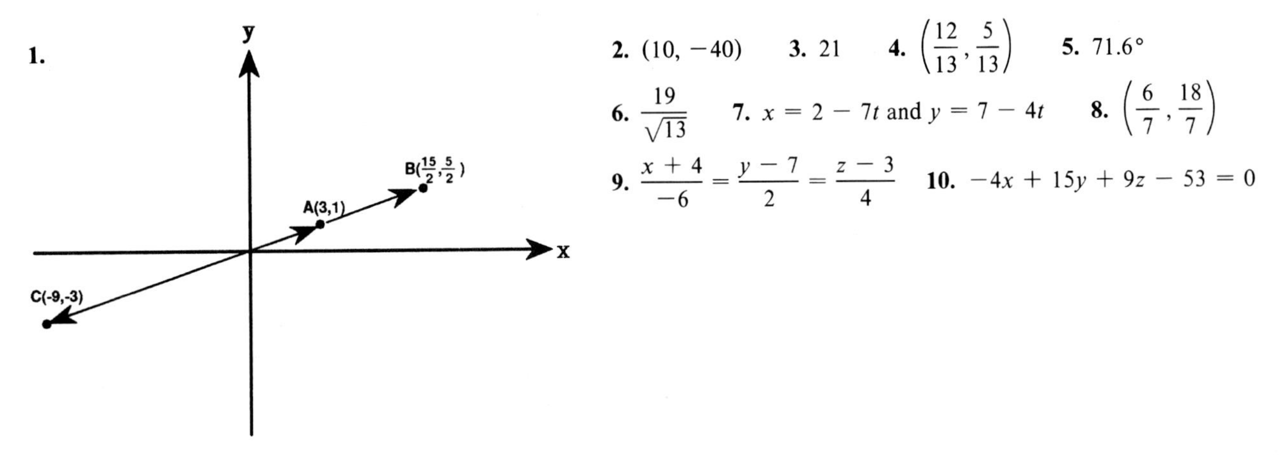

2. $(10, -40)$ **3.** 21 **4.** $\left(\frac{12}{13}, \frac{5}{13}\right)$ **5.** 71.6°

6. $\frac{19}{\sqrt{13}}$ **7.** $x = 2 - 7t$ and $y = 7 - 4t$ **8.** $\left(\frac{6}{7}, \frac{18}{7}\right)$

9. $\frac{x + 4}{-6} = \frac{y - 7}{2} = \frac{z - 3}{4}$ **10.** $-4x + 15y + 9z - 53 = 0$

Chapter 12 Test 2

1. 63.4° **2.** $\frac{17\sqrt{5}}{5}$ **3.** $x = 2 + 8t$ and $y = -5t$ **4.** $\left(-\frac{3}{11}, \frac{28}{11}\right)$ **5.** $(2, 2)$ **6.** $\frac{x + 2}{-5} = \frac{y - 5}{9} = \frac{z}{2}$

7. $-80x - 43y + 62z = 0$ **8.** $x + 4y + (-8 - 3\sqrt{17}) = 0$ or $x + 4 + (-8 + 3\sqrt{17}) = 0$

9. $(x, y) = (6, -2) + t(4, 7)$ **10.** $\frac{x - 2}{-4} = y - 8 = \frac{z + 3}{2}$

Section 13.1

1. Prove: $1 + 3 + 5 + \cdots + (2n - 1) = n^2$

S_1: $2(1) - 1 = 1^2$

$1 = 1$ so S_1 is true

Assume S_k is true; that is, assume $1 + 3 + 5 + \cdots + (2k - 1) = k^2$. Show S_{k+1} is true; that is, show:

$$1 + 3 + 5 + \cdots + (2k - 1) + (2(k + 1) - 1) = (k + 1)^2$$

Start with S_k.

$$\begin{aligned} 1 + 3 + 5 + \cdots + (2k - 1) &= k^2 \\ 1 + 3 + 5 + \cdots + (2k - 1) + (2(k + 1) - 1) &= k^2 + 2(k + 1) - 1 \\ &= k^2 + 2k + 2 - 1 \\ &= k^2 + 2k + 1 \\ &= (k + 1)^2 \end{aligned}$$

3. $1^2 + 2^2 + 3^3 + \cdots + n^2 = \dfrac{n(n+1)(2n+1)}{6}$

S_1: $1^2 = \dfrac{1(1+1)(2\cdot 1+1)}{6}$

$1 = \dfrac{6}{6} = 1$

S_k: $1^2 + 2^2 + 3^2 + \cdots + k^2 = \dfrac{k(k+1)(2k+1)}{6}$

$$\begin{aligned}
1^2 + 2^2 + 3^2 + \cdots + k^2 + (k+1)^2 &= \frac{k(k+1)(2k+1)}{6} + (k+1)^2 \\
&= \frac{k(k+1)(2k+1) + 6(k+1)^2}{6} \\
&= \frac{(k+1)(2k^2 + k + 6k + 6)}{6} \\
&= \frac{(k+1)(2k^2 + 7k + 6)}{6} \\
&= \frac{(k+1)(k+2)(2k+3)}{6} \\
&= \frac{(k+1)[(k+1)+1][2(k+1)+1]}{6}
\end{aligned}$$

Thus, $S_k \rightarrow S_{k+1}$.

5. $1\cdot 3 + 2\cdot 4 + 3\cdot 5 + \cdots + n(n+2) = \dfrac{n(n+1)(2n+7)}{6}$

S_1: $1\cdot 3 = \dfrac{1(1+1)(2\cdot 1+7)}{6}$

$3 = \dfrac{1\cdot 2\cdot 9}{6} = 3$

S_k: $1\cdot 3 + 2\cdot 4 + 3\cdot 5 + \cdots + k(k+2) = \dfrac{k(k+1)(2k+7)}{6}$

$$\begin{aligned}
1\cdot 3 + 2\cdot 4 + 3\cdot 5 + \cdots + k(k+2) + (k+1)(k+3) &= \frac{k(k+1)(2k+7)}{6} + (k+1)(k+3) \\
&= \frac{k(k+1)(2k+7) + 6(k+1)(k+3)}{6} \\
&= \frac{(k+1)[k(2k+7) + 6(k+3)]}{6} \\
&= \frac{(k+1)(2k^2 + 7k + 6k + 18)}{6} \\
&= \frac{(k+1)(2k^2 + 13k + 18)}{6} \\
&= \frac{(k+1)(k+2)(2k+9)}{6} \\
&= \frac{(k+1)(k+2)[2(k+1)+7]}{6}
\end{aligned}$$

Thus $S_k \rightarrow S_{k+1}$.

7. $(ab)^n = a^n b^n$

S_1: $(ab)^1 = a^1 b^1$

$ab = ab$

S_k: $(ab)^k = a^k b^k$

$(ab)^k \cdot (ab) = a^k b^k \cdot ab$

$(ab)^{k+1} = a^k \cdot a \cdot b^k \cdot b$

$(ab)^{k+1} = a^{k+1} \cdot b^{k+1}$

Thus $S_k \rightarrow S_{k+1}$

9. $\sin(\theta + n\pi) = (-1)^n \sin\theta$

S_1: $\sin(\theta + \pi) = -1(\sin\theta)$

$$\begin{aligned}\sin(\theta + \pi) &= \sin\theta\cos\pi + \cos\theta\sin\pi \\ &= \sin\theta(-1) + \cos\theta(0) \\ &= -\sin\theta\end{aligned}$$

S_k: $\sin(\theta + k\pi) = (-1)^k \sin\theta$

$$\begin{aligned}\sin(\theta + (k+1)\pi) &= \sin\theta\cos(k+1)\pi + \cos\theta\sin(k+1)\pi \\ &= \sin\theta\cos(k\pi + \pi) + \cos\theta\sin(k\pi + \pi) \\ &= \sin\theta(\cos k\pi\cos\pi - \sin k\pi\sin\pi) + \cos\theta(\sin k\pi\cos\pi + \cos k\pi\sin\pi) \\ &= \sin\theta((-1)^k(-1) - 0) + \cos\theta((0)(-1) + (-1)^k(0)) \\ &= \sin\theta(-1)^{k+1} + \cos\theta(0) \\ &= (-1)^{k+1} \cdot \sin\theta\end{aligned}$$

11. The statement is true if $h = 0$ because $1^n \geq 1$. Thus, in the induction proof we assume $h \neq 0$.

$$(1 + h)^n \geq 1 + nh \quad \text{if } h > -1 \text{ and } n \in N$$

S_1: $(1 + h)^1 \geq 1 + 1 \cdot h$

$1 + h = 1 + h$

S_k: $(1 + h)^k \geq 1 + kh$

Prove S_{k+1}: $(1 + h)^{k+1} \geq 1 + (k+1)h$

$$\begin{aligned}(1 + h)^k(1 + h) &\geq (1 + kh)(1 + h) \\ (1 + h)^{k+1} &\geq 1 + kh + h + kh^2 \\ &\geq 1 + (k+1)h + kh^2 \\ &\geq 1 + (k+1)h, \text{ since } kh^2 > 0\end{aligned}$$

13. $1 + 2n < 3^{n-1}$ if $n > 2$, $n \in N$

S_3: $1 + 2(3) < 3^{3-1}$

$1 + 6 < 3^2$

$7 < 9$

S_k: $1 + 2k < 3^{k-1}$, $k > 3$

Prove S_{k+1}: $1 + 2(k+1) < 3^k$

$$\begin{aligned}(1 + 2k) + 2 &< 3^{k-1} + 2 \\ &< 3^{k-1} \cdot 3 \\ &= 3^k\end{aligned}$$

Therefore $1 + 2(k+1) < 3^k$

15. $2 + 7 + 12 + \cdots + (5n - 3) = \dfrac{n(5n - 1)}{2}$

S_1: $5(1) - 3 = \dfrac{1(5 \cdot 1 + 1)}{2}$

$$2 = \frac{1(4)}{2} = 2$$

S_k: $2 + 7 + 12 + \cdots + (5k - 3) = \dfrac{k(5k - 1)}{2}$

$$\begin{aligned} 2 + 7 + 12 + \cdots + (5k - 3) + [5(k + 1) - 3] &= \frac{k(5k - 1)}{2} + 5(k + 1) - 3 \\ &= \frac{k(5k - 1) + 10(k + 1) - 6}{2} \\ &= \frac{5k^2 - k + 10k + 10 - 6}{2} \\ &= \frac{5k^2 + 9k + 4}{2} \\ &= \frac{(k + 1)(5k + 4)}{2} \\ &= \frac{(k + 1)[5(k + 1) - 1]}{2} \end{aligned}$$

17. $\left(1 + \frac{1}{1}\right)\left(1 + \frac{1}{2}\right)\left(1 + \frac{1}{3}\right) \cdot \ldots \cdot \left(1 + \frac{1}{n}\right) = n + 1$

S_1: $\left(1 + \frac{1}{1}\right) = 1 + 1$

$$1 + 1 = 2$$

S_k: $\left(1 + \frac{1}{1}\right)\left(1 + \frac{1}{2}\right)\left(1 + \frac{1}{3}\right) \cdot \ldots \cdot \left(1 + \frac{1}{k}\right) = k + 1$

$$\begin{aligned} \left(1 + \frac{1}{1}\right)\left(1 + \frac{1}{2}\right)\left(1 + \frac{1}{3}\right) \cdot \ldots \cdot \left(1 + \frac{1}{k}\right)\left(1 + \frac{1}{k + 1}\right) &= (k + 1)\left(1 + \frac{1}{k + 1}\right) \\ &= (k + 1)\left(\frac{k + 2}{k + 1}\right) \\ &= k + 2 \end{aligned}$$

19. n points produce $\dfrac{n(n - 1)}{2}$ lines, $n \geq 2$, $n \varepsilon N$

S_2: 2 points produce $\dfrac{2(1)}{2} = 1$ line

S_k: k points produce $\dfrac{k(k - 1)}{2}$ lines

Prove S_{k+1}: $k + 1$ points produce $\dfrac{(k + 1)(k)}{2}$ lines

Proof: k points produce $\dfrac{k(k - 1)}{2}$ lines. If one additional point is added, lines must be drawn from it to each of the original k points, making k new lines. So we now have $\dfrac{k(k - 1)}{2} + k$ lines.

$$\frac{k(k - 1) + 2k}{2} = \frac{k^2 - k + 2k}{2} = \frac{k^2 + k}{2} = \frac{(k + 1)k}{2} \text{ lines}$$

Section 13.2

1. $2^0, 2^1, 2^2, 2^3, \ldots, 2^{n-1}$ **3.** $\frac{1}{3}, \frac{1}{9}, \frac{1}{27}, \frac{1}{81}, \ldots, \frac{1}{3^n}$ **5.** $-1, 4, -36, 512, \ldots, (-1)^n(2n)^{n-1}$

7. $\frac{2}{1}, \frac{3}{2}, \frac{4}{3}, \frac{5}{4}, \frac{6}{5}$ **9.** $1, 0, -\frac{1}{3}, \frac{1}{2}, -\frac{3}{5}, \frac{2}{3}$ **11.** $\frac{5}{1}, \frac{5}{2}, \frac{5}{3}, \frac{5}{4}, \frac{5}{5}$ **13.** $\frac{2}{1} + \frac{3}{2} + \frac{4}{3} + \frac{5}{4} + \frac{6}{5}$

15. $2^0 + 2^1 + 2^2 + \ldots + 2^{n-1} + \ldots$ **17.** $\frac{5}{1} + \frac{5}{2} + \frac{5}{3} + \frac{5}{4} + \frac{5}{5}$

19. $\sum_{k=1}^{8} (2k + 5) = 7 + 9 + 11 + 13 + 15 + 17 + 19 + 21$

21. $\sum_{j=1}^{9} \left(\frac{j}{2} - 3\right) = -\frac{5}{2} - 2 - \frac{3}{2} - 1 - \frac{1}{2} + 0 + \frac{1}{2} + 1 + \frac{3}{2}$

23. $\sum_{i=1}^{5} [(i + 2)(i - 1)] = 0 + 4 + 10 + 18 + 21$ **25.** 112 **27.** $-\frac{9}{2}$ **29.** 60 **31.** $\sum_{i=1}^{4} 4i$

33. $\sum_{i=1}^{4} 2(3^{i-1})$ **35.** 2, 7, 12, 17 **37.** $\frac{1}{2}, \frac{1}{4}, \frac{1}{8}, \frac{1}{16}$ **39.** **(a)** 3, 6, 9, 12, 15, 18 **(b)** 3, 6, 9, 6, −9, −42

(c) no **41.** $\frac{1}{n}$ **43.** $\frac{nx}{n+1}$ **45.** $a_1 = 6, a_{n+1} = a_n + 1$ **47.** $a_1 = \frac{1}{2}, a_{n+1} = \frac{a_n}{2}$ **49.** $a_n = \frac{1}{n}$

51. $a_n = 4n - 1$

53. Prove: $\sum_{i=1}^{n} ca_i = c \sum_{i=1}^{n} a_i$

S_1: $\sum_{i=1}^{1} ca_i = ca_1 = c \sum_{i=1}^{1} a_i$

Assume S_k: $\sum_{i=1}^{k} ca_i = c \sum_{i=1}^{k} a_i$

$$\begin{aligned}\sum_{i=1}^{k+1} ca_i &= \sum_{i=1}^{k} ca_i + ca_{k+1} \\ &= c \sum_{i=1}^{k} a_i + ca_{k+1} \\ &= c\left[\sum_{i=1}^{k} a_i + a_{k+1}\right] \\ &= c \sum_{i=1}^{k+1} a_i\end{aligned}$$

55. $\sum_{i=4}^{9} \frac{i}{(i-3)} = \frac{4}{1} + \frac{5}{2} + \frac{6}{3} + \frac{7}{4} + \frac{8}{5} + \frac{9}{6}$

$\sum_{i=1}^{6} \frac{(i+3)}{i} = \frac{4}{1} + \frac{5}{2} + \frac{6}{3} + \frac{7}{4} + \frac{8}{5} + \frac{9}{6}$

57. Prove: Sum of first n Fibonacci numbers is $a_{n+2} - 1$

$$a_0 = 0$$
$$a_1 = 1$$
$$a_2 = 1$$
$$a_n = a_{n-2} + a_{n-1}$$

$S_1 = a_1 = a_3 - 1$
$1 = 2 - 1$
$1 = 1$

$S_k = \text{sum of first } k \text{ terms} = a_{k+2} - 1$

Show $S_{k+1} = a_{k+3} - 1$

$$S_{k+1} = S_k + a_{k+1}$$
$$= a_{k+2} - 1 + a_{k+1}$$
$$= a_{k+3} - 1 \text{ since } a_{k+1} + a_{k+2} = a_{k+3}$$

Section 13.3

1. arithmetic; 84; $7n$ **3.** geometric; 512; 2^n **5.** arithmetic; -3; $-\frac{2}{5}n + \frac{17}{5}$

7. geometric; $\frac{19{,}683}{1{,}024}$; $\frac{3^{n-1}}{2^n}$ **9.** geometric; $-\frac{128}{243}$; $\frac{(-2)^{n-1}}{3^{n-3}}$ **11.** arithmetic; $12x + 11$; $n(x + 1) - 1$

13. 112 **15.** $\frac{121}{729}$ **17.** $\frac{21}{128}$ **19.** -6 **21.** -480 **23.** $\frac{1{,}533}{32}$ **25.** 2,132 **27.** $\frac{125}{3}$

29. Sum does not exist. **31.** 3 **33.** 1 **35.** Sum does not exist. **37.** 3 **39.** $\frac{4}{9}$ **41.** $\frac{1}{2}$ **43.** $\frac{3{,}923{,}073}{999{,}999}$

45. The forty-second term **47.** $n = 14$

49. $2 + 4 + 6 + \cdots + 2n: = n(n + 1)$

$a_1 = 2,\ d = 2,\ a_n = 2n$

$$S = \frac{n}{2}(a_1 + a_n)$$
$$= \frac{n}{2}(2 + 2n)$$
$$= n(n + 1)$$

51. $x = \frac{24}{5}$ feet

53. Prove: $a_n = a_1 + (n - 1)d$

S_1: $a_1 = a_1 + (1 - 1)d$
$a_1 = a_1$

Assume S_k: $a_k = a_1 + (k - 1)d$

$$a_{k+1} = a_k + d$$
$$= a_1 + (k - 1)d + d$$
$$= a_1 + kd$$

55. \$342 **57.** $1.645\ P$

59. $P\left(1+\frac{r}{12}\right)^{12n-1}+P\left(1+\frac{r}{12}\right)^{12n-2}+\ldots+P=P\left[\left(1+\frac{r}{12}\right)^{12n-1}+\left(1+\frac{r}{12}\right)^{12n-2}+\ldots+1\right]$

The expression in the bracket is a geometric series with $a=1$ and $\left(1+\frac{r}{12}\right)$ the common multiplier. The series has $12n$ terms.

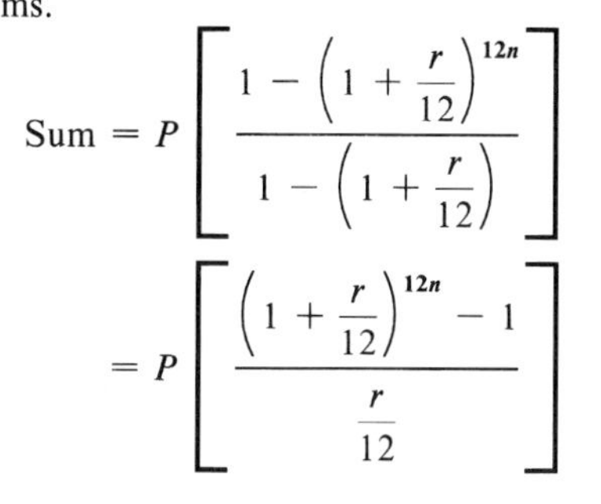

$$\text{Sum}=P\left[\frac{1-\left(1+\frac{r}{12}\right)^{12n}}{1-\left(1+\frac{r}{12}\right)}\right]$$

$$=P\left[\frac{\left(1+\frac{r}{12}\right)^{12n}-1}{\frac{r}{12}}\right]$$

61. \$316.01

Section 13.4

1. 720 **3.** 2,520 **5.** 10 **7.** 504 **9.** 120 **11.** 495 **13.** (**a**) 120 (**b**) 3,125 **15.** (**a**) 15,120 (**b**) 59,049 **17.** 336 **19.** 252 **21.** 56 **23.** 24

25. 1,716 ways to select six books from thirteen. 330 ways to select six books from thirteen if two certain books must be selected. 462 ways to select six books from thirteen if two certain books must not be chosen.

27. $x^6+18x^5+135x^4+540x^3+1215x^2+1458x+729$

29. $x^7+14x^6y+84x^5y^2+280x^4y^3+560x^3y^4+672x^2y^5+448xy^6+128y^7$

31. $729x^6-2{,}916x^5y+4{,}860x^4y^2-4{,}320x^3y^3+2{,}160x^2y^4-576xy^5+64y^6$

33. $16x^4-\frac{32}{3}x^3y+\frac{8}{3}x^2y^2-\frac{8}{27}xy^3+\frac{1}{81}y^4$ **35.** $x^{48}+96x^{47}y+4{,}512x^{46}y^2+138{,}368x^{45}y^3$

37. $x^{36}-18\sqrt{2}x^{34}+306x^{32}-1632\sqrt{2}x^{30}$ **39.** 875 **41.** (**a**) 80 (**b**) 200 **43.** 720

45. (**a**) 288 (**b**) 400

47. $\binom{n+1}{r+1}=\frac{(n+1)!}{(r+1)!(n+1-r-1)!}=\frac{(n+1)!}{(r+1)!(n-r)!}$

$$\binom{n}{r+1}+\binom{n}{r}=\frac{n!}{(r+1)!(n-r-1)!}+\frac{n!}{r!(n-r)!}$$

$$=\frac{n!r!(n-r)!+n!(r+1)!(n-r-1)!}{(r+1)!(n-r-1)!r!(n-r)!}$$

$$=\frac{n!r!(n-r)(n-r-1)!+n!(r+1)r!(n-r-1)!}{(r+1)!(n-r-1)!r!(n-r)!}$$

$$=\frac{n!(n-r)+n!(r+1)}{(r+1)!(n-r)!}$$

$$=\frac{n!(n-r+r+1)}{(r+1)!(n-r)!}$$

$$=\frac{n!(n+1)}{(r+1)!(n-r)!}$$

$$=\frac{(n+1)!}{(r+1)!(n-r)!}$$

49. (a) 16 (b) 4,845

51. From exercise 50, $2^n = \binom{n}{0} + \binom{n}{1} + \cdots + \binom{n}{n} \cdot \binom{n}{0} = 1$ is the number of ways of choosing 0 of n things.

$\binom{n}{1} + \binom{n}{2} + \cdots + \binom{n}{n}$ is the number of ways of choosing at least one of n things.

$$2^n = 1 + \binom{n}{1} + \binom{n}{2} + \cdots + \binom{n}{n}$$

$$2^n - 1 = \binom{n}{1} + \binom{n}{2} + \cdots + \binom{n}{n}$$

53. 5,148

55.
1 6 15 20 15 6 1
1 7 21 35 35 21 7 1

Each number in a row may be obtained by adding the two numbers in the row above it, one to the left and one to the right.

Chapter 13 Review Exercises

1. 4, 7, 10, 13 **3.** $\frac{1}{2}, \frac{1}{4}, \frac{1}{8}, \frac{1}{16}, \ldots, \frac{1}{2^n}$ **5.** $3 + (-6) + 9 + (-12) + \cdots + (-1)^{n+1}3n + \cdots$

7. $\sum_{i=1}^{8} (3i - 1) = 2 + 5 + 8 + 11 + 14 + 17 + 20 + 23$

9. $\sum_{j=1}^{6} (j + 3)(j - 1) = 4 \cdot 0 + 5 \cdot 1 + 6 \cdot 2 + 7 \cdot 3 + 8 \cdot 4 + 9 \cdot 5$ **11.** $\sum_{i=1}^{6} 3 \cdot 2^i$ **13.** $a_n = \frac{(-1)^{n-1}}{2n - 1}$

15. $-1, -5, -13, -29$ **17.** arithmetic; 102; $12n - 6$ **19.** geometric; $\frac{1{,}048{,}576}{6{,}561}$; $\frac{4^n}{3^{n-2}}$ **21.** 117

23. $-\frac{543}{1{,}024}$ **25.** 3,209 **27.** 2 **29.** $\frac{32}{99}$ **31.** 5,040 **33.** 126

35. $x^7 + 14x^6 + 84x^5 + 280x^4 + 560x^3 + 672x^2 + 448x + 128$ **37.** $x^5 - \frac{5}{2}x^4 + \frac{5}{2}x^3 - \frac{5}{4}x^2 + \frac{5}{16}x - \frac{1}{32}$

39. 24 **41.** 11,232,000 **43.** 120

45. $\frac{1}{2} - \frac{1}{4} - \frac{1}{8} - \frac{1}{16} - \cdots - \frac{1}{2^n} = \frac{1}{2^n}$

S_1: $\frac{1}{2} = \frac{1}{2^1}$

Assume S_k: $\frac{1}{2} - \frac{1}{4} - \frac{1}{8} - \cdots - \frac{1}{2^k} = \frac{1}{2^k}$

$$\frac{1}{2} - \frac{1}{4} - \frac{1}{8} - \cdots - \frac{1}{2^k} - \frac{1}{2^{k+1}} = \frac{1}{2^k} - \frac{1}{2^{k+1}}$$
$$= \frac{2}{2^k 2} - \frac{1}{2^{k+1}}$$
$$= \frac{2}{2^{k+1}} - \frac{1}{2^{k+1}}$$
$$= \frac{1}{2^{k+1}}$$

47. $a_1 = 2$, $a_{n+1} = a_n + 3$ **49.** The thirty-second term **51.** $a_1 = 6$ **53.** $2,400 **55.** 141 **57.** 560

Chapter 13 Test 1

1. $-1, 1, 3, 5, 7; -1 + 1 + 3 + 5 + 7$ **2.** $a_n = 2 \cdot 4^{n-1}$ **3.** 390 **4.** 1,093 **5.** $\frac{4}{9}$ **6. (a)** 40,320 **(b)** 28 **7.** $x^7 + 21x^6 + 189x^5 + 945x^4 + 2{,}835x^3 + 5{,}103x^2 + 5{,}103x + 2{,}187$ **8.** 60 **9.** 126 **10.** $0.528\ C$

Chapter 13 Test 2

1. $-\frac{12}{5}$ **2.** 3,926 **3.** $\frac{559}{990}$

4. $6{,}561x^8 - 8{,}748x^7y + 5{,}103x^6y^2 - 1{,}701x^5y^3 + \frac{2{,}835}{8}x^4y^4 - \frac{189}{4}x^3y^5 + \frac{63}{16}x^2y^6 - \frac{3}{16}xy^7 + \frac{1}{256}y^8$

5. 30 numbers **6.** 455 **7.** $\frac{306}{5}$ **8.** $a_1 = 5, a_{n+1} = a_n + 2$ **9.** 0.107 V is left. **10.** $825

Index

A

Abscissa, 76
Absolute value, 57
  definition, 57
  in measurement, 63
  notation, 57
  properties, 58
Amplitude modulation (AM), 374
"Analytic Geometry," 407
Angle(s)
  central, 267, 279
  of incidence, 458
  of inclination, 459
  initial side, 267
  between lines, 459, 478
  negative, 267
  positive, 267
  of reflection, 458
  in standard position, 267
  terminal side, 267
Angular velocity, 344
Apollonius, 406
Arc length, 266
Area of a circular sector, 269 (ex. 33)
Asymptote(s)
  horizontal, 208, 304, 305
  oblique, 216
  vertical, 206, 290, 294, 303

B

Basic cosine curve, 288
Basic sine curve, 287
Basic tangent curve, 289
Beats, 373
Binomial theorem, 524, 530

C

Cardioid, 389
Cauchy-Schwartz Inequality, 482
Circle(s), 408
  center, 408
  general equation, 408
  radius, 408
  standard equation, 408
Combinations, 528
Common logarithm, 242
Completing the square, 34
Complex numbers, 28
  addition, 30
  argument of, 399
  complex conjugates, 30
  division, 31
  equality of, 29
  imaginary part, 29
  imaginary unit *i*, 29
  modulus of, 399
  multiplication, 30
  polar form, 399
  real part, 29
  subtraction, 30
Compound interest, 237, 250
Conic sections, 406
  eccentricity, 449
  focus-directrix definition, 450
  polar coordinate form, 454
  reflecting property, 463
  tangents, 460
Consistency tests for systems of linear equations, 133
Consistent system of linear equations, 132, 158
Coordinate planes, 492
Cosine double angle formulas, 321
Cosine half angle formula, 324
Cosine sum and difference formulas, 316
Cycloid, 397

D

Damping function, 240 (ex. 43), 252 (ex. 40)
Decibel, 247
deMoivre's theorem, 399
Dependent variable, 72
Descartes, 407
Descartes' Rule of Signs, 184
Difference quotient, 101
Direction numbers of a line, 496
Discriminant, 37
Distance between two points, 91
Distance from a point to a line, 137, 410, 480
Division algorithm for polynomials, 171

E

Elementary transformations for matrices, 148
Ellipse(s), 423
  foci, 423
  general equation, 428
  major and minor axes, 423
  parametric equation, 427
  standard equations, 426, 428
  vertices, 424
Ellipsis symbol, 1
Ellipsoid, 464

Equation(s)
- with absolute value, 57
- equivalent, 3
- exponential, 248
- general second degree polynomial, 407
- with inverse trigonometric functions, 332
- linear, 5, 495
- linear systems ($m \times n$), 156
- linear systems (2×2), 131
- linear systems (3×3), 141
- logarithmic, 247
- of a plane, 497
- quadratic, 33
- in quadratic form, 52
- with radicals, 52
- with rational exponents, 54
- trigonometric, 328

Exponent(s)
- integral, 12
- irrational, 230
- natural, 11
- rational, 16
- zero, 12

F

Factorial, 526
Factoring, 22, 176, 329
- completely factored form, 22
- difference of squares, 25
- Foil method, 23
- by grouping, 26
- monomial factors, 22
- sum and difference of cubes, 25

Four-leaf rose, 393
Frequency modulation (FM), 374
Function(s), 71
- composition of, 116
- continuous, 77, 178
- decreasing, 78, 227
- difference of, 116
- discontinuous, 205
- domain, 72, 107
- exponential, 226
- graph of, 76, 93
- greatest integer, 94
- increasing, 78, 227
- inverse, 119, 122
- linear, 83
- logarithmic, 241
- one-to-one, 119
- periodic, 256, 282, 283
- polynomial, 169
- product of, 116
- quadratic, 97
- quotient of, 116
- range, 72, 107
- rational, 205
- sum of, 115
- trigonometric, 256

Fundamental Theorem of Algebra, 175

G

General form of the equation of a line, 87
Graphing techniques, 104
- contractions, 112, 291
- reflections, 109, 292, 299
- rotations, 440, 442
- stretchings, 112, 291
- symmetry tests, 105, 288, 390
- translations, 111, 293, 416, 426, 435, 442

H

Hertz (HZ), 373
Horizontal line equation, 88
Horizontal line test, 120, 298
Hyperbola(s), 430
- asymptotes, 433
- center, 432
- foci, 430
- general equation, 438
- standard equation, 435, 438
- transverse axis, 432
- vertices, 432

Hypocycloid, 399 (ex. 32)

I

Identifying conics, 447
Identity, 2, 280, 312
Inconsistent system of linear equations, 132, 158
Independent variable, 74
Inequalities, 42
- absolute value, 59
- equivalent, 43

Intercepts of a graph, 106
Interval(s), 46
- closed, 46
- half open, 46
- open, 46

Inverse trigonometric functions, 298, 325
Inverse trigonometric functions, 298, 325
Involute of a circle, 399 (ex. 31)
Irrational roots of polynomial equations, 197
Iteration, 197, 337

K

Kirchhoff's Laws, 159

L

Law of Cosines, 360
Law of Sines, 356
Leading coefficient of a polynomial, 169
Learning curve, 240 (ex. 44), 252 (ex. 41)
Lemniscate, 394
Linear interpolation, 338
Linear velocity, 344
Lines in two and three dimensions, 83, 494
Logistic curve, 241 (ex. 45), 252 (ex. 42)
Lower Bound Theorem, 184

M

Marginal cost, 103
Mathematical induction, 504
Mathematical models, 5
Matrices, 150, 157
- augmented, 151
- coefficient, 151
- echelon form, 152

Method of linear combinations, 135, 142
Midpoint of a segment, 92

N

Natural base (e), 234
Natural logarithm, 242
Newton, Issac, 524
Newton's Law of Cooling, 252 (ex. 39)

O

Octants, 492
Ordinate, 76

P

Pappus, 406
Parabola(s)
- axis, 414
- directrix, 414
- focus, 414

general equation, 420
standard equation, 420
vertex, 98, 414
Paraboloid, 463
Parallel lines, 90
Parameter, 134, 395, 485
Parametric equations, 395, 413, 438, 486, 495
Partial fractions, 219
Pascal, Blaise, 524, 532 (ex. 55)
Pencil of lines, 134
Perfect square trinomial, 34
Permutations, 525
Perpendicular lines, 90
Planes in three dimensions, 497
Point of subdivision of a line, 488
Point-slope form of the equation of a line, 85
Polar axis, 383
Polar coordinate system, 382
Pole, 383
Polynomial(s), 22, 169
binomial, 22
monomial, 22
trinomial, 22
Population growth, 235
Principal n^{th} root, 14

Q

Quadratic formula, 36

R

Radian measure, 264
Radical notation, 17
Radioactive decay, 237, 249
Rationalize the denominator, 19
Rational Roots Theorem, 181
Reciprocal trigonometric functions, 278, 283
Relations, 72
Remainder Theorem, 175
Richter scale, 251 (ex. 32)
Roots of polynomial equations, 179, 181, 197
multiplicity of, 179
quadratic equations, 36

S

Scalar, 364
Sequence, 510
arithmetic, 515
common difference, 515
common ratio, 517
finite, 511
geometric, 517
infinite, 511
n^{th} term, 511, 516, 517
recursion formula, 514
Series, 511
arithmetic, 516
finite, 511
geometric, 518
infinite, 511, 520
sum, 517, 518, 521
Set(s), 1
empty (null) set, 2
natural number set, 2
proper subset, N, 2
real number set, R, 2
replacement set, 2
set intersection, 46
set union, 48
solution set, 2
subset, 2
subsets of C, 31
Set builder notation, 1
Sigma (Σ) notation, 512
Sign graphs, 47
Sine double angle formula, 316
Sine half angle formula, 324
Sine sum and difference formulas, 316
Slope-intercept form of the equation of a line, 85
Slope of a line, 83
Square root, 14
Synthetic division, 171

T

Tangent sum formula, 318
Traffic flow, 161
Transformation equations
rotations, 416
translations, 440
Triangle inequality, 59, 482
Trigonometric cofunctions, 284
Trigonometric function definitions, 270, 272, 350
Trigonometric identities, 278, 280, 312, 320

U

Upper Bound Theorem, 183

V

Vector(s), 364, 470
addition, 366, 473
algebra, 470
angle between, 478
basis, 473
components, 364
inner (dot) product, 477, 482
linear combination, 473
magnitude, 365, 480, 493
normal, 480, 498
perpendicular, 479
position, 485
resultant, 370
scalar multiplication, 471, 473
in three dimensions, 493
unit, 472
Vector parametric equation of a line, 485
Vertical line equation, 88
Vertical line test, 78

Y

Y-intercept, 85

Z

Zeros of polynomial functions, 174